Union List of Geologic Field Trip Guidebooks of North America

Union List of Geologic Field Trip Guidebooks of North America

Fifth Edition

Compiled and edited by the
Geoscience Information Society
Guidebooks Committee
Charlotte Derksen, Chair

PUBLISHED BY THE
AMERICAN GEOLOGICAL INSTITUTE
IN COOPERATION WITH THE
GEOSCIENCE INFORMATION SOCIETY

Earlier editions:

Geologic Field Trip Guidebooks of North America
A Union List Incorporating Monographic Titles
Copyright © 1968 Phil Wilson, Publisher, Houston, Texas

Geologic Field Trip Guidebooks of North America
A Union List Incorporating Monographic Titles
Copyright © 1971 Phil Wilson, Publisher, Houston, Texas

Union List of Geologic Field Trip Guidebooks of North America
Copyright © 1978 American Geological Institute

Union List of Geologic Field Trip Guidebooks of North America
Copyright © 1986 American Geological Institute

Union List of Geologic Field Trip Guidebooks of North America
Copyright © 1989 American Geological Institute
All rights reserved

Library of Congress Catalog Card Number 88-082744
International Standard Book Number 0-913312-97-5
Printed in the United States of America

American Geological Institute
4220 King Street
Alexandria, Virginia 22302

Price: $60.00, includes postage and handling in the U.S.

TABLE OF CONTENTS

Introduction vii

Plan of the volume ix

Samples of title entries and holdings ix

Cross references ix

Holdings statements x

Plan of the geographic index x

Plan of the stratigraphic index xi

Directory of countries xiii

Guidelines for Authors/Publishers xv

Participating libraries xvii

Directory of library services xxiii

Union list of geologic field trip guidebooks 1

Geographic index 183

Stratigraphic index 219

Corrections form 223

INTRODUCTION

The Guidebook Committee for the fifth edition of the *Union List of Geologic Field Trip Guidebooks of North America* greatly appreciates the efforts of all of the librarians who submitted holdings lists and new titles. Many new reporting libraries and many new guidebook titles have been added to the database. We thank everyone who contributed and answered our inquiries. The reward is an ever more complete and accurate work.

The fifth edition of the Union List builds upon the previous four editions published in 1968, 1971, 1978, and 1986 respectively. The motivation for continuing this work persists. Guidebooks, despite their usefulness and importance to the geoscience community, can go rapidly out-of print, remain difficult to obtain, and be confusing to adequately control bibliographically. The Guidelines for Authors and Publishers on page xv should be consulted for recommended ways of preparing and distributing guidebooks. Adherance to these recommendations would help make guidebooks easier to find when needed.

As in earlier editions, the fifth edition is restricted geographically to North America, that is, the area north of the Panama Canal and including Bermuda and the West Indies. To be included in this list, a guidebook needed to have either a road log or a clear description of the field covered, so that the trip could be replicated. Guidebook titles for field trips scheduled through 1985 are included in this edition.

Guidebooks are compiled for field conferences, meetings, class trips, reporting research results, and general information. Publishers may be societies, sponsors of conferences, educational institutions, geological surveys, other government agencies, or commercial.

A guidebook prepared for a field conference or meeting is entered under the name of the society or organization holding the meeting or conference. In the case of a joint meeting, the organization covering the largest geographic area or the broadest subject area is considered the major organization. For example, the American Association of Petroleum Geologists (AAPG) Pacific Section is considered the major organization when meeting with the Coast Geological Society. In a case where two societies of equal size met jointly, the guidebooks are entered under the name of the society listed first on the title page. All guidebooks for a single meeting or conference are entered under the same main entry. See references are included from entries for all societies, local hosts, surveys, or schools which participated in a conference or meeting to the main entry for the meeting. See references are not included for societies or institutions which functioned solely as the publisher of the guidebook.

All trip titles and all holding information for any meeting are placed solely under the main entry. No library holdings information is included under SEE references.

A guidebook which was compiled for a field conference or meeting and also appears in a numbered series is placed under the entry for the society or organization holding the conference or meeting with a cross reference from the numbered series. An attempt was made to make cross-references between identical guidebooks used for two difference conferences or meetings.

A guidebook which was not prepared for a conference or a meeting has its main entry under the series in which it was published. A guidebook which was prepared for a class trip, to report research results, or for general information rather than for a conference or a meeting, is placed under the entry for the institution, agency or society responsible for its preparation.

Guidebooks prepared by individuals and published privately or commercially are listed by author under "Individual Authors (272)".

SYMBOLS OF AMERICAN LIBRARIES, 13th ed., Washington, DC Library of Congress, 1985, is the primary source for assigning symbols to the participating libraries. For several libraries not listed in SYMBOLS OF AMERICAN LIBRARIES, symbols were assigned by the Library of Congress and the National Library of Canada upon request for inclusion in this work. For the other libraries, appropriate similar symbols were devised; these are shown in brackets.

For this edition the geographic indexing was completely redone for all entries. This effort was funded by the Geoscience Information Society. Guidelines for the geographic indexing were drawn up by Louise Zipp of the University of Iowa and Richard Spohn of the University of Cincinnati. Geologists John Black and Karen Grove, assisted by Randy Vogel, Stanford University, were responsible for the geographic indexing.

A new feature of the fifth edition is a Stratigraphic Index. Cathy Pasterzyk, Sandia National Laboratories, and Dena Stepp, STN, examined all entries in the Union List for inclusion in the Stratigraphic Index.

Special thanks are due to Joan Sandoz of the American Geological Institute for her invaluable work in overseeing the transition to the new computer and inputting the data. Thanks are also due to John Mulvihill of GeoRef, and William Sanders of Stanford University for special assistance.

Please send corrections and suggestions to the Chair, Guidebooks Committee, Geoscience Information Society, c/o American Geological Institute, 4220 King Street, Alexandria, VA 22302.

Charlotte R.M. Derksen, Chair
Head, Branner Earth Sciences Library, Stanford University, Stanford, CA

Brigida Cobb
(Former GeoRef Librarian)

Claren M. Kidd
Geology Librarian, University of Oklahoma, Norman, OK

Michael Noga
Head, Geology/Geophysics Library, Univ. of California, Los Angeles, CA

Catherine E. Pasterczyk
Technical Information Specialist, Sandia National Laboratories, Albuquerque, NM

Joan Sandoz
Librarian, GeoRef, Alexandria, VA

Richard A. Spohn
Head, Geology Library, University of Cincinnati, Cincinnati, OH

Dena F. Stepp
Columbus, OH

Nancy Thurston
Mines Librarian, Ontario Geological Survey, Toronto, ON

Julia H. Triplehorn
Librarian, University of Alaska, Fairbanks, AK

Louise S. Zipp
Geology Librarian, University of Iowa, Iowa City, IA

PLAN OF THE VOLUME

SAMPLE ENTRIES WITH TITLES AND LIBRARY HOLDINGS

(208) GEOLOGICAL SOCIETY OF IOWA. FIELD TRIP GUIDEBOOK. **(Title varies)**
1976 Sept. Geologic points of interest in the Fort Dodge area. (332)
1983 Apr. No. 39 New stratigraphic interpretations of the middle Devonian rocks of Winneshiek and Fayette counties, northeastern Iowa.
     Oct. No. 40 Karstification on the Silurian escarpment in Fayettenian County, northeastern Iowa.
    I-GS   1983(40)
    IaCfT   983(39)
    IaU   1976, 83

(287) INTERNATIONAL CONFERENCE ON PERMAFROST. GUIDEBOOK.
1983 4th. No. 1. Guidebook to permafrost and Quaternary geology along the Richardson and Glenn Highways between Fairbanks and Anchorage, Alaska. (8)
        No. 3. Guidebook to permafrost and related features of the northern Yukon territory and Mackenzie Delta, Canada. (8,295)
        No. 4. Guidebook to permafrost and related features along the Elliott and Dalton highways, Fox to Prudhoe Bay, Alaska. (8,295)
        No. 5. Guidebook to permafrost and related features, Prudhoe Bay, Alaska. (8,295)
    CMenUG   1983(1,5)
    CPT   1983

(191) GEOLOGICAL ASSOCIATION OF CANADA. CORDILLERAN SECTION. [GUIDEBOOK]
1973   Vancouver geology; a short guide. (Revised in 1977).
1982 Apr. Copper Mountain-Phoenix tour, southern British Columbia.
1985   See Geological Association of Canada. (190)
    CaAEU   1977, 82
    CoDGS   1982

Society and organization names are arranged in alphabetical order and numbered consecutively. Under each society or organization entry, the titles of trips are listed chronologically. The listing includes the year and, when applicable, the month, a series number, and a cross reference number. When the society has several trips per year and at different times during the year, months have been used to distinguish the trips, as in sample entry (208) for 1983. When several trips are held for a meeting, the trips are indicated as assigned in the society or organization program. See entry (287) for 1983. Contrived information is bracketed as in entry (191).

Revised or second editions are included in the listings, See entry (191) for 1977.

CROSS REFERENCES

A standard "see" reference is used to refer from entries for an organization or society which prepared a guidebook or was a joint sponsor of the meeting to the society or organization holding the meeting for which the guidebook was compiled. See entry (191) for 1985.

A second type of reference is used in the lists of guidebook titles. A number in parentheses following a guidebook title refers to another society or organization which participated in preparation or publication of the guidebook, as in entry (208) for 1976.

Cross references are included for alternate forms of agency, society or series names. Cross references are also included for variant titles of conferences or symposia.

HOLDINGS STATEMENTS

Library symbols, such as I-GS, IaCfT, and IaU, follow the lists of guidebook titles. The dates of the guidebooks owned by each library are listed after its library symbol. When a library holds all the guidebooks listed for a year, only the year is shown (see entry 287 for library CPT). When holdings for a year are incom plete, the year is followed by parentheses enclosing the parts held, which may be indicated by months, series numbers (see entry 208 for library I-GS), or trip numbers (see entry 287 for library CMenUG).

Consult the "Directory of Library Services" on page xxiii and "Participating Libraries" on page xvii to decipher the library symbols and to find out how guidebooks from these institutions can be accessed. In the holding statements as well as in the directories library symbols are arranged alphabetically, with capital letters preceding lower case letters.

DIRECTORIES

LIBRARY SYMBOLS AND SERVICES

PARTICIPATING LIBRARIES

LIBRARIES ARRANGED BY COUNTRY, STATE/PROVINCE

Body of the Union List

PLAN OF THE GEOGRAPHIC INDEX

Indexing is based almost exclusively on the titles of the guide books. The index is divided into three main geographic regions: Caribbean, Central America, and North America. The names of the Countries included in this union list are listed alphabetically on page xiii.

The countries are subdivided into general, directional, named areas, and states/provinces. States or provinces also are subdivided into general, directional, and named areas.

As an example, guidebooks to the Rolla, Missouri, area are entered under:

Missouri
Rolla 74-1970(1); 87-1972; 380-1960

which leads to

(74) ASSOCIATION OF AMERICAN STATE GEOLOGISTS. GUIDE BOOKS AND ROAD LOGS.
1970 62nd Trip 1. Guidebook to Ozark carbonate terrane, Rolla-Devils Elbow area, Missouri
Trip 2. Guide to the southeast Missouri mineral dis trict

and

(87) ASSOCIATION OF MISSOURI GEOLOGISTS. GUIDEBOOK TO THE ANNUAL FIELD TRIP.
1972   19th Guidebook to the karst features and strati graphy of the Rolla area.

and

(380) MIDWEST GROUNDWATER CONFERENCE. GUIDEBOOK.
1960   5th   Guidebook to the geology of the Rolla area emphasizing solution phenomena. (57,397)

This index was compiled by John Black and Karen Grove, Department of Geology, Stanford University, Stanford, California.

STRATIGRAPHIC INDEX

PLAN OF THE STRATIGRAPHIC INDEX

What follows is an index to the stratigraphic (rock layer) units including members, formations, groups, and supergroups, for the guidebooks in this Union List. Indexing was done from titles of the guidebooks only. Each entry was verified and edited to conform to the form of entry as listed in the *GeoRef Thesaurus and Guide to Indexing*. 4th edition or, when not found there, the Lexicon of Geologic Names of the United States, USGS Bulletins 896, 1056A, 1200, 1350, or 1520. Some cross references are included to guide the reader.

This index was compiled by Dena F. Stepp, Columbus, Ohio, and Catherine E. Pasterczyk, Sandia National Laboratories, Albuquerque, New Mexico.

xii

DIRECTORY OF COUNTRIES

	Page
Bahamas	183
Barbados	183
Belize	183
Bermuda	183
Canada	183
Caribbean Islands	183
Cayman Islands	188
Cuba	188
Dominican Republic	188
Guatemala	188
Haiti	188
Jamaica	188
Mexico	188
Netherlands Antilles	190
Panama	190
Puerto Rico	190
United States	190
Virgin Islands	217

GUIDELINES FOR GEOLOGIC FIELD TRIP GUIDEBOOK AUTHORS/PUBLISHERS

Those who prepare guidebooks for field trips make a significant contribution to the geoscience community. The guidebooks for the field trips are recommended to colleagues and students by field trip participants. Librarians have found that guidebooks are difficult to acquire and once obtained may be difficult to catalog and identify. The searcher to whom someone has recommended a guidebook usually wants to find a guidebook associated with a particular meeting or area. Difficulties in locating guidebooks led the Geoscience Information Society to begin publication of the *Union List of Geologic Field Trip Guidebooks of North America*. Those who prepare guidebooks could contribute to the identification and control of the guidebook literature by applying the following guidelines:

Title page should include:

- Specific geographic area as part of a descriptive title, eg. county, state, or province.

- Clearly indicated subtitle.

- Name and place of meeting when the field trip is held in conjunction with a meeting. If it is an annual meeting, specify the number of the annual meeting and the number of the field trip.

- Day(s), month, and year that the field trip is conducted or the date of publication if guidebook is not compiled for a specific field trip.

- Name of the organization(s) sponsoring the field trip.

- Name and number of the consistently phrased publication series, when applicable.

- Name of field trip leader.

- Title on cover and title page should be identical.

- If reprinted, list the original publication series, guidebook number and year of publication

Verso of the title page:

- Name and address of the publisher.

- Name and address of the distributor.

- Price of the publication

General recommendations:

- Use good quality paper, printing, and binding (preferably not spiral binding.)

- Print more copies of the guidebook than are needed for the field trip participants.

- Send publication announcements containing all information that appears on the title page and its verso to *Geotimes* and *EPISODES*. If possible, send announcements to all libraries listed in the *Union List of Geologic Field Trip Guidebooks of North America* and to Geoscience Information Society members.

- Deposit a copy of the guidebook in the USGS Library in Reston, VA, and a copy in the nearest library listed in the Union List.

- Number pages consecutively.

- Identify all illustrations.

- List all unbound illustrative material in a table at the front of the guidebook and include a pocket to hold all these pieces in the back of the publication.

Prepared by the Ad Hoc Committee to Write Guidelines for Geologic Field Trip Guidebooks. Members: Mary Ansari (University of Nevada-Reno), Beatrice Lukens (University of California-Berkeley), Dorothy McGarry (University of California-Los Angeles, and Claren Kidd, Chair, (University of Oklahoma). 1983

PARTICIPATING LIBRARIES

By Alphabetical Arrangement

Library	Code
American Association of Petroleum Geologists, Energy Resources Library, Tulsa, OK	[OkTA]
Amherst College, Library, Amherst, MA	MA
Amoco Canada Petroleum Co. Ltd., Library/Information Centre, Calgary, AB	[CaACAM]
Arizona State University, D.E. Noble Science & Engineering, Tempe, AZ	AzTeS
Arkansas Geological Commission, Geological Library, Little Rock, AR	[Ar-GS]
Atlantic Richfield Company, Technical Library, Dallas, TX	TxDaAR-T
Ball State University, Braken Library, Muncie, IN	InMuB
Baylor University, F. Roemer Geological Library, Waco, TX	TxWB
Beloit College, Kohler Science Library, Beloit, WI	WBB
Boise State University, Library, Boise, ID	IdBB
Boston College, Weston Observatory, Catherine O'Connor Library, Weston, MA	MChB-WO
Boston College, Weston Observatory, Catherine B. O'Connor Library, Weston, MA	MWesB
Boston University, Science and Engineering Library, Boston, MA	MBU
British Museum (Natural History), Paleontology/Mineralogy Library, London	[Uk-Pal]
Bryn Mawr College, Geology Library, Bryn Mawr, PA	PBm
Bucknell University, Ellen Clarke Bertrand Library, Lewisburg, PA	PBU
Byrd Polar Research Center, Ohio State University, Columbus, OH	[OU (Gold)]
Calgary Public Library, Business & Science Dept., Calgary, AB	CaAC
California Division of Mines & Geology, Library, 367 Civic Drive, Suite 16, Pleasant Hill, CA	CSfCSM
California Institute of Technology, Geology and Planetary Sciences Library, Pasadena, CA	CPT
California State University, Chico, Meriam Library, Chico, CA	CChiS
Canada Institute for Scientific & Technical Information (CISTI), National Research Council Canada, Ottawa, ON	CaOON
Carleton University, Carleton University Library, Ottawa, ON	CaOOCC
Carnegie Library of Pittsburgh, Science and Technology Department, Pittsburgh, PA	PPi
Castleton State College, Library, Castleton, VT	VtCasT
Central Connecticut State University, Elihu Burrit Library, New Britain, CT	CtNbC
Chevron Oil Field Research Company, Technical Information Services, Library, La Habra, Ca	CLhC
Chevron USA, Inc., Northern Region Technical Library, Denver, CO	[CoDCh]
Cleveland Public Library, Science and Technology Dept., Cleveland, OH	OCl
Colorado Geological Survey, Library, Denver, CO	[Co-GS]
Colorado School of Mines, Arthur Lakes Library, Golden, CO	CoG
Colorado State University, Morgan Library, Fort Collins, CO	CoFS
Columbia University, Geology/Geoscience Library, New York, NY	NNC
Concordia University, Science and Engineering Library, Montreal, PQ	CaQMG
Cornell University, Engineering Library, Ithaca, NY	NIC
Dalhousie University, Macdonald Science Library, Halifax, NS	CaNSHD
Dartmouth College, Kresge Physical Sciences Library, Hanover, NH	NhD-K
DeGolyer and MacNaughton Library, 400 One Energy Square, Dallas, TX	TxDaDM
Denver Museum of Natural History, Denver, CO	[CoDMNH]
Denver Public Library, 1357 Broadway, Denver, CO	CoD
Detroit Public Library, 5201 Woodward Ave., Detroit, MI	MiD
Devereaux Library, South Dakota School of Mines & Technology, Rapid City, SD	SdRM
Earlham College, Wildman Science Library, Richmond, IN	InRE
Eastern Illinois University, Evangeline Booth Library, Charleston, IL	ICharE

Institution	Code
Eastern New Mexico University, Golden Library, Portales, NM	NmPE
Esso Resources Canada Ltd., Library Information Center, Calgary, AB	CaACERI
Federal Energy Regulatory Commission, Library, Room 8502, Washington, DC	DFPC
Field Museum of Natural History, Library, Chicago, IL	ICF
Florida Bureau of Geology, Florida Geological Survey, Tallahassee, FL	[F-GS]
Fort Lewis College, John F. Reed Library, Durango, CO	CoDuF
Geological Information Library of Dallas, Suite 100, One Energy Square, Dallas, TX	TxDaGI
Geological Society of the Oregon Country, Portland, OR	[OrPGS]
Geological Survey Alabama, P.O. Box O, Tuscaloosa, AL	[ATuGS]
Geological Survey of Canada, Library, Room 350, Ottawa, ON	CaOOG
Geology Library, Virginia Polytechnic Institute & State University, Blacksburg, VA	ViBlbV
Harold B. Lee Library, Brigham Young University, Provo, UT	UPB
Harvard University, Bernhard Hummel Library of the Geological Sciences, Cambridge, MA	MH-GS
Hofstra University, Axinn Library, Hempstead, NY	NIH
I.D. Weeks Library, University of South Dakota, Vermillion, SD	SdU
Idaho State University, Eli M. Oboler Library, Pocatello, ID	IdPI
Illinois State Geological Survey, Library, Champaign, IL	I-GS
Illinois State Geological Survey, Library, Champaign, IL	IChamIG
Illinois State University, Milner Library - Map Room, Normal, IL	INS
Indiana State University, Science Library, Terre Haute, IN	InTI
Indiana University, Geology Library, Bloomington, IN	InU
Iowa State University, Library, Ames, IA	IaAS
John Hopkins University, Dept. of Earth & Planetary Sciences, Baltimore, MD	MdBJ
Kent State University, Library, Kent, OH	OKentC
Lakehead University, The Chancellor Paterson Library, Thunder Bay, ON	CaOPAL
Lamar University at Beaumont, Mary and John Gray Library, Beaumont, TX	TxBeaL
Lehigh University, E.W. Fairchild/Martindale Library, Bethelehem, PA	PBL
Lindgren Library, Massachusetts Institute of Technology, Cambridge, MA	MCM
Lock Haven University of Pennsylvania, Stevenson Library, Lock Haven, PA	PLhS
Louisiana State University, Troy H. Middleton Library, Baton Rouge, LA	LU
Louisiana Tech University, Prescott Memorial Library, Ruston, LA	LRuL
Maine Geological Survey, Library, Agusta, ME	[ME-GS]
Marathon Oil Company, Littleton, CO	CoLiM
McGill University, Physical Sciences & Engineering Library, Montreal, PQ	CaQMME
McMaster University, Thode Library of Science & Engineering, Hamilton, ON	CaOHM
Memorial University of Newfoundland, Queen Elizabeth II Library, St. John's, NF	CaNfSM
Memphis State University, Engineering Library, Memphis, TN	TMM-E
Michigan Technological University, Library—Interlibrary Loan, Houghton, MI	MiHM
Midland Country Public Library, Petroleum Department, Midland, TX	TxMM
Ministere de L'Energie et des Ressources du Quebec, Centre de Documentation-Mines, Quebec City, PQ	CaQQERM
Mississippi Bureau of Geology, Library, Jackson, MS	MsJG
Mobil Oil Canada, Ltd., Library, Room 1420 T.1, Calgary, AB	CaACM
Mobil Producing Texas & New Mexico, Information Resource Center, Houston, TX	TxHMP
Mobil Research & Development Corp., Dallas Research Laboratory, Dallas, TX	TxDaSM
Montana College of Mineral Science and Technology, Library, Butte, MT	MtBuM
Montana State University, Roland R. Renne Library, Bozeman, MT	MtBC
NM Institute of Mining and Technology, Spears Memorial Library, Socorro, NM	NMSoI
National Museums of Canada, Branch Library, Ottawa, ON	CaOONM
New Mexico Institute of Mining and Technology, Martin Speare Memorial Library, Socorro, NM	NmSoI

Institution	Code
New Mexico State University, University Library, Las Cruces, NM	NmLcU
New York Public Library, Research Library, Science/Technology Division, New York, NY	NN
North Dakota State University, North Dakota State University Library, Fargo, ND	NdFA
Northern Arizona Univ. Lib., Flagstaff, AZ	AzFU
Northern College of Applied Arts & Technology, Haileybury School of Mines, Haileybury, ON	CaOHAINC
Northern Michigan University, Olson Library, Marquette, MI	MiMarqN
Northwestern University, Geology Library, Evanston, IL	IEN
Oklahoma City Geological Society, Inc., Library, Oklahoma City, OK	OkOkCGS
Oklahoma City University, Dulaney-Browne Library, Oklahoma City, OK	OkOkU
Oklahoma State University, Edmon Low Library, Stillwater, OK	OkS
Ontario Geological Survey, Ontario Ministry of Natural Resources, Toronto, ON	CaOTDM
Oregon Department of Geology and Mineral Industries, Library, Portland, OR	[Or-G]
Oregon State University, William Jasper Kerr Library, Corvallis, OR	OrCS
Pennsylvania Dept. of Environmental Resources, Bureau of Topographic and Geologic Survey Library, Harrisburg, PA	PHarER-T
Princeton University, Geology Library, Princeton, NJ	NjP
Purdue University, Purdue Earth and Atmospheric Sciences Library, West Lafayette, IN	InLP
Queen's University, Geology Library, Kingston, ON	CaOKQGS
Reference Library, Milwaukee Public Museum, Milwaukee, WI	WMMus
Royal Ontario Museum, Main Library, Toronto, ON	CaOTRM
SOESSO Resources Canada Limited, Calgary, AB	CaACI
SUNY at Stony Brook, Earth and Space Sciences Library, Stony Brook, NY	NSbSU
San Diego Natural History Museum, Scientific Library, San Diego, CA	SDSNH
San Diego State University, Library, Science Department, San Diego, CA	CSdS
Schlumberger Well Services, Houston Downhole Sensors - Library, Houston, TX	TxHSW
Science Bibliographer, University of California, Santa Cruz, Santa Cruz, CA	CU-SC
Sciences Library, Brown University, Providence, RI	RPB
Shell Development Company, Bellaire Research Center Library, Houston, TX	TxHSD
Smith College, Science Library, Northampton, MS	MNS
South Carolina Geological Survey, Harbison Forest Road, Columbia, SC	[SC-GS]
Southern Illinois University at Carbondale, Science Division, Morris Library, Carbondale, IL	ICarbS
Southern Methodist University, Science/Engineering Library, Dallas, Tx	TxDaM-SE
St. Cloud State University, Lerning Resource Center, St. Cloud, MN	MnStclS
Stanford University, Branner Earth Sciences Library, Stanford, CA	CSt-ES
State University College, James M. Milne Library, Oneonta, NY	NOneoU
State University College of Arts and Sciences, Geneseo, Milne Library, ILL, Geneseo, NY	NGenoU
State University of New York, Sojourner Truth Library, New Paltz, NY	NNepaSU
State University of New York at Albany, University Libraries, Albany, NY	NAlU
State University of New York at Binghamton, Binghamton, NY	NBiSU
State University of New York at Fredonia, Daniel A. Reed Library, Fredonia, NY	NFredU
Subject Cataloging Division, Library of Congress, Washington, DC	DLC
Syracuse University, Geology Library, Syracuse, NY	NSyU
Texaco Canada Resources, Library, Calgary, AB	ACTCR
Texas Tech University, Library, Lubbock, TX	TxLT
The Ohio State University, Columbus, Orton Memorial Library of Geology, Columbus, OH	OU
The Pennsylvania State University, Earth & Mineral Sciences Library, University Park, PA	PSt
The Science Museum of Minnesota, Louis S. Headley Memorial Library, St. Paul, MN	MnSSM
The University of Texas at Austin, The Joseph C. Walter, Jr. and Elizabeth C. Walter Geology Library, Austin, TX	TxU
Tulsa City County Library, Business & Technology Dept., Tulsa, OK	OkT
Tyrrell Museum of Palaeontology, Library, Drumheller, AB	CaADTMP
U.S. Bureau of Mines, Henderson Memorial Library, Denver, CO	CoDBM

Institution	Code
U.S. Geological Survey, Library (Western Region), Menlo Park, CA	CMenUG
U.S. Geological Survey, Denver Branch Library, Denver, CO	CoDGS
U.S. Geological Survey, Library, Reston, VA	DI-GS
Universite Laval, Bibliotheque Scientifique, Quebec City, PQ	CaQQLaS
Universite de Montreal, Geology Library, Montreal, PQ	CaQMUGL
University of Alberta, Science & Technology Library, Edmonton, AB	CaAEU
University of Arizona, Science-Engineering Library, Tucson, AZ	AzU
University of British Columbia, Main Library, Vancouver, BA	CaBVaU
University of Calgary, Gallagher Library of Geology and Geophysics, Calgary, AB	CaACU
University of California, Berkeley, Earth Sciences Library, Berkeley, CA	CU-EART
University of California, Davis, Physical Sciences Library, Davis, CA	CU-A
University of California, Davis, Physical Sciences Library, Davis, CA	Cu-A
University of California, Los Angeles, Geology-Geophysics Library, Los Angeles, Ca	CLU-G/G
University of California, Riverside, Physical Sciences Library, Riverside, CA	CU-RivP
University of California, Santa Barbara, Sciences-Engineering Library, Santa Barbara, CA	CU-SB
University of Cincinnati, Geology/Physics Library, Cincinnati, OH	OCU-Geo
University of Colorado, Earth Science Library, Boulder, CO	CoU
University of Colorado at Denver, Auraria Libraries, Denver, CO	CoU-DA
University of Denver, Denver, CO	[CoDU]
University of Georgia, Science Library, Athens, GA	GU
University of Hawaii, Manoa, Hawaii Institute of Geophysics, Honolulu, HI	[UH-M]
University of Houston, University Park, M.D. Anderson Library, Houston, TX	TxHU
University of Idaho, Science Library, Moscow, ID	IdU
University of Illinois at Chicago, Science Library, Chicago, IL	ICIU-S
University of Illinois at Urbana-Champaign, Geology Library, Urbana, IL	IU
University of Iowa, G.F. Kay Memorial Library, Iowa City, IA	IaU
University of Kansas, Kansas Geological Survey Library, Lawrence, KS	KLG
University of Kentucky, Pirtle Geology Library, Lexington, KY	KyU-Ge
University of Maryland, College Park, Engineering and Physical Sciences, College Park, MD	MdU
University of Massachusetts, Amherst, Biological Sciences Library, Amherst, MA	MU
University of Michigan, Natural Science Library, Ann Arbor, MI	MiU
University of Minnesota, Science & Engineering Library, Minneapolis, MN	MnU
University of Minnesota, Duluth, Geology Library, Duluth, MN	MnDuU
University of Missouri, Columbia, Geology Library, Columbia, MO	MoU
University of Missouri, Rolla, Curtis Laws Wilson Library, Rolla, MD	MoRM
University of Montana, Mansfield Library, Missoula, MT	MtU
University of Nebraska, Geology Library, Lincoln, NE	NbU
University of Nevada, Las Vegas, James R. Dickinson Library, Las Vegas, NV	NvLN
University of Nevada, Reno, Mines Library, Reno, NV	NvU
University of New Brunswick, Science Library, Fredericton, AB	CaNBFU
University of New Mexico, Zimmerman Library, Albuquerque, NM	NmU
University of New Orleans, Earl K. Long Library, New Orleans, LA	LNU
University of North Carolina at Chapel Hill, Geology Library, Chapel Hill, NC	NcU
University of North Dakota, Geology Branch Library, Grand Forks, ND	NdU
University of Northern Colorado, James A. Michener Library, Greeley, CO	CoGrU
University of Northern Iowa, Cedar Falls, IA	[IaCfT]
University of Oklahoma, Geology Library, Norman, OK	OkU
University of Oregon, Science Library, Eugene, OR	OrU-S
University of Rochester, Geology/Map Library, Rochester, NY	NRU
University of South Carolina, Thomas Cooper Library, Columbia, SC	ScU

University of Southern Colorado, USC Library, Pueblo, CO	CoPs
University of Tennessee, Knoxville, James D. Hoskins Library, Knoxcille, TN	TU
University of Texas at Austin, Bureau of Economic Geology, Austin, TX	[BEG]
University of Texas at Dallas, Eugene McDermott Library, Richardson, TX	TxU-Da
University of Toronto, Geology Library, Toronto, ON	CaOTUG
University of Utah, Science and Engineering Library, Salt Lake City, UT	UU
University of Washington, Natural Sciences Library, Seattle, WA	WaU
University of Waterloo, Engineering Mathematics & Science (EMS) Library, Waterloo, ON	CaOWtU
University of Western Ontario, Sciences Library, London, ON	CaOLU
University of Wisconsin, Madison, Geology-Geophysics Library, Madison, WI	WU
University of Wisconsin, Milwaukee, American Geographical Society Collection, Milwaukee, WI	WMUW
University of Wisconsin, Parkside, Wyllie Library/Learning Center, Kenosha, WI	WKenU
University of Wisconsin, Platteville, Karrmann Library, Platterville, WI	WPlaU
University of Wisconsin-Green Bay, University of Wisconsin-Green Bay, Library, Green Bay, WI	WGrU
University of Wyoming, Geology Library, Laramie, WY	WyU
Virginia Division of Mineral Resources, Library, Charlottesville, VA	[Vi-MR]
Washington Department of Natural Resources, Division of Geology and Earth Resources, Olympia, WA	[Wa-NR]
Washington State University, Owen Science & Engineering Library, Pullman, WA	WaPS
Washington University, Earth and Planetary Sciences Library, St. Louis, MO	MoSW
Wayne State University, Science & Engineering Library, Detroit, MI	MiDW-S
West Texas State University, Cornette Library, Canyon, TX	TxCaW
Western Michigan University, Physical Sciences Library, Kalamazoo, MI	MiKW
Weston Geophysical Corp., Library, Westboro, MA	[MWWg]
Wright State University, Library, Dayton, OH	ODaWU
Yale University, Kline Science Library, New Haven, CT	CtY-KS

DIRECTORY OF LIBRARY SERVICES

By Symbol Arrangement

ACTCR **Texaco Canada Resources**
Library
P.O. Box 3333, Station M
Calgary, AB T2P 2P8
403-267-0681
Lends: Yes
Copying services: No charge to maximum of 20 pages.
Use of materials on premises: No

ATuGS **Geological Survey Alabama**
P.O. Box O
University Stn.
Tuscaloosa, AL 35486
205-349-2852 EXT. 215/16
Lends: No
Copying services: Yes, but limited.
Use of materials on premises: Yes

Ar-GS **Arkansas Geological Commission**
Geological Library
3815 West Roosevelt Road
Little Rock, AR 72204
501-371-1488 ext. 23
Lends: Yes
Copying services: Xerox $0.10 per copy plus postage.
Use of materials on premises: Yes

AzFU **Northern Arizona Univ. Lib.**
C.U. Box 6022
Flagstaff, AZ 86011
602-523-2171
Lends: Yes
Copying services: Full range of duplication services, including self-service copiers, reduction, enlargement, and transparencies.
Use of materials on premises: Yes

AzTeS **Arizona State University**
D.E. Noble Science & Engineering
Library
Tempe, AZ 85287
602-965-7607
Lends: Yes (not all titles)
Copying services: Photocopies ($0.10 per page).
Use of materials on premises: Yes

AzU **University of Arizona**
Science-Engineering Library
Tucson, AZ 85721
602-621-6394
Lends: Yes
Copying services: Yes
Use of materials on premises: Yes

BEG **University of Texas at Austin**
Bureau of Economic Geology
Reading Room/Data Center
University Station Box X
Austin, TX 78713-7508
512-471-1534 ext. 142
Lends: No
Copying services: Limited
Use of materials on premises: Yes

CChiS **California State University, Chico**
Meriam Library
Chico, CA 95929
916-895-6479
Lends: Yes
Copying services: 1-10 pages - $2.00 plus $0.10 per page 10-20, subject to change.
Use of materials on premises: Yes

CLU-G/G **University of California, Los Angeles**
Geology-Geophysics Library
4697 Geology Bldg.
Los Angeles, Ca 90024
213-825-1055
Lends: Yes
Copying services: Two photocopy machines, 8 1/2" x 14" sheet; one microfiche reader/printer.
Use of materials on premises: Yes

CLhC **Chevron Oil Field Research Company**
Technical Information Services, Library
3282 Beach Boulevard
La Habra, Ca 90631
213-694-7500
Lends: Yes
Copying services: Photocopy, limited to 10 pages, no charge.
Use of materials on premises: Yes

CMenUG **U.S. Geological Survey**
Library (Western Region)
345 Middlefield Road MS 955
Menlo Park, CA 94025

CPT **California Institute of Technology**
Geology and Planetary Sciences Library
210 North Mudd
Caltech
Pasadena, CA 91125

CSdS   **San Diego State University**
Library, Science Department
San Diego, CA 92182-0511
619-265-6715
Lends: No
Copying services: Photocopying available through Interlibrary Loan.
Use of materials on premises: Yes

CSfCSM   **California Division of Mines & Geology, Library**
367 Civic Drive, Suite 16
Pleasant Hill, CA 94523-1997
415-671-4941
Lends: Yes
Copying services: Free up to 20 pages.
Use of materials on premises: Yes

CSt-ES   **Stanford University**
Branner Earth Sciences Library
Mitchell Bldg.
Stanford, CA 94305-2210
415-723-2746
Lends: Yes, through RLG ILL service
Copying services: Machines on premises for patron use: photocopy, fiche-to-fiche and fiche-to-paper.
Use of materials on premises: Yes

CU-A   **University of California, Davis**
Physical Sciences Library
Davis, CA 95616
916-752-0459
Lends: Yes, via ILL
Copying services: Yes, 8 1/2" x 11"; $0.10 per page.
Use of materials on premises: Yes

CU-EART   **University of California, Berkeley**
Earth Sciences Library
230 Earth Science Bldg.
Dept. of Geology & Geophysics
Berkeley, CA 94720
415-642-2997
Lends: Yes, through Ill Dept., General Library
Copying services: Information availabe from: Library Photographic Service, General Library, Univ. of Calif., Berkely, CA 94720
Use of materials on premises: Yes

CU-RivP   **University of California, Riverside**
Physical Sciences Library
Riverside, CA 92517
714-787-3511

CU-SB   **University of California, Santa Barbara**
Sciences-Engineering Library
Santa Barbara, CA 93106

CU-SC   **Science Bibliographer**
University of Califomia, Santa Cruz
University Library
Santa Cruz, CA 95064

CaAC   **Calgary Public Library**
Business & Science Dept.
616 MacLeod Trail S.E.
Calgary, AB T2G 2M2
403-266-4606
Lends: Yes. Circulating materials only
Copying services: $0.30 per page, $3.00 minimum charge per request
Use of materials on premises: Yes

CaACAM   **Amoco Canada Petroleum Co. Ltd.**
Library/Information Centre
Room 1012
444 7th Avenue S.W.
Calgary, AB T2P OY2
403-233-1451
Lends: Yes
Copying services: Yes
Use of materials on premises: Yes

CaACERI   **Esso Resources Canada Ltd.**
Library Information Center
237 Fourth Avenue SW
Calgary, AB T2P OH6
403-233-1451
Lends: Yes, ILL only.
Copying services: Yes, limited.
Use of materials on premises: Yes, with special permission only.

CaACI   **SOESSO Resources Canada Limited**
Calgary, AB

CaACM   **Mobil Oil Canada, Ltd.**
Library, Room 1420 T.1
P.O. Box 800
Calgary, AB T2P 2J7
403-268-7785
Lends: Yes
Copying services: Yes
Use of materials on premises: Yes

CaACU   **University of Calgary**
Gallagher Library of Geology and Geophysics
180 Earth Sciences Bldg.
Calgary, AB T2N 1N4
403-220-6043
Lends: Yes
Copying services: Self-service $0.10 per page. ILL $0.25 per page plus $5.00 service charge.
Use of materials on premises: Yes

CaADTMP   **Tyrrell Museum of Palaeontology**
Library
Box 7500
Drumheller, AB T0J 0Y0
403-823-7707
Lends: Yes, via ILL
Copying services: Yes
Use of materials on premises: Yes

CaAEU University of Alberta
Science & Technology Library
2-10 Cameron Library
University Library
Edmonton, AB   T6G 2J8
403-432-2728
Lends: Yes
Copying services: $0.25 per exposure plus $5.00 service fee.
Use of materials on premises: Yes

CaBVaU University of British Columbia
Main Library
1956 Main Mall
Vancouver, BA   V6T 1Y3
604-228-3115 (CIRCULATION) 604-228-4430 (ILL)
Lends: Yes ($11.00 ILL charge)
Copying services: Photocopy for a fee, $11.00 minimum.
Use of materials on premises: Yes

CaNBFU University of New Brunswick
Science Library
Box 7500
Fredericton, AB   E3B 5H5
506-453-4601 or 4602
Lends: Yes
Copying services: Photocopy
Use of materials on premises: Yes

CaNSHD Dalhousie University
Macdonald Science Library
Halifax, NS   B3H 4J3
902-424-2059
Lends: Yes
Copying services: $0.10 per page on site, ILL charges: $3.00 minimum charge; 10 pages supplied; $0.30 per page thereafter.
Use of materials on premises: Yes

CaNfSM Memorial University of Newfoundland
Queen Elizabeth II Library
St. John's, NF   A1B 3Y1
709-737-7424
Lends: Yes
Copying services: Yes
Use of materials on premises: Yes

CaOHAINC Northern College of Applied Arts & Technology
Haileybury School of Mines
Library
P.O. Box 'A'
Haileybury, ON   P0J 1K0
705-672-3376, ext. 56
Lends: No
Copying services: Photocopies available
Use of materials on premises: Yes

CaOHM McMaster University
Thode Library of Science & Engineering
1280 Main Street West
Hamilton, ON   L8S 4P5
416-525-9140, ext. 4252
Lends: Yes
Copying services: Interlibrary Loan, $2.00 minimum, $0.20 per page.
Use of materials on premises: Yes

CaOKQGS Queen's University
Geology Library
Bruce Wing
Kingston, ON   K7L 3N6
613-545-2840
Lends: Yes
Copying services: Through ILL photocopying service, Douglas Library.
Use of materials on premises: Yes

CaOLU University of Western Ontario
Sciences Library
Natural Sciences Centre
London, ON   N6A 5B7
519-679-6174/6175
Lends: Yes
Copying services: Photocopying for Interlibrary Loans; coin-operated photocopiers for users.
Use of materials on premises: Yes

CaOOCC Carleton University
Carleton University Library
Colonel By Drive
Ottawa, ON   K1S 5J7
613-231-2683
Lends: Yes, via Interlibrary Loan
Copying services: Yes
Use of materials on premises: Yes

CaOOG Geological Survey of Canada
Library, Room 350
601 Booth Street
Ottawa, ON   K1A 0E8
613-995-4151/4142
Lends: Yes
Copying services: $0.10 per page; limit of 50 pages copied.
Use of materials on premises: Yes

CaOON Canada Institute for Scientific & Technical Information (CISTI)
National Research Council Canada
Bldg. M-55, Montreal Road
Ottawa, ON   K1A 0S2
613-993-1585
Lends: No
Copying services: One copy of an article from a periodical or a copy of a conference paper. Charges: $6.80, 10 pages and under; $0.68 per page for longer items.
Use of materials on premises: Yes

CaOONM National Museums of Canada
Branch Library
2086 Walkley Road
Ottawa, ON K1A 0M8
613-998-3923
Lends: Yes
Copying services: Photocopying services for staff and ILL requests.
Use of materials on premises: Yes

CaOPAL Lakehead University
The Chancellor Paterson Library
Oliver Road
Thunder Bay, ON P7B 5E1
807-345-2121, ext. 205
Lends: Yes
Copying services: $0.20 per page.
Use of materials on premises: Yes

CaOTDM Ontario Geological Survey
Ontario Ministry of Natural Resources
Mines Library
Boom 812-77 Grenville Street
Toronto, ON M5S 1B3
416-965-1352
Lends: Yes, through ILL
Copying services: Yes
Use of materials on premises: Yes

CaOTRM Royal Ontario Museum
Main Library
100 Queen's Park
Toronto, ON M5S 2C6
416-586-5595
Lends: No (except ILL)
Copying services: Photocopier in reading room.
Use of materials on premises: Yes

CaOTUG University of Toronto
Geology Library
Mining Bldg. Rm. 316
170 College St.
Toronto, ON M5S 1A1
416-978-3024
Lends: To holders of U of T library cards.
Copying services: Self-service.
Use of materials on premises: Yes

CaOWtU University of Waterloo
Engineering Mathematics & Science (EMS) Library
Waterloo, ON N2L 3G1
519-885-1211
Lends: Yes, ILL charges: $8.00 per request.
Copying services: Photocopies: $8.00 per request filled up to 30 exposures, after which a charge of $0.20 per exposure will apply.
Use of materials on premises: Yes

CaQMG Concordia University
Science and Engineering Library
1455 de Maisonneuve St. W.
Montreal, PQ H3G 1M8
514-848-7723
Lends: Yes
Copying services: Yes
Use of materials on premises: Yes

CaQMME McGill University
Physical Sciences & Engineering Library
809 Sherbrook Street West
Montreal, PQ H3A 2K6
514-392-5914
Lends: Yes, via interlibrary loans.
Copying services: Photocopy rates: non-profit institutions, $0.34 per page; profit-making organizations, $0.43 per page, $4.30 per request.
Use of materials on premises: Yes

CaQMUGL Universite de Montreal
Geology Library
P.O. Box 6128
2900 Blvd. Edouard-Montpetit
Montreal, PQ H3C 3J7
514-343-6831
Lends: Yes
Copying services: Photocopy for a fee.
Use of materials on premises: Yes

CaQQERM Ministere de L'Energie et des Ressources du Quebec
Centre de Documentation-Mines
1620 Boul. de L'Entente
Local I-16
Quebec City, PQ G1S 4NG
418-643-4624
Lends: Yes
Copying services: Yes
Use of materials on premises: Yes

CaQQLaS Universite Laval
Bibliotheque Scientifique
Quebec City, PQ G1K 7P4
418-656-3967
Lends: Yes
Copying services: For ILL, $0.15 per page (minimum $5.00).
Use of materials on premises: Yes

Co-GS Colorado Geological Survey
Library
1313 Sherman Street
Room 715
Denver, CO 80203
303-866-2611
Lends: No
Copying services: Yes.
Use of materials on premises: Yes

CoD **Denver Public Library**
1357 Broadway
Denver, CO 80203
303-571-2122
Lends: No
Copying services: Yes
Use of materials on premises: Yes

CoDBM **U.S. Bureau of Mines**
Henderson Memorial Library
Bldg. 20, Federal Center
Denver, CO 80255
303-236-0474
Lends: Yes
Copying services: Up to 20 pages, no charge.
Use of materials on premises: Yes

CoDCh **Chevron USA, Inc.**
Northern Region Technical Library
700S. Colorado Blvd. Rm. 845
Denver, CO 80222
303-691-7577
Lends: Yes
Copying services: For ILL requests.
Use of materials on premises: With prior arrangement.

CoDGS **U.S. Geological Survey**
Denver Branch Library
Box 25046, Mail Stop 914
Denver Federal Center
Denver, CO 80225
303-236-1000
Lends: Yes
Copying services: Library will copy up to 15 pages. Will loan if over 15 pages.
Use of materials on premises: Yes

[CoDMNH] **Denver Museum of Natural History**
City Park
Denver, CO 80205
303-370-6347

[CoDU] **University of Denver**
Denver, CO 80210

CoDuF **Fort Lewis College**
John F. Reed Library
Durango, CO 81301
303-247-7738
Lends: Yes
Copying services: Photocopy; $0.10 per page.
Use of materials on premises: Yes

CoFS **Colorado State University**
Morgan Library
Fort Collins, CO 80523
303-491-5911
Lends: Yes
Copying services: Standard Interlibrary Loan.
Use of materials on premises: Yes

CoG **Colorado School of Mines**
Arthur Lakes Library
14th & Illinois
Golden, CO 80401
303-273-3687
Lends: Yes
Copying services: Through ILL.
Use of materials on premises: Yes

CoGrU **University of Northern Colorado**
James A. Michener Library
Greeley, CO 80639
303-351-2562

CoLiM **Marathon Oil Company**
P.O. Box 269
Littleton, CO 80160-0269
303-794-2601

CoPs **University of Southern Colorado**
USC Library
Pueblo, CO 81001
303-549-2451
Lends: Yes
Copying services: $0.10 per page.
Use of materials on premises: Yes

CoU **University of Colorado**
Earth Science Library
University Libraries
Campus Box 184
Boulder, CO 80309
303-482-6133
Lends: Yes
Copying services: Campus Coin Copies. Copy machine also on premises.
Use of materials on premises: Yes

CoU-DA **University of Colorado at Denver**
Auraria Libraries
Denver, CO 80204

CtNbC **Central Connecticut State University**
Elihu Burrit Library
1615 Stanley Street
New Britain, CT 06050
203-827-7528
Lends: Yes
Copying services: Through ILL: $3.00 minimum; $0.20 per page.
Use of materials on premises: Yes

CtY-KS **Yale University**
Kline Science Library
210 Whitney Ave.
P.O. Box 6666
New Haven, CT 06511-8130
203-432-3157
Lends: No
Copying services: Yes, $0.10 per page.
Use of materials on premises: Yes

Cu-A	**University of California, Davis** Physical Sciences Library Davis, CA 95616 916-752-0459 *Lends:* Yes *Copying services:* Yes, 8 1/2" x 11" *Use of materials on premises:* Yes
DFPC	**Federal Energy Regulatory Commission** Library, Room 8502 825 North Capitol Street Washington, DC 20426 202-357-5479 *Lends:* Yes *Copying services:* $0.10 per page; ILL form needed. *Use of materials on premises:* Yes
DI-GS	**U.S. Geological Survey** Library National Center, Mail Stop 950 12201 Sunrise Valley Drive Reston, VA 22092 703-860-6671; FTS 928-6671 *Lends:* Yes *Copying services:* Free photocopy up to 20 pages. *Use of materials on premises:* Yes
DLC	**Subject Cataloging Division** Library of Congress 10 First Street SE Washington, DC 20540 202-287-5000; ext. 75639 (Science Reference Desk), ext. 75450 (Interlibrary Loan) *Lends:* Yes *Copying services:* Yes *Use of materials on premises:* Yes
F-GS	**Florida Bureau of Geology** Florida Geological Survey Library 903 W. Tennessee Street Tallahassee, FL 32301 904-488-9380 *Lends:* Yes *Copying services:* In-house, $0.20 per copy; also through prepaid written request. *Use of materials on premises:* Yes
GU	**University of Georgia** Science Library Athens, GA 30602 404-542-4535 *Lends:* Yes *Copying services:* Yes, $0.10 per page. *Use of materials on premises:* Yes
I-GS	**Illinois State Geological Survey** Library 469 Natural Resources Building 615 E. Peabody Champaign, IL 61820 217-333-5110 *Lends:* Only thorugh Interlibrary Loan. *Copying services:* No. *Use of materials on premises:* Yes
ICF	**Field Museum of Natural History** Library Roosevelt Road at Lake Shore Dr. Chicago, IL 60605 312-922-9410, ext. 282 *Lends:* Yes *Copying services:* $0.15 per page if done in person in library; $0.30 per page for ILL requests, plus $2.00 service charge. *Use of materials on premises:* Yes
ICIU-S	**University of Illinois at Chicago** Science Library Box 7565 Chicago, IL 60680 312-996-5396 *Lends:* Yes, through ILL (not periodicals). *Copying services:* No, copies of articles through ILL. *Use of materials on premises:* Yes
ICarbS	**Southern Illinois University at Carbondale** Science Division, Morris Library Carbondale, IL 62901 618-453-2700 *Lends:* Yes *Copying services:* Self-service machines for in-house use and photocopying for external use when proper ILL form has been submitted. *Use of materials on premises:* Yes
IChamIG	**Illinois State Geological Survey** Library 469 Natural Resources Building 615 E. Peabody Champaign, IL 61820 217-333-5110 *Lends:* Only through Interlibrary Loan. *Copying services:* None *Use of materials on premises:* Yes
ICharE	**Eastern Illinois University** Evangeline Booth Library Charleston, IL 61920 217-581-6072 *Lends:* Yes *Copying services:* Charge is $0.05 per page unless ILL. *Use of materials on premises:* Yes
IEN	**Northwestern University** Geology Library Locy Hall Evanston, IL 60201 312-491-5525 *Lends:* Yes *Copying services:* Photocopy ($0.10 per page) *Use of materials on premises:* Yes
INS	**Illinois State University** Milner Library - Map Room Normal, IL 61761 309-438-3486 *Lends:* Yes *Copying services:* $0.05 per page. *Use of materials on premises:* Yes

IU	**University of Illinois at Urbana-Champaign** Geology Library 223 Natural History Bldg. 1301 W. Green Street Urbana, IL 61801 217-333-1266 *Lends:* Yes *Copying services:* Photocopies for a fee. *Use of materials on premises:* Yes	InTI	**Indiana State University** Science Library Terre Haute, IN 47809 812-237-2060 *Lends:* Yes *Copying services:* For a fee. *Use of materials on premises:* Yes
IaAS	**Iowa State University** Library Ames, IA 50011	InU	**Indiana University** Geology Library 1005 East 10th Street Bloomington, IN 47405 812-335-7170 *Lends:* Yes *Copying services:* Yes *Use of materials on premises:* Yes
IaCfT	**University of Northern Iowa** Cedar Falls, IA 50613	KLG	**University of Kansas** Kansas Geological Survey Library 1930 Constant Ave. Lawrence, KS 66046 913-864-3965 ext. 425 *Lends:* No *Copying services:* $0.10 per copy. *Use of materials on premises:* Yes
IaU	**University of Iowa** G.F. Kay Memorial Library Geology Library Iowa City, IA 52242		
IdBB	**Boise State University** Library 1910 University Dr. Boise, ID 83725	KyU-Ge	**University of Kentucky** Pirtle Geology Library 100 Bowman Hall Lexington, KY 40506
IdPI	**Idaho State University** Eli M. Oboler Library Box 8089 Pocatello, ID 83209-0009	LNU	**University of New Orleans** Earl K. Long Library Lakefront New Orleans, LA 70148 504-286-6546 *Lends:* Yes *Copying services:* Interlibrary Loan; $0.10 per page, $1.00 minimum; self-service copiers $0.10 per page. *Use of materials on premises:* Yes
IdU	**University of Idaho** Science Library Moscow, ID 83843		
InLP	**Purdue University** Purdue Earth and Atmospheric Sciences Library West Lafayette, IN 47907 317-494-3264 *Lends:* Yes *Copying services:* Through ILL; photocopies $0.05 per page. *Use of materials on premises:* Yes	LRuL	**Louisiana Tech University** Prescott Memorial Library P.O. Box 10408 Ruston, LA 71272 318-257-4357 *Lends:* Yes *Copying services:* Yes *Use of materials on premises:* Yes
InMuB	**Ball State University** Braken Library University Libraries Muncie, IN 47306 317-285-8033 *Lends:* Yes *Copying services:* Yes, through ILL or public photocopy machine. *Use of materials on premises:* Yes	LU	**Louisiana State University** Troy H. Middleton Library Baton Rouge, LA 70803
		MA	**Amherst College** Library Amherst, MA 01002 413-542-2372 *Lends:* Yes *Copying services:* In Main Library only. *Use of materials on premises:* Yes
InRE	**Earlham College** Wildman Science Library Richmond, IN 47374 317-962-6561 *Lends:* Yes *Copying services:* Yes *Use of materials on premises:* Yes		

MBU Boston University
Science and Engineering Library
38 Cummington Street
Boston, MA   02215
617-353-9475
Lends: Yes
Copying services: Coin-operated machine on premises. ILL through Mugar Memorial Library (771 Commonwealth Ave.).
Use of materials on premises: Yes

MCM Lindgren Library
Massachusetts Institute of Technology
Rm. 54-200
Cambridge, MA   02139
617-253-5679
Lends: Yes
Copying services: Yes. ILL Xerox requests should be sent to Microreproduction Laboratory, 14-0551.
Use of materials on premises: Yes

MChB-WO Boston College, Weston Observatory
Catherine O'Connor Library
Weston, MA   02193

ME-GS Maine Geological Survey
Library
State House Station #22
Agusta, ME   04333
207-289-2801
Lends: Yes
Copying services: $0.10 per page.
Use of materials on premises: Yes

MH-GS Harvard University
Bernhard Hummel Library of the Geological Sciences
24 Oxford Street
Cambridge, MA   02138
617-495-2029
Lends: Yes, some
Copying services: $15.00 per request, plus $0.10 per page.
Use of materials on premises: Yes

MNS Smith College
Science Library
Clark Science Center
Northampton, MS   01063
413-584-2700 ext. 2952
Lends: Yes
Copying services: No charge to Nelinet reciprocal agreement cosignees. Others: $4.00 for first 20 pages, plus $0.10 per page thereafter.
Use of materials on premises: Yes

MU University of Massachusetts, Amherst
Biological Sciences Library
214 Morrill Science Center
Amherst, MA   01003
413-545-2674
Lends: Yes
Copying services: Two photocopiers--cash and carry basis. Change provided.
Use of materials on premises: Yes

MWWg Weston Geophysical Corp.
Library
P.O. Box 550
Westboro, MA   01581
617-366-9191 ext. 241
Lends: Yes
Copying services: Photocopier and microfiche/film reader/printer.
Use of materials on premises: Yes

MWesB Boston College, Weston Observatory
Catherine B. O'Connor Library
Weston, MA   02193
617-899-0950 Ext. 8321
Lends: Yes
Copying services: Yes
Use of materials on premises: Yes

MdBJ John Hopkins University
Dept. of Earth & Planetary Sciences
Singewald Reading Room
Baltimore, MD   21218

MdU University of Maryland, College Park
Engineering and Physical Sciences Library
Campus Drive
College Park, MD   20742
301-454-3037
Lends: Yes
Copying services: Yes, $0.10 per page.
Use of materials on premises: Yes

MiD Detroit Public Library
5201 Woodward Ave.
Detroit, MI   48202
313-833-1450
Lends: Yes
Copying services: Photoduplication machines; overhead camera.
Use of materials on premises: Yes

MiDW-S Wayne State University
Science & Engineering Library
Detroit, MI   48202

MiHM Michigan Technological University
Library--Interlibrary Loan
Houghton, MI   49931
906-487-2507
Lends: Yes
Copying services: $0.10 per page, $5.00 minimum.
Use of materials on premises: Yes

MiKW Western Michigan University
Physical Sciences Library
Kalamazoo, MI   49008
616-383-4943
Lends: Yes
Copying services: $0.10 per page.
Use of materials on premises: Yes

MiMarqN	**Northern Michigan University** Olson Library Marquette, MI 49855-5376 906-227-2260 (Circulation/ILL); 906-227-2268 (Stephen H. Peters) *Lends:* Yes *Copying services:* ILL policies on OCLC, in house- $0.10 per page *Use of materials on premises:* Yes	MtBC	**Montana State University** Roland R. Renne Library Bozeman, MT 59717-0022
		MtBuM	**Montana College of Mineral Science and Technology** Library Butte, MT 59701
MiU	**University of Michigan** Natural Science Library 3140 Natural Science Bldg. Ann Arbor, MI 48109	MtU	**University of Montana** Mansfield Library Interlibrary Loans Missoula, MT 59812
MnDuU	**University of Minnesota, Duluth** Geology Library Room 224, Math-Geology Bldg. Duluth, MN 55812	NAlU	**State University of New York at Albany** University Libraries 1400 Washington Avenue Albany, NY 12222 518-442-3330 ext.3591 *Lends:* Yes *Copying services:* Self-service. *Use of materials on premises:* Yes
MnSSM	**The Science Museum of Minnesota** Louis S. Headley Memorial Library 30 East 10th Street St. Paul, MN 55101 612-221-9430 *Lends:* Under some circumstances *Copying services:* Yes *Use of materials on premises:* Yes	NBiSU	**State University of New York at Binghamton** Binghamton, NY 13901
		NFredU	**State University of New York at Fredonia** Daniel A. Reed Library Fredonia, NY 14063 716-673-3183 *Lends:* Yes *Copying services:* Yes *Use of materials on premises:* Yes
MnStclS	**St. Cloud State University** Leming Resource Center Centennial Hall St. Cloud, MN 56301 612-255-2084 *Lends:* Yes *Copying services:* No charge for reasonable requests. *Use of materials on premises:* Yes	NGenoU	**State University College of Arts and Sciences, Geneseo** Milne Library, ILL Geneseo, NY 14454 716-245-5595 *Lends:* Yes *Copying services:* Free up to 24 pages. *Use of materials on premises:* Yes
MnU	**University of Minnesota** Science & Engineering Library Walter Library Bldg. 117 Pleasant Street S.E. Minneapolis, MN 55455	NIC	**Cornell University** Engineering Library Carpenter Hall Ithaca, NY 14853 607-255-5936 *Lends:* Yes *Copying services:* Through ILL. *Use of materials on premises:* Yes
MoRM	**University of Missouri, Rolla** Curtis Laws Wilson Library Rolla, MD 65401		
MoSW	**Washington University** Earth and Planetary Sciences Library Room 214, Wilson Hall St. Louis, MO 63130	NIH	**Hofstra University** Axinn Library Hempstead, NY 11550 516-560-5940 *Lends:* Yes *Copying services:* Yes *Use of materials on premises:* Yes
MoU	**University of Missouri, Columbia** Geology Library Columbia, MO 65211		
MsJG	**Mississippi Bureau of Geology** Library 2525 North West Street P.O. Box 5348 Jackson, MS 39211 601-3546228 *Lends:* Yes *Copying services:* $0.10 per page. *Use of materials on premises:* Yes	NMSoI	**NM Institute of Mining and Technology** Spears Memorial Library Campus Station Socorro, NM 87801 505-835-5614

NN	**New York Public Library** Research Library, Science/Technology Division 5th Ave. & 42nd Street New York, NY 10018 212-930-0574/75 *Lends:* Yes *Copying services:* $0.25 per exposure *Use of materials on premises:* Yes	NcU	**University of North Carolina at Chapel Hill** Geology Library Mitchell Hall 029-A Chapel Hill, NC 27514 919-962-2386 *Lends:* Yes *Copying services:* Self-service copiers, $0.05 per page. *Use of materials on premises:* Yes
NNC	**Columbia University** Geology/Geoscience Library 601 Schermerhorn Bldg. New York, NY 10027 212-854-4713 *Lends:* Yes *Copying services:* Yes, $0.10 per page. *Use of materials on premises:* Yes	NdFA	**North Dakota State University** North Dakota State University Library P.O. Box 5599 State University Station Fargo, ND 58105 701-237-8876 *Lends:* Yes *Copying services:* Yes, fee for service. *Use of materials on premises:* Yes
NNepaSU	**State University of New York** Sojourner Truth Library College at New Paltz New Paltz, NY 12561 914-257-2209 *Lends:* Yes *Copying services:* $0.10 per page. *Use of materials on premises:* Yes	NdU	**University of North Dakota** Geology Branch Library Box 8068, University Station Grand Forks, ND 58202
NOneoU	**State University College** James M. Milne Library Oneonta, NY 13820-1383	NhD-K	**Dartmouth College** Kresge Physical Sciences Library Fairchild Building Hanover, NH 03755 603-646-3563 *Lends:* Yes *Copying services:* Photocopies $0.10 per page. Microform prints $0.10 per page. *Use of materials on premises:* Yes
NRU	**University of Rochester** Geology/Map Library Rochester, NY 14627 716-275-4487 *Lends:* Not journals or maps. *Copying services:* Yes *Use of materials on premises:* Yes	NjP	**Princeton University** Geology Library Guyot Hall Princeton, NJ 08544
NSbSU	**SUNY at Stony Brook** Earth and Space Sciences Library Stony Brook, NY 11794-2199 516-632-7146 *Lends:* Yes *Copying services:* Yes, $0.10 per page. *Use of materials on premises:* Yes	NmLcU	**New Mexico State University** University Library P.O. Box 3475 Las Cruces, NM 88003 505-646-2932 *Lends:* Yes *Copying services:* Self-service, $0.05 per page; Interlibrary Loan. *Use of materials on premises:* Yes
NSyU	**Syracuse University** Geology Library 300 Heroy Geology Laboratory Syracuse, NY 13244-1070 315-423-3337 *Lends:* Yes *Copying services:* Coin-operated copy machine. *Use of materials on premises:* Yes	NmPE	**Eastern New Mexico University** Golden Library Portales, NM 88130
NbU	**University of Nebraska** Geology Library 10 Bessy Hall Lincoln, NE 68588-0410	NmSoI	**New Mexico Institute of Mining and Technology** Martin Speare Memorial Library Capus Station Socorro, NM 87801 505-835-5614 *Lends:* Yes *Copying services:* Yes *Use of materials on premises:* Yes

NmU **University of New Mexico**
Zimmerman Library
Albuquerque, NM 87131
505-277-5761
Lends: Yes
Copying services: Yes
Use of materials on premises: Yes

NvLN **University of Nevada, Las Vegas**
James R. Dickinson Library
4505 Maryland Parkway
Las Vegas, NV 89154

NvU **University of Nevada, Reno**
Mines Library
Reno, NV 89557-0044

OCU-Geo **University of Cincinnati**
Geology/Physics Library
Braunstein Hall
Cincinnati, OH 45221
513-556-1324
Lends: Yes
Copying services: Via ILL Office, (480 Langsam Library). $0.10 per exposure; $0.05 per page on premises.
Use of materials on premises: Yes

OCl **Cleveland Public Library**
Science and Technology Dept.
325 Superior Avenue
Cleveland, OH 44114-1271
216-623-2932
Lends: Most of the material in this directory would not be loanable.
Copying services: Prepaid copying service available at 216-623-2901.
Use of materials on premises: Yes

ODaWU **Wright State University**
Library
Colonel Glenn Highway
Dayton, OH 45435
413-873-2925
Lends: Yes
Copying services: Yes
Use of materials on premises: Yes

OKentC **Kent State University**
Library
Rm. 406 McGilvrey Hall
Kent, OH 44242
216-672-2017
Lends: Yes
Copying services: Yes
Use of materials on premises: Yes

OU **The Ohio State University, Columbus**
Orton Memorial Library of Geology
155 South Oval Drive
Columbus, OH 43210
614-422-2428
Lends: Yes
Copying services: $8.00 minimum charge (1st 25 pages $8.00; $0.15 for each additional page). Microfiche: $8.00.
Use of materials on premises: Yes

OU (Gold) **Byrd Polar Research Center**
Ohio State University
Goldthwait Polar Library
125 South Oval Mall
Columbus, OH 43210-1308
614-292-6715
Lends: Yes
Copying services: Interlibrary Loans: $0.20 per page. Copiers available for in-house use: $0.05 per page.
Use of materials on premises: Yes

OkOkCGS **Oklahoma City Geological Society, Inc.**
Library
227-B Park Avenue
Oklahoma City, OK 73102
405-235-3648
Lends: To members only, this is a private library.
Copying services: For members only.
Use of materials on premises: Yes

OkOkU **Oklahoma City University**
Dulaney-Browne Library
2501 North Blackwelder
Oklahoma City, OK 73106
405-521-5072
Lends: Yes
Copying services: Yes
Use of materials on premises: Yes

OkS **Oklahoma State University**
Edmon Low Library
Stillwater, OK 74078
405-624-6314
Lends: Yes
Copying services: Yes, within copyright limitations.
Use of materials on premises: Yes

OkT **Tulsa City County Library**
Business & Technology Dept.
400 Civic Center
Tulsa, OK 74103
9188-592-7988
Lends: Yes
Copying services: Photocopy $0.10 per page.
Use of materials on premises: Yes

OkTA	**American Association of Petroleum Geologists** Energy Resources Library P. O. Box 979 1444 S. Boulder Tulsa, OK 74101 918-584-2555 Ext.220 *Lends:* No *Copying services:* $3.00 per item plus $0.15 per page. *Use of materials on premises:* Yes
OkU	**University of Oklahoma** Geology Library 830 Van Vleet Oval, Rm. 103 Norman, OK 73019
Or-G	**Oregon Department of Geology and Mineral Industries** Library 910 State Office Building 1400 S.W. Fifth Avenue Portland, OR 97201
OrCS	**Oregon State University** William Jasper Kerr Library Corvallis, OR 97331
OrPGS	**Geological Society of the Oregon Country** P.O. Box 8579 Portland, OR 97207
OrU-S	**University of Oregon** Science Library Eugene, OR 97403
PBL	**Lehigh University** E.W. Fairchild/Martindale Library Bldg 8A Bethelehem, PA 18015
PBU	**Bucknell University** Ellen Clarke Bertrand Library Lewisburg, PA 17837 717-524-3310 *Lends:* Yes *Copying services:* Yes, $0.10 per page. *Use of materials on premises:* Yes
PBm	**Bryn Mawr College** Geology Library Bryn Mawr, PA 19010 215-645-5118/5117 *Lends:* Limited *Copying services:* Yes, Through Interlibrary Loan. Cost is $0.10 per page. *Use of materials on premises:* Yes
PHarER-T	**Pennsylvania Dept. of Environmental Resources** Bureau of Topographic and Geologic Survey Library P.O. Box 2357 Harrisburg, PA 17120 717-787-2169 *Lends:* Yes, Through Interlibrary Loan. *Copying services:* Xerox and microfilm. *Use of materials on premises:* Yes
PLhS	**Lock Haven University of Pennsylvania** Stevenson Library Lock Haven, PA 17745
PPi	**Carnegie Library of Pittsburgh** Science and Technology Department 4400 Forbes Avenue Pittsburgh, PA 15213 412-622-3138 *Lends:* No *Copying services:* Xerox copies: $0.50 per page; $3.00 minimum charge, plus tax and postage. *Use of materials on premises:* Yes
PSt	**The Pennsylvania State University** Earth & Mineral Sciences Library 105 Deike Bldg. University Park, PA 16802 814-865-9517 *Lends:* Yes *Copying services:* Yes *Use of materials on premises:* Yes
RPB	**Sciences Library** Brown University Box I Providence, RI 02912
SC-GS	**South Carolina Geological Survey** Harbison Forest Road Columbia, SC 29210 803-737-9440 *Lends:* No *Copying services:* Yes *Use of materials on premises:* Yes
SDSNH	**San Diego Natural History Museum** Scientific Library P.O. Box 1390 San Diego, CA 92112
ScU	**University of South Carolina** Thomas Cooper Library Columbia, SC 29208-0103 803-777-3151 *Lends:* Only to members of U.S.C. community. *Copying services:* Self-service copiers; $0.05 per page. *Use of materials on premises:* Yes
SdRM	**Devereaux Library** South Dakota School of Mines & Technology 500 E. Saint Joseph St. Rapid City, SD 57701-3995
SdU	**I.D. Weeks Library** University of South Dakota Vermillion, SD 57069
TMM-E	**Memphis State University** Engineering Library Memphis, TN 38152 901-454-2179 *Lends:* Yes *Copying services:* $0.10 per page. *Use of materials on premises:* Yes

TU   University of Tennessee, Knoxville
           James D. Hoskins Library
           Interlibrary Loan Dept.
           Knoxcille, TN 37996-1000

TxBeaL   **Lamar University at Beaumont**
           Mary and John Gray Library
           Box 10021, Lamar Station
           Beaumont, TX 77710

TxCaW   **West Texas State University**
           Cornette Library
           P.O. Box 748, WT Station
           Canyon, TX 79016
           806-656-2761
           Lends: Yes
           Copying services: Photocopies; self-service, $0.10 per
               page; library service to individuals, $0.20 per page;
               library service to businesses, $0.25 per page.
           Use of materials on premises: Yes

TxDaAR-T   **Atlantic Richfield Company**
           Technical Library
           Dallas, TX 75221

TxDaDM   **DeGolyer and MacNaughton Library**
           400 One Energy Square
           4925 Greenville Avenue
           Dallas, TX 75206
           214-368-6391
           Lends: Yes
           Copying services: Xerox $0.15 per page.
           Use of materials on premises: Yes

TxDaGI   **Geological Information Library of Dallas**
           Suite 100, One Energy Square
           4925 Greenville Avenue
           Dallas, TX 75206
           214-363-1078
           Lends: Yes, to members only.
           Copying services: Yes
           Use of materials on premises: Yes

TxDaM-SE   **Southern Methodist University**
           Science/Engineering Library
           6425 Airline Road
           Science Information Center
           Dallas, Tx 75275
           214-692-2276
           Lends: Yes
           Copying services: Through Interlibrary Loan.
           Use of materials on premises: Yes

TxDaSM   **Mobil Research & Development Corp.**
           Dallas Research Laboratory
           ATTN: Library
           13777 Midway Road
           Dallas, TX 75234

TxHMP   **Mobil Producing Texas & New Mexico**
           Information Resource Center
           Nine Greenway Plaza - Suite 2700
           Houston, TX 77046
           713-235-1024
           Lends: Yes
           Copying services: Yes
           Use of materials on premises: No

TxHSD   **Shell Development Company**
           Bellaire Research Center Library
           P.O. Box 481
           Houston, TX 77001

TxHSW   **Schlumberger Well Services**
           Houston Downhole Sensors - Library
           P.O. Box 2175
           Houston, TX 77252-2175

TxHU   **University of Houston, University Park**
           M.D. Anderson Library
           4800 Calhoun
           Houston, TX 77004

TxLT   **Texas Tech University**
           Library
           Lubbock, TX 79409-0002

TxMM   **Midland Country Public Library**
           Petroleum Department
           P.O. Box 1191
           Midland, TX 79702
           915-683-2708 Ext.20
           Lends: Yes
           Copying services: Photocopies $0.20 per page; $0.25 per
               page from film.
           Use of materials on premises: Yes

TxU   **The University of Texas at Austin**
           The Joseph C. Walter, Jr. and Elizabeth C. Walter Geology
               Library
           Geology Building 302
           Austin, TX 78712-7330

TxU-Da   **University of Texas at Dallas**
           Eugene McDermott Library
           P.O. Box 830643
           Richardson, TX 75083-0643
           214-690-2955
           Lends: Yes
           Copying services: Yes, adhering to copyright laws.
           Use of materials on premises: Yes

TxWB   **Baylor University**
           F. Roemer Geological Library
           Geology Department CSB 367
           Waco, TX 76798
           817-755-2361
           Lends: Through Moody Library of Baylor University.
           Copying services: $0.10 per page.
           Use of materials on premises: Yes

UH-M **University of Hawaii, Manoa**
Hawaii Institute of Geophysics
Library, Rm. 252
2525 Correa Road
Honolulu, HI 96822

UPB **Harold B. Lee Library**
Brigham Young University
Provo, UT 84602

UU **University of Utah**
Science and Engineering Library
Marriott Library
159 Science and Engineering
Salt Lake City, UT 84112

Uk-Pal **British Museum (Natural History)**
Paleontology/Mineralogy Library
Cromwell Road
London SW7 5BK

Vi-MR **Virginia Division of Mineral Resources**
Library
P.O. Box 3667
Charlottesville, VA 22903
804-293-5121
Lends: No
Copying services: Yes
Use of materials on premises: Yes

ViBlbV **Geology Library**
Virginia Polytechnic Institute & State University
3040 Derring Hall
Blacksburg, VA 24061

VtCasT **Castleton State College**
Library
Castleton, VT 05735

WBB **Beloit College**
Kohler Science Library
Beloit, WI 53511
608-365-3391 ext.567
Lends: Yes
Copying services: Through ILL at $0.10 per page.
Use of materials on premises: Yes

WGrU **University of Wisconsin-Green Bay**
University of Wisconsin-Green Bay, Library
2420 Nicolet Drive
Green Bay, WI 54301-7001

WKenU **University of Wisconsin, Parkside**
Wyllie Library/Learning Center
Box 2000
Kenosha, WI 53141-2000
414-553-2595
Lends: Yes
Copying services: Minimum charge for out of state and corporation libraries $0.15 per page for copying over 20 pages.
Use of materials on premises: Yes

WMMus **Reference Library**
Milwaukee Public Museum
800 W. Wells Street
Milwaukee, WI 53233

WMUW **University of Wisconsin, Milwaukee**
American Geographical Society Collection
P.O. Box 399
Milwaukee, WI 53201
414-963-6282
Lends: Yes
Copying services: Photocopying for a fee.
Use of materials on premises: Yes

WPlaU **University of Wisconsin, Platteville**
Karrmann Library
Platterville, WI 53818

WU **University of Wisconsin, Madison**
Geology-Geophysics Library
440 Weeks Hall
1215 W. Dayton St.
Madison, WI 53706
608-262-8956
Lends: Through ILL, Memorial Library, U.W. Madison
Copying services: Self-service, $0.05 per page.
Use of materials on premises: Yes

Wa-NR **Washington Department of Natural Resources**
Division of Geology and Earth Resources
Library
PY 12
Olympia, WA 98504
206-459-6373
Lends: No
Copying services: 5 pages no charge; $0.20 per page thereafter (large jobs discouraged). ILL requests also honored if materials not otherwise available.
Use of materials on premises: Yes

WaPS **Washington State University**
Owen Science & Engineering Library
Pullman, WA 99164-3200

WaU **University of Washington**
Natural Sciences Library
University of Washington FM-25
Seattle, WA 98195
206-543-1243
Lends: Yes
Copying services: Yes
Use of materials on premises: Yes

WyU **University of Wyoming**
Geology Library
University Station
Box 3006
Laramie, WY 82071
307-766-3374
Lends: Yes
Copying services: Photocopy machine.
Use of materials on premises: Yes

UNION LIST OF GEOLOGIC FIELD TRIP GUIDEBOOKS OF NORTH AMERICA

AAG See ASSOCIATION OF AMERICAN GEOGRAPHERS. GREAT PLAINS DIVISION. (70) and ASSOCIATION OF AMERICAN GEOGRAPHERS. ROCKY MOUNTAIN DIVISION. (71) and ASSOCIATION OF AMERICAN GEOGRAPHERS. ANNUAL MEETING. (69) and ASSOCIATION OF AMERICAN GEOGRAPHERS. SOUTHEASTERN DIVISION. (72)

(1) AAPG-SEG PETROLEUM EXPLORATION SCHOOL. FIELD TRIP.
  1977 Dec. See American Association of Petroleum Geologists. Department of Education. School. 24.

AAPG. See AMERICAN ASSOCIATION OF PETROLEUM GEOLOGISTS. NATIONAL ENERGY MINERALS DIVISION. (TITLE VARIES) (27)

(2) ABILENE GEOLOGICAL SOCIETY. FIELD TRIP GUIDEBOOK. (TITLE VARIES)
  1946 Nov. Whitehorse-El Reno-Clearfork groups, Abilene, Sweetwater, Rotan Hamlin, Anson.
  1948 Jan. Road log and instructions for lower Paleozoic field trip, Llano Uplift, Texas.
  — June Study of Lower Permian and Upper Pennsylvanian rocks in Brazos and Colorado River valleys of west-central Texas, particularly from Coleman Junction to Home Creek limestones, Abilene-Eastland-Colorado River.
  1949 Subsurface studies and field trip (Brownwood, Ranger, and Mineral Wells districts to and from Abilene).
  1950 Nov. Road log from Brownwood to type localities of Strawn and older rocks of Pennsylvanian and Mississippian systems, north and west flanks of the Llano Uplift. (Cover title: Strawn and older rocks of Pennsylvanian and Mississippian systems of Brown, San Saba, McCulloch, Mason, and Kimble counties, Texas)
  1951 Middle-Upper Permian and Cretaceous; Abilene, Sweetwater, Bronte, Paint Rock, and San Angelo.
  1952 Nov. [1] Abilene, Coleman, Ballinger, Abilene, Fall field excursion.
  [2] [Geological road logs of west-central Texas]
  1954 Nov. Facies study of the Strawn-Canyon series in the Brazos River area, north-central Texas.
  1955 Dec. Study of the Lower Permian and Upper Pennsylvanian rocks in the Brazos and Colorado River valleys of west-central Texas, together with a preliminary report on the structural development of west-central Texas.
  1957 Study of Lower Pennsylvanian and Mississippian rocks of the Northeast Llano Uplift. Paleozoic stratigraphy of the Fort Worth Basin. Regional correlations of the Mississippian and early Pennsylvanian. (178)
  1961 Sept. A study of Pennsylvanian and Permian sedimentation in the Colorado River valley of west-central Texas.

BEG	1957
CLU-G/G	1952(2), 54, 57, 61
CLhC	1961
CMenUG	1957
CPT	1950, 54-55, 57, 61
CSdS	1946, 50
CSt-ES	1948, 50, 54-55, 57
CU-EART	1950, 52(2), 54-55, 57, 61
CU-SB	1954-55, 57, 61
CaACI	1954-55, 57
CaACU	1952(2), 54-55, 57, 61
CaOKQGS	1954-55, 57, 61
CoDBM	1957
CoDCh	1948, 49, 50
CoDGS	1954-55, 57
CoDU	1954-55, 61
CoG	1950, 54-55, 57, 61
DFPC	1961
DI-GS	1946, 48-51, 52(2), 54-55, 57, 61
DLC	1954-55, 57
I-GS	1957
ICF	1954-55, 57, 61
IEN	1948-49, 61
IU	1948-49, 54-55, 57, 61
IaU	1954-55, 57
InLP	1952(2), 54-55, 57, 61
InU	1957, 61
KyU-Ge	1952(2), 54-55, 57, 61, 85
LNU	1957
LU	1949-50, 54-55, 57, 61
MiDW-S	1954-55, 57, 61
MnU	1950, 54-55, 57, 61
MoU	1950
MtU	1954-55, 57, 61
NNC	1950, 54-55, 57, 61
NbU	1946, 48-50, 55, 57, 61
NcU	1952(2), 54-55, 57, 61
NjP	1961
NmPE	1952(2), 54-55, 61
NmSoI	1957
NvU	1954-55, 57
OCU-Geo	1952(2), 54-55, 61
OU	1950, 52(2), 54-55, 57, 61
OkT	1949, 50, 55, 57
OkTA	1957
OkU	1946, 48, 50, 52(2), 54-55, 57, 61
PSt	1954, 57
SdRM	1957, 61
TxBeaL	1948, 54-55, 57, 61
TxDaAR-T	1950, 54-55, 57
TxDaDM	1954-55, 57, 61
TxDaGI	1950, 54-55, 57, 61
TxDaM-SE	1954-55, 57, 61
TxDaSM	1948-49, 52(1), 55, 57, 61
TxHMP	1957
TxHSD	1948, 50, 52(2), 55
TxHU	1949, 54-55, 57, 61
TxLT	1950, 54-55, 61
TxMM	1946, 48-50, 54-55, 57, 61
TxU	1946, 48-51, 54-55, 57, 61
TxWB	1948, 52, 54-55
UU	1957
ViBlbV	1954-55, 57, 61
WU	1950, 52, 54, 55, 57, 61

(3) ADVENTURES IN EARTH SCIENCE SERIES.
  1977 No. 17 See University of British Columbia. Department of Geological Sciences. 676.(No.1)
  — No. 22 See University of British Columbia. Department of Geological Sciences. 676.(No.2)

AIBS See AMERICAN INSTITUTE OF BIOLOGICAL SCIENCES. (40)

AIME See AMERICAN INSTITUTE OF MINING, METALLURGICAL AND PETROLEUM ENGINEERS. ANNUAL MEETING. (41) and AMERICAN INSTITUTE OF MINING, METALLURGICAL AND PETROLEUM ENGINEERS. SOUTHWESTERN NEW MEXICO SECTION. (47) and AMERICAN INSTITUTE OF MINING, METALLURGICAL AND PETROLEUM ENGINEERS. SOUTHWEST MINERAL INDUSTRY CONFERENCE. (46) and AMERICAN INSTITUTE OF MINING, METALLURGICAL AND PETROLEUM ENGINEERS. INDUSTRIAL MINERALS DIVISION. (42) and AMERICAN INSTITUTE OF MINING, METALLURGICAL AND PETROLEUM ENGINEERS. MICHIGAN TECH STUDENT CHAPTER. (43) and AMERICAN INSTITUTE OF MINING, METALLURGICAL AND PETROLEUM ENGINEERS. SOUTH TEXAS MINERALS SECTION.

(44) and AMERICAN INSTITUTE OF MINING, METALLURGICAL AND PETROLEUM ENGINEERS. SOUTHERN CALIFORNIA PETROLEUM SECTION. (45) and AMERICAN INSTITUTE OF MINING, METALLURGICAL, AND PETROLEUM ENGINEERS. VIRGINIA SECTION. (48) and SOCIETY OF MINING ENGINEERS OF AIME. COLORADO PLATEAU CHAPTER. (603)

AIPG See AMERICAN INSTITUTE OF PROFESSIONAL GEOLOGISTS (AIPG). (49)

(4) ALABAMA ACADEMY OF SCIENCE. GUIDEBOOK. (TITLE VARIES)
- 1962 39th   A guidebook; field trip to selected brown iron ore deposits [in] Pike County, Alabama. (223)
- 1965 Apr.   Russellville brown iron ore district. Road log for field trip. (5)

ATuGS	1962
CLU-G/G	1962, 65
CMenUG	1962
CSt-ES	1962, 65
CU-EART	1965
CaOLU	1962
CaOOG	1965
CaOWtU	1962
CoDGS	1962
DI-GS	1962, 65
DLC	1962, 65
ICF	1962
ICIU-S	1965
ICarbS	1962
IEN	1962
IU	1962
IaU	1962, 65
InLP	1962, 65
InU	1962, 65
KyU-Ge	1962, 65
MNS	1962
MdBJ	1962
MiDW-S	1962
MiHM	1962
MiU	1962
MnU	1962
MoU	1962, 65
MtBuM	1962
NNC	1962
NSyU	1962
NbU	1965
NcU	1962, 65
NdU	1962
NhD-K	1962
NjP	1962
OU	1962
OkU	1962, 65
PSt	1962
TMM-E	1962
TxBeaL	1965
TxDaAR-T	1965
TxDaDM	1965
TxDaM-SE	1965
TxU	1962, 65
UU	1962
ViBlbV	1962
WU	1962, 65

(5) ALABAMA GEOLOGICAL SOCIETY. GUIDEBOOK FOR THE ANNUAL FIELD TRIP. (TITLE VARIES)
- 1964 1st   Pottsville Formation in Blount and Jefferson counties, Alabama.
- — 2nd   Alabama Piedmont geology.
- 1965 Apr.   See Alabama Academy of Science. 4.
- — 3rd   Structural development of the southernmost Appalachians.
- 1966 4th   Facies changes in the Alabama Tertiary.
- 1967 5th   A field guide to Mississippian sediments in northern Alabama and south-central Tennessee.
- 1968 6th   Facies changes in the Selma Group in central and eastern Alabama.
- 1969 7th   The Appalachian structural front in Alabama.
- 1970 8th   Geology of the Brevard fault zone and related rocks of the inner Piedmont of Alabama.
- 1971 9th   The Middle and Upper Ordovician of the Alabama Appalachians.
- 1972 10th   Recent sedimentation along the Alabama coast.
- 1973 11th   Talladega metamorphic front.
- 1974 12th   The Coosa deformed belt in the Alabama Appalachians.
- 1975 13th   Geologic profiles of the northern Alabama Piedmont.
- 1976 14th   Cretaceous and Tertiary faults in southwestern Alabama.
- 1977 15th   Cambrian and Devonian stratigraphic problems of eastern Alabama.
- 1978 16th   Stratigraphy and structure of the Birmingham area, Jefferson County, Alabama.
- 1979 17th   The Hillabee metavolcanic complex and associated rock sequences.
- 1980   See Geological Society of America. Southeastern Section. 206.
- 1981 18th   Contrasts in tectonic style between the inner Piedmont terrane and the Pine Mountain Window in eastern Alabama and adjacent Georgia.
- 1982 19th   Depositional setting of the Pottsville Formation in the Black Warrior Basin.
- 1983 20th   Current studies of Cretaceous formations in eastern Alabama and Columbus, Georgia.
- 1984 21st   Appalachian thrust belt in Alabama.
- 1985 22nd   Early evolution of the Appalachian Miogeocline; Upper Precambrian-Lower Paleozoic stratigraphy of the Talladega slate belt.

ATuGS	1964-79, 81, 83, 85
AzTeS	1964, 65(3), 66-68, 73
CLU-G/G	1964, 65(3), 66-68, 70-71, 73-79, 81-84
CLhC	1966
CMenUG	1964-79, 81-85
CPT	1969
CSdS	1964, 65(3), 66-68, 73
CSt-ES	1964, 65(3), 66-76
CU-A	1969
CU-EART	1964, 65(3), 66-79, 81-85
CU-SB	1972, 76, 81-84
CaACU	1974-77
CaBVaU	1966, 70
CaOHM	1964, 65(3), 66-74
CaOKQGS	1966
CaOOG	1964, 65(3), 66-69, 78-79
CaOWtU	1967, 69-79, 81-84
CoDBM	1981-84
CoDCh	1974, 75, 77, 78
CoDGS	1963, 65, 74, 84
CoG	1964-75
CoLiM	1980
CoU	1967, 69-74
DI-GS	1964, 65(3), 66-79, 81-85
DLC	1965(3), 66-67, 69-72, 74-76
GU	1969, 70
I-GS	1969
ICIU-S	1964, 65(3), 66-69
ICarbS	1973, 74, 75
IEN	1964, 65(3), 66-74
IU	1964, 65(3), 66-67, 70-74, 76-78
IaU	1964, 65(3), 66-79, 81-85
InLP	1964, 65(3), 66-79, 81-82, 84
InU	1964, 65(3), 66-79, 82
KyU-Ge	1964, 65(3), 66-79, 81-83
LNU	1964, 65(3), 67-75
MH-GS	1969, 81, 83
MiHM	1971, 74
MiKW	1967
MnU	1964, 65(3), 66-79, 81-83
MoSW	1964, 65(3), 66-67
MoU	1964, 65(3), 66-71
NBiSU	1965(3), 69-70
NNC	1964, 65(3), 66-77, 82, 83
NOneoU	1968
NSyU	1964-68, 70-77
NbU	1965(3)
NcU	1964, 65(3), 66-76, 81-84
NhD-K	1971, 73-74, 81
NjP	1964, 65(3), 66-74
OCU-Geo	1964, 65(3), 66-76
OU	1964, 65(3), 66-79, 81-85
OkU	1964, 65(3), 66-79, 81-82, 84
TU	1964, 65(3), 66
TxBeaL	1964, 65(3), 66-75

TxDaAR-T	1964, 65(3), 66-74	CMenUG	1981
TxDaDM	1964, 65(3), 66-77	CU-A	1981
TxDaDM-SE	1964, 65(3)	CU-EART	1981
TxDaM-SE	1964, 65(3)	CaACU	1981
TxDaSM	1964(2), 65(3), 66, 76	DI-GS	1981
TxHU	1965(3), 66-77	IaU	1981
TxLT	1970	MnU	1981
TxMM	1965(3)	TxU	1981
TxU	1964, 65(3), 66-74, 76, 81-85	WU	1981
TxWB	1964, 66, 68, 70-79, 81		
ViBlbV	1964, 65(3), 66-74, 76-77		
WU	1964, 65(3), 66-79, 81-84, 85		
WaPS	1973-75		

ALABAMA. GEOLOGICAL SURVEY. See GEOLOGICAL SURVEY OF ALABAMA. (222) and GEOLOGICAL SURVEY OF ALABAMA. (223)

(6) ALASKA GEOLOGICAL SOCIETY. GUIDEBOOK. (TITLE VARIES)
- 1963 Anchorage to Sutton.
- 1964 Oil fields, earthquake, geology. Anchorage to Sutton--1963; Sutton to Caribou Creek--1964. (1963 reprinted and included in 1964)
- 1970 Oil and gas fields in the Cook Inlet basin, Alaska.
- 1973 Road log and guide; geology and hydrology for planning, Anchorage area.
- 1984 See Geological Society of America. Cordilleran Section. 199.
- 1985 See American Association of Petroleum Geologists. Pacific Section. 28.

AzTeS	1963-64
AzU	1963-64
CLU-G/G	1963-64, 70
CLhC	1963-64
CMenUG	1963-64, 70, 73
CPT	1963-64, 70, 73
CSdS	1964, 70, 73
CSt-ES	1963-64, 70, 73
CU-A	1963, 70, 73
CU-EART	1963-64, 73
CU-SB	1964
CaACU	1964, 70, 73
CaOHM	1973
CaOKQGS	1964
CaOOG	1984
CaOTRM	1963-64
CaOWtU	1970
CoDGS	1963-64, 73
CoG	1964
CoU	1963-64
DI-GS	1964-64, 70, 73
DLC	1970
IU	1963-64, 73
IaU	1963-64, 73
InLP	1984
KyU-Ge	1973
MH-GS	1963-64
MnU	1964, 70
MoSW	1963-64
NIC	1984
NNC	1963, 64, 84
NSyU	1984
NdU	1964, 70
OCU-Geo	1963, 70, 73
TxBeaL	1970, 73
TxDaAR-T	1963-64, 70, 73
TxDaDM	1963-64
TxDaSM	1963-64
TxLT	1963-64
TxMM	1984
TxU	1964, 73
TxU-Da	1970
ViBlbV	1963-64, 70, 73

(7) ALASKA GEOLOGICAL SOCIETY. PUBLICATION.
- 1981 No. 1. Guide to the bedrock geology along the Seward Highway, north of Turnagain Arm.

(8) ALASKA. DIVISION OF GEOLOGICAL AND GEOPHYSICAL SURVEYS. GUIDEBOOK
- 1983 1. See International Conference on Permafrost. 287.
- — 2. See International Conference on Permafrost. 287.
- — 3. See International Conference on Permafrost. 287.
- — 4. See International Conference on Permafrost. 287.
- — 5. See International Conference on Permafrost. 287.
- — 6. See International Conference on Permafrost. 287.

(9) ALBERTA GEOLOGICAL SURVEY. OPEN-FILE REPORT.
- 1984 83-12 See Canadian Society of Petroleum Geologists. 125.

(10) ALBERTA INSTITUTE OF PEDOLOGY.
- 1981 No. M-80-2. See Alberta Research Council. 12.

(11) ALBERTA RESEARCH COUNCIL.
- 1980 See International Symposium on Water-Rock Interaction. 321.

(12) ALBERTA RESEARCH COUNCIL. EARTH SCIENCES REPORT.
- 1979 79-4. See Geological Society of America. 197.(15)
- 1981 No. 81-1. Guidebook for use with soil survey reports of Alberta provincial parks and recreation areas. (10)

CLU-G/G	1981
CSfCSM	1981
CSt-ES	1981
CU-EART	1981
CaAEU	1981
CaOKQGS	1981
CaOOG	1981
CaOTDM	1981
DI-GS	1981
KLG	1981
KU	1981
MnU	1981
NIC	1981
NSbSU	1981
NdU	1981
OCl	1981
OU	1981
PSt	1981
TxU	1981
WU	1981

(13) ALBERTA RESEARCH COUNCIL. INFORMATION SERIES.
- 1973 No.65. See Canadian Society of Petroleum Geologists. 125.

ALBERTA RESEARCH COUNCIL. OPEN FILE REPORT. See ALBERTA RESEARCH COUNCIL. (11)

(14) ALBERTA SOCIETY OF PETROLEUM GEOLOGISTS.
- 1950 See Geological Association of Canada. 190.
- 1953 See Canadian Society of Petroleum Geologists. 125.
- 1954 See Canadian Society of Petroleum Geologists. 125.
- 1955 See Canadian Society of Petroleum Geologists. 125.
- 1956 See Canadian Society of Petroleum Geologists. 125.
- 1957 See Canadian Society of Petroleum Geologists. 125.
- 1958 See Canadian Society of Petroleum Geologists. 125.
- 1959 See Canadian Society of Petroleum Geologists. 125.
- 1960 See Geological Association of Canada. 190.
- 1961 See Canadian Society of Petroleum Geologists. 125.
- 1962 See Canadian Society of Petroleum Geologists. 125.

1967 See International Symposium on the Devonian System. 318.
1970 See American Association of Petroleum Geologists. 18.
1971 See Canadian Society of Petroleum Geologists. 125.
1982 See American Association of Petroleum Geologists. 18.

(15) AMERICAN ASSOCIATION FOR THE ADVANCEMENT OF SCIENCE. COMMITTEE ON ARID LANDS.
1979 See International Conference on Arid Lands in a Changing World. Guidebook 285.

(16) AMERICAN ASSOCIATION FOR THE ADVANCEMENT OF SCIENCE. PACIFIC COAST COMMITTEE. [GUIDEBOOK]
1915 Nature and science on the Pacific coast; a guidebook for scientific travelers in the west.

CSt-ES	1915
CaOWtU	1915
DI-GS	1915
DLC	1915
IEN	1915

(17) AMERICAN ASSOCIATION FOR THE ADVANCEMENT OF SCIENCE. PACIFIC DIVISION. GEOLOGIC FIELD TRIP.
1983 Sept. Guidebook and road log to the St. Maries River (Clarkia) fossil area of northern Idaho. (258)
— Dec. Geologic section and road log across the Idaho batholith. (258)

CMenUG	1983
CPT	1983
CSt-ES	1983
CU-EART	1983
CaOWtU	1983
CoDBM	1983
DI-GS	1983
IU	1983
IaU	1983
IdU	1979
InLP	1983
InU	1983
KyU-Ge	1983
MnU	1983
NvU	1983
UU	1983
WU	1983

(18) AMERICAN ASSOCIATION OF PETROLEUM GEOLOGISTS. GUIDEBOOK FOR THE FIELD CONFERENCE HELD IN CONNECTION WITH THE ANNUAL CONVENTION (TITLE VARIES)
1924 9th Palestine, [Texas].
1929 [14th] Vicinity of Fort Worth, Texas. (585)
1932 17th Highway geology of Oklahoma. (504, 585, 599)
  [1] Oklahoma City field trip.
  [2] [Ardmore field trip (Arbuckle Mountains)]
1933 [18th] [2] Sugarland oil field, Fort Bend County, [Texas]. (255, 585, 599)
1936 [21st] [2] Geologic road log: Tulsa to Sulphur, Oklahoma. Log from Tulsa to Stroud modified after Oklahoma Geological Society. (585, 599, 662)
1937 22nd Los Angeles Basin, San Joaquin Valley, Ventura County, [California]. (585, 599)
1939 24th [1] Anadarko Basin field trip. (504, 585, 599)
  [2] Arbuckle Mountains-Ardmore Basin field trip. (504, 585, 599)
1940 [25th] 1. Pre-convention field trip. [Chicago area] (262, 585, 599)
  2. Post-convention field trip at La Salle, Illinois. (262, 585, 599)
  [3] The Kentland Uplift and the coral reefs in the Silurian rocks of northern Indiana. (262, 585, 599)
1941 26th [Gulf Coast fields] (255, 585, 599)
1947 32nd [California oil fields and their geologic features] (28, 585, 599)
1948 33rd Geology of central Colorado. (Also in Quarterly of the Colorado School of Mines, v. 43, no.2, Apr. 1948) (549, 585, 599)
  1. Denver, Boulder, Golden, Morrison.
  2. Denver, Colorado Springs, Canon City.
  3. Denver, Glenwood Springs, Rifle, Leadville, South Park.
1949 34th Southeastern Missouri and southwestern Illinois. (262, 585, 599)
1950 [35th] Niagaran reefs in the Chicago area. (265, 585, 599)
1951 36th Southeastern Missouri, Cape Girardeau and Gulf Embayment areas. (585, 599)
1952 37th Oil fields; geology [California]. (28, 585, 599)
  1. Palos Verdes Hills and San Pedro.
  2. Los Angeles Basin including Palos Verdes Hills and Willington.
  3. California Institute of Technology and United Geophysical Company laboratories.
  4. Los Angeles - Ojai - Cuyama - San Andres Fault.
  5. San Andreas Fault - Maricopa - Bakers-Field.
  8. San Joaquin Valley.
  10. San Francisco Bay area.
1953 [38th] Oil fields; geology. (255, 585, 599)
1954 [39th] Guide to the structure and Paleozoic stratigraphy along the Lincoln Fold in central western Illinois. (265, 585, 599)
1955 [40th] [1] Field guidebook of Appalachian geology; Pittsburgh to New York. (541, 585, 599)
1956 [41st] Niagaran reef at Thornton, Illinois. (265, 585)
1957 42nd Paleozoic section in the St. Louis area. (585)
1958 [43rd] A guide to the geology and oil fields of the Los Angeles and Ventura regions. (28, 585)
1959 [44th] Geology of the Ouachita Mountains. Symposium and field trip guidebook. (62, 156, 585)
1960 [45th] Guidebooks.
  1. Geology of north-central part of the New Jersey coastal plain. (339, 585)
  2. Geology of the region between Roanoke and Winchester in the Appalachian Valley of western Virginia. (339, 585)
  3. Lower Paleozoic carbonate rocks in Maryland and Pennsylvania. (339, 585)
  4. & 5. Structural geology of South Mountain and Appalachians in Maryland. With a section on geology and quarrying, Warm Spring Ridge, West Virginia. (Prepared for the field trip of the Pennsylvania Field Conference, May 1958) (339(No.17), 585, 172(1958))
1961 [46th] See Geological Society of America. 197.
1962 [47th] Geologic guide to the gas and oil fields of Northern California. (28, 112, 487, 585, 593, 601)
  Trip 1. Sacramento Valley.
  Trip 2. Santa Cruz Mountains.
  Trip 3. Point Reyes Peninsula and San Andreas Fault.
  Trip [4] Geologic guide to the Merced Canyon and Yosemite Valley, California. (28, 112, 585, 593, 601)
  Trip 5. San Francisco Peninsula.
  Trip 6. North flank of Mount Diablo.
1963 [48th] Stratigraphic study, Pleistocene to Middle Eocene, Houston, Texas. (255, 585, 588, 600)
1964 [49th] Geology of central Ontario, Toronto, Canada. (190, 377, 383, 585)
1965 [50th] [1] Airplane field trip; Louisiana delta complex. (466, 585)
  2. Salt Dome field trip. (466, 585)
  [3] Molluscan fauna of Mississippi River delta, Mudlump province, Mudlump Island field trip. (466, 585)
  4. Yucatan field trip. (466, 585)
1966 [51st] [1] Sedimentary structures and morphology of late Paleozoic sand bodies in southern Illinois. (262, 265, 271, 585)
  [2] Middle Ordovician and Mississippian strata, St. Louis and St. Charles counties, Missouri. (262, 271, 399, 585)
1967 52nd [A] Pre-convention field trips. (28, 412, 585, 593, 601, 408)
  2. Pliocene Seaknoll, South Mountain, Ventura County, California.
  4. Stream injection, Wilmington Field, Los Angeles area, California.
  5. North-central Los Angeles Basin and Whittier oil field.
  6. Baldwin Hills and Palos Verdes Hills, Los Angeles.
  7. Central Santa Monica Mountains stratigraphy and structure.
  8. Structural complexities, eastern Ventura Basin.
  [B] Post-convention field trips. (28, 412, 585, 593, 601, 408)
  1. Do-it-yourself road log, Los Angeles to Death Valley.
  2. Santa Catalina Island.
  3. Ventura Basin stratigraphic field trip. Preface to road log for AAPG field trip to Hall Canyon and Wheeler Canyon [and] Road log.
  4. Underwater field trip.

1968 53rd [1] A guidebook to the geology of the Bluejacket-Bartlesville Sandstone, Oklahoma. (504, 585)

[2] A guidebook to the geology of the western Arkoma Basin and Ouachita Mountains, Oklahoma. (504, 585)

[3] Regional geology of the Arbuckle Mountains, Oklahoma. (508, 585)

1969 [54th] 1. Guidebook to the late Pennsylvanian shelf sediments, north-central Texas. (156, 585)

[4] A guidebook to the depositional environments and depositional history, Lower Cretaceous shallow shelf carbonate sequence, west-central Texas. (156, 585)

[5] A guidebook to the stratigraphy, sedimentary structures and origin of flysch and pre-flysch rocks, Marathon Basin, Texas. (156, 585)

1970 [55th] 1. A geological guide along the highways between Drumheller-Calgary-Lake Louise. (30, 125, 585, 14)

2. Red Deer River badlands. (30, 125, 585, 14)

1971 [56th] 1. NASA field trip. No guidebook published.

2. Recent alluvial deltaic and barrier island sediments, Southeast Texas. (255, 585)

[3] Uranium geology and mines, South Texas. (255, 585, 713(No.12))

[4] SEPM field trip. Trace fossils; a field guide to selected localities in Pennsylvanian, Permian, Cretaceous, and Tertiary rocks of Texas and related papers. (255, 585, 361)

1972 [57th] [A] Energy and mineral resources of the Southern Rocky Mountains. (549, 585, 405)

Field trip 1. Geology of the Denver Mountain area.

Field trip 2. Tertiary and Cretaceous energy resources of the Southern Rocky Mountains.

Field trip 3. Structure and ore deposits of central Colorado.

[B] Environments of sandstone, carbonate and evaporite deposition. (549, 585, 405)

SEPM field trip 1. Environments of sandstone deposition, Colorado Front Range.

SEPM field trip 2. Carbonate and evaporite facies of the Paradox Basin.

1973 58th Trip 1. Metropolitan oil fields and their environmental impact. (28, 585, 593, 601)

Trip 2. Imperial Valley regional geology and geothermal exploration. (28, 585, 593, 601)

Trip 3. Santa Barbara Channel region revisited. (28, 585, 593, 601)

SEG trip 4. A profile of Southern California geology and seismicity of Los Angeles basin. (28, 585, 593, 601)

SEPM field trip 1. Miocene sedimentary environments and biofacies, southeastern Los Angeles Basin. (28, 585, 593, 601)

SEPM field trip 2. Sedimentary facies changes in Tertiary rocks-California transverse and southern coast ranges. (28, 585, 593, 601)

1974 [59th] 1. Geology of the Big Bend region, Texas. (247, 585, 612)

[2] Stratigraphy of the Edwards Group and equivalents, eastern Edwards Plateau, Texas. (247, 585, 612)

[3] Hydrogeology of the Edwards Limestone aquifer, San Antonio area, Texas. (247, 585, 612)

[4] Mexico City and environs. (247, 585, 612)

SEPM field trip 1. Precambrian and Paleozoic rocks of central mineral region, Texas. (247, 585, 612)

SEPM trip 2:1. Aspects of Trinity Division geology. A symposium on the stratigraphy, sedimentary environments, and fauna of the Comanche Cretaceous Trinity Division (Aptian and Albian) of Texas and northern Mexico. (247, 585, 612, 241)

SEPM trip 2:2. A field guide: shallow marine sediments of Early Cretaceous (Trinity) platform in central Texas. (247, 585, 612, 361)

1975 [60th] Trip 1. Regional geology of Arbuckle Mountains, Oklahoma. (156, 585, 31)

Trip 2. Relationships in continuous deposition of clastic (deltaic) and calcareous (bank) facies of Missourian (Canyon) age, north-central Texas. (156, 585, 31)

Trip 3. Edwards (Lower Cretaceous) reef complex and associated sediments in central Texas. (156, 585, 31)

SEPM field trip 1. A guidebook to the sedimentology of Paleozoic flysch and associated deposits, Ouachita Mountains-Arkoma Basin, Oklahoma. (156, 31, 585)

[SEPM field trip 2] A guidebook to the Mississippian shelf-edge and basin facies carbonates, Sacramento Mountains and southern New Mexico region. (31, 156, 585)

1976 [61st] [1] Classic Tertiary and Quaternary localities and historic highlights of Jackson-Vicksburg-Natchez area. (466, 585)

[2] Louisiana delta plain and its salt domes with a visit to Weeks and Avery islands and Morton Salt Company Mine. (466, 585)

[3] Modern Mississippi Delta - depositional environments and processes. (466, 585)

[4] Field guide to Carboniferous littoral deposits in the Warrior Basin. (466, 585)

[5] Carbonate rocks and hydrogeology of the Yucatan Peninsula, Mexico. (Revised in 1978 for Gulf Coast Association of Geological Societies No.2) (466, 585, 247)

[6] SEPM-GCAGS 1974 Guatemala field-trip guidebook [with] addendum ... for AAPG 1976 post convention field trip to Guatemala. (466, 585)

1977 [62nd] [2] A field guide to proposed Pennsylvanian System stratotype West Virginia. (585, 751)

[3] A field guide to Cretaceous and lower Tertiary beds of the Raritan and Salisbury embayments, New Jersey, Delaware, and Maryland. (585, 751)

[4] A field guide to thin-skinned tectonics in the Central Appalachians. (585, 751)

[6] A geologic and environmental guide to northern Fairfax County, Virginia. (585, 751)

1978 [63rd] 1. Regional geology of the Arbuckle Mountains, Oklahoma. (Prepared originally for The Geological Society of America 1973 Annual Meeting Dallas, Texas, Guidebook for Field Trip No. 5) (504, 505, 585)

2. Desmoinesian coal deposits in part of the Arkoma Basin, eastern Oklahoma. (504, 585, 27)

[3] Field guide to structure and stratigraphy of the Ouachita Mountains and the Arkoma Basin.

[4] Guidebook to uranium mieralization in sedimentary and igneous rocks of Wichita Mountains region, southwestern Oklahoma. (504, 585)

SEPM trip 1. A guidebook to the trace fossils and paleoecology of the Ouachita geosyncline. (504, 585)

SEPM trip 2. Field guide to Upper Pennsylvanian evolothemie limestone facies in eastern Kansas; algal sparite facies in Captain Creek Limestone. (504, 585, 635)

[SEPM trip 3] Guidebook, Upper Cretaceous stratigraphy and depositional environments of western Kansas. (504, 585, 635)

1979 [64th] [1] Oil fields and their relation to subsidence and active surface faulting in the Houston area. (Cover title: Oil fields, subsidence and surface faulting in the Houston area) (255, 585)

[2] Claiborne sediments of the Brazos Valley, Southeast Texas. (Cover title: Lower Tertiary of the Brazos River Valley) (255, 585)

[3] Damon Mound field trip guidebook. (Cover title: Damon Mound, Texas). (255, 585)

[4] South Texas uranium province--geologic perspective. (255, 585, 713)

[5] Lignite resources in east-central Texas (Cover title). (255, 585)

SEPM trip 2. A field guide to Lower Cretaceous carbonate strata in the Moffatt Mound area neaqr Lake Bleton, Bell County, Texas. (255, 585, 588)

1980 June SEPM trip 1. Trace fossils of nearshore environments of Cretaceous and Ordovician rocks, Front Range, Colorado. (585)

SEPM trip 3. Depositional environments of the Dakota Sandstones in southeastern Colorado. (585)

1981 May San Francisco, California.

Field guide to the Mesozoic-Cenozoic convergent margin of Northern California. (28, 487)

1. Upper Mesozoic Franciscan rocks and Great Valley sequence, central coast ranges, California. (SEPM Field Trips 1 and 4) (585, 593)

2. Geology of the central and northern Diablo Range, California (SEPM Trips 2 and 5). (585, 593)

3. Modern and ancient biogenic structures, Bodega Bay, California (SEPM Trip 3). (585, 593)

4. Upper Cretaceous and Paleocene turbidites, central California coast (SEPM Trip 6). (585, 593)

AAPG field trip #2: The Geysers geothermal field.

(18) American Association of Petroleum Geologists.

The Franciscan Complex and the San Andreas Fault from the Golden Gate to Point Reyes, California. (28, 487)
Guide to the Monterey Formation in the California coastal area, Ventura to San Luis Obispo. (28, 487)
— Oct.   See Atlantic Margin Energy Symposium. 90.

1982 June   Calgary, Alberta. (125, 14)
CSPG trip no. 1. Depositional environments of the McMurray Formation oil sands, with reference to surface mining technology.
CSPG field trip no. 2. Geology of the Waterton area, Alberta.
CSPG trip no. 3. Geology of the Southern Rocky Mountains; Calgary-Lake Louise-Radium-Fernie-Lundbreck-Calgary.
CSPG trip no. 4. Central and Southern Rocky Mountains of Alberta and British Columbia.
CSPG trip no. 5. Cambrian and Devonian geology of the Grassi Lakes and Big Hill area, Alberta.
CSPG trip no. 6. Facies relationships and paleoenvironments of a Late Cretaceous tide-dominated delta, Drumheller, Alberta.
CSPG trip no. 7. Hummingbird Reef complex.
CSPG trip no. 8. Trap Creek-Lundbreck area; sedimentology of Upper Cretaceous, fluviatile, deltaic and shoreline deposits.
CSPG trip no. 9. Geology of the Plateau Mountain area.
CSPG trip no. 10. Upper Cretaceous-Paleocene stratigraphy, micropaleontology and palynology of the Bow Valley area, Alberta.
CSPG trip no. 11. Macro-paleo, Bearpaw Formation, southern Alberta.
CSPG trip no. 12. Calgary to Eisenhower Junction, structural and stratigraphic overview.
CSPG trip no. 13. Burnt Timber Creek.
[14] Ladies' field trip; Calgary to Banff.
[15] Ladies' field trip; Drumheller.

1983 Apr.   Dallas, Texas.
[1] Sedimentation and diagenesis of mid-Cretaceous platform margin, east-central Mexico with accompanying field guide. (156, 585)
[2] Stratigraphic and structural overview of Upper Cretaceous rocks exposed in the Dallas vicinity. (Based on the "Introduction" page information, mentioning AAPG Annual Meeting that this guidebook is listed here) (156, 585)

1984 May   San Antonio, Texas.
[1] Meteor impact site, Anacacho asphalt deposits. (612)
[2] Structure and Mesozoic stratigraphy of Northeast Mexico. (152, 585, 612)
[3] Upper Jurassic and Lower Cretaceous carbonate platform and basin systems. (585, 588)
SEPM trip no. 3. Lower Cretaceous carbonate strata in the Moffatt Mound area near Lake Belton, Bell County, Texas. (585, 588)
[4] Modern depositional systems, Texas coastal plain. (612)
[5] The Cretaceous-Tertiary boundary and Lower Tertiary of the Brazos River Valley.
[6] Stratigraphy and structure of the Maverick Basin and Devils River Trend, Lower Cretaceous, Southwest Texas; a field guide and related papers. (585, 555)
[7] Stratigraphy of the El Paso border region; Texas and New Mexico. (167)

1985 Mar.   New Orleans, Louisiana.
[1] Tertiary and Upper Cretaceous depositional environments of central Mississippi. (466)
[2] Coastal evolution; Louisiana to Northwest Florida. (466)
[3] A tour of selected Tertiary and Quaternary localities and landscapes of the Jackson-Vicksburg-Natchez-Old River areas, Mississippi-Louisiana; pointing out the influence of geology on human use (abuse?) and exploitation of the landscape for 12,000 years. (466)
[n.d.] See Baylor Geological Society. 94.

ACTCR	1982
ATuGS	1939, 66(1)
AzFU	1955, 59, 62
AzTeS	1948, 50, 59, 62, 66(1), 67(A2, 4-8, B1-4), 68(1-2), 69(1, 4-5), 70(1), 71, 73(1-3, SEPM 1), 78(SEPM 3)
AzU	1953, 59, 63, 64, 69(1, 4-5), 71(3), 76(2, 5), 78(1)
BEG	1983-84
CChiS	1958
CLU-G/G	1932, 37, 47-48, 52-55, 58-60, 63, 66, 67(A2, 4-5, 7-8), 68(1-2), 69(1, 4-5), 70, 71(2-4), 72-74(1-3 SEPM 2:2), 75(3, SEPM 1), 76(1-5), 78(1, SEPM 1-3), 79(4-5), 80, 81(SEPM 1+4, 2+5, 3, 6), 82(1-9, 11-12)
CLhC	1952-53, 58-59, 62, 66(1), 69(1, 4-5), 71(3), 72, 73(SEPM 1, SEG 1), 79(4), 80(2 SEPM 2), 80, 81, 83
CMenUG	1939, 47, 52-54, 56-60, 62-64, 66-68(2), 69-70, 71(4), 72-74(2), 75(1), 77, 84
CPT	1932, 47, 49, 52, 55, 58, 62, 66, 67(A2, 4-5, 7-8, B1-3), 72(SEPM 1-2), 73, 74(SEPM 2:1), 79, 80, (SEPM 2), 81
CSdS	1947, 52-53, 55, 58, 62, 73, 81, 82(1-4)
CSfCSM	1937, 68(3), 70, 73(1-2, SEG 1, SEPM 1-2), 79(4), 81
CSt-ES	1929, 32, 40(2), 41, 47-50, 52-58, 60, 62, 66, 67(A2, 4-8, B1-4), 68(2-3), 69(1, 4-5), 70(1), 71(3-4), 72-73, 74(SEPM 2:1), 75(1, SEPM 1-2), 76(1-5), 78(1-2, SEPM 2-3), 79(1-5), 80(2), 81-82, 84
CU-A	1947, 52, 58-59, 62, 66(1), 67(A2, 4-5, 7-8, B1-4), 70, 71(3), 73(1-3, SEPM 1), 76(2), 78(SEPM 1), 80, 82, 84([2]-[7])
CU-EART	1932, 41, 47, 52-53, 55, 58-60, 62-63, 65(2), 66, 67(A2, 4-5, 7-8, B1-4), 68, 69(1, 4-5), 70(1), 71(3-4), 72(SEPM 2:1), 75(1-3, SEPM 1-2), 76(1-5), 78(SEPM 1), 79(1-2, 4-5), 80(2), 81(1-4), 84
CU-SB	1939, 41, 48-49, 52, 55, 57, 63, 67-69, 75, 78-80, 81-82(3, 5, 7, 8, 11) 84
CU-SC	1981, 83
CaAC	1953, 70
CaACAM	1953, 64, 69(1, 4-5), 70, 72, 82
CaACI	1950, 62-64, 65(3), 66, 67(A2, 4-8, B1-4), 68(1-2), 69(1, 4-5), 70
CaACM	1970, 72(SEPM 1-2)
CaACU	1941, 47-48, 52, 56, 58, 62, 68(1-2), 69, 70, 71(3, 4), 72, 73, 74(SEPM 2:1), 75(1, SEPM 1-2), 76(1-5), 78(1, 3-4 SEPM 1-3), 79(2, 4), 82
CaAEU	1962, 69(4), 70(1), 71(4)
CaBVaU	1953, 56, 62, 66(1), 70-71, 79(2, 4), 80(2), 82(1-9, 11-12, [14]-[15])
CaOHM	1959-60, 62-63, 66(1), 67(A2, 4-8, B1-4), 68, 69(1, 4-5), 70, 71(3), 73(2)
CaOKQGS	1932, 55, 59, 62(4, 6) 66(1), 68(3), 69(1, 4-5), 81-82
CaOLU	1948, 55, 68, 70, 71(3)
CaOOG	1937, 50, 55, 60, 64, 66, 68(1, 3), 69(1, 4-5), 70, 71(3-4), 72, 78(SEPM 2-3), 82
CaOTDM	1964, 81(1-4), 82
CaOTRM	1953, 62, 66(1)
CaOWtU	1957, 62, 63, 66(1), 67(B4), 68(2), 69, 70(1), 71(3-4), 72, 74(SEPM 2:1), 75(3), 78(1, SEPM 1-3), 79(4, SEPM 2)
CaQQLaS	1962
CoDCh	1949, 50, 52, 54, 56-69, 60(Pt.1, 3), 62(Pt.2) 64, 67(Pt.8), 68(Pt.1-3), 69(Pt.5), 74(Pt.2, SEPM 1-2), 75(SEPM 2), 77(Pt.4), 78(Pt.3, SEPM 1), 79(Pt.5, SEPM 2), 80(SEPM 1), 81(1, 4), 82(Pt.3)
CoDGS	1950, 54, 56, 64, 66, 71, 75, 79(4), 80, 81(SEPM 1-6), 82(1-3, 5-9), 84
CoDU	1968, 70
CoFS	1956, 66(1), 68(3)
CoG	1932, 39, 47-48, 52-53, 55, 58-59, 60(1-3), 62-63, 66, 67(A2, 4-8, B2-4), 68, 69(1, 5), 70, 71(2-4), 72, 76
CoLiM	1980(5, SEPM 3), 81(1, 4), 83, 84(2, 3, 5, SEPM 3), 85
CoU	1948, 55, 57, 59, 62, 66, 68, 69(1, 4-5), 71(2-3)
CoU-DA	1981(1, 4)
DFPC	1959
DI-GS	1932, 39, 41, 47-60, 62-64, 65(1, 3), 66, 67(A2, 5, 7-8, B1-4), 68, 69(1, 4-5), 70(1), 71(3), 72(1-3), 73(1-2, SEPM 1-2), 74(2), 75(SEPM 2), 76, 77(2-4, 6), 78, 79(1-5, SEPM 2), 80-82, 84
DLC	1948, 52-53, 55, 62, 66, 68(3), 71(4), 72(A), 74(SEPM 2:1), 75(1), 76(5), 78(1), 79(4)
GU	1974(SEPM 2:1)
I-GS	1924, 32, 39, 40(1-2), 49, 50(1), 53, 54(3), 55-58, 60, 62, 66, 68(2), 71(3), 80
ICF	1932, 49, 55-56, 59-60, 62-63, 66, 68(1-2), 72, 74(SEPM 2:1), 79(4)
ICIU-S	1964, 66, 68(2-3), 69(1, 4-5)
ICarbS	1954-56, 59-60, 62 (pt. 1), 63, 66, 68, 70, 72(1, SEPM 2), 71(3), 79[4], 84(Aug., SEPM 2)
IEN	1941, 47, 49, 52, 54-56, 59-60, 62(4, 6), 66, 68(2-3), 69(1, 4-5), 71(3), 73
IU	1939-41, 49-50, 53-56, 59-60, 62-63, 66, 67(A2, 4-8, B1-4), 68, 69(1, 4-5), 71(3), 72-73, 74(2-3, SEPM 1-2), 75(1-3, SEPM 1), 76(5), 77(4), 78, 79(1-2, 4, SEPM 2), 80, 81(1-4), 82-84
IaU	1932, 48, 53-55, 57-60, 62, 64, 66, 68, 69(1, 4-5), 70(1), 71(3-4), 72, 73(1-2, SEG 1, SEPM 1-2), 74(2, SEPM 2), 75(1, SEPM 1-2), 76(1-2, 4-5), 78(1-2, 4, SEPM 1-3), 79(4,SEPM 2), 80(2), 81, 82(4,8), 84(4)
IdBB	1959, 62, 66(1), 67(A2, 4-8, B2-4), 68(3), 69(1, 4-5)

IdPI	1953, 59, 62
IdU	1962(1-4)
InLP	1948, 54-57, 59-60, 62-64, 66, 68, 69(1, 4-5), 70(1), 71(3-4), 72, 73(1-3, SEPM 1, SEG 1), 74(2-3, SEPM 2:1), 75(1, SEPM 1-2), 76(1-5), 78(1-4, SEPM 1-3), 79(1-5, SEPM 2), 80, 81, 82, 84
InRE	1966(2), 68(3), 71(3), 78(1)
InU	1948-50, 54-56, 58-60, 62, 64, 66(1), 68, 69(1, 4-5), 70, 71(2-4), 72, 75, 76(1, 4), 79(1-3, 5, SEPM 2), 82-84
KLG	1954, 71, 78, 79(4)
KyU-Ge	1948, 50, 52-56, 58-60, 62-63, 66(2), 68(1-2), 69(1, 4-5), 70(1), 71(3-4), 72-73, 74(1-3, SEPM 2:2), 76(pt. 1-4), 78(1-2, SEPM 1), 79(1-5, SEPM 2), 81(SEPM 1-6), 82-85
LNU	1953, 59, 62, 65(2-3), 66(2), 68(1-2), 69(1, 4-5), 71(4), 74(SEPM 2:1), 76(3-5), 77(4), 79(4)
LU	1941, 49, 52-53, 55-56, 59, 62, 66(1), 68(3)
MCM	1982
MH-GS	1959, 66(1)
MNS	1955-56, 58-60, 66, 68(3), 69(1, 4-5), 71(4)
MU	1981
MdBJ	1960
MiDW-S	1955, 62, 66, 68(3), 71(3)
MiHM	1948, 60(1, 3), 62, 68(3), 69(1, 4-5), 71(3), 79(4)
MiKW	1953, 68(2), 69(4-5)
MiU	1941, 52, 67(B2), 68, 71(3)
MnU	1948-60, 62-64, 65(2), 66, 68, 69(1, 4-5), 70(1), 71(3), 72, 73(1-3, SEPM 1), 75(SEPM 1-2), 76(1-2), 78(2-4, SEPM 1), 79(1-5), 81-82, 84
MoRM	1955
MoSW	1941, 48-49, 51-56, 59, 62, 66, 68(3), 69(1, 4-5), 71(3-4), 72, 74(SEPM 2:1)
MoU	1932, 49, 52-54, 58-60, 62, 66, 68, 69(1, 4-5), 71(3-4)
Ms-GS	1963
MsJG	1963
MtBuM	1947-48, 62, 66, 71(3), 72(1-3), 78(SEPM 1), 79(4), 82(3)
NBiSU	1955, 62, 66(1), 68(2-3), 69(1, 4-5), 71(3)
NIC	1980, 81
NNC	1941, 47-50, 52-59, 60(1-3), 63-66, 68, 69(1, 4-5), 70, 71(3)
NOneoU	1969(5)
NRU	1968(1-2), 69(1, 4-5), 82
NSbSU	1960, 62-63, 66(1), 69, 74(SEPM 2), 75, 79-82
NSyU	1955, 58, 60, 62, 66, 68(3), 69(1, 4-5), 72, 74(SEPM 2:1), 79(4)
NbU	1948, 53, 55-56, 60, 62, 64, 66, 68(1-2), 69(1, 4-5), 71(4)
NcU	1932, 55, 59, 60, 62, 64, 66, 67(A7), 68, 69(1, 4-5), 70(1), 71(2-4), 73-74, 75(SEPM 2), 76(1), 77(2-4, 6), 80(SEPM 1, 3), 81(2-4)
NdU	1932, 47-48, 59, 62, 66(2), 68(1-2), 69(1, 4-5), 71(2, 4), 72, 76(2), 79(5), 82
NhD-K	1948, 55, 57, 59-60, 62, 64, 66, 68(1-2), 69(1, 4-5), 71(4), 72, 74(SEPM 2:1), 75-76, 77(2), 79(1-3, 5), 80, 82
NjP	1932, 39, 47-49, 52, 55, 58-60, 62, 66, 68, 69(1, 4-5), 70(1), 71(3), 72, 75(1-2)
NmPE	1959, 68
NvU	1947, 52, 58, 59, 62, 64, 66, 67(A2, 4-8, B1-4), 73, 81(1-4)
OCU-Geo	1941, 47, 49, 54, 56, 59, 60, 62-63, 66, 67(A8, B1), 68, 69(1, 4-5), 70, 71(4), 75(1-3, SEPM 1-2), 77, 78(1), 79(3, 4), 80(2), 83
OKentC	1981
OU	1932, 41, 47-49, 53-54, 56, 58-60, 62, 65-66, 67(A1-5, 7, 8, B2, 4), 68, 69(1, 4-5), 70, 71(3-4), 72, 78(1-4, SEPM 2-3), 79(1-5, SEPM 2)
OkOkCGS	1968(1, 3)
OkOkU	1968(3)
OkT	1929, 39, 47, 52-53, 55, 58, 59, 62-64, 68, 78(1, 4)
OkTA	1980, 81(Monterrey, Mesozoic), 82, 83[1], 84, 85
OkU	1929, 32, 36, 39, 41, 47-49, 52-53, 55, 59, 62-63, 65-66, 68, 69(1, 4), 71(3-4), 72, 73(1-2, SEG 1, SEPM 1-2), 74(SEPM 2:1), 75(SEPM 1-2), 76(1-5), 78, 80(2, 5), 81-84, 85[pt.2-3]
OrCS	1972, 79(4)
OrU-S	1955, 59, 62, 66(2), 68(3), 71(3), 72, 74(SEPM 2:1)
PBL	1955, 63, 68(1-2), 71(3)
PBm	1949, 52, 55, 60(1-3), 66(1), 68(3)
PHarER-T	1955(1), 60(3), 77(4)
PSt	1952, 55, 62, 66(2), 71(3), 79(4), 80, 84
RPB	1956, 62, 66
SdRM	1962, 66, 68(1, 2), 69(1, 4-5)
TMM-E	1959, 66(2), 69(1, 4-5)
TU	1959
TxBeaL	1948, 53, 59, 63, 66, 68(1-2), 69(1, 4-5), 70, 71(2-4), 72, 76, 78(1, 3-4), 79(1-5, SEPM 2)
TxCaW	1969(1, 4-5), 71(3), 75(SEPM 2), 79(1, 5)
TxDaAR-T	1932, 37, 39, 41, 47-48, 52-53, 58-60, 62-66, 67(A2, 4-8, B1-4), 68(2), 69(1, 4-5), 70, 71(2-4), 72, 73(1-2, SEG 1, SEPM 1-2), 74, 75(2, SEPM 2)
TxDaDM	1932, 39, 41, 53, 55, 59, 60(1-3), 62, 66(2), 68, 69(1, 4-5), 70, 72, 73(1-3, SEG 1, SEPM 1), 79(2, 4-5)
TxDaGI	1929, 32, 39, 41, 49, 52-53, 58-59, 60(1-3), 62(4), 64, 66, 69(1, 4-5), 70(1), 71(3), 75(SEPM 1), 78(2, 4)
TxDaM-SE	1939, 47, 52-53, 60, 62, 72-73, 80, 81(1-4), 82-84
TxDaSM	1932, 48, 53-55, 57-59, 60(3-4), 62, 66(2), 67(A7, B3), 68(1-2), 69(1, 4-5), 71(4), 72-73, 75(SEPM 1), 76(3-4), 78(3, SEPM 1), 79(3)
TxHSD	1932, 47, 52-53, 59, 62-63, 65, 66(1), 67(A2), 68(1-2), 69(1, 4-5), 71(3), 74(2-3, SEPM 2:2), 75(1, SEPM 1-2), 78, 79(4)
TxHU	1929, 41, 48, 51-54, 56, 58, 60, 62, 64, 66, 68(2-3), 71(3)
TxLT	1932, 39, 41, 47, 50, 52-53, 55, 59, 66, 69(1, 4-5), 71(3)
TxMM	1929, 32-33, 39, 41, 47, 52-53, 58-59, 62, 64(2-3), 66, 68, 69(1, 4-5), 79(2-3), 81(AAPG 2), 82-83
TxU	1924, 29, 32, 37, 39, 41, 49, 52-56, 58-60, 62-64, 66, 68, 69(1, 4-5), 70, 71(3-4), 72, 73(1-3), 75(SEPM 1-2), 76(3-5), 79(3-5, SEPM 2), 80, 81, 82(1-2, 4-9), 83-84
TxU-Da	1932, 41, 47, 52-53, 56, 60, 68(3), 69(1, 4-5), 71(3), 73(1-2, SEPM 1-2)
TxWB	1939, 41, 49, 52, 55, 60, 65, 68, 76, 78, 84
UU	1955, 62(4), 66, 68(3), 72(1-3), 78(SEPM 1)
ViBlbV	1953, 55, 58-59, 60(2), 62, 66, 67(A5-8, B2-3), 68, 69(1, 4-5), 70(1), 71(3-4), 72(SEPM 1-2), 73, 75(SEPM 1-2), 76(1-2)
WU	1948, 55, 56, 57-59, 60(3), 62, 66, 68(2-3), 69(1, 4-5), 71(4), 72(A:1-3), 75(SEPM 1-2), 76(1-3, 5), 78(1, SEPM 1-3), 79(1-5), 80, 81(1-4, 5 SEPM 6), 82, 83, 84
WaU	1971(3), 79(4), 81(SEPM 1, 3, 4, 6)
WyU	1947, 81(2-3)

(19) AMERICAN ASSOCIATION OF PETROLEUM GEOLOGISTS. AAPG RESEARCH COMMITTEE CONFERENCE. (TITLE VARIES)

1984　Geology and hydrocarbon deposits of the Santa Maria, Cuyama, Taft-Mckittrick and Edna oil districts, Coast Ranges, California.

OkTA	1984

(20) AMERICAN ASSOCIATION OF PETROLEUM GEOLOGISTS. AAPG STRUCTURAL SCHOOL. [FIELD TRIP GUIDEBOOK]

1981 May　Guide to the geology and structure of the Mecca Hills, Salton Trough.

1983 Aug.　Field trip guide; northern Wyoming Thrust Belt and Teton Pass. Field trip guide; Rendezvous Peak and Jackson Hole.

OCU-Geo	1981
OkTA	1983

(21) AMERICAN ASSOCIATION OF PETROLEUM GEOLOGISTS. AAPG STUDENT CHAPTER. OKLAHOMA STATE UNIVERSITY. FIELD TRIP. [GUIDEBOOK]

1983 Apr.　The geology of the Blue Creek Canyon area in the Slick Hills of southwestern Oklahoma.

OkTA	1983
OkU	1983

(22) AMERICAN ASSOCIATION OF PETROLEUM GEOLOGISTS. AAPG TRUSTEES. FIELD TRIP GUIDE.

1984 Apr.　Field trip for AAPG trustees to eastern Greenbrier County and southern Pocahontas County, West Virginia.

OkTA	1984

(23) AMERICAN ASSOCIATION OF PETROLEUM GEOLOGISTS. DEPARTMENT OF EDUCATION. FIELD SEMINAR. GUIDEBOOK. (TITLE VARIES)

1977 Oct.　Deep water oil sand reservoirs of the Monterey Formation, Fillmore-Piru area, Ventura County, California. (134)

1978 Apr.　Carboniferous depositional environments; eastern Kentucky and southern West Virginia. (707)

1979 Oct.　See American Association of Petroleum Geologists. Eastern Section. 25.

1982 May　Paleozoic stratigraphy and Appalachian exploration trends.

(24) American Association of Petroleum Geologists. Department of Education. School.

CSfCSM	1977
CSt-ES	1977
CU-SB	1978
CaACU	1977
DI-GS	1977
IU	1977
InLP	1982
InU	1978
KyU-Ge	1977
NcU	1978
OCU-Geo	1978, 82
OkU	1982
TxBeaL	1978
ViBlbV	1978

(24) AMERICAN ASSOCIATION OF PETROLEUM GEOLOGISTS. DEPARTMENT OF EDUCATION. SCHOOL. FIELD TRIP GUIDEBOOK.

1977 Dec.   Guide to San Diego area stratigraphy. (1)

CSdS	1977
CSfCSM	1977
CSt-ES	1977
CU-A	1977
CU-EART	1977
DI-GS	1977
IU	1977
IaU	1977

(25) AMERICAN ASSOCIATION OF PETROLEUM GEOLOGISTS. EASTERN SECTION. FIELD TRIP GUIDEBOOK.

1956   One-day field trip in the New York City area.
1966   Guide for field trip. [Newark, New Jersey area]
1972   1st   No. 1. Geology of Silurian rocks, northwestern Ohio. (494)
   No. 2. Pennsylvanian deltas in Ohio and northern West Virginia. (58)
1974   3rd   Conemaugh (Glenshaw) marine events. (541, 586)
1976   5th   Stratigraphic evidence for Late Paleozoic tectonism in northeastern Kentucky. (347, 209(1977))
1979   Oct.   [3] Field trip guidebook for Carboniferous coal short course. (767, 23, 762)
   [4] Devonian clastics in West Virginia and Maryland.
1980   Structure and depositional environments of some Carboniferous rocks of western Kentucky. (406)
1981   See American Association of Petroleum Geologists. National Energy Minerals Division. (Title varies) 27.
1982   11th   Amherst, New York. (Reprinted for the Clay Minerals Society) (469(1983))
   Geology of the northern Appalachian Basin western New York.
   Guidebook for field trips in western New York, northern Pennsylvania and adjacent, southern Ontario.
   A-1. Upper Moscow-Genesee stratigraphic relationships in western New York; evidence for regional erosive beveling in the late Devonian.
   A-2. Carbonate facies of the Onondaga and Bois Blanc formations, Niagara Peninsula, Ontario.
   A-3. Eurypterids, stratigraphy, late Silurian-early Devonian of western New York state and Ontario, Canada.
   A-4. Glacial geology of the Erie Lowland and adjoining Allegheny Plateau, northwestern New York.
   A-5. Recent oil and gas development on public lands in western New York state.
   A-6. Geologic and engineering history of Presque Isle Peninsula, PA.
   B-1. Stratigraphy and facies variation of the Rochester Shale (Silurian; Clinton Group) along Niagara Gorge.
   B-2. Glacial and engineering geology aspects of the Niagara Falls and Gorge.
   B-3. Quaternary stratigraphy and bluff erosion, western Lake Ontario, New York.
   B-4. Sedimentologic and geomorphic processes and evolution of Buttermilk Valley, West Valley, New York.
   B-5. Devonian black shales of western New York.
   [B-6] Subsurface expression and gas production of Devonian black shales in western New York.
1983   Oct.   Carbondale, Illinois. (622)
   Field trip no. 1. Geology of the no. 5 and no. 6 coals of southern Illinois.
   Field trip no. 2. Surface exposures of selected oil producing horizons in the Illinois Basin.
   Field trip no. 3. Loading docks (Cora) and Illinois power generating plant (Baldwin).
   Field trip no. 4. Hydrocarbon operations in the Illinois Basin, Raleigh Field, Saline County, Illinois.
1984   Oct.   Application of the PAC hypothesis to limestones of the Helderberg Group.
   Trenton Limestone fracture reservoirs in Lee County, southwestern Virginia.
1985   Nov.   Williamsburg, Virginia.
   Field trip 1. The Late Cenozoic geology of southeastern Virginia and the Great Dismal Swamp.
   Field trip 2. Middle Pleistocene depositional environments and facies sequences, Accomack Barrier complex, southern Delmarva Peninsula, Virginia.
   Field trip 3. Disturbed zones, lateral ramps and their relationship to the structural framework of the central Appalachians.
   Field trip 5. Stratigraphy of outcropping Tertiary beds along the Pamunkey River - central Virginia coastal plain.

CLU-G/G	1972(1), 76
CLhC	1979(3)
CMenUG	1956, 76
CPT	1979(3)
CSfCSM	1976
CSt-ES	1976, 79(3), 82
CU-A	1982(2)
CU-EART	1956, 76, 79(3-4), 82
CU-SB	1972, 76, 79, 84
CaACU	1972, 79, 82
CaOKQGS	1972(1), 82
CaOOG	1982
CaOTDM	1982
CaOWtU	1972(1), 76, 82
CoDCh	1972(Pt.1-2), 76, 82, 85
CoDGS	1976, 79
CoLiM	1982
CtY-KS	1976
DI-GS	1956, 72, 74, 76, 79(3-4), 80, 82, 84
I-GS	1980
ICF	1972(1)
ICarbS	1979(Oct.3)
IU	1972, 80
IaU	1974, 76, 79(3-4), 80, 82, 84
IdU	1976
InLP	1972, 76, 79(3-4), 80, 82-83
InU	1956, 76
KyU-Ge	1972, 76, 80, 82-83, 85(1-3, 5)
LNU	1972(2), 76
MNS	1956
MdBJ	1979
MnU	1972, 76, 79(3-4), 80, 83
MoSW	1972(1)
NRU	1982, 84
NSyU	1974
NcU	1972, 74, 79(4)
NdU	1976
NhD-K	1972, 76, 82, 84
NjP	1972(1)
OCU-Geo	1972(1), 79(3), 80, 82
OU	1972(1), 76-77, 79
OkU	1956, 72, 76, 79(4)
PBm	1979(4), 84
PSt	1974
TxBeaL	1976
TxDaAR-T	1972
TxDaDM	1981
TxDaGI	1966
TxDaM-SE	1980
TxHSD	1976
TxU	1956, 72, 76, 82, 84
TxWB	1956
ViBlbV	1972
WU	1972, 76, 79(3), 80, 82
WyU	1980

(26) AMERICAN ASSOCIATION OF PETROLEUM GEOLOGISTS. MID-CONTINENT SECTION. GUIDEBOOK FOR THE FIELD TRIP HELD AT THE ANNUAL MEETING.

1975 Sept.   Upper Pennsylvanian limestone facies in southeastern Kansas. (341)

1979 Oct.   The Desmoinesian coal cycles and associated sands of east central Oklahoma. (662)

1981 Sept.   Oklahoma City, Oklahoma.

  Field trip no. 2. Type areas of the Seminole and Holdenville formations western Arkoma Basin.

1983   Geology of the Plattsmouth (Oread Formation, Pennsylvanian System) Marine Bank in Chautauqua County, Kansas, and Osage County, Oklahoma. (341, 779)

CPT	1975
CSt-ES	1979
CU-EART	1975
CU-SB	1975
CaACU	1975, 79
CoG	1975
DI-GS	1975
IEN	1975
IU	1975, 79
IaU	1975
InLP	1975
InU	1975
KyU-Ge	1952, 75, 79, 83
MiU	1975
MnU	1975, 79
NRU	1975
NhD-K	1975
NmLcU	1975
OU	1975-76
OkT	1979
OkTA	1981(2)
OkU	1975, 81
TxHSD	1975
ViBlbV	1975
WU	1975

(27) AMERICAN ASSOCIATION OF PETROLEUM GEOLOGISTS. NATIONAL ENERGY MINERALS DIVISION. (TITLE VARIES)

1978   See American Association of Petroleum Geologists. 18. (no. 2)

1979 [28th]   Guidebook issue, AAPG-SEPM-EMD Rocky Mountain Section meeting. (596, 30, 782)

  [1] Casper Mountain.

  [2] Upper Cretaceous sandstones-Casper-Salt Creek-Tisdale Mountain.

  [3] Powder River uranium mines.

  [4] Eastern Powder River Basin coal fields.

1980   See American Association of Petroleum Geologists. 18.

1981

  — Oct.   Atlantic City, New Jersey. (18, 25, 90, 586)

  [Trip 1] Cretaceous and Tertiary sediments of the New Jersey coastal plain.

  [Trip 2] Rift basins of the Passive margin: tectonics, organic-rich lacustrine sediments, basin analyses.

  [Trip 3] Extinct and active continental margin deposits and their tectonic-switchover products: Appalachian orogen ("Eastern Overthrust Belt") - Catskill Plateau-Newark Basin-Atlantic coastal plain.

  [Trip 4] Field guide to the anthracite coal basins of eastern Pennsylvania. (539)

  [Trip 5] Field guide to the geology of the Paleozoic, Mesozoic, and Tertiary rocks of New Jersey and the central Hudson Valley. (539)

1985   See American Association of Petroleum Geologists. Rocky Mountain Section. 30.

  —Apr.   See Oil Shale Symposium. 501.

AzTeS	1979
CLU-G/G	1979
CPT	1979
CSt-ES	1979
CU-EART	1979
CaACU	1979
CoG	1979
CoU	1979
CtY-KS	1981
DI-GS	1979
IU	1979
IaU	1979
InLP	1979
KLG	1979
KyU-Ge	1981
NSbSU	1979, 81
NdU	1979
NhD-K	1979
NvU	1979
OCU-Geo	1981
OU	1979, 81(4,5)
OkU	1979
TxDaDM	1979
TxU	1979
ViBlbV	1979
WU	1979
WyU	1979

(28) AMERICAN ASSOCIATION OF PETROLEUM GEOLOGISTS. PACIFIC SECTION. GUIDEBOOK FOR THE FIELD TRIP.

1944 Nov.   Type locality of Sycamore Canyon Formation, Whittier Hills, Los Angeles County, California. (593)

1947 Sept.   Gaviota Pass-Refugio Pass areas, Santa Barbara County, California. (593)

1950 May   North Mt. Diablo monocline, Contra Costa County, California. (593)

1951 May   Cuyama District, California. (557, 593)

1952 Mar.   See American Association of Petroleum Geologists. 18.

1953 Apr.   Ventura-Ojai-Santa Paula area. (593)

1954 May 7-8   Capay Valley-Wilbur Springs, Westside Sacramento Valley. (487)

— May 15   San Marcos Pass to Jalama Creek. (593)

1955 May   Devils Den-McLure Valley area. (557, 593)

1958 Mar.   See American Association of Petroleum Geologists. 18.

— May   Imperial Valley, California. (593)

1960 June   San Gabriel Fault.

1961 May   Geology and paleontology of the southern border of San Joaquin Valley, Kern County, California. (557, 593, 601)

— June   Rincon Island-Casitas-Ventura Avenue field areas.

1962 Mar.   See American Association of Petroleum Geologists. 18.

— June   Geology of southeastern Ventura Basin including Aliso Canyon, Cascade, Horse Meadows, Mission oil fields.

— Oct.   Geology of Carrizo Plains and San Andreas Fault. (557, 593)

1963 May   Geology of Salinas Valley; production, stratigraphy, structure and the San Andreas Fault. (557, 593)

1964 Oct.   San Andreas fault zone from the Temblor Mountains to Antelope Valley, Southern California. (557, 593)

1965 June   Placerita-Soledad-Vasquez rocks area, Soledad Basin, Los Angeles County, California.

1966 Mar.   A tour of the coastal oil fields of the Los Angeles Basin in and adjacent to San Pedro Bay, California. (593, 601)

— June   Santa Susana Mountains. (593)

1967 Apr.   See American Association of Petroleum Geologists. 18.

— June   Geology of the Big Mountain oil field and the nearby area, including notes on the trip from Piru to Big Mountain, Ventura County, California.

— Oct.   Gabilan Range and adjacent San Andreas Fault. (593)

1968 Mar.   Geology and oil fields, west side, southern San Joaquin Valley. (593, 601)

— June   Tehachapi Mountains crossing of the California aqueduct, Kern and Los Angeles counties, Southern California.

— Oct.   Field trip guide to Santa Rosa Island. (593)

1969 Mar.   Geology and oil fields of coastal areas, Ventura and Los Angeles basins, California. (593, 601)

— June   Geology of the central part of the Fillmore Quadrangle, Ventura County, California.

1970 Mar.   Southeastern rim of the Los Angeles Basin, Orange County, California; Newport Lagoon-San Joaquin Hills-Santa Ana Mountains. (593, 601)

— June   Ventura Avenue & San Miguelito oil fields.

— Nov.   Pacific slope geology of northern Baja California and adjacent Alta California. (593, 601)

(28) American Association of Petroleum Geologists. Pacific Section.

1971 Feb. Summary field trip guide, San Andreas Fault-San Francisco Peninsula.
1972 Mar. Geology and oil fields, west side central San Joaquin Valley. (593, 601)
— June Central Santa Ynez Mountains, Santa Barbara County, California. (593)
1973 May See American Association of Petroleum Geologists. 18.
— June Traverse of Castaic-Ridge basins and basement complex north of Valencia, California, with tour of west branch, California aqueduct system, Antelope Valley to Castaic.
1974 Apr. A guidebook to the geology of peninsular California. (593, 601)
1975 Apr. A tour of the oil fields of the Whittier fault zone, Los Angeles Basin, California. (593, 601)
— June Geology of the Torrey Canyon, Oakridge, Santa Susana, South Tapo, and Tapo Ridge oil fields, and the nearby area.
1976 Apr. A tour of the reservoir rocks of the western Sacramento Delta. (593, 601)
— June Pliocene geology and the Santa Paula oil field areas; northeast part of Rancho San Buenaventura, Ventura Basin, California. (134)
1977 [Apr.] Late Miocene geology and new oil fields of the southern San Joaquin Valley. (593, 601, 557)
— June San Cayetano Fault field trip (with apologies to the San Andreas).
1978 Apr. Field trip guidebook: Castle steam field, Great Valley sequence. (593, 601)
1979 Mar. Geologic guide to San Onofre nuclear generating station and adjacent regions of southern California. (Guidebook 46) (593, 601)
— June Geology of the Lake Casitas area, western Ojai Valley, Ventura County, California. (Cover title: Geology of the Lake Casitas area, Ventura County, California) (134)
1980 May (1) Geologic guide to the stratigraphy & structure of the Topanga Group; central Santa Monica Mountains, Southern California. (356)
(2) Kern River oilfield. (593, 584)
1981 May See American Association of Petroleum Geologists. 18.
1982 Apr. Anaheim, California. Cenozoic nonmarine deposits of California and Arizona. (593)
(2) Late Cretaceous depositional environments and paleogeography, Santa Ana Mountains, Southern California. (593, 601, 199(8))
(3) Geologic history of Ridge Basin, Southern California. (593, 601, 199(1))
— June Geologic guide of the central Santa Clara Valley, Sespe and Oak Ridge Trend oil fields, Ventura County, California. (134)
1984 Apr. Miocene and Cretaceous depositional environments, northwestern Baja California, Mexico.
— June San Andreas Fault-Cajon Pass to Wrightwood.
1985 May Anchorage, Alaska. (593, 601, 6)
1. Geologic guide to the Fairbanks-Livengood area, east-central Alaska.
2. Guide to the geology of the Kenai Peninsula, Alaska.
Guide to the engineering geology of the Anchorage area. (593, 601)

AzTeS	1951, 54(May 15), 56, 58(May), 59-61, 62(Oct.), 63-72, 73(June), 74(Apr.), 77, 78, 79(Mar.), 80, 81, 82(June), 84(Apr.), 85(1)
AzU	1963, 70(Nov.)
CLU-G/G	1947, 50, 53(Apr.), 54-55, 58(May), 61, 62(Oct.), 63, 65-72, 73(June), 74(Apr.), 75, 77-78, 79, 80, 82(Apr.), 84(Apr., June)
CLhC	1962(June, Oct.), 63-64, 65(Apr.), 66(Mar.), 67(Oct.), 68(Mar.), 69(Mar., June), 70(Mar., Nov.), 72(Mar.), 74(Apr.), 76(June), 82(Apr., June)
CMcnUG	1947, 51, 54, 61(May)
CPT	1947, 50-51, 54, 61(May), 62(Oct.), 63-66, 67(June, Oct.), 68-71, 72(Mar.), 73(June), 74-75, 77-82, 85
CSdS	1950, 54, 58(May), 66(Mar.), 67(Oct.), 68-70, 72, 73(June), 74(Apr.), 75, 76(Apr.), 77-80, 82(Apr., June), 84(Apr.)
CSfCSM	1950, 54(May 7-8), 62(June, Oct.), 63-65, 66(Mar.), 67(Oct.), 68(Mar., June), 69(Mar.), 70(Apr.), 71, 72(Mar., June), 73(June), 75(June), 76(June), 77(Apr., June), 78, 79(Mar.), 80, 82(Apr.)
CSt-ES	1954(May 7-8), 60, 61(May), 62(Oct.), 63-72, 73(June), 75, 76(June), 77-80, 82, 85
CU-A	1944, 47, 50-51, 53(Apr.), 54-55, 58(May), 61, 62(June, Oct.), 63-71, 72(Mar.), 74(Apr.), 77, 79(Mar.), 80, 81(2), 82, 84(Apr., June)
CU-EART	1947, 50-51, 53(Apr.), 54-55, 58(May), 60-61, 62(June, Oct.), 63-66, 67(Oct.), 68-72, 73(June), 74(Apr.), 75(Apr.), 76(June), 78-80, 82(Apr.), 84
CU-SB	1947, 50-54, 56-66, 67(4, 7), 68-75, 77-80, 82, 84(Apr.)
CU-SC	1980, 82(Apr., June)
CaACU	1965(Apr.), 67(June, Oct.), 68(Mar., June), 69(Mar.), 70, 72(June), 76(June), 77(Apr.), 78
CaBVaU	1964, 67(Oct.)
CaOHM	1965, 75
CaOKQGS	1967-69, 80, 82
CaOLU	1963, 65, 66(June), 67-68
CaOTRM	1958(May), 63, 66(Mar.), 67(Oct.)
CaOWtU	1961(May), 62(Oct.), 64, 65(Apr.), 67(Oct.), 68(Oct.), 70(Mar., Nov.)
CoDCh	1972(Mar.)
CoDGS	1965
CoFS	1962(Oct.), 66(Mar.), 67, 68(Mar.), 69(Mar., June)
CoG	1947, 50, 54(May 7-8), 58(May), 60, 61(May), 62(Oct.), 64(Oct.), 65, 66(Mar.), 67-70
CoLiM	1980, 1982
CoU	1968-70
CoU-DA	1980, 82
DI-GS	1947, 50-51, 53(Apr.), 54-55, 58(May), 61(May), 62(Oct.), 63-66, 67(Oct.), 68-72, 73(June), 74(Apr.), 75, 76(June), 77-80, 82, 84
DLC	1962(June, Oct.), 67(Oct.), 68(Mar.)
ICF	1954(May 7-8), 62(June, Oct.), 68(Mar.)
IU	1947, 54, 58(May), 61, 62(June, Oct.), 63-64, 65(June), 66-72, 74(Apr.), 75, 77-78, 79(Mar.), 84(55), 80
IaU	1958(May), 63-64, 66-68, 69(Mar.), 70(Mar., Nov.), 71-72, 73(June), 74(Apr.), 75, 76(June), 77(Apr.), 78, 79(Mar.), 82, 84
IdBB	1962(June, Oct.), 64, 67(Oct.), 68, 70-71
IdPI	1963
InLP	1965(Apr.), 66(Mar.), 67, 68(Mar.), 70(Mar., Nov.), 72(Oct.), 74, 75(Apr.), 77(June), 78-80, 82, 85
InU	1950, 54, 62(Oct.), 68(Mar.), 69, 70(Nov.), 72(Mar., June)
KyU-Ge	1947, 51, 54(May 15), 61(May), 62(Oct.), 63, 64-67, 68(June, Oct.), 69-72, 73(June), 74(Apr.), 75, 76(June), 77-78, 82, 84, 85(1, 2)
MH-GS	1970(Nov.)
MNS	1958(May)
MarLiCO	1980
MiU	1961(June), 67(June), 69(Mar.), 70(June, Nov.)
MnU	1944, 50-51, 53(Apr.), 54-55, 58(May), 61, 62(June, Oct.), 63-72, 73(June), 74(Apr.), 75, 77(Apr.), 78, 80, 82
MoSW	1963, 68(Mar.)
MoU	1966(Mar.), 67, 68(Mar., June), 69(Mar., June), 70(Nov.), 72(June)
NNC	1961(May), 62(June, Oct.), 63, 65-67, 68(Oct.), 69
NOncoU	1967(Oct.), 68(Mar.), 69-70
NRU	1970(Mar., Nov.)
NSbSU	1970
NSyU	1958(May), 72(Mar.), 85(60)
NbU	1962(June, Oct.)
NcU	1962(June, Oct.), 64-65, 67, 68(Mar., June), 69-72, 74(Apr.), 82
NdU	1961(May), 68(June), 70(Nov.), 72
NhD-K	1970(Nov.)
NjP	1958(May), 66-68, 69(Mar.), 70(Nov.)
NvU	1947, 51, 54, 60-61, 62(Oct.), 63(June), 64-65, 66(June), 67-68, 69(Mar.), 70-72, 73(June), 74(Apr.), 75(Apr.), 77, 81, 82(Apr.)
OCU-Geo	1947, 61(May), 62(Oct.), 63(June), 64-71
OKentC	1980
OU	1947, 55, 61(May), 62(Oct.), 63-72, 75(June), 76(Apr.)
OkTA	1980, 82(June), 84
OkU	1954(May 7-8), 63, 66(Mar.), 67(Oct.), 68, 69(Mar.), 70(Nov.), 71, 72(June), 73(June), 74(Apr.), 75(Apr.), 80, 82(Apr.)
OrU-S	1964, 65(June)
PSt	1970
SdSNH	1984(Apr.)
TxBeaL	1950, 54(May 7-8), 65(Apr.), 66-69, 70(Mar., Nov.), 71
TxDaAR-T	1961(May), 62(June, Oct.), 63-64, 65(June), 66-72, 73(June), 74(Apr.), 75(Apr.)
TxDaDM	1947, 61, 62(June, Oct.), 63(June), 64, 65(June), 66-71, 72(June), 73(June), 74(Apr.), 75, 77-78
TxDaGI	1968(Mar.), 70(Nov.), 75(June)
TxDaM-SE	1954(May 7-8), 58(May), 61(May), 62, 63(June), 65-66, 82
TxDaSM	1954, 61(May), 62(June, Oct.), 63-65, 68(Mar., Oct.), 69(Mar.), 70(Mar., Nov.), 72, 74(Apr.), 75(Apr.), 76(June), 77(Apr.), 78
TxHSD	1954(May 7-8), 61(May), 62(Oct.), 63-64, 66(Mar.), 67(Oct.), 68(Mar., June), 69(Mar.), 72, 79(Mar.)
TxMM	1961(May)
WaU	1974, 80, 82(2)

(29) AMERICAN ASSOCIATION OF PETROLEUM GEOLOGISTS. REGIONAL MEETING. [GUIDEBOOK] (TITLE VARIES)
1937   Southwestern Pennsylvania, West Virginia, western Virginia, western Maryland. [Midyear meeting]
1938   Field trips, midyear meeting, El Paso, Texas. (756)
   No. 1. Alamogordo trip.
   No. 2. Glass Mountains-Marathon Basin trip.
   No. 3. Carlsbad; oil field trip.
   No. 4. Malone Mountain-Green Valley trip.
   No. 5. Silver City trip.
1947 Nov.   San Antonio-George West-Beeville-Falls City. (612)
1948 Oct.   Geology of the northern portion of the Appalachian Basin. (541)
1950 Sept.   See Geological Association of Canada. 190.
1951 Oct.   Paleozoic and Cretaceous of eastern Llano Uplift. (612)
1957 Nov.   Eastern Oklahoma. (662)
1958 Oct.   See Southwestern Federation of Geological Societies. 627.
1959 Oct.   South-central Kansas. (341, 634)
   — Nov.   See Gulf Coast Association of Geological Societies. 247.
1961 Oct.   Palo Duro Canyon State Park. (528)
1965 Oct.   See Gulf Coast Association of Geological Societies. 247.

AzTeS	1947, 51, 57, 59(Oct.)
CLU-G/G	1948, 57, 59(Oct.)
CPT	1948, 59(Oct.)
CSt-ES	1947-48, 51
CU-EART	1947, 59(Oct.)
CU-SB	1948
CaACAM	1948
CaACU	1959(Oct.)
CaBVaU	1948
CaOHM	1947, 57, 59(Oct.)
CaOLU	1959(Oct.)
CaOWtU	1957
CoDCh	1957, 59(Oct.)
CoG	1947-48, 51, 59(Oct.)
CoU	1957, 59(Oct.)
DFPC	1957
DI-GS	1937, 47-48, 51, 57, 59(Oct.)
DLC	1951
I-GS	1948, 51, 59(Oct.)
ICF	1948, 59(Oct.)
ICIU-S	1957
IEN	1948, 57
IU	1947-48, 57, 59(Oct.)
IaU	1948, 59(Oct.)
InLP	1957, 59(Oct.)
InU	1948, 59(Oct.)
KLG	1959
KyU-Ge	1947, 59(Oct.)
LNU	1947, 59(Oct.)
LU	1947-48, 51, 57, 59(Oct.)
MH-GS	1948
MNS	1948
MiDW-S	1948, 59(Oct.)
MiU	1959(Oct.)
MnU	1947-48, 51, 57-59(Oct.)
MoSW	1948
MoU	1947, 59(Oct.)
MtBuM	1948
MtU	1959(Oct.)
NBiSU	1948
NNC	1948, 59(Oct.)
NSyU	1948
NbU	1947, 51, 57, 59(Oct.)
NcU	1947, 51, 57, 59(Oct.)
NhD-K	1959(Oct.)
NjP	1948, 59(Oct.)
NmPE	1947, 57
OCU-Geo	1937, 47-48
OU	1948, 57, 59(Oct.)
OkT	1948, 57, 59(Oct.)
OkU	1947-48, 51, 57, 59(Oct.)
PBL	1948
PBm	1948
PHarER-T	1948
RPB	1948
TMM-E	1957
TU	1948
TxBeaL	1947, 59(Oct.)
TxDaAR-T	1948, 51, 57, 59(Oct.)
TxDaDM	1947, 51, 57, 59(Oct.)
TxDaGI	1947-48
TxDaM-SE	1947, 51, 57, 59(Oct.)
TxDaSM	1947, 51
TxHSD	1947, 59(Oct.)
TxHU	1947-48, 51, 57, 59(Oct.)
TxLT	1947
TxMM	1938, 47, 51
TxU	1938, 47-48, 51, 59(Oct.), 61
TxWB	1947-48, 51
ViBlbV	1947, 51, 57, 59(Oct.)
WU	1948, 59(Oct.)

(30) AMERICAN ASSOCIATION OF PETROLEUM GEOLOGISTS. ROCKY MOUNTAIN SECTION. GUIDEBOOK FOR THE ANNUAL MEETING AND FIELD TRIP.
1964   Durango-Silverton. (180)
1967   Road log - Casper-Casper Mountain-Bates Hole and Alcova via Circle Drive and Highway 220.
1970   See American Association of Petroleum Geologists. 18.
1974   [1] Parkman Delta in central Wyoming;
   [2] Upper Cretaceous field trip log;
   [3] Pratt Ranch First Frontier Sandstone outcrop;
   [4] Geology and geochemistry of the Highland uranium deposit;
   [5] Dave Johnston coal strip mine road log. (596, 782)
1975   Field trips to central New Mexico.
1984 Aug.   Thrust belt and Wasatch-Uinta intersect area.
1985   Rocky Mountain section; field trip guide. (27, 549, 596)

ATuGS	1964
AzTeS	1964, 73(5)
CLU-G/G	1985
CLhC	1964
CMenUG	1985
CPT	1974(2)
CSt-ES	1974
CU-EART	1964, 67, 74
CaACI	1964
CaACU	1974
CaOHM	1964
CaOOG	1964
CoD	1985
CoDCh	1964, 85
CoDGS	1985
CoDuF	1964
CoG	1964
CoU	1974
DI-GS	1964, 74-75
ICIU-S	1964
ICarbS	1964
IEN	1964
IU	1964, 85
IaU	1964
IdBB	1964
InLP	1974, 85
KyU-Ge	1964, 84
LU	1964
MiDW-S	1964
MnU	1964
NBiSU	1964
NIC	1985
NNC	1964
NcU	1964, 85
NdU	1974-75
NmPE	1964
NmU	1985
NvU	1982(Apr.)
OU	1964, 85
OkU	1964
PBL	1964
TxDaAR-T	1964
TxDaDM	1964
TxDaGI	1964
TxDaM-SE	1964
TxDaSM	1964, 75
TxHSD	1964
TxHU	1964
TxLT	1964

(31) American Association of Petroleum Geologists. Southwest Section.

TxMM	1964, 85
TxU	1964, 67
UU	1974
WU	1974(5)
WyU	1974, 85

(31) AMERICAN ASSOCIATION OF PETROLEUM GEOLOGISTS. SOUTHWEST SECTION. ANNUAL MEETING. GUIDEBOOK. (TITLE VARIES)

1974 Feb. Sierra de Juarez field trip. (Supplement to the El Paso Geological Society Sixth Annual Field Trip Guidebook. 1972) (167, 594)
1975 See American Association of Petroleum Geologists. 18.
1980 Feb. El Paso, Sierra de Juarez Chihuahua Mexico. Structure and stratigraphy. (167)
1981 Apr. West central Texas field trip. (555)
1985 Feb. Middle & Upper Pennsylvanian cratonic basin facies models, north-central Texas.

CLU-G/G	1980
CSt-ES	1974, 80
CU-A	1980
CaOKQGS	1985
CoDCh	1980
CoDGS	1980, 85
DI-GS	1980, 85
IaU	1980, 85
InLP	1985
MU	1985
MnU	1980
NmU	1980
OCU-Geo	1985
OkTA	1980, 85
OkU	1980-81, 85
TxDaM-SE	1980
TxHMP	1985
TxMM	1980, 85
TxU	1981
TxWB	1985

(32) AMERICAN ASSOCIATION OF STRATIGRAPHIC PALYNOLOGISTS. FIELD TRIP GUIDEBOOK.

1968 1st Lower Mississippi Valley, Baton Rouge to Vicksburg.
1969 2nd Sedimentation in the Carboniferous of western Pennsylvania. (Cover title: Field trip to the Allegheny Front and Appalachian Plateau of central Pennsylvania)
1970 3rd Niagara Escarpment field trip.
1971 4th [1] Geology along Arizona Highways 77 and 87 between Tucson, Holbrook, and Phoenix. Field guidebook from Tucson, Arizona to the Petrified Forest National Park and the Meteor Crater.
  [2] Lehner early man-mammoth site.
1972 5th Field guide to geology of the Cape Cod National Seashore.
1973 6th [1] Guidebook [to] coastal area geology at Torrey Pines State Preserve, and La Jolla, San Diego County, California.
  [2] Field guidebook for Sequoia and Kings Canyon national parks.
1977 10th Field trip; Tulsa area.
1978 11th Paleontology, stratigraphy and vegetation in east-central Arizona.
1979 12th Lower Cretaceous shallow marine environments in the Glen Rose Formation: Dinosaur tracks and plants.
1980 13th Keystone to Denver.
1981 [14th] Alluvial valley and upper deltaic plain.
1983 16th San Francisco, California.
  The Geysers Geothermal Field.
1984 [17th] Cretaceous and Tertiary stratigraphy, paleontology, and structure, southwestern Maryland and northeastern Virginia.

AzTeS	1980-81
AzU	1971
CLU-G/G	1969, 80, 84
CLhC	1969, 80, 84
CMenUG	1971-73, 77-80, 83
CSdS	1969
CSt-ES	1969, 72-73, 77-80, 83-84
CU-A	1980, 84
CU-EART	1971(2), 73, 77-81, 83-84
CU-SB	1967, 71-73, 77-81, 84
CaACU	1979
CaOLU	1968
CaOOG	1968, 73, 77, 80, 84
CaOWtU	1969, 71-73, 79
CoDGS	1977, 80, 83-84
DI-GS	1968-69, 71-73, 77-80, 83-84
ICarbS	1969
ICharE	1984
IU	1971, 80
IaU	1969, 72-73, 77-81, 84
InLP	1971(2), 73, 77-80
InU	1972
KyU-Ge	1969, 71, 73, 77-81, 83-84
MH-GS	1972
MnU	1969-71, 73, 77-80
MoSW	1969
MoU	1972
NN	1985
NNC	1980, 84
NRU	1972
NcU	1972
NdU	1971
NhD-K	1972
NjP	1969, 72
NmU	1984
OU	1968-69, 71(2), 73, 77-79
OkU	1968-69, 71(2), 73(1), 77-80, 83-84
PHarER-T	1969
PSt	1969, 84
TxBeaL	1972, 77
TxDaAR-T	1968-69, 71-73
TxDaDM	1969
TxDaSM	1969, 71
TxU	1968-69, 73(1), 77
WU	1973(1-2), 77

(33) AMERICAN CRYSTALLOGRAPHIC ASSOCIATION. GUIDEBOOK FOR FIELD TRIPS.

1965 Gatlinburg, Tennessee [southern Appalachians]. (384)
  1. East Tennessee marble district.
  2. Ducktown, Tennessee: Part I: Introduction and road guide. Part II: The Ducktown copper district.
  3. The Mascot-Jefferson City zinc district.
  4. Corundum Hill, North Carolina.

CSt-ES	1965
CU-EART	1965
CoG	1965
DI-GS	1965
IEN	1965
IU	1965
IaU	1965
InLP	1965
KyU-Ge	1965
LNU	1965
NBiSU	1965
NNC	1965
NOneoU	1965
NcU	1965
NdU	1965
NhD-K	1965
NjP	1965
NmPE	1965
OU	1965
OkU	1965
PBm	1965
PSt	1965
TMM-E	1965
TxDaDM	1965
ViBlbV	1965
WaU	1965

(34) AMERICAN GEOGRAPHICAL ASSOCIATION OF NEW YORK. [GUIDEBOOK]

1912 Guidebook for the transcontinental excursion.

CSt-ES	1912
CoG	1912
DI-GS	1912
IaU	1912
MtBuM	1912
NdU	1912

(35) **AMERICAN GEOLOGICAL INSTITUTE. AGI SELECTED GUIDEBOOK SERIES.**
1979 1. See International Congress of Carboniferous Stratigraphy and Geology. 290.
— 2. See International Congress of Carboniferous Stratigraphy and Geology. 290.
— 3. See International Congress of Carboniferous Stratigraphy and Geology. 290.

(36) **AMERICAN GEOLOGICAL INSTITUTE. INTERNATIONAL FIELD INSTITUTE. GUIDEBOOK. (TITLE VARIES)**
1970 Guidebook to the Caribbean Island-arc system.

CLU-G/G	1970
CLhC	1970
CSt-ES	1970
CaBVaU	1970
CaOHM	1970
CaOKQGS	1970
CaOWtU	1970
CoU	1970
DI-GS	1970
IU	1970
InLP	1970
KyU-Ge	1970
LNU	1970
MH-GS	1970
MoRM	1970
NBiSU	1970
NRU	1970
NcU	1966, 70
NdU	1970
NhD-K	1970
PSt	1970
TxBeaL	1970
TxU	1970
ViBlbV	1970

(37) **AMERICAN GEOMORPHOLOGICAL FIELD GROUP. FIELD TRIP GUIDEBOOK.**
1982 Sept. Pinedale, Wyoming.
1983 Chaco Canyon country; a field guide to the geomorphology, Quaternary geology, paleoecology, and environmental geology of northwestern New Mexico.
1985 Redwood country; northwestern California.

CLU-G/G	1983, 85
CMenUG	1982-83
CPT	1983
CSt-ES	1983
CU-A	1982, 83
CU-EART	1982-83, 85
CU-SC	1985
CaOWtU	1983
CoDGS	1982-83
DI-GS	1982
IdU	1983
KyU-Ge	1985
NmLcU	1983
NmSoI	1982
NmU	1982-83
OCU-Geo	1983
TxU	1983
WyU	1982

(38) **AMERICAN GEOPHYSICAL UNION. FALL ANNUAL MEETING. FIELD TRIP.**
1976 Dec. Field trip via BART to the Hayward Fault to see evidence of fault movement and demonstration of portable seismometers.
1983 Dec. Guide to the AGU field trip. [This trip is held in conjunction with the nano-plate tectonic symposium on Friday afternoon, to illustrate some of the distinctive terranes in the Franciscan assemblage of the San Francisco Bay region]
1984 A streetcar to subduction and other plate tectonic trips by Public Transport in San Francisco.

CSt-ES	1983, 84
CU-EART	1976
DI-GS	1983
NSbSU	1984

(39) **AMERICAN GEOPHYSICAL UNION. PACIFIC NORTHWEST MEETING. GUIDEBOOK.**
1979 Sept. Guides to some volcanic terranes in Washington, Idaho, Oregon, and Northern California. (218(1981))
1. Guide to geologic field trip between Lewiston, Idaho, and Kimberly, Oregon, emphasizing the Columbia River Basalt Group.
2. Guide to geologic field trip between Kimberly and Bend, Oregon with emphasis on the John Day Formation.
3. Central High Cascade roadside geology.
4. Newberry Volcano, Oregon.
5. High Lava Plains, Brothers fault zone to Harney Basin, Oregon.
6. A field trip to the maar volcanoes of the Fort Rock-Christmas Lake Valley basin, Oregon.
7. Roadlog for field trip to Medicine Lake Highland.
1981 [See Sept., guidebook field trips 1-7. This guidebook actually refers to the "series of field trips held in conjunction with the Pacific Northwest, American Geophysical Union meeting held in Bend, Oregon, September 1979] (218)

AzTeS	1981
BEG	1981
CLU-G/G	1981
CLhC	1981
CMenUG	1979, 81
CPT	1979, 81
CSdS	1981
CSfCSM	1981
CSt-ES	1981
CU-EART	1979, 81
CU-SB	1981
CaACU	1979, 81
CaABU	1981
CaOOG	1981
CaOTDM	1981
CaQQERM	1981
CoD	1981
CoDBM	1981
CoDGS	1981
DI-GS	1981
DLC	1979
ICarbS	1979
IaCfT	1981
IaU	1979, 81
IdBB	1981
IdU	1979, 81
InLP	1979
KLG	1981
MA	1981
MiD	1981
MnU	1979, 81
NIC	1981
NMU	1981
NNC	1981
NRU	1981
NSbSU	1979, 81
NcU	1981
NdFA	1981
NdU	1981
NhD-K	1981
NmLcU	1981
NvU	1981
OCU-Geo	1979, 81
OU	1981
OkT	1981
OrU-S	1981
PPi	1981
TxCaW	1981
TxDaM-SE	1981
TxU	1981
UU	1981
WU	1979, 81
WyU	1981

AMERICAN INSTITUTE OF BIOLOGICAL SCIENCES. See BOTANICAL SOCIETY OF AMERICA. PALEOBOTANICAL SECTION. (101)

(40) AMERICAN INSTITUTE OF BIOLOGICAL SCIENCES. AIBS ANNUAL MEETING. GUIDEBOOK FOR FIELD TRIP.
1950 1st Annual field trip, paleobotanical section of Botanical Society of America.
Pennsylvanian of southeastern Ohio.
1958 Guidebook for paleobotanical field trip (Bloomington, IN).
1960 [11th] Guidebook for paleobotanical field trip (Stillwater, Oklahoma). (101)
1961 [12th] Guidebook for paleobotanical field trip (Lafayette, Indiana). (101)
1970 21st Guidebook to Eocene plant localities of western Kentucky and Tennessee. (101)
1974 25th Guidebook to Devonian, Permian and Triassic plant localities, east-central Arizona. (101)

AzU	1974
CU-EART	1970, 74
CaOOG	1974
CoDGS	1974
DI-GS	1961, 74
OU	1950, 58, 60-61, 70
OkU	1960
TxDaSM	1970
TxU	1970

(41) AMERICAN INSTITUTE OF MINING, METALLURGICAL AND PETROLEUM ENGINEERS. ANNUAL MEETING. GUIDEBOOK.
1976 Guidebook: Las Vegas to Death Valley and return. 1975. (452)

AzTeS	1976
CPT	1976
CSfCSM	1976
CSt-ES	1976
CU-EART	1976
CaACU	1976
CaBVaU	1976
CaOWtU	1976
CoDBM	1976
CoU	1976
DI-GS	1976
IaU	1976
InLP	1976
InU	1976
KyU-Ge	1976
MtBuM	1976
NhD-K	1976
NvU	1976
OU	1976
TxDaDM	1976

(42) AMERICAN INSTITUTE OF MINING, METALLURGICAL AND PETROLEUM ENGINEERS. INDUSTRIAL MINERALS DIVISION. [GUIDEBOOK FOR FIELD TRIP]
1950 Guidebook of field trip in the Arbuckle Mountains.
1951 Morgantown, West Virginia.
Logs of field trips.
1952 Industrial sand deposits of the Indiana dunes.

IU	1950
InLP	1951
OU	1952
OkU	1950
TxLT	1950
TxU	1950

(43) AMERICAN INSTITUTE OF MINING, METALLURGICAL AND PETROLEUM ENGINEERS. MICHIGAN TECH STUDENT CHAPTER. FIELD TRIP.
1969 Spring Western field trip.
1970 Spring Canadian field trip.
1971 Spring Southern field trip.

MiHM	1969-71

(44) AMERICAN INSTITUTE OF MINING, METALLURGICAL AND PETROLEUM ENGINEERS. SOUTH TEXAS MINERALS SECTION.
1977 See Uranium in Situ Symposium. 728.

(45) AMERICAN INSTITUTE OF MINING, METALLURGICAL AND PETROLEUM ENGINEERS. SOUTHERN CALIFORNIA PETROLEUM SECTION.
1956 Oct. Ventura Avenue and San Miguelito fields. (Cover title: Ventura and San Miguelito fields)

AzTeS	1956
CMenUG	1956
CPT	1956
CSfCSM	1956
CU-EART	1956
CU-SB	1956
CoG	1956
DI-GS	1956
KyU-Ge	1956
MnU	1956
NvU	1956
OCU-Geo	1956
OU	1956
TxDaDM	1956
TxU	1956
ViBlbV	1956

(46) AMERICAN INSTITUTE OF MINING, METALLURGICAL AND PETROLEUM ENGINEERS. SOUTHWEST MINERAL INDUSTRY CONFERENCE. [FIELD TRIP GUIDEBOOK]
1965 May A.I.M.E. Pacific Southwest Mineral Industry Conference.
1. Metals.
2. Industrial minerals.

CSt-ES	1965
CoG	1965
DI-GS	1965
NvU	1965

(47) AMERICAN INSTITUTE OF MINING, METALLURGICAL AND PETROLEUM ENGINEERS. SOUTHWESTERN NEW MEXICO SECTION.
1949 See Geological Society of America. 197.

(48) AMERICAN INSTITUTE OF MINING, METALLURGICAL, AND PETROLEUM ENGINEERS. VIRGINIA SECTION. ANNUAL MEETING. FIELD TRIP.
1981 Geologic investigations in the Willis Mountain and Andersonville quadrangles. (744(No.29))

CLU-G/G	1981
CMenUG	1981
CSfCSM	1981
CSt-ES	1981
CU-EART	1981
CaOOG	1981
DI-GS	1981
IaU	1981
InLP	1981
MnU	1981
NmSoI	1981
OkU	1981
TxU	1981
WU	1981

(49) AMERICAN INSTITUTE OF PROFESSIONAL GEOLOGISTS (AIPG). FIELD TRIP [GUIDEBOOK].
1983 Mar. Field trip to Arbuckle Mountains. (662)

IU	1983
OkT	1983
TxMM	1983
TxU	1983

AMERICAN INSTITUTE OF PROFESSIONAL GEOLOGISTS (AIPG). COLORADO SECTION. See AMERICAN INSTITUTE OF PROFESSIONAL GEOLOGISTS (AIPG). (49)

(50) AMERICAN INSTITUTE OF PROFESSIONAL GEOLOGISTS. CALIFORNIA SECTION.
1977 Geologic guide to the San Andreas fault zone between Bolinas Lagoon and Tomales Bay, Marin County, California.
KyU-Ge 1977

(51) AMERICAN MUSEUM OF NATURAL HISTORY. [GUIDEBOOK]
1968 Geology of New York City and environs.
A. Geologic features in Fort Tryon and Indwood Hill Parks.
B. The geology seen on a cross-Bronx to Manhattan field trip.
C. The geology of the northern Palisade and Watchung ridges.
D. Geologic features in the northern part of the New Jersey lowland.
E. Some geologic features on Staten Island (Borough of Richmond), New York City.
F. The Precambrian and Paleozoic geology of the Hudson Highlands and vicinity.
G. The geology seen on a trip to the Delaware Water Gap.
H. Some geologic features seen in the southern part of the Connecticut Valley lowland.

AzTeS	1968
AzU	1968
CChiS	1968
CLU-G/G	1968
CSdS	1968
CSt-ES	1968
CU-A	1968
CU-EART	1968
CaAC	1968
CaACU	1968
CaAEU	1968
CaBVaU	1968
CaOOG	1968
CaOWtU	1968
CoG	1968
DI-GS	1968
DLC	1968
ICF	1968
ICarbS	1968(A)
IaU	1968
InLP	1968
InRE	1968
InU	1968
MiHM	1968
MnU	1968
MoSW	1968
NSyU	1968
NbU	1968
NcU	1968
NdU	1968
NhD-K	1968
NjP	1968
NmPE	1968
OU	1968
OkOkCGS	1968
OkT	1968
OrU-S	1968
PSt	1968
TMM-E	1968
TxBeaL	1968
TxMM	1968
TxU	1968
UU	1968
ViBlbV	1968
WU	1968

(52) AMERICAN PETROLEUM INSTITUTE. [GUIDEBOOK]
1948 Spring White Sulphur Springs, West Virginia, log of field trip Anthony Creek Gap & return.

CSt-ES	1948
InLP	1948
InU	1948
OkU	1948

(53) AMERICAN PETROLEUM INSTITUTE. NORTHERN CALIFORNIA CHAPTER. DIVISION OF PRODUCTION. [GUIDEBOOK]
1970 June Field trip to the Geysers geothermal field, Lake and Sonoma counties.
1972 May Field trip to the Geysers geothermal field, Lake and Sonoma counties. Rev. ed.

CMenUG	1970
CU-A	1972
DI-GS	1970
TxHSD	1970

(54) AMERICAN PETROLEUM INSTITUTE. ONSHORE EXPLORATION COMMITTEE
1976 Appalacian field trip, Pittsburgh, PA to Washington, DC.
CoDCh 1976

(55) AMERICAN QUATERNARY ASSOCIATION. AMQUA FIELD CONFERENCE. (TITLE VARIES)
1970 1st C. Natural processes and ecological relationships of the east flank of the Bridger Range-Bangtail Ridge area, 18 miles northeast of Bozeman, Montana. (Revised in 1976.)
D. Quaternary aspects of the Yellowstone Valley south of Livingston, Montana.
1972 2nd Field trip no. 1. Windley's Key Quarry (Key Largo Limestone-late Pleistocene).
Field trip no. 2. Holocene sedimentation in the Everglades and saline tidal lands.
1974 3rd Late Quaternary environments of Wisconsin.
1. Glacial geology and pedology in central and northeastern Wisconsin.
2. Late Quaternary environments of the Driftless Area: southwestern Wisconsin.
1976 4th San Francisco Peaks; a guidebook to the geology. 2d ed., 1976. 1st ed., 1970. See Friends of the Pleistocene. Rocky Mountain Cell (Formerly: Rocky Mountain Group). 185.(1970)
1978 5th Edmonton, Alberta.
— Aug. [1] Pre-conference field trip, Medicine Hat - Edmonton, Aug. 31 - Sept. 1, 1978.
— Sept. [2] Post conference field trip, Edmonton - Rocky Mountains - Edmonton, Sept. 4-6, 1978.
1980 Trips A & D. Glaciomarine geology of the eastern coastal zone.
Trip B. Marine adaptations and sea level rise.
Trips C & E. Mt. Katahdin: modern vegetational sequence and glacial geology.
Trip F. Paleo-Indian to ceramic period archaeology.
Trip G. Maritime plant and fire ecology; glacial and glaciomarine geology--Acadia National Park.
1984 8th Boulder, Colorado.
Late Cenozoic history and soil development, northern Bighorn Basin, Wyoming and Montana. (185)
Field trip 2. Paleo-Indian sites from the Colorado Piedmont to the Sand Hills, northeastern Colorado.
Field trip 3. Geology of a late Quaternary lake and Pinedale glacial history, Front Range, Colorado.
Field trip 4. Niwot Ridge soils, weathering and periglacial features.
Field trip 5. Front Range glacial deposits.
Field trip 8. Historic and prehistoric floods and debris flows, east- central Front Range.
Field trip 11. Sedimentation and palynology in high-level lakes, Front Range.
Post-conference field trip [1]. (Aug. 16, 1984) Guidebook to Holocene deposits in Arapaho Cirque, Colorado Front Range.
Post-conference field trip [2]. (Aug. 16-18, 1984) Guidebook to the glacial geology, archeology, and natural history of the Arapaho Pass area, Colorado Front Range.
Quaternary stratigraphy of the Upper Arkansas Valley.

AzTeS	1976, 84(5)
AzU	1976
CLU-G/G	1974
CSt-ES	1976
CaACU	1978
CoDGS	1984(5)
DI-GS	1984

(56) American Society of Agronomy.

I-GS	1972
ICarbS	1976
IU	1976
IaCfT	1984(5)
IaU	1974, 78, 84
InU	1970
LNU	1984(5)
ME-GS	1980
MeGS	1980
MnU	1974
NcU	1976, 84(5)
NdU	1974
NhD-K	1984
NvU	1976
OU	1970, 74, 78, 84(3, 4, 5, 8, 11)
PBU	1984
PBm	1984
UU	1984(5)
ViBlbV	1976
WU	1974, 76

(56) AMERICAN SOCIETY OF AGRONOMY. GUIDEBOOK.

1977 Nov. Soil development, geomorphology, and Cenozoic history of the northeastern San Joaquin Valley and adjacent areas, California: a guidebook for the joint field session of the American Society of Agronomy, Soil Science Society of America and the Geological Society of America. (606)

CMenUG	1977
CSt-ES	1977
CU-A	1977
CU-EART	1977

AMERICAN STATE GEOLOGISTS. See **ASSOCIATION OF AMERICAN STATE GEOLOGISTS. (74)**

AMQUA FIELD CONFERENCE. See **AMERICAN QUATERNARY ASSOCIATION. (55)**

(57) ANNUAL MIDWEST GROUNDWATER CONFERENCE.
1960 5th See Midwest Groundwater Conference. 380.

(58) APPALACHIAN GEOLOGICAL SOCIETY. FIELD TRIP [GUIDEBOOK]

1949 May Log of field trip; Morgantown, West Virginia to Terra Alta gas field, Salt Lick Creek.
1952 May Development of new gas fields in eastern West Virginia and western Maryland; West Virginia field trip log book.
1953 June Elkins, West Virginia to Clifton Forge, Virginia to White Sulphur Springs, West Virginia. (763)
1955 May Harrisonburg area, Virginia. (763)
1957 Apr. Some stratigraphic and structural features of the Middlesboro Basin. (209, 347)
— Oct. Blackwater Falls State Park, West Virginia. Blackwater Falls to Mouth of Seneca, West Virginia. (541, 763)
1958 Oct. See Field Conference of Pennsylvania Geologists. 172.
1959 Oct. Cacapon State Park, Berkeley Springs, West Virginia. (541)
1961 Oct. Blackwater Falls State Park, West Virginia. (541, 763)
1963 Sept. Tectonics and Cambrian-Ordovician stratigraphy in the Central Appalachians of Pennsylvania. (541)
1964 Oct. The Great Valley in West Virginia. (541, 763)
1970 Apr. Silurian stratigraphy, central Appalachian Basin.
1972 May See American Association of Petroleum Geologists. Eastern Section. 25.
1982 Apr. The structural development and deformation of the Allegheny frontal zone and Wills Mountain anticlinorium, the central Eastern Overthrust Belt.
Cumberland, Maryland to Petersburg, West Virginia.
Petersburg, West Virginia to Franklin, West Virginia.
Franklin, West Virginia to Cumberland, Maryland.
1984 Oct. The structural development and deformation of the Allegheny frontal zone and Wills Mountain anticlinorium; the central-eastern Overthrust Belt. (Cover title: Golden anniversary field trip guide book)

AzTeS	1953, 55, 57, 63-64
CLU-G/G	1952-53, 55, 57, 61, 64, 70, 82
CLhC	1970
CMenUG	1955, 57 (Oct.)
CPT	1953, 57 (Apr.), 64, 70
CSt-ES	1949, 53, 64, 82
CU-EART	1955, 63
CU-SB	1982
CaACU	1959, 70, 82
CaNSHD	1982
CaOKQGS	1963, 70
CaOOG	1953, 57 (Apr.), 63
CaOWtU	1953, 61, 64, 70
CoDCh	1953, 55, 57 (Apr.), 61, 64
CoDGS	1957(Apr.)
CoG	1952, 55, 57(Apr.), 59, 61, 70
CoLiM	1982
CoU	1970
DI-GS	1953, 55, 57, 59, 61, 63, 70, 82, 84
DLC	1953
I-GS	1957(Apr.)
ICF	1957(Apr.), 59, 61
ICarbS	1955
IEN	1970
IU	1957(Apr.), 59, 61, 64
IaU	1953, 57(Apr.), 63-64, 70, 82
InLP	1953, 55, 57, 59, 61, 63-64, 70, 82, 84
InU	1949, 53, 57(Apr.), 63
KyU-Ge	1953, 55, 57(Apr.), 63-64, 70
LNU	1959, 70
MH-GS	1970
MnU	1953, 55, 57, 59, 61, 63-64, 70, 82
MoSW	1953
MoU	1957(Apr.), 70
NBiSU	1953, 57
NIC	1982
NNC	1949, 52-53, 57, 59, 61, 63-64
NSyU	1970
NbU	1959, 61
NcU	1953, 55, 57, 59, 63-64, 70, 82
NhD-K	1970, 82
NjP	1957(Apr.), 63
OCU-Geo	1957, 64, 70, 82
ODaWU	1984
OU	1949, 52-53, 55, 57(Apr.), 59, 70
OkT	1955
OkTA	1982
OkU	1949, 52-53, 55, 57, 59, 61, 70, 82
PBU	1982
PBm	1963
PHarER-T	1963
PSt	1953, 63-64, 70
RPB	1953
TMM-E	1955, 57(Apr.)
TxBeaL	1953
TxDaAR-T	1955, 59, 70
TxDaDM	1952-53, 55, 57(Apr.), 59, 61, 63-64, 70
TxDaM-SE	1957(Apr.), 59, 61
TxDaSM	1970
TxHSD	1970
TxHU	1959
TxMM	1953, 63
TxU	1957(Apr.), 59, 61, 63-64, 82, 84
TxWB	1982
ViBlbV	1953, 55, 57(Apr.), 59, 64, 70
WU	1959, 63, 70, 84
WyU	1982

(59) APPALACHIAN-CALEDONIAN OROGEN, PROJECT 27 OF THE INTERNATIONAL GEOLOGICAL CORRELATION PROGRAMME. CANADIAN COMMITTEE. [GUIDEBOOK].
1982 See International Geological Correlation Programme (IGCP). Project 27: The Caledonide Orogen. 302.

(60) APPLACHIAN PETROLEUM GEOLOGY SYMPOSIUM.
1984 15th Upper Paleozoic oil-bearing clastics and carbonates of northern West Virginia.
OCU-Geo 1984

(61) ARCHEAN GEOCHEMISTRY FIELD CONFERENCE. FIELD GUIDE. (TITLE VARIES)
1978 See International Geological Correlation Programme (IGCP) [Project] 92. Archean geochemistry 297.
1979 See International Geological Correlation Programme (IGCP) [Project] 92. Archean geochemistry 297.
1983 Aug. See International Geological Correlation Programme (IGCP) [Project] 92. Archean geochemistry 297.

(62) ARDMORE GEOLOGICAL SOCIETY. FIELD CONFERENCE GUIDEBOOK.
1936 Mar. Pennsylvanian of the Ardmore Basin.
— Apr. 4 Pre-Pennsylvanian of the Ardmore area.
— Apr. 18 Ardmore to the Wichita Mountains.
— May 5 The Pennsylvanian System of the Ardmore Basin.
— June Ardmore to the Ouachita Mountains, Oklahoma.
— Dec. Study of the Pennsylvanian outcrops of Palo Pinto, Parker, Eastland, Brown and Coleman counties, Texas.
1937 Mar. The Hoxbar and upper Deese formations south of Ardmore.
— Apr. Study of Lower Pennsylvanian, Mississippian, and Ordovician formations on north and west sides of the Llano Uplift, central Texas.
— May Structure of the Criner Hills.
1938 Apr. The Lower Pennsylvanian of the Berwyn and Baum areas.
1940 11th Washita Valley fault system and adjacent structures.
1946 12th A structural and stratigraphic consideration of the Arbuckle Mountains and the Criner Hills.
1948 13th A study of Pennsylvanian formations, Ardmore area, Oklahoma.
1950 14th Study of structure and stratigraphy in the Arbuckle Mountains and related structures in Carter, Murray and Johnston counties, Oklahoma.
1952 15th Study of Paleozoic structure and stratigraphy in Arbuckle and Ouachita Mountains, Johnston and Atoka counties, Oklahoma.
1954 16th Southern part of the Oklahoma Coal Basin.
1955 17th Geology of the Arbuckle Mountain region. (508)
1956 18th Ouachita Mountain field conference, southeastern Oklahoma.
1957 19th Criner Hills field conference, Lake Murray area, southern Oklahoma.
1959 See American Association of Petroleum Geologists. 18.
1963 Oct. Basement rocks and structural evolution of southern Oklahoma.
1966 Oct. Pennsylvanian of the Ardmore Basin, southern Oklahoma.
1969 Oct. Geology of the Arbuckle Mountains along Interstate 35, Carter and Murray counties, Oklahoma.

ATuGS	1950
AzFU	1954-56, 66
AzTeS	1956, 66
AzU	1955
CLU-G/G	1950, 52, 54-57, 66, 69
CLhC	1955, 66, 69
CMenUG	1948, 50, 54-57
CPT	1956-57, 63, 66, 69
CSt-ES	1936(Apr., May, June), 37(Mar.), 40, 48, 50, 52, 54-56, 63, 69
CU-EART	1936(Apr., June, Dec.), 37, 38, 54, 66, 69
CaAC	1956
CaACU	1948, 55, 69
CaAEU	1955
CaBVaU	1955, 57, 63
CaOHM	1966, 69
CaOOG	1955
CoDCh	1937, 40, 46, 48, 52, 54, 57, 63, 66, 69
CoDGS	1948, 50, 54, 56-57, 66, 69
CoDU	1969
CoG	1940, 48, 50, 55-57, 63, 66, 69
CoU	1955, 63
DI-GS	1936(May, Dec.), 37(Mar.), 48, 50, 54-57, 63, 66, 69
DLC	1955-56
I-GS	1940
ICF	1950, 54-57
ICIU-S	1955-56
ICarbS	1955
IEN	1940, 48, 55
IU	1948, 50, 54-57, 63, 66
IaU	1957, 63, 66, 69
IdPI	1957
InLP	1950, 55, 57, 63, 69
InU	1950, 54-57, 63, 69
KyU-Ge	1948, 54-55, 57, 66, 69
LNU	1956
LU	1955
MH-GS	1955
MNS	1955
MiDW-S	1955
MiU	1955
MnU	1950, 54-57, 63, 66, 69
MoSW	1955, 63, 66, 69
MoU	1955, 63, 66
NNC	1940, 48, 50, 54-57, 63, 66
NSyU	1955, 63
NbU	1950, 56-57, 63
NcU	1955, 63, 66
NdU	1955, 63
NhD-K	1955, 69
NjP	1955
NmPE	1955, 66, 69
NvU	1955
OCU-Geo	1952, 63
OU	1936(Mar., June, Dec.), 37(Mar., May), 48, 50, 52, 54-55, 57, 63
OkOkCGS	1955
OkOkU	1955, 57, 63
OkT	1936(Mar., Apr. 4), 37(Mar.), 38, 48, 50, 54-57, 63
OkU	1936(Mar., Apr., June), 37-38, 40, 46, 48, 50, 52, 54-57, 63, 66, 69,
OrU-S	1955
PBL	1956
PBm	1955
PSt	1955, 63
RPB	1955
TMM-E	1966, 69
TU	1955
TxBeaL	1948, 50, 52, 63, 66, 69
TxCaW	1955, 69
TxDaAR-T	1940, 48, 50, 52, 54-57, 63, 66, 69
TxDaDM	1950, 54-57, 63, 69
TxDaGI	1937(Apr.), 52, 55, 66
TxDaM-SE	1940, 46, 50, 52, 54-57, 63, 66
TxDaSM	1948, 52, 56-57, 63, 66, 69
TxHSD	1936, 55, 57, 66
TxHU	1948, 50, 52, 54-56, 63, 66, 69
TxLT	1936(Apr. 18), 46, 50, 54-57
TxMM	1936(Apr. 4, Dec.), 37(Apr.), 38, 40, 46, 48, 50, 54-57, 63, 66
TxU	1936(Mar., Apr., May 5, June), 37-38, 48, 50, 52, 54-57, 63, 66, 69
TxWB	1936-37, 48, 50, 52, 54, 56-57, 63, 66, 69
UU	1946, 56
ViBlbV	1950, 54-56, 66, 69
WU	1948, 55-57, 63, 66, 69
WyU	1955, 63

(63) ARDMORE GEOLOGICAL SOCIETY. PUBLICATIONS.
1977 Fossil collecting in the Ardmore area.
IaU 1977
OkT 1977

(64) ARIZONA GEOLOGICAL SOCIETY DIGEST.
1980 V. 12 See Arizona Geological Society. 65.(1979)
1981 Guide to the geology of the Salt River Canyon region, Arizona.
CSt-ES 1981
CU-SB 1981
CoDGS 1981
KyU-Ge 1981

(65) ARIZONA GEOLOGICAL SOCIETY. GUIDEBOOK. (TITLE VARIES)
1952 See Geological Society of America. Cordilleran Section. 199.
1958 See New Mexico Geological Society. 458.
1959 See Geological Society of America. Cordilleran Section. 199.
1962 See New Mexico Geological Society. 458.
1966 Road log for southern Santa Rita Mountains, Santa Cruz and Pima counties, Arizona. (220)
1968 See Geological Society of America. Cordilleran Section. 199.
1978 See New Mexico Geological Society. 458.
1979 May Tucson-Yuma-Quartzsite-Buckeye. With side trips to the Marine Corps Gunnery Range and the Silver mining district. (Cover title: Studies in Western Arizona) (64(1980))

(66) Arizona State University. Center for Meteorite Studies.

1981 Trip no. 8. Hardshell Silver, Base-Metal, Manganese Oxide Deposit, Patagonia Mountains, Santa Cruz County, Arizona: A field trip guide. See Arizona Geological Society Digest. 64.
1982 Sept. Field guide (with road logs) to selected parts of the western and central San Francisco volcanic field, Coconino County, Arizona.
1983 Fall Field trip to Northern Plomosa Mountains, Granite Wash Mountians, and western Harquahala Mountains. (674)

AzTeS	1966
AzU	1966
CMenUG	1952, 58, 59, 62, 66, 68
CSt-ES	1979, 81, 83
CU-EART	1979
CU-SB	1966
CaBVaU	1979
CoDGS	1979
DI-GS	1966
KyU-Ge	1981
MnU	1966
OCU-Geo	1979, 82

(66) ARIZONA STATE UNIVERSITY. CENTER FOR METEORITE STUDIES. PUBLICATION.
1978 Jan. See Planetary Geology Field Conference. 542.
1979 17. See Meteoritical Society. Annual Meeting. 373.

ARIZONA. BUREAU OF GEOLOGY AND MINERAL TECHNOLOGY. See UNIVERSITY OF ARIZONA. BUREAU OF GEOLOGY AND MINERAL TECHNOLOGY. (673) and UNIVERSITY OF ARIZONA. BUREAU OF GEOLOGY AND MINERAL TECHNOLOGY. (672)

(67) ARKANSAS. GEOLOGICAL COMMISSION. GUIDEBOOK. (TITLE VARIES)
1951 See Shreveport Geological Society. 566.
1955 Guidebook to the Paleozoic rocks of Northwest Arkansas, 1951 (Reprinted 1955).
1956 See Kansas Geological Society. 341.
1963 63-1 See Fort Smith Geological Society. Regional Field Conference. 177.
1967 67-1 See Geological Society of America. 197.
1973 73-1 (revised in 1983) See Geological Society of America. South Central Section. 205.
— 73-2 See Geological Society of America. South Central Section. 205.(2)
— 73-3 See Geological Society of America. South Central Section. 205.(3)
— 73-3 See Geological Society of America. South Central Section. 205.(4)
1974 Feb. See Society of Economic Geologists. 583.
1975 [74-1] Arkansas-Texas: economic geology field trip, February, 1974 (Reprinted 1975).
1977 Apr. (Revised in 1980) See Tulsa Geological Society. 662.
1978 [Feb.] See Geological Society of America. South Central Section. 205.
1979 [79-1] See Geological Society of America. South Central Section. 205.
— [79-3] See Association of American State Geologists. 74.(2)
— [79-4] See Association of American State Geologists. 74.(1)
1980 [77-1] A guidebook to the geology of the Arkansas Paleozoic area (Ozark Mountains, Arkansas Valley, and Ouachita Mountains), 1977 (Revised 1980).
— 80-1. A guidebook to southwestern Arkansas and Lake Ouachita.
— 80-2. A guidebook to southwestern Arkansas. (Revised in 1982)
1981 81-1. See Geological Society of America. 197.(3) and Houston Geological Society. 255.
1982 [82-1] (Reprinted in 1985) See Fort Smith Geological Society. Regional Field Conference. 177. and Houston Geological Society. 255.
— [82-2] See Houston Geological Society. 255.
1983 A guidebook to the geology of the Ouachita Mountains, Arkansas, 1973 (Revised 1983).
1984 Oct. [84-1] See Society of Economic Paleontologists and Mineralogists. Mid-Continent Section. 589. (2nd)
84-2. A guidebook to the geology of the central and southern Ouachita mountains, Arkansas.

AGC	1979-80, 82-84
BEG	1984
CSt-ES	1980, 82-83
CU-EART	1980
CaOOG	1979-81, 83
CoDGS	1979, 80-84
CoLiM	1984
DI-GS	1975, 79, 82, 84
ICarbS	1951, 55
IU	1979(Apr.)
IaU	1975
InLP	1980-82, 84
InU	1984
MdBJ	1955
MnU	1963, 75, 84
NhD-K	1981
OCU-Geo	1975
OkTA	1980
OkU	1955, 80, 84(pt.2)
PSt	1981
TxDaM-SE	1980, 82
TxU	1979-80, 83
ViBlbV	1963
WU	1963, 79-83

ARKANSAS. RESOURCES AND DEVELOPMENT COMMISSION. DIVISION OF GEOLOGY. [GUIDEBOOK] See ARKANSAS. GEOLOGICAL COMMISSION. (67)

(68) ASOCIACION MEXICANA DE GEOLOGOS PETROLEROS.
1961 See Southwestern Federation of Geological Societies. 627.

(69) ASSOCIATION OF AMERICAN GEOGRAPHERS. ANNUAL MEETING. FIELD TRIP GUIDE. (TITLE VARIES)
1968 [A] Nine geographical field trips in the Washington, D.C. area.
  1. Washington, D.C. and environs.
  2. The Potomac Valley, Washington to Great Falls.
  3. Geographic landscape: Washington to Mt. Vernon.
  4. Reston, Virginia.
  5. Greenbelt and Columbia, Maryland.
  6. The coastal plain of Maryland.
  7. Geographic landscape: Washington to the Shenandoah.
  8. Civil War battlefields.
  9. Development and industrialization of Baltimore Harbor.
  [B] Geographical reconnaissance of the Potomac River tidewater fringe of Virginia from Arlington Memorial Bridge to Mount Vernon.
1970 70th Seattle, Washington
  Views of Washington State.
1971 Guide to outer Cape Cod and the Cape Cod National Seashore.
1972 Geographical setting of the Missouri River valley between Kansas City and St. Joseph.
1975 Landscapes of Wisconsin; a field guide.
1982 San Antonio, Texas.
  (A) San Antonio de Bexar: a walking tour of the Alamo, the River Walk, and La Villita.
  Trip 1. Mustang and Padre islands: problems in the Texas coastal zone.
  Trip 2. Texas Golden Triangle.
  Trip 3. Contemporary urban issues in the Sunbelt: Dallas on the way.
  Trip 4. The missions of San Antonio.
  Trip 5. Enchanted Rock exfoliation dome.
  Trip 6. Aspects of the physical, economic, and cultural geography of the Texas hill country.
  Trip 7. Remote sensing and the San Antonio area.
  Trip 8. Art museums of San Antonio.
  Trip 9. Historical San Antonio.
  Trip 10. The hill country Germans of Texas.
  Trip 11. Contemporary issues in the urban environment of San Antonio.
  Trip 12. Overlooking a Texas fault: the Balcones Escarpment; settlement and land use in the New Braunfels region.
  Trip 13. Physical geography of the Balcones Escarpment.
  Trip 14. Mexico: cultural collision and border identity.
  Trip 15. Big Bend National Park.
1985 Quaternary geomorphology of southeastern Michigan.

CU-EART	1982
CaOWtU	1968
DI-GS	1975
DLC	1968, 75
ICarbS	1975
InLP	1970
KLG	1972
MH-GS	1971
NcU	1968, 75
NhD-K	1971
OCU-Geo	1975
WGrU	1975
WU	1975

(70) ASSOCIATION OF AMERICAN GEOGRAPHERS. GREAT PLAINS DIVISION.
1984   See Association of American Geographers. West Lakes Division. 73.

(71) ASSOCIATION OF AMERICAN GEOGRAPHERS. ROCKY MOUNTAIN DIVISION.
1984   See Association of American Geographers. West Lakes Division. 73.

(72) ASSOCIATION OF AMERICAN GEOGRAPHERS. SOUTHEASTERN DIVISION. GUIDEBOOK FOR THE ANNUAL MEETING.
1980   35th   Environmental implications of karst and other geomorphic features of the New River Valley.
    OkU   1980

(73) ASSOCIATION OF AMERICAN GEOGRAPHERS. WEST LAKES DIVISION.
1984   Joint Meeting, field trip guide. (71, 70)
    KLG   1984

(74) ASSOCIATION OF AMERICAN STATE GEOLOGISTS. GUIDEBOOKS AND ROAD LOGS.
1927   Program and itinerary of field excursion in northern Illinois.
1946   40th   Field conference, Black Hills of South Dakota and Wyoming.
1948   Alabama meeting.
1949   June   Late Cenozoic geology of Mississippi Valley, southeastern Iowa to central Louisiana.
1952   44th   A summary of the geology of Florida and a guidebook to the Cenozoic exposures of a portion of the state.
1953   Feb.   Geologic log for Ventura Basin field trip, Los Angeles to Wheeler Springs and return via coast route. (113)
— Sept.   Hartford, Connecticut. Trip A-B. (Bibliography separate)
1955   Southeastern New Mexico-Socorro to Carlsbad, Carlsbad Caverns, potash mining district. (Published by New Mexico Bureau of Mines and Mineral Resources, New Mexico Institute of Mining and Technology, Socorro)
1956   Selected features of Kentucky geology.
1957   Keweenaw copper range, Marquette iron range; Houghton, Michigan.
1958   Mar./Apr.   Field excursion, eastern Llano region. (713)
1959   Kansas field conference. Geological understanding of cyclic sedimentation represented by Pennsylvanian and Permian rocks of northern midcontinent region.
1960   Apr.   Guidebook, southeastern Pennsylvania.
1961   Guidebook to the geology of the Coeur D'Alene mining district. (257)
1963   Centennial field trip: Appalachian Mountains of West Virginia. (763)
1964   Ardmore Basin.
1966   May   Excursions in Indiana geology. (268)
1967   59th   Centennial guidebook to the geology of southeastern Nebraska. (449)
1968   May   Geology of the Alabama coastal plain.
1970   62nd   Trip 1. Guidebook to Ozark carbonate terrane, Rolla-Devils Elbow area, Missouri.
Trip 2. Guide to the Southeast Missouri mineral district.
1972   T-10. Some geologic hazards and environmental impact of development in the San Diego area. (Reprinted, 1973, under the auspices of the Far Western Section, National Association of Geology Teachers. [Also] reprinted 1980) (412)
1973   Geologic log for Ventura Basin field trip, Los Angeles to Wheeler Springs and return via coast route. (108)
1975   May   Field guidebook to the geology of the Central Blue Ridge of North Carolina and the Spruce Pine mining district.
1977   June   Geologic field trips in Delaware. (157)
  [1] Geology and water resources of the western portion of the Delaware Piedmont.
  [2] Environmental geology, northern Delaware.
  [3] Field trip to the Chesapeake and Delaware Canal.
  [4] Newark to Rehoboth Beach with notes on points of geologic interest.
  [5] Field trip - Delaware coastal area.
1979   June   [1] A guidebook to the second geological excursion on Lake Ouachita. (67(No.79-4))
  [2] Field trip guide to three major mines in central Arkansas. (67(No.79-3))
1982   74th   Harrisburg, Pennsylvania.
Geology of the southern part of the anthracite coal fields of Pennsylvania.

ATuGS	1966
AzTeS	1956, 58, 61, 63
AzU	1958, 61, 75
CLU-G/G	1956, 58, 61, 63, 66-68
CLhC	1958-59
CPT	1953(Feb.), 56, 63, 66-67
CSfCSM	1958
CSt-ES	1953(Feb.), 56, 58, 61, 63, 66-67, 73, 77, 79
CU-EART	1958, 61, 66
CU-SB	1979
CaACU	1958, 67
CaAEU	1961, 67
CaBVaU	1961
CaOKQGS	1961, 66
CaOLU	1961
CaOOG	1958, 67, 79
CaOWtU	1963, 66-67, 77
CoDCh	1959, 66, 67, 79
CoDGS	1956, 58-59, 63, 67
CoG	1958, 61, 66
CoU	1958, 61, 67
DI-GS	1952, 53(Sept.), 56-61, 63, 66-67, 70(1), 73, 75
DLC	1958-61
I-GS	1927, 46, 48, 52, 56-58, 61, 66-68, 70(1)
ICF	1958, 61, 66
ICIU-S	1966
ICarbS	1953, 61, 66
IEN	1958, 61
IU	1957-59, 61, 63, 66-67, 75, 77
IaU	1956, 58-61, 63, 66-68, 75, 77, 79
IdBB	1961
IdPI	1958, 61
IdU	1958, 61, 66
InLP	1953, 56, 58, 61, 63, 66-67, 73, 75, 77, 79
InRE	1966
InU	1946, 49, 52, 56, 58, 61, 66-67, 79(1)
KLG	1956, 56-60, 63, 66
KyU-Ge	1953, 56, 58, 63, 66-67, 75(May), 77, 79
LNU	1958-59, 75
LU	1958, 61, 66
MNS	1958, 61, 66
MiDW-S	1958, 61, 66
MiHM	1953(Feb.), 57-58, 66
MiKW	1966
MiU	1958, 66
MnU	1946, 48, 53, 56, 58-59, 61, 63, 66-67, 75, 79(1)
MoSW	1958, 61, 66-67
MoU	1956, 58
Ms-GS	1952
MsJG	1952
MtBuM	1958, 61, 66-67
NBiSU	1958
NNC	1956, 58-59, 63, 66, 70, 73, 75
NSyU	1958, 61
NbU	1953, 58, 61, 66-67
NcU	1956, 58, 66, 75
NdU	1948, 53, 56-59, 61, 63, 66-68
NhD-K	1958, 61, 66
NjP	1956, 58, 66
NvU	1958-59, 61, 66-67
OCU-Geo	1958, 63, 66-67, 79(1)

(75) Association of Earth Science Editors.

OU	1953, 55-58, 61, 66-68
OkT	1958, 59, 61
OkU	1927, 46, 53, 55-61, 63, 66-68, 70, 73, 75
OrU-S	1958, 61
PBL	1958, 61
PSt	1958, 61, 63, 66, 82
RPB	1958, 61, 66
SdRM	1961
TMM-E	1956, 58
TxBeaL	1958-59
TxDaAR-T	1959, 66
TxDaM-SE	1953(Feb.), 56, 58, 61, 66
TxDaSM	1956, 58
TxHSD	1958
TxHU	1958, 61, 66
TxLT	1958
TxMM	1958, 61
TxU	1948, 53(Feb.), 56-58, 61, 64, 66-67
TxU-Da	1958
TxWB	1961
UU	1953, 57-58, 60-61, 66-67
ViBlbV	1956, 58, 63, 66, 75
WU	1958, 61, 66-67, 79(1, 2), 82
WaU	1958, 66

(75) ASSOCIATION OF EARTH SCIENCE EDITORS. [GUIDEBOOK]
1973 Ottawa region.
1980 Field trip notes and fantasy.
1982 Guidebook to the Late Cenozoic geology and economic geology of the lower York-James Peninsula. (Adapted from guidebook 3 of National Association of Geology Teachers (N.A.G.T.). Eastern Section.) (136)
1985 Lawrence, Kansas.
Lawrence to Kansas City road log, including a discussion of the downtown underground limestone mine in Kansas City, Missouri. (637)

CU-EART	1982
CaOOG	1973
DI-GS	1973, 80, 82
KLG	1985
OU	1982
OkTA	1982
OkU	1982

(76) ASSOCIATION OF ENGINEERING GEOLOGISTS. FIELD TRIP GUIDEBOOK.
1968 [11th] Seattle, Washington, annual meeting.
  1-5. Various areas of Washington State.
  [6] Geologic exploration of the Guayanes River multipurpose dam, Puerto Rico.
1969 [12th] San Francisco, California, annual meeting.
  1. California state water project.
  2A and 2B. Hayward and Calaveras fault zones.
  3. San Francisco peninsula-Stanford linear accelerator.
  4. Bay area rapid transit system.
  5. San Francisco peninsular trip for planners and public officials.
  6. The San Francisco Bay and Delta model.
1970 [13th] Washington, DC, annual meeting.
  1., pt. 1. Engineering geology of the Raystown Dam.
  1., pt. 2. Pennsylvania glass sand corporation plant, Berkeley Springs, West Virginia.
  2. Engineering geology in northeastern Pennsylvania and New Jersey.
  5. Historical engineering geology of the Chesapeake and Ohio canal.
  6. Engineering geology in Puerto Rico.
1971 [14th] Portland, Oregon, annual meeting.
  1. Columbia Gorge cruise.
  2. Oregon coast.
  [3] Mount Hood Loop.
  [4] Portland metropolitan area.
1972 [15th] Kansas City, annual meeting.
  1. Proposed salt mine repository for radioactive wastes.
  2. Underground mines in Northeast Kansas.
  3. Two-tier occupancy of space-surface and subsurface in the Kansas City area.
  4. Highway construction in the Kansas City area.
  5. Engineering geology and utilization of underground space in the Kansas City area.
1973 16th North Hollywood, California, annual meeting.
  1. Malibu Coast-Santa Monica Mountains.
  2. Point Fermin/Portuguese Bend landslide.
  3. The Sierra Madre-Cucamonga-Raymond Fault.
  4. LaBrea Tar Pits and L.A. County Museum of Natural History.
  5. Los Angeles/Long Beach harbor areas.
  6. San Fernando earthquake zone.
  7. South coastal zone, Los Angeles to southern Orange County, San Onofre nuclear power plant.
  8. Castaic Dam and Pyramid Dam, California aqueduct system.
1975 18th Lake Tahoe, California, annual meeting.
  [1] Geologic guide to Lake Tahoe, Mother Lode, Comstock Lode and Auburn Dam, California.
1976 [19th] Philadelphia, Pennsylvania, annual meeting.
  1. Coastal plain, Delaware and New Jersey.
  2. Piedmont, Reading Prong, Appalachian Ridge and Valley, Pennsylvania.
  3. Piedmont, New England, Appalachian Ridge and Valley, Pennsylvania and New Jersey.
1977 [20th] Seattle, Washington, annual meeting.
  1. Dams and associated landslides.
  [1-A] Southern British Columbia.
  [1-B] Chief Joseph Dam and Bridgport Slide.
  [1-C] Grand Coulee and Bacon Siphon.
  2. Teton Dam failure.
  3. Problems of recent landslides, Cascade Mountain front, Snoqualmie underground power plant.
  4. Geology and foundation excavation at the Satsop nuclear power plant.
  5. Coastal engineering geology northern Olympic Peninsula.
1978 21st Hershey, Pennsylvania, annual meeting.
  1. Lower Susquehanna River Power Complex.
  2. Ridge and valley physiographic province of Pennsylvania.
  3. Anthracite coal region, northeastern Pennsylvania field trip route.
1979 [22nd] Chicago, Illinois, annual meeting.
  Engineering geology of the greater Chicago area and the south shore of Lake Michigan.
1980 [23rd] Dallas, Texas.
  Planning; a geological perspective.
  — Oct. Lignite field trip.
1981 24th Portland, Oregon.
  Engineering geology in the Pacific Northwest.
1982 25th Montreal, Quebec. (120)
  Field trip guidebook: 25th annual meeting of the Association of Engineering Geologists jointly with the 35th technical conference of the Canadian Geotechnical Society.
  Field trip 1. Beauharnois-Carillon.
  Field trip 2. Montreal subway construction.
  Field trip 3. Active construction sites in Montreal and 1976 Olympic facilities.
  Field trip 4. Visit to LG-2, James Bay.
1983 Field guide to selected dams and reservoirs, San Diego County, California.
1985 28th Winston-Salem, North Carolina.
Oct. 5-6. Appalachian Mountains field trip.
Oct. 11-13. Coastal plain field trip.

AzFU	1969
AzTeS	1968-69
CLU-G/G	1966, 69, 70(1-2, 5-6), 73
CMenUG	1968, 70-73, 75-82
CPT	1969, 73, 75, 77, 81
CSdS	1973(1)
CSfCSM	1968, 73
CSt-ES	1968(1-5), 70(2, 6), 71, 73, 75, 77, 80-81
CU-A	1970(1:1, 2, 6), 71(1, 4), 72
CU-EART	1975
CU-SB	1973, 75-81
CaACU	1970(1:1, 2, 5-6), 71(1), 72, 76-77
CaBVaU	1982
CaOOG	1968-69, 75
CaOWtU	1968(1-5), 71(1-2, 4)
CoDGS	1969, 71-73, 76-77, 81-82

CoLiM	1983
CoU	1968
DI-GS	1968-69, 70(1:1, 2, 5-6), 71-73, 75-79, 80(23), 81-82
IU	1973, 77, 80
IdU	1969-70, 71(1), 72-73, 75-79, 80-82
InLP	1968(1-5), 69-70, 71(1-2, 4), 72-73, 75-80
InU	1968
KyU-Ge	1966, 69, 70(1-2, 5-6), 71(2-4), 72-73, 75, 76-79, 80(Oct.), 81-82, 84, 85
MNS	1969
MU	1980
MnU	1968(1-5), 76
MoU	1969
MtBuM	1966
NSyU	1973
NcU	1969, 77
NdU	1969
NjP	1968, 73
OCU-Geo	1968, 69(1), 70(1-2), 71(1-2, 4), 72
OU	1969-70, 71(1, [4]), 72-73, 78-81
OkU	1977, 80-82
PHarER-T	1970(1, 5)
PSt	1970(1, 2, 5)
TMM-E	1969
TxBeaL	1970(5)
TxU	1968, 77-78, 80-81
ViBlbV	1968-69, 70(2, 6), 71-72
WaU	1981

(77) **ASSOCIATION OF ENGINEERING GEOLOGISTS. CALIFORNIA SECTION. ANNUAL CONFERENCE. GEOLOGIC GUIDE.**
1978 1st   Geologic guide and engineering geology case histories, Los Angeles metropolitan area.

CSfCSM	1978
CSt-ES	1978
DI-GS	1978
InLP	1978

(78) **ASSOCIATION OF ENGINEERING GEOLOGISTS. DENVER SECTION. PROCEEDINGS AND FIELD TRIP GUIDE.**
1969   See Governor's Conference on Environmental Geology. 245.
1981 June   Denver, Colorado.
     Colorado tectonics, seismicity and earthquake hazards. (141)

CLU-G/G	1981
CMenUG	1981
CPT	1981
CSt-ES	1981
CU-EART	1981
CU-SB	1981
CaACU	1981
CaAEU	1981
CaOKQGS	1981
CaOOG	1981
CaOWtU	1981
CoDBM	1981
CoDGS	1981
CoG	1981
CoU-DA	1981
IaU	1981
InLP	1981
KyU-Ge	1981
MnU	1981
NRU	1981
NSbSU	1981
NvU	1981
OU	1981
TxDaM-SE	1981
TxU	1981
UU	1981

(79) **ASSOCIATION OF ENGINEERING GEOLOGISTS. LOWER MISSISSIPPI VALLEY SECTION.**
1984   Hattiesburg, Mississippi. Geologic excursion, Natchez, Mississippi to St. Francisville, Louisiana. (621)

MnU	1984

(80) **ASSOCIATION OF ENGINEERING GEOLOGISTS. NEW YORK SECTION.**
1976   Annual Meeting. Field trip guidebook. (81)

CaACU	1976

(81) **ASSOCIATION OF ENGINEERING GEOLOGISTS. PHILADELPHIA SECTION.**
1976   See Association of Engineering Geologists. New York Section. 80.

(82) **ASSOCIATION OF ENGINEERING GEOLOGISTS. SACRAMENTO SECTION. [FIELD TRIP]**
1976 May   The Geysers geothermal field. (83)

CSt-ES	1976
CaACU	1976
InLP	1976

(83) **ASSOCIATION OF ENGINEERING GEOLOGISTS. SAN FRANCISCO SECTION. [FIELD TRIP].**
1976   See Association of Engineering Geologists. Sacramento Section. [Field trip] 82.

(84) **ASSOCIATION OF ENGINEERING GEOLOGISTS. SOUTHERN CALIFORNIA SECTION. ANNUAL SPRING FIELD TRIP.**
1973 May   See San Diego Association of Geologists. 556.
1975 Sept.   Sycamore Canyon Fault, Verdugo Fault, York Boulevard Fault, Raymond Fault, and Sierra Madre fault zone.
1976 May   Geologic guide to the San Bernardino Mountains, Southern California.
1977 May   Field trip guide book to Santa Catalina Island.
1978 May   Geologic guide to the San Bernardino Mountains, Southern California.
1979 May   Field guide to selected engineering geologic features: Santa Monica Mountains.

CLU-G/G	1976, 79
CSfCSM	1976
CSt-ES	1976, 78
CU-A	1976, 78
CU-EART	1976
CU-SB	1979
CaACU	1976
DI-GS	1976
InLP	1975, 79
KyU-Ge	1975-77
ViBlbV	1976
WU	1976

(85) **ASSOCIATION OF EXPLORATION GEOCHEMISTS. SYMPOSIUM. FIELD TRIP GUIDEBOOK.**
1984 Mar.   Exploration for ore deposits of the North American Cordillera.
     Field trip 1. Sediment-hosted gold deposits.
     Field trip 2. Sediment-hosted precious metal deposits.
     Field trip 3. Precious metal districts in southern and western Nevada.
     Field trip 4. Precious metal districts in west-central Nevada.
     Field trip 5. Cancelled, replaced by trip 7.
     Field trip 6. Massive sulfide deposit: the western world massive sulfide deposit, Yuba City, California.
     Field trip 7. Porphyry molybdenum deposits.
     Field trip 8. Sulphur mining district.
     Field trip 9. Bedded barite deposits.
     Field trip 10. Skarn deposits, Lyon County.
     Field trip 11. Virginia City mining district, Storey County.
     Field trip 12. Gooseberry Mine.

CLU-G/G	1984
CMenUG	1984
CPT	1984
CSt-ES	1984
CU-A	1984
CU-EART	1984
CoDGS	1984
CoG	1984
CtY-KS	1984
DI-GS	1984
IaU	1984

(86) Association of Iowa Archeologists.

IdU	1984
InLP	1984
KyU-Ge	1984
MnU	1984
NcU	1984
NvU	1984
OkU	1984
PSt	1984
TxU	1984

(86) ASSOCIATION OF IOWA ARCHEOLOGISTS. FIELD TRIP.

1982 Apr.   Interrelations of cultural and fluvial deposits in northwest Iowa.

IaU	1982

(87) ASSOCIATION OF MISSOURI GEOLOGISTS. GUIDEBOOK TO THE ANNUAL FIELD TRIP.

1954 1st   Southeast Missouri lead belt area, sponsored by the St. Joseph Lead Company.
1955 2nd   Western Missouri; Desmoinesian section, cyclic sedimentation. (399)
1956 3rd   Central Missouri.
1957 4th   Eastern Kansas, from Kansas City to Manhattan, Kansas via the Turnpike, U.S. Highway 40 and Kansas Highway 13.
1958 5th   Mexico to Cave Hill.
1959 6th   [Basal relations of the Mississippian in central Missouri]
1960 7th   Middle Mississippian and Pennsylvanian stratigraphy of St. Louis and St. Louis County, Missouri.
1961 8th   Guidebook to the geology of the St. Francois Mountain area. (399)
1962 9th   Geology in the vicinity of Cape Girardeau, Missouri, including Crowleys Ridge.
1963 10th   Geology in the vicinity of Joplin, Missouri, including Westside-Webber Mine, Oklahoma.
1964 11th   Cryptovolcanic structures of south-central Missouri.
1965 12th   No guidebook published.
1966 13th   Engineering geology of Kaysinger Bluff and Stockton dams, west-central Missouri.
1967 14th   Middle Devonian of central Missouri.
1968 15th   Guidebook to Pleistocene and Pennsylvanian formations in the St. Joseph area, Missouri.
1969 16th   Bonne Terre, Missouri.
1970 17th   Guidebook to the highway geology of Missouri, Route 79 TR, Hannibal to Clarksville.
1971 18th   Guidebook to the geology and utilization of underground space in the Kansas City area, Missouri.
1972 19th   Guidebook of the karst features and stratigraphy of the Rolla area.
1973 20th   Engineering and environmental geology of the Springfield urban area.
1974 21st   Geology of east-central Missouri with emphasis on Pennsylvanian fire clay and Pleistocene deposition.
1975 22nd   Pennsylvanian-Pleistocene channel fill and Quaternary geomorphology near Warrensburg, Missouri.
1976 23rd   Studies in Precambrian geology of Missouri with a guide to selected parts of the St. Francois Mountains, Missouri. (399, 395)
1977 24th   Geology in the area of the Eureka-House Springs Anticline with emphasis on stratigraphy-structure-economics.
1978 25th   Energy, environment, geology in Bates County.
1981 28th   Cape Girardeau, Missouri.
1984   Guidebook to the geology of the Bolivar - Mansfield Fault Zone.

ATuGS	1955, 61
AzTeS	1961, 63
CLU-G/G	1955, 61, 76
CLhC	1955, 61
CPT	1955, 61, 76
CSt-ES	1955, 61, 76
CU-EART	1955, 61, 63, 67, 76
CaACU	1961
CaBVaU	1961
CaOLU	1961
CaOOG	1955, 61
CaOWtU	1961
CoDCh	1961, 76
CoDGS	1962
CoG	1955, 61, 67-68
CoU	1955, 61, 63, 67, 76
DFPC	1961
DI-GS	1954-64, 66-78
DLC	1955, 61, 76
I-GS	1955, 61-63, 68
ICF	1955, 61
ICIU-S	1955, 61-62
ICarbS	1955, 61, 74, 76
IEN	1955, 61
IU	1955, 57-58, 60-63, 66-67, 71, 73-78
IaU	1955, 61, 63, 76
InLP	1955, 61, 63, 76
InRE	1955, 61
InU	1955, 61, 66-69, 71
KLG	1955, 57, 61, 68, 76, 84
KyU-Ge	1955, 61, 67-70
LNU	1961, 67, 76
LU	1961
MNS	1955, 61
MWesB	1961
MiDW-S	1961
MiKW	1961-62
MnU	1955, 61-63, 66-68, 70-72, 76
MoSW	1954-55, 61-63, 66-67, 71, 73
MoU	1955, 59, 61-63, 66-73
MtBuM	1955, 61, 76
NNC	1962-63
NRU	1955, 61-63
NSyU	1955, 61
NbU	1955, 61
NcU	1961
NdU	1961
NhD-K	1955, 61-63, 76
NjP	1961
NvU	1955, 61
OCU-Geo	1955, 58, 61, 67-71, 76, 81
OU	1955, 61-62
OkT	1955, 61
OkU	1954-56, 61, 76
OrU-S	1955, 61
PBL	1971, 76
PSt	1955, 61, 66
RPB	1955, 61
SdRM	1955, 61, 76
TMM-E	1967
TxBeaL	1955
TxDaAR-T	1955, 67-68, 70
TxDaDM	1955, 59-64, 66-70, 76
TxDaM-SE	1955, 61
TxHSD	1955, 61, 63
TxHU	1955, 61
TxLT	1955, 61
TxMM	1955
TxU	1954-64, 66-71, 75
UU	1955, 61, 76
ViBlbV	1955, 61
WU	1955, 61, 72, 76

(88) ASSOCIATION QUEBECOISE POUR L'ETUDE DU QUATERNAIRE. [LIVRET-GUIDE]

1976   Stratigraphie du Wisconsinien dans la region de Trois Rivieres-Shawinigan.
1979   Hudson Bay field meeting = rencontre geologique de la Baie d'Hudson, juin 1979. (546)
  Holocene stratigraphy & sea level changes in southeastern Hudson Bay, Canada = stratigraphie de l'Holocene et evolution des lignes de rivage au sud-est de la Baie d'Hudson, Canada. (Rev. ed. of 1979 field trip sponsored by "Association Quebecoise pour l'Etude du Quaternaire".) (546)
1980   Holocene stratigraphy & sea level changes in southeastern Hudson Bay, Canada = stratigraphie de l'Holocene et evolution des lignes de rivage au sud-est de la Baie d'Hudson, Canada. (Rev. ed. of 1979 field trip sponsored by "Association Quebecoise pour l'Etude du Quaternaire".) (546)
1984 5e   Le Quaternaire de Quebec meriodional: aspects stratigraphiques et geomorphologiques. (187)

CSt-ES	1980
CaAEU	1980
CaBVaU	1980
CaOKQGS	1980
CaOOG	1980
CaOWtU	1980
DI-GS	1980
IaU	1984

(89) ATLANTIC COASTAL PLAIN GEOLOGICAL ASSOCIATION. GUIDEBOOK FOR THE ANNUAL FIELD CONFERENCE.

1960 1st Stratigraphic problems of the latest Cretaceous and earliest Tertiary sediments in New Jersey.
1961 2nd Some remarks pertaining to geology of Atlantic Coastal Plain of New Jersey, Delaware, and Maryland. Non-marine Cretaceous sediments.
1962 3rd Guidebook to the coastal plain of Virginia north of the James River. (743)
1963 4th Geology of northeastern North Carolina.
1964 5th The Cretaceous formations along the Cape Fear River, North Carolina.
1965 6th Terrace sediment complexes in central South Carolina.
1966 7th See Southeastern Geological Society. 614.
1967 8th Delaware guidebook.
1968 9th Coastal plain geology of southern Maryland. (370)
1969 10th Geology of the York-James Peninsula and south bank of the James River. (136, 747)
1970 11th Part 1. Geology of the outer coastal plain, southeastern Virginia, Chesapeake-Norfolk-Virginia Beach.
Part 2. Geology of the upland gravels near Midlothian, Virginia.
1971 12th Neogene stratigraphy of the lower coastal plain of the Carolinas.
1972 13th [The geology of the coastal plain from the sounds near New Bern to the Piedmont of Wake County] (from Introduction) (129)
1973 14th A guide to the geology of Delaware's coastal environments. (Originally prepared for field trip for the Geological Society of America annual meeting, 1971)
1974 15th Environmental geology and stratigraphy of the Richmond, Virginia area. (747)
1976 Plio-Pleistocene faunas of the central Carolina coastal plain. 1975.
1979 Oct. Structural and stratigraphic framework for the coastal plain of North Carolina. (129)
1980 17th Guidebook to the late Cenozoic geology of the lower York-James Peninsula, Virginia. (Revised in 1981) (136(1981), 747)
1982 Apr. New Bern, North Carolina.
Structural-stratigraphic framework and geomorphic signature of the Graingers Wrench Zone, North Carolina coastal plain.
1983 Oct. Martha's Vineyard, Massachusetts.
The autochthonous and allochthonous coastal plain deposits of Martha's Vineyard and the Marshfield-Scituate area, southeastern Massachusetts.
1984 [19th] Stratigraphy and paleontology of the outcropping Tertiary beds in the Pamunkey River region, central Virginia coastal plain. (591(1985))

ATuGS	1962
CLU-G/G	1962, 65, 68, 71
CPT	1962
CSt-ES	1962, 65, 68, 71-72, 74, 80, 82, 84
CU-EART	1962-63, 65, 67-70, 72-74, 76, 79-80
CU-SB	1968, 70
CaAEU	1962
CaBVaU	1962
CaOOG	1968, 79
CaOWtU	1968, 71
CoDCh	1963
CoG	1962, 65, 67-71
CoU	1962
DI-GS	1960-65, 67-74, 80, 82-84
DLC	1965
GU	1971
I-GS	1962, 72
ICF	1962
ICIU-S	1962, 65, 67-72, 74
ICarbS	1962
IU	1961-65, 67, 69, 72, 76, 79, 80
IaU	1962, 65, 68-69, 71, 74, 80, 82
InLP	1962, 65, 71-72, 82-84
InU	1961-62, 65, 67-71
KyU-Ge	1962, 65, 67-72, 74, 76, 79, 80, 83-84
LNU	1971-72
MH-GS	1983
MdBJ	1968
MiHM	1962, 68
MiKW	1968
MiU	1965, 67
MnU	1961-62, 65, 67-72, 79-80, 82-84
MoSW	1962, 65, 68, 71
MtBuM	1962
NSyU	1962
NbU	1962
NcU	1960-65, 67-72, 79, 82
NdU	1962, 76
NhD-K	1965, 71, 82
NjP	1960-65, 67-72
NvU	1962, 65
OCU-Geo	1962, 64-65, 67-69, 71, 74
OU	1962, 65, 68, 71
OkU	1962, 65, 68, 71, 82-83
PBL	1970
PBm	1961, 63, 68
PSt	1962, 65, 67-74
RPB	1962
SdRM	1962
TxBeaL	1962, 74
TxDaAR-T	1963, 68, 70-72, 74
TxDaDM	1960-65, 67-68
TxDaM-SE	1962
TxHSD	1971
TxHU	1965
TxU	1960-65, 67-69, 71-72, 74
TxWB	1968
ViBlbV	1960-63, 69, 71-72
WU	1965, 71, 80

(90) ATLANTIC MARGIN ENERGY SYMPOSIUM. FIELD GUIDE.
1981 See American Association of Petroleum Geologists. Eastern Section. 25.

(91) AUGUSTANA COLLEGE, ROCK ISLAND, ILLINOIS. DEPARTMENT OF GEOLOGY. GUIDEBOOK FOR THE ANNUAL SPRING FIELD TRIP.
1953 St. Francois Mountains and mineral district, southeastern Missouri.
1954 Ouachita Mountains Magnet Cove, Bauxite, Arkansas.
1970 41st Upper Peninsula Michigan.

IaU	1970
TxU	1953-54

(92) AUSTIN GEOLOGICAL SOCIETY FIELD TRIP GUIDEBOOK.
1973 [1] Urban flooding and slope stability in Austin, Texas.
1974 Apr. [2] Geomorphic and hydrologic features of central Texas hill country.
1979 Dec. 3. Urban hydrology and other environmental aspects of the Austin area.
1981 Spring 4. Cretaceous volcanism in the Austin area, Texas. (revised in 1982)
1982 Fall (This is a revised edition of Guidebook 4, 1981)
1983 See Geological Society of America. South Central Section. 205.(1983(2))
1984 5. Geology of the Precambrian rocks of the Llano Uplift, central Texas; field trip notes. (Reprinted from Geological Society of America, South-Central Section, 1983 field trip; Central mineral region crystalline rocks, Llano Uplift, central Texas) (205(1983(5)))
6. Hydrogeology of the Edwards Aquifer; Barton Springs segment; Travis and Hays counties, Texas.
1985 7. Austin chalk in its type area; stratigraphy and structure. (205(1983 no. 3))
8. Edwards Aquifer - northern segment; Travis, Wiliamson, and Bell counties, Texas.

CPT	1984
CSt-ES	1981
CU-A	1981, 84
CU-SB	1973, 81
CaOWtU	1973, 81, 84
DI-GS	1981-82, 84, 85(7)
IU	1973, 81, 84(5), 85(8)
IaU	1981
InLP	1973, 81-82, 84-85
KyU-Ge	1973, 81, 84(6)
NSbSU	1982
NmU	1981
OU	1981
OkU	1973-74, 79, 82, 84, 85(7)
TxWB	1982, 84-85
WU	1981, 84(4, 5b), 85(7, 8)

(93) BASIN AND RANGE GEOLOGY FIELD CONFERENCE.

1965 1st Guidebook.
1969 2nd Guidebook.

AzTeS	1969
CMenUG	1965, 69
CSt-ES	1965, 69
CU-A	1969
CU-EART	1965, 69
CaACU	1969
CoDCh	1969
CoDGS	1969
DI-GS	1965, 69
DLC	1965
IdBB	1969
InLP	1969
KyU-Ge	1969
MnU	1969
NdU	1969
NvU	1965, 69
UU	1965, 69

(94) BAYLOR GEOLOGICAL SOCIETY. FIELD CONFERENCE GUIDEBOOK.

1958 1st Mid-Cretaceous geology of central Texas.
1959 2nd Layman's guide to the geology of central Texas.
1960 5th See Southwestern Association of Student Geological Societies (SASGS). 626.
1961 Central Texas field trip.
1962 Feb. Upper Cretaceous and lower Tertiary rocks in east-central Texas.
— Mar. Precambrian rocks of the Wichita Mountains, Oklahoma.
1963 Apr. See Southwestern Association of Student Geological Societies (SASGS). 626.
—Nov. See University of Texas at Austin. University Student Geological Society 716.
1964 Oct. Shale environments of the mid-Cretaceous section, central Texas.
— Dec. See Texas Academy of Science. 654.
1966 Precambrian and Paleozoic rocks of the eastern part of the Llano Uplift, central Texas.
1967 Valley of the giants.
1968 [1] The Hog Creek watershed.
[2] The Waco region.
1969 1. The Bosque watershed.
2. Mound Valley.
1970 1. The middle Bosque Basin.
2. Field conference guide. "Lampasas Cut Plain."
3. Field conference guide.
— Fall See Southwestern Association of Student Geological Societies (SASGS). 626.
1971 The Walnut Prairie field conference guidebook.
1972 [1] The hill country (Revised edition of a 1963 guidebook).
[2] See Southwestern Association of Student Geological Societies (SASGS). 626.
1973 [1] Exploration of a delta - 65 million years ago.
[2] "Geology in the city," urban geology of the I-35 growth corridor, central Texas.
[3] "Valley of the Giants" around the Paluxy River basin.
1974 [1] "The Black and Grand prairies."
[2] Bosque Watershed (Revised edition of 1969 guidebook).
[3] "Whitney Reservoir."
1975 Fall See Southwestern Association of Student Geological Societies (SASGS). 626.
1976 Tertiary-Cretaceous border.
1978 [1] The Paluxy Basin.
[2] Streams of central Texas.
[3] Urban development along the White Rock Escarpment: Dallas, Texas.
1979 Spring See Southwestern Association of Student Geological Societies (SASGS). 626.
— [Fall] [1] Geology of urban growth.
[2] The geomorphic evolution of the Grand Prairie.
[3] The Leon River Valley.
1981 See Southwestern Association of Student Geological Societies (SASGS). 626.
1983 Southeastern Llano country. (415)
— Fall Geologic section of the Cretaceous rocks of central Texas. (415)
1985 Tectonism and sedimentation in the Arbuckle Mountain region, southern Oklahoma Aulacogen. (96)
[n.d.] A day in the Cretaceous.

AzTeS	1968(1-2)
CLU-G/G	1958, 64(Oct.), 66-69, 85
CLhC	1958
CSt-ES	1973(3), 74(1)
CU-EART	1968
CaAC	1970(3)
CoDGS	1958, 60, 74, 78-79
CoU	1968
DI-GS	1958, 62, 64(Oct.), 66-71, 72(1), 73-74, 76, 78-79, 83, 85, [n.d]
IEN	1958
IU	1958, 62, 63(Apr.), 64(Oct.), 66, 68-71, 73(2-3), 74(1, 3), 76
IaU	1962, 70(3)
InLP	1974(1), 78(1, 3), 79(1, 2)
KyU-Ge	1961
LNU	1958
LU	1958
MNS	1958
MnU	1958, 62(Feb.), 64(Oct.), 66-68
NNC	1958, 62, 64(Oct.), 68-69
NjP	1968
OCU-Geo	1963(Apr.), 64(Oct.), 68, 70
OkT	1962
OkU	1958, 62, 85
TxBeaL	1958, 62(Mar.), 68(2)
TxDaAR-T	1958, 62(Feb.), 64(Oct.)
TxDaDM	1958, 62, 64(Oct.)
TxDaGI	1962(Mar.)
TxDaM-SE	1958-59
TxDaSM	1958
TxHSD	1962(Mar.)
TxHU	1958, 61-62, 63(Apr.), 64(Oct.), 68, 70
TxLT	1964(Oct.)
TxMM	1958, 60, 70(3)
TxU	1958, 61, 62(Mar.), 63(Apr.), 64(Oct.), 67-69, 78(1)
TxU-Da	1959
TxWB	1959, 61-62, 64, 66-74, 76, 78-79, 83, 85
UU	1958
ViBlbV	1974(1, 3)
WU	1962(Feb.)

(95) BAYLOR GEOLOGICAL SOCIETY. POPULAR GEOLOGY. NON-TECHNICAL FIELD CONFERENCE.

1960 4th Popular geology of central Texas; west-central McLennan County.
1961 5th Bosque County. Popular geology of central Texas, Bosque County.
— 6th Popular geology of central Texas, northwestern McLennan County.
1962 7th Southwestern McLennan County and eastern Correll County.
1963 8th Hill Country.

CLU-G/G	1963
DI-GS	1961-63
IU	1961-63
IaU	1961-63
MnU	1961-63
NNC	1961(6)
OCU-Geo	1963
OU	1960

OkU	1961-62
PBL	1962
TxBeaL	1961
TxDaDM	1961, 63
TxHU	1961(5), 62-63
TxLT	1963
TxMM	1961
TxU	1961-63
TxU-Da	1963

(96) BAYLOR UNIVERSITY. AAPG STUDENT CHAPTER.
1985 See Baylor Geological Society. 94.

(97) BELT SYMPOSIUM. FIELD TRIPS.
1973 I Belt Symposium, Volume 1. (260)
  Trip no.1. Belt rocks in the Clark Fork region of northern Idaho and the Purcell Mountains of British Columbia.
  Trip no.2. Stratigraphy and sedimentary features of the Missoula Group.
  Trip no.3. A geological field trip in Benewah and Whitman counties Idaho and Washington, respectively.
  Trip no.4. St. Joe field trip road log.
1983 II Guide to Field Trips Belt Symposium II
  Trip no.1. & 7. Stratigraphy of the Eastern Facies of the Ravalli Group, Helena Formation, and Missoula Group Between Missoula and Helena, Montana.

CMenUG	1973
CPT	1973
CSt-ES	1973
CU-EART	1973
CaACU	1973
CoDGS	1973
IU	1973
IdU	1973
InLP	1973
MtBuM	1973, 83(1, 7)
NvU	1973
OCU-Geo	1973

(98) BERMUDA BIOLOGICAL STATION FOR RESEARCH. SPECIAL PUBLICATION.
1970 No. 4. Field guide to Bermuda geology.

AzTeS	1970
CSt-ES	1970
CaACU	1970
DLC	1970
NSbSU	1970
NSyU	1970

(99) BIG RIVERS AREA GEOLOGICAL SOCIETY. FIELD GUIDE.
1975 2nd A field guide to the Precambrian geology of the St. Francois Mountains, Missouri. (613)
1978 Fall The stratigraphy of the outcropping Upper Cretaceous in the southern portion of west Tennessee.

CU-EART	1975
CoDGS	1975
DI-GS	1975
OCU-Geo	1978
ViBlbV	1975

BILLINGS GEOLOGICAL SOCIETY. See MONTANA GEOLOGICAL SOCIETY. (402)

(100) BOSTON COLLEGE. DEPARTMENT OF GEOLOGY AND GEOPHYSICS.
1980 Oct. See New England Intercollegiate Geological Conference. 456.

(101) BOTANICAL SOCIETY OF AMERICA. PALEOBOTANICAL SECTION.
1960 Aug. See American Institute of Biological Sciences. 40.
1961 Aug. See American Institute of Biological Sciences. 40.
1970   See American Institute of Biological Sciences. 40.
1974 June See American Institute of Biological Sciences. 40.

(102) BOY SCOUTS OF AMERICA. GEOLOGICAL FIELD TRIP. GUIDEBOOK.
1956 Aug. Geology and landscape of Camp Naish area, Wyandotte County; notes on counties in administrative area, Kaw Council. (634)
1957 Oct. Kansas.
  First field conference in southwestern Kansas.
  — Oct. North Dakota.
  1. Valley City area. (480(no. 1))
  2. Minot area. (480(no. 2))
  3. Devils Lake area. (480(no. 3))
  4. Bismarck-Mandan area. (480(no. 4))
  5. Dickinson area. (480(no. 5))
  6. Williston area. (480(no. 6))
  7. Jamestown area. (480(no. 7))
  8. Fargo to Valley City. (480(no. 8))
  9. Grand Forks to Park River. (480(no. 9))
1959 July Geology...landscape...mineral resources of the Mo-Kan Council area.
1961   Cimarron River valley, Black Mesa area.
1963 Apr. Tulsa County & vicinity. (Mimeographed, stapled handout)
1967   Meade County field trip.
1968   Black Mesa area of Cimarron County, Oklahoma and Union County, northern Missouri.

ATuGS	1957
CLU-G/G	1957(2-3, 5-9)
CPT	1957
CSt-ES	1957
CU-EART	1957
CaOLU	1957
CaOOG	1957
CaOWtU	1957(2-3, 5-9)
CoDCh	1957(4, 6), 68
CoG	1957(1-7)
CoU	1957
DI-GS	1957
DLC	1957
I-GS	1957
IEN	1957
IU	1957, 59, 61, 67-68
IaU	1957
InLP	1957
KLG	1956, 57, 67
KyU-Ge	1957
LU	1957
MNS	1957
MiDW-S	1957
MiHM	1957
MnDuU	1957(1-2, 4-7)
MnU	1957, 67
MoSW	1957
MtBuM	1957
NSyU	1957
NbU	1957
NcU	1957
NdU	1957
NjP	1957
NvU	1957
OU	1956
OkU	1957, 63
OrU-S	1957
PBL	1957
PSt	1957
SdRM	1957
TxBeaL	1957
TxDaAR-T	1957(2-3, 5, 7-9), 67
TxDaDM	1957, 67
TxDaM-SE	1957
TxHU	1957
TxLT	1957
TxMM	1957
TxU	1957
UU	1957
ViBlbV	1957
WU	1957

(103) BRIGHAM YOUNG UNIVERSITY. DEPARTMENT OF GEOLOGY. GEOLOGY STUDIES.
1968 V. 15 Pt. 2. Guide to the geology of the Wasatch Mountain Front, between Provo Canyon and Y Mountain, northeast of Provo, Utah. (Studies for Students No. 1)
Pt. 3. Guide to the geology and scenery of Spanish Fork Canyon along U.S. highways 50 and 6 through the southern Wasatch Mountains, Utah. (Studies for Students No. 2)
Pt. 4. Bonneville--an ice-age lake. (Studies for Students No. 3)
Pt. 5. Guidebook to the Colorado River, Part 1: Lee's Ferry to Phantom Ranch in Grand Canyon National Park. (Studies for Students No. 4)
1969 V. 16 Pt. 2. Guidebook to the Colorado River, Part 2: Phantom Ranch in Grand Canyon National Park to Lake Mead, Arizona-Nevada. (Studies for Students No. 5)
1971 V. 18 Pt. 2. Guidebook to the Colorado River, Part 3: Moab to Hite, Utah, through Canyonlands National Park. (Studies for Students No. 6)
1973 V. 20 Pt. 2. Geologic road logs of western Utah and eastern Nevada. (Studies for Students No. 7)
1974 V. 21 Pt. 2. See Geological Society of America. 197.(1975(10))
1975 V. 22 Pt. 2. See Geological Society of America. 197.(2)
1980 V. 27:3 Studies for students, #10, geologic guide to Provo Canyon and Weber Canyon, Central Wasatch Mountains, Utah.

AzFU	1968(5)
AzTeS	1968-69, 71
AzU	1968-69, 71, 73
CLU-G/G	1980
CMenUG	1968-69, 71, 73
CPT	1968-69, 71, 73, 80
CSdS	1968-69, 71, 73
CSt-ES	1968-69, 71, 73, 80
CU-EART	1968-69, 71, 73, 80
CU-SB	1980
CaACI	1968(3-4)
CaACU	1968-69, 71, 73, 80
CaAEU	1968-69, 71, 73, 80
CaBVaU	1968-69, 71, 73
CaNSHD	1980
CaOOG	1968-69, 80
CaOWtU	1968-69, 71, 73
CoDCh	1980
CoDGS	1980
CoU	1968-69, 71, 73
DI-GS	1973, 80
DLC	1968(2-3, 5), 71, 73
F-GS	1980
FBG	1980
I-GS	1973, 80
ICF	1968-69, 71, 73
ICIU-S	1968-69, 71, 73
ICarbS	1968-69, 71, 73
IEN	1968-69, 71
IU	1968-69, 71, 73, 80
IaU	1968-69, 71, 73, 80
IdBB	1968(5), 80
IdPI	1968(3-5), 69, 71, 73
IdU	1980
InLP	1968-69, 71, 73, 80
InRE	1969, 71, 73
InU	1968(5), 71
KyU-Ge	1968(5)
MiHM	1968-69, 71, 73
MnU	1968-69, 71, 73, 80
MoSW	1968-69
MtBuM	1968-69, 71, 73
NBiSU	1968-69, 71, 73
NMSoI	1980
NNC	1980
NSbSU	1968-69, 71, 73, 80
NcU	1961-81
NdU	1968-69, 71, 73, 80
NhD-K	1968-69, 80
NjP	1968-69, 71, 73
NmU	1980
NvU	1968(5), 69, 71, 73, 80
OCU-Geo	1968-69, 71, 73, 80
OU	1968-69, 71, 73, 80
OkU	1968-69, 71, 73, 80
OrU-S	1968-69, 71, 73
PBL	1968-69, 71, 73
PSt	1968-69, 71, 73
SdRM	1968-69, 71, 73, 80
TxMM	1968(5)
TxU	1968-69, 71, 73
TxU-Da	1968(5), 69
TxWB	1980
UPB	1971, 73
UU	1968-69, 71, 73, 80
ViBlbV	1968-69, 71, 73
WU	1968-69, 71, 73, 80
WyU	1980

(104) BRIGHAM YOUNG UNIVERSITY. DEPARTMENT OF GEOLOGY. GEOLOGY STUDIES. SPECIAL PUBLICATION.
1969 1. Grand Canyon perspective, a guide to the Canyon scenery by means of interpretive panoramas.
1978 See Geological Society of America. Rocky Mountain Section. 204.

CMenUG	1969
CSt-ES	1969
CU-A	1969
CU-EART	1969
CaACU	1969
CaAEU	1969
CaOWtU	1969
CoDGS	1969
DLC	1969
IU	1969
InLP	1969
MiHM	1969
NvU	1969

(105) BRIGHAM YOUNG UNIVERSITY. DEPARTMENT OF GEOLOGY. GUIDEBOOK FOR THE GEOLOGY FIELD TRIP.
1957 Provo to Bryce Canyon and Zion national parks, Utah.
1959 Provo to Bryce Canyon and Zion national parks.

CLU-G/G	1957, 59
CMenUG	1957
CaBVaU	1959
CoDGS	1957
DI-GS	1957, 59
OU	1959
TxDaSM	1957
TxMM	1959
UU	1957

(106) BRIGHAM YOUNG UNIVERSITY. DEPARTMENT OF GEOLOGY. STUDIES FOR STUDENTS.
1968 No. 1 See Brigham Young University. Department of Geology. 103.(Pt. 2)
— No. 2 See Brigham Young University. Department of Geology. 103.(Pt. 3)
— No. 3 See Brigham Young University. Department of Geology. 103.(Pt. 4)
— No. 4 See Brigham Young University. Department of Geology. 103.(Pt. 5)
1969 No. 5 See Brigham Young University. Department of Geology. 103.
1971 No. 6 See Brigham Young University. Department of Geology. 103.
1973 No. 7 See Brigham Young University. Department of Geology. 103.
1974 No. 9 See Brigham Young University. Department of Geology. 103.
1980 No. 10 See Brigham Young University. Department of Geology. 103.

BRITISH COLUMBIA. UNIVERSITY. See UNIVERSITY OF BRITISH COLUMBIA. DEPARTMENT OF GEOLOGICAL SCIENCES. (676)

CALGARY UNIVERSITY. See UNIVERSITY OF CALGARY, DEPARTMENT OF GEOLOGY AND GEOPHYSICS. (677)

(107) CALIFORNIA ASSOCIATION OF ENGINEERING GEOLOGISTS. SACRAMENTO SECTION. GUIDEBOOK.
1959 East side San Joaquin Valley field trip.
1961 June Guidebook; annual field trip: Northern California.

[1] [West of Orland including stops at Black Butte Dam, Glenn Reservoir site, Stony Gorge Dam, Oroville]
[2] [Oroville-Wyandotte Irrigation District construction project on the South Fork of the Feather River ending at Miners Ranch dam site]
1962 See Geological Society of Sacramento. 215.

CMenUG	1959, 61
CSfCSM	1959
CSt-ES	1959
CoDGS	1959
IU	1961

(108) CALIFORNIA GEOLOGY.
1973 V. 26. No. 2, supplement. See Association of American State Geologists. 74.
1975 V. 28. No. 1. The Stanislaus River - A study in Sierra Nevada Geology.
— V. 28. No. 5. See Geological Society of America. Cordilleran Section. 199.(No. 4:2)
1978 V. 31. No. 5. Klamath River geology, Curly Jack Camp to Ti Bar Siskiyou County, California.
1984 V. 37. No. 5. Field guide; Courtright Intrusive Zone, Sierra National Forest, Fresno County, California.
1985 V. 38. No. 9. Guide to Titus Canyon, Death Valley National Monument.

CLU-G/G	1985
CPT	1975
CSt-ES	1975, 78, 84-85
CU-EART	1978
IaU	1978, 84-85
OCU-Geo	1985
OkU	1985

(109) CALIFORNIA STATE UNIVERSITY, CHICO. GEOLOGY DEPARTMENT. FIELD TRIPS.
1968 Northern California. (Note: Covers field trips from 1959 thru 1968 in one volume plus ephemeral sheet for each trip)

CChiS	1968
DI-GS	1968

(110) CALIFORNIA STATE UNIVERSITY, SAN DIEGO. DEPARTMENT OF GEOLOGICAL SCIENCES.
1979 Nov. See Geological Society of America. 197.

(111) CALIFORNIA. DEPARTMENT OF EDUCATION. [GUIDEBOOK]
1964 Oct. Geology field trips in northern California. A guide for improvement of instruction in geology in California junior colleges.

CU-EART	1964

(112) CALIFORNIA. DIVISION OF MINES AND GEOLOGY. BULLETIN.
1948 No. 141. Geologic guidebook along Highway 49, Sierran gold belt, the Mother Lode country.
1951 No. 154. Geologic guidebook of the San Francisco Bay counties.
1954 No. 170. See Geological Society of America. 197.
1962 No. 181. See American Association of Petroleum Geologists. 18.
— No. 182. See American Association of Petroleum Geologists. 18.
1966 No. 190. See Geological Society of America. 197.

ATuGS	1948, 51
AzFU	1951
AzTeS	1951
CLU-G/G	1948, 51
CLhC	1948, 51
CMenUG	1948, 51
CPT	1948, 51
CSdS	1948, 51
CSt-ES	1948, 51, 54, 62, 66
CU-A	1948, 51
CU-EART	1948, 51
CaACU	1948, 51, 54, 62, 66
CaAEU	1951
CaBVaU	1948, 51
CaOHM	1951
CaOKQGS	1948, 51
CaOTRM	1948, 51
CaQQLaS	1948, 51
CoDuF	1948
CoG	1948, 51
CoU	1948, 51
DFPC	1951
DI-GS	1948, 51
DLC	1948, 51
ICF	1948, 51
ICarbS	1948, 51
IEN	1948, 51
IU	1948, 51
IaU	1948, 51
IdPI	1948, 51
IdU	1948, 51
InLP	1948, 51
InU	1948, 51
KyU-Ge	1948, 51
LNU	1948
LU	1948, 51
MH-GS	1951
MiDW-S	1948, 51
MiHM	1948, 51
MnU	1948, 51
MoSW	1948, 51
MoU	1948, 51
MtBuM	1948, 51
NBiSU	1948, 51
NSbSU	1951
NbU	1948, 51
NcU	1948, 51
NdU	1948, 51
NhD-K	1948, 51
NjP	1951
NvU	1948, 51
OCU-Geo	1948, 51
OU	1948, 51
OkT	1948, 51
OkU	1948, 51
OrU-S	1948, 51
PSt	1948, 51
RPB	1948, 51
SdRM	1948, 51
TxDaAR-T	1951
TxDaDM	1948, 51
TxDaGI	1948
TxDaM-SE	1948, 51
TxDaSM	1951
TxHSD	1948, 51
TxHU	1948, 51
TxMM	1948, 51
TxU	1948, 51
ViBlbV	1948, 51
WU	1948, 51

CALIFORNIA. DIVISION OF MINES AND GEOLOGY. CALIFORNIA GEOLOGY. See CALIFORNIA GEOLOGY. (108)

(113) CALIFORNIA. DIVISION OF MINES AND GEOLOGY. MINERAL INFORMATION SERVICE. (NAME CHANGED TO: CALIFORNIA GEOLOGY, EFFECTIVE V. 24, NO. 1 JANUARY 1971 ISSUE)
1953 V. 6. No. 2, supplement. See Association of American State Geologists. 74.

(114) CALIFORNIA. DIVISION OF MINES AND GEOLOGY. SPECIAL PUBLICATION.
1969 35. Preliminary report and geologic guide to Franciscan melanges of the Morro Bay-San Simeon area, California.

CMenUG	1969
CPT	1969
CSdS	1969
CSt-ES	1969
CU-A	1969
CU-EART	1969
InLP	1969
NvU	1969
PBm	1969

(115) CALIFORNIA. DIVISION OF MINES AND GEOLOGY. SPECIAL REPORT.
1975 No. 118. See Geological Society of America. Cordilleran Section. 199.(No. 1)

(116) CALIFORNIA. NATURAL HISTORY FOUNDATION OF ORANGE COUNTY.
1980 Nov. See Southern California Paleontological Society. 619.

CALIFORNIA. UNIVERSITY. See UNIVERSITY OF CALIFORNIA, SAN DIEGO, GEOLOGY DEPARTMENT. (679) and UNIVERSITY OF CALIFORNIA, RIVERSIDE. (678) and UNIVERSITY OF CALIFORNIA, SANTA BARBARA. DEPARTMENT OF GEOLOGICAL SCIENCES. (680) and EARTH AND SPACE SCIENCE STUDENT ORGANIZATION, U.C.L.A. (161)

(117) CANADA WIDE SCIENCE FAIR. FIELD TRIP.
1974 13th Geology of the Banff area; a student's guide. (125)
CU-EART	1974
CaACAM	1974
CaACU	1974
CaAEU	1974
CaBVaU	1974
CaOWtU	1974
DI-GS	1974
DLC	1974
TxU	1974
ViBlbV	1974
WU	1974

CANADA. GEOLOGICAL SURVEY. See GEOLOGICAL SURVEY OF CANADA. (224) and GEOLOGICAL SURVEY OF CANADA. (225) and GEOLOGICAL SURVEY OF CANADA. (226)

(118) CANADIAN GEOPHYSICAL UNION. FIELD TRIP GUIDEBOOK.
1977 See Geological Association of Canada. 190.
1981 See Geological Association of Canada. 190.
1983 May See Geological Association of Canada. 190.

(119) CANADIAN GEOSCIENCE COUNCIL. EDGEO CONFERENCE. FIELD TRIP GUIDE.
1985 May Field trip guide for EdGEO Conference, NAGT Eastern Section Meeting, OAGEE Spring Conference. (512, 411)
Geology, physiography and land use: Toronto to Madoc.
Silurian and Devonian rocks of the Niagara Peninsula, Ontario.
Niagara Gorge, prepared for secondary school students, field trip.
CaOTDM	1985
KyU-Ge	1985

(120) CANADIAN GEOTECHNICAL SOCIETY. GEOTECHNICAL CONFERENCE. FIELD TRIP GUIDEBOOK.
1982 35th Montreal, Quebec. See Association of Engineering Geologists. 76.

(121) CANADIAN INSTITUTE OF MINING AND METALLURGY. CIM GEOLOGY DIVISION. EXCURSION GUIDEBOOK. (TITLE VARIES)
1964 The Matagami Area, northwestern Quebec: guidebook for geological field excursion.
1967 C.I.M.M. Centennial field excursion; northwestern Quebec and northern Ontario.
1980 Sept. Gold symposium and field excursion; Val D'Or-Kirkland Lake-Timmins.
1981 Sept. Saskatchewan uranium field trip guide. (Cover title: Uranium field excursion guidebook)
— Sept./Oct. "The St. Honore and Crevier niobium-tantalum deposits and related alkaline complexes, Lac St. Jean, Quebec".
1983 Sept. Metallogeny of the southern province Sudbury-Elliot Lake area, Ontario.
1984 Oct. Gold deposits in the Meguma terrane of Nova Scotia.
2. Hemlo - Manitouwadge - Winston Lake metallogenesis of hibghly metamorphosed Archean gold-base metal terrain. (Updated version printed in 1985)
4. The geology and slope stability problems in the metropolitan Toronto region.
1985 Gold and copper-zinc metallogeny within metamorphosed greenstone terrain, Hemlo-Manitouwadge-Winston Lake, Ontario, Canada. (192)
CLU-G/G	1981, 84
CMenUG	1984
CPT	1980
CSt-ES	1967, 81, 84, 85
CU-SB	1981, 83, 84
CaACU	1967, 80-81, 84
CaAEU	1981
CaBVaU	1981
CaNSHD	1980, 84
CaOHaHa	1967
CaOKQGS	1981, 84
CaOOG	1967, 81, 84
CaOTDM	1980-81, 83-84
CaOWtU	1964
CoG	1981
DI-GS	1967, 81, 84
DLC	1967
IU	1967
IaU	1985
InU	1967
MiHM	1967
MnU	1981, 84
NBiSU	1967
NN	1981
NOneoU	1967
NjP	1967
NvU	1981
PSt	1967
TxU	1967

(122) CANADIAN PALEONTOLOGY AND BIOSTRATIGRAPHY SEMINAR. FIELD EXCURSIONS GUIDEBOOK.
1981 Oct. Atlantic City, New Jersey.
Field guide to the anthracite coal basins of eastern Pennsylvania. (18, 25, 27, 539, 585)
Field guide to the geology of the Paleozoic, Mesozoic, and Tertiary rocks of New Jersey and the central Hudson Valley. (539)
[Trip 1] Cretaceous and Tertiary sediments of the New Jersey coastal plain.
[Trip 2] Rift basins of the Passive margin: tectonics, organic-rich lacustrine sediments, basin analyses.
[Trip 3] Extinct and active continental margin deposits and their tectonic-switchover products: Appalachian orogen ("Eastern Overthrust Belt") - Catskill Plateau-Newark Basin-Atlantic coastal plain.
1985 Sept. 1. The Trenton Group of the Quebec City area.
2. The Citadel Formation: its age on the basis of trilobites, graptolites and brachiopods.
3. The Levis Formation: passion margin slope process aand dynamic stratigraphy in the western area.
IaU	1985
NvU	1985

(123) CANADIAN SOCIETY OF EXPLORATION GEOPHYSICISTS.
1975 See Canadian Society of Petroleum Geologists. 125.
1984 See Canadian Society of Petroleum Geologists. 125.

(124) CANADIAN SOCIETY OF PETROLEUM GEOLOGISTS. CSPG-U OF C SHORT COURSE [NOTES].
1983 May Clastic diagenesis.
MnU	1983

(125) CANADIAN SOCIETY OF PETROLEUM GEOLOGISTS. GUIDEBOOK FOR THE ANNUAL FIELD CONFERENCE.
1950 See Geological Association of Canada. 190.
1952 2nd Kananaskis Valley. (Also contains additional information on Highwood Valley east to Calgary via Turner Valley oil field)
1953 3rd Calgary and Crowsnest Pass, southern Alberta.
1954 4th Banff-Golden-Radium; southern Canadian Rocky Mountains.

1955	5th	Jasper National Park.
1956	6th	Bow Valley, Alberta.
1957	7th	Waterton, Alberta.
1958	8th	Nordegg, Alberta.
1959	9th	Drumheller-Moose Mountain, Alberta. (14)
1960	10th	Map 1 geological. Rocky Mountains, Banff to Golden; structure and stratigraphy.

Map 2 geological. Banff-Minnewanka-Canmore area. (Maps only for these field trips were published)

1961 11th Turner Valley-Savannah Creek, Kananaskis. (14)
1962 12th [1] Coleman-Cranbrook-Radium. (Special guidebook issue, Journal of the Alberta Society of Petroleum Geologists, v. 10:7, July-Aug. 1962) (14)

[2] Road log: Crowsnest-Cranbrook-Windermere.

1963 13th Ghost River area.
1964 14th Flathead Valley, Southeast British Columbia. (Bulletin of Canadian Petroleum Geology, v. 12, Special Issue, Aug. 1964)
1965 15th Cypress Hills Plateau, Alberta and Saskatchewan.
1967 See International Symposium on the Devonian System. 318.
1968 16th Canadian Rockies-Bow Valley to North Saskatchewan River.
1970 See American Association of Petroleum Geologists. 18.
1971 A guide to the geology of the Eastern Cordillera along the Trans-Canada Highway between Calgary, Alberta, and Revelstoke, British Columbia. (14)
1973 Oil Sands symposium. Guide to the Athabasca Oil Sands area. (13)
1974 See Canada Wide Science Fair. 117.

Lower Carboniferous stratigraphy, biostratigraphy, and sedimentology, Moose Dome, Rocky Mountain Foothills.

1975 May Structural geology of the foothills between Savanna Creek and Panther River, southwestern Alberta, Canada. (123)
— [Sept.] Guidebook to selected sedimentary environments in southwestern Canada.
1977 [May] Jura Creek field trip guidebook.
— June Lake Minnewanka field trip guidebook.
— [July] Kicking Horse Pass field trip guidebook.
— [Sept.] Geological guide for the CSPG 1977 Waterton-Glacier Park field conference.
— [Oct.] Burnt Timber Creek guidebook.
1978 May Moose Mountain field trip guidebook.
— June International conference: Facts and principles of world oil occurrence.

[1] Geological and geographical guide to the Mackenzie Delta area.
[2] Geological guide to the central foothills and Rocky Mountains of Alberta.
[4] Field guide to rock formations of southern Alberta (stratigraphic sections guidebook)
— Aug. Lake Minnewanka field trip.
— Sept. Big Hill section field trip.
1979 (1) Plateau Mountain field trip.
(2) Cypress Hills Plateau field trip.
1980 May Grassi Lakes-Whiteman Gap field trip.
— June Trap Creek-Lundbreck field trip. Sedimentology of Upper Cretaceous fluviatile, deltaic and shoreline deposits.
— Aug. Mount Wilson field trip.
— Sept. Rocky Mountain overview field trip. Calgary-Lake Louise-Columbia Icefields-Saskatchewan Crossing-Nordegg.
1981 Upper Devonian Hummingbird Reef field trip.
— Aug. Plateau Mountain field trip.
— Sept. Seebe-Jura Creek field trip.
1982 June See American Association of Petroleum Geologists. 18.
1983 May Calgary, Alberta.

The Mesozoic of middle North America.

1. Sedimentology of the Upper Cretaceous Judith River (Belly River) Formation, Dinosaur Provincial Park, Alberta.
2. Facies relationships and paleoenvironments of a Late Cretaceous tide-dominated delta, Drumheller, Alberta.
3. Sedimentology of Upper Cretaceous fluviatile, deltaic and shoreline deposits, southwestern Alberta foothills.
4. Structure, stratigraphy, sedimentary environments and coal deposits of Jura-Cretaceous Kootenay Group, Crowsnest Pass area, Alberta and British Columbia.
7. Sedimentology of Jurassic and Upper Cretaceous marine and nonmarine sandstones, Bow Valley.
8. Biogenic structures in Upper Cretaceous outcrops and cores.
9. Structure, stratigraphy and sedimentary facies of the Paleocene and Lower Cretaceous coal-bearing strata in the Coalspur and Grande Cache areas, Alberta.
— Aug. Calgary, Alberta.
[10] Precambrian Miette conglomerates, Lower Cambrian Gog quartzites and modern braided outwash deposits, Kicking Horse Pass area.
[11] Stratigraphy and sedimentary environments of the Jurassic-Cretaceous Kootenay Group and adjacent strata, Highwood Pass - Kananaskis Country area, Alberta.
[12] The sedimentology of the Blood Reserve Sandstone in southern Alberta.
— Sept. [13] Late Paleozoic shelf deposits - the new precision in stratigraphy; field trip notes.
1984 Aug. Depositional cycles and facies relationships within the Upper Cretaceous Wapiabi and Belly River formations of west central Alberta.
Trip 1. Calgary to Nordegg; stratigraphy and structural overview.
Trip 1A. Central Rocky Mountains; Calgary-Forestry Trunk Road-Nordegg-Calgary. (123)
[2] Sedimentology of a foreland coastal plain: Upper Cretaceous Judith River Formation at Dinosaur Provincial Park. (9)
Geological guidebook; Lake Minnewanka trip.
1985 CSPG Geologic guidebook; Lake Minnewahka Trip.

ACTCR	1983-84
AzFU	1955
AzTeS	1956-57, 62(2), 65, 68, 71
CLU-G/G	1953-59, 62-65, 68, 71, 78(June)
CLhC	1954-59, 62, 64
CMenUG	1975, 78, 82-83
CPT	1954-59, 61-64, 73
CSdS	1978(4)
CSt-ES	1952, 54-60, 62-65, 68, 71, 73, 75, 77(Sept.), 78(June), 83, 84(Aug.)
CU-A	1963-64, 73, 75(Sept.), 77(Sept.), 78(June 1-2), 84(4)
CU-EART	1953-59, 61-65, 68, 71, 73, 75, 77(Sept.), 78(June), 83
CU-SB	1978, 84(4)
CaAC	1953-59, 61, 64-65, 68, 71, 73, 75(Sept.), 77(Sept.), 78(June) (Circulate only 1954-56, 58, 62, 64)
CaACAM	1954-59, 61-65, 68, 71, 73, 75, 77-78, 83
CaACI	1952-59, 61-65, 68
CaACM	1953-59, 61-65, 68, 71, 75, 77(Sept.), 78(June 1-2))
CaACU	1953-59, 61-65, 68, 71, 73, 74, 75(May), 77-81, 83-84
CaADTMP	1984(2)
CaAEU	1954-56, 58-59, 61-65, 68, 71, 73, 75(May), 77(Sept.), 78(June 1-2), 83
CaBVaU	1953-59, 61-62, 65, 68, 71, 73, 75(Sept.), 77(Sept.), 78(June (1-2)), 81(10-12), 84(1, 1A)
CaDTMP	1084(2)
CaNBFU	1954
CaOHM	1953
CaOKQGS	1953-59, 61-65
CaOLU	1953-59, 62-63, 73
CaOOCC	1954-60, 73
CaOOG	1952-65, 68, 71, 73, 77(Sept.), 78(June 1-2), 83
CaOPAL	1955-56, 59
CaOTRM	1954-59, 61-65
CaOWtU	1954-59, 61-64, 68, 71, 73, 75(May)
CaQMU	1954-59, 61-64
CoDGS	1973, 82(3, 12)
CoDU	1956, 68
CoG	1953-59, 62-65, 68
CoLiM	1984
CoU	1953-59, 61-65, 73
CtY-KS	1978(4)
Cu-A	1978[3]
DI-GS	1953-59, 62-65, 68, 71, 73, 75, 77-78, 80-83
DLC	1954-59, 73, 75(Sept.), 78(June 2)
I-GS	1954-59
ICF	1955-59, 61
IEN	1954-59, 62-64
IU	1954-59, 61-65, 68, 71, 75, 77(Sept.), 78(June)
IaU	1954-59, 61-65, 75
IdBB	1956-57, 62, 68
IdPI	1959
IdU	1977(Sept.), 1978(June), 1983(11)
InLP	1954-59, 61-65, 71, 75, 77(Sept.), 78(June [3], 4), 83

(126) Canadian Society of Petroleum Geologists. Paleontology Division.

InU	1954-59, 61-62, 65, 68, 75(Sept.), 77(Sept.), 78(June)	DLC	1959
KyU-Ge	1954-59, 62, 64-65, 68, 71, 75(May, Sept.), 77(Sept.), 78(June)	ICF	1959
LNU	1955-56, 59	IU	1959, 68
LU	1952-59	IaU	1977, 80
MH-GS	1954	KyU-Ge	1971
MNS	1954-56	MdBJ	1977
MiDW-S	1965	NSbSU	1977
MiHM	1973	NcU	1968
MnU	1954-56, 61-62, 64-65, 75, 78(June (2, [3], 4), 83	TxDaAR-T	1968
MoSW	1975(May)		
MtBuM	1953-60, 62, 64, 78(2, 4)		
NBiSU	1971		
NNC	1953-59, 62-65, 71, 75(May), 77(May), 78(May)		
NSyU	1954-59, 62-64, 73		
NbU	1955-59, 62-63		
NcU	1954-59, 61-65, 68, 71, 75(Sept.)		
NdU	1954, 56, 61-62, 64, 73, 75(Sept.)		
NhD-K	1954-59, 62-65, 68, 73		
NjP	1954-64		
NvU	1954-59, 61-63		
OCU-Geo	1953-57, 62, 65, 68, 71, 73, 75(May), 78(May)		
ODaWU	1983		
OU	1954-59, 61, 63, 65, 68, 71		
OkT	1953-56, 58-59		
OkTA	1983(3, [11])		
OkU	1953-59, 61, 64, 75(May), 77(Sept.), 78		
OrU-S	1956-59, 62-64		
PSt	1954-59, 62, 64-65		
RPB	1964		
SdRM	1954-59, 62, 64-65		
TU	1954-59, 62		
TxBeaL	1953, 55-58, 62-63, 75(May)		
TxDaAR-T	1953-59, 62, 64-65, 71, 75		
TxDaDM	1953-59, 61, 64, 68		
TxDaGI	1953-57, 59		
TxDaM-SE	1953-59, 62-65		
TxDaSM	1953-59, 64-65, 71, 75		
TxHSD	1953-55, 59, 64, 73, 75, 78(June (1))		
TxHU	1953-59, 62-65, 75		
TxLT	1953, 55		
TxMM	1953-59, 62, 64-65, 83		
TxU	1952-65, 71, 73, 75(Sept.), 77(Sept.)		
TxU-Da	1953-59, 62-63		
TxWB	1956		
UU	1955-59, 61, 64, 75(May), 77(Sept.), 78(June (2))		
ViBlbV	1956, 68, 71, 73, 75		
WGrU	1975(May)		
WU	1954-59. 62(1), 64, 65, 75(Sept.), 77(Sept.), 78(June), 84		
WaPS	1975(May)		

(126) CANADIAN SOCIETY OF PETROLEUM GEOLOGISTS. PALEONTOLOGY DIVISION.
1979   Cretaceous-Tertiary, Gleichen and Bassand districts.
    CaACU   1979

(127) CARIBBEAN GEOLOGICAL CONFERENCE. FIELD GUIDE.
1959 2nd Roadlog and guide for a geologic field trip through central and western Puerto Rico. (234)
1968 5th Field guide to the geology of the Virgin Islands.
    1. Guide to the geology of St. Thomas and St. John, Virgin Islands.
    2. Field guide to the geology of St. Croix, U.S. Virgin Islands.
1971 6th PC-4:1. Excursion en lancha a lo largo de la costa norte Venezolona entre Cumana y Pertigalete.
1977 8th Guide to the field excursions on Curacao, Bonaire and Aruba, Netherlands.
    [1] Field trip of general nature to Bonaire.
    [2] Field trip to salinas of Bonaire.
    [3] Field trip of general nature to Auraba.
    [4] Field trip to the Eocene of Cer'i Cueba (Curacao)
    [5] Field trip to Late Senonian Knip Group.
1980 9th Field trip to some submerged Pleistocene reef terraces. vol. 2.

CLU-G/G	1959
CSt-ES	1977
CU-EART	1977
CaACU	1959
CoDGS	1959
DI-GS	1959, 71

(128) CAROLINA COAL GROUP.
1979 See Geological Society of America. Southeastern Section. 206.

(129) CAROLINA GEOLOGICAL SOCIETY. GUIDEBOOK OF EXCURSIONS. (TITLE VARIES)
1952 The Great Smoky Mountains.
1953 Oct. Road log for the annual excursion of the Carolina Geological Society. [Relations among granites and gneisses along the boundary between the Shelby and Lincolnton quadrangles, North Carolina] (From the introduction)
1955 The coastal plain of North Carolina.
1957 Guidebook for the South Carolina coastal plain field trip.
1958 Lake Murray, South Carolina, and surrounding area.
1959 Geology of the Albemarle and Denton quadrangles, North Carolina; stratigraphy and structure in the Carolina volcanic-sedimentary group. (607)
1960 Road log of the Grandfather Mountain area, North Carolina.
1961 Relationships between the Carolina Slate Belt and the Charlotte Belt in Newberry County, South Carolina. (608)
1962 Road log of the geology of Moore County, North Carolina.
1963 Guide to the geology of Pickens and Oconee counties, South Carolina. (608)
1964 Nov. Road log of the Chatham, Randolph and Orange County areas, North Carolina.
1965 Guide to the geology of York County, South Carolina. (608)
1966 Excursion in Cabarrus County, North Carolina.
1967 Guide to the geology of Mount Rogers area, Virginia, North Carolina, Tennessee.
1968 Stratigraphy, structure and petrology of the Piedmont in central South Carolina. (608)
1969 A guide to the geology of northwestern South Carolina. (608)
1970 Stratigraphy, sedimentology and economic geology of Dan River basin, N.C.
1971 Stratigraphy and structure of the Murphy Belt, North Carolina.
1972 See Atlantic Coastal Plain Geological Association. 89.
1973 34th Granitic plutons of the central and eastern Piedmont of South Carolina.
1974 Geology of the Piedmont and coastal plain near Pageland, South Carolina and Wadesboro, North Carolina.
1975 Nov. Guide to the geology of the Blue Ridge south of the Great Smoky Mountains, North Carolina.
1976 Oct. Introduction to the geology of the eastern Blue Ridge of the Carolinas and nearby Georgia.
1977 Field guide to the geology of the Durham Triassic Basin North Carolina.
1978 Oct. Geological investigations of the eastern Piedmont, Southern Appalachians. (With a field trip guide on the bedrock geology of central South Carolina)
1979 Oct. See Atlantic Coastal Plain Geological Association. 89.
1981 Oct. Geological investigations of the Kings Mountain Belt and adjacent areas in the Carolinas.
1982 Oct. Geological investigations related to the stratigraphy in the Kaolin mining district, Aiken County, South Carolina.
1983 Oct. Geologic investigations in the Blue Ridge of northwestern North Carolina.
    Deformational history of the region between the Grandfather Mountain and Mountain City windows, North Carolina and Tennessee.
1984 Oct. A stratigrapher's view of the Carolina slate belt, south central North Carolina.
1985 Nov. The Virgilina deformation: implications of stratigrahic correlation in the Carolina slate belt.
    CLU-G/G   1957, 61, 63, 65, 68-69, 82

CMenUG	1952, 55, 57, 59-60, 67, 71, 76, 81-83
CPT	1957, 81-82
CSt-ES	1961, 63, 65, 67-69, 72, 73, 76-78, 81-82
CU-EART	1981-83
CaBVaU	1982
CaOOG	1981-82
CaOTDM	1981-82
CaOWtU	1965, 68-69, 82
CoDCh	1955, 70, 78, 80
CoDGS	1952, 55, 57-60, 62, 67, 81-82
CoG	1957
CoU	1957, 61, 63, 65, 68-69
DI-GS	1952-53, 55, 57-69, 73-74, 76-78, 81-83
DLC	1983
I-GS	1957, 78, 81, 82
IEN	1957
IU	1953, 57, 64-65, 67-71, 73, 76-78
IaU	1957, 59-60, 62, 64-71, 73, 75-78, 81-83
InLP	1957-71, 73-78, 81-83
InU	1961, 63, 66-69, 82
KyU-Ge	1952, 57-61, 63-71, 73-74, 78, 81
LNU	1957, 59-60, 62, 64, 66-67, 70, 74, 77
MH-GS	1952, 67
MdBJ	1962
MnU	1957, 59-68, 70-71, 73, 76-77, 81-82
MoSW	1957-58, 81
NIC	1981, 82, 84
NNC	1952, 57, 68
NbU	1961
NcU	1952-53, 55, 57-60, 62, 64, 66-67, 70-73, 78, 80-82
NdU	1968-69
NhD-K	1958, 73, 76, 81
NjP	1959-60, 62, 64, 66-69
OCU-Geo	1952-53, 66, 76, 81
OU	1957, 61, 63, 65, 68-69, 76, 81
OkU	1957, 59-61, 63, 65, 68-69, 72, 73, 75-77, 80-82
PSt	1957
SC-GS	1981-82
SCgs	1981-82
ScU	1981
SdRM	1957
TxBeaL	1976
TxDaAR-T	1959, 61, 63, 65-71, 73-74
TxDaDM	1963-65, 68-69
TxDaM-SE	1957, 81-82
TxDaSM	1955
TxHSD	1957
TxU	1957, 59, 61, 63, 65-66, 68, 76, 81-83
ViBlbV	1959-63, 65-71, 76
WU	1973-74, 76, 78, 81-82

(130) CAVE MAN EXPEDITION.
1962 6th A guide to the caverns and geology of Cave River Valley Park and vicinity, Washington County, Indiana.
    InLP   1962

CIM GEOLOGY DIVISION. See CANADIAN INSTITUTE OF MINING AND METALLURGY. CIM GEOLOGY DIVISION. (121)

CIRCUM-PACIFIC JURASSIC RESEARCH GROUP. FIELD CONFERENCE. GUIDEBOOK. See INTERNATIONAL GEOLOGICAL CORRELATION PROGRAMME (IGCP). [PROJECT] 171. CIRCUM-PACIFIC JURASSIC RESEARCH GROUP. (301)

(131) CIRCUM-PACIFIC TERRANE CONFERENCE. PROCEEDINGS.
1983 Tectonostratigraphic Terranes of the North San Francisco Bay Region. (Field Trip guide). (630)
    CSt-ES   1983
    IaU   1983

(132) CLAY MINERALS CONFERENCE. PROCEEDINGS. GUIDEBOOK TO THE FIELD EXCURSIONS. (TITLE VARIES)
1952 1st No guidebook.
1953 2nd [Clays of east-central Missouri in the vicinity of Columbia, McCredie, Auxvasse, Mexico, and Hermann] (419, 407)
1954 3rd No guidebook.
1955 4th Clay minerals in sedimentary rocks. (419, 407)
1956 5th [Early Tazewell moraines (Illinois)] (419)
1957 6th Ione clay area. (419)
1958 7th Northeastern Maryland and northern Delaware. (422)
1959 8th Wichita Mountain area, southwestern Oklahoma. (419)
1960 9th [West Lafayette to High Bridge (Indiana)] (419)
1961 10th Field excursion, central Texas; bentonites, uranium-bearing rocks, vermiculites. (713)
1962 11th Gatineau area, Quebec, Canada. (419)
1963 12th Attapulgite fuller's earth localities in Georgia and Florida. (419)
1964 13th Field trips in southern Wisconsin. (776, 133)
    No. 1. South-central Wisconsin.
    No. 2. East-central Wisconsin.
1965 14th Field trips in central California. (133)
    No. 1. Lincoln plant of the Interpace Corp.; Solano and San Joaquin soils.
    No. 2. Rock alteration in the North San Francisco Bay area.
1969 18th Field excursion: East Texas; clay, glauconite, ironstone deposits. (713, 197(1973-No. 11))
1976 25th [1] Geology and soils. Field trip to Oregon Coast and Coast Range. (133)
    [2] U.S.-Japan seminar [on] amorphous and poorly crystalline clays. Field tour, Oregon Cascades: Crater Lake and Mazama deposits, August 6, 7, and 8, 1976. (133)
1979 [28th] Field conference on kaolin, bauxite, and Fuller's earth. (133)
1980 29th Waco, Texas.
Soils guidebook for the ... annual Clay Minerals Conference.
Landscape and land use in Central Texas.

AzTeS	1953, 55-57, 59-63, 69
AzU	1961, 69
CLU-G/G	1953, 55-63, 69
CLhC	1961
CMenUG	1980
CPT	1953, 55-57, 59-60, 62-63
CSdS	1956-65
CSt-ES	1961, 69
CU-EART	1956-63, 65, 69
CU-SB	1953, 55-65
CaACU	1953, 55-63, 69
CaBVaU	1953, 55-63
CaOHM	1953, 55-63, 69
CaOKQGS	1953, 55-63
CaOLU	1953, 55-63
CaOOG	1953, 55-63, 69
CaOWtU	1953, 55-57, 59-60, 62-63, 69
CoDGS	1961, 69
CoG	1961, 69
CoU	1953, 55-63, 69
DI-GS	1953, 55-63, 69, 79-80
DLC	1961, 69
I-GS	1953, 55-61, 63, 80
ICF	1953, 56, 59-61, 69
ICarbS	1953, 55-57, 59-60, 62, 63, 69
IEN	1958, 61, 69
IU	1953, 55-63, 69
IaU	1953, 55-63, 69, 79
IdBB	1960, 62
InLP	1953, 55-57, 59-65, 69, 76
InU	1953, 55-63, 69
KyU-Ge	1953, 55-63, 69, 72-73, 78, 80
LNU	1961, 69
LU	1953, 55-63
MNS	1957, 61, 69
MiDW-S	1953, 55-63, 69
MiHM	1953, 55-57, 59-63, 69
MiU	1961
MnU	1961, 69
MoSW	1953, 55-63, 69
MoU	1953, 55-63, 69
MtBuM	1953, 55-61, 63, 69
NBiSU	1961, 69
NNC	1961, 69
NSyU	1961, 69
NbU	1953, 55-63
NcU	1953, 55-63, 69
NdU	1953, 55-57, 59-60, 62-63

(133) Clay Minerals Society.

NhD-K	1953, 55-63, 69
NjP	1953, 55-63, 69
NvU	1953, 55-63, 69
OCU-Geo	1953, 55-63, 69
OU	1953, 55-63, 69
OkT	1953, 55, 57-62, 69
OkU	1961-62, 64, 69
OrCS	1976
OrU-S	1953, 55-63, 69
PBL	1953, 55-63
PSt	1953, 55-63, 69
RPB	1953, 55-63
TMM-E	1955, 57-62, 69
TxBeaL	1953, 55-63, 69
TxCaW	1957-63, 69
TxDaAR-T	1953, 55-63
TxDaGl	1961, 69
TxDaM-SE	1961
TxDaSM	1953, 55, 57-63
TxHSD	1969
TxHU	1953, 55-63, 69
TxLT	1961, 69
TxMM	1959, 61-62, 69
TxU	1958, 61, 69
TxU-Da	1961, 69
UU	1961
ViBlbV	1961, 69
WU	1955-56, 58, 61-64, 69, 76, 79
WaU	1961, 69

(133) CLAY MINERALS SOCIETY. ANNUAL MEETING FIELD TRIP.
1964 1st   See Clay Minerals Conference. 132.
1965 2nd   See Clay Minerals Conference. 132.
1976 13th   See Clay Minerals Conference. 132.
1979   See Clay Minerals Conference. 132.
1981 18th   Clay mineralogy of Pleistocene and Pennsylvanian sediments in east-central Illinois.
1983 Oct.   Niagara Falls field trip. (Reprinted from 54th Annual Meeting, Field Trips Guidebook for New York State Geological Association) See New York State Geological Association. 469.
1985 July-Aug.   See International Clay Conference. 282.
   CU-EART   1981
   I-GS   1981, 83

(134) COAST GEOLOGICAL SOCIETY. GUIDEBOOK.
1965 Oct.   See Society of Economic Paleontologists and Mineralogists. Pacific Section. 593.
1976 June   See American Association of Petroleum Geologists. Pacific Section. 28.
1977 Oct.   See American Association of Petroleum Geologists. Department of Education. 23.
1979 June   See American Association of Petroleum Geologists. Pacific Section. 28.
1982 June   See American Association of Petroleum Geologists. Pacific Section. 28.

(135) [COLLEGE CENTER OF THE FINGER LAKES] BAHAMIAN FIELD STATION.
1981   Field guide to the geology of San Salvador. (2d ed. December 1981)
1983   Field guide to the geology of San Salvador. (3rd ed. 1983)
1984   See Symposium on the Geology of the Bahamas. 645.
   Geology field trip workbook, San Salvador, Bahamas.
1985 Oct.   See Geological Society of America. 197.
   CSt-ES   1984
   CU-EART   1983
   KyU-Ge   1981
   NBiSU   1984
   OU   1983
   OkU   1983, 84

(136) COLLEGE OF WILLIAM AND MARY. DEPARTMENT OF GEOLOGY. GUIDEBOOK.
1969 No. 1.   See Atlantic Coastal Plain Geological Association. 89.
1981 No. 2.   See Atlantic Coastal Plain Geological Association. 89.
1982 No. 3.   See National Association of Geology Teachers. Eastern Section. 411.
— No. 4.   See Association of Earth Science Editors. 75.
   CUBG   1981

(137) COLLEGE OF WOOSTER, GEOLOGY FIELD TRIP GUIDEBOOK.
1976   Geology of central Pennsylvania.
   PSt   1976

(138) COLLOQUE SUR LE QUATERNAIRE DU QUEBEC.
1973 2nd   Aspects du Quaternaire dans le region au Nord de Joliette, Montreal, Quebec.
   CaOOG   1973

(139) COLORADO GEOLOGICAL SURVEY. BULLETIN.
1972 No. 32.   Prairie peak and plateau; a guide to the geology of Colorado.
   CSt-ES   1972
   DLC   1972
   IdU   1972

(140) COLORADO GEOLOGICAL SURVEY. DEPARTMENT OF NATURAL RESOURCES. RESOURCE SERIES.
1980 No. 8.   See Forum on the Geology of Industrial Minerals. 179.(1979)
— No. 10.   See Symposium on the Geology of Rocky Mountain Coal (ROMOCOAL). 644.

(141) COLORADO GEOLOGICAL SURVEY. DEPARTMENT OF NATURAL RESOURCES. SPECIAL PUBLICATION.
1969 No. 1.   See Governor's Conference on Environmental Geology. 245.
1981 No. 19.   See Association of Engineering Geologists. Denver Section. 78.
1985 No. 27.   Scenic trips into Colorado geology (Uncompahgre Plateau).
   CoDGS   1985
   IaU   1985

(142) COLORADO GROUND-WATER ASSOCIATION.
1985 4th   Water and energy resources of west central Colorado.
   CSt-ES   1985

(143) COLORADO SCHOOL OF MINES. PROFESSIONAL CONTRIBUTIONS.
1976 Vol. 8.   See Geological Society of America. 197.(No.15)
No. 9. See Geological Society of America. 197.(No.3)
No. 15. See Geological Society of America. 197.(No.17)
No. 17. See Geological Society of America. 197.(No.18)
No. 18. See Geological Society of America. 197.(No.12,18)
No. 24. See Geological Society of America. 197.(No.1)
No. 26. See Geological Society of America. 197.(No.21)
No. 27. See Geological Society of America. 197.(No.6)
No. 28. See Geological Society of America. 197.(No.8)
No. 30. See Geological Society of America. 197.(No.16)
No. 34. See Geological Society of America. 197.(No.19)
No. 35. See Geological Society of America. 197.(No.9)
No. 36. See Geological Society of America. 197.(No.7)
No. 37. See Geological Society of America. 197.(No.7,20)
No. 38. See Geological Society of America. 197.(No.22)
No. 39. See Geological Society of America. 197.(No.5)
No. 40. See Geological Society of America. 197.(No.10)
No. 41. See Geological Society of America. 197.(No.2)
1978 Vol. 9.   See International Association on the Genesis of Ore Deposits. 280.
1980 Vol. 10.   See American Association of Petroleum Geologists. 18.
1983 Vol. 11.   See International Symposium on Fossil Algae. 311.

(144) COLORADO STATE UNIVERSITY. DEPARTMENT OF EARTH RESOURCES.
1979   See Geological Society of America. Rocky Mountain Section. 204.

COLORADO. UNIVERSITY. See UNIVERSITY OF COLORADO. MUSEUM. (682)

(145) COLUMBIA RIVER BASALT SYMPOSIUM. FIELD TRIP. [GUIDEBOOK]
1969 2nd Proceedings of the second Columbia River Basalt Symposium. Basalt symposium field trip: [Cheney, Odessa, Ephrata, Royal City, Cheney]. (491)

CU-EART	1969
MtBuM	1969

(146) COLUMBIA UNIVERSITY. NEW YORK GANDER CONFERENCE. (TITLE VARIES)
1967 Geology along the North Atlantic.
Field trip no. 1. St. John's area.
Field trip no. 2. St. John's-Random Island-Gander.
Field trip no. 3. Gander-Carmanville-Change Islands.
Field trip no. 4. Gander-Boyds Cove-New World Island.
Field trip no. 5. Gander-Baie Verte.
Field trip no. 6. Deer Lake-Bonne Bay-St. Anthony.
Field trip no. 7. Corner Brook-Humber Arm Stephenville-Port au Port Peninsula.

CaOWtU	1967
DI-GS	1967
DLC	1967
MH-GS	1967
MnDuU	1967
MnU	1967
OU	1967
TxHSD	1967
WU	1967

(147) COMMISSION FOR FLORIDA STRATIGRAPHY.
1976 Nov. See Southeastern Geological Society. 614.

(148) COMMONWEALTH MINING AND METALLURGICAL CONGRESS. FIELD CONFERENCE. [GUIDEBOOK]
1957 6th Banff, Alberta.
1. Field conference; a study of the stratigraphy and structure of the Canadian Rockies, as exposed in the Banff-Lake Louise area, Alberta, Canada.
[2] Geology and mineral deposits of the Sudbury area, Ontario.

AzTeS	1957
CSt-ES	1957 (1)
CaBVaU	1957
CaOKQGS	1957
CaOLU	1957
DI-GS	1957(2)
InU	1957
OkT	1957
PBL	1957

(149) CONFERENCE ON CENOZOIC GEOLOGY OF THE TRANS-PECOS VOLCANIC FIELD OF TEXAS. FIELD GUIDE.
1978 May Alpine, Texas. Cenozoic geology of the Trans-Pecos volcanic field of Texas. Conference proceedings and field guide. (713(1979-No.17))

CSt-ES	1978
DI-GS	1978
DLC	1978
NcU	1978
NhD-K	1978
PSt	1978
TxBeaL	1978
TxCaW	1978
TxU	1978

(150) CONFERENCE ON GEOLOGIC PROBLEMS OF THE SAN ANDREAS FAULT SYSTEM, STANFORD UNIVERSITY.
1967 Self-guiding field trip map. (630)

AzTeS	1967
CLU-G/G	1967
CSdS	1967
CSt-ES	1967
CU-A	1967
CU-EART	1967
CaBVaU	1967
CaOOG	1967
CoG	1967
DLC	1967
ICF	1967
ICIU-S	1967
InU	1967
KyU-Ge	1967
MoU	1967
NSbSU	1967
NSyU	1967
NbU	1967
NcU	1967
NdU	1967
NhD-K	1967
PBL	1967
TMM-E	1967
TxU	1967
UU	1967

(151) CONFERENCE ON THE HISTORY OF GEOLOGY, UNIVERSITY OF NEVADA, RENO.
1964 Aug. (1) Sierran trip; Reno, Steamboat Springs, Carson City, Glenbrook, Bijou, Emerald Bay, Tahoe City, Truckee, Reno.
[2] Basin and Range trip; Reno, Fallon, Fairview Peak, Fallon, Virginia City, Reno.

CU-EART	1964
CoDGS	1964
CoU	1964
DI-GS	1964
DLC	1964
IU	1964
NhD-K	1964
NvU	1964

CONNECTICUT. STATE GEOLOGICAL AND NATURAL HISTORY SURVEY. See STATE GEOLOGICAL AND NATURAL HISTORY SURVEY OF CONNECTICUT. (631) and STATE GEOLOGICAL AND NATURAL HISTORY SURVEY OF CONNECTICUT. (632)

(152) CORPUS CHRISTI GEOLOGICAL SOCIETY. ANNUAL FIELD TRIP.
1948 Field trip.
1949 Three Rivers, Cotulla, Artesia Wells, Carrizo Springs, Eagle Pass, Del Rio.
1950 Corpus Christi to Laredo to Rio Grande City.
1951 A trip to six selected salt dome structures in Southwest Texas.
1952 Northeastern Mexico; Reynosa to Monterrey, Mexico; Cortinas and Huasteca canyons.
1953 Quaternary (Beaumont) to Cretaceous (Fredericksburg).
1954 Quaternary (Beaumont) to Eocene (Mt. Selman).
1955 Cretaceous of Austin, Texas area.
1956 Route Corpus Christi, Laredo, Monterrey [Mexico].
1957 South Texas domes.
1958 See Gulf Coast Association of Geological Societies. 247.
1959 Geology of the upper Rio Grande Embayment and a portion of the Edwards Escarpment, Corpus Christi to Del Rio.
1960 Geology of the Chittim Arch and the area north to the Pecos River.
1961 Geology of the Pleistocene-Jurassic of northeastern Mexico, Nuevo Laredo to Monterrey to Reynosa, Mexico.
1962 Sedimentology of South Texas; Corpus Christi to Brownsville.
1963 Geology of Peregrina Canyon and Sierra de El Abra. Corpus Christi to Cds. Valles and Victoria, Mexico.
1964 See Gulf Coast Association of Geological Societies. 247.
1965 Upper Cretaceous asphalt deposits of the Rio Grande Embayment.
1966 Geology of the Austin-Llano area, central Texas. (Supplement to University of Texas. Bureau of Economic Geology. Guidebook 5, 1963)
1968 South Texas uranium.
1970 Spring Hidalgo Canyon and La Popa Valley, Nuevo Leon, Mexico. Tertiary-Mesozoic; structure-stratigraphy. Corpus Christi, Nuevo Laredo, Sabinas Hidalgo, Monterrey, China, Reynosa.
1972 See Gulf Coast Association of Geological Societies. 247.

(153) Cuba. Secretaria de Agricultura, Comercio y Trabajo.

1975 Apr. Triple energy field trip, Duval, Webb, and Zapata counties, Texas; uranium, coal, gas.
1978 May Minas de Golondrinas and Minas Rancherias. [Northwestern Mexico]
1979 Apr. Portrero Garcia and Huasteca Canyon, northeastern Mexico.
1980 Apr. Gulf Coast uranium.
1981 See Gulf Coast Association of Geological Societies. 247.
1982 Spring Geology of the Llano Uplift, central Texas and geological features in the Uvalde area.
1983 May Structure and Mesozoic stratigraphy of Northeast Mexico.
1984 Spring Big Bend National Park, Texas.
— May See American Association of Petroleum Geologists. 18.

AzFU	1950, 61-63, 65, 68, 70
AzTeS	1950, 61-63, 65, 68, 70
BEG	1984
CLU-G/G	1949-57, 59-63, 65-66, 68, 70
CLhC	1951-54, 57, 59, 61-63, 65, 82
CMenUG	1950-54, 56-57, 59-66, 68, 70, 72, 75, 78-82, 84
CPT	1950-51, 55, 57, 61-63, 78-79, 82-84
CSt-ES	1950-51, 53, 55-57, 59-63, 65-66, 68, 70, 75, 78-80
CU-A	1982
CU-EART	1952, 55-56, 61-63, 65, 68, 70
CU-SB	1968, 75, 82
CaACU	1968, 70, 75, 79
CaBVaU	1963
CaOHM	1961, 65, 68, 70
CaOKQGS	1950, 62-63, 65-66, 82
CaOOG	1978-80
CaOWtU	1968, 70, 75, 78-79, 82, 84
CoDCh	1948, 52
CoDGS	1950-57, 59-63, 65-66, 68, 70, 75, 78-80, 82, 84
CoG	1950-53, 55-57, 59-63, 65, 68, 70, 75, 78-79
CoLiM	1982, 84
CoU	1961-63, 65, 68
DFPC	1954
DI-GS	1950-56, 68, 70, 75, 78-79, 82, 84
DLC	1951-55, 65
ICF	1950, 61-63, 65
IEN	1960
IU	1950, 57, 59-63, 65, 68, 70, 83
IaU	1963, 70, 78-80, 82-84
InLP	1963, 65-66, 68, 70, 75, 78-80, 82-84
InU	1950-52, 54, 57, 60-63, 65, 68, 70, 78-79, 82
KyU-Ge	1950-51, 56-57, 61-63, 65-66, 68, 70, 75, 78(May), 79(Apr.), 82-84
LNU	1950, 61-63, 65, 70
LU	1949-57, 59-63, 65, 68, 70
MiU	1950, 56-57, 61-63, 65-66
MnU	1950-51, 57, 59-63, 65-66, 68, 70, 75, 78-79, 82
MoSW	1950, 61-63, 65
MoU	1950, 61-63, 65, 68
NNC	1950, 61, 63, 65, 68, 70, 75, 78-79
NSbSU	1968, 70, 75, 78-79
NbU	1950, 61-62
NcU	1950, 63, 65, 75
NhD-K	1978
NjP	1963, 65
NmPE	1968, 70
NmU	1984
OCU-Geo	1965, 68, 70
OU	1950, 52-53, 55, 59-63, 65, 68, 70, 75
OkT	1950-51, 57, 59-61, 63, 68
OkTA	1982, 84
OkU	1948-50, 52-55, 57, 59-60, 63, 65-66, 68, 70, 75, 78-79, 82-84
PBU	1982, 84
TxBeaL	1949-50, 52-53, 56, 59, 61, 63, 65-66, 68, 70
TxCaW	1968
TxDaAR-T	1950-54, 56-57, 59, 61-63, 65-66, 68, 70
TxDaDM	1950-51, 53, 56-57, 59-63, 65, 68, 70, 75, 78-80
TxDaGI	1953, 55, 57, 59, 63, 65, 75
TxDaM-SE	1984
TxDaSM	1949-50, 52-57, 59, 63, 65, 75
TxHMP	1982
TxHSD	1950, 60, 61-63, 65
TxHU	1950-51, 53-57, 59-63, 65, 68
TxLT	1954, 61, 63, 68
TxMM	1950-51, 57, 61, 63, 65, 68, 70, 75, 78, 80, 82
TxU	1949-57, 59-63, 65-66, 68, 70, 84
TxWB	1949-51, 56-65, 68, 70, 75, 78-79, 81-84
ViBlbV	1957, 65-66, 68, 70, 75, 78

(153) CUBA. SECRETARIA DE AGRICULTURA, COMERCIO Y TRABAJO.
1938 Field guide to geological excursion in Cuba.
DI-GS 1938

(154) CYCLOTHEM GEOLOGY CLUB, UNIVERSITY OF ILLINOIS AT URBANA-CHAMPAIGN. FIELD TRIP GUIDEBOOK.
1977 Spring Big Bend National Park and Carlsbad Caverns National Park, New Mexico; Magnet Cove, Arkansas; Palo Duro Canyon, Texas; and Guadalupe Mountains National Park, Texas.
IU 1977

(155) DAKOTERRA.
1985 V. 2. Pt. 2. See Society of Vertebrate Paleontology. 604.

(156) DALLAS GEOLOGICAL SOCIETY. FIELD GUIDE.
1955 The Washita Group in the valley of the Trinity River, Texas. (624)
1959 See American Association of Petroleum Geologists. 18.
1969 See American Association of Petroleum Geologists. 18.
1975 See American Association of Petroleum Geologists. 18.
1983 Apr. See American Association of Petroleum Geologists. 18.

CLU-G/G	1955
CoG	1955
CoU	1955
DI-GS	1955
DLC	1955
ICF	1955
IEN	1955
NmPE	1955
TxDaAR-T	1955
TxDaDM	1955
TxDaM-SE	1955
TxDaSM	1955
TxHU	1955
TxLT	1955
TxU	1955
TxU-Da	1955
ViBlbV	1955

(157) DELAWARE GEOLOGICAL SURVEY. OPEN FILE REPORT.
1976 8. See Friends of the Pleistocene. Eastern Group. 182.
1977 9. See Association of American State Geologists. 74.

(158) DENVER REGION EXPLORATION GEOLOGISTS' SOCIETY (D.R.E.G.S.). GUIDEBOOK. FIELD TRIP. (TITLE VARIES)
1980 1st Silver Cliff volcanic center and Tallahassee Creek uranium deposits.
1981 Field trip notes; Creede mining district, San Juan volcanic province, Colorado.
1982 Aspen-Grizzly Peak-Leadville, Colorado.
1983 Gunnison gold belt and Powderhorn carbonatite field trip.

CU-SB	1980
CaOOG	1980
CoD	1981-82
CoDGS	1981-82
CoG	1983
DI-GS	1980-82

(159) DESK AND DERRICK CLUB. DALLAS CHAPTER. GUIDEBOOK FOR THE FIELD TRIP.
1951 [East Texas oil fields]
TxDaSM 1951

(160) EARLHAM COLLEGE. GEOLOGY SENIOR SEMINAR.
1973 Spring Ohio and Pennsylvania geology.
InRE 1973

(161) EARTH AND SPACE SCIENCE STUDENT ORGANIZATION, U.C.L.A. ESSSO GUIDEBOOK. (TITLE VARIES)

1971 Spring Field guide to Papoose Flat in the Inyo Mountains, eastern California.
— Fall Field guide to the southern Coast Ranges of California.
1972 Spring Field guide to the Peninsular Ranges of Southern California.
1973 Spring Field guide to the San Gabriel anorthosite of Southern California.
1976 Fall GSUCLA Field Guide 5. Geologic guidebook to the Long Valley-Mono Craters region of eastern California.
1977 Fall Geologic guidebook to the Zion-Bryce Canyon region, Utah.
1978 Fall 7. Field guide to selected aspects of the geology between Los Angeles and San Diego.
— Dec. 8 (unofficial). Death Valley. (Revised in 1980)
1979 Mar. 9. Grand Canyon.
— Fall 10. Guidebook to the southern Coast Ranges; geology and plate tectonics.
1980 Spring 11. Guidebook to the Mojave Desert region. (Cover title: Geologic guidebook to the Mojave Desert)
— Fall 12. Guidebook to the geology of a portion of the eastern Sierra Nevada, Owens Valley, and White-Inyo Range.
— Dec. 8. Death Valley. (Revised 1980)
1982 Fall 13. Geologic guidebook to the western Sierra Nevada.
1983 Gunnison gold belt and Powderhorn carbonatite field trip.
1984 Oct. See Society of Economic Geologists. 583.

CLU-G/G	1971-73, 76-80, 82
CLhC	1976(Fall), 80(Fall)
CMenUG	1971-73, 76-79, 80(Fall, Dec.), 82
CPT	1976, 80, 82
CSfCSM	1976, 80(Fall), 82
CSt-ES	1976, 78(Fall), 80, 82
CU-A	1972, 76, 78, 80(Fall), 82
CU-EART	1976, 80(Fall), 82
CU-SB	1977, 82
CU-SC	1980(Fall), 82
CaOOG	1982
CoDGS	1976
DI-GS	1971(Fall), 72, 76-77, 78(Fall), 79-80, 82
DLC	1976
IU	1971(Spring), 72-73, 76-77, 78(Fall), 79(Fall), 80
IaU	1976, 80(Fall), 82
KyU-Ge	1982
MnU	1982
NvU	1976
OkU	1982
TxU	1976

(162) EAST TEXAS GEOLOGICAL SOCIETY. FIELD TRIP.

1939 Claiborne field trip, Tyler, Texas to Natchitoches, Louisiana.
1945 [Pecan Gap, Wolfe City and Annona formations in East Texas]
1951 The Woodbine and adjacent strata of the Waco area of central Texas. (624)
1959 The Edwards Formation of central Texas. (Guide book to accompany Symposium on Edwards Limestone in central Texas; Texas University. Bureau of Economic Geology. Publication 5905)
1960 Claiborne-Wilcox Eocene, Smith and Cherokee counties, Texas.
1984 Mar. Tyler, Texas.
  The Jurassic of east Texas. (Prepared and published by the East Texas Geological Society for the East Texas Jurassic Exploration Conference)

AzU	1959
CLU-G/G	1951, 84
CMenUG	1984
CSt-ES	1984
CU-A	1984
CU-EART	1951, 84
CU-SB	1984
CaOOG	1984
CoDGS	1984
CoG	1945, 60
CoU	1951
DI-GS	1945, 51, 60, 84
ICF	1951
IEN	1945, 51, 59-60
IU	1951
IaU	1951
InLP	1951, 59, 84
InU	1951, 59
KyU-Ge	1945
LNU	1960
LU	1951
MnU	1984
MoSW	1951
NNC	1945, 51
NSbSU	1984
NSyU	1951
NcU	1951
NhD-K	1984
OkT	1951, 84
OkTA	1984
OkU	1939, 45
PSt	1951
RPB	1951, 59
TxBeaL	1945, 51, 59
TxDaAR-T	1951, 59
TxDaDM	1951
TxDaGI	1951, 59
TxDaM-SE	1945, 84
TxDaSM	1945, 51
TxHMP	1984
TxHSD	1945, 51
TxHU	1945, 51
TxLT	1945
TxMM	1951, 84
TxU	1945, 59-60, 84
TxU-Da	1951, 59-60
TxWB	1945, 51, 59-60
WU	1984

(163) EAST TEXAS STATE UNIVERSITY GEOLOGICAL SOCIETY.

1972 Spring See Southwestern Association of Student Geological Societies (SASGS). 626.
1977 Fall See Southwestern Association of Student Geological Societies (SASGS). 626.

(164) EASTERN MICHIGAN UNIVERSITY. EARTH SCIENCE FIELD STUDY.

1959 No. 1. Glacial geology of central and western Washtenaw County; Field study 1, A traverse from Ypsilanti to Waterloo Recreation Area and return; Cary drifts of the Erie and Saginaw lobes, their genesis and surface expression.

CoDGS	1959
InU	1959

(165) EASTERN NEVADA GEOLOGICAL SOCIETY.

1960 See Intermountain Association of (Petroleum) Geologists. 276.

(166) EDMONTON GEOLOGICAL SOCIETY FIELD TRIP GUIDEBOOK.

1959 1st Cadomin area.
1960 2nd Rock Lake.
1961 3rd Jasper.
1962 4th Peace River.
1963 5th Sunwapta Pass area.
1964 6th Medicine and Maligne lakes, Jasper Park.
1965 7th David Thompson Highway, from near Nordegg to Banff-Jasper Highway.
1966 8th Cadomin, Alberta.
1967 No field trip held.
1969 Edmonton, Jasper, Ft. Nelson, Watson Lake, Pine Pt. (lead-zinc), Ft. McMurry, Tersands, Edmonton.
1970 Peace River, Pine Pass, Yellowhead, Field Conference.
1976 See Geological Association of Canada. 190.

CMenUG	1960-62, 64-66, 69, 70
CSt-ES	1963, 66, 69
CaAC	1961-64, 66, 69
CaACAM	1960-61, 63-64, 66, 69-70
CaACI	1959-66, 69-70
CaACM	1959-66, 69-70
CaACU	1959-66, 69-70
CaAEU	1959-62

(167) El Paso Geological Society.

CaBVaU	1961-64, 66, 69-70
CaOHM	1962-64, 66, 69
CaOKQGS	1962-64, 66, 69-70
CaOLU	1962-64, 66, 69
CaOOG	1959-64
CoDCh	1961
CoDGS	1959-66, 69-70
CoG	1962-64, 66, 69-70
DI-GS	1959-66, 69-70
DLC	1966
ICF	1962-64
IU	1959, 61-64, 66
IaU	1964
MiU	1962-66, 69-70
MnU	1962-64, 66, 69
NNC	1962-64, 66, 69
NhD-K	1962-66, 69-70
NjP	1962-64
OU	1963, 66, 69
OkU	1962-64, 66
TxDaAR-T	1962, 64, 70
TxDaDM	1961-64, 66, 69
TxDaM-SE	1962, 64, 66
TxDaSM	1961
TxHSD	1960, 63
TxU	1959-66, 69-70
ViBlbV	1966, 69-70

(167) EL PASO GEOLOGICAL SOCIETY. GUIDEBOOK.
1967 [1st] Precambrian rocks of the Franklin Mountains, Texas.
1968 [2nd] General geology of the Franklin Mountains, Texas.
1969 3rd See Society of Economic Paleontologists and Mineralogists. Permian Basin Section. 594.
1970 4th Cenozoic stratigraphy of the Rio Grande Valley area, Dona Ana County, New Mexico.
1971 5th A glimpse of some of the geology and mineral resources of the Sierra Blanca-Van Horn area, Hudspeth and Culberson counties, Texas. (Cover title: A glimpse of some of the geology and mineral resources, Sierra Blanca-Van Horn country, Hudspeth and Culberson counties, Texas.)
1972 6th The stratigraphy and structure of the Sierra de Juarez, Chihuahua, Mexico.
1973 7th The geology of southcentral Dona Ana County, New Mexico.
1974 8th Geology of the Florida Mountains, Luna County, New Mexico.
— Feb. See American Association of Petroleum Geologists. Southwest Section. 31.
1975 9th Exploration from the mountains to the basin.
1980 Feb. See American Association of Petroleum Geologists. Southwest Section. 31.
1981 Geology of the border; southern New Mexico-northern Chihuahua.
1982 Mar. See Symposium on the Paleoenvironmental Setting and Distribution of the Waulsortian Facies. 646.
1983 Geology and mineral resources of north-central Chihuahua.
1984 See American Association of Petroleum Geologists. 18.

AzTeS	1975
AzU	1967-68, 70-75
CLU-G/G	1970-74, 83
CMenUG	1967-69, 72-74, 81, 83
CPT	1975, 83
CSt-ES	1971-74, 83
CU-A	1972, 74, 83
CU-EART	1971-74
CaACU	1975
CaOWtU	1972, 75
CoDCh	1974, 81
CoDGS	1967-69, 72-75, 81, 83
CoG	1967-68, 70-73
CoLiM	1983
DI-GS	1967-68, 70-75, 81, 83
IU	1970-72, 74-75
IaU	1967-68, 73-75
InLP	1971-75, 83
InU	1967
KyU-Ge	1970-75, 81
MnU	1967-68, 72-74, 81, 83
NdU	1972, 74
NhD-K	1974-75, 83
NmLcU	1981
NmPE	1970-75
OCU-Geo	1967-68, 70-74, 83
OkTA	1981, 83
OkU	1971-74, 81, 83
TxBeaL	1970-75
TxDaAR-T	1967-68, 70-74
TxDaGI	1967, 70-74
TxDaM-SE	1981, 83
TxDaSM	1967-68, 70-72
TxHMP	1983
TxHSD	1972
TxLT	1971-72, 74
TxMM	1967-68, 70-75
TxU	1967-68, 70-75, 80-81, 83
TxWB	1967, 72, 74, 81
ViBlbV	1975
WU	1975, 83

(168) EMORY AND HENRY COLLEGE, EMORY, VIRGINIA. DEPARTMENT OF GEOLOGY.
1972 No. 1. See Virginia Geology Field Conference. 747.

ESSSO GUIDEBOOK. See EARTH AND SPACE SCIENCE STUDENT ORGANIZATION, U.C.L.A. (161)

(169) EXPLORATION FOR METALLIC RESOURCES IN THE SOUTHEAST. [GUIDEBOOKS FOR FIELD TRIPS]
1981 Sept. Exploration for metallic resources in the southeast.
[1] Slate belt field trip.
[2] Piedmont west of Atlanta field trip.

CU-EART	1981

(170) FAIRLEIGH DICKINSON UNIVERSITY. WEST INDIES LABORATORY. SPECIAL PUBLICATION.
1974 5. Guidebook to the geology and ecology of some marine and terrestrial environments, St. Croix, U.S. Virgin Islands.

CSt-ES	1974
CaACU	1974
CaOWtU	1974
DI-GS	1974
IU	1974
InU	1974
NSbSU	1974
NSyU	1974
NcU	1974
NdU	1974
NhD-K	1974
OkU	1974
PSt	1974
TxBeaL	1974
ViBlbV	1974

(171) FAULT FINDERS GEOLOGICAL SOCIETY. [GUIDEBOOK]
1949 Cretaceous Comanche field trip. (716)

CLU-G/G	1949
DI-GS	1949
TxBeaL	1949
TxU	1949

(172) FIELD CONFERENCE OF PENNSYLVANIA GEOLOGISTS. ANNUAL MEETING. GUIDEBOOK.
1931 [1st] Guidebook for the first field conference of Pennsylvania geologists.
Trip I. Inspection of limestone mine of the American Lime and Stone Co., Bellefonte; bentonite beds, geologic structure and Ordovician stratigraphy in the vicinity of Bellefonte, Pennsylvania.
Trip II. The Silurian and Lower Devonian section of the Bald Eagle Valley from Bellefonte to Lock Haven, Pennsylvania, including the outcrop of the Oriskany horizon.
Trip III. Stratigraphy of the Allegheny Front from Altoona to Gallitzen.
Trip IV. Trips to caves of central Pennsylvania.

Field Conference of Pennsylvania Geologists. (172)

1932 2nd   Around and near the "Forks of the Delaware," and various sundry "Gaps."
1933 [3rd] Guidebook for the third field conference of Pennsylvania geologists.
1934 [4th] Guidebook for the fourth field conference of Pennsylvania geologists.
1935 5th   In the Philadelphia area of southeastern Pennsylvania.
1936 6th   Geological inspection of anthracite field.
1937 7th   Bradford district trip.
1938 [8th] Guidebook, Virginia. (742)
1939 [9th] Log for West Virginia.
1940 10th Guidebook for the tenth field conference of Pennsylvania geologists.
1941 11th Allegheny Front trip; Blue Knob, East Freedom, Hollidaysburg, Williamsburg area, Pennsylvania.
1946 12th From the Cambrian to the Silurian near State College and Tyrone.
1947 13th Bethlehem, Pennsylvania: local geology.
1948 14th Harrisburg, Pennsylvania. Excursions.
  No. 1. South Mountain.
  No. 2. Pennsylvania Turnpike.
  No. 3. Cornwall Mines.
  No. 4. Susquehanna-Juaniata rivers.
  Supplement: Silurian sediments and relationships at Susquehanna Gap in Blue or Kittatinny Mountain, five miles north of Harrisburg, Pennsylvania.
1949 15th Lancaster, Pennsylvania.
  Trip 1. Old Mines and Mine Ridge anticline.
  Trip 2. Martic overthrust.
  Trip 3. Appalachian drainage and Pleistocene terraces.
1950 16th Pittsburgh-Pennsylvania field excursions.
  Trip 1. Aliquippa Plant, Jones and Loughlin Steel Company.
  Trip 2. Glacian "foreland" of Northwest Pennsylvania.
  Trip 3. Chestnut Ridge Anticline.
1951 17th Geology of Philadelphia area.
1952 18th Sussex County, New Jersey.
  1. Pleistocene geology.
  2A. Dikes of special petrologic interest.
  2B. Silurian and Devonian stratigraphy.
  3A. Cambro-Ordovician and Pre-Cambrian rocks.
  3B. Silurian-Devonian section at Nearpass Quarries.
1953 19th Easton, Pennsylvania field trip and guidebook with summary of regional geology and sections on Paleozoic rocks and Precambrian geology.
1954 20th Cambro-Ordovician limestones. Martinsburg Formation.
1955 21st [Geology of central Pennsylvania]
  No. 1. Stratigraphy of Ordovician limestones and dolomites of Nittany Valley from Bellefonte to Pleasant Gap.
  No. 2. Stratigraphy and structure of ridge and valley from University Park to Tyrone, Mt. Union and Lewistown.
  No. 3. Structure and stratigraphy of Pennsylvanian sediments of the plateau area near Philipsburg and Clearfield.
1956 22nd Trenton, New Jersey.
1958 23rd Structural geology of South Mountain and Appalachians in Maryland. (58, 339(No.17), 18(1960))
1959 24th Titusville, Pennsylvania. Drake Well Centennial, 1859-1959.
  A. Glacial geology of Crawford and Erie counties, Pennsylvania.
  B. Bedrock and oil geology of northwestern Pennsylvania and the great Oildorado.
  C. Erosion channel in Penn Dixie limestone mine.
1960 25th Some tectonic and structural problems of the Appalachian piedmont along the Susquehanna River.
1961 26th Structure and stratigraphy of the Reading Hills and Lehigh Valley, in Northampton and Lehigh counties, Pennsylvania.
1962 27th Stratigraphy, structure, and economic geology of southern Somerset County and adjacent parts of Bedford and Fayette counties, Pennsylvania.
1963 28th Stratigraphy and structure of Upper and Middle Devonian rocks in northeastern Pennsylvania.
1964 29th Cyclic sedimentation in the Carboniferous of western Pennsylvania.
1965 30th Stratigraphy of the Pennsylvanian and Permian rocks of Washington, Mercer, and Lawrence counties, Pennsylvania.
1966 31st [A] Comparative tectonics and stratigraphy of the Cumberland and Lebanon valleys.
[B] Revised edition of the 21st meeting held in 1955.
Trip 1. Stratigraphy of Ordovician limestones and dolomites of Nittany Valley from Bellefonte to Pleasant Gap.
Trip 2. Stratigraphy and structure of Ridge and Valley area from University Park to Tyrone, Mt. Union and Lewistown appendices, 1966.
Trip 1, appendix no. F1. Erosional benches along Bald Eagle Mountain in the area from Waddle Gap to Milesburg Gap, Pennsylvania.
Trip 2, appendix no. S1. Subsurface faults in the vicinity of Birmingham and Jacksonville, Pennsylvania.
Appendix no. S2. Comparison of geologic, magnetic and gravimetric features in parts of central Pennsylvania.
1967 32nd Geology in the region of the Delaware to Lehigh water gaps.
1968 33rd Geology of mineral deposits in south-central Pennsylvania.
1969 34th The Pocono Formation in northeastern Pennsylvania.
1970 35th New interpretations of eastern Piedmont geology of Maryland or granite and gabbro or graywacke and greenstone?
1971 36th Upper Devonian sedimentation in Susquehanna County; and, hydrology, glacial geology and environmental geology of the Wyoming-Lackawanna County.
1972 37th Stratigraphy, sedimentology, and structure of Silurian and Devonian rocks along the Allegheny Front in Bedford County, Pennsylvania, Allegheny County, Maryland, and Mineral and Grant counties, West Virginia.
1973 38th Structure and Silurian-Devonian stratigraphy of the Valley and Ridge Province, central Pennsylvania.
1974 39th Geology of the Piedmont of southeastern Pennsylvania.
1975 40th The late Wisconsinan drift border in northeastern Pennsylvania.
1976 41st Bedrock and glacial geology of northwestern Pennsylvania in Crawford, Forest and Venango counties.
1977 42nd Stratigraphy and applied geology of the Lower Paleozoic carbonates in northwestern New Jersey.
1978 43rd Uranium in Carbon, Lycoming, Sullivan and Columbia counties, Pennsylvania.
1979 44th Devonian shales in south-central Pennsylvania and Maryland. (Cover title: Devonian shales of south-central Pennsylvania and Maryland)
1980 45th Land use and abuse; the Allegheny County problem. (541)
1981 46th Geology of Tioga and Bradford counties, Pennsylvania.
1982 47th Geology of the Middle Ordovician Martinsburg Formation and related rocks in Pennsylvania.
1983 48th Silurian depositional history and Alleghanian deformation in the Pennsylvania Valley and Ridge.
1984 49th Geology of an accreted terrane: the eastern Hamburg Klippe and surrounding rocks, eastern Pennsylvania.
1985 50th Central Pennsylvania geology revisted.
Field trip no. 1. Guide to the Lock-Haven reef.
Field trip no. 2. Catskill sedimentation in central Pennsylvania.
Field trip no. 3. On the soils and geomorphic evolution of Nittany Valley.
Field trip no. 4. Cross-strike and strike-parallel deformation zones in central Pennsylvania.
Field trip no. 5. Economic and mining aspects of the coal-bearing rocks in western Pennsylvania.

ATuGS	1938
CLU-G/G	1946, 49, 55, 59-60, 62-66
CLhC	1955
CMenUG	1946, 60, 71, 83
CPT	1961
CSt-ES	1958, 73, 77, 80
CU-EART	1938, 58, 63-64, 66-67, 72, 77
CU-SB	1980
CaACU	1977-79
CaAEU	1966(B)
CaBVaU	1961
CaOHM	1958
CaOKQGS	1955, 58-60, 62-69
CaOOG	1958
CaOWtU	1960, 62-69, 77, 80-83
CoDGS	1935, 58-59, 61, 65, 71
CoG	1938-39, 48-49, 54, 58

(173) Flintstone Rock Club.

CoLiM	1980-82
DFPC	1959
DI-GS	1932, 35, 38, 47-56, 58-85
DLC	1938, 53, 59, 62, 64-71, 73-79, 82
ICF	1946, 55, 58
ICarbS	1938, 55, 58, 66(B)
IEN	1955, 58
IU	1946, 50, 55, 58-79, 82
IaU	1938, 55, 58, 60, 66, 77, 79-80
InLP	1958-59, 62, 64-65, 66(A), 67-81
InRE	1958, 70
InU	1949, 55, 58-60, 62-65, 66(A), 67-68, 79
KyU-Ge	1938, 58-60, 62, 64-85
LNU	1958, 66
MH-GS	1949, 51, 53, 74
MNS	1948-49, 58
MdBJ	1958
MiU	1962-69
MnDuU	1960
MnU	1949, 55-56, 58-60, 62-75, 77-85
MoSW	1949, 58, 71
MoU	1958
NBiSU	1958
NNC	1935, 38, 49, 51-53, 55-56, 58, 60, 62-64, 68
NRU	1958-59, 62, 64-73, 75-76, 80, 81, 83, 84
NSyU	1951, 59, 62, 64-65, 66(A), 67-75, 77
NbU	1958
NcU	1938, 58, 61, 64, 66-69, 71-74, 76-77, 83
NdU	1938
NhD-K	1953, 58
NjP	1935, 37-38, 55-56, 60, 62-68
OCU-Geo	1958-59, 62, 64-73, 77, 79, 85
OU	1946, 49-50, 60, 62-69
OkU	1958, 80
PBL	1936, 61, 64
PBU	1980, 81, 83-85
PBm	1935, 46-56, 58-60, 62, 64-65, 66(A), 67-79, 82
PHarER-T	1931-41, 46-56, 58-78, 82
PSt	1931-41, 46-50, 55, 58-60, 62-79, 82
RPB	1959-67
TxDaAR-T	1963, 67, 69-74
TxDaDM	1938, 55, 58-60, 62-69
TxDaGI	1958
TxDaM-SE	1958
TxHSD	1946
TxHU	1958
TxU	1935, 46, 48(suppl.), 50, 55, 58-68, 70, 73
ViBlbV	1938, 60
WU	1958

(173) FLINTSTONE ROCK CLUB. [GUIDEBOOK]
1964 1st Okmulgee, Oklahoma.
    OkU   1964

(174) FLORIDA BUREAU OF GEOLOGY. LEAFLET.
1982 12 A geologic guide to Suwannee River, Ichetucknee Springs, O'Lebi and Manatee Springs State Parks.
1984 13. A geologic guide to the state parks of the Florida Panhandle coast.
    InLP   1982, 84

(175) FLORIDA BUREAU OF GEOLOGY. SPECIAL PUBLICATION.
1959 5. Summary of the geology of Florida and a guidebook to the classic exposures. (Revised in 1964)
1964 5. Summary of the geology of Florida and a guidebook to the classic exposures. Revised ed.

CLhC	1959
CMenUG	1959, 64
CPT	1964
CSt-ES	1964
CU-EART	1964
CaACU	1959
CaOOG	1959, 64
CaOWtU	1964
DI-GS	1964
DLC	1959
ICarbS	1959
IaU	1964
InLP	1959
KyU-Ge	1959, 64
MNS	1964
MiHM	1959, 64
NcU	1959
NdU	1959
NhD-K	1959, 64
OU	1959, 64
OkT	1959
OkU	1959, 64
OrCS	1959, 64
OrU-S	1959
PBL	1959
PSt	1964
ViBlbV	1959, 64
WU	1959, 64

FLORIDA GEOLOGICAL SURVEY. See **FLORIDA BUREAU OF GEOLOGY. (175)** and **FLORIDA BUREAU OF GEOLOGY. (174)**

(176) FLORIDA SINKHOLE RESEARCH INSTITUTE. FSRI REPORT.
1984 See Southeastern Geological Society. 614.
1985 85-86-1 See Geological Society of America. 197.

(177) FORT SMITH GEOLOGICAL SOCIETY. REGIONAL FIELD CONFERENCE. GUIDEBOOK.
1959 1st Western portion of Arkansas Valley basin.
1961 See Tulsa Geological Society. 662.
1963 2nd Southeastern Arkansas Valley and the Ouachita and frontal Ouachita Mountains, Arkansas. (67(No.63-1))
1964 May Informal field trip; Magnet Cove, barite and bauxite mines, Ouachita Mountains.
1982 Apr. Field guide to the Magnet Cove area and selected mining operations and mineral collecting localities in central Arkansas. (67(No.82-1))

AGC	1982
CSt-ES	1963, 82
CU-EART	1959, 63
CaOOG	1982
CoDCh	1959, 63
CoDGS	1959
CoG	1959
DI-GS	1959
ICF	1963
IU	1959, 63
IaU	1959, 63
InLP	1963, 82
InU	1959
KyU-Ge	1963, 82
MnU	1959, 63, 82
MoU	1959
NNC	1963
NcU	1963
NjP	1963
OCU-Geo	1982
OkOkCGS	1959
OkT	1959, 63
OkU	1959, 63-64
TMM-E	1963
TxDaAR-T	1959, 63
TxDaDM	1959, 63
TxDaGI	1963
TxDaM-SE	1959, 63
TxDaSM	1959, 63
TxHSD	1959, 63
TxHU	1963
TxU	1963
ViBlbV	1963
WU	1963, 82

(178) FORT WORTH GEOLOGICAL SOCIETY. [GUIDEBOOK]
1939 Fall See West Texas Geological Society. 756.
1957 See Abilene Geological Society. 2.
1969 Arbuckle Mountains field trip.

CoDGS	1969
CoG	1969
DI-GS	1969
MnU	1969
OkOkCGS	1969
OkU	1969
TxBeaL	1969
TxDaAR-T	1969
TxU	1969
WU	1957

(179) FORUM ON THE GEOLOGY OF INDUSTRIAL MINERALS. FIELD TRIP GUIDEBOOK. (TITLE VARIES)

1967   Inland underwater facilities.
1971 7th The central Florida phosphate district.
1972 8th [1] Pleistocene Lake Calvin.
            [2] Underground gas storage in Iowa.
            [3] Some notes on the Osage; Raid's Nelson quarry.
            [4] Tour of U.S. Gypsum Company's Sperry Mine and plant.
1973 9th 1. A geologic excursion to fluorspar mines in Hardin and Pope counties, Illinois, Illinois-Kentucky mining district and adjacent upper Mississippi Embayment. (265)
            2. Geologic guide to a portion of the fluorspar mining district in Livingston and Crittenden counties, Kentucky.
1974 10th Field trip. Sequential land use of some pits and quarries along the Scioto River.
1975 11th Kalispell, Montana.
1979 15th Industrial minerals in Colorado and the Rocky Mountain region. (Appendix 2 title: "Aggregates and industrial minerals in the Golden Morrison, and Lyons areas.) (Cover title: Proceedings of the Fifteenth Forum on Geology of Industrial Minerals.) (140(1980))
1983 19th Field trip A: Industrial Mineral Industries (Talc and Nepheline Syenite). (515)
      Field trip B: Industrial Mineral Industries (Aggregates, Stone, and Glass). (515)

AzTeS	1973(2)
CLU-G/G	1973
CMenUG	1971-74, 79, 83
CPT	1973, 83
CSfCSM	1979
CSt-ES	1973-74, 79, 83
CU-A	1973(1)
CU-EART	1973(1), 83
CU-SB	1979, 83
CaACU	1979
CaAEU	1973(1), 79
CaBVaU	1973(1)
CaNSHD	1983
CaOHM	1973(1)
CaOKQGS	1973(1), 79, 83
CaOOG	1973, 79-80, 83
CaOTDM	1976-83
CaOTRM	1973(1)
CaOWtU	1973(1), 79
CoD	1979
CoDBM	1979
CoDCh	1973(1)
CoDGS	1973(1), 79, 83
CoG	1973(1), 83
CoLiM	1979
DI-GS	1971, 73, 83
DLC	1973(1)
I-GS	1973
ICF	1973(1)
ICIU-S	1973(1)
ICarbS	1973(1)
IEN	1973(1)
IU	1972-74
IaU	1971-73, 79, 83
IdBB	1973(1)
InLP	1973, 79
InU	1973(1)
KLG	1967, 79, 83
KU	1979, 83
KyU-Ge	1971, 73, 79
LNU	1971
LU	1973(1)
MNS	1973(1)
MiDW-S	1973(1)
MiHM	1973(1)
MiKW	1973(1)
MnU	1973(1), 79
MoSW	1973(1)
MoU	1973(1)
MtBuM	1973(1)
NBiSU	1973(1)
NIC	1979
NN	1979
NRU	1979
NSbSU	1973(1), 75
NSyU	1973(1)
NU	1973(2)
NbU	1973(1)
NcU	1973, 83
NdU	1973(1), 83
NhD-K	1973
NjP	1973
NvU	1973, 79
OCU-Geo	1973
OU	1967, 71, 72(1), 73(1), 75
OkT	1973(1)
OkU	1973, 83
PBL	1973(1)
PBm	1973(1)
PSt	1973(1)
TMM-E	1973(1)
TxBeaL	1973(1), 74
TxDaDM	1973(1)
TxHU	1973(1)
TxLT	1973(1)
TxU	1971, 73, 79, 83
UU	1973(1), 79
ViBlbV	1973(1)
WU	1972, 73(1), 78, 79, 83

(180) FOUR CORNERS GEOLOGICAL SOCIETY. GUIDEBOOK FOR THE FIELD CONFERENCE.

1955 1st Geology of parts of Paradox, Black Mesa, and San Juan basins.
1957 2nd Geology of southwestern San Juan Basin.
1960 3rd Geology of Paradox Basin fold and fault belt.
1963 4th Shelf carbonates of the Paradox Basin.
1964   See American Association of Petroleum Geologists. Rocky Mountain Section. 30.
1969 5th Geology and natural history of the Grand Canyon region; the Powell Centennial. River expedition 1969. Including separate river log. Lee's Ferry to Phantom Ranch and geologic color map of eastern Grand Canyon (Colorado River).
1971 6th Geology of Canyonlands and Cataract Canyon.
1972 Oct. Cretaceous and Tertiary rocks of the southern Colorado Plateau. Road log: Durango to Farmington via Cortez, Shiprock, Biclabito Dome, and Big Gap reservoir with selected side trips.
1973 7th Geology of the canyons of the San Juan River.
1975 8th Canyonlands country.
1979 9th Permianland.
1984   See Geological Society of America. Rocky Mountain Section. 204.

AzFU	1957, 60, 69, 71
AzTeS	1957, 60, 69, 71
AzU	1957, 60, 63, 69, 71, 73, 75, 79
CChiS	1955
CLU-G/G	1955, 57, 60, 63, 69, 71-72
CLhC	1955, 57, 60, 63, 71, 75
CPT	1955, 57, 60, 63, 69, 71-73, 75, 79
CSfCSM	1955
CSt-ES	1955, 57, 60, 63, 69, 71-73, 79
CU-A	1957, 63, 69, 71, 73, 75, 79
CU-EART	1955, 57, 60, 63, 69, 71-73, 75, 79
CU-SB	1957
CaACI	1957
CaACU	1955, 57, 60, 63, 69, 71-73, 75, 79
CaAEU	1963
CaBVaU	1955, 60, 63, 72
CaOHM	1957, 60, 63
CaOKQGS	1957, 60, 63
CaOLU	1969

CaOOG	1955, 57, 60, 63, 69, 79
CaOWtU	1957, 63, 69, 72, 75, 79
CoDCh	1955, 57, 60, 63, 69, 72, 75, 79
CoDGS	1955, 57, 60, 63, 69, 71, 73, 75, 79
CoDU	1957, 63, 71
CoDuF	1955, 57
CoG	1955, 57, 60, 63, 69, 71, 73
CoGrU	1969, 75
CoU	1955, 57, 60, 63, 69, 71, 73, 75
DFPC	1957
DI-GS	1955, 57, 60, 63, 69, 71-73, 75, 79
DLC	1955, 60, 69
ICF	1957, 60, 63, 69
ICTU-S	1957, 60, 63, 69, 71, 73, 75, 79
ICarbS	1957, 60, 63, 69, 71
IEN	1955, 57, 60, 63, 69
IU	1955, 57, 60, 63, 69, 71, 73
IaU	1955, 57, 60, 63, 69, 71, 75, 79
InLP	1957, 60, 63, 69, 71-73, 75, 79
InU	1957, 60, 63, 69, 71, 79
KyU-Ge	1955, 57, 60, 63, 69, 71, 73, 75, 79
LNU	1955
LU	1955, 57, 60, 63, 69, 71
MH-GS	1957, 60, 63, 69, 84
MNS	1971
MiHM	1963, 69, 71
MiU	1969, 71
MnU	1955, 57, 60, 63, 69, 71, 73, 75, 79
MoSW	1955, 60, 75
MoU	1955, 60, 63
MtBC	1960
MtBuM	1955, 63
NBiSU	1957, 60, 63, 71, 73
NNC	1955, 57, 60, 63, 71, 73, 75
NOneoU	1971
NRU	1972
NSyU	1957, 60, 63, 69, 71
NbU	1955, 57, 60, 63, 69, 71-73
NcU	1957, 60, 63, 69, 71-72, 75
NdU	1957, 60, 63, 69, 71-73, 79
NhD-K	1955, 57, 60, 63, 69, 71, 73, 75, 79
NjP	1957, 60, 69, 71
NmPE	1957, 60, 69, 72
NmU	1957
NvU	1955, 57, 60, 63, 73
OCU-Geo	1957, 60, 63, 69, 71, 73, 75
OU	1955, 57, 60, 63, 69, 71, 73, 75, 79
OkOkCGS	1957
OkT	1957, 60, 63, 69
OkU	1955, 57, 60, 63, 69, 71, 73, 75, 79
OrCS	1979
OrU-S	1957, 60, 63, 69, 71, 73, 75
PBL	1975
PSt	1955, 57, 60
RPB	1955
SdRM	1960, 69, 79
TMM-E	1969
TU	1957, 60, 63
TxBeaL	1955, 57, 60, 69, 75, 79
TxCaW	1969
TxDaAR-T	1955, 57, 60, 63, 69, 73
TxDaDM	1955, 57, 60, 63, 69, 71, 75
TxDaGI	1955, 57, 60, 63
TxDaM-SE	1957, 60, 63
TxDaSM	1957, 60, 63, 69, 71-72, 75
TxHSD	1955, 57, 60, 63, 69, 72-73, 75
TxHU	1955, 57, 60, 63, 69, 71
TxLT	1955, 57, 60
TxMM	1955, 57, 60, 63, 64, 69, 71, 73, 79
TxU	1955, 57, 60, 63, 69, 71, 73, 75
TxU-Da	1957, 63, 69
TxWB	1955, 57, 60, 63, 69, 71, 73
UU	1955, 57, 60, 63, 69, 71-73, 75
ViBlbV	1957, 60, 63, 69, 73
WU	1955, 57, 60, 63, 69, 71, 73, 75, 79
WaPS	1975

(181) FRIENDS OF THE GRENVILLE. [GRENVILLE PROVINCE OF QUEBEC] FIELD GUIDEBOOK. (TITLE VARIES)

- 1983 No. 46. Bedrock geology of the High Peaks region, Marcy Massif, Adirondacks, New York. (693)
- 1984 Geology of the Minden and Chandos township areas, central metasedimentary belt, Ontario.
- 1985 Precambrian potassic intrusions in the central metasedimentary belt, Grenville Province of Quebec.
- — Oct. The Montreal-Val D'or geotraverse.

CaOOG	1983
CaOTDM	1983-85
DI-GS	1983
MA	1983
MU	1983
NhD-K	1983
TxU	1983

(182) FRIENDS OF THE PLEISTOCENE. EASTERN GROUP. FIELD CONFERENCE GUIDEBOOK. (TITLE VARIES)

- 1935 2nd New England Pleistocene geologists' conference; suggested program.
- 1937 4th Hanover-Mt. Washington, New Hampshire.
- 1939 6th Annual invasion; announcement only.
- 1940 7th No guidebook.
- 1941 8th Glacial features of the Catskills.
- 1947 10th Finger Lakes.
- 1948 11th Toronto-Barrie, Ontario.
- 1950 13th Ithaca, New York.
- 1952 15th Columbus, Ohio.
- 1953 16th Glacial geology field meeting. Ayer, Massachusetts Quadrangle and adjacent areas.
- 1954 17th Wellsboro, Elmira, Towanda region, Pennsylvania-New York.
- 1955 18th Malone, New York area.
- 1956 19th Drummondville region, Quebec, Canada.
- 1959 22nd London, Ontario.
- 1960 23rd Dunkirk, New York. Glacial geology of Cattaraugus County, New York.
- 1961 24th Late Pleistocene stratigraphy and history of southwestern Maine.
- 1962 25th The Charlestown Moraine and the retreat of the last ice sheet in southern Rhode Island.
- 1963 26th Riviere-du-Loup, Quebec, Canada.
- 1964 27th The Pleistocene geology of Martha's Vineyard.
- — 28th Long Island.
- 1966 29th Annual reunion. Chesapeake, Va.
- 1967 30th Annual reunion.
- 1968 31st Cape Cod, Massachusetts.
- 1969 32nd Detailed road log and field trip notes: Sherbrooke, Quebec.
- 1970 33rd 1. Glen House, Chandler Ridge, top of Mt. Washington. 2. Pinkham Notch, Dolly Copp Road.
- 1973 36th Binghamton, New York.
- 1976 [39th] Guidebook: Columbia deposits of Delaware. (157)
- 1978 41st Quaternary deposits and soils of the central Susquehana Valley of Pennsylvania. (532)
- 1979 42nd Evidence of local glaciation, Adirondack Mountains, New York.
- 1980 43rd Late Wisconsin stratigraphy of the upper Cattaraugus Basin.
- 1981 44th Nashua River Valley, Leominster, Massachusetts. (220)
- 1982 45th Drummondville - St.- Hyacinthe, Quebec, Canada.
- 1985 48th Woodfordian deglaciation of the Great Valley, New Jersey.

CLU-G/G	1952
CMenUG	1959
CaBVaU	1959
CaOOG	1959, 78
CoDGS	1959-60
DI-GS	1978
I-GS	1952, 59, 63
IU	1948, 52, 59-61, 64, 81
IaU	1952, 76
InLP	1976
InU	1952, 59, 62, 64
MnU	1959, 61-62
NNC	1959, 61
NSyU	1960
OCU-Geo	1959

OU	1950, 52, 61-62, 64, 67-68, 79, 82
OkU	1952, 78
PHarER-T	1978
TxDaSM	1952
TxU	1970
WU	1948

(183) FRIENDS OF THE PLEISTOCENE. MIDWEST GROUP. ANNUAL FIELD CONFERENCE. GUIDEBOOK. (TITLE VARIES)

1950	1st	Wisconsin Pleistocene.
1951	2nd	Southeastern Minnesota.
1952	3rd	Tazewell drift, western Illinois and eastern Iowa.
1953	4th	Northeastern Wisconsin.
1954	5th	Central Minnesota.
1955	6th	Review of the relationships of the Pleistocene stratigraphy, geomorphology and soils in Pottawattamie, Cass, and Adair counties, southwestern Iowa.
1956	7th	The northwestern part of the southern peninsula of Michigan.
1957	8th	Field guide and road log for study of Kansan, Illinoian and early Tazewell tills, loesses and associated faunas in south-central Indiana.
1958	9th	East-central North Dakota. (480)
1959	10th	Glacial geology of west-central Wisconsin.
1960	11th	Eastern South Dakota.
1961	12th	Edmonton, Alberta, Canada.
1962	13th	Evidence for "early Wisconsin" in central and Southwest Ohio.
1963	14th	Loess stratigraphy, Wisconsinan classification and accretion-gleys in central western Illinois. (265)
1964	15th	Eastern Minnesota. Wisconsin glaciation of Minnesota. (387)
1965	16th	Iowan problem.
1966	17th	Evidence of multiple glaciation in the glacial-periglacial area of eastern Nebraska. (449)
1967	18th	Glacial geology of the Missouri Coteau and adjacent areas. (480)
1969	19th	Saskatchewan and Alberta; a reappraisal of the pre-late Ferry Paleosol.
1971	20th	Pleistocene stratigraphy of Missouri River valley along the Kansas-Missouri border. (636)
1972	21st	Pleistocene stratigraphy of east-central Illinois. (265)
1973	22nd	The Valderan problem; Lake Michigan basin.
1975	23rd	Quaternary paleoenvironmental history of the western Missouri Ozarks.
1976	24th	Stratigraphy and faunal sequence, Meade County, Kansas. (635)
1978	25th	Loess stratigraphy and Paleosols in Southwest Indiana.
1979	26th	Wisconsinan, Sangamonian, and Illinoian stratigraphy in central Illinois. (265)
1980	27th	Yarmouth revisited. (332)
1981	[28th]	Late-Wisconsinan history of northeastern lower Michigan.
1982	May	Prairie du Chien, Wisconsin.
		Quaternary history of the Driftless area, with special papers (777)
		Quaternary history of the Kickapoo and lower Wisconsin River valleys, Wisconsin.
	— Oct.	Tectonic features in the central gneiss belt, Parry Sound region, Ontario.
1983	30th	Interlobate stratigraphy of the Wabash Valley, Indiana.
1984	31st	Pleistocene history of west-central Wisconsin; discussion, roadlog, and geologic stop descriptions. (777)
1985	32nd	Illinoian and Wisconsinan stratigraphy and environments in northern Illinois: the Altonian revised. (265)

ATuGS	1963
AzTeS	1967, 76, 85
CLU-G/G	1958, 63, 66-67, 72, 76, 79, 85
CMenUG	1955, 59, 61, 63, 67, 72-76, 78-81
CPT	1959, 66, 67, 79, 85
CSdS	1967
CSt-ES	1958, 63-64, 67, 71-73, 76, 82, 84-85
CU-A	1972, 78, 85
CU-EART	1958, 64, 66-67, 72, 76, 79, 82, 85
CU-SB	1966
CU-SC	1985
CaACU	1958, 63, 66-67, 71-72, 79
CaAEU	1966, 72, 79, 85
CaBVaU	1972
CaOHM	1972
CaOKQGS	1972
CaOLU	1958
CaOOG	1958, 60, 63, 67, 78-79, 82, 83(6, 9), 85
CaOTDM	1982(Oct.), 85
CaOTRM	1972, 79
CaOWtU	1958, 67, 71-76, 79, 82, 85
CaQQERM	1985
CoDCh	1972, 79
CoDGS	1972, 79, 81, 85
CoFS	1958, 63
CoG	1958, 67, 72, 79
CoU	1966-67, 71-72
DI-GS	1950-67, 71-73, 75-76, 78-79, 81, 82(May), 84-85
DLC	1958, 72, 79
F-GS	1985
FBG	1985
I-GS	1952-60, 62-66, 72, 80, 83
ICF	1958, 67, 72, 79
ICIU-S	1958, 72, 79
ICarbS	1958, 63, 67, 72, 79, 85
IEN	1958, 63, 67, 72
IU	1951-52, 55-60, 62-67, 69, 71-73, 75-76, 78
IaU	1952-53, 55, 57-59, 62-67, 71-72, 75-76, 78-85
IdBB	1972
InLP	1955-58, 62-63, 66-67, 71-72, 76, 78-79
InU	1957-59, 63, 67, 72, 79
KLG	1966-67, 71, 76, 85
KyU-Ge	1958, 66-67, 71-73, 76, 78-79, 81-82, 85
LNU	1972
LU	1958, 63, 72
MH-GS	1963, 72
MNS	1963-64, 67, 72
MiDW-S	1958, 63, 67, 72
MiHM	1958, 67, 72, 79, 85
MiKW	1966, 72
MiU	1956, 58, 67
MnU	1954-55, 57-59, 63-64, 66-67, 69, 71-73, 75-76, 78-85
MoSW	1958, 64, 66-67, 72, 85
MoU	1958, 72
MtBuM	1958, 63, 67, 72, 79
NBiSU	1967, 72
NNC	1959-64, 72
NSbSU	1972, 79
NSyU	1958, 63, 67, 72
NbU	1958, 63, 67, 72
NcU	1958, 67, 72, 79
NdU	1954, 57-61, 63, 67, 71-73, 79
NhD-K	1964, 67, 72, 79, 85
NjP	1958, 67, 72
NmSoI	1985
NvU	1958, 66-67, 72, 79, 84-85
OCU-Geo	1957, 63, 67, 72, 76, 78-79, 85
OU	1951, 53, 56-58, 60-63, 67, 72, 76, 78-80, 82
OkT	1972
OkTA	1985
OkU	1958, 63, 67, 71, 72, 76, 79, 82, 84-85
OrU-S	1958, 67, 85
PBL	1958, 67, 72
PBm	1972, 79, 82
PHarER-T	1982
PSt	1958, 72, 79
RPB	1958, 63
SdRM	1958, 64, 67, 72, 79
TMM-E	1972
TxBeaL	1958, 67, 72, 76, 78-79
TxCaW	1976
TxDaAR-T	1958, 66-67, 71-73
TxDaDM	1958, 60, 71-72, 79
TxDaM-SE	1958, 63, 67, 82, 84
TxDaSM	1963
TxHSD	1963
TxHU	1958, 72
TxLT	1958, 67, 72
TxU	1958, 63, 67, 72, 85
UU	1958, 63, 72, 79
ViBlbV	1958, 67, 72
WBB	1985
WKenU	1982, 84
WU	1953, 56-59, 62-67, 72-73, 75-76, 79, 80-85
WyU	1985

FRIENDS OF THE PLEISTOCENE. NORTHEAST [GROUP]. GUIDEBOOK. See FRIENDS OF THE PLEISTOCENE. EASTERN GROUP. (182)

(184) FRIENDS OF THE PLEISTOCENE. PACIFIC CELL (FORMERLY: PACIFIC COAST GROUP). GUIDEBOOK.

- 1966 1st Glaciomarine environments and the Fraser glaciation in northwest Washington.
- 1967 Pleistocene geology and palynology, Searles Valley, California.
- 1969 No title, text or itinerary. It consists of copies of several preliminary USGS topographic maps, profiles, and other data. Contains Sugar House Quadrangle, Salt Lake County, Utah, and Fort Douglas Quadrangle, Utah.
- — Sept. Pleistocene geology of the east-central Cascade Range, Washington.
- 1971 Sept. Glacial and Pleistocene history of the Mammoth Lakes Sierra - a geologic guidebook. (Cover title: Glacial and Pleistocene history of the Mammoth Lakes Sierra, California - a geologic guidebook) (701)
- 1972 Progress report on the USGS Quaternary studies in the San Francisco Bay area.
- 1975 Field guide: San Diego.
- 1978 Nov. Fluvial history of Panamint Valley, California.
- 1979 Aug. Field guide to relative dating methods applied to glacial deposits in the third and fourth recesses and along the eastern Sierra Nevada, California, with supplementary notes on other Sierra Nevada localities.
- 1981 July Quaternary stratigraphy, soil geomorphology, chronology and tectonics of the Ventura, Ojai, and Santa Paula areas, western Transverse Ranges, California. (412(July))
- 1984 Oct. Holocene paleoclimatology and tephrochronology east and west of the central Sierran Crest.

CChiS	1971
CLU-G/G	1966-67, 71, 79, 84
CMenUG	1967, 69, 71-72, 75, 78, 79, 81, 84
CPT	1971
CSfCSM	1979, 81
CSt-ES	1971-72, 79, 84
CU-A	1971, 81, 84
CU-EART	1966, 71-72, 79, 81, 84
CaBVaU	1966
CaOOG	1966-67, 69, 75, 79
CoDGS	1969, 78-79
DI-GS	1967, 69, 71-72, 75, 78, 81
IU	1966, 69
InLP	1972
KyU-Ge	1971
MnU	1969
NbU	1969(maps)
NdU	1971
NvU	1978, 84
OU	1966, 71-72
PSt	1984
TxDaAR-T	1971
TxU	1967
WaU	1969

(185) FRIENDS OF THE PLEISTOCENE. ROCKY MOUNTAIN CELL (FORMERLY: ROCKY MOUNTAIN GROUP). FIELD TRIP GUIDEBOOK.

- 1952 1st Estes Park, Colorado.
- 1953 2nd Twin Lakes area, Colorado.
- 1958 4th Jackson Hole area, Wyoming.
- — 5th Not issued.
- 1960 6th Little Valley, Promontory Point, Utah, and Little Cottonwood-Draper area, near Salt Lake City, Utah.
- 1961 7th Recent geologic history of Bear Lake Valley, Utah-Idaho.
- 1962 8th Twin Falls, Glenns Ferry and Bruneau, Idaho to study Pleistocene history of Snake River canyon.
- 1963 Madison River and Yellowstone River from Hayden Valley to Pine Creek, Montana.
- 1964 10th Quaternary geology of the Duncan-Virden-Safford area, New Mexico, Arizona. (USGS Geological Map I-442 used as tour guide)
- 1965 10th Upper Gila River region, New Mexico, Arizona.
- 1966 11th Landscape evolution and soil genesis in the Rio Grande region, southern New Mexico.
- 1967 No formal guidebook.
- 1968 13th San Pedro Valley, Arizona.
- 1969 14th Pleistocene lake deposits, Jordan Valley.
- 1970 15th Guidebook to the geology of the San Francisco Peaks, Arizona. (Reprinted in 1976) See American Quaternary Association. 55.(1976)
- 1971 16th Guidebook to the Quaternary geology of the east-central Sierra Nevada.
- 1972 Canon City, Colorado, to Wetmore, Westcliffe, Texas Creek and return.
- 1974 Hebgen Lake area to Bozeman, Montana via Hwy. 191. Itinerary and road log. Excursion to Schmitt site, Three Forks, Montana.
- 1976 Road log and field guide: The southern Alberta Caper.
- 1981 Sept. Relative dating of the glacial sequences in the Roaring Fork, Lower Lake Creek, and upper Arkansas River valleys, central Colorado.
- 1982 Field trip to Little Valley and Jordan Valley, Utah. (220)
- 1983 Aug. Day 1. Jokulhlaups into the Sanpoil arm of glacial Lake Columbia, Washington.
- Day 2. Jokulhlaups near Spokane, Washington, and Lewiston, Idaho.
- Day 3. Glacial sequence near McCall, Idaho. (220)
- 1984 Aug. See American Quaternary Association. 55.

AzTeS	1971
AzU	1971
CLU-G/G	1964(map), 69-71, 76, 81-82
CLhC	1971
CMenUG	1960, 71, 81-82
CSdS	1971
CSt-ES	1964(map), 70-71, 82, 83
CU-A	1971
CU-EART	1976, 82
CaOHM	1971
CaOOG	1971
CaOWtU	1971
CoDGS	1970-71, 76, 81-82
CoDuF	1969
CoG	1971
CoU	1969, 71
DI-GS	1952-53, 58, 60-66, 68-72, 74, 76, 81, 83
DLC	1976
I-GS	1966
ICarbS	1971
IEN	1971
IU	1966, 69-72, 74, 76
IaU	1983
IdU	1983(3)
InLP	1964
InRE	1963
InU	1960
KyU-Ge	1970-71, 74
MH-GS	1971
MiHM	1964, 71
MnU	1960, 62-63, 68-69, 71, 76
MoSW	1971
MoU	1971
NNC	1960-61, 71
NRU	1971
NbU	1964(map), 69
NcU	1971
NdU	1971
NhD-K	1971, 82, 83
NjP	1971
NmPE	1971
NvU	1971
OCU-Geo	1981
OU	1968-72, 74, 76
OkU	1971
OrU-S	1971
TxBeaL	1971
TxCaW	1966
TxDaAR-T	1968-71, 74
TxDaDM	1960-61
TxDaSM	1971
TxU	1966, 68, 70-71, 82
UU	1971
WU	1982

(186) FRIENDS OF THE PLEISTOCENE. SOUTH-CENTRAL CELL. FIELD TRIP. [GUIDEBOOK]
1983   Guidebook to the central Llano Estacado. (656, 657, 521)
1985 Apr.   Loesses in Louisiana and at Vicksburg, Mississippi.
CU-EART	1983
CaAEU	1983
CaOWtU	1983
CoDGS	1983
DI-GS	1983
IaU	1983
KyU-Ge	1985
OkU	1983

(187) GEOGRAPHIE. BULLETIN DE RECHERCHE.
1984   Num. 77-78.   See Association Quebecoise pour l'Etude du Quaternaire. 88.

(188) GEOLOGIC FIELD CONFERENCE FOR HIGH SCHOOL TEACHERS.
1958 1st   Camp Wood and Elmdale Hill, Chase County, Kansas.
  KLG   1958

(189) GEOLOGIC FIELD CONFERENCE FOR KANSAS SCHOOL TEACHERS. GUIDEBOOK.
1956 1st   Geology, landscape, and mineral resources, Camp Naish area, Wyandotte County. (634)
CaACU	1956
KLG	1956
OU	1956

(190) GEOLOGICAL ASSOCIATION OF CANADA. GUIDEBOOK FOR THE ANNUAL MEETING.
1950   [1] Banff area. (29, 125, 602, 14)
  2. Banff Formation, Lac des Arc Fault, Palliser Formation, etc., along Banff-Calgary Highway. (29, 125, 602, 14)
1953   See Geological Society of America. 197.
1963 16th   [Southern Quebec] field trips; region around Montreal. (383)
  No. 1. Oka Complex.
  No. 2. Pleistocene geology, St. Lawrence lowland north of Montreal.
  No. 3. Laurentian area north of Montreal.
  No. 4. Grenville mineralization, Glen Almond, Quebec.
  No. 5. Ordovician stratigraphy, St. Lawrence lowland north of Montreal.
  No. 6. Mount Royal.
  No. 7. Pleistocene geology between Montreal and Covey Hill.
  No. 8. Mounts St. Hilaire and Johnson.
  No. 9. Western edge of Appalachians in southern Quebec.
  No. 10. Breccia localities.
1964   See American Association of Petroleum Geologists. 18.
1966 19th   Geology of parts of the Atlantic provinces. (383)
  Trip 1. Southwestern New Brunswick.
  Trip 2. Bathurst-Dalhousie area, New Brunswick.
  Trip 3. Zeolites, Nova Scotia.
  Trip 4. Surficial geology, north shore of Minas Basin.
  Trip 5. Halifax to Bay of Fundy, Nova Scotia.
  Trip 6. Walton area, Nova Scotia.
  Trip 7. Cape Breton Island, Nova Scotia.
  Trip 8. Corner Brook area, Newfoundland.
1967 20th   Geology of parts of eastern Ontario and western Quebec. (383)
  Trip 1. Sudbury, Cobalt and Bancroft areas, Ontario.
  Trip 2. Nepheline rocks, Bancroft areas, Ontario.
  Trip 3. Paleozoic stratigraphy, Kingston area, Ontario.
  Trip 4. The Grenville Province (12-day field trip).
  Trip 5. Surficial geology east and west of Kingston.
  Trip 6. Grenville structure, stratigraphy, and metamorphism, southeastern Ontario.
  Trip 7. Alkaline rocks, Oka and St. Hilaire, Province of Quebec.
  Trip 8. Tectonic intrusions, Shawinigan region, Province of Quebec.
1968   [A] Guidebook for geologic field trips in southwestern British Columbia; a revision of a guidebook prepared by the Geological Discussion Club in March, 1960. (676, 196(1960), 675(No. 6))
  1. Field trip to illustrate geology of Coast Mountains, North Vancouver, B.C.
  2. Field trip, Vancouver to Kamloops and return.
  [B] Guidebook for M.A.C. field trip to the southern interior of British Columbia. (676, 383, 675(No. 7))
1969 June   Guidebook for the geology of Monteregian Hills. (383)
  No. 1. Flysch sediments in parts of the Cambro-Ordovician sequence of the Quebec Appalachians.
  No. 4. Geologic evolution of the Grenville Province in the Shawinigan area.
  No. 5. Quaternary geology of the Lac St. Jean and Seguenay River areas.
  No. 6A. Diatreme breccia pipes and dykes and the related alnoite, kimberlite and carbonatite intrusions occurring in the Montreal and Oka areas, Quebec.
  No. 6B, pt. 1. Geology of Mount Rougemont.
  Pt. 2. Geology of Mount Johnson.
  No. 6C. Geology of Brome and Shefford Mountains.
  No. 7. Structural geology of the Sherbrooke area.
  No. 8A. Stratigraphy of the St. Lawrence Lowlands (Covey Hill). The Cambrian sandstones of the St. Lawrence Lowlands.
  No. 9. Mineralogy at Mount St. Hilaire.
1970 23rd   Field trip no. 1. Geology of the Moak-Setting Lakes area, Manitoba (Manitoba nickel belt). (383)
  Field trip no. 2. Comparative geology and mineral deposits of the Flin Flon-Snow Lake and Lynn Lake-Fox Lake areas.
  Field trip no. 4. Geology of the Bisset-Lac du Bonnet area, Manitoba.
  Field trip no. 5. Lower Paleozoic of the interlake area, Manitoba.
  Field trip no. 6. Paleozoic and Mesozoic of the Dawson Bay area and Manitoba escarpment, Manitoba.
  Field trip no. 7. Surficial geology and Pleistocene stratigraphy of an area between the Red and Winnipeg rivers.
1973 [25th]   An excursion guide to the geology of Saskatchewan. (383, 559, 560)
  (1) Geological road log of the Winnipeg and Interlake areas.
  (2) Geological road log of the lakes Winnipegosis and Manitoba areas, Manitoba.
  (3) Geological road log of the Hanson Lake Road-Flin Flon area, Saskatchewan-Manitoba.
  (4) Geological road log of the Montreal Lake-Wapawekka Hills area.
  (5) Tour of the Estevan coalfield.
  (6) Geological road log of the Avonlea-Big Muddy Valley area.
  (7) Geological road log of the Cypress Hills-Milk River area, southeastern Alberta.
  (8) Upper Cretaceous and Tertiary stratigraphy of the Swift Current-Cypress Hills area.
  (9) Geological road log for Highway 2; La Ronge to Reindeer Lake.
  (10) Northern Saskatchewan uranium tour: La Ronge-Rabbit Lake-Beaverlodge-Cluff Lake.
  (11) Geology and ores of the Hanson Lake, Flin Flon and Snow Lake areas of Saskatchewan and Manitoba.
  Trip C. Quaternary geology and its application to engineering practice in the Saskatoon-Regina-Watrous area, Saskatchewan.
1974 [26th]   St. John's, Newfoundland. Field trip manual. (383)
  A-1. A cross-section of the Newfoundland Appalachians.
  A-2. The regional setting and structure of the West Newfoundland ophiolites.
  A-3(see B-3).
  A-4. Aspects of the Pleistocene geology of the eastern Avalon Peninsula.
  A-5. Environmental geology of a rural area near St. John's. Conception Bay south.
  B-2. The regional setting and structure of the West Newfoundland ophiolites.
  B-3. Ore deposits and their setting in the central mobile belt of Northeast Newfoundland. (Cover title: Ore deposits and their tectonic setting in the central mobile belt of Northeast Newfoundland)
  B-3a. Advocate asbestos mine.
  B-3b. Ore deposits and their tectonic setting in the central Paleozoic mobile belt of northern Newfoundland.
  B-3c. The consolidated rambler deposits.
  B-3d. Geology of Tilt Cove.
  B-3e. Volcanogenic copper deposits in probable ophiolitic rocks, Springdale Peninsula.
  B-3f; B-3g. Geology of the Pilleys Island area.
  B-3h. Geology, geochemistry and ore deposits of the Buchans area.
  B-3i. Geology of the Betts Cove area.

Geological Association of Canada.

B-4. Fossils and sedimentology of the Paleozoic strata of Port au Port Peninsula. (Cover title: Lower Paleozoic stratigraphy of the Port au Port area, West Newfoundland)
B-5. Igneous rocks of the Avalon Platform.
B-6. Late Precambrian and Cambrian sedimentary sequences of southeastern Newfoundland.
B-7. Polydeformed metamorphic belts on the southeastern and northwestern margins of the Newfoundland central mobile belt. (Cover title: Geology of the Gander Lake and Fleur de Lys belts)
B-8. Applications of refraction seismology and resistivity surveys in environmental and engineering geology.
B-9. St. Lawrence, Canada's only fluorspar producing area.
B-10. Geology and mineralization of the Newfoundland Carboniferous. (Cover title: Geology and industrial minerals of the Newfoundland Carboniferous)
B-11. The Cape Ray Fault; a possible cryptic suture in S.W. Newfoundland.
S-1. Aerial field trip over eastern Avalon Platform.
S-2. The pyrophyllite mine south of Foxtrap, Conception Bay. (Cover title: Pyrophyllite mine, Foxtrap)

1975 [27th] Waterloo, Ontario. Waterloo '75; field excursions guidebook. (200, 383)
Part A. Precambrian geology.
Trip 1. Granitoid rocks of the Madoc-Bancroft-Haliburton area of the Grenville Province.
Trip 2. Grenville gneisses in the Madawaska highlands (eastern Ontario).
Trip 3. Precambrian economic geology of the Timmins area.
Trip 12. Mineralogy and economic geology of the Cobalt silver deposits.
Part B. Phanerozoic geology.
Trip 4, 5. Ordovician to Devonian stratigraphy and conodont biostratigraphy of southern Ontario.
Trip 6. Quaternary stratigraphy of the Toronto area.
Trip 7. Late Quaternary stratigraphy of the Waterloo-Lake Huron area, southwestern Ontario.
Part C. Environmental geology.
Trip 8. Industrial minerals of the Paris-Hamilton District, Ontario.
Trip 9. Engineering geology of the Niagara Peninsula.
Trip 10. Environmental geology, Kitchener-Guelph area. (200)

1976 29th Edmonton, Alberta. Edmonton, '76. (166, 383)
A-1. Uranium deposits of northern Saskatchewan.
A-2. Mineral deposits, Great Slave Lake region, N.W.T.
A-3. Coal deposits of west central Alberta.
A-4. Engineering and environmental geology around Edmonton.
A-5. Mesozoic stratigraphy in central Alberta Foothills and near Drumheller.
A-6. Upper Cretaceous and Paleocene vertebrate paleontology in Alberta.
C-7. Paleozoic stratigraphy in Central Rocky Mountains.
C-8. Geology of the Edmonton area, Alberta.
C-9. Guide to the Athabasca Oil Sands area. (Reprinted 1975)
C-10. Structural geology in central Rocky Mountains.
C-11. Geology of Canadian Cordillera between Edmonton and Vancouver.

1977 30th Vancouver, British Columbia. (383, 583, 118)
Field trip no. 1. Lead-zinc deposits of southeastern British Columbia. (583)
Field trip no. 2. Porphyry copper and molybdenum deposits of north central British Columbia. (583)
Field trip no. 3. Guichon Creek Batholith and mineral deposits. (583)
Field trip no. 4. Coal deposits of the East Kootenay coalfields. (583)
Field trip no. 5. Nicola volcanics, plutons, and mineral deposits. (583)
Field trip no. 6. Volcanic suites of southern British Columbia. (583)
Field trip no. 7. Geology of Vancouver Island. (383)
Field trip no. 8. A guide to the geology of the southern Canadian Cordillera - Calgary-Monte Creek-Vernon-Osoyoos-Princeton-Vancouver. (Cover title: Southern Canadian Cordillera - Calgary to Vancouver) (383, 583, 118)
Field trip no. 9. Southern end of the coast Plutonic complex. (383, 583, 118)
Field trip no. 10. Quaternary stratigraphy of [the] Fraser Lowland. (383, 583, 118)
Field trip no. 11. Vancouver field trip. (Cover title: Vancouver geology) (383, 583, 118)
Field trip no. 12. Fraser Delta field trip. (Cover title: Fraser Delta) (383, 583, 118)
Field trip no. 13. Engineering geology field trip. (Cover title: Engineering geology) (383, 583, 118)
Field trip no. 14. Hydrogeology of Vancouver and Fraser Lowland.
Field trip no. 16. Volcano-fly by. (383, 583, 118)

1978 See Geological Society of America. 197.

1979 Quebec '79.
A-1. Volcanology and sedimentology of Rouyn-Noranda area, Quebec. (383)
A-4. Lithostratigraphy and ultramafic rocks of the Val D'Or-Amos region.
A-5. Geology and tectonics of the Magdalen Islands salt domes.
A-6. Cambro-Ordovician submarine channels and fans, L'Islet to Sainte-Anne-Des-Monts, Quebec.
A-7. Paleozoic stratigraphy of the lowland of Quebec.
A-8. Charlevoix Astrobleme, St-Urbain anorthosite and stratigraphy.
A-9. Structure and stratigraphy of platform and Appalachian sequences near Quebec city.
A-11. Quaternary geology south of Quebec city (Thetford Mines, Beauce, Notre-Dame Mountains).
A-12. Cambro-Ordovician stratigraphy and tectonics south of Sherbrooke area.
A-13. Thetford Mines ophiolite complex.
A-14. Interference patterns in the Oak Hill Group in the Quebec Appalachians.
A-16. Saint-Dominique Slice, county of Bagot: Sedimentological aspects of the Ordovician rocks and tectonic features of the4 Appalachian Front (Quebec).
A-18. Hydrological case studies in the Montreal area.
B-1. Stratigraphy and me4tallogeny in the Chibougamau area.
B-2. Facies of the Silurian West Point reef complex.
B-3. The Jeffrey Asbestos Mine and the ophiolitic complex at Asbestos, Quebec.
B-10. Structural setting of the Thetford Mines ophiolite complex.
B-11. Sensitive clays, unstable slopes, corrective works and slides in the Quebec and Shawinigan area.
B-12. The Montauban-Les-Mines mineralized zone; Portneuf County, Quebec.

1980 May Halifax, Nova Scotia. (383)
Trip 1. Modern and Pleistocene carbonates of Bermuda.
Trip 2. Stratigraphy and volcanology of the western Avalon Zone of Newfoundland, from proterozoic oceanic crust to carboniferous ignimbrites.
Trip 3. The Northern Appalachian geotraverse: Quebec-New Brunswick-Nova Scotia.
Trip 5. Stratigraphy and paleoenvironment of the Windsor Group in New Brunswick.
Trip 6. Mineral deposits and mineralogenic provinces of Nova Scotia.
Trip 7. Geology of carboniferous coal deposits in Nova Scotia.
Trip 8. Mesozoic vulcanism and structure - northern Bay of Fundy region, Nova Scotia.
Trip 9. Quaternary stratigraphy of southwestern Nova Scotia: glacial events and sea-level changes.
Trip 10. Evolution of Sable Island.
Trip 11. Magmatic stoping at Meguma-granite contact near Halifax, Nova Scotia.
Trip 12. Rocks and fossils of Late Precambrian and Early Paleozoic age (including Late Precambrian glacial sequence), Avalon Peninsula, Newfoundland.
Trip 13. Cambro-Ordovician of West Newfoundland - sediments and faunas.
Trip 14. A cross-section through the Appalachian Orogen in Newfoundland.
Trip 16. Geology and massive sulphides of the Bathurst area, New Brunswick.
Trip 17. Volcanism and mineralization in southwestern New Brunswick.
Trip 18. Zeolites in the North Mountain Basalt, Nova Scotia.
Trip 19. Structure and stratigraphy of the Cobequid Highlands, Nova Scotia.
Trip 20. Paleozoic history of Nova Scotia - a time trip to Africa (or South America?)
Trip 21. Igneous and metamorphic geology of southern Nova Scotia.
Trip 22. Stratigraphy, sedimentology, and mineralization of the Carboniferous Windsor Group, Nova Scotia.
Trip 23. Geomorphology and sedimentology of the Bay of Fundy.
Trip 24. The Maritime hydrogeologic field trip.

1981 Calgary, Alberta

Field guides to geology and mineral deposits. (383, 118)
- (1) Structure, stratigraphy, sedimentary environments and coal deposits of the Jura-Cretaceous Kootenay Group, Crowsnest Pass area, Alberta and British Columbia.
- (2) Lead-zinc and copper-zinc deposits in southeastern British Columbia.
- (3) Basal type uranium deposits, Okanagan region, south central British Columbia.
- (4) Quaternary stratigraphy of the region between Calgary and the Porcupine Hills.
- (5) Structural geology of the foothills and front ranges west of Calgary, Alberta.
- (6) Saskatchewan uranium field trip guide.
- (7) Oil sands geology and technology.
- (8) Facies relationships and paleoenvironments of a Late Cretaceous tide-dominated delta, Drumheller, Alberta.
- (9) Mississippian stratigraphy and sedimentology, Canyon Creek (Moose Mountain), Alberta.
- (10) Devonian stratigraphy and sedimentation, Southern Rocky Mountains.
- (11) The Quaternary stratigraphy and geomorphology of Southwest/central Alberta.
- (12) Cordilleran cross-section -- Calgary to Victoria.
- (13) Southern Cordilleran cross-section -- Cranbrook to Kamloops.
- (14) Metamorphism and its relation to structure within the core zone west of the Southern Rocky Mountains.
- (15) Engineering geology of the Mica and Revelstoke dams.

1982 May   Winnipeg, Manitoba.
Trip 1. Till stratigraphy and proglacial lacustrine deposits in the Winnipeg area. (383, 194)
Trip 2. Stratigraphy of the Lynn Lake Greenstone Belt and surface geology at Ruttan and Fox Mine. (383)
Trip 3. Stratigraphy and structure of the western Wabigoon subprovince and its margins, northwestern Ontario. (383, 194)
Trip 4. Proterozoic geology of the northern Lake Superior area. (383)
Trip 5. Quaternary stratigraphy and geomorphology of a part of the lower Nelson River. (383)
Trip 6. Flin Flon volcanic belt: geology and ore deposits at Flin Flon and Snow Lake, Manitoba. (383)
Trip 7. Environmental geology of Winnipeg area. (383, 194)
Trip 8. Volcanic stratigraphy, alteration and gold deposits of the Red Lake area, northwestern Ontario.
Trip 9. Bird River Greenstone Belt, Southeast Manitoba; geology and mineral deposits.
Trip 10. Paleozoic stratigraphy of southwestern Manitoba.
Trip 11. Industrial minerals of the Pembina Mountain and Interlake area, Manitoba. (383)
Trip 12. Granitic pegmatites of the Black Hills, South Dakota and Front Range, Colorado. (Cover title: Granite pegmatites of the Black Hills, South Dakota and Front Range, Colorado) (383, 194)
Granitic pegmatites of the Black Hills, South Dakota and Front Range, Colorado, 2nd printing, April, 1983. (383, 194)
Trip 14. Geological setting of the mineral deposits at Ruttan, Thompson, Snow Lake and Flin Flon. (383)
Trip 15. Geology of the underground research laboratory site, Lac du Bonnet Batholith, southeastern Manitoba.

1983 May   Victoria, British Columbia. (118, 383)
- V. 1.   1. Volcanology structure, coal and mineral resources of early Tertiary outliers in south-central British Columbia.
2. Copper, molybdenum and silver deposits of north central British Columbia.
4. Some gold deposits in the western Canadian Cordillera.
5. Pre-Tertiary geology of San Juan Islands, Washington and Southeast Vancouver Island, British Columbia.
6. Late Quaternary geology of southwestern British Columbia.
7. Coastal environments of southern Vancouver Island.
8. Geology and tectonic history of the Queen Charlotte Islands.
- V. 2.   9. Mineral deposits of Vancouver Island: Westmin resources (Au-Ag-Cu-Pb-Zn), Island copper (Cu-Au-Mo), Argonaut (Fe).
10. Porphyry deposits of southern British Columbia.
11. Quaternary geology of southern Vancouver Island.
12. The Tertiary Olympic terrane, Southwest Vancouver Island and Northwest Washington.
13. Stratabound base metal deposits in southeastern British Columbia and northwestern Montana.
14. Metamorphism and structure of the coast plutonic complex and adjacent belts, Prince Rupert and Terrace areas, British Columbia.
15. Slope stability and mountain torrents, Fraser Lowlands and southern Coast Mountains, British Columbia.
16. Geology of the Hawaiian Islands: Hawaii, Maui and Oahu.

1984 May   London, Ontario. (383)
Field trip 1. Paleozoic stratigraphy of southwestern Ontario.
Field trip 2. Industrial minerals of southwestern Ontario.
Field trip 3. Geology of the Sudbury and Elliot Lake mining areas.
Field trip 4. Geology, silver, and gold deposits: cobalt and Kirkland Lake.
Field trip 5. Proterozoic geology Lake Superior, south shore.
Field trip 8. Hydrogeology and groundwater resources of the Kitchener-Waterloo area.
Field trip 9A/10A. Grenville traverse A; cross-sections of parts of the central metasedimentary belt.
Field trip 12. Glacial stratigraphy and sedimentology - central north shore area, Lake Erie, Ontario.
Field trip 13. Carbonate rocks and coral reefs; Bonaire, Netherlands Antilles.
Field trip 14. Surface geology and volcanogenic base metal massive sulphide deposits and gold deposits of Noranda and Timmins.
Field trip 16. Petroleum production and transportation facilities in southwestern Ontario.
Field trip 17. Cross-section of archean crust in the Wawa-Chapleau-Timmins region.

1985   Fredericton, New Brunswick. (383)
- V. 1-3.   Field excursions. Excursion 1. Appalachian geotraverse (Canadian mainland).
Excursion 2. Massive sulphide deposits of the Bathurst-Newcastle area, New Brunswick, Canada.
Excursion 3. Repeated orogeny, faulting, and stratigraphy in the Cobequid Highlands, Avalon terrain of northern Nova Scotia.
Excursion 4. Stratigraphy and structure of the Ordovician, Silurian and Devonian of northern New Brunswick.
Excursion 5. Stratigraphy and paleoenvironmental setting of the Windsor Group potash deposits, southeastern New Brunswick.
Excursion 6. The Albert Formation - oil shales, lakes, fans and deltas.
Excursion 7. Geomorphic features associated with deglaciation and marine submergence in Southwest New Brunswick.
Excursion 8. Gold deposits of Fundy coastal zone.
Excursion 9. Structure in Southwest New Brunswick.
Excursion 10. Igneous and metamorphic geology of the Cape Breton highlands.
Excursion 11. Turbidite-hosted gold deposits, Meguma domain, eastern Nova Scotia.
Excursion 13. Tungsten, molybdenum, and tin deposits in New Brunswick.
Excursion 14. Lithostratigraphic, physical diagenetic and economic aspects of the Pennsylvanian Permian transition sequence of Prince Edward Island and Nova Scotia.
— May   See Geological Society of America. Cordilleran Section. 199.

CLU-G/G	1966-67, 77(2, 8, 9, 13-16), 82(2, 11), 83
CLhC	1981
CMenUG	1966, 77(2, 4. 7-9, 12-14, 16), 80(1, 3, 5-8, 10-14, 17-24), 81, 82(1-12, 14-15), 83(V.1, no.1-8)(V.2, no.9-16), 1984(13), 1985(3, 10-11, 14)
CPT	1963, 69(1), 83
CSt-ES	1963, 66-68, 70(4-7), 73(B), 75, 81-83
CU-A	1963, 75
CU-EART	1963, 66-67, 69(1), 70(5-6), 81, 82(3, 7, 12-13)
CU-SB	1967, 68
CaAC	1973(B), 76(A3), 1977(2-3, 12, 16)
CaACAM	1966-67, 70(5-6), 82
CaACI	1950, 66-67
CaACM	1950, 66-67, 69(1, 4-5, 7, 8A), 73(B), 76(A1-2, 4-6, C7-8, 10-11), 81
CaACU	1950, 67, 68(A), 69-70, 73, 75, 76(A1-A6, C7-C8, C10), 77(1-6, 8-11, 13-14), 79-83, 85(3, 10, 14)
CaAEU	1975, 80-83

(191) Geological Association of Canada. Cordilleran Section.

CaBVaU	1967-68, 69(1, 4-5, 7, 8A), 73(B), 1977(1-14, 16), 83(2-9, 15-16), 85(1, 3, 10, 14)
CaNSHD	1980, 85(1-2, 9)
CaOHM	1963, 66, 74
CaOKQGS	1963, 67, 69, 70(4-5, 7), 80-83
CaOLU	1963, 67
CaOOG	1963, 66-68, 69(1, 4-5, 6A-C, 7, 8A, 9), 73(B), 74(A2, 4-5, B2-10, S-2), 75, 76(A1-6, C7-8, 10-11), 77(1-14, 16), 80-83, 85
CaOONM	1967, 70(4-7)
CaOTDM	1963, 67, 74(A1, 3), 75, 84(1-5, 8-10A, 12, 14, 16-17), 82, 84-85
CaOTRM	1966-67
CaOWtU	1963, 66-67, 68(A), 69(1), 70, 73(B), 75, 77(2, 4, 9, 13-14), 82, 83, 85(V.1)
CoDGS	1966, 68, 77, 80, 82-83, 84(13), 85(3, 10, 11, 14)
CoU	1967, 69
DI-GS	1963, 66-69, 70(1-2), 73(B), 74(A1-2, 4, B3, 6-9, S2), 75, 80-82, 83(1-2, 9), 84
DLC	1966-67, 75
ICF	1966-67, 69(1, 4-5, 7, 8A), 70(4-7)
ICarbS	1963, 67
IU	1963, 66-67, 69(4-5, 7, 8A), 70(4-7), 73(B), 75
IaU	1950, 63, 67, 73(B), 75, 77, 79(A-1), 80(1-3, 5-11, 13-14, 16-24) 81-82
IdU	1983(5, 8, 12, 14)
InLP	1963, 67, 70(5-6), 75, 81, 82(1-12, 14-15)
InRE	1963
InU	1966, 67-68, 73(B), 75
KyU-Ge	1963, 67, 70(4-7), 75, 76(A1-6, C7-11), 77(4, 9, 13, 14), 82, 85
LNU	1975
LU	1950
MH-GS	1963, 67
MNS	1967
MiDW-S	1963
MiHM	1963, 67, 75
MnU	1966-67, 70(1-2, 4-7), 75, 81, 83
MoSW	1967, 69(1), 82, 83
MoU	1963
MtBC	1963
NBiSU	1963, 67
NIC	1982
NNC	1963, 67-68
NOncoU	1967
NRU	1963, 67, 75
NSbSU	1975, 77, 85
NSyU	1963
NbU	1968(A)
NcU	1963, 67, 75, 80, 82
NdU	1963, 67, 69(1, 4-5, 7, 8A), 70, 73(C), 75, 82
NhD-K	1967, 69, 70(4-7), 75, 82
NjP	1950, 63, 66-67, 75
NvU	1963, 81
OCU-Geo	1950(1), 63, 67, 73(B), 75, 80
OU	1963, 67, 69(4-9), 70(1, 4-7), 74(A1-2), 75, 80, 84(4-5, 8, 13-14, 16), 85(4)
OkT	1950
OkU	1963, 67, 70(5-6), 75, 82-83, 85
PBL	1975
PBU	1983
PBm	1967, 75
PSt	1963, 67, 69, 81, 85
RPB	1967
TxBeaL	1950(1), 63, 69(1, 4-5, 7, 8A), 75
TxDaAR-T	1950, 67, 69(1, 4-5, 7, 8A), 70(1-2), 75
TxDaDM	1966-67, 69(4-5, 7, 8A), 75
TxDaGl	1950(1)
TxDaSM	1969(1)
TxHSD	1950, 77(3-6, 8-9, 11, 16)
TxHU	1966
TxU	1950, 63, 66-67, 69, 70(1-2), 73(B), 83(1, 9)
TxWB	1963
UH-M	1983(16)
UU	1963, 83
ViBlbV	1967, 70(5-7), 75
WBB	1981
WU	1963, 77(4, 9, 13, 14), 79(A1), 81-83, 85
WaU	1968, 83
WyU	1981, 83

(191) GEOLOGICAL ASSOCIATION OF CANADA. CORDILLERAN SECTION. [GUIDEBOOK]

1973   Vancouver geology; a short guide. (Revised in 1977)
1975   Garibaldi geology; a popular guide to the geology of the Garibaldi Lake area.
1977   Vancouver geology; a short guide. 1973. (Revised [ed.] 1977).
1982 Apr.   Copper Mountain-Phoenix tour, southern British Columbia.
1985   See Geological Association of Canada. 190.

CLU-G/G	1982-83
CMenUG	1982
CSt-ES	1977
CaACU	1975, 77
CaAEU	1975, 77, 82
CaBVaU	1973
CaOOG	1973, 82
CaOWtU	1973, 75, 77, 82
CoDGS	1982
DI-GS	1982
DLC	1975
IaU	1982
OkU	1982
WU	1982

(192) GEOLOGICAL ASSOCIATION OF CANADA. MINERAL DEPOSITS DIVISION.

1985   See Canadian Institute of Mining and Metallurgy. CIM Geology Division. 121.

(193) GEOLOGICAL ASSOCIATION OF CANADA. MINERAL DEPOSITS DIVISION. FIELD GUIDE AND REFERENCE MANUAL SERIES.

1985 No. 1.   Geology and ore deposits of the Highland Valley Camp.
   CaOTDM   1985

(194) GEOLOGICAL ASSOCIATION OF CANADA. WINNIPEG SECTION. FIELD TRIP.

1982 May   See Geological Association of Canada. 190.

(195) GEOLOGICAL ASSOCIATION OF NEW JERSEY. ANNUAL FIELD CONFERENCE.

1984 1st   Igneous Rocks of the Newark Basin: petrology, mineralogy, ore deposits and guide to field trip.
1985 2nd   Geological Investigations of the Coastal Plain of Southern New Jersey.

CU-A	1984
KyU-Ge	1984, 85

(196) GEOLOGICAL DISCUSSION CLUB, VANCOUVER, BRITISH COLUMBIA.

1960   (Revised and reprinted in 1968) See Geological Society of America. Cordilleran Section. 199.
1968   See Geological Association of Canada. 190.

(197) GEOLOGICAL SOCIETY OF AMERICA. ANNUAL MEETING. GUIDEBOOK FOR FIELD TRIPS. (TITLE VARIES)

1931 44th   Tulsa, Oklahoma. Log of G.S.A. field trip in Arbuckle Mountains from Ada to Springer, Oklahoma.
   Field trip no. 3. Ardmore Basin and western Arbuckle Mountains.
1940 53rd   Austin, Texas.
   Trip 1, 2. Cretaceous in the vicinity of Austin.
   Trip 3. Fault-line oil fields.
   Trip 4. Pre-Cambrian of the Llano region.
   Trip 5. Paleozoic of the Llano region.
   Trip 6. Lower Tertiary of the Colorado River.
   Trip 7. Oil fields of South Texas.
   Trip 8. Pennsylvanian, Permian, and Triassic of northwestern Texas.
   Trip 9. Meteor Crater of Ector County.
   Trip 10. Paleozoic of the Marathon region.
   Trip 11. Vertebrate fossils and artifacts, Bee County.
1948 61st   New York, New York.
   Excursion no. 1. Franklin Mine, Franklin, New Jersey.

Excursion no. 2. Pleistocene and Tertiary geology, Staten Island and eastern New Jersey.
Excursion no. 3. Petrology and flow structure of the Cortlandt norite, and granitization phenomena at Bear Mountain.
Excursion no. 4. Engineering geology in and near New York.
Excursion no. 5. Squier laboratory, Fort Monmouth, New Jersey.
Excursion no. 6. American Museum of Natural History, New York.
Excursion no. 7. Bell Telephone laboratory, Murray Hill, New Jersey (for American citizens only).
Excursion no. 8. Glacial and ground-water geology on Long Island.
Excursion no. 9. Detailed geology of the Palisades diabase.
Excursion no. 10. Dutchess County, New York.
Excursion no. 11. Zeolite minerals and geology of the Watchung basalt sheets.
Excursion no. 12. Engineering geology of the New York City aqueducts.

1949 [62nd] El Paso, Texas.
1. Marathon region, Big Bend region, Green Valley-Paradise Valley region, Sierra Blanca region, Texas. (756)
2. Cenozoic geology of the Llano Estacado and Rio Grande Valley. (756, 457)
3. Geology and ore deposits of Silver City region, New Mexico. (47, 756)
4. Permian rocks of the Trans-Pecos region. (756)
5. Pre-Permian rocks of the Trans-Pecos area and southern New Mexico. (756)

1950 [63rd] Washington, D.C. Guidebooks to the geology of Maryland.
1. Geology of the South Mountain Anticlinorium.
2. Geology of Bear Island, Potomac River.
3. The coastal plain geology of southern Maryland.

1951 [64th] Detroit, Michigan.
1. Study of Pleistocene features of the Huron-Saginaw ice lobes in Michigan.
2. Devonian rocks of southeastern Michigan and northwestern Ohio.

1952 65th Boston, Massachusetts. Guidebook for field trips in New England.
Field trip no. 1. Geology of the Appalachian highlands of east-central New York, southern Vermont, and southern New Hampshire.
Field trip no. 2. Outstanding pegmatites of Maine and New Hampshire.
Field trip no. 3. Geology of the "Chelmsford Granite" area.
Field trip no. 4. Glacial geology in the Buzzards Bay region and western Cape Cod.

1953 66th Toronto, Ontario. (190)
No. 1. Petrology of the nepheline and corundum rocks, Bancroft area, eastern Ontario.
No. 2. Mineral occurrences of Wilberforce, Bancroft and Craigmont-Lake Clear areas of southeastern Ontario.
No. 3. Glacial geology of the Toronto Orangeville area, Ontario.
No. 4, 5. Geology of part of the Niagara Peninsula of Ontario.
No. 6. Sir Adam Beck-Niagara generating station no. 2, Niagara Falls, Ontario; a resume of the engineering geology.
No. 7. Geology and mineral deposits, Sudbury area, Ontario.
No. 8. Geology and mineral deposits of the Kirkland-Larder mining district, Ontario.
No. 9. The Porcupine mining district, Ontario.
No. 10. Geology and mineral deposits of northwestern Quebec.

1954 67th Los Angeles, California. Geology of Southern California. (112, 199)
1. Western Mojave Desert and Death Valley region.
2. Ventura Basin.
3. Los Angeles Basin.
4. Southwestern part of the Los Angeles Basin.
5. Northern part of the Peninsular Range province.
6. Road log for Santa Catalina Island.

1955 68th New Orleans, Louisiana. Guides to southeastern geology. With supplemental field guide covering physiography and geology along U.S. Highway 90 from New Orleans to Lafayette.
No. 1. Appalachian; an outline of the geology in the segment in Tennessee, North Carolina and South Carolina. The Kings Mountain area. Geomorphology. Outline of the geology of the Great Smoky Mountains area.
No. 2. Type localities (Mississippi and Alabama).
No. 3. Alabama (coastal plain and west-central).
No. 4. Louisiana (northwestern and southern).
No. 5. South Louisiana salt domes.
[6] Guidebook. The California Company. Offshore field trip, southeastern Lousiana offshore operations for members of college and university faculties. (Cover title: California Company, New Orleans. Guidebook. Offshore field trip, southeastern Lousiana offshore operations)
Addenda to field trips:
Appalachian, inner piedmont belt. Guide to the geology of the Spruce Pine District, North Carolina.

1956 69th Minneapolis, Minnesota.
No. 1. Precambrian of northeastern Minnesota.
No. 2. Lower Paleozoic of the upper Mississippi Valley.
No. 3. Glacial geology, eastern Minnesota.

1957 [70th] Atlantic City, New Jersey.
No. 1. Cretaceous and Cenozoic of New Jersey coastal plain.
No. 2. Triassic formations in the Delaware Valley.
No. 3. Precambrian of New Jersey highlands.
No. 4. Delaware Valley Paleozoics.
No. 5. Crystalline rocks of Philadelphia area.
No. 6. Cretaceous and Tertiary geology of New Jersey, Delaware, and Maryland.
No. 7. General geology of the folded Appalachian Mountains of Pennsylvania.

1958 71st St. Louis, Missouri.
No. 1. Southeast Missouri lead belt.
No. 2. Problems of Pleistocene geology in the greater St. Louis area.
No. 3. Mississippian rocks of western Illinois.
No. 4. Onondaga Cave.
No. 5. Pennsylvanian (Desmoinesian) of Missouri. (399)

1959 72nd Pittsburgh, Pennsylvania. Guidebook for field trips.
No. 1. Structure and stratigraphy in central Pennsylvania and the anthracite region.
No. 2. The Pennsylvanian of western Pennsylvania.
No. 3. Monongahela series, Pennsylvanian system, and Washington and Green series, Permian system, of the Appalachian Basin.
No. 4. Mineral deposits of eastern Pennsylvania.
No. 5. Glacial geology of northwestern Pennsylvania.
No. 6. Engineering geology of the Pittsburgh area.

1960 73rd Denver, Colorado. Guide to the geology of Colorado. (Prepared for the 1960 Annual Meeting of the Geological Society of America and Associated Societies and the 1961 Annual Meeting of the American Association of Petroleum Geologists. 00008(1961)) (549)
A-1. Geology of west-central Colorado. Mancos Shale and Mesaverde Group of Palisade area. Colorado National Monument and adjacent areas.
A-2. Geology of south-central Colorado. Brief description of the igneous bodies of the Raton Mesa region, south-central Colorado. Permo-Pennsylvanian stratigraphy in the Sangre de Cristo Mountains, Colorado. Great sand dunes of Colorado. Geology near Orient Mine, Sangre de Cristo Mountains.
A-3. Tectonics and economic geology of central Colorado. Kokomo mining district. Tectonic relationships of central Colorado. Structure and petrology of north end of Pikes Peak Batholith, Colorado. Pre-Cambrian rocks and structure of the Platte Canyon and Kassler quadrangles, Colorado. Gravity map of the Hartsel area, South Park, Colorado. Cripple Creek District. Geologic formations and structure of Colorado Springs area, Colorado. Placers of Summit and Park counties, Colorado.
B-1. Quaternary geology of the Front Range and adjacent plains. First day: Pleistocene geology of the eastern slope of Rocky Mountain National Park, Colorado Front Range. Second day: Surficial geology of the Kassler and Littleton quadrangles near Denver, Colorado. Quaternary sequences east of the Front Range near Denver, Colorado.
B-2. Geology of the northern Front Range-Laramie Range, Colorado and Wyoming. Summary of Cenozoic history, southern Laramie Range, Wyoming and Colorado. Laramie anorthosite. Cross-lamination and local deformation in the Casper Sandstone, Southeast Wyoming. Dakota Group in northern Front Range area.
B-3. Field trip to Climax, Colorado (no road log). Geology of the Climax molybdenite deposit; a progress report.
C-1. Stratigraphy of Colorado Springs-Canon City area.
C-2. Engineering geology-distribution system of the Colorado-Big Thompson Project.

(197) Geological Society of America.

C-3. Precambrian geology of the Idaho Springs-Central City area, Colorado. Geology of the Central City-Idaho Springs area, Front Range, Colorado.
C-4. Fossil vertebrates and sedimentary rocks of the Front Range foothills, Colorado.
C-5. Geology of the Front Range foothills, Boulder-Lyons-Loveland area, Colorado.
C-6. Geology of Mountain Front west of Denver.

1961 74th Cincinnati, Ohio.
No. 1. Grand Appalachian field excursion with a description of the Pennsylvanian rocks along the West Virginia Turnpike. (749)
No. 2. Geology from Chicago to Cincinnati.
No. 3. Pleistocene geology of the Cincinnati region (Kentucky, Ohio, and Indiana).
No. 4. Pennsylvanian geology of eastern Ohio.
No. 5. Engineering geology of flood control and navigation structures in the Ohio River Valley.
No. 6. A guide to the hydrogeology of the Mill Creek and Miami River valleys, Ohio.
No. 7. Sunday all-day field excursion in the Cincinnati region.
No. 8. Examination of Ordovician through Devonian stratigraphy and the Serpent Mound chaotic structure area.
No. 9. Field excursion to the Falls of the Ohio.
No. 10. Geology of the St. Francois Mountain area.

1962 75th Houston, Texas. Geology of the Gulf Coast and central Texas and guidebook of excursions. (255)
No. 1. Geology of Llano region and Austin area. (Revised in 1972 by the University of Texas at Austin, Bureau of Economic Geology) (713)
No. 2. Tertiary and uppermost Cretaceous of the Brazos River valley, southeastern Texas.
No. 3. Recent and Pleistocene geology of Southeast Texas.
No. 4. Tertiary stratigraphy and uranium mines of the Southeast Texas coastal plain, Houston to San Antonio, via Goliad.
No. 5. Active faults, subsidence and foundation problems in the Houston, Texas area.
No. 6. Palestine and Grand Saline salt domes, eastern Texas.
No. 7. Sulphur mine at Boling Dome.
No. 8. Fresh water from sea water ... magnesium from sea water.
No. 9. Coastal Louisiana swamps and marshlands.
No. 10. Tertiary formations between Austin and Houston, with special emphasis on the Miocene and Pliocene.
No. 11. Hydrogeology of the Edwards and associated limestones.
No. 12. Engineering geology of Canyon Dam, Guadalupe River, Comal County, Texas.

1963 76th New York City, New York.
No. 1. Stratigraphy, facies changes and paleoecology of the Lower Devonian Helderberg limestones and the Middle Devonian Onondaga limestones.
No. 2. Stratigraphy, structure, and petrology of Rhode Island and southeastern Connecticut.
No. 3. Stratigraphy, structure, sedimentation and paleontology of the southern Taconic region, eastern New York.
No. 4. Coal group. Geology of the southern part of the eastern Pennsylvania anthracite region. (198)
No. 5. Postglacial stratigraphy and morphology of coastal Connecticut.
No. 6. Atlantic Cement Company at Ravena, New York. (Not published)
No. 7. Hydrologic problems of central Long Island. (Not published)
No. 8. Engineering geology of urban area. (Not published)

1964 77th Miami Beach, Florida.
No. 1. South Florida carbonate sediments. (Reprinted in 1972. Used for University of Iowa field trip in 1978.) (688)
No. 2. Carbonate sediments, Great Bahama Bank.
No. 3. Living and fossil reef types of South Florida.
No. 4. Geology and hydrology of southeastern Florida.
No. 5. [Not held]
No. 6. The geology and geochemistry of the Bone Valley Formation and its phosphate deposits, west-central Florida.
No. 7. Littoral marine life of southern Florida.
No. 8. [Not held]
No. 9. Guidebook for field trip in Puerto Rico.
No. 10. Environments of coal formation in southern Florida.

1965 78th Kansas City, Missouri.
No. 1. Pennsylvanian marine banks in southeastern Kansas.
No. 2. Upper Cretaceous stratigraphy, paleontology, and paleoecology of western Kansas.
No. 3. Upper Pennsylvanian cyclothems in the Kansas River valley.
No. 4. Cryptoexplosive structures in Missouri. (399)
No. 5. Geology of the Kansas City Group at Kansas City. (399)
No. 6. Pennsylvanian fossil plants from Kansas coal balls.
No. 7. Hydrogeology of the lower Kansas River valley.
No. 8. Engineering geology of Kaysinger Bluff and Stockton dams, west-central Missouri.

1966 79th San Francisco, California.
[A] Central Utah coals. (730)
[B] Geology of Northern California. (112)
No. 1. Point Reyes Peninsula and San Andreas fault zone.
No. 2. San Francisco Peninsula.
No. 3. San Andreas Fault from San Francisco to Hollister.
No. 4. Hydrogeology field trip East Bay area and northern Santa Clara Valley.
No. 5. Sacramento Valley and northern coast ranges.
No. 6. Yosemite Valley and Sierra Nevada Batholith.
No. 7. Mineralogy of the Laytonville quarry, Mendocino County, California.
[C] A walker's guide to the geology of San Francisco. (Special supplement to Mineral Information, November 1966, v. 19, no. 11) (113)

1967 80th New Orleans, Louisiana.
No. 1. Geology of the coastal plain of Alabama.
[No. 2] Lower Mississippi alluvial valley and terrace.
[No. 3] Central Arkansas; economic geology and petrology. (67)
[No. 4] Cancelled. (Arkansas-Louisiana coastal plain)
[No. 5] Coal Division. Carboniferous detrital rocks in northern Alabama. (198)
[No. 6] Five Islands salt domes and the Mississippi deltaic plain.
No. 7. Yucatan field trip: Peninsula of Yucatan. (2nd ed.) (466)
[No. 8] Guatemala. (274)
Local excursion A. Aerial tour of Mississippi River deltaic plain.
Local excursion B. New Orleans and vicinity.

1968 [81st] Mexico, D.F.
No. 1. Sabinas coal region guidebook. Geology of the Sabinas coal basin, Coahuila.
No. 2. Geology of the northern part of the Valley of Mexico and of the Pachuca-Real del Monte mining district.
No. 3. General geology of the Sierra Madre Oriental between Tulancingo and Poza Rica.
No. 4. General, ground water and engineering geology of the metropolitan area of Mexico City and immediate (sic) surroundings.
No. 5. General geology of the Morelos Basin and adjacent areas.
No. 6. Geology and utilization of geothermic energy at Pathe, State of Hidalgo.
No. 7. Paleozoology of the marine Upper Jurassic in the Petlalcingo area, State of Puebla, and paleobotany of the continental Middle Jurassic of the Tezoatlan area, State of Oaxaca.
No. 8. Volcanology and geomorphology of the southeast corner of Mexico Basin, west side of Ixtacihuatl and north side of Popocatepetl volcanoes, Mexico.
No. 9. General geology of south-central Mexico, between Mexico City and Acapulco.

1969 [82nd] Atlantic City, New Jersey.
[A] Pre-convention. Coal Division. Some Appalachian coals and carbonates; models of ancient shallow-water deposition. (198)
[B] Geology of selected areas in New Jersey and eastern Pennsylvania and guidebooks of excursions.
No. 1A. Precambrian and Lower Paleozoic geology of the Delaware Valley, New Jersey-Pennsylvania.
No. 1B. Geology of the Valley and Ridge Province between Delaware Water Gap and Lehigh Gap, Pennsylvania. Structural control of wind gaps and water gaps and of stream capture in the Stroudsburg area, Pennsylvania, and New Jersey.
No. 1C. Sedimentology of some Mississippian and Pleistocene deposits of northeastern Pennsylvania.
No. 2. Shelf and deltaic paleoenvironments in the Cretaceous-Tertiary formations of the New Jersey coastal plain.

No. 3. Quaternary geology of part of northern New Jersey and the Trenton area.
No. 4. Late Triassic Newark Group, north-central New Jersey and adjacent New York and Pennsylvania.
No. 5. Engineering geology of the Yards Creek hydro-electric pumped storage project. Geology of Tocks Island area and its engineering significance.
No. 6. Ilmenite deposits of the New Jersey coastal plain.

1970 [83rd] Milwaukee, Wisconsin.
No. 1. Cambrian-Ordovician geology of western Wisconsin. (778)
No. 2. Geology of the Baraboo District, Wisconsin; a description and field guide incorporating structural analysis of the Precambrian rocks and sedimentologic studies of the Paleozoic strata. (587, 778)
No. 3. The Mississippian and Devonian of Iowa. (332)
No. 4. Pleistocene geology of southern Wisconsin. (778)
No. 5. Marquette Range field trip guidebook. (Cover title: Marquette Iron Range, Michigan)
No. 6. Hydrogeology of the Rock-Fox River Basin of southeastern Wisconsin. (778)
No. 7. Upper Mississippi Valley base metal district. (778)
No. 8. Coal Geology Division. Depositional environments in parts of the Carbondale Formation; western and northern Illinois, Francis Creek Shale and associated strata and Mazon Creek biota. (265, 198)
No. 9. Glacial geology of Two Creeks forest bed, Valderan type locality, and northern Kettle Moraine State Forest. (778)

1971 [84th] Washington, D.C.
1. A guide to the geology of Delaware's coastal environments. (University of Delaware. College of Marine Studies. Publication 2GL039)
2. Guidebook to contrast in style in deformation of the southern and central Appalachians of Virginia. (749)
3. New interpretations of the eastern Piedmont geology of Maryland, or granite and gabbro or graywacke and greenstone. (370)
4. The Piedmont crystalline rocks at Bear Island, Potomac River, Maryland. (370)
5. Environmental history of Maryland Miocene. (370)
6. Environmental geology in the Pittsburgh area.
7. Historical engineering geology of the Chesapeake and Ohio Canal. 1970.
8. Alkalic complex and related rocks of the central Shenandoah Valley (Virginia), Devonian Tioga tuff, and Eocene felsites.
[1] Field trip to igneous rocks of Augusta, Rockingham, Highland and Bath counties, Virginia. (743)
[2] Geologic setting of Triassic-Jurassic-Eocene dike swarm in west-central Virginia, and adjacent parts of West Virginia.
9. Depositional environments of eastern Kentucky coals. (209)
10. Slope stability and denudational processes; central Appalachians.
11. Hydrogeology and geochemistry of folded and faulted carbonate rocks of the central Appalachian type and related land use problems.
12. No guidebook.

1972 [85th] Minneapolis, Minnesota.
1. Field trip guidebook for lower Precambrian volcanic-sedimentary rocks of the Vermilion District, northeastern Minnesota. (388)
2. Field trip guidebook for Precambrian North Shore Volcanic Group, northeastern Minnesota. (388)
3. Field trip guidebook for Paleozoic and Mesozoic rocks of southeastern Minnesota. (388)
4. Field trip guidebook for Precambrian migmatitic terrane of the Minnesota River valley. (388)
5. Field trip guidebook for Precambrian geology of northwestern Cook County, Minnesota. (388)
6. Field trip guidebook for geomorphology and Quaternary stratigraphy of western Minnesota and eastern South Dakota. (388)
7. Field trip guidebook for hydrogeology of the Twin Cities artesian basin. (388)
8. Coal Geology Division. Depositional environments of the lignite-bearing strata in western North Dakota. (480, 703, 200)

1973 [86th] Dallas, Texas.
1. The Edwards Reef Complex and associated sedimentation in central Texas. (713(No. 15))
4. Hydrogeology of the Edwards Limestone Aquifer.
5. Regional geology of the Arbuckle Mountains, Oklahoma. (505, 18(1978))
6. Igneous geology of the Wichita Mountains and economic geology of Permian rocks in Southwest Oklahoma.
7. Environmental geologic atlas of the Texas coastal zone, Galveston-Houston area. 1972.
8. Pennsylvanian depositional systems in north-central Texas. A guide for interpreting terrigenous clastic facies in a cratonic basin. (713(No. 14))
10. Lignite geology, mining, and reclamation at Big Brown Steam Plant near Fairfield, Texas.
11. Mineral resources of East Texas. (Texas. Univ. Bureau of Economic Geology Guidebook 9, "Field excursion, East Texas: Clay glauconite, ironstone deposits", 1969, supplemented by a new road log) See University of Texas at Austin. Bureau of Economic Geology. 713.(1969)

1974 [87th] Miami Beach, Florida.
No. 1. Guidebook to the geology and ecology of some marine and terrestrial environments.
No. 2. Field seminar on water and carbonate rocks of the Yucatan Peninsula, Mexico. (466)
No. 3. Modern Bahaman platform environments.
No. 4. Field trip cancelled; no guidebook published.
No. 5. Sabellariid reef, beach erosion and environmental problems of the Barrier Island-Lagoon system of the lower east Florida coast.
No. 6. The comparative study of the Okefenokee Swamp and the Everglades-Mangrove swamp-marsh complex of southern Florida.
No. 7, 8. No guidebook. (Used "Land from the sea;" the geologic story of south Florida)
No. 9. Field trip cancelled; no guidebook published.
No. 10. Same guidebook as 1974, no. 3. Modern Bahamian platform environments.
No. 11. Field guide to selected Jamaican geological localities. (337)
No. 12. Field guide to some carbonate rock environments, Florida Keys and western Bahamas.

1975 [88th] Salt Lake City, Utah.
[2] Coal geology division. Field guide and road log to the western Book Cliffs, Castle Valley, and parts of the Wasatch Plateau. (103, 198)
[4] Environmental geology of the middle Wasatch Front, Utah.
[10] Geologic guide to the northwestern Colorado Plateau. (Studies for Students No. 9) (103(1974))
11. Guidebook to the Cenozoic structural and volcanic evolution of the southern Marysvale volcanic center, Iron Springs mining district and adjacent areas, southwestern Utah. (577)
[13] Field trip guide, Great Salt Lake and deposits of Lake Bonneville at Little Valley, Utah.

1976 89th Denver, Colorado.
A. Studies in Colorado field geology.
1. Precambrian and Cenozoic geology, south-central colorado. (143(No.8 #24))
2. Hydrogeology of the San Luis Valley, Colorado. (143(No.8 #41))
3. Paleozoic depositional environments, northern Front Range, Colorado and Wyoming. (143(No.8 #9))
5. Urban geology of the greater Denver area. (143(No.8 #39))
6. Lacustrine and related nonmarine depositional environments in Tertiary rocks. (143(No.8 #27))
7. Economic geology, central Front Range. (143(No.8 #36,37))
8. Mechanism of deformation along the northeastern Front Range. (143(No.8 #28))
9. Geology of the Schwartzwalder Uranium Mine, Jefferson County. (143(No.8 #35))
10. Geology, landuse and resource development, Denver area (for non-geologists). (143(No.8 #40))
11. Mountain front geology, Denver.
12. Sedimentology of Paleozoic and Mesozoic strata, Morrison-Golden area. (143(No.8 #18))
13. No field trip guide.
14. No field trip guide.
15. Precambrian geology, Front Range. (143(No.8 #1))
16. Structure, volcanism, and geothermal features, Rio Grande Rift, Colorado. (143(No.8 #30))
17. Paleozoic tectonics and sedimentation and Tertiary volcanism in the western San Juan Mountains. (143(No.8 #15))

18. Cretaceous tectonics, sedimentation and energy resources, western Denver Basin. (143(No.8 #17,18))
19. Energy resou8rces of Northwest Colorado. (143(No.8 #34))
20. Henderson Molybdenite Deposit, Front Range. (143(No.8 #37))
21. Quaternary soil sequence, Goulden-Boulder area. (143(No.8 #26))
22. Land use and engineering geology, Front Range. (143(No.8 #38))
B. Coal geology of northwestern New Mexico and southwestern Colorado. (461, 198)

1977 [90th]   Seattle, Washington. Geological excursions in the Pacific Northwest.
Field trip no. 1. Bedrock geology of the North Cascades.
Field trip no. 2. General geology of the southern Olympic coast.
Field trip no. 3. Tertiary stratigraphy of the central Cascades Mountains, Washington state.
Field trip no. 4. Volcanic stratigraphy and structure of the southern Cascade Range, Washington.
Field trip no. 5. The stratigraphy and structure of Orcas Island, San Juan Islands.
Field trip no. 6. Quaternary geology of the Fraser Lowland.
Field trip no. 9. Cenozoic stratigraphy of southwestern Washington.
Field trip no. 10. Structure, stratigraphy, plutonism, and volcanism of the central Cascades, Washington.
Field trip no. 11. Geology of the southern San Juan Islands.
Field trip no. 12. Stratigraphy of the Yakima basalts and structural evolution of the Yakima Ridges in the western Columbia Plateau.
Field trip no. 13:1. The Okanogan lobe of the Vashon continental glacier.
13:2. Lake Missoula flooding and the Channeled Scabland.
13:[3] Guidebook to Quaternary geology of the Columbia, Wenatchee, Peshastin, and upper Yakima valleys, west-central Washington. (220)
[Field trip no. 14] Geologic hazards in Seattle.
[Field trip no. 15] Archaeological geology of Birch Creek Valley and the eastern Snake River Plain, Idaho.

1978 [91st]   Toronto, Ontario. Field trips guidebook. (190, 383, 522, 384, 583, 408, 604)
No. 1. Metallogeny of Archean and Proterozoic rocks of the Superior and southern provinces of the Canadian Shield.
No. 2. Volcanic stratigraphy and geochemistry in the Timmins mining area.
No. 3. PreCambrian stratigraphy and uranium deposits, Elliot Lake area, Ontario.
No. 4. The English River subprovince, an Archean gneissic belt.
No. 5. Volcanology and mineral deposits of the Uchi-Confederation Lakes area, northwestern Ontario.
No. 6. Geology of the Grenville front tectonic zone in Ontario.
No. 7. GSA Coal Geology Division. Geology of Carboniferous coal deposits in Nova Scotia. See Geological Society of America. Coal Division. 198.
No. 8. Alkalic rocks of the Haliburton-Bancroft region.
No. 9. Middle Wisconsinan stratigraphy in southern Ontario.
No. 10. Geology and engineering phenomena of Champlain Sea clays.
No. 11. Archeological sites: Pittsburgh to Toronto.
No. 12. Engineering geology at Niagara hydroelectric plants.
No. 13. Geology of the greater Toronto region.
No. 14. Nuclear power plants, underground space and engineering geology.
No. 15. Airborne tour of the geomorphology and urban geology of the western Lake Ontario drainage basin.
No. 17. Archean komatiitic, tholeiitic, calc-alkalic and alkalic volcanic sequences in the Kirkland Lake area.
No. 18. Silurian stratigraphy of the Niagara Escarpment, Niagara Falls to the Bruce Peninsula.
No. 19. Geology and mineral deposits of the Sudbury area.
No. 20. Isograds around the Hastings metamorphic "Low."
No. 21. The Kidd Creek Mine, Timmins.
No. 22. Same as No. 14.
No. 23. Southeastern Ontario, a geological overview.
No. 24. Hydrogeology and subsurface waste disposal in the Toronto region.
No. 25. GAC Structural Geology Division. Structure and lithology of Muskoka-southern Georgian Bay region, central Ontario.
No. 26. Glacial geology of the Toronto-Owen Sound area.

1979 [92nd]   San Diego, California. (384, 522, 583)
[Field trip nos. 1, 4, 8, 11, 16] Geological excursions in the Southern California area. (110)
Field trip no. 1. Quaternary terraces and crustal deformation in Southern California.
Field trip no. 4. (NAGT trip) Geomorphology of the Salton Basin. See National Association of Geology Teachers. 408.
Field trip no. 8. Some prehistoric earthquakes on the San Andreas Fault, Los Angeles area.
Field trip no. 11. Upper Cretaceous deep-sea fan deposits, San Diego.
Field trip no. 16. Regional Miocene detachment faulting and early Tertiary (?) mylonitic terranes in the Colorado River trough, southeastern California and western Arizona.
Field trip no. [3] Guidebook for roundtrip flight to Colorado Plateau. (110)
[Field trip nos. 5, 20-21, 24] Mesozoic crystalline rocks: Peninsular Ranges Batholith and pegmatites, Point Sal ophiolite.
Field trip nos. 5 and 20. Gem-bearing pegmatites in San Diego County.
Field trip no. 21. Peninsular Ranges Batholith, San Diego and Imperial counties.
Field trip no. 24. Point Sal ophiolite.
Field trip no. 7. Geology and geothermics of the Salton Trough. (678)
[Field trip nos. 10, 12-13, and 26] Baja California geology; field guides and papers.
Field trip no. 10. Geology of Isla Cedros-Vizcaino Peninsula, Baja California Sur, Mexico.
Field trip no. 12. Geologic transect along Mexican Highway 1, La Paz to Tijuana, Baja California.
Field trip nos. 13 and 26. Geology of northern Baja California, Mexico.
Field trip no. 15. Coal Geology Division. Phoenix-Black Mesa-Page, Arizona. (198, 674)
[Field trip no. 18] Eocene depositional systems, San Diego, California. (593)
[Field trip no. 19] Earthquakes and other perils, San Diego region. (556)
[Field trip no. 22] Tectonics of the juncture between the San Andreas fault system and the Salton Trough, southeastern California. (680)
[Field trip no. 23] A guidebook to Miocene lithofacies and depositional environments, coastal Southern California and northwestern Baja California. (593)
Field trip no. 27. Geology of northern Sonora.

1980 [93rd]   Atlanta, Georgia.
Field trip no. 4. Archaelogical Geology Division. The archaeology-geology of the Georgia coast. (237(20))
— V. 1.   [Field trip nos. 1-3, 5, 7-13] Excursions in southeastern geology.
Field trip no. 1. Engineering Geology Division. Geology related to construction of the Rocky Mountain pumped storage project.
Field trip no. 2. Superimposed folding and its bearing on geologic history of the Atlanta, Georgia, area.
Field trip no. 3. Petrology and structure of the Stone Mountain granite and Mount Arabia migmatite, Lithonia, Georgia.
Field trip no. 5. Geology of the eastern Piedmont fault system in South Carolina and eastern Georgia.
Field trip no. 6. Transect of the Southern Appalachians.
Field trip no. 7. Coal Geology Division. Depositional environments of a part of the Southern Appalachian coal fields. (198)
Field trip no. 8. Society of Economic Geologists. Volcanogenic ore deposits of the Carolina slate belt.
Field trip no. 9. Cenozoic biostratigraphy of the Carolina outer coastal plain.
Field trip no. 10. Middle Ordovician carbonate shelf to deep water basin deposition in the Southern Appalachians. (709, 525)
Field trip no. 11. Geohydrology of the Chattahoochee River.
Field trip no. 12. Guide to geology along a traverse through the Blue Ridge and Piedmont provinces of North Georgia.
Field trip no. 13. Carboniferous paleodepositional environments of the Chattanooga area.
— V. 2.   [Field trip nos. 14-23] Excursions in southeastern geology.
Field trip no. 14. Coastal environments of Georgia and South Carolina.
Field trip no. 15. Kaolin deposits and the Cretaceous-Tertiary boundary in east-central Georgia.
Field trip no. 16. Depositional facies in Middle-Upper Ordovician and Silurian rocks of Alabama and Georgia.
Field trip no. 17. Repeat of field trip no. 3.
Field trip no. 18. Petrology and structural setting of post-metamorphic granites of Georgia.

Field trip no. 19. Society of Economic Geologists. Barite deposits in the Cartersville District, Bartow County, Georgia.
Field trip no. 20. Upper Cretaceous and lower Tertiary geology of the Chattahoochee River valley, western Georgia and eastern Alabama.
Field trip no. 21. Tectonic framework of the Appalachian orogen in Alabama.
Field trip no. 22. Surficial deposits, weathering processes, and evolution of an inner coastal plain landscape, Augusta, Georgia.
Field trip no. 23. Structural and stratigraphic setting of arc volcanism in the Talladega slate belt of Alabama.
Field trip no. 24. Coral World of the Virgin Islands.
1981 [94th] Cincinnati, Ohio. (384, 522, 583, 408)
— V. 1. [Field trip nos. 11-12, 1-2, 13, 4] Stratigraphy, sedimentology.
Field trip 11. Paleoenvironmental interpretation of the Middle Ordovician high bridge group in central Kentucky.
Field trip 12. Stratigraphy, sedimentology, and paleoecology of the Cincinnatian Series (Upper Ordovician) in the vicinity of Cincinnati, Ohio.
Field trip 1. Lithostratigraphy, cyclic sedimentation and paleoecology of the Cincinnatian Series in southwestern Ohio and southeastern Indiana.
Field trip 2. Devonian and early Mississippian smaller foraminiferans of southern Indiana and northwestern Kentucky.
Field trip 13. Early Mississippian deltaic sedimentation in central and northeastern Ohio.
Field trip 4. Mississippian-Pennsylvanian boundary in the central part of the Appalachian Basin.
— V. 2. [Field trip nos. 3, 19, 10, 6/16, 15] Economic geology, structure.
Field trip 3. Chattanooga and Ohio shales of the southern Appalachian Basin.
Field trip 19. Structure and stratigraphy of the Pine Mountain thrust sheet.
Field trip 10. Stratigraphic and structural geology of the Hot Springs traverse, Tennessee-North Carolina.
Field trip 6/16. The Serpent Mound cryptoexplosion structure, southwestern Ohio.
Field trip 15. Geology and ore deposits of the St. Francois Mountains, Missouri. (395, 399(no.67))
— V. 3. [Field trip nos. 5, 9, 7-8, 17-18] Geomorphology, hydrogeology, geoarcheology, engineering geology.
Field trip 5. Hydrogeology Division. Geohydrology of the Ohio River alluvial aquifer. [Abstracts]
Field trip 9. Quaternary Geology and Geomorphology Division. Quaternary deposits of southwestern Ohio.
Field trip 7. Coastal geomorphology and geology of the Ohio shore of Lake Erie.
Field trip 8. Friends of the Karst. Hydrogeology of the Mammoth Cave region, Kentucky.
Field trip 17. Archaeological Geology Division. Geoarcheology of the Flint Mammoth Cave system and the Green River, western Kentucky.
Field trip 18. Engineering Geology Division. Engineering geology of the Cincinnati area.
Field trip 14. Coal Geology Division. Coal and coal-bearing rocks of eastern Kentucky. (198, 347)
1982 [95th] New Orleans, Louisiana. (384, 408, 522, 583)
[1] Coal Geology Division. Lignite of the Gulf Coastal Plain, Texas and Louisiana.
[2] Mississippi River and delta depositional environments.
[3] Quaternary Geology and Geomorphology Division. A field guide to the Paleozoic rocks of the Ouachita Mountain and Arkansas Valley provinces, Arkansas. (67(1981 No.8 1.1))
[4] Archaeological Geology Division. Cultural and morphological changes in the upper Barataria Basin, ca. 900-1700 AD.
[5] Reefs and associated sediments of Grand Cayman Island, B.W.L: recent carbonate sedimentation.
[6] Mississippian-Pennsylvania shelf to basin transition, Ozark and Ouachita regions Arkansas.
[7] Transgressive depositional environments of the Mississippi River delta plain. (357)
[8] Upper Cretaceous in the Lower Mississippi embayment of Tennessee and: lithostratigraphy and biostratigraphy.
Field trip no. 10. Hydrogeology Division. Geology and hydrogeology of carbonate rocks of the northeastern Yucatan Peninsula; road log and supplement to 1978 guidebook, geology and hydrogeology of northeastern Yucatan. (466)
[11] Hydrogeology Division. Engineering geology of the Jefferson Island event -- November 20, 1980. (357)
[12] Sedimentary processes and environments along the Louisiana-Texas coast. (466)
13. Appalachian thrust belt in Alabama: tectonics and sedimentation.
[14] Hydrogeology Division. Geopressured-geothermal energy resource appraisal: hydrogeology and well testing determine producibility. (357)
SEG Field trip. See Society of Economic Geologists. 583.
1983 [96th] Indianapolis, Indiana. (268, 270, 384, 408, 583, 522)
— V.1. [Field trip nos. 1-6, 12-13, 11] Field trips in midwestern geology.
Field trip 1. The Paleozoic systemic boundaries of the southern Indiana-adjacent Kentucky area and their relations to depositional and erosional patterns.
Field trip 2. Paleontology Division. Paleontology and stratigraphy of the Borden Delta of southern Indiana and northern Kentucky.
Field trip 3. Quaternary Geology and Geomorphology Division. Stratigraphy of the Wedron and Trafalgar formations (Wisconsinan) in east-central Illinois and west-central Indiana.
Field trip 4. Structural Geology Division. Precambrian geology south of Lake Superior.
Field trip 5. Geology of the Kentland Dome structurally complex anomaly, north-western Indiana.
Field trip 6. Quaternary Geology and Geomorphology Division. Shoreline processes and geomorphology, southwestern Lake Michigan.
Field trip 12. Silurian reef and interreef strata as responses to a cyclical succession of environments, southern Great Lakes area.
Field trip 13. Quaternary Geology and Geomorphology. History of Pleistocene alluviation of the middle and upper Wabash Valley.
Field trip 11. History of geology. The New Harmony geologic legacy.
— V. 2. [Field trip nos. 7-10, 14-16] Field trips in midwestern geology.
Field trip 7. Quaternary Geology and Geomorphology Division. Ground water hydrology and geomorphology of the Mammoth Cave region, Kentucky, and of the Mitchell Plain, Indiana.
Field trip 8. Engineering Geology Division. Urban and engineering geology of the Indianapolis area.
Field trip 9. Coal Geology Division. Origin and economic geology of the Springfield Coal member in the Illinois Basin. (198)
Field trip 10. Society of Economic Geologists. Metalliferous shales of the Illinois Basin.
Field trip 14. Archeological Geology Division. Archaeological geology of the Wyandotte Cave region, south-central Indiana.
Field trip 15. Lithostratigraphy, mineralogy, and geochemistry of the New Albany Shale (Devonian and Mississippian) in southeastern Indiana.
Field trip 16. Society of Economic Geologists. The Salem Limestone in the Indiana building-stone district.
1984 97th Reno, Nevada. (454, 384, 408, 522, 583)
—V. 1. [Field trip nos. 1-2, 13-14, 14(cor. & rev.)] Western geological excursions.
Field trip 1. The Mississippian-Pennsylvania boundary in the eastern Great Basin.
Field trip 2. Biomeres and biomere boundaries.
Field trip 13. Archeological Geology Division. Quaternary stratigraphy and archaeology of the Lake Lahontan area.
Field trip 14. Quaternary geology of the eastern Mojave Desert, eastern California.
Field trip 14. Surficial geology of the eastern Mojave Desert, California. (Corrected and revised field trip 14: Quaternary geology of the eastern Mojave Desert, California, which is pages 101 through 251 of volume 1, "Western Geological Excursions")
— V. 2. [Field trip nos. 10-12, 23, Trip A] Western geological excursions.
Field trip 10. Geology of the Nevada Test Site.
Field trip 11. Mono craters, Long Valley caldera: Seismicity, volcanism and engineering geology.
Field trip 12. Engineering Geology Division. Engineering geology of the slide mountain rock slide and waterflood-debris flow.
Field trip 23. Structure and dynamics of the nearshore sedimentological systems in Lake Tahoe and the effects of manmade structures.
Field trip A. Magma mixing in some Sierran plutonic rocks.
— V. 3. [Field trip nos. 3-4, 15, 20-22] Western geological excursions.

(197) Geological Society of America.

Field trip 3. Coal Division. Coal deposits, stratigraphy, and the Cretaceous-Tertiary boundary southern Raton Basin, New Mexico and Colorado. (198)

Field trip 4. Hydrogeology Division. The hydrogeology of the Carson and Truckee River basins, Nevada.

Field trip 15. Mineral deposits of central Nevada. (453)

Field trip 20. Remote sensing imagery for mineral exploration.

Field trip 21. Hydrogeology of the Nevada Test Site.

Field trip 22. Geothermal areas of western Nevada.

— V. 4. [Field trip nos. 5-9, 16-19] Western geological excursions.

Field trip 5. Structural Geology and Tectonics Division. Tectonic development of the northern Sierra Nevada: An accreted late Paleozoic island arc and its basement.

Field trip 6. Structural Geology and Tectonics Division. Paleozoic and Mesozoic continental margin collision zone features: Mina to Candelaria, Nevada, traverse.

Field trip 7. Paleozoic stratigraphy and tectonics of the southwestern Great Basin.

Field trip 8. Tertiary tectonics of west-central Nevada: Yerington to Gabbs Valley.

Field trip 9. Geology of the Ruby Mountains-East Humboldt range: A Cordilleran metamorphic core complex.

Field trip 16. Mesozoic-Cenozoic convergent margin of Northern California.

Field trip 17. Structural Geology and Tectonics Division. The pre-Cordilleran active overthrust belt, San Juan Province, Argentina.

Field trip 18. Engineering Geology Division. Neotectonics of western Nevada.

Field trip 19. Tertiary extensional tectonics in the Sevier Belt of southern Nevada.

SEG Field trip. See Society of Economic Geologists. 583.

1985 96th Orlando, Florida. (384, 408, 522, 583)

[1] Geology of the Barrier Island and marsh-dominated coasts, west-central Florida.

Field trip no. 2. Pleistocene and Holocene carbonate environments on San Salvador Island, Bahamas. (135)

[3] Karst hydrogeology of central and northern Florida. (176)

Field trip no. 4. Coastal geology and the occurrence of Beachrock: central Florida Atlantic coast.

[5] Geology of the southwestern Piedmont of Georgia.

[6] Society of Economic Geologists. Volcanic-hosted gold and high alumina rocks of the Carolina slate belt. (583)

Field trip no. 7. Coal Division field trip. Characteristics of the Mississippian-Pennsylvania boundary and associated coal-bearing rocks in the southern Appalachians.

[8] A self-guided field trip to the Winter Park sinkhole.

9. Florida land-pebble phosphate district.

10. Canceled.

11. Geology of Haiti.

[Field trip no. 12] Cenozoic geology of the Apalachicola River area, northwest Florida. (614(1983 no.25))

13. Coastal morphology of Southwest Florida and its relevance to past human occupation of that coast.

AGC	1982(3)
ATuGS	1949, 55(1-5), 57-58, 67, 82
AzFU	1954(1-5), 60, 66(A)
AzTeS	1949, 54(1-5), 55(1-5), 56-62, 63(5), 64-66, 67(1, 5, 7), 68(1), 69, 70(8), 71(2, 4, 9), 73(8), 75(2, 11), 77(1-6, 9-13:2), 78
AzU	1940, 49-50, 55(1-5), 58-60, 66(A), 67(7), 70(4), 73(1), 75(2, 4, 10), 79(1, 4, 8, 11, 16, 22)
BEG	1980(2)
CChiS	1969
CLU-G/G	1940, 48, 49(3, 5), 50, 52, 53(1-10), 54(1-6), 55(1-5), 56-62, 63(3-4), 64(1-4, 6, 10), 65(1-7), 66(A-B), 67, 69(A), 70, 71(1-8:1, 9-11), 72, 73(1, 8, 10), 74(1-2, 5-8, 11-12), 77(1-6, 9-13:2, 14), 78, 79(1, 3-5, 7-8, 10-13, 16, 18, 20-24, 26-27), 80(V.1, V.2, 10), 81(V.1-3), 83(V.1, 2), 84, 85(no.2, [6], [7])
CLhC	1949(1-2, 4-5), 50, 54(1-5), 55(1-5), 56-62, 64(1-3), 65(1), 66(B), 67(6), 71(2), 73(1, 8), 76, 79(18, 23), 81, 83-84
CMenUG	1957(1), 60, 62, 63(3-5), 64(1, 2, 6), 68(2), 70(1, 2, 4-9), 71, 72, 73(1, 4-11), 74(1-3, 5, 6, 11, 12), 75(2, 10, 11, 13), 77(1, 4, 5, 8-14), 78, 79(1, 3-5, 7-8, 10-13, 15-24, 26-27), 80, 81, 82(3-8, 11-14), 83, 84, 85(1-9, 12)
CPT	1949, 51, 53(1, 2, 4, 5, 7-10), 54-62, 63(3), 65(1-7), 66, 69, 70(1, 2, 4, 6-9), 71(3, 8:1), 72(8), 75(2, 10, 11), 76, 78, 79(1, 3-5, 7-8, 11, 16, 18-24), 80(1-5, 7-10, 12-23), 81(1-4, 6-13, 16-19), 82(3, 11, 14), 83(1, 2, 4-16), 84, 85(1, 2, 5, 7)
CSdS	1949(1-2, 4-5), 50, 53, 54(1-5), 55(1-5), 56, 59, 61-62, 64, 66(B), 67, 73(1), 75(2, 10), 79, 82(Apr.)
CSfCSM	1954(1-5), 57-59, 61-62, 79(1, 4-5, 7-8, 10-13, 16, 20-21, 24, 26), 81(15)
CSt-ES	1940, 48-51, 53, 54(1-5), 55(1-5), 56-62, 63(3-5), 64(1, 3, 6, 10), 65-66, 67(1, 3, 5-8, 13), 69(A, B1(A-B)), 70, 71(1-8:1, 9), 72, 73(1, 5-6, 8, 11), 74(2-3, 5-8, 11-12), 75(2, 10), 76(B), 77(1-6, 9-13:2), 78, 79(1, 3-5, 7-8, 10-13, 16, 20-22, 24, 26-27), 80(1), 81, 82(3, 11, 12, 14), 83-84, 85(1-9)
CU-A	1949(1, 3), 50, 54(1-5), 55(1-5), 56, 61-62, 66, 69, 70(1-2, 4, 6-9), 73(1, 8), 77(1-6, 9-13:2), 78, 79(1, 3-5, 8, 10-13, 16, 19-24, 26), 80(1), 82(3, [8]), 84(1, 3)
CU-EART	1948, 49(2-5), 50, 52, 53(4-5), 54(1-5), 55(1-5), 56-60, 61(1-9), 62, 64(1-3, 6-7, 10), 65-66, 67(1-3, 5-8, A-B), 68-70, 71(1-8:1, 10-11), 72, 73(1, 5-8, 10), 74(3,9), 75(2, 4, 10, 11, 13), 76, 77(1-6, 9-13:2), 78, 79(1, 3-5, 7-8, 10-13, 15-16, 18-22, 24, 26-27), 80(1, 10), 81(1-2), 82-84, 85(2, 3, 4, 9)
CU-SB	1949, 56, 58-59, 61, 64, 67-69, 71(2-5, 7), 73(6), 75, 77-78, 79(7), 80(1), 81(11), 84
CaAC	1957-59, 61, 71(3)
CaACI	1959-60, 64(1-3, 6-7, 9-10)
CaACM	1949(5), 56(1-3), 57(1-7), 59(1-6), 61(1-9), 65(4), 71(3), 73(1, 5-7)
CaACU	1953(3-5), 54, 55(1-5, Addenda), 56-57, 59-60, 62, 64(1-3, 6), 66(B), 67(1), 68(9), 69, 70(8), 73(1, 5, 7-8), 74(3, 11-12), 75(2, 10), 76, 78, 84
CaAEU	1954(1-5), 56-59, 61-62, 63(3), 64(1-3, 6-7, 9-10), 65(1-7), 66(B), 67(1, 3, 5, 7-8), 69(B), 74, 75(2), 77(1-6, 9-13:2), 79(5, 20-21, 24), 81, 83
CaBVaU	1953(1, 5, 8-10), 54(1-5), 56(1-3), 57-61, 64(1-3), 65(1-5), 66(B), 69, 70(8), 71(2-3, 5-6, 8:1), 74, 75(2, 10), 76, 77(1-3), 79(3), 81
CaOHM	1953(1-3, 7-10), 54(1-5), 55(1-5), 57(1-2), 59, 61-62, 63(3-5), 64(2-3), 67(1, 5, 7), 69, 70(8), 71(2-4), 73(8)
CaOKQGS	1940, 50, 53(1-5, 7-10), 54(1-5), 55(1-5), 56-61, 66(B), 67(7), 69, 83
CaOLU	1948, 53, 55(1-5), 56-57, 59-62, 64(10), 70(8), 73(1, 8)
CaOOG	1940, 49(1-3, 5), 53, 55(1-5), 56-61, 63(4), 64(1-3), 65(4-5, 8), 66(C), 67(B), 68-69, 70(5), 71(1, 11), 72(8), 73(1, 8), 76(B), 78, 79(3, 7, 10, 12-13, 26), 80(1), 81-84
CaOONM	1953(1-9)
CaOTDM	1953(1-9), 77(1-6, 9-13:2), 78, 83
CaOTRM	1952-53, 54(1-5), 55(1-5), 56-59, 61-62, 66(A-B), 70(8)
CaOWtU	1954(1-5), 60, 63(3-4), 64(3, 10), 65(1-3, 5-7), 66(B), 67(1, 3, 5), 69, 70(1-2, 4, 6-9), 71(1-5, 8:1, 9, 11), 72, 73(1, 5, 8), 74, 75(2, 10, 11), 76, 78, 79(1, 3-4, 7-8, 10-13, 16, 18, 23, 26), 80-81, 82(11-12), 83(1-2), 85
CaQMME	1950
CaQQLaS	1954(1-5), 66(B)
CoDCh	1955-61, 64(1), 65(1-2, 4-5), 67(3), 70(8), 71(5, 9), 72(8), 73(5-6), 76(A), 81, 82(3), 83
CoDGS	1960, 70, 71, 73(1, 8, 11), 74(11), 77, 80(1), 81-82, 84, 85(2, 8)
CoDU	1960
CoDuF	1957, 60
CoFS	1956-57, 60-61, 65, 69
CoG	1940, 48-50, 53(1-9), 54(1-5), 55(1-5), 57(1-5), 58-60, 61(1-9), 62, 63(1-3, 5), 64(1-2, 6, 10), 65(4-5), 66(A-B), 67(1, 5, 7), 69, 70(1-2, 4, 6-9), 71(2, 6-7, 10), 72(8), 75(2, 10), 80-81, 85(3)
CoGrU	1960, 63(3), 76(A), 82([8], 3)
CoLiM	1980(1, 10), 81
CoU	1940, 49-50, 53(1, 6, 8), 54(1-5), 55(1-5), 56-62, 63(3-5), 64(1-3, 6-7, 10), 65, 66(A-B), 67(1-3, 5-8, A-B), 68-69, 70(1-2, 4-9), 71(2-3, 6-7, 8:1, 9, 11), 72, 73(1, 8), 74(3, 9, 11), 75(10), 76
CogrU	1960, 63(3), 76(A), 82([8], 3)
DI-GS	1940, 48-53, 54(1-5), 55-62, 63(1-5), 64-66, 67(1, 3-8, A-B), 68-70, 71(1-7, 8:1, 9, 11), 72, 73(1, 4-8, 10-11), 74, 75(2, 10), 76(B), 77(1-6, 9-13:2, 14), 78-85
DLC	1940, 49(2, 4), 50, 51(2), 54(1-5), 55(1-5), 58, 60-61, 62(1), 63(4), 65(1-7), 66, 67(1, 3, 5, 7, B), 69, 70(1-2, 4, 6-9), 71(1, 9), 73(1, 5, 8), 74(5, 9, 12), 75(2, 10), 76, 77(1-6, 9-13:2, 14-15), 83
F-GS	1985(9)
FBG	1985(9)
I-GS	1940, 48, 49(2, 5), 50, 51(1), 52, 53(3-5), 55(1-5), 56-60, 61(1-9), 62, 63(1, 4), 64(1-2, 6-7, 9-10), 65(3-6), 67(2), 70(6, 8), 71(3-5, 11), 72(11), 73(1, 8), 83
ICF	1950-51, 54(1-5), 55(1-5), 56-62, 64(1-3, 6-7), 65(3-5), 66, 67(1-2, 5-8), 69, 70(1-2, 4, 6-9), 71(8:1), 72(8), 74(9), 75(2, 10), 78

ICIU-S	1948, 55(1-5), 58(5), 59, 61, 65(4-5), 66(A), 69, 70(1-2, 4, 6-9), 72(8), 75(2, 10)
ICarbS	1950, 54(1-5), 55(1-5), 56-61, 62, 65(4, 5), 66[A], 69, 70(1-2, 4, 6-9), 71(1, 8(1)), 72(8), 73(18), 74(9), 75(2, 10), 76(A), 76[B], 79(15), 81
IEN	1948-50, 52, 53(1-5, 7-10), 54(1-5), 55(1-5), 56-60, 61(1-9), 62, 63(3-4), 64(1-3, 6, 9, 10), 65-66, 69-72, 73(1, 5, 6, 8, 10), 74(1-2, 6-7), 74(10)
INS	1983
IU	1931, 40, 48-50, 52-53, 55(1-5), 56-63, 64(1-3, 6-7, 10), 65(1-5), 66-74, 75(2, 10), 77(1-6, 9-13:2, 14-15), 78-79, 80, 82
IaAS	1940, 49, 56-60
IaCfT	1980(1)
IaU	1940(1-10), 48-50, 52, 54(1-5), 55(1-5), 56-60, 61(1-9), 62, 63(1-5), 64(1-3, 6-7, 9-10), 65-66, 67(1-3, 5-8, A-B), 69-70, 71(1-7, 9-11), 72(1, 7-8), 73(1, 6, 8), 74(3, 9), 75(2, 10, 11), 76, 77(1-6, 9-13, 15), 78, 79(1, 3-5, 7-8, 10-13, 15-16, 18, 20-24, 26-27), 80-84, 85(1-3, 5, 7)
IdBB	1954(1-5), 56, 59-60, 61(1-9), 66(A), 70(8)
IdPI	1964(3), 70(8), 74(9), 75(2, 10)
IdU	1954, 57-59, 61, 76(A), 77, 80-81, 83-84
InLP	1950, 53(1, 4-5, 7, 9-10), 54(1-5), 55(1-5), 56-57, 58(5), 59-62, 63(3-4), 64(3), 65-67, 70, 71(2-8:1, 9), 72(1, 6-8), 73(1, 5-8, 11), 74(2-3, 5-12), 75(2, 10, 11), 77(1-6, 9-13:2), 78, 79(3, 5, 7, 18, 20-24, 27), 80, 81(V.1-3, 14, 15), 82-84, 85(2-3, 5, 8)
InRE	1956-59, 61, 65(1-7), 69-70, 71(1, 7), 72(1-7), 77, 82(4-5, 8, 11-14), 83-84, 85(2-3, 5, 8, 12)
InTI	1983
InU	1948-52, 53(1-5, 7-9), 54(1-5), 55(1-5), 56-57, 58(5), 59-62, 63(4-5), 64(1-3, 6-7, 10), 65, 66(A-B), 67(1, 3, 5, 7), 69, 70(1-2, 5-9), 71(2, 10), 72(8), 74(2, 9), 75(9, 10, 11), 79(7), 82-84
KLG	1955, 65(1-7), 73(8), 81(15)
KyU-Ge	1949-51, 54(1-5), 55(1-5), 56-62, 63(3-5), 64(1-3, 6-7, 9-10), 65(1, 4-5), 66(A-B), 67(A, 1, 3, 5, 7-8), 69, 70(1-9), 71(1-7, 9-11), 72, 73(1, 5-8, 10-11), 74(1, 3, 5-7, 11-12), 75(2, 11), 76(A), 77(1-6, 9-13:2, 14), 78, 79(1, 3-5, 7-8, 10-13, 16, 18, 20-22, 24, 26-27), 80(1-3, 5-10, 12-16, 18-23), 81(1-4, 6-19), 82-84, 85(1-5, 7-9, 12)
LNU	1940, 50(3), 51(1), 53(1-2, 7-10), 54(1-5), 55(1-5), 56(1), 57, 58(5), 59, 61, 64(1), 65(1-7), 66(7), 67(1, 3, 5), 69, 70(7-8), 72(3, 6), 73(1, 5, 8), 74(2), 77(1-6, 9-13:2), 81(15), 82
LU	1949, 54(1-5), 55(1-5), 56-63, 66(A-B), 69, 70(8)
MH-GS	1949, 52, 53(1, 3, 7), 55(1-5), 56(1), 57-62, 63(3), 64(1-2, 10), 66(B), 68(9), 69, 71(6), 72(8), 80
MNS	1950, 53(1-5, 7-8, 10), 55(1-5), 57-61, 63(3), 64(1-3, 6-7), 65(4-5), 70(1-2, 4, 6-9), 71(1), 73(1, 8), 77(1-6, 9-13:2)
MU	1981
MWesB	1958(5), 64(4-5), 62
MdBJ	1950, 53, 55(1-5), 56-61, 67(3), 71
MdU	1981
MiDW-S	1950, 52-53, 54(1-5), 55(1-5), 56-61, 63, 65(1-7), 66(A-B), 67, 69, 70(8), 72(8), 73(1, 8)
MiHM	1950, 52, 53(1-5, 7-10), 54(1-5), 55(1-5), 56-59, 61(1-9), 62, 63(3), 65(1-3, 6-8), 66, 67(8), 69, 70(1-2, 4-9), 71(8:1), 72, 73(1, 7-8), 78
MiKW	1951, 56-57, 59, 61, 67(7), 70(8), 71(3-5), 72(1-7)
MiMU	1969
MiU	1949, 54(1-5), 55(1-5), 56-57, 59, 61, 63, 69, 72(2, 4-8)
MnDuU	1948, 53(2), 56(1), 58, 66(A), 70(1-2, 4-7, 9), 72(7)
MnU	1940, 48, 49(2, 4-5), 50, 51(1-2), 52(1-4), 53(1-5, 7-10), 54(1-5), 55(1-5), 56-60, 61(1-2), 62, 63(1-5), 64(1-3, 6-7, 9-10), 65-67, 70, 71(1, 3-6, 8:1, 10-11), 72, 73(1, 6-8), 74(9), 75(2), 76(A), 77(1-6, 9-13:2, 14-15), 78, 79(1, 3-5, 7-8, 10-13, 16, 18, 20-24, 26-27), 80(1), 81-82
MoRM	1949, 55(1-5), 56-62, 71(2)
MoSW	1940, 48-51, 53(7), 54(1-5), 55(1-5), 56-59, 61-62, 63(4), 65(2-7), 66(A-B), 67(1-3, 5-8, A-B), 70(8), 71(1-2, 6-7, 8:1, 9-10), 72(8), 73(1, 5-6, 8), 80(1)
MoU	1948, 49(2, 4-5), 50, 53(1-5, 7-10), 54(1-5), 55(1-5), 56-62, 63(3-4), 64(3, 10), 65(1-3), 66(A), 67(1, 3, 5), 69, 70(1-2, 4, 6-9), 71(8:1), 72(8), 76(B)
MtBC	1958-61, 65(1-2), 68
MtBuM	1949(2, 5), 54(1-5), 55(1-5), 57-60, 61(1-9), 65(4-5), 66(A-B), 70(1-2, 4, 6-9), 71(8:1), 72(8), 73(1, 8), 75(2)
MtU	1949, 58-59, 61
NBiSU	1950-51, 53(10), 54(1-5), 55(1-5), 56(1-3), 57, 59-60, 61(2-9), 62, 63(3, 4), 64(1, 6, 9-10), 65(1-5), 66(A-B), 67(1, 3, 5-6, A-B), 69, 70(8), 71(1-5), 73(1, 8)
NCSB	1982(11)
NIC	1980, 82
NNC	1940, 48, 50-53, 55(1-5), 56-62, 63(5), 64(1-3), 65(1-7), 67(7), 73(1, 8), 81
NOneoU	1958, 60, 63(3), 69, 71(4)
NRU	1949, 53(1-7), 56, 61-62, 69, 71(3, 5), 75(11), 83
NSbSU	1948, 50, 52, 55-57, 59-61, 64, 66(B), 69, 70(8), 74, 75(2), 76(A), 78-85
NSyU	1940, 50(1-3) 53(1), 54(1-5), 56-57, 58(5), 59-60, 61, 62, 63(5), 64(1), 65(1-7), 66(A-B), 67(A), 69, 70(3, 8, 10), 71(10), 73(1, 8), 74(10) 75(11), 77(1-6, 9-13:2), 80, 83
NbU	1949-50, 53, 54(1-5), 55(1-5), 56-60, 61(1-9), 62, 65(1-7), 66(A-B), 70(8)
NcU	1949(5), 50, 51(1), 54(1-5), 55(1-5), 56-62, 63(3, 5), 64(1-3, 6-7, 10), 65(2-7), 66(A-B), 67(1-3, 5, 7-8), 69-70, 71(1-6, 8:1, 9), 72(1, 5-8), 73(1, 5-6, 8), 74(2-3, 5-6), 75(2, 11), 76(B), 77(1-6, 9-13:2), 80(1), 81(15), 85(2)
NdU	1954(1-5), 55(1-5), 56-59, 61-62, 63(3-5), 64(1-3, 6-7), 65(1-7), 66(A-B), 67(1, 3, 7-8, A-B), 69, 70(1-2, 4, 6-9), 71(1-7, 8:1, 9-10), 72, 73(1, 8), 74(1-3, 10-11), 75(2), 77(1-6, 9-13:2), 78, 79(1, 3-5, 7-8, 11, 16, 20-22, 24, 27), 80(1), 83-84
NhD-K	1948-51, 53, 54(1-5), 55(1-5), 56-59, 61(1), 62, 64(3, 6), 65(4-5), 66(A-B), 67(7), 69, 70(2, 4, 6-9), 71(1), 72(8), 73(1, 8), 74(2), 75(2, 4, 11), 76(B), 77(1-6, 9-13:2), 79(1, 3-5, 7-8, 11, 16, 20-22, 24, 27), 80-82
NjP	1940, 49-53, 54(1-5), 55(1-5), 56-62, 63(4-5), 64(1-3, 6, 10), 65, 66(A-B), 67(1, 3, 5, 7), 68(7), 69, 70(1-2, 4, 6-9), 71(1-3, 5-6), 72(8), 73(5-6, 14-15), 75(2)
NmPE	1954(1-5), 57-61, 66(B), 69, 73(5), 76
NmU	1949(5)
NvLN	1966(A), 69
NvU	1949, 54(1-5), 55(1-5), 56-61, 65(4-5), 66(B), 67(5, 7), 70(5, 7-8), 71(8:1), 73(5, 6), 75(2, 11), 79(18, 23), 80, 82-84
OCU-Geo	1940, 48, 49(3), 50, 52-53, 54(1-5), 55(1-5), 56-60, 61(1-9), 62, 64(1-3, 6-7, 9-10), 65(1-7), 66(B, C), 67(2-3, 7-8), 69, 70(1-4, 7-9), 71(1-5, 8-9), 72(3, 5-6, 8), 73(1, 8), 75(2, 10, 11), 76(B), 77(1-6, 9-13:2, 14-15), 78, 79(1, 3-5, 7-8, 10-13, 15-16, 18-24, 26-27), 80, 81, 82(5), 83-84
OCl	1982
ODaWU	1980, 81, 82(8)
OU	1948-52, 53(1-5, 10), 54(1-5), 55(1-5), 56-60, 61(1-9), 62, 63(4-5), 64(1-3, 6-7, 9-10), 65(1-7), 66(A-B), 67(3, 7), 68(1, 4, 7, 9), 69, 70(1-2, 4, 6-9), 71(1, 8-10), 72(8), 73(1:8, 5), 74(3), 75(2, 11), 76, 79(1, 3-5, 8, 10-13, 15-16, 18, 20-24, 26-27), 80-82
OkT	1949, 54(1-5), 55(1-5), 57-59, 61-62, 63(3), 65(4-5), 66(B), 67(3), 70(8), 81(15)
OkTA	1981(15), 82(3, 5, 8, 12), 85(1-2)
OkU	1931, 48-51, 53, 54(1-5), 56-59, 61-62, 64(1-4, 6-7, 9-10), 65(1-7), 66(A-B), 67(1-3, 5, 7), 69, 70(1-2, 4, 6-9), 71(2-5, 9), 73(1, 5-6, 8), 74(2), 75(11), 76, 77(15), 78, 80(1), 81, 82-85
OrCS	1949(4)
OrU-S	1948, 49(2, 4), 54(1-5), 55(1-5), 56-60, 61(1-9), 65(4-5), 66(A-B), 72(8), 73(1, 8), 75(2)
PBL	1955(1-5), 57-59, 66(A), 69, 70(1, 2, 5-7, 9), 75(11)
PBU	1980
PBm	1948, 51(1), 55(1-5), 56-59, 61, 70(8), 71(1-7, 10), 77(1-6, 9-13:2, 14-15), 81
PHarER-T	1982, 83
PSt	1940, 48, 52, 53(6), 54(1-5), 55(1-5), 56-62, 63(3-5), 64(1-3, 6-7, 9-10), 65-66, 67(1-7), 69, 70(1-7), 71(1-8:1, 9-11), 72, 73(1, 6-8, 11), 74(1-3, 6, 12), 75(2), 76(B), 78, 79(1, 3-5, 7-8, 10-13, 16, 19-22, 24, 26-27), 82, 85(1-3, 5, 7)
RPB	1951-53, 54(1-5), 55(1-5), 56-59, 61(1-9), 63(2-5), 64(4, 10), 65(1-7), 66, 67(1, 3, 5-8, A-B), 69
SdRM	1950, 54, 70, 71([1]), 72(8), 75(2, 10)
SdU	1958, 60, 61
TMM-E	1949(1), 59. 61, 67(3, 5), 69, 70(8), 71(2, 8), 73(5)
TU	1949(2-4), 58-59, 61, 65(1-7)
TxBeaL	1949, 55(1-5), 56-59, 61-62, 63(5), 64(1-3, 6, 10), 65(6-7), 66(B-C), 67(1, 5), 69(B), 70(8), 71(3, 5, 7, 8:1, 9-10), 72, 73(1, 6-8, 10-11), 74(1-3, 5, 7-8, 12), 77(1-6, 9-13:2), 78, 79(1, 3-5, 7-8, 10-13, 15-16, 19-22, 24, 26-27)
TxCaW	1973(1, 8)
TxDaAR-T	1940, 49, 54(1-5), 55(1-5), 56-62, 63(3), 64(1-3, 6-7, 10), 65(1-7), 66(A-B), 67(1, 3, 5, 7), 69-72, 73(1, 6-8, 10-11), 74(2-3, 6, 10)
TxDaDM	1940, 48, 49(4-5), 53(1-5, 9), 54(1-5), 55(1-5), 56-62, 64(1), 65(2, 4, 8), 66(A-B), 67(1-2, 6-8, B), 69(A), 70(1, 4, 6-7, 9), 71(3, 5, 8:1, 9), 72(8), 73(1, 8), 75(2), 79(10, 12-13, 26)
TxDaGI	1940(1-11), 48, 49(1-5), 50(1-3), 51-52, 53(4-5), 55(1-5), 57-60, 61(1-9), 63(3), 64(6), 67(1, 5-6), 73(8)

TxDaM-SE	1940, 49(4-5), 50, 55(1-5), 56-62, 63(3-4), 64, 65(4-5), 66(B), 67(7), 80(1), 79(10, 12, 13, 26), 82(3), 83
TxDaSM	1949(2, 4-5), 54(1-5), 55(1-5), 56-62, 65(3), 66(B), 67(1-2, 5, 7), 69, 71(1), 72(8), 73(1, 5-6, 8), 74(2), 75(2), 77(1-6, 9-13:2), 79(18, 22)
TxHMP	1980(1)
TxHSD	1948-49, 54(1-5), 55(1-5), 58-60, 62, 63(3-4), 64(1-3, 6-7, 9-10), 65(1-2, 4-5), 66(A-B), 67(3), 69-70, 71(1), 73(1, 5-6, 8), 74(11), 77(1-6, 9-13:2), 79(1, 4, 8, 10-13, 16, 18, 22-23, 26-27)
TxHU	1940, 49(1), 50, 54(1-5), 55(1-5), 56-62, 63(3-5), 65(1-5), 66(A-B), 67(7), 69, 70(8), 73(1, 8)
TxLT	1949(2-5), 50, 53(1-5, 7-10), 57-61, 62(1), 65(4-5), 69, 70(8), 72(8), 73(8)
TxMM	1940, 49, 53(1-5, 7-10), 55(1-5), 56-60, 62, 64(1-3, 6-7, 9), 65(1-7)
TxU	1940, 48-50, 52-53, 54(1-5), 55(1-5), 56-71, 72(4-5, 8), 73(1, 5-8, 10-11), 74, 75(2), 77(1-6, 9-13:2, 14-15), 78, 79(1, 4, 7-8, 10-13, 16, 26), 80(1), 81-84
TxU-Da	1949(1-2, 4-5), 54(1-5), 55(1-5), 57-58, 60, 64(1-3), 65(4), 67(B)
TxWB	1980(1), 81-82, 85
UU	1949(4), 53, 54(1-5), 57-60, 65(4-5), 66(A-B), 67(3), 69(B), 70(1-2, 4, 6-9), 72(8), 73(1), 75(2, 4, 11), 78
ViBIbV	1949(2, 5), 50, 54(1-5), 55(1-5), 57-59, 61(1-9), 62, 64(1), 65(4-5), 66(A-B), 67(3), 69, 70(8), 71(2-5, 8:1, 10), 72(7-8), 73(1, 8), 74(2, 5-6, 12), 75(2), 79(1, 3-5, 7-8, 11, 16, 19-22, 24, 27)
WGrU	1970(1-2, 4, 6-7)
WKenU	1980(1)
WU	1940, 49(1-2, 4-5), 50, 53(1-5, 7-9), 54(1-5), 55(1-5), 56-60, 61(1-9), 62, 63(2, 4-5), 64(2-3, 10), 65(1-7), 66, 67(1, 3, 5-8, A-B), 68(1-6), 69, 70(1-2, 4-9), 71(2), 72, 73(1, 5-8, 11), 74(2-3, 11-12), 75(2, 11, 13), 76(1, 2), 77(1-6, 9-13:2), 78, 79(1, 3-5, 7-8, 10-13, 15-16, 19-22, 24-27), 80(1-10), 81-84, 85(1, 2, 5, 7, 10)
WaPS	1971
WaU	1950, 73(1, 8), 77, 79([3], 23), 84(4)
WyU	1949-50, 56-61, 80(1), 83

(198) GEOLOGICAL SOCIETY OF AMERICA. COAL DIVISION. FIELD TRIP.

1963 See Geological Society of America. 197.
1967 See Geological Society of America. 197.([5])
1969 See Geological Society of America. 197.([A])
1970 See Geological Society of America. 197.(8)
1972 See Geological Society of America. 197.(8)
1975 See Geological Society of America. 197.([2])
1976 See Geological Society of America. 197.(B)
1978 See Geological Society of America. 197.(7)
1979 See Geological Society of America. 197.(15)
1980 See Geological Society of America. 197. (7)
1981 See Geological Society of America. 197.(14)
1983 See Geological Society of America. 197. (9)
1984 See Geological Society of America. 197. (3)
1985 See Geological Society of America. 197. (7)

(199) GEOLOGICAL SOCIETY OF AMERICA. CORDILLERAN SECTION. GUIDEBOOK FOR THE ANNUAL MEETING.

1952 [48th] Tucson, Arizona. Guidebook for field trip excursions in southern Arizona. (65)
  Trip 1. Ground water problems of Queen Creek area.
  Trip 2. Paleozoic and Cretaceous stratigraphy of the Tucson Mountains.
  Trip 3. Santa Catalina Mountains metamorphic area.
  Trip 4. Economic geology; Ajo porphyry copper.
  Trip 5. Stratigraphy, structure, and economic geology typical of southern Arizona.
1954 See Geological Society of America. 197.
1955 [51st] Berkeley, California.
  Trip 1. Petrology: Sonoma-Petaluma area.
  Trip 2. Stratigraphy: Oakland-Mt. Diablo area.
1956 [52nd] Guidebook for the December 16, 1954, Dixie Valley-Fairview Peak earthquake area field trip.
1958 [54th] Eugene, Oregon. Columbia River Gorge: Portland to The Dalles; Eugene to Coos Bay field trip; Willamette Valley field trip.
1959 55th Tucson, Arizona. Southern Arizona guidebook II. (65)
  1. Structure and ore deposits of the east Sierrita area, Arizona.
  2. Stratigraphy of the Waterman and Silver Bell Mountains.
  3. Geology of the Santa Catalina Mountains.
  4. Chaotic breccias in the Tucson Mountains.
  5. General geology of southeastern Arizona.
  6. Volcanic craters of the Pinacate Mountains, Sonora, Mexico.
1960 [56th] Vancouver, British Columbia. Guidebook for geological field trips in southwestern British Columbia. (196)
  [1] A field trip to illustrate geology of Coast Mountains, North Vancouver, British Columbia.
  [2] Engineering geology, North Vancouver.
  [3] Field trip; Vancouver to Kamloops and return.
1961 57th San Diego, California. San Diego County-1961 (Cover title). (679)
  1. Geology and geomorphology of eastern San Diego County.
  2. Southwestern San Diego County.
  3. Notes on the Pala and Mesa Grande pegmatite districts, San Diego County.
  4. Cruise on motor vessel "Horizon."
  5. Geology of northwestern Baja California, Mexico.
1963 59th Berkeley, California. Guidebook to field trips in Alameda and Contra Costa counties, California.
  1. The Berkeley Hills and Livermore Valley, Alameda and Contra Costa counties, California.
  2. San Pablo Syncline, Martinez and Mount Diablo.
1964 60th Seattle, Washington.
  No. 1. Tertiary stratigraphy of the Port Angeles, Lake Crescent, Olympic Peninsula area.
  No. 2. A geologic trip guide along the northern Olympic Peninsula highways.
1965 61st Fresno, California.
  No. 1. Mercury, jadeite, and asbestos regions near Panoche Pass, California.
  No. 2. The alluvial fans of western Fresno County, California.
  No. 3. Geology of the Sierran foothills in eastern Fresno and Madera counties, California.
  [4] Engineering geology of the San Luis project.
1966 62nd Reno, Nevada. Guidebook for field trip excursions in northern Nevada.
  1. Dixie Valley-Fairview Peak earthquake areas.
  2. Mining areas in western Nevada.
  3. Cenozoic geology of the lower Truckee Canyon.
  4. Tertiary and Quaternary geology along the Truckee River with emphasis on the correlation of Sierra Nevada glaciation with fluctuation of Lake Lahontan.
  5. Geology of the western Pershing County.
  6-6a. Geology of the Paleozoic Thrust Complex in north-central Nevada. Geology of the Antler Peak Quadrangle.
1967 63rd Santa Barbara, California.
  No. 1. The South Mountain area, Ventura County, California.
  No. 2. The central Santa Ynez Mountains, Santa Barbara County, California.
  No. 3. The San Luis Obispo-Nipomo area, San Luis Obispo County, California.
1968 64th Tucson, Arizona. Southern Arizona guidebook III. (65)
  1. Volcanic geology, southwestern New Mexico and southeastern Arizona.
  2. Mesozoic stratigraphy and Laramide tectonics of part of the Santa Rita and Empire Mountains, southeast of Tucson, Arizona.
  3. Engineering geology, Tucson and Benson areas.
  4. Structure and ore deposits of the Pima mining district.
  5. Stratigraphic and volcanic geology, Tucson Mountains.
  6. Quaternary geology of the San Pedro River valley.
1969 65th Eugene, Oregon.
  No. 1. Road log, northern Klamath Mountains field trip.
  No. 2. Andesite conference guidebook. Volcaniclastic rocks of central Oregon. (Eugene and Bend, Oregon) 1968. (517)
  No. 3. Geology of the Newport area, Oregon. (518)
1970 [66th] Hayward, California.
  Field trip no. 1. Central California coast ranges.
  Field trip no. 4. Geology of the Diablo Canyon nuclear power plant site, San Luis Obispo County, California.
  Field trip no. 5. Hayward-Hollister field trip. Active slippage of the Calaveras, Hayward and San Andreas faults.
1971 [67th] Riverside, California. Geological excursions in Southern California. (678(No.1))

Field trip no. 1. Clark Mountain thrust complex in the Cordillera of southeastern California.
Field trip no. 2. Structural geology and tectonics of the Salton Trough, Southern California.
Field trip no. 3. Vertebrate paleontology of the northern Mojave Desert, Southern California.
Field trip no. 4. Geology of the northern peninsular ranges, Southern California.
Field trip no. 5. Contact metamorphic minerals at Crestmore Quarry, Riverside, California.
Field trip no. 6. Geological engineering problems posed by the San Jacinto Fault.
Field trip no. 7. The San Andreas Fault between San Bernardino and Palmdale, California.
Field trip no. 8. Stratigraphy and structure of the area between Oceanside and San Diego, California.
Field trip no. 9. Non-marine turbidites and the San Andreas Fault, San Bernardino Mountains, California.
1972 68th Honolulu, Hawaii. Road guide to points of geologic interest in the Hawaiian Islands.
1973 [69th] Portland, Oregon. Geologic field trips in northern Oregon and southern Washington. (517, 524)
Trip 1. Cretaceous and Cenozoic stratigraphy of north-central Oregon.
Trip 2. Volcanoes and intrusive rocks of the central part of the Oregon Coast Range.
Trip 3. Cenozoic stratigraphy of northwestern Oregon and adjacent southwestern Washington.
Trip 4. Columbia River Gorge; basalt stratigraphy, ancient lava dams and landslide dams.
Trip 5. Urban environmental geology and planning, Portland, Oregon.
Trip 6. Stratigraphy and structure of Yakima Basalt in Pasco Basin, Washington.
Trip 7. Geological field trip guide, Mt. St. Helen's lava tubes.
1974 70th Las Vegas, Nevada.
Field trip no. 1. Guidebook: Death Valley region, California and Nevada.
[Field trip no. 2.] Black Mountain volcanic center. (452)
Field trip no. 3. Guidebook to the geology of four Tertiary volcanic centers in central Nevada. A road log and trip guide to the collapse calderas at Northumberland Canyon in the Toquima Range and at Black Mountain in Pahute Mesa and to the precious-metal mining districts of Tonopah and Goldfield:
Field trip no. 4. Interbasin ground-water flow in southern Nevada. (452)
Austin-Northumberland caldera-Carver station;
Carver station-Tonopah District;
Goldfield District;
1975 71st Los Angeles, California.
No. 1. San Andreas Fault in Southern California, a guide to San Andreas Fault from Mexico to Carrizo Plain. (115)
No. 2. Peninsular ranges.
No. 3. Eocene sedimentation and paleocurrents San Nicolas Island, California.
No. 4:1. Guidebook to the Quaternary geology along the western flank of the Truckee Meadows, Washoe County, Nevada. (452(No. 22))
No. 4:2. A field guide to Cenozoic deformation along the Sierra Nevada Province and Basin and Range boundary. (108)
No. 5. Preliminary report and geologic guide to the Jurassic ophiolite near Point Sal, Southern California coast.
1976 72nd Pullman, Washington.
Field guide no. 1. Guide to field trip between Pasco and Pullman, Washington; emphasizing stratigraphy, vent areas, and intracanyon flows of Yakima basalt.
Field guide no. 2. Channeled scablands of southeastern Washington; a roadlog via Spokane, Coulee City, Vantage, Washtucna, Lewiston [and] Pullman.
Field guide no. 3. Hydrology and engineering geology of the Columbia Basin.
Field guide no. 4. The Idaho batholith and related subduction complex.
Field guide no. 5. Geologic guide to Hells Canyon, Snake River.
1977 73rd Sacramento, California.
[1] Guidebook: Paleozoic-Mesozoic rocks of the northern Sierra Nevada.
[2] Field trip guide to the Geysers-Clear Lake area.
[3] Geological road log for the California Mother Lode country along highway 49 between Mariposa and Grass Valley.
[4] Tectonics and stratigraphy of the Calaveras complex, central Sierra Nevada foothills.
[5] Guidebook; San Andreas Fault in Marin County with section across the Coast Ranges from Sacramento to San Francisco.
[6] Guidebook to the geology of the Klamath Mountains, northern California.
[7] Fieldtrip guide to the Kings-Kaweah suture, southwestern Sierra Nevada foothills, California.
[8] Field guide: Great Valley sequence, Sacramento Valley.
[9] Plate tectonic history of the Yolla Bolly junction, Northern California.
[10] Field trip around the Sacramento-San Joaquin delta.
1978 74th Tempe, Arizona. Guidebook to the geology of central Arizona. (673, 524)
1979 [75th] San Jose, California.
[1] Geology of the central Diablo Range between Hollister and New Idria, California.
[2] Recent deformation along the Hayward, Calaveras, and other fault zones, eastern San Francisco Bay region, California.
[3] Engineering and environmental geology of the Santa Clara Valley, Santa Clara County, California - a field guide.
[5] The Calaveras fault zone field trip.
[6] Tertiary paleontology and stratigraphy of the central Santa Cruz Mountains, California coast ranges.
[7] Geology of the Santa Cruz Mountains, California.
[8] Coastal tectonics and coastal geologic hazards in Santa Cruz and San Mateo Counties, California.
[9] The evolution of a Late Paleocene submarine canyon and fan system: the Meganos Formation, southern Sacramento Basin, California.
[10] Modern and ancient coastal sedimentary facies, Monterey Bay, California.
[11] Geologic guide to the Kings Canyon Highway, central Sierra Nevada, California.
[12] Field trip to observe natural and resource management-related erosion in Franciscan terrane of northwestern California.
1980 [76th] Corvallis, Oregon.
Geologic field trips in western Oregon and southwestern Washington. (517)
[4] Geologic field trip guide through the north-central Klamath Mountains. (518(V.42 no.2))
1981 77th Sonora, Mexico.
Geology of northwestern Mexico and southern Arizona. (Geologia del noroeste de Mexico y del sur de Arizona) (670)
Field trip no. 12. Mesozoic through early sedimentational and tectonic patterns of Northeast Sonora and Southeast Arizona. (674)
1982 78th Anaheim, California. (524)
— [1] See American Association of Petroleum Geologists. Pacific Section. 28.(3)
Mesozoic-Cenozoic tectonic evolution of the Colorado River region; Anderson-Hamilton Volume.
Field trip nos. 2, 7, 13. Geologic excursions in the California desert.
Field trip no. 2. Late Cenozoic tectonic and magmatic evolution of the central Mojave Desert, California.
Field trip no. 7. Comparison of Mesozoic tectonics with mid-Tertiary detachment faulting in the Colorado River area, California, Arizona, and Nevada.
Field trip no. 13. Geology and mineral deposits of the central Mojave Desert.
Field trip nos. 3, 4, 14. Neotectonics in Southern California.
Field trip no. 3. Neotectonics of the Ventura Basin.
Field trip no. 4. Geologic hazards along the San Andreas fault system, San Bernardino-Helmet-Chino, California.
Field trip no. 14. Surficial structure and geomorphology of the San Andreas Fault, western portion of the Big Bend.
Field trip nos. 5, 6, 11. Geologic excursions in the Traverse Ranges, Southern California.
Field trip no. 5. Geology and structural setting of the San Gabriel anorthosite-syenite body and adjacent rocks of the western San Gabriel Mountains, Los Angeles County, California.
Field trip no. 6. Late Cenozoic stratigraphy and structure of the San Bernardino Mountains.

(199) Geological Society of America. Cordilleran Section.

Field trip no. 11. Crystalline basement terranes in the southern eastern Transverse Ranges, California.

Field trip no. 9. Geology of selected areas in the San Bernardino Mountains, western Mojave Desert, and southern Great Basin, California.

Field trip no. 10. Landslides and landslide abatement, Palos Verdes Peninsula, Southern California.

Field trip no. 12. Late Quaternary pedogenesis and alluvial chronologies of the Los Angeles and San Gabriel mountains areas, Southern California and Holocene faulting and alluvial stratigraphy within the Cucamonga fault zone; a preliminary view.

Field trip no. 8. See American Association of Petroleum Geologists. Pacific Section. 28.(2)

1983 [79th] Salt Lake City, Utah. (204, 524, 733)

Part 1. Field trips: 1, 2, 6. Geological excursions in the overthrust belt and metamorphic core complexes of the intermountain region.

Field trip 1. Geology of the Albion-Raft River-Grouse Creek Mountains area.

Field trip 2. Mesozoic and early Tertiary structure and sedimentology of the central Wasatch Mountains and Uinta Basin.

Field trip 6. Style of mid-Tertiary extension in east-central Nevada.

Part 2. Field trips: 3, 7, 8. Geologic excursions in stratigraphy and tectonics; from southeastern Idaho to the southern Inyo Mountains, California, via Canyonlands and Arches national parks, Utah.

Field trip 3. Upper Proterozoic diamictites and volcanic rocks of the Pocatello Formation and correlative units, southeastern Idaho and northern Utah.

Field trip 7. Evolution of Early Mesozoic tectonostratigraphic environments - southwestern Colorado Plateau to southern Inyo Mountains.

Field trip 8. The geology in and near Canyonlands and Arches national parks, Utah.

Part 3. Field trips: 4, 9. Geologic excursions in volcanology; eastern Snake River plain (Idaho) and southwestern Utah.

Field trip 4. Holocene basaltic volcanism along the Great Rift, central and eastern Snake River plain, Idaho.

Field trip 9. Mid-Tertiary history of the central Pioche-Marysvale igneous belt, southwestern Utah.

Part 4. Field trips: 5, 10, 11, 12. Geologic excursions in neotectonics and engineering geology in Utah.

Field trip 5. Paleoseismicity along the Wasatch Front and adjacent areas, central Utah.

Field trip 10. Lake Bonneville stratigraphy, geomorphology, and isostatic deformation in west-central Utah.

Field trip 11. Geologic aspects of upper Stillwater Damsite, Bonneville Unit, central Utah Project, Utah.

Field trip 12. Engineering geologic problems along Utah's urban corridor.

1984 80th Anchorage, Alaska. (524, 6)

[1] Guide to the glacial geology of Glacier Bay, southeastern Alaska.

[2] Guide to the bedrock geology of a traverse of the Chugach Mountains from Anchorage to Cape Resurrection. (6)

[3] Guide to surficial geology and glacial stratigraphy in the Upper Cook Inlet Basin.

[4] Guide to the engineering geology of the Anchorage area.

[5] Guide to Late Pleistocene and Holocene deposits of Turnagain Arm.

[6] Guide to the Willow Creek gold mining district, with stops at the Castle Mountain Fault and the Alaska Tsunami warning center.

[8] Guide to the bedrock and glacial geology of the Glenn Highway, Anchorage to the Matanuska Glacier and the Matanuska coal mining district.

[9] Geology of the Yukon Tanana Upland Fairbanks to Livergood area. Fairbanks gold mining district and subartic periglacial features.

1985 81st Vancouver, British Columbia.

— May Field guides to geology and mineral deposits in the southern Canadian Cordillera. (524, 191)

Trip 1. Westmin Resources' massive sulphide deposits, Vancouver Island.

Trip 3. Cordilleran cross-section Calgary to Vancouver.

Trip 4. Slope hazards in the southern Coast Mountains of British Columbia.

Trip 5. Volcanology and structure of early Tertiary outliers in south central British Columbia.

Trip 6. Stratigraphy and sedimentology of the Eocene Chuckanut Formation on Bellingham Bay, Washington.

Trip 7. Mesozoic melange of the Pacific Rim Complex, western Vancouver Island.

Trip 8. LITHOPROBE profile across southern Vancouver Island: geology and tectonics.

Trip 9. Precious metal mineralization in southwestern British Columbia.

Trip 10. Gulf Islands boat cruise; geology and scenery.

Trip 11. Stratabound base metal deposits in southeastern British Columbia.

Trip 12. Metamorphic complexes and extensional tectonics, southern Shuswap Complex, southeastern British Columbia.

Trip 13. Tethyan exotic terranes in southern British Columbia.

Trip 14. Thrust and strike-slip faults bounding tectonostratigraphic terranes, central British Columbia.

Trip 15. Late Quaternary geology of the Fraser Lowland, southwestern British Columbia.

Trip 16. Geology and mineral deposits of the Adams Plateau-Clearwater region.

AzFU	1952, 59, 68
AzTeS	1952, 55, 59-60, 63-68, 69(3), 71, 74(2), 75(1, 3, 4:1)
AzU	1952, 59, 68, 73, 74(3-4), 75(4:1)
CLU-G/G	1960-61, 67-68, 69(2-3), 70(1, 4-5), 71, 73, 74(1, 3-4), 75(1, 3), 76, 77(3-5, 7, 9), 78, 79(2, 5-11), 82, 85
CLhC	1952, 59, 68, 71, 74(1), 75(1, 3), 79(1-3, 5-11), 81, 82(2-7, 9, 11-13), 83 (pt.1)
CMenUG	1952, 55, 59-61, 63-64, 65(2-3), 67, 70(1,4), 71, 73-74, 75(1-3, 5), 76-79, 81(1), 82(2-5, 9-10, 12, 14), 84(1-6, 8), 85
CPT	1955, 61, 67(2), 71, 73-74, 75(1, 3-4), 77-78, 79(1-11), 80, 81, 82(3-6, 9-11, 14), 84(1-3,6), 85
CSdS	1952, 59, 61, 68, 71, 73, 75(1, 3, 4:2), 79(1-3, 5-11), 82(2, 7, 13)
CSfCSM	1952, 55, 58-61, 63, 65(2), 68, 71, 77, 79(1-3, 5-11), 82(3-6, 9-12, 14)
CSt-ES	1952, 55, 59-61, 65, 66(1), 67-68, 69(2-3), 70(1, 4-5), 71, 73, 74(1, 3-4), 75(1, 3-5), 76, 77(1-9), 78, 79(1, 5-11), 80-85
CU-A	1961, 68, 71, 73, 74(1), 75(3), 76(5), 77(1-9), 79(1-3, 5-11), 80-81, 82(2, 5-7, 9, 11-13), 83
CU-EART	1955-56, 58-61, 63, 65-68, 69(2-3), 71, 73, 74(1, 3-4), 75(1, 3-5), 77-78, 79(1-3, 5-12), 80-81, 82(2, 7, 9, 13), 83-85
CU-SB	1952, 63, 67, 74-78, 79(2-3, 5, 7-11), 82(2-4, 9, 12)
CU-SC	1982(3-6, 9-12, 14)
CaACAM	1960
CaACM	1973
CaACU	1960-61, 68, 71, 74(1), 75(1, 3, 4:2), 77-78, 79(1-3, 5-11), 84
CaAEU	1969(1), 74(3-4), 80-81
CaBVaU	1952, 59-60, 74(1-3), 75(1-4), 78, 84(1-6, 8)
CaOHM	1968
CaOKQGS	1964, 68
CaOLU	1959, 68
CaOOG	1959-60, 68, 70(1, 5), 73, 74(1), 77, 81, 84
CaOONM	1973
CaOTDM	1973
CaOWtU	1970(5), 71, 73, 74(1), 75(1, 4), 78, 80, 82(3-6, 9-12, 14), 83-84
CoDBM	1984
CoDCh	1952, 59, 78
CoDGS	1952, 55, 59, 61, 63, 66, 68-71, 73-79, 81-82, 84-85
CoDuF	1968
CoG	1952, 59-60, 68, 70(4), 73, 74(1, 3-4), 75(4:1), 79(1-3, 5-10), 82
CoLiM	1981, 82(2, 7, 10, 13)
CoU	1952, 59, 68, 69(3), 73, 74(1, 3-4), 75(4:1)
DI-GS	1952, 55, 58-61, 63, 65-69, 70(1, 4-5), 71-73, 74(1, 3-4), 75-78, 79(1-3, 5-12), 80-85
DLC	1959, 68, 73, 75(1), 83
I-GS	1955, 60, 73, 74(3-4)
ICF	1960, 68, 75(1)
ICarbS	1959, 74(3-4), 75(4), 85
IEN	1959, 68
IU	1952, 55, 59-61, 63, 67-68, 70-73, 74(1, 3-4), 75(1), 76-78, 79(1-3, 5-12)
IaU	1959, 68, 73, 74(1, 3), 75(3), 76-78, 79(1-3, 5-11), 81, 82-84
IdBB	1973
IdPI	1968
IdU	1968, 76, 78, 80, 84(2, 5-6, 8)
InLP	1959, 68-69, 71, 73, 74(1, 3-4), 75(1, 4), 76-78, 79(1-3, 5-11), 80, 83, 84(1-2, 4-5, 8), 85(1)
InRE	1968, 74(1)
InU	1952, 55, 59-60, 68-69, 74(1, 3-4), 75, 82
KyU-Ge	1959, 68, 70(1, 4-5), 71, 75l-4), 76-78, 79(1-3, 5-11), 80, 82-83, 85
LNU	1968, 71
LU	1959
MH-GS	1958-59, 68, 74(3), 82
MNS	1952, 59
MWesB	1959

MiDW-S	1952, 59, 68
MiHM	1968, 75(4:2)
MiKW	1974(1)
MiU	1968, 74(3-4)
MnDuU	1973
MnSSM	1968
MnU	1952, 59-60, 63, 67, 69(1), 71-73, 74(3-4), 75(1, 4), 76(5), 78, 79(1-3, 5-11), 81-82, 84
MoSW	1952, 59, 68, 73, 74(3-4), 75(1, 4:1)
MoU	1968, 74(3)
MtBuM	1952, 73, 75(1)
NBiSU	1968
NNC	1952, 55, 59-61, 67-68
NRU	1959, 68,
NSbSU	1959, 70, 75-76, 81-82
NSyU	1959, 68, 74(1)
NbU	1968
NcU	1952, 59, 68, 71, 73, 74(1), 75(1), 77(2, 5-6, 8), 79(1-3, 5-11), 80, 82(5-7, 9, 11, 13)
NdU	1959, 68, 69(2), 71, 73, 74(1, 3-4), 75(4:1) 79(1-3, 5-11)
NhD-K	1961, 66, 68, 71, 73, 74(3-4), 75(1, 4:1), 78, 82,
NjP	1952, 60, 68, 69(2), 71, 73, 74(3)
NmPE	1968
NmSoI	1983
NvLN	1952, 68
NvU	1952, 58-59, 61, 66, 68, 71, 74(1, 3-4), 75(1, 3, 4:1), 77, 79(1-3, 5-11), 81-82
OCU-Geo	1952, 59, 67(1, 3), 68, 71, 74(1), 78, 82
ODaWU	1980
OU	1959, 68, 69(3), 71, 74(1), 75(4:1), 77(2), 78, 79(1-3, 5-11), 82(10)
OkS	1971
OkT	1959, 68
OkU	1952, 70(5), 71(4), 73, 74(1, 3-4), 75(4:1), 79(3, 5-7, 9, 11), 81, 83, 84(1, 4-6, 8), 85
OrCS	1980
OrHS	1980
OrU-S	1952, 59-60, 68
PBL	1959
PBU	1982
PBm	1959, 71, 78, 82
PSt	1959, 68, 71, 74(1), 78, 79(1-3, 5-11)
RPB	1959, 68
SDSNH	1980
SdRM	1961, 74(2, 4), 75(1)
TxBeaL	1955, 59, 68, 74(1), 78
TxCaW	1968
TxDaAR-T	1952, 59, 67(3), 68-69, 70(1, 4-5), 71-73
TxDaDM	1952, 55, 59-61, 67(1-3), 68-69, 73, 74(3), 75
TxDaGI	1952
TxDaM-SE	1959, 82(3-4, 14), 83
TxDaSM	1952, 59, 71, 74(1), 77(1, 4, 6, 8-10), 79(1, 6-7, 9-10)
TxHSD	1959, 66, 68, 73, 74(1, 3), 75(3, 5), 78
TxLT	1959, 68, 76(4)
TxMM	1952, 59, 63, 68
TxU	1952, 59-61, 68-69, 70(1, 4-5), 74(3-4), 77(2-3, 6-9), 78, 79(1-3, 5-11), 81, 82(2, 7, 13), 83
TxWB	1971, 77, 79, 82(3-4, 14)
UU	1952, 59, 73, 74(3-4)
ViBlbV	1959, 68, 69(2), 71, 73, 74(1, 3-4), 75(2-5), 76(1-2), 77(3-5, 7, 9), 78, 79(1-3, 5-11)
WGrU	1968
WU	1968, 73, 74(1, 3-4), 75(1, 4:1), 76(5), 77-78, 80, 81, 82(2-4, 7, 13, 14), 84, 85(1-16)
WaPS	1973
WaU	1984(1-3, 5, 7-8)
WyU	1982(2, 10, 12))

(200) **GEOLOGICAL SOCIETY OF AMERICA. NORTH CENTRAL SECTION. FIELD TRIP GUIDEBOOK.**
1967 1st Bloomington, Indiana. Geologic tales along Hoosier trails.
No. 1. Karst geomorphology of south-central Indiana.
No. 2. Silurian and Devonian stratigraphy of southeastern Indiana.
No. 3. Pleistocene stratigraphy of west-central Indiana.
No. 4. Nonmetallic mineral resources of southwestern Indiana.
1968 2nd Iowa City, Iowa.
No. 1. Niagaran bioherms in the vicinity of Iowa City (Linn and Cedar counties).
No. 2. Geology of U.S. Gypsum Company's Sperry mine.
No. 3. Mississippian (Osage and Kinderhook) stratigraphy and Mississippian-Devonian boundary problems in southeastern Iowa.
No. 4. Middle River traverse of Iowa (Pennsylvanian stratigraphy). (688)
No. 5. A review of Pleistocene Lake Calvin.
No. 6. Conodont successions.
1969 3rd Columbus, Ohio.
No. 1. Devonian outcrops in Columbus, Ohio and vicinity.
No. 2. Till stratigraphy from Columbus southwest to Highland County, Ohio.
No. 3. Ordovician conodont localities, southwestern Ohio.
No. 4. Mississippian strata of the Granville-Newark area, Ohio.
1970 4th East Lansing, Michigan. Guidebook for field trips. (Cover title: Field excursions; Pleistocene, Devonian, Precambrian) (377)
No. 1. Devonian strata of Alpena and Presque Isle counties, Michigan.
No. 2. Glacial history of the Glacial Grand Valley.
No. 3. Precambrian of the Marquette area, Michigan.
1971 [5th] Lincoln, Nebraska.
No. 1. Guidebook to selected Pleistocene paleosols in eastern Nebraska. (449)
No. 2. Guidebook for field trip on urban geology in eastern Nebraska. (449)
No. 3. Guidebook to the geology along portions of the lower Platte River valley and Weeping Water Valley of eastern Nebraska. (449)
No. 4. Guidebook to the late Pliocene and early Pleistocene of Nebraska. (449)
1972 [6th] DeKalb, Illinois.
No. 2. Geology for land-use planning, McHenry County, Illinois.
No. 3. Pennsylvanian conodont assemblages from La Salle County, northern Illinois. (265, 527)
No. 4. Pleistocene geology between DeKalb and Danville, Illinois.
1973 [7th] Columbia, Missouri.
No. 1. Field trip not held; therefore no guidebook published.
No. 2. Pleistocene and engineering geology of north-central Missouri.
No. 3. Fletcher Mine trip.
No. 4. Barite deposits of the central mineral district.
No. 5. Central Missouri fire clay deposits.
1974 [8th] Kent, Ohio. No. [1,2,3] Selected field trips in northeastern Ohio. (228)
Field trip 1. General geology of the International Salt Company Cleveland mine, Cleveland, Ohio.
Field trip 2. Sedimentary environments of the Lower Pennsylvanian Sharon Conglomerate.
Field trip 3. Engineering and Pleistocene geology of the lower Cuyahoga River valley.
Field trip 4. Natural and manmade features affecting the Ohio shore of Lake Erie.
Field trip 5. Pennsylvanian conodont localities in northeastern Ohio. (527)
Field trip 6. Energy resources Canton-Cadiz area field trip.
1975 9th See Geological Association of Canada. 190.
1976 10th Kalamazoo, Michigan.
No. 1. Geology of the Kentland structural anomaly, northwestern Indiana. (768)
No. 2. Guidebook for a field trip on some aspects of the glacial geology of the Kalamazoo area. (768)
No. 3A. Indiana portion of guidebook for a field trip on Silurian reefs, interreef facies, and faunal zones of northern Indiana and northeastern Illinois; with a section on glacial geology. (768)
No. 3B. Illinois portion of guidebook for a field trip on Silurian reefs, interreef facies, and faunal zones of northern Indiana and northeastern Illinois. Thornton Reef (Silurian) northeastern Illinois, 1976 revisitation; with a section on Trilobites of the Thornton Reef (Silurian) northeastern Illinois, 2 v. (768)
No. 4. Coastal sedimentation and stability in southern Lake Michigan. (Cover title: Field trip guidebook: coastal and environmental geology of southeastern Lake Michigan) (768)
1977 11th Carbondale, Illinois.
Volume 1. Premeeting field trip. (622)
[1] The type section of the Meramecian series.
Volume 2. Postmeeting field trips. (622)
[2] Geology of southern Illinois coal deposits.
[3] A typical stratiform replacement ore body in the Illinois-Kentucky fluorspar district.

(200) Geological Society of America. North Central Section.

[4] Alluvial fan, braided stream, and shallow marine deposits in the Lamotte Sandstone, southeastern Missouri, with a note on transport and composition of coarse detritus, St. Francis River, Southeast Missouri.
[5] Upper Mississippian conodonts and boundary relations in southern Illinois.
[6] Middle Devonian stratigraphy of southern Illinois.
[7] Quaternary geology of the Mississippi Valley in southern Illinois.

1978 [12th] Ann Arbor, Michigan. Field excursions from the University of Michigan. (410, 523)
[1] Pleistocene geology of the Thumb area of Michigan.
[2] Strata of the Middle Devonian silica formation.
[3] Patch reef development and effects of repeated subareal exposure in Silurian shelf carbonates, Maumee, Ohio. Stratigraphic facies of carbonate platform and basinal deposits, late Middle Silurian, Michigan Basin.

1979 13th Duluth, Minnesota. (273)
No. 1. Middle Precambrian geology of northern Wisconsin. (777)
[2] Field trip guidebook for stratigraphy, structure, and mineral resources of east-central Minnesota. (388(No.9))
3. Quaternary geology of the Duluth area.
[4] The Mesabi Range trip.
5. Geologic history and engineering geology of the western Lake Superior region.
6. Cambrian and Ordovician stratigraphy and paleontology of southeastern Minnesota.
[7] Field trip guidebook for the Keweenawan (Upper Precambrian), north shore volcanic group, Minnesota. (388(No.11))
[8] Field trip guidebook for the western Vermilion district, northeastern Minnesota. (388(No.10))

1980 [14th] Field trips 1980 from the Indiana University campus, Bloomington. (268, 270)
Field trip 1. Quarrying and milling of the Salem Limestone in the Bloomington-Bedford District.
Field trip 2. Platform communities and rocks of the Borden Siltstone Delta (Mississippian) along the south shore of Monroe Reservoir, Monroe County, Indiana.
Field trip 3. Silurian stratigraphy and conodont paleontology, southeastern Indiana.
Field trip 4. The origin of terra rossa in the karst of southern Indiana. (527, 587, 523)

1981 [15th] Ames, Iowa. (527, 587, 523)
1. Mississippian stratotypes. (265)
2. Depositional environments of glacial sediments and landforms on the Des Moines Lobe, Iowa. (332)
3. Cretaceous stratigraphy and sedimentation in Northwest Iowa, Northeast Nebraska, and South Dakota, Southeast. (332)
4. Cherokee sandstones and related facies of central Iowa: an examination of tectonic setting and depositional environments. (332)

1982 16th West Lafayette, Indiana. (410, 527, 523)
Field trip no. 1. Coastal evolution of southern Lake Michigan. (544)
Field trip no. 2. Geomorphology and glacial history of the Great Bend area of the Wabash Valley, Indiana. (544)
Field trip no. 4. Geology of the Kentland structural anomaly, northwestern Indiana; update, study for earth science students, field guide.
Field trip no. 5. Silurian reefs at Delphi and Pipe Creek Jr. Quarry, Indiana, with emphasis on the question of deep vs. shallow water. (544)

1983 17th Madison, Wisconsin. (527, 523, 587)
No. 1. Late glacial history and environmental geology of southeastern Wisconsin. (777)
[2] Sedimentology of Ordovician carbonates and sandstones in southwestern Wisconsin. (777)
[3] The oolitic Neda iron ore (upper Ordovician?) of eastern Wisconsin. (777)
[4] Precambrian rhyolite and granite inliers in south-central Wisconsin.
No. 5. Middle Proterozoic to Cambrian rocks in central Wisconsin: anorogenic sedimentary and igneous activity. (777)

1984 18th Richardson, Texas. See Geological Society of America. Southeastern Section. 206.

1985 19th DeKalb, Illinois. (587, 523, 410)
[1] Depositional environments and correlation problems of the Wedron Formation (Wisconsinan) in northeastern Illinois. (265)
[2] Silurian geology of the Des Plaines River valley, northeastern Illinois. (265)
[3] Early Pennsylvanian paleotopography and depositional environments, Rock Island County, Illinois. (265)

CLU-G/G	1970, 71(4), 72(3), 74(1-5), 81, 85
CLhC	1981(3)
CMenUG	1967, 79(1-3), 80, 82(1-2, 4-5), 84, 85
CPT	1971, 72(3), 85
CSt-ES	1967, 70, 71(4), 72(3), 73(2-3, 5), 74(1-3, 5), 79(1, 7-8), 81, 83(2, 5), 85
CU-A	1972(3), 85(1-2)
CU-EART	1967, 70, 71(4), 72(3), 74(1-5), 76(1-2, 3A, 4), 79(2, 3), 81, 83, 85
CU-SB	1976
CU-SC	1985
CaACU	1971, 72(3), 76-78
CaAEU	1969(1), 72(3), 76(2-4)
CaBVaU	1972(3)
CaOHM	1970, 72(3), 74(1-3, 5)
CaOKQGS	1970, 72(3), 74(1-3, 5), 81
CaOLU	1970
CaOOG	1967, 70, 74(1-3, 5), 83(3), 84-85
CaOTDM	1976
CaOTRM	1972(3)
CaOWtU	1967-68, 69(1), 70-71, 72(3), 74(1-3), 76, 78, 79(1-2, 7-8), 81(2-4)
CoDCh	1969, 71(1, 3), 72(3), 78, 79(7), 81(3-4)
CoDGS	1967, 73-76, 78-80, 81(3-4), 82
CoG	1967, 70, 72(3), 73(2, 4-5), 82(1-2, 4)
CoU	1970, 71, 72(3)
DI-GS	1967-71, 72(2, 4), 73(2-5, 7), 74, 76(1-3A, 4), 77-81, 83
DLC	1967, 70, 71(4), 72(3)
I-GS	1969, 71, 72(2-3), 74(1-3, 5), 81, 82(5), 83(4), 85
ICF	1967, 70, 72(3), 73(4), 76(1), 77(2-7), 79(6)
ICIU-S	1970, 72(3)
ICarbS	1972(3)
IEN	1970, 72(3)
INS	1985(1)
IU	1967-71, 72(2-4), 73(2-5), 74, 76-78, 79(1, 3, 6-8), 80, 85(1)
IaCfT	1981(4)
IaU	1967-71, 72(4), 74(5), 76-78, 79(1, 7), 80(1-4), 81(2-4), 82(1-2, 4-5), 83(2-3, 5, 7), 85(1-3)
IdBB	1972(3)
IdU	1980
InLP	1967, 69-71, 72(3), 74(1-5), 76-78, 79(1, 3, 7, 8), 80, 81(2-4), 82(1-2, 5) 83-85
InRE	1974(1-3, 5)
InU	1967, 68(4), 69-70, 71(3), 72(3), 74, 76(1-4), 82(2, 5), 85(1)
KLG	1971(1-4), 85(1-2)
KyU-Ge	1969-71, 72(3), 73(2-5), 74, 76(1-3A, 4), 77-78, 79(1-2, 7-8), 80-81, 82(2), 83(1, 3, 7-8), 85
LNU	1967-69
LU	1972(3)
MH-GS	1970, 84
MNS	1972(3)
MU	1985
MiDW-S	1972(3), 85(1-2)
MiHM	1967, 70, 72(3), 73(2), 76(2), 79(1), 85(1-2)
MiKW	1971, 72(3), 74(1-3, 5)
MiU	1969
MnDuU	1969
MnU	1967, 69-71, 72(2-4), 76(1-3A, 4), 77-81, 82(2, 4-5), 83(2-3, 7-8), 85
MoSW	1967, 71, 72(3), 74(1-3, 5)
MoU	1967, 68(4), 69-70, 72(3), 74(1-3), 77(2-7)
Ms-GS	1971(1-2)
MsJG	1971(1-2)
MtBuM	1971, 72(3), 74(1-3, 5)
NBiSU	1972(3)
NGenoU	1981
NMSoI	1985
NNC	1967, 70, 81, 85
NRU	1970, 76(4), 81
NSbSU	1970, 72, 76, 81
NSyU	1972(3)
NbU	1970, 72(3)
NcU	1970, 72(3), 76(1-3A), 80
NdU	1968, 71, 72(3), 76(1-3A, 4)
NhD-K	1970-71, 72(3), 76(1-3A, 4), 81, 85(1-2)
NjP	1967-70, 71(4), 72(3), 74(1-3)
NmSoI	1981
NvU	1968, 71, 72(3), 74(1-3, 5)

OCU-Geo	1971, 72(2-4), 74(1-5), 76, 78, 80, 82(4)
OCl	1985
OU	1968-71, 72(3), 74(1-3, 5), 76, 80, 81, 82(4), 83(2, 7, 8)
OkT	1972(3)
OkTA	1985
OkU	1967, 69, 71, 72(3), 74(1-3, 5), 76(3B), 81, 83(3, 7), 85(2)
OrU-S	1985(2)
PBL	1972(3)
PBm	1972(3), 76(3A-B), 85(1-2)
PHarER-T	1981, 85
PSt	1970(1), 72(3), 74(1-5), 85(1-2)
SdRM	1971, 72(3), 81(2-4), 85
TMM-E	1972(3)
TxBeaL	1972(3), 76
TxCaW	1971(4)
TxDaAR-T	1967, 70-71, 72(3-4)
TxDaDM	1968, 70, 72(2-4), 74(1-3, 5)
TxDaM-SE	1981
TxDaSM	1970, 76(3A, 4), 77(2-7)
TxHSD	1972(3)
TxHU	1970, 72(3)
TxLT	1972(3)
TxMM	1970
TxU	1967-71, 72(3), 74(1-3, 5), 77, 81(2-4), 85
TxWB	1976, 80
UU	1971, 72(3), 74(1-3, 5)
ViBlbV	1972(3), 74(1-3, 5), 76(1-3A, 4), 77
WGrU	1976(1, 3A-B)
WKenU	1983(2-3, 7-8)
WU	1970-71, 72(3), 74(1-3, 5), 76(1-3A, 3B, 4), 78, 79(1-4, 6-8), 80-81, 82(4), 83(2-4, 7-8), 85
WaU	1972(3), 81(1), 82(3, 4, 14), 85

(201) GEOLOGICAL SOCIETY OF AMERICA. NORTHEASTERN SECTION. GUIDEBOOK FOR FIELD TRIPS.

1969 No field trip.
1970 No field trip.
1971 No field trip.
1972 No field trip.
1973 [8th] Allentown, Pennsylvania.
  Field guide to the Friedensville Mine of the New Jersey Zinc Company.
1974 No field trip.
1975 No field trip.
1976 11th Arlington, Virginia. (206)
  1,4. Piedmont geology of the Fredericksburg, Virginia, area and vicinity.
  2. Selected examples of carbonate sedimentation, lower Paleozoic of Maryland. (370)
  3. Carboniferous stratigraphy of southwestern Virginia and southern West Virginia.
  5. Geology of the Blue Ridge Anticlinorium in northern Virginia.
  6. The geology of the crystalline rocks near Baltimore and its bearing on the evolution of the eastern Maryland piedmont. (No road log) (371)
  7a. Coastal plain stratigraphy of the upper Chesapeake Bay region.
  7b. Stratigraphy of the Chesapeake Group of Maryland and Virginia.
1977 No field trip.
1978 No field trips.
1979 No field trips.
1980 No field trips.
1981 No field trips.
1982 17th Washington, DC. Central Appalachian geology. (206, 586)
  [1] Tectonic control of cyclic sedimentation in the Chesapeake Group of Virginia and Maryland.
  [2] Mesozoic and Cenozoic compressional faulting along the Atlantic Coastal Plain margin, Virginia.
  [3] Geologic traverse across the Culpeper Basin (Triassic-Jurassic) of northern Virginia.
  [4] Geology of the Catoctin-Blue Ridge Anticlinorium in northern Virginia.
  [5] Anorthosite, ferrodiorite, and titanium deposits in Grenville Terrane of the Roseland District, central Virginia.
  [6] Geology of the Blue Ridge and Valley and Ridge at the junction of the Central and Southern Appalachians.
  [7] Facies of the Great American Bank in the Central Appalachians.
1983 Mar. 18th Kiamesha Lake, New York. (592, 411)

Field trip 1. Tectonic setting, chemical petrology and petrogenesis of the Cortlandt Complex and related igneous rocks of southeastern New York State.
Field trip 2. Middle and Upper Ordovician sandstone-shale sequences of the mid-Hudson region west of the Hudson River.
1984 No field trip.
1985 No field trip.

CLU-G/G	1976
CLhC	1982
CMenUG	1976
CPT	1976(2, 6)
CSt-ES	1976(2, 7a-7b)
CU-EART	1976(2, 6), 82
CU-SB	1982
CaACU	1976
CaAEU	1982
CaOOG	1976(7a), 82
CaOWtU	1976(2, 6), 82
CoDCh	1976(2, 7a)
CoDGS	1982
DI-GS	1973, 76(2-3), 82-83
ICarbS	1976(6)
IU	1973, 76(1-5, 7)
IaU	1976(2-3, 5, 7)
IdU	1982
InLP	1976(2, 6), 82
InU	1976(7a), 82
KyU-Ge	1976(2, 6), 82
LNU	1976(2)
MdBJ	1976
MnU	1976(1-5, 7), 82
NNepaSU	1983
NSbSU	1976
NcU	1976(1-4, 7), 82
NdU	1976(7a)
NhD-K	1982
OCU-Geo	1976(1-4, 7)
ODaWU	1982
OkU	1982
PBm	1976(2)
PHarER-T	1973, 82
PSt	1976(2)
SdRM	1976(6)
TxDaDM	1976(2)
TxDaM-SE	1982
TxHSD	1976(1-2, 4, 7)
TxU	1982
TxWB	1982
WU	1976(7a, 7b)

(202) GEOLOGICAL SOCIETY OF AMERICA. NORTHEASTERN SECTION. SHORT COURSE. FIELD TRIP GUIDEBOOK.

1976 Mar. 3. A geologic and environmental guide to northern Fairfax County, Virginia. (207)

CLU-G/G	1976
IU	1976
IaU	1976
NcU	1976
OCU-Geo	1976
TxHSD	1976

(203) GEOLOGICAL SOCIETY OF AMERICA. PENROSE CONFERENCE. GUIDEBOOK. (TITLE VARIES)

1978 May Structural style of the Blue Ridge. Relative chronology of thrusting in the Southern and Central Appalachians; its significance and relationships to other orogenic processes.
1979 June Near-surface batholiths, related volcanism, tectonism, sedimentation, and mineral deposition.
— Sept. Elko to Battle Mountain, Shoshone Range, and Cortez Mountains.
1983 Depositional environments and paleoclimates of the Greenhorn tectomoeustatic cycle, Rock Canyon Anticline, Pueblo, Colorado.

CMenUG	1979(Sept.)
CoLiM	1983
MtBuM	1979(June)

(204) GEOLOGICAL SOCIETY OF AMERICA. ROCKY MOUNTAIN SECTION. GUIDEBOOK FOR FIELD TRIPS.

1951 4th Rapid City, South Dakota.
   No title. (Prepared by the members of the geological staffs of the South Dakota School of Mines and the Homestake Mining Company)
1952 5th Salt Lake City, Utah.
   Geology and hydrothermal alteration of the Bingham Copper Mine.
1953 6th Butte, Montana. Guidebook of field excursions, Butte, Montana.
   Trip 1. Stratigraphy and structure in the Three Forks area.
   Trip 2. Underground trip to mines of the Anaconda Company.
   Trip 3. Dewey-Divide contact of the Boulder Batholith.
   Trip 4. Petrology and ore deposits in the northern part of the Boulder Batholith.
   Trip 5. Structure and stratigraphy in the Phillipsburg-Drummond area.
1956 [9th] Albuquerque, New Mexico.
   Road log from Albuquerque, New Mexico, to Grants Uranium District. (457)
1957 10th Logan, Utah.
   Northern central Utah and southeastern Idaho.
1958 11th Golden, Colorado.
   Precambrian field trip, Front Range. Sedimentary field trip, Golden to Deer Creek, Colorado.
1959 12th Missoula, Montana. Montana State University. Guidebook to field trips.
   Field trip 1. Stratigraphy of the Belt series.
   Field trip 2. Drummond-Helmville-Ovando area with emphasis on post-Precambrian geology.
   Field trip 3. Glacial geology of the Flathead Valley and environs.
   Field trip 4. Northern Bitterroot Range and Idaho Batholith border facies.
1960 13th Rapid City, South Dakota. Rapid City to Bell Fourche, South Dakota via Sturgis, Deadwood, and Spearfish.
1962 15th Provo, Utah. Geology of the southern Wasatch Mountains and vicinity, Utah; a symposium.
1964 17th Moscow, Idaho.
   Field trip no. 3. The Idaho Batholith.
1965 18th Fort Collins, Colorado. Resume of the geology of the Laramie anorthosite mass and road log.
   Field trip no. 1. The anorthosites of the Laramie Range.
   Field trip no. 2. Diatremes containing lower Paleozoic rocks in southern Wyoming and northern Colorado.
   Field trip no. 3. Itinerary, cement plant field trip.
   Field trip no. 4. Tertiary of the Gangplank area; trip from Fort Collins, to Wellington, Cheyenne, Granite Canyon and return via State Highway 14, U.S. Highways 87 and 30 and local roads.
   Field trip no. 5. Sandstone members of the Pierre Shale in the Fort Collins, Colorado area.
   Field trip no. 6. Big Thompson Canyon and vicinity.
1966 [19th] Las Vegas, Nevada.
   Road log; Cedar City, Utah to Las Vegas, Nevada via Caliente, Nevada.
1967 20th Golden, Colorado.
   Field trip no. 1. Precambrian basement rocks of the central Colorado Front and its 700 million year history.
   Field trip no. 4. Sedimentary structures of the formations in the Golden-Red Rocks area.
1968 [21st] Bozeman, Montana.
   No. 1. Butte mines.
   No. 2. Cretaceous rocks of the western Crazy Mountains Basin and vicinity, Montana.
   No. 3. Boulder Batholith.
   No. 4. Stillwater igneous complex.
   No. 5. Geomorphology and Cenozoic history of the Yellowstone Valley, south of Livingstone, Montana.
   No. 6. Igneous and hydrothermal geology of Yellowstone National Park.
1969 [22nd] Salt Lake City, Utah. Guidebook of northern Utah. (730)
   [1] Tintic mining district.
   [2] Geology of Bingham mining district.
   [4] Willard Thrust.
   [6] Great Salt Lake and Antelope Island.
1970 23rd Rapid City, South Dakota.
   Field trip no. 1. Road log, Rapid City to Homestake Mine via Interstate 90. Road log, Rapid City to Homestake Mine via Central Hills. The Homestake Mine.
   Field trip no. 2. Road log, Precambrian metasediments and pegmatites of the Black Hills.
   Field trip no. 3. Road log, Tertiary stratigraphy of the northern part of the Big Badlands.
   Field trip no. 4. Road log, Paleozoic and Mesozoic stratigraphy of the Rapid City area.
   Field trip no. 5. Road log, engineering geology of the Rapid City area.
1971 24th Calgary, Alberta.
   No. 1. Coal, oil, gas and industrial mineral deposits of the interior plains, foothills and Rocky Mountains of Alberta and British Columbia.
   No. 1a. Cascade and Crowsnest coal basins.
   No. 1b. A guide to the geology of the Eastern Cordillera along the Trans-Canada Highway between Calgary, Alberta and Revelstoke, British Columbia.
   No. 2a. Geological guide along the Trans-Canada Highway from Calgary to Banff.
   No. 3. Geologic guidebook to the Canadian Cordillera between Calgary and Revelstoke.
   No. 4. A guide to the Pleistocene geology.
   No. 5, 6. A guide to the geology between Calgary and Banff.
1972 [25th] Laramie, Wyoming.
   No. 1. Depositional environment of the Green River Formation, Wyoming.
   No. 2. Mining in the Hanna coal field.
   No. 3. Field trip guide and road log to northern Laramie anorthosite complex. (726)
   No. 4. Late Paleozoic rocks of the southern Laramie Basin.
1973 26th Boulder, Colorado.
   No. 1. Geomorphology, palynology, and paleomagnetic record of glacial Lake Devlin, Front Range.
   No. 2. Virginia Dale ring-dike complex.
   No. 3. No guidebook published.
   No. 4. Nature of the early Tertiary intrusives between Golden and Lyons, Colorado, and their relation to the structural development of the Front Range.
   No. 5. Lyons Sandstone.
   No. 6. Urban geology of the Boulder area, Colorado.
   No. 7. Tundra environment on Niwot Ridge, Colorado Front Range.
   No. 8. Petrologic, tectonic, and geomorphic history of central Colorado.
   No. 9. Dakota Group.
   No. 10. Environmental geology for planning Windsor study area: Fort Collins, Loveland, and Greeley.
1974 27th Flagstaff, Arizona. Geology of northern Arizona with notes on archaeology and paleoclimate; Part 1, Regional studies; Part 2, Area studies and field guides.
   [No. 1] Kaibab trail guide to the southern part of Grand Canyon, northern Arizona.
   [No. 2] Geologic resume and field guide, north-central Arizona.
   [No. 3] Field guide to the geology of the San Francisco volcanic field, Arizona.
   [No. 4] Field guide to the geology of the San Francisco Mountain, northern Arizona.
   [No. 5] Field guide for Southeast Verde Valley, northern Hackberry Mountain area, north-central Arizona.
   [No. 6] Field guide for Hopi buttes and Navajo buttes area, Arizona.
   [No. 7] Field guide for the Black Mesa-Little Colorado River area, northeastern Arizona.
   [No. 8] Economic geology and field guide for the Jerome District, Arizona.
1975 28th Boise, Idaho.
   No. 1. Engineering geology approaches to highway construction in central Idaho: Boise, Idaho to Whitebird Hill, Idaho.
   No. 2. No guidebook.
   No. 3. Geologic field guide to the Quaternary volcanics on the south-central Snake River plain, Idaho. (259)
   No. 4. Field trip guide to the Idaho-Wyoming thrust fault zone.
   No. 5. Rock alteration and slope failure, Middle Fork of the Payette River.
   No. 6. The evaluation of geologic processes in the Boise foothills that may be hazardous to urban development. (A report prepared for the Ada Council of Governments, May 1973)
   No. 7. The later Tertiary stratigraphy and paleobotany of the Weiser area, Idaho. (258(No. 28))

No. 8. The geology and scenery of the Snake River on the Idaho-Oregon border from Brownlee Dam to Hells Canyon Dam. (258(No. 28))
1976 29th   Albuquerque, New Mexico.
1. No guidebook issued.
2. No guidebook issued.
[3] Guidebook to Albuquerque Basin of the Rio Grande Rift, New Mexico. (461(No. 153))
4. No guidebook issued.
1977 30th   Missoula, Montana.
1. Mylonite detachment zone, eastern flank of Idaho Batholith. (701)
2. Big Fork-Avon environmental geology mapping project. (701)
3. Stratabound copper occurrences in green beds of the Belt Supergroup, western Montana. (701)
4. Glacial geology of Flathead Valley and catastrophic drainage of glacial Lake Missoula. (701)
5. Alluvial fan, shallow water, and sub-wave base deposits of the belt supergroup near Missoula, Montana. (701)
1978 [31st]   Provo, Utah. (527)
[4] Upper Cambrian to Middle Ordovician conodont faunas of western Utah. (625)
10. Conodont biostratigraphy and sedimentation of Devonian to Triassic rocks along the Wasatch Front and Disturbed Belt, Utah. (625)
Volume 25, Part 1. (104)
[8] Quaternary tectonics along the Intermountain Seismic Belt South of Provo, Utah.
[3] Sevier Orogenic Attenuation Faulting in the Fish Springs and House Ranges, western Utah.
[5] Geology of Volcanic rocks and mineral deposits in the southern Thomas Range, Utah: A brief summary.
[6] Tintic mining district, Utah.
[7] Mesozoic and Cenozoic sedimentary environments of the northern Colorado Plateau.
[2] Geology of the Marysvale Volcanic Field, west central Utah.
[1] Geology, geochemistry, and geophysics of the Roosevelt Hot Springs Thermal Area, Utah: A summary.
1979 32nd   Fort Collins, Colorado. Field guide: Northern Front Range and Northeast Denver Basin, Colorado (Cover title). (144)
1. Sedimentology and stratigraphy of selected Paleozoic and Mesozoic sequences: Northwest Denver Basin.
Part A. Permian stratigraphy, paleotectonics, and structural geology, Bellvue-Livermore area, Larimer County, Colorado.
Part B. Lacustrine deltaic deposition of the Jurassic Morrison Formation of north-central Colorado.
Part C. Lower Dakota Group depositional environments and stratigraphic evolution: Colorado Front Range, Larimer County, Colorado.
Part D. Depositional environments and processes, Laramie and Fox Hills formations northeast of Wellington, Colorado.
2. Precambrian structural relations, metamorphic grade, and intrusive rocks along the northeast flank of the Front Range in the Thompson Canyon, Poudre Canyon, and Virginia Dale areas.
3. Till sequence and soil development in the North St. Vrain Drainage Basin, east slope, Front Range, Colorado.
4. Field guide for the Sloan and Nix kimberlites in the southern portion of the Colorado-Wyoming State Line kimberlite district.
1980 33rd   Ogden, Utah.
1. Geology of the Albion-Raft River-Grouse Creek mountains area, northwestern Utah and southern Idaho.
2. No field guide; field trip cancelled.
3. Mining of the Phosphoria Formation in southeastern Idaho, the old and the new.
4. Late Quaternary lacustrine geology and geologic hazards along the Wasatch Front.
4A. Geologic evidence for recurrence for major earthquakes on the Wasatch fault zone, Kaysville site.
4B. Field trip guide to the Quaternary stratigraphy and faulting in the area north of the mouth of Big Cottonwood Canyon, Salt Lake County, Utah. (220)
4C. Tectonostratigraphic and morphostratigraphic hyphotheses concerning threshold-controlled Lake Bonneville shorelines, with emphasis on shorelines and associated deposits in the Long Bench-North Ogden area, Utah.
5. No field guide.
6. Precambrian rocks of the northern Wasatch.
6A. Field guide to the geology of Little Mountain.
6B. Geology of the Farmington Canyon Complex, north central Utah.
7. The Bingham mining district, Utah. (220)
1981 [34th]   Rapid City, South Dakota.
Geology of the Black Hills, South Dakota and Wyoming. (Revised in 1985.)
Field trip no. 1. Stratigraphy and depositional environments of Lower and Upper Cretaceous strata, southern Black Hills, South Dakota.
Field trip no. 2. Geology of the Tertiary intrusive province of the northern Black Hills, South Dakota and Wyoming.
Field trip no. 3a. Engineering geology of the central Black Hills, South Dakota.
Field trip no. 3b. Engineering geology of the northern Black Hills, South Dakota.
Field trip no. 4. Paleozoic and Mesozoic stratigraphy of the northern Black Hills.
Field trip no. 5. Geology of uraniferous conglomerates in the Nemo area, Black Hills, South Dakota.
Field trip no. 6. Geology and paleontology of the Badlands and Pine Ridge area, South Dakota.
1982 35th   Bozeman, Montana.
[1] Cenozoic history of the Yellowstone Valley south of Livingston, Montana. (404(No. 12))
[2] Geology of the fold and thrust belt, west-central Montana. (401(1983-V.86))
[3] Late Cretaceous volcanic and intrusive rocks near the eastern margins of the Boulder and Tobacco Root batholiths. (404(No.10))
[4] Stratigraphy, depositional environments, and paleotectonics of the Lahood Formation. (404(No.11))
1983 36th   Salt Lake City.
Field trips 1-12. See Geological Society of America. Cordilleran Section. 199.
1984 37th   Durango, Colorado.
[1] Paleotectonics - San Juan Mountains.
[3] Dolores Formation - Paleosols and depositional systems.
[4] Quaternary deposits and soils - Durango area. (180)
[5] Jurassic depositional systems - San Juan Basin.
1985 38th   Boise, Idaho.
1. Raft trip through the Snake River Birds of Prey Area to view canyon-filling volcanics and lake beds of the Pleistocene Snake River. (Cover title: Power-raft trip through the National Birds of Prey Natural Area to view canyon- filling volcanics and lake beds of the Pleistocene Snake River)
2. Tertiary volcanics and epithermal mineralization in the Owyhee Mountains, southwestern Idaho.
3. Late Cenozoic volcanic geology of the Jordan Valley-Owyhee River region, southeastern Oregon.
4. The Borah Peak earthquake of October 18, 1983: surface faulting, mass movements, and Quaternary tectonic setting of the Lost River Range. (Cover title: The Borah Peak earthquake of October 18, 1983: surface faulting, mass movements, and Quaternary tectonic setting)
5. Geologic road log: Boise-Mountain Home-Wood River Junction via Interstate Highway 84 and U.S. Highway 20. (Cover title: Thrust belt in south- central Idaho at the Fish Creek Reservoir window)
6. Repeat of trip 1.
7. Field trip guide for geology of the Boise geothermal system. (Cover title: Boise geothermal system, western Snake River Plain, Idaho)
8. Ancestral canyons of the Snake River: geology and hydrology of canyon-fill deposits in the Thousand Springs area, south-central Snake River plain, Idaho. (Cover title: Ancestral canyons of the Snake River: geology and geohydrology of canyon-fill deposits in the Thousand Springs area, south-central Snake River plain, Idaho)
Geology of the Black Hills, South Dakota and Wyoming. (Revised from 1981.)

AzFU	1966, 67(1), 69(1-2, 4, 6), 74
AzTeS	1969(1-2, 4, 6), 76(3), 79
AzU	1966, 69(1-2, 4, 6), 74, 75(3)

(205) Geological Society of America. South Central Section.

CLU-G/G	1953, 58, 68(6), 69(1-2, 4, 6), 70, 73(1-2, 4-10), 74, 81, 83
CLhC	1976, 81
CMenUG	1979, 81, 82(1, 4), 83-85
CPT	1969, 72(3), 75(3, 7-8), 76, 81-85
CSfCSM	1957, 59
CSt-ES	1953, 58, 66, 68(6), 69(1-2, 4, 6), 70, 71(1, 2a, 3, 5, 6), 73(1-4, 6-7), 74, 75(3-4, 7-8), 76(3), 78(V25.pt.1), 79, 80, 82, 83, 85
CU-A	1974, 79, 81, 83(1, 3-4), 84
CU-EART	1956, 59, 62, 65, 67(1), 68, 69(1-2, 4, 6), 75(3, 7-8), 76(3), 79, 81-83
CU-SB	1967, 79, 81, 83-84
CaACAM	1971(3), 84
CaACU	1969, 70, 71(1B, 3, 5-6), 72(3), 74, 75(7-8), 76, 77(4-5), 79, 81
CaAEU	1969(1-2, 4, 6), 81
CaBVaU	1966, 69(1-2, 4, 6), 79, 81
CaOHM	1969(1-2, 4, 6)
CaOKQGS	1981
CaOLU	1969(1-2, 4, 6)
CaOOG	1969(1-2, 4, 6), 72(2-3), 73(1-2, 4, 6-7)
CaOWtU	1971(1, 2a, 3, 5-6), 74, 79, 81
CoDCh	1956-57, 59, 68(2), 69, 75(4), 79, 81, 82, 83(1-2)
CoDGS	1953, 56(8), 72(3), 75-76, 81
CoG	1967(1), 69(1-2, 4, 6), 70, 71(3-4), 72, 74, 77, 81, 84
CoGrU	1981
CoLiM	1984
CoU	1951, 57, 69(1-2, 4, 6), 70, 71(1, 4-6), 72, 73(1-2, 4-10), 74, 75(3, 7-8), 76(3)
Dl-GS	1951-53, 57-60, 62, 64(3), 65-66, 67(4), 68, 69(1-2, 4, 6), 70, 71(1, 2a, 3-6), 72, 73(1-2, 4, 6-8, 10), 74, 75(1, 3-8), 76(3), 77, 78(4), 79, 81-85
DLC	1969(1-2, 4, 6), 75(3, 7)
ICF	1969(1-2, 4, 6), 75(3)
ICIU-S	1962, 66, 69(1-2, 4, 6)
ICarbS	1969, 72(3), 74(2)
IEN	1969(1-2, 4, 6)
IU	1959, 65-66, 67(1), 69(1-2, 4, 6), 70, 71(1, 2a, 3-6), 72, 73(1-2, 4-8), 75(1, 3-8), 76(3), 77(1-3, 5), 79, 81(1-6)
IaCfT	1981
IaU	1969(1-2, 4, 6), 75(7-8), 76(3), 77(2-5), 79, 81, 83
IdBB	1975(1, 3-8), 81, 85
IdPI	1969(1-2, 4, 6), 75(3, 7-8)
IdU	1981
InLP	1966, 69(1-2, 4, 6), 72(3), 75(3, 8), 76(3), 77, 78(10), 79-81, 82(1-3), 85
InU	1951, 59, 69(1-2, 4, 6), 70, 75(3, 7-8), 79, 82(4), 83
KLG	1983
KU	1983
KyU-Ge	1966, 68, 69(1-2, 4, 6), 70, 71(1, 2a, 3-6), 72, 73(1, 2, 4-10), 74, 75(1, 3-8), 76(3), 77-79, 80(1, 3-4, 4B, 6A, 7), 81-85
MH-GS	1958, 74, 83
MiHM	1966, 83
MiKW	1979
MnU	1966, 68, 69(1-2, 4, 6), 70, 71(2, 5-6), 75(3, 8), 76(3), 77, 79, 81-82, 82(4)
MoSW	1969(1-2, 4, 6), 73(1-2, 4, 6-8), 75(3, 7-8), 81
MoU	1966, 68(4), 69(1-2, 4, 6), 73(1-2, 4, 6), 76(3)
MtBuM	1953, 59, 68, 69(1-2, 4, 6), 71(3), 75(3, 7-8), 82(1, 3-4)
MtU	1970
NBiSU	1973(1-2, 4, 6-7)
NNC	1959, 69(1-2, 4, 6)
NRU	1981
NSbSU	1969, 81, 83
NSyU	1969(1-2, 4, 6)
NbU	1968, 69(1-2, 4, 6)
NcU	1966, 69(1-2, 4, 6), 76(3), 81
NdU	1967(1), 68(1, 3), 69(1-2, 4, 6), 70, 74, 75(3, 7-8), 79, 83(1)
NhD-K	1975(3), 76(3), 83
NjP	1959, 69(1-2, 4, 6)
NmPE	1966, 74, 76(3)
NmSoI	1979, 83
NvLN	1969(1-2, 4, 6)
NvU	1969(1-2, 4, 6), 75(3, 7-8), 79, 83(1-4), 81(85 rev.)
OCU-Geo	1970, 74, 76(3), 77(3-5), 79
ODaWU	1981
OU	1969(1-2, 4, 6), 70, 72, 73(1-2, 4, 6-7), 74, 76(3), 79, 83(1-3)
OkU	1969(1-2, 4, 6), 75, 76(3), 79, 81, 83
OrCS	1979, 81
OrU-S	1969(1-2, 4, 6)
PBL	1975(3, 7-8)
PBm	1974, 79
PSt	1969(1-2, 4, 6), 81(85 rev. ed.)
SdRM	1969-70
SdU	1970
TxBeaL	1974, 76(3)
TxDaAR-T	1969(1-2, 4, 6), 70, 71(3, 5-6), 72(1, 3), 73(1-2, 4-5, 7-9), 74, 75(3-5)
TxDaDM	1959, 60, 69(1-2, 4, 6), 75(3, 7-8), 82
TxDaM-SE	1979, 81, 83(2)
TxDaSM	1959, 72(1, 4), 74, 77(5), 79
TxHSD	1969(1-2, 4, 6), 74
TxHU	1969(1-2, 4, 6)
TxLT	1976(3)
TxMM	1969(1-2, 4, 6)
TxU	1966, 69(1-2, 4, 6), 70, 71(1, 3), 72, 74, 75(7-8), 76(3), 79, 81, 82(1, 4), 83(1), 84
TxU-Da	1960
TxWB	1974, 79
UU	1969(1-2, 4, 6), 74, 75(3, 7-8), 81, 83
ViBlbV	1974, 75(1, 3-4, 7-8)
WKenU	1981
WU	1969(1-2, 4, 6), 75(1, 3-4, 6-8), 76(3), 78(10), 79, 80(1-7), 81, 83
WaU	1979
WyU	1957, 59, 83(1)

(205) GEOLOGICAL SOCIETY OF AMERICA. SOUTH CENTRAL SECTION. FIELD TRIP GUIDEBOOK FOR THE ANNUAL MEETING.

1967 1st   Norman, Oklahoma.
The structure and igneous rocks of the Wichita Mountains, Oklahoma.

1968 2nd   Dallas, Texas.
Stratigraphy of the Woodbine Formation, Tarrant County, Texas.

1969 3rd   Lawrence, Kansas.
Pleistocene geology of Doniphan County, Kansas.

1970 4th   College Station, Texas.
Outcrops of the Claiborne Group (middle Eocene) in the Brazos Valley, Southeast Texas.

1971 5th   Lubbock, Texas.
Mesozoic and Cenozoic geology of the Lubbock, Texas region.

1972 6th   Manhattan, Kansas.
Stratigraphy and depositional environments of the Crouse Limestone (Permian) in north-central Kansas.

1973 7th   Little Rock, Arkansas
No. 1. A guidebook to the geology of the Ouachita Mountains, Arkansas. (Revised in 1983.) (67(No.73-1))
No. 2. Geological field trip excursion on Lake Ouachita. (67(No.73-2))
No. 3. Guidebook to Lower and Middle Ordovician strata of northeastern Arkansas and generalized log of route from Little Rock to Batesville, Arkansas. (67(No.73-3))
— Apr.   [4] Field trip guide to four major mines in central Arkansas. (67(No.73-4), 583)

1974 [8th]   Stillwater, Oklahoma.
No. 1. Environmental geology of metropolitan Tulsa.
No. 2. Distribution of algae and corals in Upper Pennsylvanian-Missourian rocks in northeastern Oklahoma.
No. 3. Guidebook to the depositional environments of selected Pennsylvanian sandstones and carbonates of Oklahoma.

1975 9th   Austin, Texas.
Field trip no. 1. Precambrian rocks of the southeastern Llano region, central Texas. Geologic description and road log.
Field trip no. 2. Utilization of land resources in the Austin area.
Field trip no. 3. Geology of the Llano region and Austin area. (713(1972 No.13))
[4] Stratigraphy of the Austin Chalk in the vicinity of Pilot Knob.

1976 10th   Houston, Texas.
1. Subsidence and active surface faulting in the Houston vicinity. (255)
Field trip no. 2. Plutonic igneous geology of the Wichita magmatic province, Oklahoma.
3. Recent sediments of Southeast Texas; a field guide to the Brazos alluvial and deltaic plains and the Galveston Barrier Island complex. (713(1973 or 1970-No.11))

1977 11th   El Paso, Texas.
  1. Geology of Potrillo Basalt Field, south-central New Mexico. (461(1976-No.149))
  2. No guidebook.
  3a. Geology of Cerro de Cristo Rey Uplift, Chihuahua and New Mexico. (463(1976-No.31))
1978 [12th]   Tulsa, Oklahoma.
  [1] A guidebook to the Atoka Formation in Arkansas. (67)
  [2] The Coffeyville Formation (Pennsylvanian) of northern Oklahoma: a model for an epeiric sea delta.
  3. Structural style of the Arbuckle region.
  4. Tulsa's physical environment. 1972. (661)
1979 13th   Mountain View, Arkansas.
  1. A guidebook to the Ordovician-Mississippian rocks of north-central Arkansas. (67(No.79-1))
  2. Geologic float guide on the upper Buffalo National River. (67(1978))
1980 14th   Wichita, Kansas.
  [2] Solution and collapse features in the salt near Hutchinson, Kansas. (779)
1981 [15th]   San Antonio, Texas. (418)
  [1] South Texas field trip - 1981. Meteor impact site, asphalt deposits & volcanic plugs. (612)
  [2] Environmental geology and hydrology of the greater San Antonio area.
  [3] Texas uranium belt. (612)
1982 [16th]   Norman, Oklahoma. (418)
  Field trip 1. Geology of the eastern Wichita Mountains, southwestern Oklahoma. (508(No.21))
  Field trip 2. Lower and Middle Pennsylvanian stratigraphy in south-central Oklahoma. (508 (No.20))
  Field trip 3. Structural styles of the Ouachita Mountains, southeastern Oklahoma.
1983 [17th]   College Station, Texas. (418)
  Trip no. 2. (NAGT) Geologic section of the Cretaceous rocks of central Texas. (92)
  Trip no. 3. Austin Chalk in the Austin area. (92(1985 No. 7))
  Trip no. 5. Two day Central Mineral region crystalline rocks. (92(1984-No.5))
1984 18th   Richardson, Texas. See National Association of Geology Teachers. Texas Section. 418.
  Field trip no. 1. Recent developments in the Wichita Mountains.
  Field trip no. 2. Planning - a geological perspective.
  Field trip no. 4. Comparative structural evolution of the Arbuckle and Ouachita Mountains.
1985 Apr.   Fayetteville, Arkansas.
  [3] Alkalic rocks and carboniferous sandstones, Ouachita Mountains; new perspective.

AGC	1979-84
AzTeS	1968, 74(2), 77, 78(4), 79, 80, 81(1)
BEG	1982
CLU-G/G	1967-71, 74, 76(2)
CLhC	1982(1-2)
CMenUG	1973, 74(2-3), 75(1-2, 4), 76(1-2), 78(1), 79-81, 82(1-2), 84(1, 4), 85(3)
CPT	1967-68, 77(1, 3a), 78(4), 82(1-2)
CSfCSM	1982(2)
CSt-ES	1967-72, 73(1-3), 74, 75(1), 76(2), 78(1), 79-80, 82(1-2)
CU-A	1967-70, 81
CU-EART	1967, 69-71, 74(3), 76(2), 81(1), 82(1-2)
CU-SB	1967-74, 76, 78-79, 81, 82(1)
CaACU	1967-72, 74(3), 76(2), 77(3a), 82(1-2)
CaAEU	1967-69, 82
CaOHM	1967-68
CaOLU	1967
CaOOG	1971, 74(3), 79, 82
CaOWtU	1967, 69-71, 82(1-2)
CoDCh	1967, 71-72, 73(1-3), 74(3), 76(2), 78(1), 82(1, 2), 84(1)
CoDGS	1979-80, 82(1-2), 84-85
CoLiM	1980, 81(2)
CoU	1967
DI-GS	1967-72, 73(1), 74, 75(1-2, 4), 76(1-2), 77(1, 3a), 78-81, 84(1, 4), 85
DLC	1982(1)
GU	1973(3)
I-GS	1982(1-2)
ICF	1973(3)
ICarbS	1973
IEN	1967-72, 73, 74(2)
IU	1967-68, 70-75, 76(1-2), 78(1-2), 82(1-2)
IaU	1967-72, 73(1-3), 74(2), 75(1), 76(1), 78(1-2), 79, 81, 82(1-2), 84,
IdBB	1967
IdPI	1967-68
IdU	1982(1-2)
InLP	1967-73, 74(3), 75(3), 76(2-3), 77-79, 80-82, 83(2), 84-85
InU	1967-68, 71, 82(1-2), 85
IvU	1982(2)
KLG	1980, 82(1)
KyU-Ge	1967-72, 73(1-3), 74(3), 75, 76(2-3), 77(1, 3a), 78-80, 81(2), 82(1-2), 83-84
LNU	1973(1-3), 74(3), 81
MBU	1982(1)
MH-GS	1967-68
MiHM	1982(1-2)
MnU	1967-71, 77(1, 3a), 79-80, 81(1-2), 82(1-2)
MoSW	1973(2)
MoU	1973(2, 4), 74(3)
MtBuM	1982(1-2)
NBiSU	1967-68, 70-72
NHD	1982(1)
NIC	1982(2)
NNC	1982(1-2)
NSbSU	1981
NSyU	1967-68, 79
NYU	1982(1)
NbU	1968
NcU	1967-68, 74, 76, 78(2), 79
NdU	1967-68
NhD-K	1967, 71-72, 75, 76(2), 82(2)
NhE	1982(2)
NjP	1967-72
NmPE	1973(1-2)
NvU	1973, 82(2)
OCU-Geo	1968-72, 74(3), 76(2), 78-79
OCl	1982(1-2)
OU	1967-70, 74, 82(1-2)
OkT	1967, 69-71, 74(3)
OkTA	1979
OkU	1967-71, 73-74, 76(2), 78(1-3), 79(1-2), 80, 81(1, 2), 82, 83(3), 84(1), 85
PBm	1967-68
PHarER-T	1982(1-2)
PSt	1982(1)
RPB	1967-68
SdRM	1967-68
TMM-E	1967
TxBeaL	1967-70, 76, 78(1)
TxCaW	1967-72, 82(1)
TxDaAR-T	1967-72, 73(1-3), 74(2-3)
TxDaDM	1967-72, 73(1, 3), 74(3), 76(2)
TxDaM-SE	1979, 82
TxDaSM	1967, 70, 73(1, 3), 78(1, 3)
TxHSD	1974(3)
TxLT	1968
TxU	1967-70, 73(1-2), 74, 75(3-4), 82, 84(1, 4), 85
TxWB	1967-68, 71-72, 74-75, 78, 84-85
ViBlbV	1973(1-3), 75(1-2, 4), 76(2), 78(1, 3a)
WU	1973(1, 3, 4), 78(1), 79, 81(2)

(206) GEOLOGICAL SOCIETY OF AMERICA. SOUTHEASTERN SECTION. FIELD TRIP GUIDEBOOK. (TITLE VARIES)
1953 [2nd]   Nashville, Tennessee.
  Central Tennessee phosphate district.
1954 [3rd]   Columbia, South Carolina.
  Coastal plain and Piedmont.
1955 [4th]   Durham, North Carolina.
  Coastal plain field trip.
1956 [5th]   Tallahassee, Florida.
  Panhandle Florida.
1957 [6th]   Morgantown, West Virginia.
  No. 1. Morgantown, West Virginia to Greer, West Virginia. (767)

(206) Geological Society of America. Southeastern Section.

No. 2. Morgantown, West Virginia to the Humphrey coal preparation plant. (767)
1958 [7th]   Tuscaloosa, Alabama.
Birmingham area and celebrated coastal plain fossil locations. (223)
1959 [8th]   Chapel Hill, North Carolina.
[1] Piedmont field trip featuring metamorphic facies in the Raleigh area, North Carolina.
[2] Coastal plain field trip featuring basal Cretaceous sediments of the Fayetteville area, North Carolina.
1960 [9th]   Lexington, Kentucky.
No. 1. Physiographic and stratigraphic profile in Kentucky, Lexington to the Mammoth Cave region. (209)
No. 2. Geology of the central Bluegrass area. (209)
1961 [10th]   Knoxville, Tennessee.
— [1] Geology of the Mascot-Jefferson City zinc district, Tennessee. (651)
— [2] Structural geology along the eastern Cumberland Escarpment, Tennessee. (651)
1962 [11th]   Atlanta, Georgia.
No. 1. The Georgia marble district. (237)
No. 2. Stone Mountain-Lithonia District. (237)
No. 3. Ocoee metasediments, north-central Georgia, Southeast Tennessee. (237)
1963 [12th]   Roanoke, Virginia. Geological excursions in Southwest Virginia. (749(No. 2))
1964 [13th]   Baton Rouge, Louisiana.
No. 1. Jackson-Vicksburg type sections.
No. 2. Geology of central Louisiana.
No. 3. Flood plain and terrace geomorphology; Baton Rouge fault zone.
No. 4. Five Islands and Mississippi deltaic plain.
1965 [14th]   Nashville, Tennessee. Guidebook for field trips, Nashville, Tennessee.
No. 1. Geologic structures in northern Sequatchie Valley and adjacent portions of the Cumberland Plateau, Tennessee.
No. 2. Selected features of the Wells Creek basin cryptoexplosive structure.
No. 3. Ordovician of central Tennessee.
1966 [15th]   Athens, Georgia.
No. 1. Pleistocene and Holocene sediments, Sapelo Island, Georgia. (684)
No. 2. Extrusive volcanics and associated dike swarms in central-east Georgia.
No. 3. Stratigraphy and economic geology of the coastal plain of the central Savannah River area, Georgia.
1967 [16th]   Tallahassee, Florida.
No. 1. Sedimentation and coastal features of the Alligator Point area and the area between the Fenholloway and Steinhatchee Rivers.
No. 2. Paleontology of a part of West Florida.
No. 3. Attapulgite; economic geology.
1968 [17th]   Durham, North Carolina. Guidebook for field excursions. (617)
No. 1. Sedimentation in Onslow Bay.
No. 2. Geology of the Sauratown Mountain Anticlinorium and vicinity, North Carolina.
1969 [18th]   Columbia, South Carolina.
No. 1. Paleocene stratigraphy of South Carolina and the question of Tertiary volcanism. (616, 484)
No. 2. Geology of the slate belt in central South Carolina.
No. 3. Geology of the Charlotte belt in central South Carolina. (484)
1970 [19th]   Lexington, Kentucky.
No. 1. Geologic features of southeastern Kentucky.
No. 2. Lithology and fauna of the Lexington Limestone (Ordovician) of central Kentucky.
No. 3. Borden Formation (Mississippian) in southeast-central Kentucky.
No. 4. Paleozoic section on east flank of Cincinnati arch along Interstate 64, Lexington to Olive Hill, Kentucky. (209)
Pt. 1. Lexington eastward to valley of Licking River.
Pt. 2. Valley of Licking River eastward to Olive Hill interchange.
1971 [20th]   Blacksburg, Virginia. Guidebook to Appalachian tectonics and sulfide mineralization of southwestern Virginia. (749(No.5))
Field trip no. 1. Sulfide mineralization, southwestern Virginia.
Field trip no. 2. Geology of the Blue Ridge in southwestern Virginia and adjacent North Carolina.
Field trip no. 3. Appalachian structural and topographic front between Narrows and Beckley, Virginia and West Virginia.
Field trip no. 4. Appalachian overthrust belt, Montgomery County, southwestern Virginia.
1972 21st   Tuscaloosa, Alabama.
[A] Carboniferous depositional environments in the Cumberland Plateau of southern Tennessee and northern Alabama. (651(No.33))
[B] Guide to Alabama geology.
Field trip no. 1. Meta-Paleozoic rocks, Chilton County, Alabama.
Field trip no. 2. Upper Cretaceous series in central Alabama.
Field trip no. 3. Southern Appalachian Valley and Ridge Province; structures and stratigraphy.
Field trip no. 4. Limestone hydrology and environmental geology.
1973 [22nd]   Knoxville, Tennessee. Geology of Knox County, Tennessee. (650)
Field trip nos. 1, 2. Stratigraphy and depositional environments in the Valley and Ridge at Knoxville.
Field trip no. 3. Mineral resources of Knox County, Tennessee.
1974 23rd   Atlanta, Georgia.
Field trip 1. Brevard fault zone in western Georgia and eastern Alabama. (237)
Field trip 2. Tertiary stratigraphy of the central Georgia coastal plain. (237)
1975 [24th]   Memphis, Tennessee. Field trips in West Tennessee. (651(No.36))
Field trip no. 1. Fossiliferous Silurian, Devonian, and Cretaceous formations in the vicinity of the Tennessee River.
Field trip no. 2. Environmental geology of Memphis, Tennessee.
Field trip no. 3. Geology of Reelfoot Lake and vicinity.
Field trip no. 4. The northeastern part of West Tennessee.
Field trip no. 5. Paleocene and Eocene localities in Southwest Tennessee.
1976 25th   See Geological Society of America. Northeastern Section. 201.
1977 [26th]   Winston-Salem, North Carolina. Field guides for Geological Society of America, Southeastern Section Meeting, Winston-Salem, North Carolina.
[1] Geology of the Carolina volcanic slate belt in the Asheboro, North Carolina, area.
[2] Platform and platform margin carbonate facies, Cambrian shady dolomite, Virginia.
[3] Guide to the geology of the Kings Mountain belt in the Kings Mountain area, North Carolina and South Carolina.
1978 27th   Chattanooga, Tennessee. Field trips in the Southern Appalachians. (651(No.37))
1. A structural transect in the Southern Appalachians, Tennessee, and North Carolina.
2. Chattanooga overflight.
3. Cave trip.
4. Engineering geology of the Chattanooga, Tennessee area.
5. Carboniferous depositional environments in the southern Cumberland Plateau.
1979 [28th]   Blacksburg, Virginia. Guides to field trips 1-3, Blacksburg, Virginia. (749(No.7))
1. Virginia Piedmont geology along the James River from Richmond to the Blue Ridge.
2. Lithofacies and biostratigraphy of Cambrian and Ordovician platform and basin facies carbonates and clastics, southwestern Virginia.
3. Geology of the Pulaski Overthrust near Blacksburg, Virginia.
[4] Carboniferous depositional environments in the Appalachian region. (128, 707)
1980 29th   Birmingham, Alabama. See Alabama Geological Society. 5.(1980-No.17)
1. Facies changes and paleogeographic interpretations of the Eutaw Formation (Upper Cretaceous) from western Georgia to central Alabama.
2. Depositional setting of the Mississippian Hartselle Sandstone and lower Bangor Limestone in Northwest Alabama.
3. Geology of the Pine Mountain window and adjacent terranes in the Piedmont province of Alabama and Georgia.
4. The Pleistocene vertebrate assemblage of Little Bear Cave, Colbert County, Alabama.
1981 30th   Hattiesburg, Mississippi. Field trip guidebook for southern Mississippi. (621)

Field trip no. 1. Neogene geology of southeastern Mississippi and southwestern Alabama.
Field trip no. 2. Teaching hard-rock and soft-rock geology in the Gulf Coastal Plain.
Field trip no. 3. Detailed mid-Tertiary stratigraphy along the Chickasawhay River, east central Mississippi.
Field trip no. 4. General overview of the economic geology and stratigraphy of central and east-central Mississippi.
Field trip no. 5. Miocene(?) geology of the Hattiesburg District.
1982 31st Washinton, DC. See Geological Society of America. Northeastern Section. 201.
1983 32nd Tallahassee, Florida.
[No. 1] The central Florida phosphate district.
4. See Southeastern Geological Society. 614.(1982-No.26)
5. Hydrogeology and geomorphology of the Dougherty Plain, Southwest Georgia.
1984 Lexington, Kentucky. (200, 410, 527, 587, 523)
Field trip 1. Stratigraphy and structure in the Pine Mountain Thrust area of Kentucky, Virginia, and Tennessee.
Field trip 2. Coal and coal-bearing rocks of eastern Kentucky.
Field trip 3. Devonian and Mississippian bone beds, paracontinuities, and pyroclastics, and the Silurian-Devonian paraconformity in southern Indiana and northwestern Kentucky.
Field trip 4. Hydrogeology and environmental geology of the Inner Bluegrass Karst Region, Kentucky.
Field trip 5. The Silurian stratigraphy of east-central Kentucky and adjacent Ohio.
Field trip 6. Lithostratigraphy and depositional environments of the Middle Ordovician limestones in central Kentucky.
Field trip 7. Fossil soils and subaerial crusts in the Mississippian of eastern Kentucky.
Field trip 8. Stratigraphy and structure along a geotraverse from Corbin, Kentucky to Newport, Tennessee.
1985 Knoxville, Tennessee. (527)
Field trips 1-5, 7. Field trips in the southern Appalachians. (709)
1. Zinc deposits of the Mascot-Jefferson City and Copper Ridge districts, east Tennessee.
2. Lower Pennsylvanian sediments and associated coals in the Walden Ridge south.
3. Structural transect in SW Virginia and NE Tennessee.
4. Ophiolites (?) of the southern Appalachian Blue Ridge.
5. Massive sulfide deposits of the Ducktown District Tennessee.
7. Dynamic Quaternary landscapes of east Tennessee: an integration of paleoecology, geomorphology, and archeology.
Field trip 6. The geologic history of the Thorn Hill Paleozoic section (Cambrian-Mississippian), eastern Tennessee. (709)

ATuGS	1958, 60, 62, 65(1), 72, 80
AzTeS	1960-61, 64-65, 68, 71-72
CLU-G/G	1960-63, 68, 70-75, 80(1-4), 84, 85
CLhC	1958
CMenUG	1957, 59-60, 62, 65(1), 66, 68, 70, 72-73, 79(1-3), 80(2), 82, 83(5), 84(1)
CPT	1981
CSdS	1964(1)
CSt-ES	1960-63, 65, 68(2), 69(1, 3), 70(1-3), 71-75, 78(1, 4-5), 83-84
CU-A	1971, 72(B)
CU-EART	1961-62, 64-66, 68, 72-73, 75, 78(1, 4-5), 79(1-3), 81, 84, 85(1-7)
CU-SB	1971-72, 80, 82, 85
CaACU	1960(1), 62, 71, 73
CaBVaU	1961(2)
CaOHM	1972
CaOKQGS	1960(2), 65, 66(3), 81
CaOLU	1972(A)
CaOOG	1960, 68(1), 69, 71, 73, 81, 84(2)
CaOWtU	1960(1), 61, 62(1-2), 64, 69(1), 71, 72(A), 73, 75, 79, 80(1-4), 82, 85
CoDCh	1959(2), 65, 68, 77, 80
CoDGS	1957-59, 70-71, 72(A), 73-76, 78-79, 83, 84, 85(6)
CoG	1960-62, 65-66, 70-72, 75, 81
CoLiM	1980(1-3)
CoU	1956-57, 61, 65-66, 68, 71, 73
CtY-KS	1979(4)
DI-GS	1953-66, 67(1-2), 68, 69(1, 3), 70-75, 77, 78(1, 4-5), 79, 81, 83-85,
DLC	1958, 61-62, 71-73, 75, 85
F-GS	1983
FBG	1983
GU	1974
I-GS	1956, 72-73, 75
ICF	1960(2), 61-62, 71-73, 75, 78(1, 4-5)
ICIU-S	1961, 72-73, 75
ICarbS	1958(7), 61, 72(A), 73
IEN	1958, 63, 71
IU	1959-62, 64-65, 66(1), 67-75, 77, 78(1, 4-5), 80, 85(1-5, 7)
IaU	1959, 60(1), 61, 64, 66(1, 3), 68-70, 72-73, 75, 77, 78(1, 4-5), 79, 83-84, 85(6)
IdU	1961, 72(A), 75
InLP	1955, 58-63, 65(1), 67(1-2), 68(2), 69(1-3), 70-75, 77, 78(1, 4-5), 79, 80(1-2), 81, 83, 84, 85(1-5,7)
InU	1953, 58, 60-63, 66, 68(2), 69, 70-72, 75, 80(1-3)
KyU-Ge	1952, 54, 57-65, 66(2-3), 67-68, 70-75, 77, 78(1, 4-5), 79(1-3), 80(1-4), 81, 83-84, 85(6)
LNU	1959, 61-62, 66(2-3), 72(B), 77, 81
MH-GS	1958-59, 84
MNS	1958, 61, 66, 72
MarLiCO	1980(1-3)
MdBJ	1958, 61, 63, 66, 68-69, 72-73, 75, 78
MiDW-S	1962
MiHM	1961, 68, 71-73, 75
MiKW	1964, 70
MiU	1958, 68, 72
MnU	1958-64, 65(1), 66(1), 70-75, 77, 78(1, 4-5), 79(1-3), 80-81, 85
MoSW	1961, 63, 66(1), 68(2), 70, 74-75
MoU	1960, 62(2), 63, 71
MtBuM	1961, 72-73, 75
NBiSU	1962
NNC	1958, 60-63, 65-66, 69-70, 72
NRU	1975
NSbSU	1968, 75, 85
NSyU	1961, 69, 72-73
NbU	1961
NcU	1955, 59-64, 65(1), 66(1), 67-75, 77, 80, 83, 85
NdU	1961-62, 69(1, 3), 72(A), 73, 75
NhD-K	1961, 68(2), 71-73, 75, 78(1, 4-5), 79(4), 81
NjP	1960-63, 65(1), 66, 68-70, 72, 79(4)
NvU	1961-62, 72-73, 75
OCU-Geo	1953, 56-57, 60-62, 63, 65, 66(2-3), 68(2), 69-73, 75, 78, 79(1-3), 85(1-5, 7)
OU	1953, 56-58, 60-62, 66(3), 68, 69(1, 3), 70, 72-74, 77, 80, 84-85(1-5, 7)
OkTA	1980
OkU	1960-65, 66(1, 3), 70-73, 75, 77, 79(4), 80-81, 85(1-5, 7)
OrU-S	1972(A), 75
PBm	1972(A)
PSt	1958, 61-62, 66, 68
RPB	1961
SdRM	1961, 66(2-3), 72-73, 75, 78
TMM-E	1958, 60-61, 65, 68, 71, 73
TxBeaL	1961, 68(2), 69(1), 72, 75, 78(1, 4-5)
TxDaAR-T	1958, 60, 62-63, 66, 67(3), 68, 71-75
TxDaDM	1958, 60(2), 61-62, 64-65, 66(1), 70, 72-75
TxDaM-SE	1958, 60, 66(1, 3), 80-81
TxDaSM	1964, 66(1)
TxHSD	1965(1), 72
TxHU	1961-62, 68
TxLT	1961, 72
TxMM	1965(1)
TxU	1957-58, 60-71, 77, 80(1-3), 84
TxWB	1964, 66, 72, 80, 85
UU	1961, 72-73, 75
ViBlbV	1952, 58, 60(1), 61-63, 68, 70-71, 73, 75, 77, 78(1, 4-5), 79(1-3)
WU	1961-63, 71-75, 78(1, 4-5), 79(1-3), 80(1-4), 81, 84, 85
WaPS	1971
WaU	1968
WyU	1985

(207) GEOLOGICAL SOCIETY OF AMERICA. SOUTHEASTERN SECTION. SHORT COURSE. FIELD TRIP GUIDEBOOK.
1976 Mar. See Geological Society of America. Northeastern Section. 202.

(208) GEOLOGICAL SOCIETY OF IOWA. FIELD TRIP GUIDEBOOK. (TITLE VARIES)
1959 Apr. [Study of the Meramec Series]

(209) Geological Society of Kentucky.

1. Kaser's Selma Quarry.
2. Douds Stone Co. Mine near Douds.
3. Henry County Quarry.
1960 Apr. Field trip to River Products Company Quarry (in conjunction with Iowa Academy of Science for the Jr. Academy of Science).
— Sept. [Are the Iowa Falls Dolomite and Eagle City Limestone the lateral equivalents of the Gilmore City Limestone?]
1961 Aug. The Pennsylvanian between Muscatine and Davenport.
— Dec. [Malcolm Stone Company Mines - Osage-Kinderhook section]
1962 Mar. [United States Gypsum Company Mine, Sperry, Iowa]
— May Skvor-Hartl area, Southeast Linn County, Iowa.
— July Maquoketa of Northeast Iowa.
1963 May Silurian bioherms of eastern Iowa.
— July Upper Devonian in Mason City and Garner areas.
1964 Aug. Southwestern Iowa field trip (Pennsylvanian).
1965 May Pre-Cedar Valley, post-Maquoketa sediments of Northeast Iowa.
1967 Field trip 2. Emphasis on industry; plant tours, Concrete Materials Division of Martin Marietta Corporation, Cedar Rapids and tour of quarries.
— June 2 United States Gypsum Company Mine, Sperry, Iowa.
— June 3 Osage and Kinderhook series, Des Moines County, Iowa.
— Fall Middle River traverse.
1968 March Field trip; Middle River traverse.
— June Field trip (Mississippian: Maynes Creek, Chapin, Prospect Hill, McCraney; Devonian: English River, Maple Mill, Aplington, Sheffield, Lime Creek).
1969 Geology of upper Mississippi Valley zinc-lead district: New Jersey Zinc Co., Elmo Mine, Wisconsin.
1970 Geology and geohydrology of Red Rock Dam, Marion County, Iowa.
1972 April Pleistocene Lake Calvin-Landforms and landuse of an ancient lake basin in Southeast Iowa.
— May Revision of galena stratigraphy.
— July General geology of Black Hawk County and adjacent areas.
1975 Karst topography along the Silurian escarpment in southern Clayton County, Iowa
1976 Sept. Geologic points of interest in the Fort Dodge area. (332)
1977 Oct. [Stratigraphy and lithofacies of the Spergen of Southeast Iowa]
1978 April Pints Quarry.
1979 Oct. Field trip guidebook to the Cambrian stratigraphy of Allamakee County.
1980 Apr. No. 33. Field guide to Upper Pennsylvanian cyclothems in south-central Iowa; a field trip along the Middle River traverse, Madison County, Iowa.
— Sept. No. 34. Geomorphic history of the Little Sioux River Valley.
1981 Apr. No. 35. Silurian stratigraphy of eastern Linn and western Jones counties, Iowa; a field trip investigation of carbonate mound and intermound facies of the Scotch Grove (new) and Gower formations.
— Sept. No. 36. Glacial sedimentation and the Algona Moraine in Iowa.
1982 Apr. No. 37. Mississippian biofacies-lithofacies trends - north central Iowa; a new look at an old attraction.
— Oct. No. 38. Cretaceous stratigraphy and depositional systems in Guthrie County, Iowa; with comments on the Pennsylvanian sequence.
1983 Apr. No. 39. New stratigraphic interpretations of the middle Devonian rocks of Winneshiek and Fayette counties, northeastern Iowa.
— Oct. No. 40. Karstification on the Silurian escarpment in Fayette County, northeastern Iowa.
1984 Apr. No. 41. Underburden-overburden; an examination of Paleozoic and Quaternary strata at the Conklin Quarry near Iowa City.
No. 42. The Cedar Valley Formation (Devonian), Black Hawk and Buchanan counties; carbonate facies and mineralization.
1985 Nov. No. 43. After the great flood; exposures in the emergency spillway, Saylorville Dam.

CLU-G/G	1964
CSt-ES	1980
CU-EART	1981
CoDCh	1984(42)
DI-GS	1960, 62-65, 67(2, June), 68, 72(July), 76, 79-85
DLC	1972(July)
I-GS	1963, 65, 80(34), 81, 83(40)
ICF	1972(May)
IU	1959, 63-65, 67-72, 75, 76(Sept), 79(Oct), 80-82, 83(40)
IaCfT	1983(39), 84
IaU	1959, 60-65, 67-68, 70, 72, 76-77, 79-85
InLP	1959-65, 67-70, 72, 76-85
InU	1968
KyU-Ge	1960
MnU	1960, 63(July), 65, 83(39)
NNC	1964
NhD-K	1981
TxDaAR-T	1963, 67
TxU	1962(May), 63(July), 64-65, 68

(209) GEOLOGICAL SOCIETY OF KENTUCKY. GUIDEBOOK FOR THE FIELD TRIP. (TITLE VARIES)

1941 Itinerary of field conference held April 25-26, 1941, starting at Litchfield, Kentucky.
1942 May Pennsylvanian stratigraphy of Laurel, Clay, Perry, Leslie, McCreary, Whitley, Knox, Bell, and Harlan counties, southeastern Kentucky.
1946 May Southeastern Kentucky and southwestern Virginia.
1950 Southwestern Virginia.
1952 Chester field excursion: outcrop of the Chester formations of Crawford and Perry counties, Indiana, and Breckinridge County, Kentucky.
Since 1952 these guidebooks have been prepared in cooperation with the Kentucky Geological Survey.
1953 Some Pennsylvanian sections in Morgan, Magoffin, and Breathitt counties, Kentucky.
1954 Geology of the Mammoth Cave region; Barren, Edmonson, and Hart counties, Kentucky.
1955 Exposures of producing formations of northeastern Kentucky.
1956 Selected geologic features of southwestern Kentucky.
1957 See Appalachian Geological Society. 58.
1958 Sedimentation and stratigraphy of Silurian and Devonian rocks in the Louisville area, Kentucky.
1959 Stratigraphy of Nelson County and adjacent areas.
1960 See Geological Society of America. Southeastern Section. 206.
1961 Apr. Itinerary: geologic features of the Cumberland Gap area, Kentucky, Tennessee, and Virginia.
1962 Selected features of the Kentucky fluorspar district and the Barkley Dam site.
1963 Geologic features of the Mississippian plateau, south-central Kentucky.
1964 May Itinerary: geologic features of the Mississippian plateaus in the Mammoth Cave and Elizabethtown areas, Kentucky.
1965 No. 1. Lithostratigraphy of the Ordovician Lexington Limestone and the Clays Ferry Formation of the central Bluegrass area near Lexington, Kentucky.
No. 2. Excursion to the cryptoexplosive structure near Versailles, Kentucky.
1966 Geologic features of selected Pennsylvanian and Mississippian channel deposits along the eastern rim of the western Kentucky coal basin.
1967 Some aspects of the stratigraphy of the Pine Mountain front near Elkhorn City, Kentucky with notes on pertinent structural features.
1968 May Geologic aspects of the Maysville-Portsmouth region, southern Ohio and northeastern Kentucky. (494)
1969 Middle and Upper Pennsylvanian strata in Hopkins and Webster counties, Kentucky.
1970 See Geological Society of America. Southeastern Section. 206.
1971 Apr. Carboniferous depositional environments in northeastern Kentucky.
1972 Geology of the Jackson Purchase region, Kentucky.
1973 See Geological Society of America. 197.(1971)(9))
1974 Apr. Late Cenozoic geologic features of the middle Ohio River valley.
1975 Oct. Selected structural features and associated dolostone occurrences in the vicinity of the Kentucky River fault system.
1977 Stratigraphic evidence for Late Paleozoic tectonism in northeastern Kentucky. (Used American Association of Petroleum Geologists. Eastern Sections 1976 guidebook.) See American Association of Petroleum Geologists. Eastern Section. 25.(1976)
1978 Oct. Surface rocks in the western Lake Cumberland area, Clinton, Russell, and Wayne counties, Kentucky.
1979 Oct. Depositional environments of Pennsylvanian rocks in western Kentucky.
1980 Oct. Stratigraphy, trace fossil associations, and depositional environments in the Borden Formation (Mississippian), northeastern Kentucky.

1981 Apr. Energy resources of Devonian-Mississippian shales of eastern Kentucky.
1984 Sept. Alluvial processes and sedimentation of the Mississippi River.
1985 Oct. Stratigraphy along and adjacent to the Bluegrass Parkway.

AzTeS	1952-56, 58-59, 61-67, 69, 71, 74
AzU	1974-75
CLU-G/G	1952-56, 59, 61-69, 71-72, 74-75, 78-81, 84-85
CMenUG	1950, 52-56, 58-59, 61-62, 64-65, 68, 71-72, 74-75, 79-81
CPT	1952, 55-56, 58-59, 61-69
CSfCSM	1980-81
CSt-ES	1954, 58, 63, 65, 68, 71, 74-75, 78-80
CU-EART	1971, 74-75, 78-80
CU-SB	1975
CaACU	1979-81
CaOKQGS	1952, 55, 58-59, 65
CaOLU	1954, 61-62
CaOOG	1952-56, 58-59, 61-62
CaOWtU	1981
CoDCh	1953, 58-59, 61, 67-68, 71-72, 75
CoDGS	1952, 54-59, 80-81
CoG	1958-59, 61-69
DI-GS	1952-56, 58-59, 61-69, 71-72, 74-75, 78-81
DLC	1954, 58, 65, 68, 71
I-GS	1958, 64, 67, 69, 71, 74, 80
ICF	1952, 54, 56, 58-59, 61-64
IEN	1950
IU	1941, 52-56, 58-59, 61-65, 75, 78-80
IaU	1955-56, 59, 61-62, 64-68, 71-72, 74-75, 78-81
InLP	1952-56, 58-59, 61-69, 71-72, 74-75, 78-79, 80-81, 85
InU	1952-55, 58-59, 61-62, 64-65, 67-69, 71-72, 74-75, 79
KyU-Ge	1942, 50, 52-56, 58-59, 61-69, 71-72, 74-75, 77-81, 84-85
LU	1952-53
MiDW-S	1955-56, 61, 63-64, 66-67, 71-72
MnU	1952-56, 58-59, 61-69, 71-72, 74-75, 78-79
MoSW	1954, 67, 74-75
MoU	1952, 55, 58-59, 61-66
NBiSU	1971, 74
NIC	1980
NNC	1950, 52-56, 58-59, 61-69, 71, 74-75
NRU	1975, 80, 81
NSbSU	1975, 80
NbU	1954-55, 58-59, 61-62
NcU	1952-56, 59, 61-69, 71-72, 75, 78-79
NdU	1964, 74-75, 78
NhD-K	1971, 75, 78, 80
NjP	1952, 55-56, 58-59, 61-69, 75
NvU	1974-75, 78
OCU-Geo	1950, 55-56, 59, 60-68, 71-72, 78-79, 80
OU	1942, 53-56, 58-59, 61-69, 71-72, 74-75, 78-79, 81
OkOkCGS	1954-55
OkU	1942, 46, 52-56, 58-59, 61-69, 74, 75, 78, 80-81
PBL	1952-55
PBm	1979
PPi	1980
PSt	1941, 55
RPB	1954
TMM-E	1952, 55-56, 59, 61-66, 68-69
TxDaAR-T	1958-59, 69, 71-72, 74
TxDaDM	1952-56, 58-59, 61-69, 71-72, 74-75
TxDaGI	1954
TxDaM-SE	1952-56, 58-59, 61-66, 80
TxDaSM	1954-56, 64
TxHSD	1968
TxLT	1968
TxMM	1942
TxU	1952, 54-56, 58-59, 61-68, 71, 75, 78, 81
ViBlbV	1950, 52, 54-56, 58-59, 62-63, 65-69, 79
WU	1950, 59, 67, 71, 74-75, 78-79, 81

(210) GEOLOGICAL SOCIETY OF MAINE. GUIDEBOOK FOR FIELD TRIPS.
1977 1 and 2. Bedrock geology; Machias-Eastport-Calais-Wesley area. (Reprinted in 1983)
— 1. Silurian-lower Devonian volcanic rocks of the Machias-Eastport area, Maine.
— 2. Geology of the inland rocks of the Calais-Wesley area, Maine.
1978 3 and 4. Bedrock geology; Casco Bay Group-Portland area. Coopers Mills-Liberty area. (Reprinted in 1983)
1980 Field trip 5. Bedrock and surficial geology of the upper St. John River area, northwestern Aroostook County, Maine. (Reprinted in 1983)
1983 July Field trips of the Geological Society of Maine, 1978-1983. (Field trips 1 thru 5 are reprints of the 1977 and 1980 field trips) (368)
Field trip 6. Major structural features of the Gardiner and Wiscasset quadrangles, Maine.
Field trip 7. The Waldoboro Moraine and related glaciomarine deposits, Lincoln and Knox counties, Maine.
Field trip 8. Post Acadian brittle fracture in the Norumbega fault zone, Brooks quadrangle, Maine.
Field trip 9. Plutonism and post-Acadian faulting in east-central Maine.
Field trip 10. Late Wisconsinan features of southeastern Maine between the Pineo Ridge system and the upper Narraguagus River.
Field trip 11. Bedrock geology of the Androscoggin Lake igneous complex, Wayne and Leeds, Maine.
Field trip 12. Glaciomarine deltas and moraines, Sebago Lake region, Maine.

CMenUG	1983
CaOKQGS	1983
CoDGS	1978, 80, 83
DI-GS	1977-78, 80, 83
IaU	1983
InLP	1983
KyU-Ge	1977-78, 80, 83
M	1983
MA	1983
MH-GS	1977, 83
Me	1980, 83
NIC	1983
NhD-K	1983
TxU	1983
ViBlbV	1978
WU	1983(1, 12)

(211) GEOLOGICAL SOCIETY OF NEVADA. MEETING AND FIELD TRIP ROAD LOG.
1962 See Nevada Geological Society. 453.
1984 Pinson Mine, Florida Canyon Deposit, Rochester District, Relief Canyon Deposit. (212)
1985 Alligator Ridge Mine, Little Bald Mountain Mine, Buckhorn Mine. (212)

NvU	1984, 85

(212) GEOLOGICAL SOCIETY OF NEVADA. SPECIAL PUBLICATION.
1984 1. See Geological Society of Nevada. 211.
1985 3. See Geological Society of Nevada. 211.

(213) GEOLOGICAL SOCIETY OF NEW JERSEY. [GUIDEBOOK]
1959 Stokes Forest.

DI-GS	1959
InLP	1959
NcU	1959
NjP	1959
TxU	1959

(214) GEOLOGICAL SOCIETY OF PUERTO RICO. [GUIDEBOOK]
1966 Field trip on Puerto Rico. [No guidebook published]
1968 Karst field trip; stratigraphy and geomorphology.

TxU	1968

(215) GEOLOGICAL SOCIETY OF SACRAMENTO. ANNUAL FIELD TRIP GUIDEBOOK.
1955 [Itinerary of the spring field trip]
1956 Indian Valley region, Plumas County, California.
1957 The Cretaceous and associated formations of the Redding area, Shasta County, California.
1958 East side Sacramento Valley-Mother Lode area, California.
1959 Coast Ranges; Livermore Valley to Hollister area.
1960 Northwestern California; a traverse of the Klamath Uplift, northern Coast Ranges, and Eel River basin.
1961 East-central Sacramento Valley: Marysville (Sutter) Butte, Chico Creek, and Oroville.

(216) Geological Society of the Oregon Country.

1962 U.S. Highway 40; Sacramento to Reno, Dixie Valley and Sand Springs Range, Nevada. (107, 453)
1963 Central portion of Great Valley of California, San Juan Bautista to Yosemite Valley.
1964 Mount Diablo.
1965 La Porte to the summit of the Grizzly Mountains, Plumas County, California.
1966 East-central front of the Sierra Nevada.
1967 Quaternary geology of northern Sacramento County, California.
1968 Geological studies in the Lake Tahoe area, California and Nevada.
1969 Geologic guide to the Lassen Peak, Burney Falls and Lake Shasta area, California.
1970 Geologic guide to the Death Valley area, California.
1971 Geologic guide to the northern Coast Ranges, Point Reyes region, California.
1972 Geologic guide to the northern Coast Ranges: Lake, Mendocino and Sonoma counties, California.
1973 Environmental geology; a field trip to eastern Sacramento County and western El Dorado County.
1974 Geologic guide to the southern Klamath Mountains.
1975 Stanislaus River guide, Camp Nine to Melones.
1978 Geologic guide to the Northern California coast ranges, Sacramento to Bodega Bay.
1980 Geologic guide to the Modoc Plateau and the Warner Mountains.

AzFU	1970-73
AzTeS	1967-70, 72-73
AzU	1956, 63, 65-69, 72-75
CChiS	1956, 58, 62-63, 65-67, 69-73
CLU-G/G	1956-59, 61, 63, 65-75, 78
CLhC	1963, 65, 68, 70-72
CMenUG	1980
CPT	1960-63, 65, 67-69, 71, 73, 75, 78, 80
CSfCSM	1956-74, 78, 80
CSt-ES	1955, 57, 59-75, 78, 80
CU-A	1956-57, 63-73, 80
CU-EART	1955-64, 66, 68-74, 78, 80
CU-SB	1957-59, 65, 76-80
CaACU	1970, 72, 78
CaBVaU	1968-70
CaOHM	1967-68
CaOLU	1958, 68
CaOOG	1963
CaOTRM	1965-66
CaOWtU	1968-70
CoDGS	1957-73, 75, 78, 80
CoG	1959-60, 62, 65-73
CoU	1963, 65-67
DI-GS	1956-75, 78, 80
ICF	1959-60
ICIU-S	1965, 67-74
IU	1956-62, 65-75, 78, 80
IaU	1965-75, 78, 80
IdBB	1968, 70
InLP	1965, 67-68, 70-75, 78, 80
InU	1969, 70
KyU-Ge	1961-62, 65, 67-71
MH-GS	1968-70
MiDW-S	1968-69
MiHM	1968
MiKW	1962-63
MiU	1965-73
MnU	1959-63, 65-75, 80
MoSW	1963, 65-67
MoU	1964, 65, 68-73
NNC	1955-56, 58-63, 65-75, 78
NSyU	1970
NbU	1955, 58, 62-75
NcU	1968-70, 72-75
NdU	1967, 68, 71
NhD-K	1962-74, 78, 80
NjP	1965, 67-74
NvU	1960-63, 66-75, 78, 80
OCU-Geo	1963-64, 66-73
OU	1958-60, 63, 65-72, 74-75, 78, 80
OkU	1956, 58, 63, 65-68, 71-75
OrCS	1978, 80
OrU-S	1968-75
PBL	1969-70
PBU	1980
PBm	1968-69
PSt	1968-69
SdRM	1963, 65-67
TMM-E	1968-70
TxBeaL	1967, 69, 70
TxDaAR-T	1968, 70-72, 74
TxDaDM	1958-61, 65, 67-68, 70-71, 73-74
TxDaM-SE	1960, 65
TxDaSM	1962
TxLT	1968-75
TxU	1956-70, 72
UU	1963, 65, 67-68, 70
ViBlbV	1959-60, 62-63, 66, 68-70, 72-74

(216) GEOLOGICAL SOCIETY OF THE OREGON COUNTRY. FIELD GUIDE TO PRESIDENT'S CAMPOUT. (TITLE VARIES)

1964 Geological trip log through the eastern foothills of the Oregon Coast Range between Vernonia and banks on the Vernonia, South Park and Sunset Steam Railroad.
1965 No. 1. Geological guidebook for central Oregon.
 No. 2. The Columbia River Gorge; geological notes from Cascade Locks to Bingen.
1967 Columbia River Gorge and "Grand Canyon" of the Deschutes River. (Cover title: Geological trip log along the Columbia and Deschutes Rivers) 5th rev. ed.
 Field trip no. 1. Columbia River section.
 Field trip no. 2. Deschutes River section.
1968 No. 1. President's campout. Our central Oregon "Moon Country", 2nd ed.
 No. 2. Deschutes Canyon field trip.
1969 Geological trips in the Mitchell-John Day area.
 Field trip no. 1. To Dayville and the South Fork of the John Day River.
 Field trip no. 2. To Ochoco Summit area and cuts to the east as far as the Eocene-Cretaceous contact.
 Field trip no. 3. To Mitchell Mesozoic rock and fossil areas.
 Field trip no. 4. To Twickenham area and the Painted Hills.
1970 Sept. [No. 1] A glimpse in the Quaternary and Tertiary of central Oregon (Maupin to Smith Rock State Park and Cover Park).
 No. 2. Condon's first island; geological trips in the Siskiyous and along the Rogue.
 Field trip no. 1. Dow Rogue River to old Benton Mine.
 Field trip no. 2. Up Illinois River as far as the Waldo-Takilma gold mining area.
 Field trip no. 3. Up the Applegate River via Jacksonville.
 Field trip no. 4. Up through Medford area to Mt. Ashland area.
1971 Condon's second island; a guidebook issue.
1972 Geological trips in Wallowa County.
 Field trip no. 1. Elgin to Wallowa Lake.
 Field trip no. 2. Wallowa Lake to Mount Howard.
 Field trip no. 3. Lostine Valley.
 Field trip no. 4. Wallowa Lake to Hat Point.
 Field trip no. 5. Joseph to Black Marble Quarry.
 Field trip no. 6. Joseha Imnaha Loop.
1973 July Geologic excursions through Douglas County.
1974 A geological guide book with selected field trips along the Oregon coast in Lincoln County.
1977 Selective field trips investigating the geology of the North Cascades; President's campout, 1977.
1981 Gold country!
1982 Central Oregon's volcanic wonderland and how it came to be.
1984 Golden anniversary campout; Lewiston, Idaho, Geological Society of the Oregon Country.
1985 July John Day country; areas in the Oregon counties of Grant and Harney. (516)

CMenUG	1965-73, 77, 81-82, 84
CU-A	1974
CU-SB	1965, 71, 82
DI-GS	1964-65, 67, 68(2), 69-73, 81-82, 84-85
IU	1964-65, 67-74, 77
InLP	1965, 67, 68(2), 69-73, 77
NhD-K	1982

GEOLOGICAL SOCIETY OF U.C.L.A. (NAME OF ORGANIZATION CHANGED TO: EARTH AND SPACE SCIENCE STUDENT ORGANIZATION, U.C.L.A.) See EARTH AND SPACE SCIENCE STUDENT ORGANIZATION, U.C.L.A. (161)

(217) GEOLOGICAL SURVEY (U.S.). BULLETIN.
1915 No. 611. Guidebook of the western U.S. Part A. The Northern Pacific Route.
— No. 612. Guidebook of the western U.S. Part B. The Overland Route with a side trip to Yellowstone Park.
— No. 613. Guidebook of the western U.S. Part C. The Santa Fe Route with a trip to the Grand Canyon of the Colorado.
— No. 614. Guidebook of the western U.S. Part D. The Shasta Route and Coast Line.
1922 No. 707. Guidebook of the western U.S. Part E. The Denver and Rio Grande Western Route.
1933 No. 845. Guidebook of the western U.S. Part F. The Southern Pacific lines, New Orleans to Los Angeles.
1974 No. 1327. The geologic story of Canyonlands.
1975 No. 1393. The geologic story of Arches National Park.

AzFU	1915, 22, 33
AzTeS	1915, 22, 33
AzU	1915, 22, 33
CLU-G/G	1915, 22, 33
CLhC	1915, 74
CMenUG	1915, 22, 33
CPT	1915, 22, 33
CSdS	1915, 22, 33
CSt-ES	1915, 22, 33, 74, 75
CU-EART	1915, 22, 33
CaACU	1915, 22, 33
CaAEU	1915, 22, 33
CaBVaU	1915, 22, 33
CaOHM	1915, 33
CaOKQGS	1915, 22, 33
CaOLU	1915, 22
CaOWtU	1915, 22, 33
CoG	1915, 22, 33
CoU	1915, 22, 33
DFPC	1922
DI-GS	1915, 22, 33
DLC	1915, 22, 33, 74-75
ICF	1915, 22, 33
ICarbS	1915, 22, 33
IEN	1915, 22, 33
IU	1915, 22, 33
IaU	1915
IdBB	1915(613-614)
IdU	1915, 22, 33
InLP	1915, 22, 33
InU	1915, 22, 33
KyU-Ge	1915, 22, 33
LNU	1915, 22, 33
MNS	1915, 22, 33
MiHM	1915, 22, 33
MiMarqN	1915, 22, 33
MnU	1915, 22, 33
MoSW	1915, 22, 33
MoU	1915, 22, 33
MtBuM	1915, 22, 33
NBiSU	1915, 22, 33
NSbSU	1915, 22, 33
NSyU	1915, 22, 33
NbU	1915, 22, 33
NcU	1915, 22, 33
NdU	1915, 22, 33
NhD-K	1915, 22, 33
NjP	1915, 22, 33
NmPE	1915, 22, 33
NvLN	1915, 22, 33
NvU	1915, 22, 33
OCU-Geo	1915, 22, 33
OU	1915, 22, 33
OkT	1915, 22, 33
OkU	1915, 22, 33, 74-75
OrU-S	1915, 22, 33
PBL	1915, 22, 33
PSt	1915, 22, 33
SdRM	1915(611)
TMM-E	1915, 22, 33
TxBeaL	1915
TxDaGI	1915, 33
TxDaSM	1915, 22, 33
TxU	1915, 22, 33
UU	1915, 22, 33
ViBlbV	1915, 22, 33
WU	1915, 22, 33

(218) GEOLOGICAL SURVEY (U.S.). CIRCULAR.
1981 838. See American Geophysical Union. Pacific Northwest Meeting. 39.

(219) GEOLOGICAL SURVEY (U.S.). [INFORMATION LEAFLET]
1967 67-12. Mountains and plains, Denver's geologic setting.
CPT   1967
IaU   1967

(220) GEOLOGICAL SURVEY (U.S.). OPEN-FILE REPORT.
1966 66-827. See Arizona Geological Society. 65.
1969 69-38. Guidebook for past field trips to the Nevada Test Site. (Prepared by R. L. Chritstiansen)
1977 77-753. See Geological Society of America. 197.
1978 78-446. See International Geological Correlation Programme (IGCP). Project 114: Biostratigraphic Datum-planes of the Pacific Neogene. 298.
— 78-1068. See International Geological Correlation Programme (IGCP). Project 60: Correlation of Caledonian Stratabound Sulphides. 303.
1980 80-823. See Geological Society of America. Rocky Mountain Section. 204.(1980(7))
1981 81-651. See Friends of the Pleistocene. Eastern Group. 182.
— 81-773. See Geological Society of America. Rocky Mountain Section. 204.(1980(4B))
— 81-1291. Road log and documentary photographs for 15 significant biostratigraphic sites in Miocene-Pliocene limestone, Kingshill Seaway, St. Croix, U.S. Virgin Islands.
1982 82-845. See Friends of the Pleistocene. Rocky Mountain Cell (Formerly: Rocky Mountain Group). 185.
— 82-850. Guidebook to the late Cenozoic geology of the Beaver Basin, south-central Utah.
1984 84-98. Field trip guide to deposition and diagenesis of the Monterey Formation, Santa Barbara and Santa Maria areas, California.
1985 85-290-B. See National Earthquake Prediction and Hazard Programs. 420.

CSt-ES	1981-82, 84
CaOOG	1982
CoDCh	1982
CoDGS	1982
CoG	1982
DI-GS	1984
IaU	1981, 82
KLG	1982
KU	1982
NcU	1982
NmLcU	1982
NvU	1969
OkU	1982, 84
TxDaM-SE	1982

(221) GEOLOGICAL SURVEY (U.S.). WRD DISTRICT CHIEFS' CONFERENCE. GUIDE TO A FIELD TRIP.
1975 Sept. Guide to a field trip through part of the Rio Grande Valley and Jemez Mountains of northern New Mexico.
DI-GS   1975

(222) GEOLOGICAL SURVEY OF ALABAMA. CIRCULAR.
1968 No. 47. Geology of the Alabama coastal plain; a guidebook.
1973 No. 90. A field guide to mineral deposits in south Alabama.

(223) Geological Survey of Alabama.

AzTeS	1968, 73
CLU-G/G	1968, 73
CMenUG	1968, 73
CPT	1968
CSt-ES	1968, 73
CU-EART	1968, 73
CaAEU	1973
CaOWtU	1968, 73
CoDGS	1968, 73
CoG	1968
CoU	1968, 73
DI-GS	1968, 73
DLC	1968, 73
GU	1973
ICF	1968, 73
ICIU-S	1968, 73
ICarbS	1973
IEN	1968
IU	1968, 73
InLP	1968, 73
InU	1968, 73
LNU	1968, 73
MNS	1968
MiHM	1968, 73
MnU	1968, 73
MoSW	1968
MoU	1968
MtBuM	1968, 73
NBiSU	1968
NSyU	1968, 73
NbU	1968
NcU	1968, 73
NdU	1968, 73
NhD-K	1968
NjP	1968, 73
NvU	1968, 73
OCU-Geo	1968, 73
OU	1968
OkT	1968, 73
OkU	1968
OrU-S	1968
PSt	1968
TMM-E	1968
TxBeaL	1968
TxDaAR-T	1968
TxDaGI	1968
TxHSD	1968
TxU	1968, 73
UU	1968, 73
ViBlbV	1968, 73
WU	1968, 73

(223) GEOLOGICAL SURVEY OF ALABAMA. INFORMATION SERIES.

1958 No. 13. See Geological Society of America. Southeastern Section. 206.
1960 No. 18. Selected outcrops of the Eutaw Formation and Selma Group near Montgomery, Alabama.
1962 No. 26. See Alabama Academy of Science. 4.

ATuGS	1960
AzTeS	1960
CLU-G/G	1960
CSt-ES	1960
CoDGS	1960
DI-GS	1960
DLC	1960
I-GS	1960
ICF	1960
ICarbS	1960
IEN	1960
IU	1960
InLP	1960
InU	1960
KyU-Ge	1960
MNS	1960
MdBJ	1960
MiDW-S	1960
MiHM	1960
MiU	1960
MnU	1960
MtBuM	1960
NNC	1960
NSyU	1960
NdU	1960
OCU-Geo	1960
OU	1960
OkU	1960
PSt	1960
TxDaDM	1960
TxU	1960
UU	1960
WU	1960

(224) GEOLOGICAL SURVEY OF CANADA. GUIDEBOOK.
1913 See International Geological Congress. 296.

(225) GEOLOGICAL SURVEY OF CANADA. MISCELLANEOUS REPORT. (ONLY THOSE WITH ROAD LOGS ARE INCLUDED)

1960 [1] Story of the mountains, Banff National Park.
1962 2. Rocks and scenery of Fundy National Park.
— 3. Prince Edward Island National Park; the living sands.
— 4. Yoho National Park; the mountains, the rocks, the scenery.
— 5. Cape Breton Highlands National Park; where the mountains meet the sea.
1963 6. Jasper National Park; behind the mountains and glaciers.
1964 9. Kootenay National Park; wild mountains and great valleys.
— 10. Waterton Lakes National Park; lakes amid the mountains.
1965 11. Glacier and Mount Revelstoke National Parks; where rivers are born.
1966 12. Rocks and scenery of Terra Nova National Park.
1967 13. Banff National Park; how nature carved its splendour.
1968 15. Guide to the geology and scenery of the National Capital area.
1974 20. Guide to the geology of Riding Mountain National Park and its vicinity; history of its upland and other scenery.
1975 24. The geology of Long Beach Segment, Pacific Rim National Park, and its approaches; the story of the beaches, mountains and other scenery.
1977 Banff National Park, how nature carved its splendour.
1983 36. See International Geological Correlation Programme (IGCP). Project 60: Correlation of Caledonian Stratabound Sulphides. 303.
— 37. Classic mineral collecting localities in Ontario and Quebec. (296)

CMenUG	1960, 62-68, 74-75
CPT	1960, 62-68, 74-75
CSt-ES	1962-68, 74-75
CU-EART	1960, 62-68, 74-75
CaACI	1960, 62(2-3, 5), 64-68
CaACU	1962-68, 74-75
CaOTDM	1960-75
CaOWtU	1962(2-3, 5), 63, 64(10), 65-68, 75
IaU	1963, 65-66, 74, 83
KyU-Ge	1960, 62-68, 74-75
MdBJ	1962, 64-65
MiHM	1962-66, 68, 74
MiMarqN	1968
MnU	1962(4-5), 63-68, 74
NcU	1968
OCU-Geo	1962-68, 74-75
WU	1962-68, 74-75

(226) GEOLOGICAL SURVEY OF CANADA. PAPER. (ONLY THOSE WITH ROAD LOGS ARE INCLUDED)

1971 70-67. Quaternary geology road logs, Banff area, Alberta.
1973 72-32. Rocks and minerals for the collector: the Alaska Highway; Dawson Creek, British Columbia to Yukon/Alaska border.

CMenUG	1973
CPT	1971, 73
CSdS	1971, 73
CSt-ES	1971, 73
CU-EART	1971, 73
CaACU	1971, 73
CaAEU	1973
CaBVaU	1973
CaOOG	1973
CaOWtU	1971, 73
ICarbS	1973

IU	1973
IaU	1971
InLP	1973
MdBJ	1971, 73
MiHM	1973
MnU	1973
NSbSU	1971, 73
NhD-K	1973
OCU-Geo	1971, 73
WU	1971, 73

(227) GEOLOGICAL SURVEY OF OHIO. EDUCATIONAL LEAFLET.
1979 No. 11. Guide to the geology along U.S. Route 23 between Columbus and Portsmouth.
1982 No. 13. Guide to the geology along Interstate 75 between Toledo and Cincinnati.

CMenUG	1979
CPT	1979, 82
CSfCSM	1982
CoDGS	1979, 82
DI-GS	1982
IaU	1979
InLP	1979, 82
OU	1982

(228) GEOLOGICAL SURVEY OF OHIO. GUIDEBOOK.
1973 No. 1. Natural and man-made features affecting the Ohio shore of Lake Erie.
1974 No. 2. See Geological Society of America. North Central Section. 200.
— No. 3. See Geological Society of America. North Central Section. 200.
1975 No. 4. Geology of the Hocking Hills State Park region.

CLU-G/G	1973, 75
CMenUG	1973-75
CSt-ES	1973, 75
CU-EART	1973, 75
CU-SB	1973-75
CaOHM	1973, 75
CaOKQGS	1973
CaOOG	1973, 75
CaOWtU	1973
CoDGS	1973, 75
DI-GS	1973, 75
DLC	1973, 75
I-GS	1973
IU	1973
IaU	1973, 75
InLP	1973, 75
InRE	1973
InU	1973, 75
KyU-Ge	1973, 75
MiHW	1973
MiKW	1973
MoSW	1973, 75
MtBuM	1973, 75
NcU	1973
NvU	1973, 75
OCU-Geo	1973, 75
OU	1973, 75
OkU	1973, 75
PSt	1973, 75
TxBeaL	1973
UU	1973, 75
ViBlbV	1973, 75
WU	1973, 75

(229) GEOLOGICAL SURVEY OF OHIO. INFORMATION CIRCULAR.
1955 16. The geology along Route 40 in Ohio.

CMenUG	1955
CoDCh	1955
CoDGS	1955
ICarbS	1955
IaU	1955
WU	1955

(230) GEOLOGICAL SURVEY OF WYOMING.
1937 See Rocky Mountain Association of Geologists. 549.
1979 See Wyoming Geological Association. 783.

(231) GEOLOGICAL SURVEY OF WYOMING. BULLETIN.
1971 No. 55. Traveler's guide to the geology of Wyoming.

CLhC	1971
CMenUG	1971
CPT	1971
CSt-ES	1971
CaACU	1971
CaAEU	1971
CaOLU	1971
CaOWtU	1971
CoDGS	1971
DI-GS	1971
DLC	1971
ICF	1971
ICarbS	1971
IU	1971
InLP	1971
InU	1971
MnU	1971
MoSW	1971
MtBuM	1971
NdU	1971
NvU	1971
OCU-Geo	1971
OU	1971
OkU	1971
PBL	1971
TxBeaL	1971
TxU	1971
UU	1971
ViBlbV	1971
WU	1971

(232) GEOLOGICAL SURVEY OF WYOMING. PUBLIC INFORMATION CIRCULAR.
1980 No. 14. Guidebook to the coal geology of the Powder River Coal Basin, Wyoming.
1983 No. 20. See Wyoming Geological Association. 781.
1984 No. 21. Self-guided tour of the geology of a portion of southeastern Wyoming.
— No. 23. Tour guide to the geology and mining history of the South Pass Gold mining district, Fremont County, Wyoming.

CMenUG	1980, 84
CSfCSM	1984(23)
CSt-ES	1984(21)
CU-SB	1984(21)
CaOWtU	1984
CaQQERM	1984
CoDCh	1980, 84(21)
CoDGS	1980, 84
CoG	1980, 84
DI-GS	1980, 84
ICarbS	1980, 84
IU	1980, 84
IaU	1980, 84
IdU	1984(23)
InLP	1980, 84
KLG	1984
KU	1984
MBU	1980
MnU	1980, 84
NNC	1980, 84
NdFA	1980, 84
NdU	1980, 84
NvU	1980, 84
OCU-Geo	1980, 84
OU	1984
OkU	1984
PHarER-T	1980, 84
PSt	1980
TxU	1984(21)
UU	1980, 84(21)
WU	1980, 84
WyU	1980, 84

(233) GEOLOGICAL SURVEY OF WYOMING. REPRINT
1978 45. See Wyoming Field Science Foundation. 780.
    InLP   1978

(234) GEOLOGY CLUB OF PUERTO RICO. BULLETIN.
1959 No. 1. See Caribbean Geological Conference. 127.

(235) GEOLOGY TODAY.
1985 1. Geological tours of the south-western United States; follow the footsteps of the pioneers from Denver to San Francisco.

CSt-ES	1985
CaOKQGS	1985
CoDGS	1985
IaU	1985
OkTA	1985
TxDaM-SE	1985
WU	1985

(236) GEORGIA GEOLOGIC SURVEY. GEOLOGIC GUIDE.
1977 1. Geologic guide to Sweetwater Creek State Park.
   — 2. Geologic guide to Panola Mountain State Park - rock outcrop trail.
   — 3. Geologic guide to Panola Mountain State Park - watershed trail.
1980 4. Geologic guide to Stone Mountain Park.
1982 6. Geologic guide to Cumberland Island national seashore.
1983 7. Geologic guide to Cloudland Canyon State Park.

DI-GS	1980, 82, 83
IaU	1980
InLP	1977, 80, 83
NIC	1980, 83
NcU	1980
NvU	1980
PHarER-T	1980, 83
TxU	1977

(237) GEORGIA GEOLOGIC SURVEY. GUIDEBOOK.
1962 1. See Geological Society of America. Southeastern Section. 206.
   — 2. See Geological Society of America. Southeastern Section. 206.
   — 3. See Geological Society of America. Southeastern Section. 206.
1966 4. See Georgia Geological Society. 240.
   — 5. See Southeastern Geological Society. 614.
1967 6. See Georgia Geological Society. 240.
1968 7. See Georgia Geological Society. 240.
1969 8. See Georgia Geological Society. 240.
1970 [9] See Georgia Geological Society. 240.
1971 [10] See Georgia Geological Society. 240.
1972 11. See Georgia Geological Society. 240.
1974 12. See Geological Society of America. Southeastern Section. 206.
   — 13. See Georgia Geological Society. 240.
   — 13-A. See Georgia Geological Society. 240.
   — 14. See Society of Economic Geologists. 583.
1975 15. See Georgia Geological Society. 240.
1976 16. See Georgia Geological Society. 240.
1977 16-A. See Georgia Geological Society. 240.
1978 17. See Georgia Geological Society. 240.
1979 18. See Georgia Geological Society. 240.
1980 19. See Georgia Geological Society. 240.
   — 20. See Geological Society of America. 197.

(238) GEORGIA GEOLOGIC SURVEY. MISCELLANEOUS PUBLICATIONS.
1980 6 Urban Geology Field Trip Atlanta, Georgia.

IaU	1980
InLP	1980

(239) GEORGIA GEOLOGIC SURVEY. OPEN FILE.
1979 80-1. See Georgia Geological Society. 240.

(240) GEORGIA GEOLOGICAL SOCIETY. ANNUAL FIELD TRIP.
1966 1st The Cartersville Fault problem. (237(No. 4))
1967 2nd The geology of the Barnesville area and the Towaliga Fault, Lamar County, Georgia. (237)
1968 3rd Late Tertiary stratigraphy of eastern Georgia. (237)
1969 4th A guide to the stratigraphy of the Chickamauga Supergroup in its type area. (237)
1970 5th Stratigraphic and structural features between the Cartersville and Brevard fault zones. (237)
1971 6th [Economic geology of the Georgia Fall Line area] (237)
   [1] Norite intrusives in western Jasper County and eastern Monroe County, Georgia.
   [2] Lithostratigraphy and biostratigraphy of the north-central Georgia coastal plain.
   [3] The mining methods utilized by Freeport Kaolin Company at their mines near Gordon, Georgia.
   [4] Stratigraphy and paleontology of Huber Kaolin Company, Pit 22.
1972 7th Sedimentary environments in the Paleozoic rocks of Northwest Georgia. (237)
1973 8th The Neogene of the Georgia coast.
1974 9th [1] The Lake Chatuge sill outlining the Brasstown Antiform. (237(no.13))
   [2] An introduction to the Blue Ridge tectonic history of Northeast Georgia. (237(no.13A))
1975 [10th] A guide to selected Upper Cretaceous and lower Tertiary outcrops in the lower Chattahoochee River valley of Georgia. (237)
1976 11th Stratigraphy, structure, and seismicity in slate belt rocks along the Savannah River. (237)
1977 12th Stratigraphy and economic geology of Cambrian and Ordovician rocks in Bartow and Polk counties, Georgia. (237)
1978 [13th] Stratigraphy, structure, and metamorphism east of the Murphy Syncline: Georgia-North Carolina. (237)
1979 14th The stratigraphy of the Barnwell Group of Georgia. (239, 237)
1980 15th Geological, geochemical, and geophysical studies of the Elberton Batholith, eastern Georgia. (237(19))
1982 17th Geology of Late Precambrian and Early Paleozoic rocks in and near the Cartersville District, Georgia.

CLU-G/G	1966, 74-75
CMenUG	1980
CPT	1966, 74-78, 80
CSt-ES	1966, 68, 70-77, 80
CU-A	1973
CU-EART	1966-68, 70-72, 74-78, 80
CaACU	1973
CaAEU	1973
CaOHM	1973
CaOLU	1973
CaOOG	1973, 80
CaOWtU	1966, 67, 73
CoDCh	1977
CoG	1967
CoU	1974, 76
DI-GS	1966-68, 70-77, 80
DLC	1966, 73
GU	1966, 68, 70, 72-73, 74(2)
ICF	1966-69, 72
IU	1966, 67, 69, 74-77, 80
IaU	1966, 67, 73-80
InLP	1966-67, 73-74, 76-80
InU	1966-68
KyU-Ge	1966-69, 72-78, 80, 82
LNU	1966-68, 75
MH-GS	1968
MdBJ	1966, 69, 72, 74
MiDW-S	1966-70, 72
MnU	1966-69, 71, 73, 74(2), 75-78, 80
MoSW	1971-75
NBiSU	1966, 69, 72
NOneoU	1973
NRU	1973
NSyU	1973
NcU	1966-70, 74-76, 80
NdU	1966
NhD-K	1973
NjP	1966-73
NvU	1966-67
OCU-Geo	1966-69, 74-76
OU	1966, 73, 75
OkU	1966, 74-75
PBL	1975

PBm	1974-76
PSt	1966, 71-72, 74-75
TMM-E	1970
TxBeaL	1967, 72-73, 76
TxDaAR-T	1966-67, 72-74
TxDaDM	1966-68, 74(2), 75-76
TxDaM-SE	1966
TxDaSM	1973
TxHU	1966-68, 72, 74
TxU	1967-70, 73
UU	1974(2), 75-76
ViBlbV	1966-68, 73-76
WU	1966, 70, 74-79, 80

GEORGIA. DEPARTMENT OF MINES, MINING AND GEOLOGY. See GEORGIA GEOLOGIC SURVEY. (237) and GEORGIA GEOLOGIC SURVEY. (236) and GEORGIA GEOLOGIC SURVEY. (239)

GEORGIA. DEPARTMENT OF NATURAL RESOURCES. EARTH AND WATER DIVISION. GEOLOGICAL SURVEY. See GEORGIA GEOLOGIC SURVEY. (237) and GEORGIA GEOLOGIC SURVEY. (236) and GEORGIA GEOLOGIC SURVEY. (239) and GEORGIA GEOLOGIC SURVEY. (238)

GEORGIA. GEOLOGIC & WATER RESOURCES DIVISION. See GEORGIA GEOLOGIC SURVEY. (237) and GEORGIA GEOLOGIC SURVEY. (236) and GEORGIA GEOLOGIC SURVEY. (239) and GEORGIA GEOLOGIC SURVEY. (238)

GEORGIA. UNIVERSITY. See UNIVERSITY OF GEORGIA. MARINE INSTITUTE, SAPELO ISLAND. (684)

(241) GEOSCIENCE AND MAN.
1974 No. 8. See American Association of Petroleum Geologists. 18.

(242) GEOSCIENCE WISCONSIN.
1978 V. 2. See Institute on Lake Superior Geology. 273.

(243) GEOTHERMAL RESOURCES COUNCIL. FIELD TRIP GUIDEBOOK.
1978 Feb. Direct utilization of geothermal energy in the Imperial Valley, California; field trip guidebook.
1983 Oct. Portland, Oregon. (518)
A field trip guide to the central Oregon Cascades.
First day: Mount Hood-Deschutes Basin.
Second day: Santiam Pass-Belknap Hot Springs-Breitenbush Hot Springs.

AzTeS	1983
CLU-G/G	1978
CMenUG	1983
CPT	1983
CSfCSM	1978, 83
CSt-ES	1983
CU-EART	1983
CU-SB	1983
CaAEU	1983
CaOOG	1983
DI-GS	1978, 83
IaU	1983
IdBB	1983
IdU	1983
InLP	1983
MnU	1983
NdU	1983
OkU	1983
OrU-S	1983
TxDaM-SE	1983
TxU	1983
UU	1983
WU	1983
WyU	1983

(244) GONDWANA SYMPOSIUM. FIELD EXCURSION. GUIDEBOOK.
1985 6th No. 2. Geology of the southern Appalachian Mountains.
No. 3. Glacial geology of central Ohio. (500)
No. 6. Carboniferous of eastern Kentucky. (500)
No. 7. Quaternary and Proterozoic glacial deposits. (500)
No. 4 & 9. Lower Carboniferous clastic sequence of central Ohio. (500)
No. 10. Lower and Middle Paleozoic geology of southern Ohio.
OU   1985

(245) GOVERNOR'S CONFERENCE ON ENVIRONMENTAL GEOLOGY.
1969 An environmental geology field trip: road log. (78, 141)
KyU-Ge 1969
OkU 1969

(246) GRAND JUNCTION GEOLOGICAL SOCIETY. GUIDEBOOK.
1960 July Guidebook to northern San Juan Mountains.
1961 July Guidebook to western San Juan Mountains.
1962 July Field trip to Dunton and Rico areas.
1978 Sept. Scenic geology of the San Juan Mountains, Colorado.
1982 Sept. Southeastern Piceance Basin, western Colorado.
1983 Oct. Northern Paradox Basin - Uncompahgre Uplift.
1984 Sept. Road log from Grand Junction to Resource Enterprises' Deep Seam Project (coal bed methane) east of Collbran, Colorado.
1985 Apr. See Oil Shale Symposium. 501.

CLU-G/G	1982, 83
CMenUG	1982-84
CSt-ES	1960, 83
CU-EART	1982, 83
CU-SB	1960, 82, 83
CoD	1982-83
CoDBM	1983
CoDCh	1982
CoDGS	1982, 84
CoG	1982
CoLiM	1982, 83
CoU-DA	1982, 83
DI-GS	1960, 62, 82, 84
IU	1960, 62, 84
IaU	1982-83
InLP	1960, 62, 78, 82-84
KyU-Ge	1960, 62, 82, 83
MnU	1982-83
NRU	1982
NdU	1983
OkTA	1982, 83
OkU	1978, 83
TxDaDM	1961
TxDaM-SE	1982, 83
TxMM	1982, 83
TxU	1982-83
UU	1982
WU	1982, 83
WyU	1982

GSUCLA FIELD GUIDE. See EARTH AND SPACE SCIENCE STUDENT ORGANIZATION, U.C.L.A. (161)

(247) GULF COAST ASSOCIATION OF GEOLOGICAL SOCIETIES. ANNUAL MEETING. GUIDEBOOK FOR THE FIELD TRIPS.
1956 6th [No. 1] Lower Claiborne. (588, 612)
[No. 2] Lower Cretaceous. (588, 612)
1958 8th Sedimentology of South Texas. (152)
1959 9th Recent sediments of the north-central Gulf Coastal Plain. (29, 588)
1960 10th Cenozoic field trips. (588)
[No. 1] Recent sedimentation on Horn Island, Mississippi.
[No. 2] Stratigraphy of the Quaternary and upper Tertiary of the Pascagoula Valley, Mississippi.
1961 11th Southern Edwards Plateau. (612)
1962 12th Little Stave Creek, Salt Mountain, Jackson, Alabama. (588)
1964 14th Depositional environments, south-central Texas coast. (152)
1965 15th Deltaic coastal plain. (29, 255)
1966 16th Guidebook for field trips.

(248) Gulf Geological Subcommittee.

No. 1. Lafayette-Atchafalaya-Five Islands flight.
No. 2. Belle Isle salt dome trip.
1967 17th [1] San Antonio, Uvalde, Carrizo Springs, Laredo, Freer, Encinal, Pearsall, San Antonio, Texas.
[2] San Antonio, Hondo, Eagle Pass, Laredo. (612)
1968 18th A field guide to Cretaceous and Tertiary exposures in west-central Alabama.
1969 19th No. 1. Field guide to some carbonate rock environments, Florida Keys and western Bahamas. (588)
No. 3. Geological field guide to Neogene sections in Jamaica, West Indies. (588)
No. 4. Late Cenozoic stratigraphy of southwestern Florida. (588)
1970 20th [1] Guidebook to north-central Louisiana salt domes. (588)
[2] Geology of Lone Star, Texas, area, specifically Lone Star Steel's open pit iron mines, and visit to Lone Star Steel's iron processing plants (Cover title: Lone Star, Texas iron deposits).
1971 21st The southern shelf of British Honduras. (466, 588)
1972 22nd Padre Island National Seashore field guide. (152)
1973 23rd [2] Lower Cretaceous strata in central Texas; a field guide to the Moffatt Mound area near Lake Belton, Bell County, Texas. (255)
[3] A field trip to northeastern coast of Yucatan. 2d ed., revised. (255)
1974 24th Lafayette, Louisiana. (588)
1. Avery Island Salt Mine, Louisiana.
2. Atchafalaya Basin, Chenier trend, Five Island salt domes flight.
3. SEPM-GCAGS 1974 Guatemala field trip.
1975 25th Field trip no. 1. Thomasville field sour gas plant, Rankin County, Mississippi.
Field trip no. 2. U.S. Army Engineers waterways experiment station, Corps of Engineers, Vicksburg, Mississippi.
Field trip no. 3. Tertiary localities of east-central Mississippi.
1976 [26th] A study of paleozoic rocks in Arbuckle and western Ouachita Mountains of southern Oklahoma. (566)
1977 27th [1] Lower Cretaceous carbonate tidal facies of central Texas (Cover title: Lower Cretaceous carbonate tidal facies). (714)
[2] Lignite resources in east central Texas. (588)
1978 [28th] 2. Geology and hydrogeology of northeastern Yucatan. (Revision of Carbonate rocks and hydrogeology of the Yucatan Peninsula. AAPG Annual Meeting 1976) (466, 588, 18(No.5))
1979 [29th] No. 1. Guidebook for the stratigraphy of Edwards Group and equivalents, eastern Edwards Plateau, Texas. (588, 612)
1980 30th Lafayette, Louisiana.
The sedimentary environments of the Louisiana coastal plain. (588, 350)
Field trip 1. Cote Blanche Island Salt Mine, Iberia Parish, Louisiana.
Field trip 2. Upper Delta Plain sediments of early Miocene age, Cane River diversion canal, Rapides Parish, Louisiana.
Field trip 3. Overflight of the Mississippi Delta and southeastern Louisiana coast.
Field trip 4. Stratigraphy and coastal processes of the Louisiana Chenier Plain microfossil associations.
1981 [31st] [Corpus Christi, Texas]
[1] Modern depositional environments of sands in South Texas. (152)
[2] Geology of Peregrina & Novillo canyons, Ciudad Victoria, Mexico. (152)
1983 [33rd] Jackson, Mississippi.
Tertiary and Upper Cretaceous depositional environments, central Mississippi and west-central Alabama.
1985 Geology and hydrogeology of the Yucatan and Quaternary geology of Northeastern Yucatan Peninsula. (Also used for New Orleans Geological Society, 1985.)

ATuGS	1962, 68
AzTeS	1956, 59, 61, 67-68, 69(1, 3-4)
AzU	1956, 58-59, 60(1), 61-62, 64
BEG	1980
CLU-G/G	1958-62, 64-66, 68, 69(1, 3), 71-72, 74(3), 78(2)
CLhC	1958-60, 64
CMenUG	1959-62, 64-65, 67-69, 71-72, 75-77, 80-81, 83
CPT	1958-62, 64-66, 72, 74(3), 81
CSdS	1959, 61-62, 64-65
CSt-ES	1956, 58-62, 64-69, 71-72, 76, 78(2), 81, 85
CU-A	1958-61, 64-66, 72, 73(3), 85
CU-EART	1956, 58-62, 64-68, 69(1, 3-4), 70(2), 76, 77(1), 78(2), 81, 85
CU-SB	1959, 61, 64-65, 69-70, 72, 76, 81
CaACAM	1980, 83
CaACU	1956, 67(1), 69(1), 72, 76, 78
CaBVaU	1958
CaOHM	1956, 58-61, 64-65, 67-68, 69(1, 3-4), 70(2)
CaOKQGS	1959-62
CaOOG	1959-61, 64-65, 68, 70(2), 76, 81
CaOWtU	1960, 72, 74(3), 75(3), 81, 83
CoDGS	1956, 58-61, 64-65, 67, 69, 72, 76-77, 79, 81, 83
CoG	1956, 58-62, 64-66, 68, 69(3), 71-72
CoLiM	1981, 83
CoU	1958-60, 64-66
DI-GS	1956, 58-62, 64-69, 69(1, 3-4), 70-72, 73(2-3), 75-77, 80-81, 83
F-GS	1980-81, 83
FBG	1980-81, 83
I-GS	1958, 64
ICF	1958-62
ICIU-S	1959, 61, 64-66, 72, 75
IEN	1958-62, 70(2), 72
IU	1956, 58-61, 64-66, 69(1, 3-4), 70(2), 72, 74(2-3), 76, 77(1), 79(1), 80
IaU	1958-62, 64-68, 69(1, 3-4), 72, 74(2-3), 75, 77(1), 79(1), 81, 83
InLP	1956, 60, 64, 69(3), 72, 74(3), 75-77, 79-83
InU	1956(1), 58, 64, 66, 68, 69(1, 3-4), 71, 76, 81
KyU-Ge	1956, 59, 60(1-2), 61-62, 65-68, 71-72, 75(1-3), 76, 78(2), 79(1), 81, 83, 85
LNU	1958-60, 64, 68, 71, 76, 78(2)
LU	1956, 58-62, 64-65
MH-GS	1969(1, 3-4), 70(2), 83
MNS	1956
MiDW-S	1964-66, 71
MiU	1958-62, 64-68, 72
MnU	1956, 58-62, 64-66, 68, 69(1, 3-4), 70-72, 78(2), 81, 83
MoSW	1959, 62, 64-66, 69(3)
MoU	1956, 59, 61, 64-67
Ms-GS	1956, 58, 60, 75
MsJG	1956, 58, 60, 75
MtU	1958-61, 64-65
NBiSU	1971-72
NNC	1956, 58-62, 64-66, 68, 69(1, 3-4), 72, 81, 83
NOneoU	1964
NSbSU	1956, 58-62, 64-81, 83
NSyU	1968
NbU	1956, 62
NcU	1958-62, 64-66, 68, 69(1, 3-4), 70(2)
NdU	1958-60, 62, 64-66, 72
NhD-K	1971
NjP	1958-62, 64-69
NmPE	1956
OCU-Geo	1956, 58-62, 64-68, 69(1, 3-4), 70(2), 71-72, 81
OU	1958-62, 64-66, 72, 81
OkTA	1980, 81, 83
OkU	1956, 58-59, 60(1-2), 61, 74, 76, 79(1), 81, 83
OrCS	1961
OrU-S	1960
PBL	1959, 61, 64-65
PSt	1958-62, 64-66, 72
TMM-E	1959-62, 64-66
TxBeaL	1956(1), 58-59, 62, 64-65, 68, 69(1, 3-4), 71, 74(3), 76
TxCaW	1972
TxDaAR-T	1956, 58-62, 64-68, 69(1, 3-4), 71-72, 73(2-3), 74(2)
TxDaDM	1956, 58-62, 64-66, 68, 70(2), 71-72, 78(2), 81, 83
TxDaGI	1956, 59, 64, 76
TxDaM-SE	1956, 59-62, 64-65
TxDaSM	1956, 58, 61, 64-65, 72, 76, 78(2)
TxHMP	1981
TxHSD	1956, 58-59, 61, 64-65, 66(2), 67, 70-71, 76, 77(1), 79(1)
TxHU	1956, 58-62, 64-66, 72
TxMM	1956, 58-59, 61-62, 64-65, 72, 83
TxU	1956, 58-62, 64-68, 71-72, 74(3), 77(1), 80-81, 83
TxU-Da	1951, 61, 76
TxWB	1956, 58-60, 64, 67, 69, 72, 75-76, 80-81
UU	1958-62, 64-66, 72
ViBlbV	1956, 58, 68, 72, 74(3)
WU	1958-59, 72, 76, 78, 79(1), 81

(248) GULF GEOLOGICAL SUBCOMMITTEE.
1953 May Geology of a portion of the Northern Appalachian Basin: Log of 1953 Field Trip.

1955 April   Pittsburg.
Field conference on Mechanics of rock failure in the folded Appalachians od West Virginia and Virginia.
CoG   1953, 55

(249) HARDIN SIMMONS UNIVERSITY GEOLOGICAL SOCIETY. FIELD CONFERENCE. GUIDEBOOK.
1963   See Southwestern Association of Student Geological Societies (SASGS). 626.
1969   See Southwestern Association of Student Geological Societies (SASGS). 626.

(250) HARRISBURG AREA GEOLOGICAL SOCIETY. ANNUAL FIELD TRIP. (TITLE VARIES)
1982 Apr.   Geology in the South Mountain area, Pennsylvania; field excursion guide.
1983 Apr.   Geology along the Susquehanna River, south central Pennsylvania; field excursion guide.
1984 3rd   Stratigraphy, structural style, and economic geology of the York-Hanover Valley.
1985 4th   Pennsylvania's polygenetic landscape.
PHarER-T   1982-85

(251) HAWAII INSTITUTE OF GEOPHYSICS. SPECIAL PUBLICATION.
1979 July   Field trip guide to the Hawaiian Islands.
DI-GS   1979
NhD-K   1979

(252) HAWAII. DEPARTMENT OF PUBLIC LANDS. DIVISION OF HYDROGRAPHY. BULLETIN.
1939 2 Geologic map & guide of the Island of Oahu, Hawaii.
CPT   1939
CSt-ES   1939

(253) HIGHWAY GEOLOGY SYMPOSIUM. FIELD TRIP GUIDEBOOK. ANNUAL.
1967 18th   Field trip guidebook for 18th annual highway geology symposium, vicinity of Lafayette, Indiana.
1969 20th   Annual field trip near East St. Louis, Illinois.
1970 21st   Geo-engineering in northeastern Kansas; greater Kansas City area.
1971 22nd   Highway geology in the Arbuckle Mountains and Ardmore area, southern Oklahoma.
1975 26th   Idaho-Montana loop trip; Coeur d'Alene Mine trip.
1977 28th   Engineering geology of central and northern Black Hills, South Dakota.
Trip A: Central Black Hills (morning).
Trip B: Northern Black Hills (afternoon).
1978 29th   Annapolis, Maryland.
1979 30th   Geology and scenery of the lower Columbia River Gorge and northern Cascade Range, Oregon.
1984 35th   Annual highway geology symposium and field trip.
CMenUG   1967, 69-71, 75, 77-79
CU-A   1975
CU-EART   1971
CaOOG   1967, 70
CaOWtU   1970
CoDCh   1971
DI-GS   1970-71, 78
I-GS   1969
IU   1969-71
IaU   1971, 77-78, 84
InLP   1971, 78-79
InU   1970
KLG   1970
MoSW   1971
OkU   1971
TxU   1967, 69-71

(254) HOBBS GEOLOGICAL SOCIETY. GUIDEBOOK.
1962   See West Texas Geological Society. 756.

1968   See New Mexico Geological Society. 458.

(255) HOUSTON GEOLOGICAL SOCIETY. GUIDEBOOK.
1933   See American Association of Petroleum Geologists. 18.
1938   Road log of the Jackson-Claiborne field trip of the Houston Geological Society.
1941   See American Association of Petroleum Geologists. 18.
1952   Geologic strip maps: U.S. Highway 77, Texas-Oklahoma state line to Dallas; U.S. Highway 75, Dallas to Galveston.
1953 Mar.   See American Association of Petroleum Geologists. 18.
1958 Dec.   See Society of Economic Paleontologists and Mineralogists. Gulf Coast Section. 588.
1959   No. 1. Boling Field, Fort Bend and Wharton counties, Texas.
No. 2. See Society of Economic Paleontologists and Mineralogists. Gulf Coast Section. 588.
No. 3. Geologic strip maps: U.S. Highway 80, Texas-New Mexico state line to Van Horn; U.S. Highway 90, Van Horn to Texas-Louisiana state line.
1960 May   See Society of Economic Paleontologists and Mineralogists. Gulf Coast Section. 588.
1961   Geology of Houston and vicinity, Texas, with appended guides to fossil, mineral, and rock collecting localities.
1962   See Geological Society of America. 197.
1963 Mar.   See American Association of Petroleum Geologists. 18.
1964   Houston and vicinity geological field trip for earth science teachers secondary schools.
1965   See Gulf Coast Association of Geological Societies. 247.
1968 Oct.   Environments of deposition, Wilcox Group, Texas Gulf Coast.
1971   See American Association of Petroleum Geologists. 18.
1972 Feb.   A field trip to northeastern coast of Yucatan.
1973 Oct./Nov. See Gulf Coast Association of Geological Societies. 247.
1976   See Geological Society of America. South Central Section. 205.
1978 Apr.   Field trip to Damon Mound.
— Nov. The Chenier Plain and modern coastal environments, southwestern Louisiana; and geomorphology of the Pleistocene Beaumont Trinity River Delta Plain. (Cover title: Louisiana Chenier Plain and Southeast Texas geomorphology)
1979   See American Association of Petroleum Geologists. 18.
1980   Economic geology of south central Texas. (South Texas Geological Society guidebook 1976, reprinted for Houston Geological Society.) See South Texas Geological Society. 612.(1976)
1981   See Geological Society of America. 197.(No.3) and Arkansas. Geological Commission. 67.(81-1)
1982 Apr.   Guidebook to the geology of the eastern Ouachita Mountains, Arkansas. (67(No.82-2))
AGC   1982
AzTeS   1968-69
AzU   1969
CChiS   1969
CLU-G/G   1952, 59(1, 3), 61, 68-69
CMenUG   1961, 80, 82
CPT   1978
CSt-ES   1968, 78, 82
CU-A   1969
CU-EART   1952, 59(3), 61, 68, 78(Nov.)
CU-SB   1976, 78-79
CaACU   1959(3)
CaBVaU   1978
CaOHM   1968
CaOKQGS   1959(1, 3), 68
CaOOG   1982
CaOTRM   1959(1, 3)
CaOWtU   1969, 82
CoG   1968
CoLiM   1980
DI-GS   1952, 59(3), 61, 68-69, 78(Nov.)
DLC   1952
ICarbS   1959, 61
IEN   1969
IU   1959(1, 3), 68, 78(Apr.)
IaU   1959(1, 3), 61, 68, 82
InLP   1961, 64, 78, 80, 82
InU   1952, 59(3), 69, 78(Nov.)
KyU-Ge   1959(1, 3), 61, 68, 78(Nov.), 82
LNU   1972

MiU	1959(1, 3)
MnU	1959(1, 3), 68, 82
MoSW	1959(1, 3)
NNC	1952, 59(1, 3), 68
NSbSU	1980
NSyU	1969
NbU	1969
NcU	1969
NdU	1969, 78(Apr.)
NhD-K	1969, 78
OCU-Geo	1959(1, 3), 61
OU	1978(Nov.)
OkU	1952, 58, 59(3), 60, 68-69
PBL	1969
TMM-E	1969
TxBeaL	1952, 59(1, 3), 64, 68-69, 78(Nov.)
TxCaW	1978(Nov.)
TxDaAR-T	1968
TxDaDM	1959(1, 3)
TxDaSM	1968, 78(Nov.)
TxHSD	1938, 52, 59(3), 61, 68, 78(Apr.)
TxHU	1961, 64, 69
TxU	1959(1, 3), 68, 78
TxWB	1959, 68
ViBlbV	1968-69
WU	1978, 82

HOUSTON, TEXAS. UNIVERSITY. See UNIVERSITY OF HOUSTON. DEPARTMENT OF GEOLOGY. (685)

(256) IDAHO ASSOCIATION OF PROFESSIONAL GEOLOGISTS. FIELD TRIP GUIDE.
1982 May   Field trip guide to the Boise foothills, Idaho.

IdBB	1982

(257) IDAHO BUREAU OF MINES AND GEOLOGY. BULLETIN.
1961   No. 16. See Association of American State Geologists. 74.

(258) IDAHO BUREAU OF MINES AND GEOLOGY. INFORMATION CIRCULAR.
1975   28. See Geological Society of America. Rocky Mountain Section. 204.
1979   33. See American Association for the Advancement of Science. Pacific Division. 17.(1983)
—   34. See American Association for the Advancement of Science. Pacific Division. 17.(1983)

(259) IDAHO BUREAU OF MINES AND GEOLOGY. PAMPHLET.
1963   No. 130.   Geology along U.S. Highway 93 in Idaho.
1975   No. 160.   See Geological Society of America. Rocky Mountain Section. 204.

AzU	1963
CLU-G/G	1963
CMenUG	1963
CPT	1963
CSt-ES	1963
CU-EART	1963
CaACU	1963
CaAEU	1963
CoDGS	1963
CoG	1963
DI-GS	1963
DLC	1963
ICF	1963
ICarbS	1963
IU	1963
IdBB	1963
IdU	1963
InLP	1963
InU	1963
MnU	1963
MoSW	1963
MtBuM	1963
NhD-K	1963
NvU	1963
OkU	1963
SdRM	1963
TxHSD	1963
UU	1963
ViBlbV	1963
WU	1963

(260) IDAHO BUREAU OF MINES AND GEOLOGY. SPECIAL REPORT.
1973   2. See Belt Symposium. 97.

(261) ILLINOIS EARTH SCIENCE ASSOCIATION.
1980 Spring   Guidebook to geologic features and resources of west-central Illinois.

I-GS	1980

(262) ILLINOIS GEOLOGICAL SOCIETY. GUIDEBOOK OF THE FIELD CONFERENCE.
1938   Field conference on Chester Series and Ste. Genevieve Formation.
1939   Urbana, Illinois to Madison, Wisconsin.
1940   See American Association of Petroleum Geologists. 18.
1946   Field conference on Chester stratigraphy.
1949   See American Association of Petroleum Geologists. 18.
1953   Wisconsin stratigraphy of the Wabash Valley and west-central Indiana. See Indiana Geological Survey. 268.
     Basis of subdivision of Wisconsin glacial stage in northeastern Illinois. (268)
1956   Southern Illinois, lower Chester rocks of southwestern Illinois.
1957   Ordovician, Silurian, Devonian and Mississippian rocks of western Illinois.
1959   Extreme southeastern Illinois.
1965   Mineral resources of Southeast Missouri.
1966   See American Association of Petroleum Geologists. 18.
1973   Guidebook to the Cambro-Ordovician rocks of eastern Ozarks.
1982   Delta environments of the lower Chesterian (Mississippian) in southern Illinois. (265)

CLU-G/G	1957
CSt-ES	1939, 53, 82
CU-SB	1956, 59, 65, 73, 82
CaACU	1957, 73
CoDCh	1956, 59
CoDGS	1956-57, 59
CoG	1956-57, 59, 65
DI-GS	1956-57, 59, 65, 73, 82
I-GS	1938, 46, 56-57, 82
ICF	1953, 56-57, 59, 65, 73
ICarbS	1953, 65
IEN	1946, 59
IU	1946, 53, 59
InLP	1946, 53, 56, 59, 65
InU	1946, 56-57, 82
KLG	1953
MA	1982
MiHM	1959
MnU	1956-57, 59, 65
NcU	1956-57, 59, 65, 73
NjP	1965
OCU-Geo	1959, 73, 82
OCl	1982
OU	1957, 82
OkT	1956-57, 59
OkTA	1982
OkU	1946, 56-57, 82
PHarER-T	1982
TxBeaL	1956-57, 59, 73
TxDaAR-T	1956-57, 59
TxDaM-SE	1982
TxDaSM	1956-57, 59, 65
TxU	1956
TxWB	1956-57, 59
ViBlbV	1956-57, 59, 65, 73
WaU	1982

(263) ILLINOIS STATE ACADEMY OF SCIENCE. [GUIDEBOOK FOR GEOLOGIC FIELD TRIPS]
1950   Guide leaflet ... geological field trip, May 6, 1950, Augustana College, Rock Island, Illinois. [Rock Island area]

Most of the guidebooks for geologic field trips held by this organization are included as issues of the series: Illinois State Geological Survey. Field Trip Guide Leaflet.
OU   1950

(264) ILLINOIS STATE GEOLOGICAL SURVEY. FIELD TRIP GUIDE LEAFLET.
1950 Guide leaflet ... geological field trip, May 6, 1950, Augustana College, Rock Island, Illinois. [Rock Island area]

Most of the guidebooks for geologic field trips held by this organization are included as issues of the series: Illinois State Geological Survey. Field Trip Guide Leaflet.

The Illinois State Geological Survey maintains a supply of the latest revision of each Guide Leaflet. Occasionally, specific titles will be withdrawn from the series and others added. Those titles showing no date of publication were published before 1960. The Survey suggests that the desired leaflet(s) be requested by current title as listed below:

[Listed guidebooks titles below have been shortened to reflect field trip areas for the given year, omitting from the complete titles "A guide to the geology of the" to avoid repetition for each area]

[1] Alton
[2] Anna-Jonesboro
[3] Apple River Canyon
[4] Barrington-Fox Lake
[5] Belvidere
[6] Benton
[7] Bloomington-Normal
[8] Cairo
[9] Canton
[10] Carbondale
[11] Carlinville
[12] Carrollton
[13] Casey
[14] Cave in Rock
[15] Champaign-Urbana
[16] Charleston
[17] Chester
[18] Chicago Heights
[19] Dallas City
[20] Danville
[21] De Kalb-Byron
[22] Des Plaines
[23] Dixon
[24] Downers Grove
[25] Eldorado
[26] Elgin
[27] Elizabeth
[28] Eureka
[29] Fairbury
[30] Fairfield
[31] Freeport
[32] Fulton
[33] Galena
[34] Galesburg
[35] Grafton
[36] Grand Tower
[37] Greenup
[38] Hardin
[39] Harrisburg
[40] Homer
[41] Hoopeston
[42] Jacksonville
[43] Joliet
[44] Kankakee
[45] Kewanee
[46] Lake region-Crystal Lake
[47] Lawrenceville
[48] Marion
[49] Macomb
[50] Marseilles-Ottawa
[51] Marshall
[52] Metropolis
[53] Millstadt
[54] Moline
[55] Monmouth
[56] Mt. Carroll
[57] Mt. Sterling
[58] Murphysboro
[59] Naperville
[60] Nashville
[61] Newton
[62] North Shore
[63] Olney
[64] Oregon
[65] Palos Park
[66] Paris
[67] Pecatonica
[68] Pekin
[69] Peoria
[70] Pere Marquette
[71] Pere Marquette State Park
[72] Petersburg
[73] Pinckneyville
[74] Pine Hills
[75] Pittsfield
[76] Pontiac-Streator
[77] Port Byron
[78] Princeville
[79] Quincy
[80] Red Bud
[81] Rock Island-Moline
[82] Rockford
[83] Rosiclare
[84] Salem
[85] Savanna
[86] Shawneetown
[87] Shelbyville
[88] Springfield
[89] Stockton
[90] Streator-Pontiac
[91] Thebes
[92] Vienna
[93] Warsaw
[94] Waterloo
[95] Watseka
[96] Waukegan
[97] West Chicago
[98] Wilmington
[99] Wyoming
[100] Yorkville

1960 [1] Grafton
[2] Harrisburg
[3] Milan-Rock Island
[4] Pana
[5] Quincy
[6] Rosiclare
[7] Salem
[8] Sparta
[9] Woodstock

1961 [1] Charleston
[2] Elgin
[3] Georgetown
[4] Hamilton
[5] Lena
[6] Morris
[7] Sparta
[8] Valmeyer

1962 [1] Amboy
[2] Golconda
[3] Greenville

(264) Illinois State Geological Survey.

    [4] Neoga
    [5] Peoria
    [6] Pittsfield
    [7] Starved Rock
    [8] Wheaton
1963   [1] Belvidere
    [2] Dupo
    [3] Farmington
    [4] Freeport
    [5] Marion
    [6] Pine Hills
    [7] Pontiac
    [8] Savanna
1964   [1] Bloomington
    [2] Chester
    [3] Colchester
    [4] Edinburg
    [5] Jonesboro
    [6] Morrison
    [7] Rochelle
1965   [1] Alto Pass
    [2] Beardstown
    [3] Carrier Mills
    [4] Galena
    [5] Shelbyville
    [6] Yorkville
1966   [1] Byron
    [2] Freeport
    [3] Paris
    [4] Quincy
    [5] Steeleville
    [6] Vienna
1967   [1] Bourbonnais
    [2] Dixon
    [3] Elizabethtown-Cave in Rock
    [4] Oakwood
    [5] Petersburg
1968   [1] Barry
    [2] Dixon
    [3] Princeton
    [4] Saint Elmo
    [5] Thebes
1969   [1] Equality
    [2] Havana
    [3] Monticello-Mahomet
    [4] Mt. Carroll
    [5] Princeton
1970   [1] Freeport
    [2] Hamilton-Warsaw
    [3] Millstadt-Dupo
    [4] Mt. Carroll
    [5] Winchester
1971   [1] Galena
    [2] Hamilton-Warsaw
    [3] LaSalle
    [4] Makanda
    [5] Mt. Sterling
    [6] Palos Hills
1972   [1] Carlock
    [2] Danville
    [3] LaSalle
    [4] Potomac-Danville
    [5] Red Bud
    [6] Stockton
1973   [1] Knoxville
    [2] Robinson
    [3] Rockton
    [4] Stockton
1974   [1] Breese
    [2] Milan
    [3] Rockford
    [4] Rockton
1975   [1] Carrollton
    [2] Metropolis
    [3] Milan
    [4] St. Anne-Momence
1976   [1] Carrollton
    [2] Mt. Carmel
    [3] St. Anne-Momence
1977   [1] Champaign-Urbana
    [2] Monmouth
    [3] Mt. Carmel
1978   A. Springfield
    B. Elmhurst-Naperville
    C. Middle Illinois Valley
    D. Marion
1979   A. Westfield-Casey
    B. Farmer City
    C. Evergreen Park-Thornton
    D. Carlinville
1980   A. Equality area.
    B. Hillsdale area.
    C. Dundee area.
    D. Quincy north area.
1981   A. Cairo area.
    B. Danville area.
    C. Zion-Lake Bluff area.
    D. Waterloo-Valmeyer area.
1982   A. Decatur area.
    B. Capron-Rockford area.
    C. Danvers-Normal area.
    D. Cave in Rock-Rosiclare area.
1983   D. Golconda area.
1984   A. Greenville area.
    B. Morris area.
    C. Pontiac-Streator area.
1985   A. Salem area.
    B. Elizabeth area.
    C. Pekin geological science field trip.
    D. Lawrenceville geological science field trip.

AzTeS	1964(2, 4, 6), 65-66, 67(2-5), 68(1, 3-5), 69(1-4)
CLhC	1971(3-5)
CSt-ES	1960(1, 3, 8, 13, 17, 18, 23, 42, 68, 82, 92)
CU-EART	1976(1-2), 77(3)
DI-GS	Pre 1960(9, 15, 26, 32, 37-38, 51-52, 58, 60, 64, 88, 98), 60(1-7, 9), 61, 62(1-4, 6-8), 63(1-3, 5-8), 64-67, 68(1, 3-5), 69(1-4), 70-71, 72(1, 3-6), 73, 74(1-2, 4), 75-79, 81-83
I-GS	Pre 1960(1-5, 7-14, 16-36, 38-48, 50-73, 75, 77-80, 82-89, 91-100), 60(1-4, 6-7, 9), 61(1, 3-8), 62, 63(1-3, 5, 7-8), 64-70, 71(1-5), 72(1, 4-6), 73, 74(1-2, 4), 75-76, 77(1, 3), 78-80, 82(A), 85(A)
ICarbS	Pre 1960, 60-85
INS	1980(C, D), 81, 82(A), 84, 85(A)
IU	Pre 1960(3-4, 7-15, 18-19, 21, 24, 26-27, 29-32, 34, 37-38, 40-41, 43, 47, 49-53, 55-58, 60-61, 63-64, 67, 73-74, 76-77, 80, 82, 86, 88, 90-91, 96, 98-99), 60(1, 3, 6-9), 61(1, 3-8), 62(1-6, 8), 63(1-5, 7-8), 64(1-7), 65-66, 67(1-3, 5), 68-70, 71(1-5), 72-73, 74(1-2, 4), 75-78, 79(A, B, D)
IaU	Pre 1960(3-4, 13, 24, 26, 29, 32, 34, 38, 40-41, 43, 50, 55, 60, 67, 73, 76, 83, 99), 60(1, 7, 9), 61(1, 6, 8), 62(1-4, 6-7), 63(1, 3, 7-8), 64(2-7), 65(1, 3, 5-6), 66(1, 3-6), 67-71, 72(1, 3-6), 73, 74(1-2, 4) 75(2-4), 76, 77(1, 3), 78-79, 81-84
IdU	1985
InLP	Pre 1960(1, 4, 10, 16, 20, 25, 31, 34, 42, 49, 55, 61, 64, 67, 77, 79-81, 86, 88, 92, 96-97), 61(1, 6), 62(1-2, 4-5), 63(1-5), 64(2-6), 65(1, 3, 5-6), 66(1-3, 5-6), 67(2, 5), 68(1-3, 5), 69(2, 4), 70, 71(1, 3-6), 72, 73(1, 2, 4), 74(1, 4), 75(1, 3-4), 76, 77(1-2), 78, 79, 80, 81(A, C), 82(B-D), 83(D), 84(B, C), 85
InU	Pre 1960(19, 38, 53), 60(1-2, 5, 7, 9), 61(4, 7-8), 62(2, 6), 63(2, 5), 64(2, 5), 65(1, 3, 6), 66(5-6), 67(3), 68(1, 5), 69(1), 70(3), 71(4), 72(3, 5)

MiHM	Pre 1960(3-4, 9-10, 12-13, 15, 18-19, 24, 26-27, 29, 32, 34, 38, 40-41, 43, 50-51, 55, 57, 60-61, 67, 73, 76, 80, 83, 90, 99), 60(1, 3, 7, 9), 61(1, 3, 5-8), 62(1-7), 63(1, 3, 5, 7-8), 64(2-7), 65(1, 3, 5-6), 66(1, 3-6), 67, 68(1, 3-5), 69(1-4), 70(1-3, 5), 71(1, 3-6), 72(1, 4-6), 73, 74(1-2, 4), 75(2-4), 76-79
MnU	1960(1-4, 7, 9), 61(3, 8), 62(1-4, 6-7), 63(1, 3, 5, 7-8), 64(2-7), 65-68, 69(1-4), 70(1-2, 4-5), 71(1-5), 72(1, 3-6), 73-76, 77(1, 3), 78-85
NNC	1961(1), 64(1)
OU	Pre 1960(1, 3-4, 6, 13-14, 17, 20, 35-36, 38-39, 43, 48, 66, 69, 76, 79, 85, 87, 93-94, 96-98), 71(3-4)
OkU	Pre 1960(55), 66-70, 71(1-5), 72(1, 2, 4-6), 73, 74, 75, 76(1, 3, 4), 77(2, 4), 78-81, 82(B, C)
PHarER-T	1981-83

(265) ILLINOIS STATE GEOLOGICAL SURVEY. GUIDEBOOK SERIES. (TITLE VARIES)

1950 1. See American Association of Petroleum Geologists. 18.
1952 2. See Tri-State Geological Field Conference. 659.
1954 3. See American Association of Petroleum Geologists. 18.
1956 4. See American Association of Petroleum Geologists. 18.
1963 5. See Tri-State Geological Field Conference. 659.
1964 6. See Tri-State Geological Field Conference. 659.
1966 7. See American Association of Petroleum Geologists. 18.
1970 8. See Geological Society of America. 197.
1972 9. See Friends of the Pleistocene. Midwest Group. 183.
— 10. See Geological Society of America. North Central Section. 200.
1973 11. See Forum on the Geology of Industrial Minerals. 179.
1974 12. See Society of Economic Paleontologists and Mineralogists. Great Lakes Section. 587.
1979 13. See Friends of the Pleistocene. Midwest Group. 183.
— 14. See Tri-State Geological Field Conference. 659.
— 15. See International Congress of Carboniferous Stratigraphy and Geology. 290.
1981 See Geological Society of America. North Central Section. 200.
1982 See Illinois Geological Society. 262.
1985 16. See Geological Society of America. North Central Section. 200.
— 17. See Geological Society of America. North Central Section. 200.
— 18. See Geological Society of America. North Central Section. 200.
— 19. See Friends of the Pleistocene. Midwest Group. 183.

(266) ILLINOIS STATE MUSEUM. GUIDEBOOKLET. TITLE VARIES

1977 Nautral history tour to Horseshoe Lake and Shawnee National Forest, October 15-16, 1977.
1979 4. See International Congress of Carboniferous Stratigraphy and Geology. 290.

ICarbS	1977(Oct.), 80(2)
IaU	1980(2)

(267) INDIANA ACADEMY OF SCIENCE. GUIDEBOOK FOR GEOLOGIC FIELD TRIP.

1954 May Guidebook to the glacial geology of northern Steuben County, Indiana.
1955 May Guidebook to some interesting features of the geology and soils of Brown County, Indiana.
1959 May Guide to some geological features of McCormick's Creek State Park and vicinity.
1962 Spring Some features of karst topography in Indiana.
1965 May Glacial geology and soils of the area around Lake Maxinkuckee.

AzTeS	1965
CLU-G/G	1962
CPT	1965
CoDCh	1962
DI-GS	1962, 65
ICF	1965
IU	1965
IaU	1955
InLP	1962, 65
InU	1954, 55, 59, 65
MoSW	1962
TxU	1962

INDIANA GEOLOGICAL SOCIETY. See INDIANA GEOLOGICAL SURVEY. (268)

(268) INDIANA GEOLOGICAL SURVEY. FIELD CONFERENCE GUIDEBOOKS.

1947 1st Silurian and Devonian formations in southeastern Indiana. (270)
1948 2nd Upper and Middle Mississippian formations of southern Indiana. (270)
1949 3rd Silurian formations and reef structures of northern Indiana. (270)
1950 4th Stratigraphy along the Mississippian-Pennsylvanian unconformity of western Indiana. (270)
1951 5th Pennsylvanian geology and mineral resources of west central Indiana. (270)
1953 6th Ordovician stratigraphy and the physiography of part of southeastern Indiana. (270)
— June See Illinois Geological Society. 262.
1954 7th Salem Limestone and associated formations in south-central Indiana. (270)
1955 8th Sedimentation and stratigraphy of the Devonian rocks of southeastern Indiana. (270)
1957 9th Rocks associated with the Mississippian-Pennsylvanian unconformity of southwestern Indiana. (270)
1961 10th Stratigraphy of the Silurian rocks of northern Indiana. (270)
1965 11th Geomorphology and groundwater hydrology of the Mitchell Plain and Crawford Upland in southern Indiana.
1966 12th See Association of American State Geologists. 74.
1972 13th See Indiana-Kentucky Geological Society. 271.
1973 [1] Guidebook to an environmental geology field trip of the Terre Haute-Brazil, Indiana, area.
[2] Geology for the public: a field guide to the Lake Michigan shore in Indiana.
[4] Guidebook to the geology of some Ice Age features and bedrock formations in the Fort Wayne, Indiana, area.
[5a] Guidebook to the geology of the New Albany-Jeffersonville area of southern Indiana.
[5b] Road log and stop description.
1980 See Geological Society of America. North Central Section. 200.
1983 See Geological Society of America. 197.

ATuGS	1947-51, 53, 54-55, 57, 61, 65
CLU-G/G	1947-51, 53, 54-55, 57, 61, 65
CLhC	1949-50, 53, 54-55, 57
CMenUG	1947-51, 53-55, 57, 61, 65
CPT	1949-51, 53, 54-55, 57, 61, 65, 73(1, 4, 5A)
CSt-ES	1947-51, 53, 54-55, 57, 61, 65
CU-EART	1947-51, 53, 54-55, 57, 61, 65
CU-SB	1947-51, 57, 61, 65
CaACU	1961
CaOKQGS	1947-51, 53, 54-55, 57, 61, 65
CaOLU	1947, 49, 53, 55, 61
CaOOG	1947-51, 53, 54-55, 57, 61, 65
CaOTRM	1947-51, 53, 54
CaOWtU	1948-51, 53, 54-55, 57, 61, 65
CoDCh	1948-51, 53-55, 57, 61
CoG	1947-51, 53, 54-55, 57, 61, 65
CoU	1954
DI-GS	1947-51, 53, 54-55, 57, 61, 65, 73(1, 4)
DLC	1947-51, 61, 65
I-GS	1947-51, 53, 54-55, 57, 61, 65
ICF	1947-51, 53, 54-55, 57, 61, 65
ICarbS	1947-51, 54-55, 57, 61, 65
IEN	1947-51, 53, 54-55, 57, 61, 65
IU	1947-51, 53, 55, 57, 61, 65, 73(1, 4-5)
IaU	1947-51, 53, 54-55, 57, 61, 65, 73(4)
IdU	1947-51, 53, 54-55, 57, 61, 67
InLP	1947-51, 53, 54-55, 57, 61, 65, 73(2, 4)
InRE	1947-48, 50-51, 53, 54-55, 57, 61, 65
InTI	1947, 71-72
InU	1947-51, 53, 54-55, 57, 61, 65, 73(2, 4-5)
KyU-Ge	1947-51, 53, 54-55, 57, 61, 65, 73(1, 4-5)
LU	1947-51, 53, 54-55, 57, 61, 65
MH-GS	1954
MNS	1947-51, 53, 55, 57, 61, 65
MdBJ	1947-51, 53-55, 57, 61, 65, 66, 72
MiDW-S	1947-51, 53, 54-55, 57, 61, 65
MiHM	1947-51, 55, 57, 61, 65

(269) Indiana University-Purdue University at Indianapolis. Geology Department.

MiKW	1947-51, 57, 61, 65
MiU	1947-51, 53, 54-55, 57, 61, 65
MnU	1947-51, 53, 54-55, 57, 61, 65
MoSW	1947-51, 53, 54-55, 57, 61, 65
MoU	1947-51, 53, 54-55, 57
MtBuM	1949-51, 53, 54-55, 57, 61, 65
NBiSU	1949-50
NNC	1947-51, 53, 54, 57, 61, 65
NSyU	1949-51, 53, 54-55, 57, 61, 65
NbU	1947-51, 53, 54-55, 57, 61, 65
NcU	1947-51, 53, 54-55, 57, 61, 65
NdU	1947-51, 53, 54-55, 57, 61, 65
NhD-K	1949-51, 53, 54-55, 57, 61, 65
NjP	1948-51, 53, 54-55, 57, 61, 65
NvU	1947-51, 53, 54-55, 57, 61, 65
OCU-Geo	1947-51, 53, 54-55, 57, 61, 65, 73(1-2, 4-5)
OU	1947-51, 53, 54-55, 57, 61, 65
OkT	1948-51, 53, 54-55, 57, 61
OkU	1947-51, 53, 54-55, 57, 61, 65
PBm	1947-51, 53, 54-55, 57, 61
PSt	1947-51, 53, 54-55, 57, 61, 65
RPB	1948-51, 53, 54-55, 57, 61, 65
TU	1948-51, 53, 54-55, 57
TxBeaL	1961
TxDaAR-T	1947-51, 53, 54-55, 57, 61, 65
TxDaDM	1947-51, 53, 54-55, 57, 61, 65
TxDaGI	1949-51, 53, 54-55, 57, 61
TxDaM-SE	1948-51, 53, 54-55, 57, 61
TxHSD	1954, 57, 61, 65
TxHU	1947-51, 53, 54-55, 57, 61, 65
TxLT	1948, 50, 54, 57, 61
TxU	1947-51, 53, 54-55, 57, 61, 65
TxU-Da	1955
TxWB	1947-55, 57, 61
UU	1947-51, 53, 54-55, 57, 65
ViBlbV	1947-51, 53, 54-55, 57, 61, 65
WMMus	1951, 54
WU	1947-51, 53, 54-55, 57, 61, 65
WaU	1947-51, 53, 54-55, 57, 61, 65

(269) INDIANA UNIVERSITY-PURDUE UNIVERSITY AT INDIANAPOLIS. GEOLOGY DEPARTMENT. GUIDEBOOK.

1982 May Western and central New York State. (544)
1983 May G420 regional geology field trip to Grand Canyon, Arizona and surrounding areas.

InLP	1982, 83

(270) INDIANA UNIVERSITY. DEPARTMENT OF GEOLOGY. [ITINERARY & GUIDEBOOK]

1935 Nov. Itinerary and guide book for trip to the Ozark Highlands.
1947 Apr. See Indiana Geological Survey. 268.
1948 May See Indiana Geological Survey. 268.
1949 May See Indiana Geological Survey. 268.
— June Geology enroute Bloomington, Indiana to Jefferson Island, Montana and return. Rev. ed.
1950 May See Indiana Geological Survey. 268.
1951 May [1] Ozarks field trip.
[2] See Indiana Geological Survey. 268.
1953 May See Indiana Geological Survey. 268.
1954 May See Indiana Geological Survey. 268.
1955 Sept. See Indiana Geological Survey. 268.
1957 Oct. See Indiana Geological Survey. 268.
1961 May See Indiana Geological Survey. 268.
1962 June Guidebook. Bloomington, Indiana to Indiana geologic field station, Cardwell, Montana. Prepared for course, Geology G429, Field Geology in the Rocky Mountains.
1965 See Sigma Gamma Epsilon. Rho Chapter, Indiana University. 578.
1966 See Sigma Gamma Epsilon. Rho Chapter, Indiana University. 578.
Sedimentation and stratigraphy: Cambrian thru Devonian.
1968 Apr. Appalachian field trip.
— June Road logs and geological descriptions of the Black Hills and the Rocky Mountains; prepared for the course Geology G429, Field Geology in the Rocky Mountains.
1969 Mar./Apr. Regional geologic field trip to the Ozarks and Ouachitas, G420.
1970 June Road logs and geological descriptions of the Black Hills and the Rocky Mountains. Prepared for the course Geology G429, Field Geology in the Rocky Mountains. Rev. ed.
— Aug. G420. Regional geology field trip to the Lake Superior region.
1972 May G420. Regional geology field trip; Central and Southern Appalachians.
1973 May G420. Regional geologic field trip to the Ozarks, Ouachitas and Arbuckles.
— Oct. See Society of Economic Paleontologists and Mineralogists. Great Lakes Section. 587.
1980 See Geological Society of America. North Central Section. 200. Manual for field study of geology of the Northern Rocky Mountains.
1983 See Geological Society of America. 197.

CaOTRM	1949(June), 51(1)
CoDCh	1962
DI-GS	1962
IU	1935
IaU	1962
InLP	1980
InU	1935, 62, 68-70, 72, 73(May)
KyU-Ge	1973(May)
MnDuU	1968
NcU	1962
OCU-Geo	1962, 66, 68-69
TxWB	1972, 80

(271) INDIANA-KENTUCKY GEOLOGICAL SOCIETY. GUIDEBOOK FOR FIELD TRIP. (TITLE VARIES)

1940 Itinerary; outcrop of the Chester series of southern Indiana.
1966 See American Association of Petroleum Geologists. 18.
1972 May A field guide to the Mt. Carmel Fault of southern Indiana. (268)
1973 Oct. Geologic features of the Rough Creek Fault, Grayson and Ohio counties, Kentucky.
1975 May Field guide to stratigraphy of Stephensport group (Middle Chester) in southern Indiana.

CLU-G/G	1972-73
CPT	1972
CSt-ES	1972
CU-EART	1972
CaOKQGS	1972
CaOWtU	1972, 78(1B, 3A-B)
CoDCh	1973
CoG	1972
DI-GS	1972-73
DLC	1972
I-GS	1940, 72-73
ICF	1972
ICIU-S	1972
ICarbS	1972
IEN	1972
IU	1940, 72-73
IaU	1972-73
IdU	1972
InLP	1972-73
InRE	1972
InU	1972-73, 75
KyU-Ge	1972-73
LU	1972
MiDW-S	1972
MiHM	1972
MiKW	1972
MiU	1972
MnDuU	1972
MnU	1972-73
MtBuM	1972
NNC	1972
NcU	1973
NdU	1972
NhD-K	1972
NjP	1972
NvU	1972-73
OCU-Geo	1972-73
OU	1972
OkU	1972-73
PBL	1972
TxDaDM	1973
TxHU	1972

TxU	1972-73
UU	1972
ViBlbV	1972-73
WU	1972-73
WaU	1972

(272) INDIVIDUAL AUTHORS

1. Adams, Arthur G. The Hudson; a guidebook to the river. Albany, NY, State University of New York Press, 1981. 424 p. WU.
2. Adams, Virginia. Illustrated guide to Yosemite - the valley, the rim, and the central Yosemite Sierra, by Virginia and Ansel Adams. San Francisco, CA, Sierra Club, 1963. ICarbS.
3. Adams, Virginia. Illustrated guide to Yosemite Valley, by Virginia and Ansel Adams. 1952. 128 p.
4. Allen, John Eliot. Geologic field guide to the Northwest Oregon coast, by John Eliot Allen and Robert VanAtta. Portland, OR, Portland State College, 1964. 39 p.
5. Allied Chemical Corporation. Welcome to the Jamesville Quarry. [Jamesville, New York], published by Allied Chemical Corporation, n.d., 8 p.
6. Alt, David D. Roadside geology of Oregon. Missoula, MT, Mountain Press Publishing Co., 1978. 268 p. CSt-ES.
7. Alt, David D. Roadside geology of the Northern Rockies. Missoula, MT, Mountain Press Publishing Company, 1972. 280 p. AzU, CSt-ES, CU-A, CU-EART, CaACU, CaOOG, DLC, IU, InLP, KyU-Ge, MtBuM, NcU, NhD-K, WU, WaPS.
8. Alt, David D. Rocks, ice and water; the geology of Waterton-Glacier Park. Missoula, MT, Mountain Press Publishing Company, 1973. CLU-G/G, CSt-ES, CU-EART, CaACU, CaBVaU, CaOWtU, MnU, MtBuM, NcU, NmPE, WU, WaPS.
9. Alt, David D. and Donald W. Hyndman. Roadside geology of Northern California. Missoula, MT, Mountain Press Publishing Co., 1975. 244 p. IU, InLP, OkU.
10. Alt, David D. [and] Donald W. Hyndman. Roadside geology of Oregon. Missoula, MT, Mountain Press Publishing Company, 1981. 2d ed. MtBuM.
11. Alt, David D. [and] Donald W. Hyndman. Roadside geology of Washington. Missoula, MT, Mountain Press Publishing Company, 1984. 282 p. CSdS.
12. Armstrong, James Brackston. Field trip guide to a Colorado River excursion: Lee's Ferry to Phantom Ranch. Austin, TX, the University of Texas at Austin, 1971. TxU.
13. Audet, Sonia et. al. Sites geologiques touristiques en Abitibi-Temiscamingue. Rouyn, PQ, College de l'Abitibi-Temiscamingue. Cahiers du Departement d'histoire et de geographie, 1984. CaOWtU.
14. Bacon, Edwin M. Boston, Massachusetts, a guidebook to the city and vicinity by Edwin M. Bacon, revised by LeRoy Phillips. Boston, MA, Ginn & Company, 1903. CaBVaU, CaOOG, CaOWtU.
15. Bacon, Edwin M. Boston: A guide book to the city and vicinity. Boston, MA, Ginn and Company, 1928. IdU.
16. Bacon, Edwin M. Boston: A guide book to the city and vicinity. Revised by LeRoy Phillips. Boston, MA, Ginn and Company, c1922. ICarbS.
17. Baird, David M. A guide to geology for visitors to Canada's national parks. [s.l.], Macmillan of Canada/Toronto in cooperation with Indian and Northern Affairs [and] Parks Canada, 1974.
18. Bezy, John V. A guide to the desert geology of the Lake Mead National Recreation Area. Globe, AZ, Southwest Parks and Monuments Association, 1978. 68 p. NcU, NmPE.
19. Bohakel, Charles A. A guidebook to: Mt. Diablo, the "Devil" mountain of California, by Charles A. Bohakel. Rev. ed., 1973. 20 p. CU-EART.
20. Bohakel, Charles A. A pictorial guidebook to Mount Diablo, the "Devil" mountain of California. 2d rev. ed. Antioch, CA, 1975. 20 p. CU-A.
21. Bolles, William H. Pennsylvania Interstate 80 geologic guide [by] William H. Bolles [and] Alan R. Geyer. [Harrisburg, PA], Pennsylvania Department of Education, [1975]. 32 p.
22. Bolles, William H. Pennsylvania Interstate 81 geologic guide by William H. Bolles [and] Alan R. Geyer. [Harrisburg, PA], Pennsylvania Department of Education, 1976. 35 p. DI-GS, PHarER-T.
23. Buchanan, Rex. Kansas geology; an introduction to landscapes, rocks, minerals, and fossils. Lawrence, KS, University Press of Kansas, 1984. CMenUG, CU-SB, CoDGS, IaCfT, IdU, KLG, KU, NmU, OkT, TxDaM-SE.
24. Buck, Marcia Castle [and] Patricia Castle Smith. Gold rush nuggets; a gold mine of information about ... ten counties in California's Mother Lode area. Pasadena, CA, Castle Ventures, 1984. 191 p. CMenUG.
25. Burchett, R. Guidebook to the geology along the Missouri River bluffs of southeastern Nebraska and adjacent area. Lincoln, NE, University of Nebraska Conservation and Survey Division, 1978. 21 p. CSt-ES.
26. Camsell, Charles. Guide to the geology of the Canadian National Park on the Canadian Pacific railway between Calgary and Revelstoke. Department of the Interior, Ottawa, Ontario. (Material taken largely from the guidebooks by D. B. Dowling, J. A. Allan and R. A. Daly for the excursions of the 20th International Geological Congress and published by the Survey.) 1914. CSt-ES, CaACU, CaAEU, CaBVaU, CaOWtU, OkU, WU.
27. Chapman, Carleton A. The geology of Acadia National Park. Chatham Press, 1970. 128p. CSt-ES.
28. Chronic, Halka. Pages of stone; geology of western national parks & monuments, [volume] 1: Rocky Mountains & western Great Plains. Seattle, WA, The Mountaineers, 1984. CSt-ES, IdU.
29. Chronic, Halka. Roadside geology of Arizona. Missoula, MT, Mountain Press Publishing Company, 1983. 314 p. CMenUG, CPT, CSdS, CU-SB, CoDGS, IaCfT, IdBB, IdU, KLG, KU, NmLcU, NmU, OkT, OkTA.
30. Chronic, Halka. Roadside geology of Colorado. Missoula, MT, Mountain Press Publishing Company, 1980. CoU-DA CSt-ES, IdU, InTI, MBU, MiD, NmU.
31. Chronic, Halka. Time, rocks, and the Rockies; a geologic guide to roads and trails of Rocky Mountain National Park. Missoula, MT, Mountain Press Publishing Company, 1984. 120 p. CMenUG, CPT, CU-SB, CoDGS, IdU, KLG, KU, NmLcU, TxDaM-SE.
32. Clark, K.A. Guide to the Alberta Oil Sand area along the Athabasca River between McMurray and Bitumount and to the Oil Sand Separation Plant of the government of Alberta at Bitumount. 1951. CLhC.
33. Conkin, James E. Guide to the rocks and fossils of Jefferson County, Kentucky, southern Indiana, and adjacent areas by James E. Conkin and Barbara M. Conkin. 2d rev. ed. Louisville, KY, University of Louisville Reproduction Services, 1976. 238 p. WU.
34. Conkin, James E. Supplement and index to guide to the rocks and fossils of Jefferson County, Kentucky, southern Indiana and adjacent areas, by James E. Conkin and Barbara M. Conkin. Louisville, KY, University of Louisville, 1974. 49 p. CU-EART, IaU.
35. Conkin, James Elvin and Barbara M. Conkin. Guide to the rocks and fossils of Jefferson County, Kentucky, southern Indiana and adjacent areas. [Louisville, KY, Univ. of Louisville Printing Services, 1972] 331 p. CSt-ES, DLC, IU, InLP, MnU, NcU, NhD-K.
36. DeWindt, J. Thomas. Geology of the Great Smoky Mountains, Tennessee and North Carolina, with road log for field excursion, Knoxville-Clingmans Dome-Maryville, by J. Thomas DeWindt. Printed in The Compass, Sigma Gamma Epsilon, 1975, p. 73-129. CSt-ES, CU-EART, IU, IaU, InLP, NhD-K.
37. Feldmann, Rodney M. Field guide [to] the Black Hills. Dubuque, IA, Kendall/Hunt Publishing Company, 1980. MBU, NGenoU, NdFA, ODaWU.
38. Feldmann, Rodney M. Field guide, southern Great Lakes [by] Rodney M. Feldmann, Alan H. Coogan, Richard A. Heimlich. Dubuque, IA, Kendall/Hunt Publishing Company, 1977. 241 p. CSt-ES, CU-A, CU-EART, CaAEU, CaOWtU, DI-GS, DLC, IaU, InLP, InU, KyU-Ge, NSyU, NcU, NdU, OkU, PBm, TxU, ViBlbV, WU.
39. Finlay, George Irving. A guidebook describing the rock formations in the vicinity of Colorado Springs. Colorado Springs, CO, The Out West Company. DLC, WU.
40. Fowkes, E.J. An educational guidebook to the geologic resources of the Coalinga District, California. Coalinga, CA, West Hills College, 1982. 260 p. CMenUG.
41. Fritz, William J. Roadside geology of the Yellowstone country. Missoula, MT, Mountain Press Publishing Company, 1985. CSdS, CSfCSM, CU-SB, CoDGS, IaCfT, IdBB, IdU, NdFA, NmU, OkT, TxDaM-SE.

42 Gaines, David and the Mono Lake Committee. Mono Lake guidebook. CSt-ES, CU-A, CU-SB, DLC, IaU.

43 Genest, Claude. Livret-guide geiologique de la region de trois-Rivieres de Saint-Roch-de-Mekinac a Notre-Dame-du-Bon-Conseil. Sillery, Quebec, Quebec Science Editeur Presses de l'Universite du Quebec, 1985. IaU.

44 Gentile, Richard J. Guidebook to the field trips held in conjunction with the Symposium on the development and utilization of underground space. Kansas City, Missouri; sponsored by the Department of Geosciences, University of Missouri-Kansas City, and supported by local underground industry in Greater Kansas City and the National Science Foundation, Washington, D.C. [Kansas City MO, Department of Geosciences, University of Missouri-Kansas City, 1975] IU.

45 Gibbs, Alan Kendrick. Nat Sci 11 1977 field trip guidebook. Compiled by A.K. Gibbs, L. Brush, S. Maxwell, and R. Loucks. [s.l.: s.n.], 1977. 49 leaves. MH-GS.

46 Gilbert, Wyatt G. A geologic guide to Mount McKinley National Park. [s.l.], Alaska Natural History Association in cooperation with National Park Service, U.S. Department of the Interior, 1979. 52 p. NcU.

47 Goff, Fraser E. [and] Stephen L. Bolivar. Field trip guide to the Valles Caldera and its geothermal systems. Los Alamos, NM, Los Alamos National Laboratory, 1983.

48 Goldman, A. Lawrence, Jr. A field guide and sourcebook to geology of the eastern Upper Peninsula of Michigan; intended primarily for secondary earth science teachers. [s.l.], Western Michigan University, Department of Geology, 1976. (Publication E-S 1976.) 88 p. CU-EART, DI-GS.

49 Hamblin, W. Kenneth. Roadside geology of U. S. Interstate 80 between Salt Lake City and San Francisco, the meaning behind the landscape, by W. Kenneth Hamblin and J. Keith Rigby, John L. Snyder, William H. Matthews, III. Sponsored by the American Geological Institute. Van Nuys, CA, Varna Enterprises, 1974. 51 p. CLU-G/G, CSt-ES, CU-A, CU-EART, CaACU, CaOOG, CaOWtU, DLC, IU, IaU, InLP, MoSW, NcU, NdU, NhD-K, NvU, OCU-Geo, OU, OkU, TxHSD, WU.

50 Hamilton, Wayne L. The sculpturing of Zion; guide to the geology of Zion National Park. Diamond Jubilee Edition. Springdale, UT, Zion Natural History Association, 1984. CU-A, CU-EART, DLC, IaU, MnU, NhD-K.

51 Hanley, Thomas and M.M. Graff. Rock trails in Central Park. New York, NY, Greensward Foundation, Inc., 1976. CU-A, DLC.

52 Harbaugh, John Warvelle. Field guide; Northern California, Dubuque, IA, Kendall/Hunt Publishing Company, 1975. (K/H Geology Field Guide Series) 123 p. CU-EART, CaBVaU, InLP, TxU.

53 Harbaugh, John Warvelle. Geology field guide to Northern California [by] John W. Harbaugh. Dubuque, IA, W. C. Brown Co. [1974]. (The Regional Geology Series) 123 p. CSt-ES, CU-A, CU-EART, CaAEU, CaOWtU, DLC, ICIU-S, IU, InRE, InU, LNU, MnU, MoSW, NSyU, NcU, NhD-K, NmPE, OU, OkU, PBm, UU, WU.

54 Harris, Albert D. Klamath River geology; Curly Jack Camp to Ti Bar Siskiyou County, CA. (In California Geology, v. 31, No. 5, May 1978, p. 108- 114) CU-EART, IaU.

55 Harris, Ann G. [and] Esther Tuttle. Geology of national parks. 3rd ed. Dubuque, IA, Kendall/Hunt Publishing Company, c. 1983 554 p. MA.

56 Harris, David V. The geologic story of the National Parks and monuments. 4th ed. New York, NY, etc., John Wiley and Sons, 1985. MiD.

57 Harris, Stanley E., Jr. [and others]. Exploring the land and rocks of southern Illinois; a geological guide. Carbondale, IL, Southern Illinois University Press, 1977. CSt-ES, CU-A, CU-EART, CU-SB, MnU, NBiSU, NmU.

58 Hayes, Philip T. River runner's guide to the canyons of the Green and Colorado Rivers, with emphasis on geologic features, vol. 1: From Flaming Gorge Dam through Dinosaur Canyon to Ouray, by Philip T. Hayes and Elmer S. Santos. Denver, CO, Powell Society Ltd., [1969]. (Powell Centennial vol. 1) 40 p. WaPS.

59 Hayes, Philip T., and Simmons, George C. River runners' guide to Dinosaur National Monument and vicinity with emphasis on geologic features. Denver, CO, Powell Society, Ltd., 1973. CSt-ES, DI-GS, DLC, InU, ViBlbV.

60 Haynes, Jack Ellis. Haynes guide: handbook of Yellowstone National Park, Park, by Jack Ellis Haynes. 51st ed., 191 p. CSt-ES, MtBuM, OkU.

61 Heilprin, Angelo. Town geology: the lesson of the Philadelphia rocks; studies of nature along the highways and among the byways of a metropolitan town. Philadelphia, PA. Published by the author, Academy of Natural Sciences, 1885. 142 p. MiHM.

62 Hiller, John, Jr. Connecticut mines and minerals. Shelton, CT, John Hiller, Jr., c. 1971. 61 p. CaOOG, DI-GS, MH-GS, NcU.

63 Hiller, John, Jr. Massachusetts mines and minerals. Stratford, CT, John Hiller, Jr., c. 1974. 51 p. CU-A, CaOOG, CaOWtU, DI-GS, InLP, MH-GS, OU, ViBlbV.

64 Hogberg, Rudolph K. Field trip guide for the environmental geology of Olmsted County, Minnesota. St. Paul, MN, Minnesota Geological Survey, 1973. 27 p. CSt-ES.

65 Jillson, Willard Rouse. Geological excursions in Kentucky; a series of twelve descriptions of localities in and about the bluegrass region exhibiting earth phenomena of unusual interest. Frankfort, KY, Roberts Print Co., 1948. 59 p. CU-EART, DLC, IU, OkU.

66 Jorgensen, Neil. A guide to New England's landscape. [Barre, MA, 1971] 256 p. CaBVaU, InLP, NhD-K, UU.

67 Kain, Joan. Rocky roots; three geology walking tours of downtown St. Paul. St. Paul, MN, Ramsey County Historical Society, 1978. 32 p. InLP, MnU.

68 Langenheim, R. Geology along the highway route from Urbana, Illinois, to San Francisco and Los Angeles, California, and return via St. Joseph, Missouri, Denver, Colorado, Carson City, Nevada, and Las Vegas, Nevada, Kingman, Arizona, Grand Canyon, Arizona, Albuquerque, New Mexico, Tulsa, Oklahoma, and St. Louis, Missouri. Compiled from guidebooks of the Rocky Mountain Association of Petroleum Geologists, the Intermountain Association of Petroleum Geologists, the California Division of Mines, the New Mexico Geological Society, the Oklahoma City Geological Society; from State Geological maps; and from other sources, by B. J. Bluck [and others] under the direction of R. L. Langenheim, Jr. 1962. NhD-K.

69 Larkin, Robert P., Paul K Grogger [and] Gary L. Peters. Field guide [to] the Southern Rocky Mountains. Dubuque, IA, Kendall/Hunt Publishing Company, 1980. CMenUG, CoDGS, CoU-DA, MA, MBU, MiD, N-GenoU, NIC, NdFA, NmU, ODaWU.

70 Lipman, Peter W. Mauna Loa southwest rift zone; field trip guide. [n.l.], Hawaiian Volcano Observatory, 1979. 26 p. CSt-ES.

71 Livingston, Alfred, Jr. Geological journeys in Southern California by Alfred Livingston, Jr. [and] William C. Putnam. Los Angeles, CA, 1933. (Los Angeles Junior College. Publication no. 1) (Geology Series no. 1) 104 p. CU-EART.

72 Livingston, Alfred, Jr. Geological journeys in Southern California. 2d ed. Dubuque, IA, Wm. C. Brown Company, rewritten and revised July 1939. 169 p. CSt-ES.

73 Lobeck, Armin Kohl. Airways of America; Guidebook No. 1. The United Airlines; a geological and geographical description of the route from New York to Chicago and San Francisco. New York, NY, the Geographical Press, Columbia University, 1933. CaAEU, CaBVaU, MiHM, MnU, NcU, NhD-K, OkU, TxCaW.

74 Lobeck, Armin Kohl. The Midland Trail in Kentucky: a physiographic and geologic guidebook to U.S. Highway No. 60. (Published as an appendix to "Devonian Rocks of Kentucky" by T. E. Savage), Frankfort, KY, the Kentucky Geological Survey, 1971. DLC, MtBuM, NhD-K.

75 Lobeck, Armin Kohl. The geology of the Midland Trail in Kentucky (with Kentucky Geological Survey, Series VI, v. 33: The Devonian rocks of Kentucky by Thomas Edmund Savage). Frankfort, KY, the Kentucky Geological Survey, 1930. [95] p. CU-EART, WU.

76 [MINOBRAS] Uranium resources of the Central and Southern Rockies, Wyoming-Colorado-New Mexico. Dana Point, CA, MINOBRAS, 1979. 9 p. CoDGS.

77 Macfarlane, James. An American geological railway guide, giving the geological formation at every railway station, with altitudes above mean tide-water, notes on interesting places on the routes and a description of each of the formations. New York, NY, D. Appleton and Company, 1890. CSt-ES, ICIU-S, IU, MiHM, MnU, NhD-K, PSt, WU.

78 Macfarlane, James. An American geological railway guide, giving the geological formation at every railway station, with altitudes above mean tide-water, notes on interesting places on the routes, and a description of each of the formations. 2d ed., rev. and enl. New York, NY, D. Appleton and Company, 1890. 426 p. CU-EART.

79 Macfarlane, James. The geologists traveling handbook. An American geological railway guide, giving the geological formation at every railway station, with notes on interesting places on the routes and a description of each of the formations. New York, NY, D. Appleton and Company, 1879. 216 p. CU-EART, MiHM.

80 Matthews, Vincent and Ralph C. Webb. Pinnacles geological trails. Globe, AZ, Southwest Parks and Monuments Association, 1972. CSt-ES.

81 McGerrigle, H.W. Tour geologique de la Gaspesie. Bibliotheque Nationale du Quebec, 1985. CaBVaU, DI-GS.

82 McKay, Robert M., Raymond R. Anderson [and] Jean C. Prior. Geology in the Johnson County area. (Cover title: An introductory guidebook to geology in the Johnson County area) Reprinted from: Johnson County environmental field trip guidebook, by Johnson County Soil Conservation District. 1980? IaU.

83 McKenzie, Garry D. Field guide to the geology of parts of the Appalachian Highlands and adjacent interior plains, by Garry D. McKenzie [and] Russell O. Utgard, Columbus, OH, The Ohio State University, Department of Geology and Mineralogy, 1985. DI-GS, OCU-Geo.

84 Mengel, J.T., Jr. Geology of the western Lake Superior region; a guide for visitors. Superior, WI, Wisconsin State University, Geology Department, 1970. MiHM.

85 Milici, Robert C. A guide to the stratigraphy of the Chickamauga Supergroup in its type area by Robert C. Milici and James W. Smith. [s.l., s.n.], 15 p. CSt-ES.

86 Multer, H. G. Field guide to some carbonate rock environments; Florida Keys and western Bahamas. New edition. Dubuque, IA, Kendall/Hunt Publishing Co., 1977. (Kendall/Hunt Geology Field Guide Series) 415 p. AzU, CSt-ES, CU-A, CaAEU, CaOOG, CaOWtU, DI-GS, IaU, InRE, InU, KyU-Ge, NSyU, NcU, NhD-K, OU, OkU, PSt, TxBeaL, TxCaW, TxU, ViBlbV, WU.

87 Mutschler, Felix E. River runner's guide to the canyons of the Green and Colorado Rivers, with emphasis on geologic features, v. 4: Desolation and Gray Canyons. Denver, CO, Powell Society, [1972]. (Powell Centennial, v. 4) 85 p. CSt-ES.

88 Mutschler, Felix E. River runner's guide to the canyons of the Green and Colorado Rivers, with emphasis on geologic features, vol. 2: Labyrinth, Stillwater, and Cataract canyons. Denver, CO, Powell Society, [1972]. (Powell Centennial, v. 2) 79 p. WaPS.

89 [North Carolina Geological Survey] Guide to geologic and other natural resources; points of interest along Interstate 40. 1984. InLP.

90 Nymeyer, Robert. Carlsbad, Caves and a camera. Teaneck, NJ, Zephyrus Press, 1978. DI-GS, DLC.

91 Oakeshott, Gordon B. California's changing landscapes. 2d ed. New York, NY, McGraw-Hill Book Company, 1978. 379 p. CU-A, CU-EART.

92 Oakeshott, Gordon B. California's changing landscapes; a guide to the geology of the state. New York, NY, McGraw-Hill, 1971. 388 p. CSt-ES, CaAEU, CaBVaU, CaOWtU, ICIU-S, IU, InLP, LNU, MiHM, MnU, NcU, NhD-K, NvU, OU, TxCaW.

93 Ojakangas, Richard W. [and] Charles L. Matsch. Minnesota's geology. Minneapolis, University of Minnesota Press, 1982. 255 p. MiD.

94 Ollerenshaw, N. C. and Hills, L. V. Field guide to rock formations of southern Alberta; stratigraphic sections guidebook. Calgary, AB, Canadian Society of Petroleum Geologists, 1978. 102 p. CSt-ES, CU-SB.

95 Orme, Antony R. The Transverse Ranges and San Andreas fault system, California, by Antony R. Ormes [and] Amalie Jo Brown. Los Angeles, CA, University of California, Department of Geography, 1979. 51 p. CLU-G/G.

96 Ostrom, Meredith E. Geology field trip; southwestern Dane County. University Extension, University of Wisconsin, 1971. 10 l. IaU, InLP, NcU, WU.

97 Pabian, Roger K. Geology along the Republican River Valley near Red Cloud, Nebraska. Lincoln, NE, University of Nebraska, Lincoln Institute of Agriculture and Natural Resources, Conservation and Survey Division, [1980] 25 p. MoSW.

98 Parsons, Willard H. Field guide; Middle Rockies and Yellowstone. Dubuque, IA, Kendall/Hunt Publishing Company, 1978. (K/H Geology Field Guide Series) 233 p. CSt-ES, CU-A, IU, IaU, InLP, InU, MnU, MtBuM, TxU, WU, WaPS.

99 Paull, Rachel K. [and] Richard A. Paull. Field guide [to] Wisconsin and upper Michigan; including parts of adjacent states highway guide. Dubuque, IA, Kendall/Hunt Publishing Company, 1980. CoU-DA, MBU, NGenoU, ODaWU, WGrU.

100 Perkins, John W. Colorado and Wyoming; a geological field guidebook. Cardiff, Britain, Department of Extra-Mural Studies, University College, 1984. 138 p. CoDGS.

101 Perkins, John W. Utah and northern Arizona; a field guidebook. A circular autotour for the layman, rockhound or naturalist, describing the geology and geography of this golden area of scenic and vacation lands. Cardiff, Great Britain, University College, Department of Extra-Mural Studies, 1981. PSt.

102 Peterson, Helen. Peterson guide to mineral collecting, Bancroft area; including locations at Bancroft, Tory Hill, Wilberforce, Gooderham, Harcourt, Haliburton, Madoc, Quadville, Marmora, Norland, Kinmount [and] Minden. Compiled by Helen Peterson. Bancroft, Ontario, Parkwood Beach Ltd., 1970. CaOOG.

103 Pewe, T. L. and Updike, R. G. San Frnacisco Peaks; a guidebook to the geology. 2nd ed., 1976. CSt-ES.

104 Pewe, Troy Lewis. Colorado River guidebook; a geologic and geographic guide from Lees Ferry to Phantom Ranch, Arizona. 2nd ed. Tempe, AZ, The Author, 1969. 78 p. CLU-G/G, CSt-ES, CaOWtU, DLC, IU, InLP, MnU, NSyU, NcU, NhD-K, OU, OkU, PBm, UU.

105 Pewe, Troy Lewis. Colorado River guidebook; a geologic and geographic guide from Lees Ferry to Phantom Ranch, Arizona. Tempe, AZ, Arizona State University, Department of Geology, 1968. 78 p. PBm.

106 Prather, Thomas. Geology of the Gunnison Country. Road logs: [1] Gunnison to Montrose; [2] Gunnison to Monarch Pass; [3] Gunnison to Crested Butte. [Gunnison, CO], Western State College Foundation, c. 1982. 149 p. CU-A, CU-EART, MnU.

107 Price, Raye Carleson. Guidebook to Canyonlands country: Arches National Park, Moab, Colorado River, Canyonlands National Park. [Pasadena, CA], W. Ritchie Press, [1974] 96 p. CaACU.

108 Raup, Omer B. [and others]. Geology along Going-to-the-Sun Road, Glacier National Park, Montana. [s.l.], Glacier National History Association, 1983. CSt-ES, CU-A, CU-EART, CoDGS, DLC, MnU, OkU.

109 Ream, Lanny R. Northwest volcanoes; a roadside geologic guide. Renton, WA, BJB Books, 1983.

110 Reiter, Martin. The Palos Verdes Peninsula: a geologic guide and more. (Contains road log.) Dubuque, IA, Kendall/Hunt Publishing Company, 1984. CSt-ES, CU-EART.

111 Reynolds, Robert E. Geologic investigations along Interstate 15 Cajon Pass to Manix Lake, California. Redlands, CA, Curator, Earth Sciences San Bernardino County Museum, 1985. CSt-ES.

112 Richmond, Gerald M. Raising the roof of the Rockies. [s.l.], Rocky Mountain Nature Association, Inc. in cooperation with the U.S. National Park Service, 1974. DLC.

113 Rigby, J. Keith. Field guide, northern Colorado Plateau. Dubuque, IA, Kendall/Hunt Publishing Company, 1976. 207 p. AzTeS, CSt-ES, CU-A, CU-EART, CaAEU, CaOOG, CaOWtU, DI-GS, DLC, IU, IaU, InLP, InRE, InU, KyU-Ge, MnU, NcU, NdU, NmPE, NvU, OU, PBL, PBm, TxBeaL, TxU, ViBlbV, WU, WaPS.

114 Rigby, J. Keith. Field guide, southern Colorado Plateau. Dubuque, IA, Kendall/Hunt Publishing Company, 1977. 148 p. AzTeS, AzU, CSt-ES, CU-A, CU-EART, CaAEU, CaOWtU, CoU, DI-GS, DLC, IU, IaU, InLP, InRE, InU, KyU-Ge, MnU, NcU, NdU, NmPE, NvU, PBL, PBm, SdRM, TxBeaL, TxU, WU, WaPS.

115 Roberts, Ralph Jackson. Guidebook-San Francisco, California to Eureka, Nevada. 1970. 52 p. NvU.

116 Roseberry, Cecil R. From Niagara to Montauk; the scenic pleasures of New York State. Albany, NY, State University of New York Press, 1982. 344 p. CU-EART, ScU.

117 Runnells, Donald D. Boulder, a sight to behold: guidebook. Boulder, CO, Donald D. Runnells, c. 1976. CoU, DI-GS.

(272) Individual Authors

118 Sansome, Constance J. Minnesota underfoot; a field guide to the state's outstanding geologic features. Bloomington, MN, Voyageur Press, 1983. MiD, NSbSU.

119 Schaffer, Jeffrey P. Lassen Volcanic National Park; a natural-history guide to Lassen Volcanic National Park, Caribou Wilderness, Thousand Lakes Wilderness, Hat Creek Valley and McArthur-Burney Falls State Park. Berkeley, CA, Wilderness Press, 1981. 216 p. CU-A, CU-EART, DLC.

120 Seguin, Maurice K. L'est du Canada. Paris; New York, Masson, 1976. 175 p. CU-EART.

121 Sevenair, John P. Trail guide to the Delta country. Baton Rouge, LA, Sierra Club, New Orleans Group, 1980.

122 Sharp, Robert P. Field guide: Southern California. Dubuque, IA, Kendall/Hunt Publishing Company, 1975. 181 p. AzTeS, AzU, CLhC, CSfCSM, CSt-ES, CU-A, CU-EART, CaAEU, CaOOG, CaOWtU, DLC, IU, IaU, InLP, InRE, InU, KyU-Ge, LNU, MiHM, MnU, NSyU, NcU, NdU, NvU, OU, PBL, PBm, ViBlbV, WU.

123 Sharp, Robert P. Field guide: Southern California. Rev. ed. Dubuque, IA, Kendall/Hunt Publishing Company, 1976. 208 p. CU-A, CaAEU, NcU, TxU.

124 Sharp, Robert P. Field guide; coastal Southern California. Dubuque, IA, Kendall/Hunt Publishing Company, 1978. 268 p. AzU, CLhC, CSt-ES, CU-A, CU-EART, CaAEU, CaBVaU, CaOWtU, DI-GS, DLC, IU, IaU, InRE, InU, KyU-Ge, MnU, NSyU, NcU, NdU, OrCS, TxBeaL, ViBlbV, WU.

125 Sharp, Robert Phillip. Geology field guide to Southern California. Dubuque, IA, Wm. C. Brown Company Publishers, 1972. (The Regional Geology Series) 181 p. CLU-G/G, CSt-ES, CU-A, CU-EART, CaACU, CaAEU, CaOOG, CaOWtU, CoU, DLC, ICIU-S, IU, InLP, KyU-Ge, LNU, MH-GS, MnU, MoSW, NmPE, NvU, OU, OkU, TMM-E, TxU, WU, WaPS.

126 Sheldon, Robert A. Roadside geology of Texas. Missoula, MT, Mountain Press Publishing Co., 1979. 180 p. CU-EART, CaAEU, IU, NcU, NmPE.

127 Sheldon, Robert A. Roadside geology of Texas. Missoula, MT, Mountain Press Publishing Company, 1982. MtBuM.

128 Sheridan, Michael F. Superstition wilderness guidebook; an introduction to the geology and trails, including a roadlog of the Apache Trail and trails from First Water and Don's Camp, by Michael F. Sheridan. [1st ed], Tempe, AZ, [1971]. 52 p. AzTeS, CSt-ES, CU-A, CaAEU, CoG, DLC, MH-GS, MoSW, NcU, NmPE, NvU, OkU, UU.

129 Shirk, William R. A guide to the geology of south central Pennsylvania. Shippensburg, PA, William R. Shirk by Robson & Kaye, Inc., 1980. CU-SB, DLC, MnU, NIC, NhD-K.

130 Simmons, George Clarke. River runner's guide to the canyons of the Green and Colorado Rivers with emphasis on geologic features, v. 3: Marble Gorge and Grand Canyon. Flagstaff, AZ, Northland Press, 1969. (Powell Centennial, v.3) 132 p. WaPS.

131 Skehan, James W. Puddingstone, drumlins, and ancient volcanoes. Boston College, 1975. 63 p. CU-EART, CaOOG, DI-GS, IaU, InLP, InU, MH-GS, TxBeaL, ViBlbV, WU.

132 Skehan, James W. Puddingstone, drumlins, and ancient volcanoes; a geologic field guide along historic trails of greater Boston. 2d., rev. ed., Weston, MA, Boston College, Department of Geology and Geophysics, 1979. 63 p. CU-EART.

133 Skinner, Hubert C. Guidebook, southern Oklahoma field trip. 1958. 20 l. OkU.

134 Smith, Genny, ed. Mammoth Lakes, Sierra; a handbook for roadside and trail. 4th ed. Palo Alto, CA, Genny Smith Book, 1976. 147 p. CU-A, CU-SB, DLC, NvU.

135 Starr, Walter A., Jr. Guide to the John Muir Trail and the High Sierra region. San Francisco, CA, Sierra Club, 1967. CLU-G/G.

136 Stearns, Harold T. Geology of the Craters of the Moon National Monument, Idaho. Arco, [ID], Craters of the Moon Natural History Association, published in cooperation with the National Park Service, 1963. CSt-ES, DLC, MnU, NcU.

137 Stearns, Harold T. Road guide to points of geologic interest in the Hawaiian Islands by Harold T. Stearns. Palo Alto, CA, Pacific Books [1966]. CaBVaU, CaOOG, CaOWtU, IU, InLP, KyU-Ge, LNU, MnU, NcU, NhD-K, WU.

138 Stearns, Harold T. Road guide to points of geologic interest in the Hawaiian Islands. 2d ed. Palo Alto, CA, Pacific Books, Publishers, 1978. 100 p. AzU, CSt-ES, CU-A, CU-EART, CaACU, DI-GS, IU, IaU, InLP, MnU, NcU, OrCS, ViBlbV, WU.

139 Stewart, G.A. [and] G.T. MacCallum. Athabasca Oil Sands guide book. Calgary, AB, Canadian Society of Petroleum Geologists, 1978. 33 p. CU-A, CU-SB, IaU, NBiSU.

140 Stoller, James Hough. Geological excursions; a guide to localities in the region of Schenectady and the Mohawk Valley and the vicinity of Saratoga Springs, 1930.

141 Stout, T. M. Guidebook to the late Pliocene and early Pleistocene of Nebraska. Lincoln, NE, University of Nebraska Conservation and Survey Division. 1971. 109 p. CSt-ES.

142 Strawn, Mary and Devlin L. Williams. The geology of Massacre Rocks State Park. Boise, ID, Idaho. Parks and Recreation Department, n.d. IdPI.

143 Tabor and Crowder. Routes and rocks in the Mt. Challenger quadrangle. Seattle, WA, The Mountaineers. 1968. CaACU.

144 Tabor, Rowland W. Guide to the geology of Olympic National Park. Seattle, University of Washington Press, 1975. 144 p. AzU, CU-A, CU-EART, CaOOG, CaOWtU, DI-GS, DLC, IU, IaU, InLP, InRE, NSyU, NcU, NhD-K, NvU, OrCS, TxBeaL, TxU, ViBlbV, WU.

145 Tesmer, Irving H. Colossal cataract; the geologic history of Niagara Fall. Albany, NY, State University of New York Press, 1981. MBU, MnU.

146 Totten, Stanley M. Wooster-Hanover geology field trip; glacial geology of northeastern Ohio. [s.l., s.n.], 1978. OCU-Geo.

147 [U.S. Geological Survey] The river and the rocks; the geologic story of Great Falls and the Potomac River Gorge. [s.l.] U.S. Geological Survey - U.S. National Park Service, 1970. DLC, NIC.

148 Untermann, G. E. and B. R. Untermann. A popular guide to the geology of Dinosaur National Monument. Dinosaur Nature Association, Dinosaur National Monument, Utah-Colorado, 1969. DLC, UU.

149 Van Diver, Bradford B. Field guide [to] Upstate New York. Dubuque, IA, Kendall/Hunt Publishing Company, 1980. CU-EART, CaAEU, NmPE, NmU.

150 Van Diver, Bradford B. Roadside geology of New York. Missoula, MT, Mountain Press Publishing Company, 1985. AzTeS, CLU-G/G, CMenUG, DLC, IU, IaCfT, IaU, IdU, InLP, InTI, MA, MBU, MCM, MU, MnU, NGenoU, NIC, NSbSU, NdFA, NmU, ODaWU, PBm, TxU.

151 Van Diver, Bradford B. Rocks and routes of the north country New York; Geological guide for: tours, minerals, rock climbing, whitewater. Geneva, NY, W. F. Humphrey Press Inc., 1976. CLU-G/G, CSt-ES, CU-EART, DI-GS, IU, IaU, InLP, NSyU, NcU, ViBlbV, WU.

152 Wahrhaftig, Clyde. A streetcar to subduction and other plate tectonic trips by public transport in San Francisco. Rev. ed. Washington, DC, American Geophysical Union, 1984. 76 p. CSt-ES, DI-GS, DLC, IU, IaU, NIC, OCU-Geo.

153 Wegemann, Carroll H. A guide to the geology of Rocky Mountain National Park, Colorado. Washington, D.C., Government Printing Office, 1944. CLU-G/G, CSt-ES, DLC, WU.

154 Wheelock, Walt. Desert Peaks guide; Part I. Glendale, CA, La Siesta Press, 1964. 40 p. CSfCSM.

155 Whitney, Stephen. A field guide to the Grand Canyon. New York, NY, William Morrow and Company, 1982. 320 p. DI-GS.

156 Wolf, Robert Charles. Fossils of Iowa; field guide to Paleozoic deposits. 1983. CaOOG, DI-GS, IaU, NdFA.

157 Wright, Terry. Guide to geology and rapids, South Fork American River. Pollock Pines, CA, Wilderness Interpretation, 1981. CU-EART.

158 Wyckoff, Jerome. The Adirondack landscape, its geology and landforms; a hiker's guide. Gabriels, N.Y., Adirondack Mountain Club [1967]. 72 p. NhD-K.

159 Young, Keith. Field trip guide for Paleontology of Texas Seminar. Houston, TX, Continuing Education-University of Houston and Houston Museum of Natural Science, 1974. 4 l. TxU.

(273) INSTITUTE ON LAKE SUPERIOR GEOLOGY. [GUIDEBOOK TO FIELD TRIPS] (TITLE VARIES)
1964 10th Field trip, Marquette iron-mining district and Republic trough.
1965 11th St. Paul, Minnesota.

1966 12th   Includes areal description and guidebooks: Ontario. (384, 583)
- [1] Sault Ste. Marie area.
- [2] Geology and mineral deposits of the Manitouwadge Lake area.
- [3] Relationship of mineralization to the Precambrian stratigraphy, Blind River area, Ontario.
- [4] Sudbury nickel irruptive tour.

1967 13th   Field trip to the Grenville of southeastern Ontario.
1968 14th   The Duluth complex, near Ely, Minnesota.
1969 15th   Central Wisconsin volcanic belt.
1970 16th   A. Proterozoic formations in the Thunder Bay area.
- B. Sturgeon River metavolcanics-metasedimentary formations in the Beardmore-Geraldton area.
- C. The Port Coldwell alkalic complex.
- D. Atikokan. (Exact title not known)

1971 17th   A. The North Shore Volcanic Group (Keweenawan).
- B. Guide to the Precambrian rocks of northwestern Cook County as exposed along the Gunflint trail.
- C. Mesabi range magnetite taconite.
- D. Geology of the Vermilion metavolcanic sedimentary belt, northeastern Minnesota.

1972 18th   A. Penokean orogeny in the central and western Gogebic region, Michigan and Wisconsin.
- B. Guide to Penokean deformational style and regional metamorphism of the western Marquette range, Michigan.

1973 19th   1. Guidebook to the Precambrian geology of northeastern and north-central Wisconsin.
- 2. Guidebook to the geology and mineral deposits of the central part of Jackson County and part of Clark County, Wisconsin.
- 3. Guidebook to the upper Mississippi Valley base-metal district. (Reprint with slight modification of Wisc. Geological and Natural History Survey Information Circular No. 16)

1974 20th   Field trip no. 1. Middle Keweenawan rocks of the Batchawana-Mamainse Point area.
- Field trip no. 3. Precambrian igneous rocks of the north shore of Lake Huron region.
- Field trip no. 4. Stratigraphy and sedimentation of the Huronian Supergroup.
- Field trip no. 5. The Michipicoten greenstone belt.

1975 21st   Field trip no. 1. Glacial geology. Cancelled, no guidebook.
- Field trip no. 2. Greenstone.
- Field trip no. 3. The Jacobsville Sandstone; evidence for a lower-middle Keweenawan age.
- Field trip no. 4. Marquette iron range.
- Field trip no. 5, 6. The Empire mine and mill, Palmer, Michigan.

1976 22nd   Proceedings.
- Field trip A. Minnesota River Valley Field Conference.
- Field trip B. Engineering and Pleistocene geology in the Twin Cities area.

1977 23rd   [A] Coldwell complex.
- [B] Proterozoic rocks of the Thunder Bay area, northwestern Ontario: field excursion guide. (Cover title: Proterozoic trip)
- [C] Archean metallogeny and stratigraphy of the South Sturgeon Lake area. (Cover title: Mattabi trip)

1978 24th   1. Southwestern Wisconsin zinc-lead district.
- 1A. Geology of Upper Mississippi Valley zinc-lead district.
- 1B. Upper Mississippi Valley base metal district. (777)
- 2. Mineral extraction and processing equipment manufacturer in the greater Milwaukee area. (No guidebook issued)
- 3. Precambrian rhyolite, granite, and quartzite inliers in south-central Wisconsin.
- 3A. Introduction, geochronology and engineering geology of Precambrian rocks in south-central Wisconsin. (242)
- 3B. Precambrian inliers in south-central Wisconsin. (777)

1979 25th See Geological Society of America. North Central Section. 200.
1980 26th   Eau Claire, Wisconsin. (720)
- Field trip 1. Precambrian geology of the Chippewa Valley, Wisconsin.
- Field trip 2. Precambrian tectonic history of the Black River Valley.
- Field trip 3. Petrology, geochemistry, and contact relations of the Wausau and Stettin syenite plutons, central Wisconsin.
- Field trip 4. The Precambrian geology & tectonics of Marathon County, Wisconsin.

1981 27th   Lansing, Michigan.
The Huronian rocks between Sault Ste. Marie and Thessalon, District of Algoma, Ontario.

1982 28th   International Falls, Minnesota.
- Field trip 1. Mineral deposits of the Fort Frances-Mine Centre area, Ontario.
- Field trip 2. Archean geology of the International Falls-Kabetogama area, Minnesota.

1983 29th   Houghton, Michigan.
- (1) Field guide to the geology of the Keweenaw Peninsula, Michigan.
- (2) Ropes Gold Mine and its geological setting.

1984 Wausau, Wisconsin.
- Field trip 1. Guide to the geology of the early Proterozoic rocks in northeastern Wisconsin.
- Field trip 2. Early Proterozoic tectonostratigraphic terranes of the southern Lake Superior region: field trip guide with summary.
- Field trip 3. The Wausau syenite complex.

1985 31st   Kenora, Ontario.
- (1) The Cameron Lake deposit.
- (2) Geologic setting and style of gold mineralization in the Lake of the Woods area.
- (3) Geological relationships in the vicinity of the Wabigoon-Winnipeg River subprovincial interface in the Kenora area.
- (4) A volcanic facies interpretation of the Berry River formation.
- (5) Granitoid related mineralization in the Dryden area.

CLU-G/G	1966, 70(A-C), 71, 74(1, 3-5), 75, 77, 78(1B, 3B), 81, 83, 84(1, 2), 85
CSt-ES	1969, 70(A-C), 73, 75-77, 78(1B) 80(1-2, 4), 83
CU-EART	1967, 69, 76
CU-SB	1973, 80(1, 4)
CaACU	1976
CaOHM	1970
CaOLU	1966
CaOOG	1969, 78(3A)
CaOONM	1967-73, 75
CaOTDM	1964, 66-67, 72(A), 74, 75(1, 5-6), 77, 80(1, 3), 81-85
CaOWtU	1972
CoU	1969
DI-GS	1964, 67, 69-73, 74(1, 3-5), 75-76, 83, 85
DLC	1975
ICF	1978(3A)
ICIU-S	1969-73
ICarbS	1978(3A, 3B)
IU	1968-73, 74(1, 3-5), 75-77, 78(1A-B, 3A-B)
IaU	1967, 69, 71, 73(1-2), 78(1B), 80(1-2, 4), 82, 84(3)
IdU	1984(3)
InLP	1965, 67-76, 78(1A, 1B, 3A-B), 80(1), 83
InU	1964, 68-70
KyU-Ge	1969-73, 76-77, 78(1A-B, 3A-B), 80(1-2, 4), 84
LNU	1969
MU	1980
MdBJ	1978
MiHM	1964, 66, 69, 70(A-C), 71-72, 73(1), 74(1, 3-5), 75-77, 80, 83, 84(1-2), 85
MiMarqN	1975, 83
MiU	1967
MnU	1964-66, 68-71, 73, 74(1, 3-5), 75-77, 78(1A-B), 3A-B), 80-83
NRU	1983
NSbSU	1978(3A)
NSyU	1969, 75
NhD-K	1978(1B, 3A-B)
NjP	1962-70
OCU-Geo	1966, 70
OU	1964, 69, 71
OkOkU	1969
PBm	1969
TxU	1968-69, 73, 75
WGrU	1980(1-2)
WU	1968-76, 77, 78(1B, 3A-B), 80-85

(274) **INSTITUTO GEOGRAFICO NACIONAL (GUATEMALA). BULLETIN.**
1967 No. 4.   See Geological Society of America. 197.

(275) **INTERAMERICAN MICROPALEONTOLOGICAL COLLOQUIUM. FIELD TRIP GUIDEBOOK.**
1970 1st   Texas. (Summary of Upper Cretaceous stratigraphy: the Gulf of Mexico province)

(276) Intermountain Association of (Petroleum) Geologists.

CLU-G/G	1970
CLhC	1970
DI-GS	1970
ICarbS	1970
MoSW	1970
TxDaAR-T	1970
TxHSD	1970
TxU	1970

(276) INTERMOUNTAIN ASSOCIATION OF (PETROLEUM) GEOLOGISTS. GUIDEBOOK FOR THE ANNUAL FIELD CONFERENCE.

1950 1st Petroleum geology of Uinta Basin. (737)
1951 2nd Geology of the Canyon, House, and Confusion Ranges, Millard County, Utah. (737)
1952 3rd Cedar City, Utah to Las Vegas, Nevada. (737)
1953 4th Geology of northern Utah and southeastern Idaho.
1954 5th Geology of portions of the high plateaus and adjacent canyon lands, central and south-central Utah.
1955 6th See Rocky Mountain Association of Geologists. 549.
1956 7th Geology and economic deposits of east-central Utah.
1957 8th Geology of the Uinta Basin.
1958 9th Geology of the Paradox Basin.
1959 10th Geology of the Wasatch and Uinta Mountains, transition area.
1960 11th Geology of east-central Nevada. (165)
1963 12th Geology of southwestern Utah.
1964 13th Geology and mineral resources of the Uinta Basin; Utah's hydrocarbon storehouse.
1965 [14th] Geology and resources of south-central Utah; resources for power. (737)
1967 15th Anatomy of the western phosphate field; a guide to geologic occurrence, exploration methods, mining engineering, recovery technology. Supplement to guidebook: Descriptive geology along the 1967 field conference route.
1969 16th Geologic guidebook of the Uinta Mountains, Utah's maverick range. (737)

AzFU	1951, 53, 56-60, 67, 69
AzTeS	1951-52, 54, 56-60, 63-65, 67, 69
AzU	1950-53, 56-60, 63-65, 67, 69
CChiS	1958-59, 63-64, 67, 69
CLU-G/G	1950-54, 56-60, 63-65, 67, 69
CLhC	1950-54, 56-60, 63-65, 67, 69
CMenUG	1950-65, 67, 69
CPT	1953-54, 56-60, 63-65, 67, 69
CSdS	1951, 53, 56-60, 63-64
CSt-ES	1950-60, 63-65, 67, 69
CU-A	1960, 64-65, 67, 69
CU-EART	1950-54, 56-60, 63-65, 67, 69
CU-SB	1950-53, 56-65, 67, 69
CaAC	1958
CaACAM	1951, 67
CaACI	1956, 58
CaACU	1957, 60, 63, 65, 67, 69
CaBVaU	1950-53, 57-59, 63-65, 67
CaOHM	1952, 56, 58-60, 63-64, 67
CaOKQGS	1952, 58-60, 63-65, 67
CaOOG	1950, 53-54, 56-60, 63-64, 69
CaOWtU	1967
CoDCh	1950-54, 56-60, 63-65
CoDGS	1950-60, 63-64, 67, 69
CoDU	1952, 65, 69
CoDuF	1957-58
CoFS	1956-60, 63-65, 67, 69
CoG	1950-54, 56-60, 63-65, 67, 69
CoGrU	1950-54, 56-65
CoU	1950-54, 56-60, 63-65, 67, 69
DFPC	1957-58, 63-65
DI-GS	1950-54, 56-60, 63-65, 67, 69
DLC	1951-54, 56-60, 63-65, 67
I-GS	1960
ICF	1950-54, 56-60, 63
ICIU-S	1952, 58-59, 63-65, 67, 69
ICarbS	1951-54, 58-60, 63-65, 67, 69
IEN	1950-54, 56-60, 63-65, 67, 69
IU	1950-54, 56-60, 63-65, 67, 69
IaU	1950-54, 56-60, 63-65, 67, 69
IdBB	1950-53, 56, 58-59, 63-65, 67
IdPI	1953, 58-60
IdU	1950-54, 56-60, 63-65, 67, 69
InLP	1951-54, 56-60, 63-65, 67, 69
InRE	1950-52
InU	1950-53, 56-60, 63-65, 67, 69
LNU	1951, 65
LU	1951-54, 56-60, 63-65, 67, 69
MH-GS	1951, 63
MNS	1952-53, 56-58, 65
MiDW-S	1950-53, 56-60, 63-65, 67, 69
MiHM	1967, 69
MiKW	1958-60, 63, 65, 67, 69
MiU	1950-54, 56-60, 63-65, 67, 69
MnDuU	1969
MnSSM	1960
MnU	1950-54, 56-60, 63-65, 67, 69
MoRM	1950-52, 65, 69
MoSW	1950-54, 56-60, 63-65, 67, 69
MoU	1950-52, 65, 69
MtBC	1953, 59
MtBuM	1953-54, 56-58, 60, 67
NBiSU	1953, 56, 58-60, 63-65, 67
NNC	1950-54, 56-60, 63-65, 67, 69
NOneoU	1958-59, 63-64
NRU	1951, 53, 56-60, 65
NSyU	1950-54, 56-60, 63, 65
NbU	1950-54, 56-60, 63-65, 67, 69
NcU	1951-54, 56-60, 63-65, 67, 69
NdU	1951-54, 56, 58-60, 63, 65, 67, 69
NhD-K	1950-54, 56-60, 63-65, 67, 69
NjP	1950-52, 65
NmU	1960
NvU	1950-54, 56-60, 64-65, 67, 69
OCU-Geo	1953, 54, 56-59, 63-65, 67, 69
OU	1950-54, 56, 58-60, 63-65, 67, 69
OkT	1950-54, 56-60, 63, 65, 67
OkU	1950-54, 56-60, 63-65, 67, 69
OrCS	1950-52, 65, 69
OrU-S	1950-54, 56-60, 63-65, 67, 69
PBL	1950-51
PBm	1950
PSt	1951-54, 56-60, 63-65, 67, 69
RPB	1950-54, 56-60, 63-65, 67, 69
SdRM	1950-53, 55-59, 65
TMM-E	1963
TU	1956-60, 63-65
TxBeaL	1951, 67, 69
TxDaAR-T	1950-54, 56-60, 63-65, 67, 69
TxDaDM	1950-54, 56-60, 63-65, 67, 69
TxDaGI	1950-54, 56-59, 63, 65
TxDaM-SE	1950-53, 56-60, 63-65
TxDaSM	1950-53, 56-60, 63-65, 67
TxHSD	1950-53, 56-60, 63-64
TxHU	1951-54, 56-60, 63-65, 67, 69
TxLT	1951, 53-54, 56-60, 63-65, 67, 69
TxMM	1950, 53-54, 56-60, 63-65, 67, 69
TxU	1950-54, 56-60, 63-65, 67, 69
TxU-Da	1952-54, 58
TxWB	1953-54, 56, 58-59, 63-64, 67
UPB	1950-52, 65
UU	1950-54, 56-60, 63-65, 67, 69
ViBlbV	1950-53, 56-60, 63-65, 67, 69
WU	1950-54, 56-60, 63-65, 67, 69
WaU	1953, 56-60
WyU	1950-53, 56-60, 63-65, 67

(277) INTERNATIONAL ASSOCIATION FOR QUATERNARY RESEARCH (INQUA). CONGRESS. GUIDEBOOKS FOR FIELD CONFERENCE.

1965 7th A. New England-New York State.
B-1. Central Atlantic coastal plain.
B-3. Mississippi delta and central Gulf Coast. (Eleven guidebooks)
C. Upper Mississippi valley.
D. Central Great Plains.
E. Northern and middle Rocky Mountains.
F. Central and south central Alaska. (Revised in 1977) (445)
G. Great Lakes-Ohio River valley.

H. Southwestern arid lands.
I. Northern Great Basin and California.
J. Pacific northwest.
K. Boulder area, Colorado.

1977   Central and south central Alaska. (Rev. ed.)

AzTeS	1965
AzU	1965
CLU-G/G	1965
CPT	1965
CSt-ES	1965
CU-EART	1965(A-H, J, G)
CU-SB	1965
CaACU	1977
CaAEU	1965
CaBVaU	1965
CaOKQGS	1965(A-J)
CaOLU	1965
CaOOG	1965(A-G, I-K) (On microfiche)
CaOONM	1965(A-J)
CaOTRM	1965
CaOWtU	1965(A-J)
CoFS	1965
CoG	1965
CoU	1965
DI-GS	1965
DLC	1965(F)
ICF	1965
ICarbS	1965
IU	1965
IaU	1965
IdBB	1965(C, E)
InLP	1965
InU	1965
KLG	1965
LU	1965(C)
MiHM	1965(C)
MiKW	1965(A-J)
MnU	1965
MoSW	1965
NSbSU	1965
NbU	1965
NcU	1965(A-J)
NdU	1965
NhD-K	1965
NjP	1965
NvU	1965
OU	1965
OkU	1965(D, K)
OrU-S	1965
RPB	1965
SdRM	1965
TxBeaL	1965
TxCaW	1965
TxDaAR-T	1965
TxDaDM	1965(H)
TxDaSM	1965
TxHU	1965
TxU	1965
UU	1965(A-C, E-K)
ViBlbV	1965
WU	1965
WaPS	1965(E, H-J)
WaU	1965

INTERNATIONAL ASSOCIATION FOR THE STUDY OF FOSSIL CNIDARIA. See **INTERNATIONAL SYMPOSIUM ON FOSSIL CNIDARIA. (312)**

(278) **INTERNATIONAL ASSOCIATION OF GEOCHEMISTRY AND COSMOCHEMISTRY. FIELD TRIP GUIDE BOOK.**
1980   See International Symposium on Water-Rock Interaction. 321.

(279) **INTERNATIONAL ASSOCIATION OF SEDIMENTOLOGISTS. INTERNATIONAL CONGRESS ON SEDIMENTOLOGY. FIELD EXCURSION GUIDE BOOK.**
1982 1A: Lower Cambrian bioherms and sandstones, southern Labrador.

— 4A: From Sabkha to Coal Swamp, the Carboniferous sediments of Nova Scotia and southern New Brunswick.
— 6A: Beach and nearshore depositional environments of the Bay of Fundy and southern Gulf of St. Lawrence.
— 10A: Weathering, soil formation and land used in south and central Ontario.
— 11A: Late Quaternary sedimentary environments of a glaciated area: southern Ontario.
— 12A: Lower Paleozoic carbonate rocks and paleoenvironments in southern Ontario.
— 14A: Precambrian sediments and environmental aspects.
— 17A: Sedimentary facies; products of sedimentary environments in a cross section of the classic Appalachian Mountains and adjoining Appalachian Basin in New York and Ontario.
— 21A: Clastic units of the front ranges, foothills and plains in the area between Field, B. C. and Drumheller, Alberta.
— 22: Athabasca oil sands, sedimentology and development technology.
— 27A: Upper Devonian Miette reef complex, Jasper National Park, Alberta.
— 30A: Late Quaternary sedimentary environments, southwestern British Columbia.
— 2B: Anatomy and evolution of a Lower Paleozoic continental margin, western Newfoundland.
— 5B: Pre-Acadian sedimentary rocks of the Meguma Zone, Nova Scotia; a passive continental margin juxtaposed against a volcanic island arc.
— 7B: Paleozoic continental margin sedimentation in the Quebec Appalachians.
— 9B: Coastal sediments and geomorphology of the Canadian lower Great Lakes.
— 13B: Depositional environments and tectonic settings in the early Proterozoic Huronian Supergroup.
— 16B: Precambrian geology of the cobalt area: southern Ontario.
— 19B: Comparative sedimentology of Paleozoic clastic wedges in the central Appalachians, U.S.A.
— 20B: Glacial and postglacial sediments Edmonton-Jasper-Banff- Calgary area, Alberta.
— 28B: Upper Devonian stratigraphy and sedimentology, southern Alberta Rocky Mountains.

CLU-G/G	1982(1A, 4A, 6A, 10A, 11A, 12A, 14A, 17A, 21A, 22, 27A, 30A, 2B, 5B, 7B, 9B, 13B, 16B, 19B)
CMenUG	1982(1A, 4A, 6A, 10A, 11A, 12A, 14A, 17A, 21A, 22, 27A, 30A, 2B, 5B, 7B, 13B, 16B, 19B)
CSt-ES	1982(1A, 4A, 6A, 10A, 11A, 12A, 14A, 22, 30A, 2B, 5B, 9B, 13B, 16B, 19B, 20B)
CU-A	1982(1A, 4A, 6A, 10A, 11A, 12A, 14A, 22, 30A, 2B, 5B, 9B, 13B, 16B, 19B)
CU-EART	1982(4A, 6A, 10A, 11A, 12A, 14A, 17A, 22, 27A, 30A, 2B, 5B, 7B, 9B, 13B, 16B, 19B)
CU-SB	1982(1A, 4A, 6A, 10A, 11A, 12A, 14A, 17A, 21A, 22, 27A, 30A, 2B, 5B, 7B, 9B, 13B, 16B, 19B)
CaACU	1982(1A, 4A, 6A, 10A, 11A, 12A, 14A, 17A, 21A, 22, 27A, 30A, 2B, 5B, 7B, 9B, 13B, 16B, 19B)
CaAEU	1982
CaNSHD	1982
CaOKQGS	1982(1A, 4A, 6A, 10A, 11A, 12A, 14A, 17A, 21A, 22, 27A, 30A, 2B, 5B, 7B, 9B, 13B, 16B, 19B)
CaOOG	1982(1A, 4A, 10A, 14A, 21A, 22, 27A, 30A, 5B, 7B, 9B, 13B, 16B, 19B)
CaOTDM	1982(1A, 4A, 6A, 10A, 11A, 12A, 14A, 17A, 21A, 22, 27A, 30A, 2B, 5B, 7B, 9B, 13B, 16B, 19B)
CaOWtU	1982(1A, 2B, 4A, 5B, 6A, 7B, 9B, 10A-12A, 13B, 14A, 16B, 17A, 19B, 21A-22A, 27A, 30A)
CoDGS	1982
CoLiM	1982(4A, 11A, 12A, 17A, 27A, 30A, 2B, 7B, 13B, 16B, 20B, 28B)
DI-GS	1982(1A, 4A, 6A, 17A, 21A, 27A, 30A, 2B, 5B, 7B, 28B)
IdU	1982(11A, 20B, 30A)
InLP	1982(1A, 4A, 6A, 10A-12A, 17A, 21A, 22, 27A, 30A, 2B, 5B, 7B, 9B, 13B, 16B, 19 B, 20B, 28B)
InU	1982
MA	1982
MH-GS	1982
MdBJ	1982
MnU	1982(1A, 4A, 6A, 10A, 11A, 12A, 14A, 17A, 21A, 22, 27A, 30A, 2B, 5B, 7B, 9B, 13B, 16B, 19B)

(280) International Association on the Genesis of Ore Deposits.

MtBuM	1982(4A, 21A)	
NIC	1982	
NRU	1982	
NSbSU	1982	
NcU	1982(1A, 6A, 10A, 11A, 12A, 14A, 17A, 21A, 22, 27A, 30A, 2B, 5B, 7B, 9B, 13B, 16B, 19B, 20B, 28B)	
NdU	1982	
NhD-K	1982	
OU	1982	
OkTA	1982	
OkU	1982	
TxDaM-SE	1982(1A, 4A, 6A, 10A, 11A, 12A, 14A, 17A, 21A, 22, 30A, 2B, 5B, 9B, 13B, 16B, 19B, 20B, 28B)	
TxU	1982(6A, 14A, 22, 27A, 30A, 2B, 5B, 7B, 9B, 13B, 19B, 28B)	
WU	1982(1A, 4A, 6A, 10A, 11A, 12A, 14A, 17A, 21A, 22, 27A, 30A, 2B, 5B, 7B, 9B, 13B, 16B, 19B)	
WaU	1982(11A, 21A)	
WyU	1982	

(280) INTERNATIONAL ASSOCIATION ON THE GENESIS OF ORE DEPOSITS. FIELD EXCURSION.

1978 5th   C-1. Guidebook to mineral deposits of the central Great Basin. (452)
C-2. Guidebook to mineral deposits of southwestern Utah. (735)
C-3/C-4. Guidebook on fossil fuels and metals, eastern Utah and western-southwestern-central Colorado. (143)

AzU	1978(C-3/C-4)
CLU-G/G	1978
CLhC	1978(C-3/C-4)
CPT	1978
CSt-ES	1978
CU-EART	1978
CU-SB	1978
CaACU	1978(C-1, C-3/C-4)
CaBVaU	1978(C-1, C-3/C-4)
CaOOG	1978(C-3/C-4)
CaOWtU	1978(C-2, C-3/C-4)
CoDBM	1978(C-3/C-4)
CoDCh	1969
CoDGS	1978
CoG	1978
CoU	1978(C-1)
DI-GS	1978
DLC	1978(C-2, C-3/C-4)
ICarbS	1978(C-2, C-3/C-4)
IU	1978(C-2)
IaU	1978
InLP	1978(C-2, C-3/C-4)
InRE	1978(C-3/C-4)
InU	1978
KyU-Ge	1978
MnU	1978
NSbSU	1978
NSyU	1978(C-2)
NhD-K	1978(C-1, C-2)
NvU	1978
OCU-Geo	1978(C-2, C-3/C-4)
OU	1978(C-2, C-3/C-4)
OkU	1978(C-2)
OrCS	1978(C-3/C-4)
PSt	1978
TxU	1978(C-2)
UU	1978(C-2, C-3/C-4)
WU	1978(C-1, C-2, C-3/C-4)

(281) INTERNATIONAL BOTANICAL CONGRESS.

1959 9th   Trip 1C. British Columbia.
23. Palaeobotanical excursion to eastern Canada. (Cover title: Palaeobotanical excursion ... to the Gaspe Peninsula, New Brunswick, and northwestern Nova Scotia.)

CSt-ES	1959(23)
DLC	1959(23)
NIC	1959(23)
OU	1959

(282) INTERNATIONAL CLAY CONFERENCE. FIELD TRIP GUIDEBOOK.

1985 July-Aug.   Denver, Colorado.
(1) Bentonite, coal and uranium deposits of the Black Hills and Powder River Basin.
(2) Clays and clay minerals, western Colorado and eastern and central Utah.
(3) The low-grade metamorphic transformation of clay minerals in sedimentary rocks within the Montana disturbed belt. (133)
(4) Clays and zeolites; Los Angeles, California to Las Vegas, Nevada.

CMenUG	1985
CoDGS	1985
DI-GS	1985(4)
I-GS	1985
KyU-Ge	1985

(283) INTERNATIONAL COMMITTEE ON GEODYNAMICS. WORKING GROUP 7.

1978 Aug.   See Lunar and Planetary Institute. Topical Conference. 364.

(284) INTERNATIONAL COMMITTEE ON NATURAL ZEOLITES. EXCURSION.

1983 6th   Zeo-trip '83; an excursion to selected zeolite deposits in eastern Oregon, southwestern Idaho, and northwestern Nevada and to the Tahoe-Truckee Water Reclamation Plant, Truckee, California.

CMenUG	1983
CSt-ES	1983
CU-A	1983
CU-EART	1983
CaOOG	1983
CoDBM	1983
CoDGS	1983
DI-GS	1983
IaU	1983
ScU	1983

(285) INTERNATIONAL CONFERENCE ON ARID LANDS IN A CHANGING WORLD. GUIDEBOOK

1969 June   Tucson, Arizona. Guidebook of northern Mexico Sonoran Desert region.

AzTeS	1969
AzU	1969
CLU-G/G	1969
CaOWtU	1969
CoDCh	1969
CoDGS	1969
CoG	1969
CoU	1969
DI-GS	1969
DLC	1969
IU	1969
IaU	1969
KyU-Ge	1969
MiU	1969
MoSW	1969
MoU	1969
NhD-K	1969
NmPE	1969
OU	1969
OkU	1969
PSt	1969
TxDaDM	1969
TxHSD	1969
TxLT	1969
TxMM	1969
TxU	1969
UU	1969

(286) INTERNATIONAL CONFERENCE ON MISSISSIPPI VALLEY TYPE LEAD-ZINC DEPOSITS. GUIDES TO FIELD TRIPS.

1982 Oct.   Rolla, Missouri.
A Guide to the geology of the Buick Mine.
Geologic guidebook to Fletcher Mine.
Magmont Mine (Cominco American Incorporated & Dresser Industries), Viburnum Trend, southeast Missouri.

Frank R. Milliken Mine of Ozark Lead Company, new lead belt, southeast Missouri.
Field trip to the St. Francois Mountains and the historic Bonne Terre Mine.
WU   1982

(287) INTERNATIONAL CONFERENCE ON PERMAFROST. GUIDEBOOK.
1978 3rd   Edmonton, Alberta.
Field trip no. 1. Central Yukon-Alaska.
Field trip no. 2. Upper Mackenzie River valley.
Field trip no. 3. Lower Mackenzie River valley.
Field trip no. 4. Northern Manitoba - District of Keewatin.
Field trip no. 5. Northern Quebec - Laborador.
Field trip no. 6. Rocky Mountains - Alberta foothills.
1983 4th   Fairbanks.
1. Guidebook to permafrost and Quaternary geology along the Richardson and Glenn Highways between Fairbanks and Anchorage, Alaska. (8)
Middle Tanana River valley.
Delta River area, Alaska Range.
Copper River basin.
Overview of the Matanuska Glacier.
Upper Cook Inlet region and the Matanuska Valley.
2. Guidebook to permafrost and related features of the Colville River Delta, Alaska. (8, 295)
3. Guidebook to permafrost and related features of the northern Yukon territory and Mackenzie Delta, Canada. (295, 8)
The Klondike and Dawson.
Dawson.
The Dempster Highway - Dawson to Eagle Plain.
The Dempster Highway - Eagle Plain to Inuvik.
Mackenzie Delta and Inuvik.
Inuvik.
Tuktoyaktuk.
4. Guidebook to permafrost and related features along the Elliott and Dalton highways, Fox to Prudhoe Bay, Alaska. (8, 295)
5. Guidebook to permafrost and related features, Prudhoe Bay, Alaska. (8, 295)
6. The Alaska Railroad between Anchorage and Fairbanks. (8)

AKU-GI	1983(2-4)
AzTeS	1983
CLU-G/G	1983
CMenUG	1983(1, 5)
CPT	1983
CSt-ES	1983(5, 6)
CU-A	1983
CU-EART	1983
CU-SB	1983
CU-SC	1983(1, 5)
CaACU	1983
CaOOG	1983(1, 5)
CaOWtU	1983
CoDGS	1983(1)
DI-GS	1983
DLC	1983
IaCfT	1983(1)
IaU	1983(1, 2, 4, 5)
InLP	1983
KLG	1983(1, 5)
KU	1983(1, 5)
KyU-Gc	1983
LNU	1983(1)
MnU	1983(1, 5)
MoSW	1983(1, 5)
MtBuM	1983(1, 4)
NRU	1983
NhD-K	1983
OkU	1983(1, 5)
OrU-S	1983(1)
TxDaM-SE	1983(1, 5)
TxU	1983(1)
WU	1983(1, 5)
WyU	1983(1, 5)

(288) INTERNATIONAL CONFERENCE ON THE NEW BASEMENT TECTONICS. FIELD GUIDE.
1976 2nd   Collection of field trips in central Pennsylvania. (Cover title: Field guide to lineaments and fractures in central Pennsylvania) Post convention field trip; transgressive structures in the crust. Central Pennsylvania-Appalachian Fold Mountains.

IU	1976
NSyU	1976
PHarER-T	1976
PSt	1976

(289) INTERNATIONAL CONFERENCE ON THE PERMIAN AND TRIASSIC SYSTEMS. GUIDEBOOK.
1971 Aug. 5. Permian and Triassic exposures of western North America; Calgary, Alberta to El Paso, Texas.

CaOOG	1971
CaOWtU	1971
CoDCh	1971
CoDGS	1971
DI-GS	1971
MnU	1971
NvU	1971

(290) INTERNATIONAL CONGRESS OF CARBONIFEROUS STRATIGRAPHY AND GEOLOGY. GUIDEBOOK.
1979 9th   May-June. Washington, DC and Urbana IL.
Field trip 1. Proposed Pennsylvanian System stratotype, Virginia and West Virginia. (35)
Field trip 2. Coal geology of northern Appalachians. (Cover title: Geology of the northern Appalachian coal field)
Field trip 4. Carboniferous geology from the Appalachian Basin to the Illinois Basin through eastern Ohio and Kentucky.
Field trip 5. Carboniferous basins of southeastern New England.
Field trip 6. Carboniferous depositional environments in the Appalachian region.
Field trip 7. Devonian-Mississippian boundary in southern Indiana-northwestern Kentucky.
Field trip 8. Stratigraphy of the Mississippian stratotype: Upper Mississippi Valley, U.S.A.
Field trip 9. Depositional and structural history of the Pennsylvanian System of the Illinois Basin. Part 1. Road log and descriptions of stops. (265)
Field trip 10. Pennsylvanian cyclic platform deposits of Kansas and Nebraska; Field guide to Pennsylvanian cyclic deposits in Kansas and Nebraska; Kansas coal resources and production; Heavy-oil-bearing sandstones of the Cherokee Group in southeastern Kansas; Paleoecology, provincialism, and substitution among late Pennsylvanian crinoids of the midcontinent United States. (635)
Field trip 11. Mississippian-Pennsylvanian shelf-to-basin transition, Ozark and Ouachita regions, Oklahoma and Arkansas. (Cover title: Ozark and Ouachita shelf to basin transition, Oklahoma-Arkansas) (508)
Field trip 13. Carboniferous stratigraphy in the Grand Canyon country, northern Arizona and southern Nevada. (35)
Field trip 15. Carboniferous of the northern Rocky Mountains. (35)
[Field trip B] Guidebook to environments of plant deposition -- coal balls, paper coals, and gray shale floras in Fountain and Parke counties, Indiana.
Field trip D. Underground coal mine-Herrin (No. 6) Coal Member: stratigraphy, deformational structures, and roof stability.
[Field trip G] Early Pennsylvanian upland compression flora - Rock Island County, Illinois. (266)
Field trip J. Surface coal mine-Major coals of the Carbondale Formation.

AzU	1979(1, 5, 11, 13, 15)
CLU-G/G	1979(1-2, 4-5, 7, 9-11, 13, 15)
CLhC	1979(1-2, 4-5, 7-11, 13, 15, B)
CMenUG	1979(11)
CPT	1979(5, 8-9, 11, 13, 15
CSfCSM	1979(9, 11)
CSt-ES	1979(1-2, 5, 7, 9-11, 13, 15)
CU-EART	1979(5, 9, 11, 15)
CU-SB	1979(7, 9, 13)
CaACAM	1979(15)
CaACU	1979(1, 4-11, 13, 15)
CaAEU	1979(5, 9, 13, 15)

(291) International Congress of Speleology.

CaBVaU	1979(1, 10)
CaOOG	1979(2, 4, 10-11)
CaOTDM	1979(9)
CaOTRM	1979(9)
CaOWtU	1979(5, 13, 15)
CoDCh	1979(1-2, 7, 10, 11, 13, 15)
CoDGS	1979(1-2, 4-5, 7-9, 13, 15, D, G)
CoG	1979(9, 11)
CoU	1979(11)
DI-GS	1979(1-2, 4-7, 9-11, 13, 15, D, J)
DLC	1979(1, 5, 13, 15)
ICF	1979(9)
ICIU-S	1979(9, 11)
ICarbS	1979(9, G)
IU	1979(1-2, 4-5, 7-11, 15, D, G, J)
IaU	1979(1-2, 4-5, 7, 9-11, 13, 15)
IdU	1979(2, 4-5, 10, 13)
InLP	1979(1-2, 4-7, 9-11, 13, 15)
InU	1979(1, 5, 9, 11, 13, 15)
KLG	1979(10)
KyU-Ge	1979(1, 4, 7, 9-11, 13, 15)
MiHM	1979(9, 11)
MnU	1979(1-2, 4-5, 7, 9-11, 13, 15)
MtBuM	1979(9, 11, 15)
NSbSU	1979(1-2, 7, 9-11, 13, 15)
NSyU	1979(5, 9, 13, 15)
NcU	1979(1, 4, 5, 7, 8, 10, 13)
NdU	1979(1, 5, 8-9, 11, 13, 15)
NhD-K	1979(1-2, 4-5, 7, 9-11, 13, 15)
NvU	1979(11, 13, 15)
OCU-Geo	1979(9)
OU	1979(1-2, 4-7, 9-11, 13, 15)
OkU	1979(1-2, 4-5, 7, 9:1, 10-11, 13, 15)
PBm	1979(1-2, 4-7, 9-11, 13, 15)
PHarER-T	1979(1)
PSt	1979(1-2, 4-5, 7-11, 13, 15)
SdRM	1979(9)
TxBeaL	1979(1-2, 4, 7, 11, 13, 15)
TxHSD	1979(4, 6)
TxMM	1979(1-2, 4-5, 7, 9, 13, 15)
TxU	1979(5, 11)
UU	1979(9, 11, 13, 15)
ViBlbV	1979(1, 4, 13, 15)
WU	1979(1-2, 4-5, 7, 9-11, 13, 15)
WaU	1979(1)

INTERNATIONAL CONGRESS OF SEDIMENTOLOGY. FIELD EXCURSION GUIDE BOOK. See INTERNATIONAL ASSOCIATION OF SEDIMENTOLOGISTS. INTERNATIONAL CONGRESS ON SEDIMENTOLOGY. (279)

INTERNATIONAL CONGRESS OF SOIL SCIENCE (EDMONTON: 11TH). TOUR. See INTERNATIONAL SOCIETY OF SOIL SCIENCE. CONGRESS (11TH: EDMONTON). (308)

(291) INTERNATIONAL CONGRESS OF SPELEOLOGY. [GUIDEBOOK]
1981 8th A guide to the historic section of Mammoth Cave.
Guidebook to the karst of the central Appalachians.

CMenUG	1981
CU-EART	1981
CoDGS	1981
DI-GS	1981
IU	1981
InLP	1981
KyU-Ge	1981
MnU	1981
NN	1981
OkU	1981
WU	1981

(292) INTERNATIONAL CONGRESS ON ROCK MECHANICS. EXPEDITION GUIDE.
1974 3rd Rock Mechanics: the American Northwest. An advance expedition report including a glimpse at the structure of the North Cascade, and Rainier, Yellowstone, Grand Teton, Glacier, National Park regions, with special reference and full descriptions concerning the mechanical problems of landslides, for citizen, emigrant and tourist. (307)
1. Big Horn Basin and adjacent highlands.
2. Absaroka Range and Heart Mountain rockslide.
3. Yellowstone Park and Hebgen Lake earthquake area.
4. Grand Tetons, Jackson Hole, Gros Ventre Valley.
5. Butte - The Berkeley Pit.
6. Northern Rocky Mountains and Libby Dam.
7. Glacier Park and Lewis Overthrust.
8. Columbia Plateau and Grand Coulee Dam.
9. Cascade Range and Mt. Rainier.

CSt-ES	1974
CU-A	1974
MdBJ	1974
MiHM	1974
NBiSU	1974
NvU	1974
TxHSD	1974

INTERNATIONAL FIELD INSTITUTE FOR COLLEGE AND UNIVERSITY GEOLOGY TEACHERS. See AMERICAN GEOLOGICAL INSTITUTE. INTERNATIONAL FIELD INSTITUTE. (36)

(293) INTERNATIONAL FLUVIAL SEDIMENTOLOGY CONFERENCE. PROCEEDINGS.
1985 3rd Fort Collins, Colorado.
Field guidebook to modern and ancient fluvial systems in the United States.
Field trip 1. The Catskill magnafacies of New York State.
Field trip 2. Upper Jurassic/lower Cretaceous and Paleocene alluvial sediments of the Bighorn Basin, northwest Wyoming.
Field trip 3. Field guide to the upper Salt Wash alluvial complex.
Field trip 4. Guide to the field study of alluvial fan and fan-delta deposits in the Fountain Formation (Pennsylvania-Permian), Colorado.
Field trip 5. Holocene braided streams of eastern Colorado and sedimentological effects of Lawn Lake Dam failure, Rocky Mountain National Park.
Field trip 6. Hydraulics and sedimentary processes in the Calamus River (Nebraska Sand Hills): a field workshop.

CMenUG	1985
CU-SB	1985
CoDGS	1985
IU	1985
InLP	1985
NmU	1985
NvU	1985

(294) INTERNATIONAL GEOBOTANY CONFERENCE. FIELD TRIP GUIDE.
1973 Mar. [1] Field conference guide for an altitudinal transect of the Great Smoky Mountains National Park.
[2] Ocoee River Gorge and Copper Basin.

DI-GS	1973

(295) INTERNATIONAL GEOGRAPHICAL UNION. COMMISSION ON THE SIGNIFICANCE OF PERIGLACIAL PHENOMENA. GUIDEBOOK.
1983 See International Conference on Permafrost. 287.

(296) INTERNATIONAL GEOLOGICAL CONGRESS. GUIDE DES EXCURSIONS. (TITLE VARIES)
1891 5th Excursion A. Geology of Washington and vicinity.
Excursion B. Geological guidebook of the Rocky Mountain excursion.
Excursion C. Excursion to Lake Superior, Pre-Cambrian geology of the Lake Superior region.
1906 10th Excursions avant le Congres.
1. De Mexico a Jalapa.
2. Excursions a Chavarrillo, Santa Maria Tatetla, Veracruz et Orizaba.

3. De Esperanza a Mexico.
4. De Mexico a Tehuacan.
5. L'archaique du Canon de Tomellin.
6. Les ruines de Mitla.
7. Excursion de Tehuacan a Zapotitlan et San Juan Raya.
8. De Mexico a Patzcuaro et Uruapam.
9. Le Xinantecatl ou Volcan Nevado de Toluca.
10. Phenomenes postparoxysmiques du San Andres.
11. Le Jorullo.
12. Les Geysers d'Ixtlan.
13. Le Volcan de Colima.
14. Les crateres d'explosion de Valle de Santiago.
15. Etude de la Sierra de Guanajuato.
16. Geologie des environs de Zacatecas.
17. Etude miniere du District de Zacatecas.
18. Le Mineral de Mapimi.
19. Excursion aux mines de Soufre de la Sierra de Banderas.
20. Excursion au Cerro de Muleros.
21. Esquisse geologique et petrographique des environs de Parral.
22. Etude miniere de la "Veta Colorada" de Minas Nueva a Hidalgo del Parral.
23. Excursions dans les environs de Parras, Coah.
24. Geologie de la Sierra de Concepcion del Oro.
25. Le Mineral d'Aranzazu.
26. Geologie de la Sierra de Mazapil et Santa Rosa.
27. Les gisements carboniferes de Coahuila.
28. Les gisements carboniferes de Coahuila.
29. Excursions dans les environs de Monterrey et Saltillo.
30. De San Luis Potosi a Tampico.
31. Excursion a l'Isthme de Tehuantepec.

1913 12th Guidebooks of excursions in Canada. (224)
1. Excursion in eastern Quebec and the Maritime Provinces.
2. Excursions in the eastern townships of Quebec and the eastern part of Ontario.
3. Excursions in the neighborhood of Montreal and Ottawa.
4. Excursions in southwestern Ontario.
5. Excursions in the western peninsula of Ontario and Manitoulin Island.
6. Excursions in vicinity of Toronto and to Muskoka and Madoc. (English edition issued by Ontario Bureau of Mines)
7. Excursions to Sudbury, Cobalt, and Porcupine. (English edition issued by Ontario Bureau of Mines)
8. Toronto to Victoria and return, via Canadian Pacific and Canadian Northern railways.
9. Toronto to Victoria and return, via Canadian Pacific and Grand Trunk Pacific and National transcontinental railways.
10. Excursions in northern British Columbia and Yukon Territory and along the North Pacific coast.

1933 16th Washington, D.C.
1. Eastern New York and western New England.
2. Mining districts of the eastern states.
3. Southern Appalachian region.
4. The Paleozoic stratigraphy of New York.
5. Chesapeake Bay region.
6. Oklahoma and Texas.
7. Geomorphology of the Central Appalachians.
8. Mineral deposits of New Jersey and eastern Pennsylvania.
9. New York City and vicinity.
9a. The Catskill region.
10. Southern Pennsylvania and Maryland.
11. Northern Virginia.
12. Southern Maryland.
13. Western Texas and Carlsbad Caverns.
14. Ore deposits of the Southwest.
15. Southern California.
16. Middle California and western Nevada.
17. The Salt Lake region.
18. Colorado Plateau region.
19. Colorado.
20. Pennsylvanian of the northern mid-continent region.
21. Central Oregon.
22. The Channeled Scablands.
23. The Butte mining district, Montana.
24. Yellowstone-Beartooth-Big Horn region
25. The Black Hills.
26. Glacial geology of the central states.
27. Lake Superior region.
28. An outline of the structural geology of the United States.
29. Stratigraphic nomenclature in the United States.
30. The Baltimore and Ohio Railroad.

1956 20th Mexico City.
Excursion A-1/C-4. Geologia minera del Noroeste de Mexico. Depositos de cobre de cananea, Sonora y de cobre y manganeso de El Boleo y Lucifer, Baja California.
Excursion A-2/A-5. Geologia a lo largo de la Carretera Panamericana entre Cuidad Juarez, Chih. y Mexico, D.F. Distritos Mineros de Santa Eulalia, Naica, Parral, San Francisco del Oro y Santa Barbara, Chih. Yacimiento de Fierro del Cerro do Mercado en Durango, Dgo. Distritos Mineros de Sombrerete, San Marin, Fresnillo y Zacatecas, Zac. y Guanajuato, Gto.
Excursion A-3/C-1. Geologia a lo largo de la carretera entre Mexico, D.F. Pachuca y Zimapan, Hgo. Distritos Mineros de Pachuca Real del Monte y de Zimapan, Hgo.
Excursion A-4/C-2. Geologia a lo largo de la carretera entre Mexico, D.F. y Taxco, Gro. Distrito Minero de Taxco. Visita a un yacimiento de Fluorita en rocas del Terciario Inferior.
Excursion A-5. See Excursion A-2.
Excursion A-6. Geologia a lo largo de la Carretera Panamericana entre Mexico, D.F. y Tehuantepec, Oax. Distritos Mineros de Natividad y Pluma Hidalgo, Oax., y visita a monumentos precoloniales de Oaxaca.
Excursion A-7. Geologia general de la parte sur de la Peninsula de Baja California. Depositos continentales y volcanicos del Cenozoico superior y marinos del inferior, asi como sedimentos marinos del Cretacico superior. Caracteristicas fisiograficas y efectos de Intemperismo en la region desertica.
Excursion A-8. Estratigrafia de las formaciones Paleozoicas y Mesozoicas de Altar-Caborca, Estado de Sonora.
Excursion A-9/C-12. Geologia a lo largo de la carretera entre Mexico, D.F. y Acapulco, Gro., via Taxco, Gro. y Chilpancingo, Gro. Geologia de los alrededores de Acapulco, Gro. Los yacimientos de dolomita de El Ocotito, Gro.
Excursion A-10/C-13. Geologia entre Mexico, D.F. y Huauchinango, Pue. Campos petroleros de Poza Rica, Ver. y la Nueve Faja de Oro, Ver.
Excursion A-11. Estratigrafia del Mesozoico y tectonica del sur del estado de Puebla; Presa de Valsequillo, Sifon de Huexotitlanapa y problemas hidrologicos de Puebla.
Excursion A-12. Estratigrafia y paleontologia del Mesozoico de la Cuenca sedimentaria de Oaxaca y Guerrero, especialmente del Jurasico Inferior y Medio.
Excursion A-13. Estratigrafia Mesozoica y tectonica de la Sierra de Chihuahua; Permico de Placer de Guadalupe, Chih.; geohidrologia de la region lagunera; estratigrafia Mesozoica y tectonica de la Sierra Madre Oriental entre Mapimi, Dgo. y Monterrey, N.L.
Excursion A-14/C-6. Estratigrafia del Cenozoico y del Mesozoico a lo largo de la carretera entre Reynosa, Tamps. y Mexico, D.F. Tectonica de la Sierra Madre Oriental. Volcanismo en el Valle de Mexico. (English edition: Stratigraphy of the Tertiary and Mesozoic along the highway between Reynosa, Tampa, and Mexico, D.F. Tectonics of the Sierra Madre Oriental. Volcanic and continental sedimentary rocks in the Valley of Mexico)
Excursion A-15. Volcanismo Terciario y Reciente del Eje Volcanico de Mexico. Formaciones andesiticas de las Sierras de Las Cruces y Ozumatian. Formaciones basalticas de las Sierras de Zitacuaro, Morelia, Paracho, y alrededores del Paricutin. Fenomenos post-paroxismales de la Sierra de San Andres y el Lago de Cuitzeo y estructura e historia del nuevo Volcan Paricutin.
Excursion A-16. Geologia a lo largo de la carretera entre Mexico, D.F. y Guadalajara, Jal., via Morelia, Mich. y entre Guadalajara, Jal. y Mexico, D.F., via Leon, Gto. Condiciones geohidrologicas de los Valles de Atemajac y Tesistan y de las zonas adyacentes a la ciudad de Guadalajara.
Excursion C-1. See Excursion A-3.

Excursion C-2. See Excursion A-4.
Excursion C-3. Geologia a lo largo de la carretera entre Mexico, D.F., y Saltillo, Coah. Distritos mineros de Guanajuato, Gto. y Avalos-Concepcion del Oro-Mazapil, Zac. Minas de carbon de Monclava y Nueva Rosita, Coah.
Excursion C-4. See Excursion A-1.
Excursion C-5. Estudio de la estratigrafia del Mesozoico y de la tectonica de la Sierra Madre Oriental entre Monterrey, N.L. y Torreon, Coah. Estudio de la cuenca carbonifera de Sabinas, Coah. Visita a las Grutas de Garcia. Morfologia tipica de bolson y observacion del tipo de pliegues en la Sierra de Parras. (For partial translation, see Southwestern Association of Student Geological Societies, 1964, 5th)
Excursion C-6. See Excursion A-14.
Excursion C-7. Geologia general de la Sierra Madre Oriental entre Mexico, D.F., y Cordoba, Ver. Depositos continentales y volcanicos del Cenozoico Superior y sedimentos marinos del Mesozoico y Cenozoico. Campos petroleros de la Cuenca de Veracruz. Obras hidraulicas del Rio Papaloapan. Campos petroleros y azufreros del Istmo de Tehuantepec. Geomorfologia de la Peninsula de Yucatan. Visitas a las zonas arqueologicas Mayas.
Excursion C-8. Estratigrafia y paleontologia del Jurasico Inferior y Medio. Marino de la region central de la Sierra Madre Oriental.
Excursion C-9. Volcanes, rocas volcanicas, sedimentos lacustres y aluviales del Pleistoceno y Plioceno; rocas clasticas y volcanicas Terciarias; yeso y caliza no marinos del Terciario Inferior; calizas y lutitas del Cretacico Superior y calizas y dolomitas del Cretacico Inferior, en el sur de la cuenca de Mexico y en el Edo. de Morelos.
Excursion C-10. Estratigrafia Mesozoica y tectonica de la Sierra Madre Oriental entre Zimapan, Hgo., y Cuidad Valles, S.L.P. Rocas igneas y sedimentarias del Cenozoico. Geologia petrolera. Plantas de refinacion y produccion de Poza Rica, Ver.
Excursion C-11. Estratigrafia del Cenozoico continental de las sierras del sur y la costa del Golfo de Mexico entre Coatzacoalcos, Ver., y Veracruz. Vulcanologia del Pleistoceno. Geologia general y paleontologia vertebrada.
Excursion C-12. See Excursion A-9.
Excursion C-13. See Excursion A-10.
Excursion C-14. Espeleologia y fenomenos carsticos de las grutas de Cacahuamilpa, Mor.; estratigrafia y geologia superficial.
Excursion C-15. Geologia del Mesozoico y estratigrafia Permica del Estado de Chiapas.
Excursion C-15A. Geologia del Mesozoico y estratigrafia del Paleozoico superior de Estado de Chiapas. Geologia sedimentaria y petrolera: campos del Istmo de Tehuantepec. Domos salinos con yacimientos de azufre y su explotacion por el metodo Frasch.
Excursion C-15B. Geologia a lo largo de la carretera entre Tuxtla Gutierrez, Chis. y Mexico, D.F., y visita a monumentos precoloniales de Oaxaca, Oax.
Excursion C-16. Visita a las localidades tipo de las formaciones del Eoceno, Oligoceno y Mioceno de la cuenca sedimentaria de Tampico-Misantla, en la llanura costera del Golfo de Mexico, entre Poza Rica, Ver., Tampico, Tamps. y Ciudad Valles, S.L.P.

1972   24th   Field excursions, Montreal.
Excursion A01/X01. Structural style of the southern Canadian Cordillera.
Excursion A02. Quaternary geology of the southern Canadian Cordillera.
Excursion A03/C03. Geology of the southern Canadian Cordillera.
Excursion A04/C04. Plutonic and associated rocks of the Coast Mountains of British Columbia.
Excursion A05/C05. Geology of Vancouver area of British Columbia.
Excursion A06/C06. Mineral deposits along the Pacific coast of Canada.
Excursion A07. Not published.
Excursion A08/C08. Engineering geology of the southern Cordillera of British Columbia.
Excursion A09/C09. Copper and molybdenum deposits of the western Cordillera.
Excursion A10. Stratigraphy and structure, Rocky Mountains and foothills, of west-central Alberta and northeastern British Columbia.
Excursion A11. Quaternary geology and geomorphology, southern and central Yukon (northern Canada).
Excursion A12. Volcanic rocks of the northern Canadian Cordillera.
Excursion A13. Not published.
Excursion A14. Lower and middle Paleozoic sediments and paleontology of Royal Creek and Peel River, Yukon, and Powell Creek, Northwest Territories.
Excursion A15/C15. The Canadian Rockies and tectonic evolution of the southeastern Canadian Cordillera.
Excursion A16. The Permian of the southeastern Cordillera.
Excursion A19. Cambrian and Ordovician biostratigraphy of the southern Canadian Rocky Mountains.
Excursion A20. The Cretaceous and Jurassic of the foothills of the Rocky Mountains of Alberta.
Excursion A21. Vertebrate paleontology, Cretaceous to Recent, interior plains, Canada.
Excursion A24/C24. Major lead-zinc deposits of western Canada.
Excursion A25/C25. Coal, oil, gas, and industrial mineral deposits of the interior plains, foothills and Rocky Mountains of Alberta and British Columbia.
Excursion A26. Hydrogeology of the Rocky Mountains and interior plains.
Excursion A27. Archean and Proterozoic geology of the Yellowknife and Great Bear areas, Northwest Territories.
Excursion A28. Archean and Proterozoic sedimentary and volcanic rocks of the Yellowknife-Great Slave Lake area, Northwest Territories.
Excursion A29. Muskox intrusion and Coppermine River lavas, Northwest Territories, Canada.
Excursion A30. Quaternary geology and geomorphology, Mackenzie delta to Hudson Bay.
Excursion A31/C31. Geology and mineral deposits of the Flin Flon, Lynn Lake and Thompson areas, Manitoba, and the Churchill-Superior front of the western Precambrian shield.
Excursion A32a/A32b. Precambrian geology of the Lake Athabasca area, Saskatchewan and Baker Lake area, Northwest Territories.
Excursion A33/C33. Archean geology and metallogenesis of the western part of the Canadian Shield.
Excursion A35/C35. The geology of the Canadian Shield between Winnipeg and Montreal.
Excursion A36/C36. Precambrian geology of the southern Canadian Shield with emphasis on the lower Proterozoic (Huronian) of the north shore of Lake Huron.
Excursion A37/C37. Not published.
Excursion A39/39b/C39. Precambrian geology and mineral deposits of the Timagami, Cobalt, Kirkland Lake and Timmins region, Ontario.
Excursion A40/C40. Precambrian volcanism of the Noranda, Kirkland Lake, Timmins, Michipicoten, and Mamainse Point areas, Quebec and Ontario.
Excursion A41/C41. Precambrian geology and mineral deposits of the Noranda-Val d'Or and Matagami-Chibougamau greenstone belts, Quebec.
Excursion A42. Quaternary stratigraphy and geomorphology of the eastern Great Lakes region of southern Ontario.
Excursion A43. The Great Lakes of Canada; Quaternary geology and limnology.
Excursion A44/C44. Quaternary geology and geomorphology, southern Quebec.
Excursion A45/C45. Stratigraphy and paleontology of the Paleozoic rocks of southern Ontario.
Excursion A46/C46. The Grenville province of the Precambrian in Quebec.
Excursion A47/C47. Classic mineral collecting localities in Ontario and Quebec.
Excursion A48. Hydrogeology of representative areas of the southern part of Ontario.
Excursion A51a. Engineering geology in Quebec-Labrador region, eastern Canada.
Excursion A53/C53. Alkalic rock complexes and carbonatites of Ontario and part of Quebec.
Excursion A54. Igneous rock of central Labrador, with emphasis on anorthositic and related intrusions.
Excursion A55. Iron ranges of Labrador and northern Quebec.
Excursion A56/C56. Appalachian structure and stratigraphy, Quebec.
Excursion A57/C57. Appalachian stratigraphy and structure of the Maritime Provinces.

Excursion A58/C58. Mineral deposits of southern Quebec and New Brunswick.
Excursion A59. Vertebrate paleontology of eastern Canada.
Excursion A60. Stratigraphy and economic geology of Carboniferous basins in the Maritime Provinces.
Excursion A61/C61. Quaternary geology, geomorphology and hydrogeology of the Atlantic provinces; Quaternary deposits and events.
Excursion A62/C62. A cross-section through the Appalachian orogen in Newfoundland.
Excursion A63/C63. Appalachian geotectonic elements of the Atlantic provinces and southern Quebec.
Excursion A64/C64. Not published.
Excursion A65. Some astroblemes, craters and cryptovolcanic structures in Ontario and Quebec.
Excursion A66. The Canadian Arctic Islands and the Mackenzie region.
Excursion A68. Eastern Canadian Cordillera and Arctic Islands; an aerial reconnaissance.
Excursion B01. Petrology and structure of the Morin anorthosite; 20 parts.
Excursion B02. Stratigraphy and tectonics of the Precambrian of the Grenville province in the Saint Paulin-Saint Boniface de Shawinigan area.
Excursion B03. Stratigraphy of the Montreal area.
Excursion B04. Pleistocene deposits northeast of Montreal.
Excursion B05. Structural geology of the Sherbrooke area.
Excursion B06. A crypto-explosion structure at Charlevoix and the St. Urbain anorthosite.
Excursion B07. Base metal deposits of southeastern Quebec.
Excursion B08. Asbestos deposits of southern Quebec.
Excursion B09. Quebec Iron and Titanium Corporation ore deposit at Lac Tio, Quebec.
Excursion B10. Monteregian Hills; diatremes, kimberlite, lamprophyres and intrusive breccias west of Montreal.
Excursion B11. The Monteregian Hills; ultra-alkaline rocks and the Oka carbonatite complex.
Excursion B12. Geology of Mount Royal.
Excursion B13. The geology of the Brome and Shefford igneous complexes.
Excursion B14. Monteregian Hills; Mounts Johnson and Rougemont.
Excursion B15. The Monteregian Hills; mineralogy of Mount St. Hilaire.
Excursion B16. Not published.
Excursion B17. Quarries in the Montreal area.
Excursion B18. Engineering geology of Montreal.
Excursion B19. Geology of the environs of Quebec City.
Excursion B20. Engineering geology in the Quebec city area.
Excursion B21. Appalachian tectonics in the eastern townships of Quebec.
Excursion B22. Not published.
Excursion B23-B27. Geology of the national capital area.
Excursion C17. Lower Carboniferous stratigraphy and sedimentology of the southern Canadian Rocky Mountains.
Excursion C18. Devonian stratigraphy and facies of the southern Rocky Mountains of Canada, and the adjacent plains.
Excursion C22. Quaternary geology and geomorphology between Winnipeg and the Rocky Mountains.
Excursion C23. Industrial and non-metallic minerals of Manitoba and Saskatchewan (Central Plains).
Excursion C34. The Precambrian rocks of the Atikokan-Thunder Bay-Marathon area.
Excursion C38. General geology of the Sudbury-Elliot Lake region.
Excursion C49. Visits to deposits of industrial minerals and building materials in Quebec and Ontario.
Excursion C50. Not published.
Excursion C51b. Engineering geology of eastern Canada-southern Ontario.
Excursion C52. Stratigraphy and structure of the St. Lawrence lowland of Quebec.
Excursion C67. Uranium deposits of Canada.

AzTeS   1956(A1 thru A7, A9 thru A-16, C-1 thru C-9, C-12 thru C-13, C-15 thru C-16), 72
CLU-G/G 1891(A), 1906, 13, 33(9-30), 56(A-1 thru A-6, A7, A11 thru A-16, C-1 thru C-2, C-4, C-6, C-8 thru C-9, C-15 thru C-16), 72
CLhC   1972(A02, A04/C04-A05/C05, A10, A12, A53/C53, A55, B13-B14, C49)
CPT   1891, 1906, 13, 33, 56(A1-A4, A7, A9-A16, C5, C8, C9, C15, C16), 72(A1-A6, A8-A12, A14-A16, A19-A21, A24-A33, A35, A36, A39-A48, A51A, A5 3-A63, A65-A66, A68, C17-C18, C22-C23, C34, C38, C49, C51b, C52, C67)
CSdS   1913, 33, 56
CSt-ES 1891, 1906, 13, 33, 56, 72
CU-A   1891, 1913
CU-EART 1906(1-2, 4-15, 17-31), 13, 33, 56(A-1 thru A-7, A-9 thru A-16, C-1 thru C-2, C-4 thru C-8, C-12 thru C-13, C-15, C-15B, C-16), 72(A01/X01-A06/C06, A08/C08-A12, A14-A16, A19-A21, A24/C24-A33/C33, A35/C35-A36/C36, A39/39b/C39-A48, 51a, A53/C53-A63/C63, A65-A66, A68, B01-B15, B17-B21)
CU-SB   1933
CaAC   1972(A02, A04/C04-A05/C05, A10, A12, A53/C53, A55, B13-B14, C49)
CaACAM 1972(A01/X01-A06/C06, A08/C08-A12, A14-A16, A19-A21, A24/C24-A33/C33, A35/C35-A36/C36, A39/39b/C39-A48, 51a, A53/C53-A63/C63, A65-A66, A68)
CaACI   1913(1, 8-10)
CaACM   1913(1, 8-10), 72(A01/X01-A06/C06, A08/C08-A12, A14-A16, A19-A21, A24/C24-A33/C33, A35/C35-A36/C36, A39/A39a/C39-A48, 51a, A53/C53-A63/C63, A65-A66, A68, C17-C18, C22-C23, C34, C38, C49, C51b, C52, C67)
CaACU   1913, 33(1-5, 7-30), 56(A-1 thru A-4, A-6 thru A-7, A-12 thru A-16, C-3, C-5, C-8, C-15, C-15A, C-15B, C-16), 72(A01/X01-A06/C06, A08/C08-A12, A14-A36, A39-A63/C63, A65-A66, A68, C17-C18, C22, C23, C34, C38, C49, C51b, C52, C67)
CaAEU   1913, 33(1-8, 9a-30)
CaBVaU 1891, 1913, 33, 56(A-1 thru A-16, C-1 thru C-8, C-12 thru C-16), 72(A01/X01-A06/C06, A08/C08-A12, A14-A16, A19-A21, A24/C24-A28, C17-C18)
CaOHM   1913, 72
CaOHaHa 1972(A39/39b/C39)
CaOKQGS 1891, 1913, 33, 56(A-1-A-4, A-7, A-11-A-16, C-8 thru C-9, C-15B, C-16)
CaOLU   1891, 1913, 33, 56, 72
CaOOG   1906, 13, 33, 56
CaOONM 1972
CaOTDM 1972(A01/X01-A06/C06, A08/C08-A12, A14-A16, A19-A21, A24/C24-A33/C33, A35/C35-A36/C36, A39/39b/C39-A48, 51a, A53/C53-A63/C63, A65-A66, A68, B01-B15, B17-B21, B23-B27)
CaOTRM 1913, 33, 56, 72
CaOWtU 1913(1-6, 8-10), 33(4), 72
CoG   1933, 56, 72
CoU   1913, 33, 56(C-9), 72
DI-GS   1891, 1906, 13, 33, 56, 72
DLC   1891, 1956(A-1 thru A-7, A-9 thru A-16, C-1 thru C-2, C-4 thru C-9, C-12 thru C-16), 72
ICF   1913(1-5, 8-10), 33(1-8, 10-27, 29-30), 56(C-9)
ICarbS 1906(1-23, 25, 27-28, 30-31), 13(1-6, 8-10), 33, 72(A01/X01-A06/C06, A08/C08-A12, A14-A16, A19-A21, A24/C24-A33/C33, A35/C35-A36/C36, A39/39b/C39-A48, 51a, A53/C53-A63/C63, B01-B06, B08-B15, B17-B21, B23-B27, C67)
IEN   1913, 33, 72
IU   1891, 1913, 33, 56, 72
IaU   1906, 13, 33, 56
InLP   1933, 72
InRE   1933(3-4, 6, 10, 13, 17, 19-21, 23-24, 28)
InU   1906, 13, 33, 56(A-1, A-16, C-4, C-9)
KyU-Ge 1891, 1906, 13, 33, 56, 72
LNU   1913(1-5, 8-10), 33, 56(A-3, C-1)
LU   1906, 13(1-5, 9-10)
MH-GS   1906, 13, 33(1-15, 17-23, 25-30), 72
MNS   1913, 33(1-29), 56(A-1 thru A-5, A-7, A-9 thru A-16, C-1 thru C-2, C-4, C-6, C-8, C-12 thru C-13, C15A thru C-16), 72(A01/X01-A06/C06, A08/C08-A12, A14-A16, A19-A21, A24/C24-A33/C33, A35/C35-A36/C36, A39/39b/C39-A48, 51a, A53/C53-A63/C63, A65-A66, A68, B01-B15, B17-B21, B23-B27)
MiDW-S 1913(1-5, 8-10), 33(1-8, 9a-30)
MiHM   1891, 1913, 33, 56(A-1 thru A-7, A-9 thru A-16, C-4 thru C-6, C-8 thru C-9, C-15A thru C-16), 72
MiKW   1913, 33(2-3, 10, 12, 26)
MiMarqN 1891(B), 1933(1, 5-6, 9-13, 15-16, 18, 20-21, 23-24)
MnU   1913(1-5, 8-10), 33, 56(A-1 thru A-7, A-9 thru A-16, C-8 thru C-9, C-11 thru C-13, C-15A thru C-16), 72
MoSW   1891, 1906, 13, 33, 56, 72

MoU	1913, 33, 56(A-1 thru A-7, A-9 thru A-16, C-4 thru C-9, C-12 thru C-13, C-15A thru C-16), 72
MtBC	1972
MtBuM	1891, 1933, 56(A-1 thru A-5, A-11 thru A-16, C-1 thru C-2, C-4, C-6, C-8, C-15B, C-16), 72(A01/X01-A06/C06, A08/C08-A12, A14-A16, A19-A21, A24/C24-A33/C33, A35/C35-A36/C36, A39/39b/C39-A48, 51a, A53/C53-A63/C63, A65-A66, A68, B01-B15, B17-B21, B23-B27, C17-C18, C22-C23, C34, C38, C49, C51b, C52, C67)
NBiSU	1933(3, 5-6, 8, 11-22, 26, 29-30)
NNC	1906, 13, 33, 56(A-1 thru A-16, C-1 thru C-16), 72(A01/X01-A06/C06, A08/C08-A12, A14-A16, A19-A21, A24/C24-A33/C33, A35/C35-A36/C36, A39/39b/C39-A48, 51a, A53/C53-A63/C63, A65-A66, A68, B01-B15, B17-B21, B23-B27, C17-C18, C22-C23, C34, C38, C49, C51b, C52, C67)
NOneoU	1972(A01/X01-A06/C06, A08/C08-A12, A14-A16, A19-A21, A24/C24-A33/C33, A35/C35-A36/C36, A39/39b/C39-A48, 51a, A53/C53-A63/C63, A65-A66, A68, B01-B02, B05-B06, B08-B15, B19, B21, B23-B27, C38, C49)
NRU	1972(A01/X01-A06/C06, A08/C08-A12, A14-A16, A19-A21, A24/C24-A33/C33, A35/C35-A36/C36, A39/39b/C39-A48, 51a, A53/C53-A63/C63, A65-A66, A68, B01-B15, B17-B21, B23-B27, C17-C18, C22-C23, C34, C38, C49, C51b, C52, C67)
NSyU	1913, 33, 56(A-1 thru A-16, C-5, C-8, C-15A thru C-16), 72
NbU	1933, 56, 72
NcU	1891, 1913, 33, 72
NdU	1913, 33(5, 10-12, 17, 20-22, 25-26, 30), 56(A-1 thru A-4, A-6, A-7, A-9 thru A-16, C-6, C-8, C-15A thru C-16), 72
NhD-K	1906, 13, 33, 56, 72
NjP	1891, 1906, 13, 33, 56, 72(A01/X01-A06/C06, A08/C08-A12, A14-A16, A19-A21, A24/C24-A33/C33, A35/C35-A36/C36, A39/39b/C39-A48, 51a, A53/C53-A63/C63, A65-A66, A68)
NvLN	1972
NvU	1913, 33, 56, 72
OCU-Geo	1891, 1913, 33, 72(A01/X01-A06/C06, A08/C08-A12, A14-A16, A19-A21, A24/C24-A33/C33, A35/C35-A36/C36, A39/39b/C39-A48, 51a, A53/C53-A63/C63, A65-A66, A68)
OU	1906, 13, 33, 56, 72
OkU	1913, 33, 56, 72(A01/X01-A06/C06, A08/C08-A12, A14-A16, A19-A21, A24/C24-A33/C33, A35/C35-A36/C36, A39/39b/C39-A48, 51a, A53/C53-A63/C63, A65-A66, A68)
OrU-S	1906, 13, 33, 56(A-3, C-1, C-9)
PBL	1913(7, 8:2, 9-10), 33, 72
PBm	1891, 1913, 1933, 72(A01/X01-A03/C03, A05/C05-A06/C06, A08/C08-A12, A14-A16, A19-A21, A24/C24-A33/C33, A35/C35-A36/C36, A39/39b/C39-A48, A51a, A53/C53-A63/C63, A65-A66, A68, C17-C18, C22-C23, C34, C38, C49, C51b, C52, C67)
PSt	1913, 33, 56(A-1 thru A-7, A-9 thru A-16, C-1 thru C-9, C-15A thru C-16), 72
RPB	1891, 1906, 13, 33, 56(A-1 thru A-7, A-9 thru A-16, C-4 thru C-9, C-11 thru C-13, C-15A thru C-16)
SdRM	1933(1-29)
TMM-E	1933
TU	1933, 56(A-1 thru A-5, A-7, A-13)
TxBeaL	1933(2, 4-6, 10-14, 17-25, 27, 30), 56, 72
TxDaAR-T	1913(1-5, 8-10), 72(A01/X01, A03/C03-A05/C05, A10, A14-A16, A19-A20, A25/C25-A26, A56/C56, A60, A62/C62-A63/C63, A66, A68, C17-C18, C52)
TxDaDM	1933, 56(A-1 thru A-7, A-9 thru A-10, A-14, C-1 thru C-8)
TxDaGI	1933, 56(A-1 thru A-16, C-4 thru C-6, C-8, C-12 thru C-14, C-15A, C-15B)
TxDaM-SE	1891, 1906, 13, 33, 56
TxDaSM	1956(A-1 thru A-3, A-5, A-11 thru A-16, C-4 thru C-6, C-8, C-15A thru C-16), 72
TxHSD	1933, 56(A-1 thru A-16, C-4 thru C-5, C-8, C-15A thru C-16), 72
TxHU	1913(2-4), 56(A-1 thru A-4, A-7, A-9 thru A-16, C-4 thru C-5, C-7, C-9, C-15A thru C-16)
TxLT	1906(1-30), 13, 33, 56(C-9)
TxMM	1913(1, 10), 33
TxU	1891, 1906, 13, 33, 56(A-14, C-6), 72
UU	1891, 1933, 56, 72
ViBlbV	1913, 33, 56, 72
WU	1891, 1906, 13, 33, 56(A-1 thru A-7, A-9 thru A-16, C-4 thru C-9, C-15 thru C-16), 72
WyU	1913(1-6, 8-10)

INTERNATIONAL GEOLOGICAL CORRELATION PROGRAM. See U.S. WORKING GROUP OF THE INTERNATIONAL GEOLOGICAL CORRELATION PROGRAM. (667)

(297) INTERNATIONAL GEOLOGICAL CORRELATION PROGRAMME (IGCP) [PROJECT] 92. ARCHEAN GEOCHEMISTRY

1978 [Aug.] [Aspects of Superior Province geology as exposed in western Ontario, Canada, and northern Minnesota, U.S.A.] (61)
1979 Aug. Guide to the Precambrian rocks of the Beartooth Mountains. (61)
1983 Aug. Workshop on a cross section of Archaean crust. (61, 363)

CSt-ES	1978
CU-SB	1978
CaACU	1978
CaAEU	1978
CaOOG	1978
CoDGS	1979, 83
Cu-A	1979
DI-GS	1979
IaU	1978
MnU	1978
MtBuM	1979
WU	1981

(298) INTERNATIONAL GEOLOGICAL CORRELATION PROGRAMME (IGCP). PROJECT 114: BIOSTRATIGRAPHIC DATUM-PLANES OF THE PACIFIC NEOGENE. FIELD CONFERENCE.

1978 June Neogene biostratigraphy of selected areas in the California Coast Ranges; field conference on the marine Neogene of California. (220(78-446))

CLU-G/G	1978
CMenUG	1978
CSfCSM	1978
CSt-ES	1978
CoDGS	1978
DI-GS	1978
NhD-K	1978
OU	1978

(299) INTERNATIONAL GEOLOGICAL CORRELATION PROGRAMME (IGCP). PROJECT 156: PHOSPHORITES. GUIDEBOOK.

1985 8th International Field Workshop and Symposium (Southeastern United States).

(300) INTERNATIONAL GEOLOGICAL CORRELATION PROGRAMME (IGCP). PROJECT 161.

1985 92. The Stillwater Complex, Montana: geology and guide. (401)

CLU-G/G	1985
CLhC	1985
CMenUG	1985
CPT	1985
CSfCSM	1985
CSt-ES	1985
CU-EART	1985
CU-SB	1985
CaACU	1985
CaOOG	1985
CaOTDM	1985
CaOWtU	1985
CaQQERM	1985
CoDBM	1985
CoDGS	1985
CoG	1985
ICarbS	1985
IaU	1985
IdU	1985
InLP	1985
MiHM	1985
MnU	1985
MtBuM	1985
NNC	1985
NdU	1985
NhD-K	1985
NmSoI	1985

NmU	1985
OCU-Geo	1985
PBU	1985
PBm	1985
PHarER-T	1985
TxDaM-SE	1985
TxMM	1985
TxWB	1985

(301) INTERNATIONAL GEOLOGICAL CORRELATION PROGRAMME (IGCP). [PROJECT] 171. CIRCUM-PACIFIC JURASSIC RESEARCH GROUP. FIELD CONFERENCE. GUIDEBOOK.

1982 1st Calgary, Alberta.
   A guidebook to the Fernie Formation of southern Alberta and British Columbia.

CMenUG	1982
CSt-ES	1982
CaOOG	1982
DI-GS	1982
MA	1982
OkU	1982

(302) INTERNATIONAL GEOLOGICAL CORRELATION PROGRAMME (IGCP). PROJECT 27: THE CALEDONIDE OROGEN. FIELD GUIDE. (TITLE VARIES)

1979 Aug.   The Caledonides in the U.S.A.
   [1] Geological excursions in the northeast Appalachians.
   [2] Guidebook for southern Appalachian field trip in the Carolinas, Tennessee, and northeastern Georgia.
1982 Aug.   Field guide for Avalon and Meguma zones. (372, 59, 442)
   Regional trends in the geology of the Appalachian-Caledonian-Hercynian-Mauritanide Orogen.
   Field trip guidebook 'A'. External and internal domains between Quebec City and Thetford Mines.
   Field trip guidebook 'B'. Sulphide deposits of the Bathurst area, N.B.
   Field trip guidebook 'C'. Tungsten-molybdenum, tin, antimony deposits in N.B.
   Field trip guidebook 'D'. Notre Dame Bay island arc, ophiolite & melange complexes.
   Field trip guidebook 'F'. Trans-Newfoundland trip.
   Field trip guidebook 'H'. Polyorogenic deformation in coastal southern N.B.

CU-A	1979(1), 82
CU-EART	1979(1)
CaACAM	1979(1)
CaACU	1979(1)
CaAEU	1979(1)
CaBVaU	1979(1)
CaNSHD	1982
CaOOG	1979(1), 82(B)
CaOWtU	1979(1), 82
CoDGS	1979
CtY-KS	1979(1), 82
DI-GS	1979
IaU	1979(1), 82
InU	1979(1)
KyU-Ge	1979(1)
MNS	1979
MdBJ	1979, 82
NSbSU	1979
NcU	1979
NhD-K	1979(1)
OU	1979(1)
OrCS	1979(1)
PSt	1979(1)
TxBeaL	1979(1)
ViBlbV	1979(1)
WU	1979(1)

(303) INTERNATIONAL GEOLOGICAL CORRELATION PROGRAMME (IGCP). PROJECT 60: CORRELATION OF CALEDONIAN STRATABOUND SULPHIDES. FIELD TRIP GUIDEBOOK. (TITLE VARIES)

1978   Massive sulfides of Virginia. (220(78-1086))
1983 Sept.   Field trip guidebook to stratabound sulphide deposits, Bathurst area, N.B., Canada and west-central New England, U.S.A. (225(No.36))

1984 May   See American Association of Petroleum Geologists. 18.

CSt-ES	1983
CU-A	1983
CaNSHD	1983
CaOWtU	1983
CoDGS	1978, 83
CtY-KS	1983
DI-GS	1983
IU	1983
IaU	1978, 83
InU	1983
KyU-Ge	1983
OkU	1983
PSt	1983
WU	1983

(304) INTERNATIONAL MINERALOGICAL ASSOCIATION. CONGRESS. [GUIDEBOOK]

1962 3rd   Washington, D.C., northern field excursion and southern field excursion. A visitor's guide to the geology of the National Capitol area. (2 separate guidebooks)

AzTeS	1962
CSt-ES	1962
CaOLU	1962
CoDGS	1962
CoG	1962
DI-GS	1962
ICIU-S	1962
IEN	1962
IU	1962
KyU-Ge	1962
LNU	1962
LU	1962(southern)
MnDuU	1962
NBiSU	1962(southern)
NOneoU	1962
NcU	1962(southern)
NdU	1962
NhD-K	1962
NjP	1962
OU	1962(southern)
OkU	1962
PBL	1962(southern)
PBm	1962(southern)
PSt	1962
TMM-E	1962
TxDaSM	1962
TxLT	1962
TxU	1962
ViBlbV	1962(southern)
WU	1962(southern)

(305) INTERNATIONAL PALYNOLOGICAL CONFERENCE.

1984 6th   Field trip no.1. Plains Region, Campanian to Paleocene.

CaACU	1984

(306) INTERNATIONAL PROTEROZOIC SYMPOSIUM. [FIELD TRIP]

1981 May   Madison, Wisconsin.
   Lake Superior field trip. (722)

CSt-ES	1981
CU-EART	1981
CaACU	1981
CaAEU	1981
CaOOG	1981
CaOTDM	1981
CoDGS	1981
IaU	1981
MU	1981
MnU	1981
NdU	1981
OkU	1981
PBU	1981
WU	1981

(307) INTERNATIONAL SOCIETY FOR ROCK MECHANICS.

1974   See International Congress on Rock Mechanics. 292.

(308) INTERNATIONAL SOCIETY OF SOIL SCIENCE. CONGRESS (11TH: EDMONTON).
1978 1, 10 Guidebook for a soil and land use tour of eastern Canada.

(309) INTERNATIONAL SYMPOSIUM ON ANTARCTIC GEOLOGY AND GEOPHYSICS. FIELD TRIP.
1977 3rd Lake Superior field trip.
AzTeS	1977
DI-GS	1977
IU	1977
KyU-Ge	1977
OU	1977
WU	1977

(310) INTERNATIONAL SYMPOSIUM ON CORAL REEFS. FIELD GUIDEBOOK.
1977 3rd [1] Field guidebook to modern and Pleistocene reef carbonates, Barbados, W.I.
[2] Field guidebook to the reefs of Belize.
[3] Field guidebook to the reefs and geology of Grand Cayman Island, B.W.I.
[4] Field guidebook to the modern and ancient reefs of Jamaica.
[5] Field guidebook to the reefs of San Blas Islands, Panama.
[6] Field guidebook to the reefs and reef communities of St. Croix, Virgin Islands.
AzU	1977
CLU-G/G	1977
CSt-ES	1977
CU-EART	1977
CU-SB	1977
CaACAM	1977
CaACU	1977
CaOWtU	1977
DI-GS	1977
IU	1977
IaU	1977(1, 3-6)
InLP	1977(1-3)
InU	1977
KyU-Ge	1977
MNS	1977 (1, 3-6)
MnU	1977
NSbSU	1977(3)
NSyU	1977(1-2, 4, 6)
NcU	1977
NhD-K	1977
OU	1977
PSt	1977
ScU	1977
TxBeaL	1977
TxU	1977
UU	1977
ViBlbV	1977(1, 5-6)
WU	1977

(311) INTERNATIONAL SYMPOSIUM ON FOSSIL ALGAE. FIELDGUIDE.
1983 3rd Precambrian and Paleozoic algal carbonates, west Texas-southern New Mexico. (143)
Field guide to selected localities of late Proterozoic, Ordovician, Pennsylvanian and Permian ages, including the Permian Reef Complex. (143)
CLhC	1983
CMenUG	1983
CPT	1983
CSt-ES	1983
CU-EART	1983
CU-SB	1983
CaACU	1983
CaAEU	1983
CaOKQGS	1983
CaOWtU	1983
CoDGS	1983
CoG	1983
DI-GS	1983
IaCfT	1983
IaU	1983
IdU	1983
NSbSU	1983
NcU	1983
NhD-K	1983
NmLcU	1983
NvU	1983
ODaWU	1983
PBU	1983
PBm	1983
TxDaM-SE	1983
TxHMP	1983
TxU	1983
WU	1983

(312) INTERNATIONAL SYMPOSIUM ON FOSSIL CNIDARIA. [GUIDEBOOK]
1981 Geologic investigations in the Willis Mountain and Andersonv quadrangles.
1983 4th Silurian and Devonian corals and stromatoporoids of New York.
CaOOG	1983
NSbSU	1983

(313) INTERNATIONAL SYMPOSIUM ON HYDROLOGIC PROBLEMS IN KARST REGIONS. GUIDEBOOK.
1976 Apr. Bowling Green, Kentucky. Hydrology of the Turnhole Spring groundwater basin and vicinity.
CSt-ES	1976
CoDGS	1976
DI-GS	1976
IaU	1976
NSyU	1976

(314) INTERNATIONAL SYMPOSIUM ON LAND SUBSIDENCE. FIELD TRIP GUIDEBOOK.
1976 2nd Land subsidence in California.
CSfCSM	1976
CSt-ES	1976
CoDGS	1976
DI-GS	1976
TxU	1976
ViBlbV	1976

(315) INTERNATIONAL SYMPOSIUM ON LANDSLIDES. FIELD TRIP GUIDE.
1984 4th The geology and slope stability problems in the metropolitan Toronto region.
CaOTDM	1984

(316) INTERNATIONAL SYMPOSIUM ON OSTRACODA. GUIDEBOOK OF EXCURSIONS.
1982 8th Texas ostracoda.
Excursion 1: Lower Tertiary and Cretaceous ostracada of the Austin area, central Texas.
Excursion 2: Classic Paleozoic ostracode localities of south-central Oklahoma.
Excursion 3: Lower Tertiary and Upper Cretaceous ostracoda of the Brazos River valley.
Excursion 4: Recent freshwater and marine ostracoda of the Houston- Galveston area, Texas Gulf Coast.
Excursion 5: Permo-Carboniferous and Cretaceous ostracoda of north-central Texas.
Excursion 6: Recent and Tertiary ostracoda of Mexico.
CLU-G/G	1982
CLhC	1982
CMenUG	1982
CU-A	1982
CU-EART	1982
CaOOG	1982
CaOWtU	1982
CoDGS	1982
DI-GS	1982
IaU	1982
InU	1982
KLG	1982
KU	1982
KyU-Ge	1982

NIC	1982
NRU	1982
NcU	1982
OKentC	1982
OU	1982
OkU	1982
TxDaM-SE	1982
TxU	1982
WU	1982
WyU	1982

(317) **INTERNATIONAL SYMPOSIUM ON THE CAMBRIAN SYSTEM. GUIDEBOOK FOR FIELD TRIP.**
   1981 2nd 1. Cambrian stratigraphy and paleontology of the Great Basin and vicinity, western United States.
      2. The Cambrian System in the southern Canadian Rocky Mountains, Alberta and British Columbia.
      3. Cambrian and lowest Ordovician lithostratigraphy and biostratigraphy of southern Oklahoma and central Texas.

CLU-G/G	1981
CMenUG	1981
CSt-ES	1981(2-3)
CoDGS	1981
DI-GS	1981
NvU	1981
OkU	1981

(318) **INTERNATIONAL SYMPOSIUM ON THE DEVONIAN SYSTEM. [GUIDEBOOK]**
   1967 Calgary. Guidebook for Canadian Cordillera field trip. (125)
     — A-11. Field guide to Devonian outcrops of southwestern Manitoba. (14)

CLU-G/G	1967(reprint 1969)
CSt-ES	1967
CU-EART	1967
CaACAM	1967
CaACI	1967
CaACM	1967
CaACU	1967(A-11)
CaAEU	1967
CaBVaU	1967
CaOKQGS	1967
CaOLU	1967
CaOOG	1967
CaOWtU	1967
CoDGS	1967
CoU	1967
DI-GS	1967
DLC	1967
IU	1967
IdBB	1967
InLP	1983
MiHM	1967
MnU	1967
MtBuM	1967
NIC	1983
NNC	1967, 83
NRU	1983
NcU	1967
NdU	1967
NhD-K	1983
OU	1967
TxDaAR-T	1967
TxHSD	1967
TxHU	1967
TxMM	1967
UU	1967

(319) **INTERNATIONAL SYMPOSIUM ON THE ORDOVICIAN SYSTEM. FIELD EXCURSION. (TITLE VARIES)**
   1977 3rd 2. Ordovician of the eastern midcontinent.
      3. The ecostratigraphy of the middle Ordovician of the Southern Appalachians (Kentucky, Tennessee, and Virginia), U.S.A. (709)

AzU	1977(3)
CLU-G/G	1977(3)
CSt-ES	1977(3)
CaACU	1977(3)
CoDGS	1977
DI-GS	1977
IU	1977(3)
IaU	1977(3)
InLP	1977(3)
InRE	1977(3)
InU	1977(3)
KyU-Ge	1977(3)
OCU-Geo	1977(3)
OU	1977
OkU	1977(3)
ScU	1977
TU	1977(3)
TxBeaL	1977(3)
TxU	1977(3)
ViBlbV	1977(3)
WU	1977(3)

(320) **INTERNATIONAL SYMPOSIUM ON THE RIO GRANDE RIFT. GUIDEBOOK.**
   1978 Guidebook to Rio Grande Rift in New Mexico and Colorado. (461)

AzTeS	1978
CLhC	1978
CPT	1978
CSt-ES	1978
CU-EART	1978
CaACU	1978
CaBVaU	1978
CaOOG	1978
CaOWtU	1978
CoDCh	1978
CoU	1978
DI-GS	1978
DLC	1978
IU	1978
IaU	1978
InLP	1978
InRE	1978
InU	1978
KyU-Ge	1978
MnU	1978
NSyU	1978
NcU	1978
NhD-K	1978
NmPE	1978
NvU	1978
OCU-Geo	1978
OU	1978
OrCS	1978
PSt	1978
TxCaW	1978
TxLT	1978
TxU	1978
UU	1978
WU	1978

(321) **INTERNATIONAL SYMPOSIUM ON WATER-ROCK INTERACTION. FIELD TRIP GUIDE BOOK.**
   1980 3rd Alberta Plains and Rocky Mountains, post-session field trip. (278, 11)

CLhC	1980
CaOOG	1980
CoDGS	1980
DI-GS	1980
IaU	1980

(322) **INTERNATIONAL UNION OF GEODESY AND GEOPHYSICS. [GUIDEBOOK]**
   1963 Aug. Guidebook for seismological study tour.

CLU-G/G	1963
CSfCSM	1963
CSt-ES	1963
CU-EART	1963
NvU	1963

(323) INTERNATIONAL UNION OF GEOLOGICAL SCIENCES (I.U.G.S.). COMMISSION ON GEOCHRONOLOGY. FIELD GUIDE.
1967 June International conference; geochronology of Precambrian stratified rocks.
[1] Field guide for the East Kootenay field trip.
[2] Field guide for the Yellowknife field trip; a field trip held June 14-16 in conjunction with the conference, geochronology of Precambrian stratified rocks, Edmonton, Alberta.

CU-EART	1967(1)
DI-GS	1967(2)
DLC	1967(1)
InU	1967(1)
NcU	1967(1)
NhD-K	1967(1)
TxU	1967(2)

(324) INTERNATIONAL UNION OF GEOLOGICAL SCIENCES (I.U.G.S.). SUBCOMMISSION ON DEVONIAN STRATIGRAPHY. [GUIDEBOOK]
1981 July Devonian biostratigraphy of New York; part 2: stop descriptions.

CSt-ES	1981(Reprinted 1986)
CoDGS	1981
DI-GS	1981
IaU	1981(Reprinted 1986)
NRU	1981
OCU-Geo	1981

(325) INTERNATIONAL UNION OF GEOLOGICAL SCIENCES (I.U.G.S.). SUBCOMMISSION ON PRECAMBRIAN STRATIGRAPHY. ANNUAL MEETING. FIELD TRIP GUIDEBOOK.
1979 5th Field trip guidebook for the Archean and Proterozoic stratigraphy of the Great Lakes area, United States and Canada. (388)

CSt-ES	1979
CU-EART	1979
CaOTDM	1979
CaOWtU	1979
DI-GS	1979
IU	1979
IaU	1979
InLP	1979
KyU-Ge	1979
MnU	1979
WU	1979

(326) INTERNATIONAL UNION OF GEOLOGICAL SCIENCES (I.U.G.S.). SUBCOMMISSION ON SILURIAN STRATIGRAPHY. ORDOVICIAN-SILURIAN BOUNDARY WORKING GROUP. GUIDEBOOK.
1981 Field meeting; Anticosti-Gaspe, Quebec, 1981; Volume 1: guidebook. Anticosti Island, Quebec.
Gaspe Peninsula.

IaU	1981
InU	1981
NPSt	1981

(327) INTERNATIONAL UPPER MANTLE COMMITTEE. GUIDEBOOK.
1968 July Andesite conference guidebook. (International Upper Mantle Project. Scientific Report 16-S) (517)
1. McKenzie Pass area.
2. Crater Lake area.
3. Newberry Caldera area.
4. Mount Hood area.
5. Andesite petrochemistry.
1970 June International symposium on mechanical properties and processes of the mantle, Flagstaff, Arizona. Field guide to Kimberlite-bearing diatremes of northern Arizona-southern Utah including stops at Hopi Buttes, Monument Valley, and Grand Canyon.

AzU	1968
CLU-G/G	1968
CPT	1968
CSdS	1968
CSt-ES	1968
CU-A	1968
CU-EART	1968
CaACU	1968
CaAEU	1968
CaBVaU	1968
CaOKQGS	1968
CaOOG	1968
CaOTDM	1968
CaOWtU	1968
CoDGS	1968
CoG	1968
CoU	1968
DI-GS	1968
DLC	1968
IEN	1968
IU	1968
IaU	1968
InLP	1968
InU	1968
MH-GS	1968
MnU	1968
MoSW	1968
MtBuM	1968
NSbSU	1968
NcU	1968, 70
NdU	1968
NhD-K	1968
NjP	1968
NvU	1968
OkU	1968
PBL	1968
TxCaW	1968
TxU	1968
UU	1968
ViBlbV	1968
WU	1968

(328) INTERNATIONAL WILLISTON BASIN SYMPOSIUM. FIELD TRIP GUIDE.
1964 3rd Third International Williston Basin Symposium. (402)
1982 4th Geology of the Little Rocky Mountains, north-central Montana. (560, 476)

AzTeS	1964
CLU-G/G	1964
CLhC	1964
CMenUG	1964
CPT	1964
CU-EART	1964
CaACAM	1964
CaACU	1964
CaOKQGS	1964
CaOOG	1964
CoDGS	1964, 82
CoG	1964
CoU	1964
DI-GS	1964, 82
ICarbS	1964
IEN	1964
IU	1964
IaU	1964
InLP	1964
InU	1964
LU	1964
MnU	1964
MtBC	1964
MtBuM	1964
MtU	1964
NNC	1964
NdU	1964
NhD-K	1964
OU	1964
OkU	1964
PSt	1964
RPB	1964
SdRM	1964
TxDaDM	1964
TxDaGI	1964
TxMM	1964
TxU	1964

TxU-Da	1964
WyU	1964

INTERNATIONAL ZEOLITE CONFERENCE. See INTERNATIONAL COMMITTEE ON NATURAL ZEOLITES. (284)

(329) INTERSTATE OIL COMPACT COMMISSION. GUIDEBOOK.
1965 June Historical trip through Pennsylvania oil fields, Drake Well Memorial Park and Museum at Titusville, Vanished City of Pithole, observing local geology. (Cover title: Field trip through Pennsylvania's historical oil fields)
Part 2. Field trip: itinerary of trip from Pittsburgh to Titusville to Pithole and return to Pittsburgh.

NdU	1965

(330) IOWA ACADEMY OF SCIENCE. GEOLOGICAL FIELD TRIP GUIDEBOOK.
1975 Oct. Field recognition of subdivision of the Galena Group within Winneshiek Co. (386, 775)
1985 Geology of the University of Iowa Campus area, Iowa City.

IaU	1975
InLP	1985

IOWA GEOLOGICAL SOCIETY [GUIDEBOOK] See GEOLOGICAL SOCIETY OF IOWA. (208)

(331) IOWA GEOLOGICAL SURVEY. EDUCATIONAL SERIES.
1967 No. 1. Fossils and rocks of eastern Iowa.
Field trip no. 1. Cambrian and Ordovician, northeastern Iowa.
Field trip no. 2. Ordovician and Silurian, northeastern and east-central Iowa.
Field trip no. 3. Silurian and Devonian, east-central Iowa.
Field trip no. 4. Middle Devonian, east-central Iowa.
Field trip no. 5. Upper Devonian, north-central Iowa.
Field trip no. 6. Lower Mississippian (Kinderhook), north-central Iowa.
Field trip no. 7. Mississippian, north-central Iowa.
Field trip no. 8. Mississippian, southeastern and south-central Iowa.
Field trip no. 9. Mississippian, southeastern Iowa.

CMenUG	1967
CPT	1967
CSt-ES	1967
CU-EART	1967
CaOWtU	1967
CoDGS	1967(1-3)
CoU	1967
DI-GS	1967
DLC	1967
ICF	1967
ICIU-S	1967
IU	1967
InLP	1967
MnU	1967
MoSW	1967
MtBuM	1967
NcU	1967
NhD-K	1967
OkT	1967
TxHSD	1967
TxLT	1967
UU	1967
ViBlbV	1967
WU	1967

(332) IOWA GEOLOGICAL SURVEY. GUIDEBOOK SERIES.
1970 See Geological Society of America. 197.
1976 See Geological Society of Iowa. 208.
1978 See Tri-State Geological Field Conference. 659.
1980 See Friends of the Pleistocene. Midwest Group. 183.
1981 No. 4. See Geological Society of America. North Central Section. 200.
  No. 5. See Geological Society of America. North Central Section. 200.
  No. 6. See Geological Society of America. North Central Section. 200.
1984 Apr. No. 7. Geology of the University of Iowa campus area, Iowa City; a walking tour along the Iowa River. (Revised in 1985)

CLU-G/G	1984
CSt-ES	1984
CaOOG	1984
CaOWtU	1981
CoDGS	1984
DI-GS	1984
IaU	1984
NNC	1981
PHarER-T	1981

(333) IOWA NATURAL HISTORY ASSOCIATION. FIELD TRIP GUIDEBOOK.
1983 Apr. No. 1. Natural history of the upper Iowa Valley between Decorah and New Albin.
1984 Apr. No. 2. Natural history of the Lake Calvin Basin of Southeast Iowa.
1985 Apr. No. 3. Carboniferous, past and present; interpreting the fossil and modern communities associated with the Pennsylvanian and Mississippian strata of the Pella-Oskaloosa area.

IaU	1983-85

(334) IOWA STATE UNIVERSITY. DEPARTMENT OF EARTH SCIENCE. PUBLICATION.
1969 No. 2. See Tri-State Geological Field Conference. 659.

(335) IOWA STATE UNIVERSITY. DEPARTMENT OF EARTH SCIENCES. ANNUAL FIELD CONFERENCE GUIDEBOOK.
1964 Mar. Northwestern Nebraska and southwestern South Dakota.

IaU	1964

IOWA. UNIVERSITY. See UNIVERSITY OF IOWA. DEPARTMENT OF GEOLOGY. (688)

(336) IVACEI INTERNATIONAL ASSOCIATION OF VOLCANOLOGY AND CHEMISTRY OF THE EARTH'S INTERIOR.
1984 Deposits and effects of devastating lithic pyroclastic density current from Mount St. Helens on 18 May 1980--Field guide for northeast radial. (220)

CSt-ES	1984

(337) JAMAICA. MINES AND GEOLOGY DIVISION. SPECIAL PUBLICATION.
1974 No. 1. See Geological Society of America. 197.

(338) JAPANESE-AMERICAN FIELD CONFERENCE. FIELD TRIP GUIDEBOOK.
1979 May A guidebook for visiting selected Southern California landslides. (Cover title: Landslides of Southern California)

CLU-G/G	1979
CSt-ES	1979
CU-A	1979
CU-EART	1979
CaACU	1979
CaOWtU	1979
DI-GS	1979
KyU-Ge	1979
TxU	1979

(339) JOHNS HOPKINS UNIVERSITY. DEPARTMENT OF GEOLOGY. STUDIES IN GEOLOGY.
1950 16th See Geological Society of America. 197.
1958 17th See Field Conference of Pennsylvania Geologists. 172.
1960 18th See American Association of Petroleum Geologists. 18.

(340) KANSAS ACADEMY OF SCIENCE. [GUIDEBOOK FOR GEOLOGY FIELD TRIP] (TITLE VARIES)
1951 83rd Geology and botany field trips in vicinity of Lawrence, Kansas.
1953 Guidebook for geology section field trip; Tuttle Creek damsite.
1958 Field trip: western Fmaklin County.

(341) Kansas Geological Society.

CoDCh	1958
I-GS	1951, 53
KLG	1951, 53, 58
NjP	1951
TxLT	1951

(341) KANSAS GEOLOGICAL SOCIETY. GUIDEBOOK FOR THE ANNUAL FIELD CONFERENCE.

- 1927 1st Iowa field trip.
- 1928 2nd Ozark Mountains of Missouri and Arkansas.
- 1929 3rd Black Hills of South Dakota and Front Range of Rocky Mountains in Wyoming and Northern Colorado.
- 1930 4th Mountains of south-central Colorado, north-central and north-eastern New Mexico and the Texas Panhandle.
- 1931 5th Wichita, Arbuckle, and the Ouachita Mountains of Oklahoma and the Ouachita Mountains of Arkansas, with geologic cross sections.
- 1932 6th Carboniferous rocks of eastern Kansas and Nebraska, and western Missouri.
- 1933 7th Older Paleozoic rocks of Missouri, Arkansas and Oklahoma.
- 1934 8th Southwestern Kansas and the adjacent parts of Colorado, New Mexico, Oklahoma and Texas.
- 1935 9th Upper Mississippi Valley; Iowa City, Iowa to Duluth, Minnesota.
- 1936 10th Pennsylvanian and Permian rocks of northeastern Kansas and northwestern Missouri.
- 1937 11th Southeastern Kansas and northeastern Oklahoma.
- 1938 12th See Rocky Mountain Association of Geologists. 549.
- 1939 13th Southwestern Illinois, southeastern Missouri.
- 1940 14th Western South Dakota, eastern Wyoming; Black Hills, Hartville Uplift, and Laramie Range.
- 1941 15th Central and northeastern Missouri and adjoining area in Illinois.
- 1946 Mar. Permian and Pennsylvanian rocks, Winfield to Sedan, Kansas.
- — June Permian rocks, Augusta to Elmdale, Kansas.
- 1947 June Southeastern Kansas-Sedan to Iola.
- 1949 Apr. Permian and Pennsylvanian rocks, Winfield to Sedan.
- — June Pennsylvanian field conference. Pennsylvanian rocks in Kansas, lower Kansas River valley.
- — Oct. Western Shawnee and eastern Wabaunsee counties, Kansas.
- 1951 Nov. Lyon County, Kansas.
- 1952 16th West-central Missouri. (399)
- 1954 17th Southeastern and south-central Missouri. (399)
- 1955 18th Southwestern Kansas.
- 1956 19th New Virgilian sections along the Kansas Turnpike, east-central Kansas.
- — 20th Northwest Arkansas and Magnet Cove. (67)
- 1957 21st Eastern Kansas, from Kansas City to Manhattan, Kansas via the Kansas Turnpike, U.S. Highway 40, and Kansas Highway 13.
- 1958 22nd South-central Colorado.
- 1959 23rd Northeastern Kansas; Pennsylvanian and Permian cyclic deposits.
- — 24th See American Association of Petroleum Geologists. Regional Meeting. 29.
- 1960 25th Northeastern Oklahoma.
- 1961 26th Northeastern Missouri and west-central Illinois. (399)
- 1962 27th Geoeconomics of the Pennsylvanian marine banks in Southeast Kansas.
- 1964 28th North-eastern Oklahoma. (704, 505)
- 1966 29th Flysch facies and structure of the Ouachita Mountains.
- 1969 30th Excursion to the Kansas Geological Survey and core party at Lawrence, Kansas.
- 1975 31st See American Association of Petroleum Geologists. Mid-continent Section. 26.
- 1977 [32nd] Carbonate field seminar, Great Bahama Bank.
- 1980 33rd North-eastern Oklahoma.
- 1982 34th Southeast Kansas.
  1. Geology of the Howard Limestone (Wabaunsee Group, Virgilian stage, Pennsylvanian) in southeastern Kansas.
  2. Upper Howard (White Cloud) limestone trend in western Sumner county, Kansas.
- 1983 35th Geology of the Plattsmouth (Oread formation, Pennsylvanian system) Marine Bank in Chautauqua County, Kansas, and Osage County, Oklahoma. (26)
- 1984 36th Environment of deposition of the Ervine Creek marine algal bank (Deer Creek formation, Shawnee Group, upper Pennsylvanian) in southeastern Kansas.

ATuGS	1952, 54, 61
AzFU	1960
AzTeS	1928, 51-52, 54, 56-57, 59(23), 60-62, 64, 66
AzU	1964, 69
CChiS	1958
CLU-G/G	1931, 41, 46(June), 47, 49(Apr.), 51-52, 54, 56(19), 57-58, 61-2, 64, 66, 69, 77, 80, 82-83
CLhC	1961-62
CMenUG	1930, 31, 37, 41, 52, 54-61, 66, 75, 82
CPT	1928, 32, 34-37, 39-41, 49, 51-52, 54-58, 60-64, 69, 77, 80, 82-84
CSdS	1931, 52
CSt-ES	1939-41, 49, 51-52, 54, 56-58, 59(23), 61-62, 64, 66, 69
CU-A	1966
CU-EART	1932, 35, 41, 49, 52, 54-58, 60-62, 64, 66, 69, 77, 80, 82-84
CU-SB	1941, 61-62, 64, 66, 69
CaACAM	1935, 39-40
CaACM	1954
CaACU	1931, 52, 54, 56, 61-62, 77
CaAEU	1932, 64, 66
CaBVaU	1961, 66
CaOHM	1954, 56-57, 61-62, 66, 69
CaOKQGS	1927-29, 32-35, 54, 57-58, 61, 69
CaOLU	1951, 54, 56(19), 57, 61-62, 66, 69
CaOOG	1935, 40-41, 46-47, 49, 51-52, 54, 61, 64, 80, 82-84
CaOWtU	1961, 66, 82-83
CoDCh	1929-37, 39-41, 49, 51-52, 54-55, 58, 61-62, 64, 66
CoDGS	1929-32, 34-64, 66, 75, 77, 80, 82
CoFS	1958, 64
CoG	1930-37, 39-41, 49, 51-52, 54, 56-58, 59(23), 60-61, 64, 66, 69, 77
CoU	1930, 35, 40, 52, 54-56, 58, 59(23), 60-62, 64, 66, 69
DFPC	1954, 60-61
DI-GS	1927-37, 39-41, 52, 54-58, 59(23), 60-62, 64, 66, 69, 77, 80, 82-84
DLC	1930-31, 34-35, 52, 54-55, 61
I-GS	1928, 31-33, 35-37, 39-40, 49(Oct.), 52, 54, 57-58, 59(23), 60-61, 66
ICF	1932, 35-37, 39-41, 52, 54, 57-58, 60-61, 66
ICIU-S	1961, 64
ICarbS	1935, 37, 39-40, 49(Oct.), 52, 54, 61-62
IEN	1931-33, 35, 39-41, 52, 54, 56-58, 60-62, 64, 66, 69
IU	1932, 35-37, 39-41, 49, 52, 54, 56-58, 59(23), 60-62, 64, 66, 80
IaU	1927-37, 39-41, 52, 54-58, 59(23), 60-62, 64, 66, 77, 80
InLP	1951-52, 54, 56(19th), 57-62, 64, 66, 69, 77, 80
InRE	1935, 61
InU	1929, 35-37, 39-41, 49(Oct.), 51-52, 54, 56(19), 57-58, 60-62, 64, 66, 69, 77
KLG	1929, 31-37, 46-47, 49, 51, 52, 54-55, 58-62, 69, 75, 77, 80, 82-84
KU	1980, 82-84
KyU-Ge	1932-37, 39-41, 46-47, 49, 52, 54-58, 59(23), 60-62, 64, 66, 69, 77, 80, 82-84
LNU	1939, 61-62
LU	1930, 35-36, 40, 52, 54-58, 59(23), 60-62, 64, 66, 69
MNS	1929, 32, 34-37, 39-41, 52, 54, 58, 61, 64
MiDW-S	1928-37, 39-41, 49(June, Oct.), 51-52, 54, 56(19), 57-58, 61-62, 64
MiHM	1935, 39-41, 49(June, Oct.)
MiKW	1961
MiU	1927-37, 39-41, 46-47, 49, 51-52, 54-58, 59(23), 60-62, 64, 66, 69
MnSSM	1952, 54
MnU	1929-36, 40-41, 46-47, 49, 51-52, 54, 56-58, 59(23), 60-62, 64, 66, 69, 77, 80, 82
MoSW	1933, 35-36, 39-41, 49(June, Oct.), 51-52, 54-55, 57, 61-62, 64, 66
MoU	1930-32, 34-37, 39-41, 46-47, 49, 51-52, 56-57, 59(23), 61-62, 64, 66
MtBuM	1929, 40, 52, 54, 61
MtU	1940, 52, 54, 56-58, 59(23), 60-62
NIC	1980, 82-84
NNC	1927-37, 39-41, 49(June, Oct.), 51-52, 54-58, 59(23), 60-62, 64, 66, 69
NRU	1929, 37, 41, 52, 54, 56-58, 59(23), 61-62, 64, 69, 80, 82-84
NSyU	1935, 52, 54, 57, 61, 69
NbU	1932, 35-37, 39-41, 54-58, 60-62, 64, 66, 69
NcU	1932, 34-37, 39-41, 49, 51-52, 54, 56-58, 61-62, 64, 66
NdU	1929, 39, 52, 54, 56(20), 58, 61-62
NhD-K	1932-37, 39-41, 49, 52, 54-58, 59(23), 60-62, 64, 66, 69, 80, 82
NjP	1929, 31-32, 34-37, 39-41, 49, 52, 54-58, 59(23), 60-62, 64, 66, 69
NmLcU	1984
NvU	1952, 54, 61, 64

OCU-Geo	1929-35, 39, 41, 52, 54, 56, 61, 64, 66
OU	1930, 32-37, 39-41, 49, 51-52, 54-58, 59(23), 60-62, 64, 66
OkOkCGS	1937, 49, 51
OkT	1928-37, 39-41, 52, 54-55, 56(20), 57-58, 61-62, 64, 66
OkTA	1980, 82-83
OkU	1927-37, 39-41, 46-47, 49, 51-52, 54-58, 60-62, 64, 66, 69, 75, 77, 80, 82-84
OrCS	1935, 61-62, 64, 66, 69
OrU-S	1952, 54, 61-62, 64
PBL	1935-37, 52, 54, 61
PBm	1940-41, 52, 61
PSt	1935, 52, 54, 61-62, 64
RPB	1935, 41, 54, 61
SdRM	1929-31, 34-41, 52, 54-55, 61
TMM-E	1952, 64
TxBeaL	1941, 49(June), 52, 54, 56(19), 57-58, 60-62, 64, 66
TxDaAR-T	1929-37, 40-41, 46, 49(June, Oct.), 51-52, 54-58, 59(23), 60-62, 66
TxDaDM	1941, 51-52, 54, 56-58, 59(23), 60-62, 64, 66, 69
TxDaGI	1930-32, 34-35, 37, 39, 41, 52, 54-58, 61
TxDaM-SE	1929-30, 32, 35-37, 39-41, 49(Oct.), 51-52, 54, 56(19), 57-58, 59(23), 60-62, 64
TxDaSM	1932, 41, 49, 51-52, 54-58, 60-61
TxHSD	1930, 32, 35, 37, 41, 49(Oct.), 51-52, 54, 57, 59(23), 60-62, 64, 66
TxHU	1930-31, 35-36, 39-41, 52, 54, 56(20), 57-58, 59(23), 60-62, 64, 66, 69
TxLT	1930-31, 35-37, 41, 49, 51-52, 54-55, 61
TxMM	1929-33, 36-41, 51-52, 58, 62, 64, 66, 75, 80
TxU	1928-37, 39-41, 46-47, 49, 51-52, 54-58, 59(23), 60-62, 64, 66, 69, 82-83
TxU-Da	1930, 32-33, 35-36, 40, 52, 54, 58, 61, 66
TxWB	1930-35, 37-38, 40-41, 49, 52, 54-59
UU	1952, 54, 58, 61
ViBlbV	1935, 51-52, 54, 56(19), 61, 64
WPlaU	1935, 54, 58
WU	1932, 35-41, 52, 54-55, 56(19), 57-58, 60-62, 64, 66, 69, 77, 80, 82-83

KANSAS STATE COLLEGE OF PITTSBURG. See PITTSBURG STATE UNIVERSITY (KANSAS). NATIONAL SCIENCE FOUNDATION CLASS. (540)

(342) **KANSAS STATE TEACHERS COLLEGE OF EMPORIA. SCIENCE TEACHERS INSTITUTE. TITLE VARIES. [GUIDEBOOK]**
- 1957 Flint Hills of Lyon, Chase, Morris and Wabaunsee counties, Kansas.
- 1958 1st Camp Wood and Elmdale4 Hill, Chase County, Kansas.
- — 2nd Geologic field conference in the Flint Hills of Lyon, Chase, Morris, and Wabaunsee counties, Kansas. (425)
- 1959 2nd Camp Wood and Elmdale Hill, Chase County, Kansas.
- 1960 4th Flint Hills-Osage Cuestas of Lyon, Chase, Butler, Greenwood, Wilson, Woodson and Allen counties, Kansas. (425)
- 1961 5th Flint Hills of Lyon, Chase, Morris, and Wabaunsee counties, Kansas. (425)
- 1962 6th Flint Hills of Lyon, Chase, Morris and Wabaunsee counties, Kansas. (425)

CaOLU	1957
CoDCh	1959
IU	1957, 60-62
KLG	1957-58, 60-62
OU	1957

(343) **KANSAS STATE UNIVERSITY. UNIVERSITY FOR MAN. [GUIDEBOOK]**
- 1976 Land of the post rock.

CaACU	1976
IU	1976

KANSAS. GEOLOGICAL SURVEY. See STATE GEOLOGICAL SURVEY OF KANSAS. (636) and STATE GEOLOGICAL SURVEY OF KANSAS. (634) and STATE GEOLOGICAL SURVEY OF KANSAS. (635)

KANSAS. UNIVERSITY. See UNIVERSITY OF KANSAS. SCIENCE AND MATHEMATICS CAMP. (689)

(344) **KENDALL/HUNT GEOLOGY FIELD GUIDE SERIES. (SEE GUIDEBOOKS UNDER INDIVIDUAL AUTHORS)**

(345) **KENT STATE GEOLOGICAL SOCIETY. FIELD TRIP.**
- 1969 See Ohio Intercollegiate Field Conference in Geology. 496.

(346) **KENT STATE GEOLOGICAL SOCIETY. PROFESSIONAL PAPER.**
- 1962 [No.] 2. See Ohio Intercollegiate Field Conference in Geology. 496.

KENTUCKY GEOLOGICAL SOCIETY. SPRING FIELD CONFERENCE. See GEOLOGICAL SOCIETY OF KENTUCKY. (209)

(347) **KENTUCKY GEOLOGICAL SURVEY.**
- 1957 Apr. See Appalachian Geological Society. 58.
- 1976 Oct. See American Association of Petroleum Geologists. Eastern Section. 25.
- 1981 Nov. See Geological Society of America. 197.(14)

(348) **KENTUCKY GEOLOGICAL SURVEY. GEOLOGIC REPORTS.**
- 1930 Vol. 33. The Midland Trail in Kentucky. [Appendix to Devonian Rocks of Kentucky.]

CSt-ES	1930
KyU-Ge	1930
MiMarqN	1930
OCU-Geo	1930

(349) **KEY GEOLOGISTS REGIONAL CONFERENCE. [GUIDEBOOK FOR FIELD TRIP]**
- 1963 Geology of DeGray damsite; field trip (Arkansas).

TxHU	1963

(350) **LAFAYETTE GEOLOGICAL SOCIETY (LOUISIANA). FIELD TRIP.**
- 1962 Tertiary of central Louisiana.
- 1965 Tertiary of Mississippi.
- 1968 Tertiary of central Louisiana.
- 1980 See Gulf Coast Association of Geological Societies. 247.

CSt-ES	1968
CU-EART	1968
CaACU	1968
CoG	1968
DI-GS	1968
IU	1965
InLP	1968
MnU	1968
NcU	1968
OU	1968
TxBeaL	1968
TxDaAR-T	1968
TxHSD	1968
TxHU	1962
TxMM	1968
TxU	1962, 68
TxWB	1968
ViBlbV	1968

LAMAR STATE COLLEGE OF TECHNOLOGY. DEPARTMENT OF GEOLOGY. See LAMAR UNIVERSITY. DEPARTMENT OF GEOLOGY. (353)

(351) **LAMAR UNIVERSITY GEOLOGICAL SOCIETY. FIELDTRIP GUIDEBOOK.**
- 1980 Apr. See Southwestern Association of Student Geological Societies (SASGS). 626.

(352) **LAMAR UNIVERSITY GEOLOGY CLUB.**
- 1976 Spring See Southwestern Association of Student Geological Societies (SASGS). 626.

(353) LAMAR UNIVERSITY. DEPARTMENT OF GEOLOGY.
1965   See Southwestern Association of Student Geological Societies (SASGS). 626.

LIBERAL GEOLOGICAL SOCIETY. ANNUAL BOY SCOUT GEOLOGY FIELD TRIP. See BOY SCOUTS OF AMERICA. (102)

(354) LOCK HAVEN STATE COLLEGE. FORUM. [GUIDEBOOK]
1981 Fall   Geological walking tour of Lock Haven State College.
OCU-Geo	1981
PLhS	1981

(355) LONG BEACH, CALIFORNIA. DEPARTMENT OF OIL PROPERTIES. [GUIDEBOOK FOR FIELD TRIP]
1967   Field trip ... to Long Beach unit, offshore operations; Wilmington oil field data.
TxU	1967

(356) LOS ANGELES BASIN GEOLOGICAL SOCIETY. GUIDEBOOK.
1971   San Fernando earthquake field trip. Road log. (609)
1980 May   See American Association of Petroleum Geologists. Pacific Section. 28.
1984 June   San Andreas Fault-Cajon Pass to Wrightwood. (593)
CMenUG	1971
CPT	1971
CU-A	1984
CU-EART	1971
CaACU	1971
DI-GS	1971, 84
KyU-Ge	1971
MnU	1971
OCU-Geo	1971
OU	1971
TxBeaL	1971
TxDaAR-T	1971
TxDaDM	1971
TxU	1971

(357) LOUISIANA GEOLOGICAL SURVEY. GUIDEBOOK SERIES.
1982 No. 1.   See Geological Society of America. 197.(11)
— No. 2.   See Geological Society of America. 197.(14)

(358) LOUISIANA GEOLOGICAL SURVEY. LOUISIANA GEOLOGICAL BULLETIN.
1966 No. 41.   See Shreveport Geological Society. 566.

(359) LOUISIANA STATE UNIVERSITY GEOLOGY CLUB. GUIDE BOOK.
1961 Apr.   Woodville, Vicksburg and Jackson areas, Mississippi.
DI-GS	1961

(360) LOUISIANA STATE UNIVERSITY. SCHOOL OF GEOSCIENCE. GUIDEBOOK.
1961 Apr.   Woodville, Vicksburg and Jackson areas, Mississippi.
1973   Guidebook to the Paleozoic Ozark Shelf of northern Arkansas. (566(1975))
CSt-ES	1973
CU-EART	1973
CaACU	1973
DI-GS	1973
InLP	1973
LNU	1973
NcU	1973
OCU-Geo	1973
OU	1973
OkU	1973
TxBeaL	1973
TxDaAR-T	1973
ViBIbV	1973

(361) LOUISIANA STATE UNIVERSITY. SCHOOL OF GEOSCIENCE. MISCELLANEOUS PUBLICATION.
1971 71-1.   See American Association of Petroleum Geologists. 18.
1974 74-1.   See American Association of Petroleum Geologists. 18.

(362) LUBBOCK GEOLOGICAL SOCIETY.
1956   See West Texas Geological Society. 756.

(363) LUNAR AND PLANETARY INSTITUTE. LPI TECHNICAL REPORT.
1982 82-01.   See Lunar and Planetary Institute. Workshop. 365.(1981)
1983 83-03.   See International Geological Correlation Programme (IGCP) [Project] 92. Archean geochemistry 297.

(364) LUNAR AND PLANETARY INSTITUTE. TOPICAL CONFERENCE. FIELD TRIP GUIDE. (TITLE VARIES)
1978 Aug.   Papers presented to the Conference on Plateau Uplift: Mode and mechanism. (283, 648)
1982 Oct.   Field trip guidebook for the Precambrian rocks of the Minnesota River Valley. (388(No.14))
CMenUG	1978, 82
CSt-ES	1978, 82
CU-EART	1978, 82
CU-SC	1978
CaOOG	1982
CoPs	1982
IaU	1978, 82
InLP	1982
KyU-Ge	1982
MnU	1982
NSbSU	1982
NdU	1982
TxDaM-SE	1982

(365) LUNAR AND PLANETARY INSTITUTE. WORKSHOP. [FIELD GUIDE]
1981 Aug.   Workshop on magmatic processes of early planetary crusts: magma oceans and stratiform layered intrusions. (363)
Field guide to the Stillwater Complex.
Informal guidebook for the Precambrian rocks of the Beartooth Mountains, Montana-Wyoming.
1983 Aug.   See International Geological Correlation Programme (IGCP) [Project] 92. Archean geochemistry 297.
The Archaean crust in the Wawa-Chapleau-Timmins region.
CMenUG	1981
CSt-ES	1981
CU-A	1983
CaAEU	1981
CaOOG	1981
CaOTDM	1983
CoDGS	1983
DI-GS	1981
LPI	1985
MnU	1981
OkU	1981
TxU	1981
WU	1983

(366) LUNAR GEOLOGICAL FIELD CONFERENCE. GUIDEBOOK.
1965   State of Oregon lunar geological field conference guide book. (468, 517, 705)
[1] Devils Hill-Broken Top area and Lava Butte area field trip.
[2] Newberry Volcano area field trip.
[3] Hole-in-the ground-Fort Rock-Devils Garden area field trip.
[4] Belknap Crater-Yapoah Crater-Collier Cone area field trip.
[5] Crater Lake area field trip.
AzTeS	1965
AzU	1965
CLU-G/G	1965
CPT	1965
CSdS	1965
CSt-ES	1965
CU-EART	1965
CaBVaU	1965

CaOOG	1965
CaOWtU	1965
CoDGS	1965
CoG	1965
CoU	1965
DI-GS	1965
DLC	1965
I-GS	1965
IEN	1965
IU	1965
IaU	1965
IdBB	1965
InLP	1965
InU	1965
KyU-Ge	1965
MH-GS	1965
MiHM	1965
MnU	1965
MoSW	1965
MtBuM	1965
NRU	1965
NcU	1965
NdU	1965
NhD-K	1965
NjP	1965
NvU	1965
OU	1965
OkU	1965
OrCS	1965
OrU-S	1965
PBL	1965
PSt	1965
RPB	1965
ScU	1965
TxDaM-SE	1965
TxLT	1965
TxU	1965
UU	1965
ViBlbV	1965
WU	1965

(367) MAINE GEOLOGICAL SURVEY. BULLETIN.
1980 No. 26. See New England Intercollegiate Geological Conference. 456.

(368) MAINE GEOLOGY. BULLETIN.
1983 No. 3. See Geological Society of Maine. 210.

(369) MAINE SEA GRANT BULLETIN.
1977 10. See Nature Conservancy. Maine Chapter. 444.

MAINE. UNIVERSITY, PRESQUE ISLE ... See UNIVERSITY OF MAINE, PRESQUE ISLE. (692)

(370) MARYLAND GEOLOGICAL SURVEY. GUIDEBOOK.
1968 No. 1. See Atlantic Coastal Plain Geological Association. 89.
1971 No. 2. See Geological Society of America. 197.
— No. 3. See Geological Society of America. 197.
— No. 4. See Geological Society of America. 197.
1976 No. 5. See Geological Society of America. Northeastern Section. 201.

(371) MARYLAND GEOLOGICAL SURVEY. REPORT OF INVESTIGATIONS.
1976 No. 27. See Geological Society of America. Northeastern Section. 201.

MASSACHUSETTS. UNIVERSITY. See UNIVERSITY OF MASSACHUSETTS AT AMHERST. DEPARTMENT OF GEOLOGY AND GEOGRAPHY. (693) and UNIVERSITY OF MASSACHUSETTS. COASTAL RESEARCH GROUP. (694)

(372) MEMORIAL UNIVERSITY OF NEWFOUNDLAND. DEPARTMENT OF EARTH SCIENCES. REPORT.
1982 No. 9. See International Geological Correlation Programme (IGCP). Project 27: The Caledonide Orogen. 302.

(373) METEORITICAL SOCIETY. ANNUAL MEETING. FIELD TRIP GUIDEBOOK.
1974 Guidebook to the geology of Meteor Crater, Arizona. (Revised in 1979)
1979 Guidebook to the geology of Meteor Crater, Arizona. Revised ed. (66(No.17))

AzTeS	1979
AzU	1974
CLU-G/G	1979
CU-EART	1979
CoDGS	1974, 79
DI-GS	1974
DLC	1979
MnU	1974
NN	1979
NvU	1979
OkTA	1979

(374) MIAMI GEOLOGICAL SOCIETY. ANNUAL FIELD TRIP.
1967 1st Field guidebook on geology and ecology of Everglades National Park.
1968 2nd Late Cenozoic stratigraphy of southern Florida; a reappraisal, with additional notes on Sunoco-Felda and Sunniland oil fields.
1969 [1] Field guide to some carbonate rock environments; Florida Keys and western Bahamas.
— 3rd [2] Late Pleistocene geology in an urban area.
1970 4th Sedimentary environments and carbonate rocks of Bimini, Bahamas.
1971 Field guide to some carbonate rock environments, Florida Keys and western Bahamas. Rev. ed.
1977 Guatemala, where plates collide; a reconnaissance guide to Guatemalan geology. (695)
1979 Jan. A field guide with road log to the Pliocene fossil reef of Southwest Florida.
1980 May Water, oil and the geology of Collier, Lee and Hendry counties.
1981 Apr. Survey of central Florida geology.
1982 Mar. Survey of the geology of Haiti; guide to the field excursions in Haiti.
1983 Apr. The Miami limestone; a guide to selected outcrops and their interpretation (with a discussion of diagenesis in the formation).

AzTeS	1977
CLU-G/G	1968-71, 77
CPT	1977
CSt-ES	1968, 70-71, 77, 79, 82
CU-A	1977
CU-EART	1968, 69(3), 77
CU-SB	1969, 82
CaACU	1968, 69(1), 71, 77, 79
CaAEU	1971
CaBVaU	1971
CaOKQGS	1969-70
CaOOG	1971
CaOWtU	1968-69(3rd), 71
CoG	1968-69, 71
CoU	1968-70
DI-GS	1967-70, 77, 79-83
F-GS	1981, 83
FBG	1981, 83
ICIU-S	1971
ICarbS	1971
IU	1968, 71, 77
IaU	1968, 69(3), 77, 79-81
IdBB	1970
InLP	1968, 77, 79-81
InU	1968-69, 77
KyU-Ge	1968-70, 77
LNU	1968, 69(3), 71
LU	1968
MH-GS	1968-69
MNS	1968
MiHM	1971
MnU	1968, 77, 79-82
MoU	1968
NBiSU	1968
NSbSU	1968-69, 77
NbU	1971
NcU	1968, 69(1), 70-71, 79

(375) Miami University, Oxford, Ohio. Department of Geology.

NdU	1968-71
NhD-K	1968-71, 77
OCU-Geo	1968-71
OU	1968-71
OkU	1968-70, 77, 80-83
PBL	1971
PSt	1971
TMM-E	1969
TxBeaL	1971
TxCaW	1971
TxDaAR-T	1968-69, 71
TxDaDM	1968-69
TxDaGI	1969
TxDaSM	1968
TxHSD	1968, 77
TxLT	1977
TxU	1967-69, 77
TxWB	1969, 77
ViBIbV	1969, 77
WU	1968-71, 77, 79

(375) MIAMI UNIVERSITY, OXFORD, OHIO. DEPARTMENT OF GEOLOGY.

1965 May Four day field trip to central Kentucky.

OU	1965

(376) MICHIGAN ACADEMY OF SCIENCE, ARTS, AND LETTERS. SECTION OF GEOLOGY AND MINERALOGY. ANNUAL GEOLOGICAL EXCURSION.

1935 5th A study of the Lucas County, Ohio-Monroe County, Michigan monocline and stratigraphy of northwestern Ohio and southeastern Michigan. (377)

1937 7th [No title given, but it is about celebrating the centennial anniversary of the founding of the Michigan Geological Survey] (377)

1938 8th [No title available] (377)

1939 9th Marquette and Menominee districts. (377)

1940 10th To Afton-Onaway District. (377)

1941 11th St. Ignace and Mackinac Island. (377)

1947 See Michigan Basin Geological Society. 377.

1948 See Michigan Basin Geological Society. 377.

CSt-ES	1947, 48
CoDCh	1937-41
CoDGS	1938, 39
DI-GS	1937-41
IU	1935, 37-39, 41
InU	1940
MiHM	1935, 37-41
OU	1938-40
OkU	1940
TxU	1941

(377) MICHIGAN BASIN GEOLOGICAL SOCIETY. GEOLOGICAL EXCURSION. (TITLE VARIES)

1935 5th See Michigan Academy of Science, Arts, and Letters. Section of Geology and Mineralogy. 376.

1937 7th See Michigan Academy of Science, Arts, and Letters. Section of Geology and Mineralogy. 376.

1938 8th See Michigan Academy of Science, Arts, and Letters. Section of Geology and Mineralogy. 376.

1939 9th See Michigan Academy of Science, Arts, and Letters. Section of Geology and Mineralogy. 376.

1940 10th See Michigan Academy of Science, Arts, and Letters. Section of Geology and Mineralogy. 376.

1941 11th See Michigan Academy of Science, Arts, and Letters. Section of Geology and Mineralogy. 376.

1944 Aug. Itinerary, Mackinac Straits field conference.

1946 Ontario geological excursion to Kettle Point, Owen Sound, Waubaushene.

1947 Michigan copper country. (376)

1948 Pleistocene and early Paleozoic of eastern part of the northern peninsula of Michigan. (376)

1949 The northern part of the southern peninsula of Michigan.

1950 The Ordovician rocks of the Escanaba-Stonington area.

1951 The Devonian and Silurian rocks of parts of Ontario, Canada and western New York.

1952 Stratigraphy and structure of the Devonian rocks in southeastern Michigan and northwestern Ohio.

1953 The Ordovician stratigraphy of Cincinnati, Ohio and Richmond, Indiana areas.

1954 The stratigraphy of Manitoulin Island, Ontario, Canada.

1955 The Niagara Escarpment of peninsular Ontario, Canada.

1956 The Devonian strata of the London-Sarnia area, southwestern Ontario, Canada.

1957 Silurian rocks of the northern peninsula of Michigan.

1958 Precambrian geology of parts of Dickinson and Iron counties, Michigan.

1959 Geology of Mackinac Island and Lower and Middle Devonian south of the Straits of Mackinac.

1960 Lower Paleozoic and Pleistocene stratigraphy across central Wisconsin.

1961 Geologic features of parts of Houghton, Keweenaw, Baraga, and Ontonagon counties, Michigan.

1962 Silurian rocks of the southern Lake Michigan area.

1963 Stratigraphy of the Silurian rocks in western Ohio.

1964 See American Association of Petroleum Geologists. 18.

1965 Geology of central Ontario. (Second edition of American Association of Petroleum Geologists-Society of Economic Paleontologists and Mineralogists-Geological Association of Canada guidebook for joint meeting, Toronto, 1964) (24, 390)

1966 Cambrian stratigraphy in western Wisconsin. (778)

1967 Correlation problems of the Cambrian and Ordovician outcrop areas, northern peninsula of Michigan.

1968 Geology of Manitoulin Island.

1969 Studies of the Precambrian of the Michigan Basin.

1970 See Geological Society of America. North Central Section. 200.

1971 Geology of the Lake Erie Islands and adjacent shores.

1972 Niagaran stratigraphy; Hamilton, Ontario.

1973 Geology and the environment; man, earth and nature in northwestern lower Michigan.

1974 Silurian reef-evaporite relationships.

1976 Devonian strata of Emmet and Charlevoix counties, Michigan. (697(1976 No. 7))

1977 May Geology of the Marquette district, a field guide.

1978 Sept. Geology of the Manitoulin area including the road log to the Michigan Basin Geological Society field trip. (378)

1980 Ordovician and Silurian geology of the northern peninsula of Michigan.

1983 Tectonics, structure, and karst in northern lower Michigan.

AzTeS	1950, 57-59, 62, 67, 69, 70, 73, 74, 76
AzU	1977
CLU-G/G	1947-48, 50-63, 66-69, 71-74, 76-77, 80, 83
CLhC	1962
CMenUG	1948, 50-63, 65-74, 76-78, 80, 83
CPT	1948, 51-53, 66, 68-69, 71-74, 76-77, 80
CSdS	1950
CSt-ES	1947-52, 54-63, 66-69, 70-74, 76, 78
CU-EART	1951-52, 54-57, 66, 71-74, 78
CU-SB	1959, 68-69, 71-78
CaAC	1956
CaACAM	1965, 68, 71-73, 78
CaACI	1968
CaACU	1959, 66, 71-74, 76-78
CaAEU	1969, 78
CaBVaU	1954, 57, 59-60, 69, 78
CaOHM	1950-52, 56-63, 67-69, 73-74
CaOKQGS	1950-63, 65, 67-69, 71, 80, 83
CaOLU	1946, 49-50, 52-63, 65-66, 68, 73
CaOOG	1948, 50-63, 65-69, 71-74, 80, 83
CaOTDM	1954-56, 59, 69, 71, 78, 80
CaOTRM	1954
CaOWtU	1950-52, 57-63, 65-66, 68-69, 71-72, 78
CoDCh	1947-49, 51, 54-55, 59, 62, 63, 68, 69, 72, 74, 77, 78, 83
CoDGS	1948, 50-63, 66-69, 71-74, 76-77, 78, 80, 83
CoG	1948-63, 65-69, 71-74, 76-77, 83
CoU	1954, 69
DI-GS	1947-63, 65-69, 71-74, 76-78, 80, 83
DLC	1952-57, 59, 66, 69, 71, 73
I-GS	1959, 66, 71, 73-74
ICF	1947-52, 54-63, 66, 68-69, 71-73, 76-78

ICtU-S	1955-56, 66, 74
ICarbS	1966, 69
IEN	1947-50, 52, 54, 56-63, 65-69, 71-74
IU	1946-47, 49-52, 54-63, 65-69, 71-74, 77-78
IaU	1950, 52, 54, 57-59, 61-63, 65-69, 71-74, 77-78, 80, 83
InLP	1947-50, 52, 55, 57-63, 66-69, 71-74, 76-78, 80, 83
InRE	1953, 55-56, 63, 78
InU	1947-52, 54-57, 59-63, 66-67, 69, 71-74, 77-78
KyU-Ge	1950-52, 57-63, 66-69, 71-73, 77-78, 80
LNU	1957
MNS	1954, 66
MdBJ	1976
MiDW-S	1950-52, 54-63, 65, 67-69, 71-73
MiHM	1944, 46-63, 65-69, 71-74, 77
MiU	1971-72
MnU	1948, 50-52, 54-63, 66-68, 71-74, 76-78, 80, 83
MoSW	1966, 69
MoU	1950, 52, 55, 57-63, 66-69, 71-74
Ms-GS	1954-55
MsJG	1954-55
MtBuM	1966
NBiSU	1969, 73
NNC	1946-63, 65-69, 72-73, 76-77, 80, 83
NOneoU	1969
NSbSU	1977-78, 83
NSyU	1969, 73, 76
NbU	1950-63, 65-69, 71-74
NcU	1950, 52, 57-60, 62-63, 66-69, 71-74, 77-78, 80
NdU	1950, 52, 57-60, 62-63, 66-69, 71-74, 77-78
NhD-K	1950-63, 65-69, 71-74, 77-78, 80, 83
NjP	1950, 52, 57-58, 68-69, 71-74
NvU	1966
OCU-Geo	1950, 52-53, 57-60, 62-63, 66-69, 71-74, 77
OU	1946-63, 65-69, 71-74, 77
OkT	1956
OkU	1948-63, 66-69, 70-74, 77, 80
PBL	1966
PBm	1955-56, 69
PSt	1954-56, 66
TMM-E	1957
TxBeaL	1950, 52-53, 56-60, 62, 67-69, 71, 73
TxDaAR-T	1949-50, 54, 56-61, 63, 66-69, 71, 73
TxDaDM	1948, 50-63, 65-69, 72, 74
TxDaGI	1957
TxDaM-SE	1947-48, 50-52, 54-63, 65-66
TxDaSM	1948, 50-63, 74
TxHSD	1957, 59, 62-63, 71, 78
TxHU	1954-60, 63, 65-69
TxLT	1956
TxMM	1955-57, 70, 74
TxU	1958-63, 65-69, 71, 74, 77, 80
TxU-Da	1950-63, 65-68
UU	1956, 66
ViBlbV	1956, 59, 71-72, 74
WU	1947, 57-63, 65-69, 71-78, 80, 83

(378) MICHIGAN BASIN GEOLOGICAL SOCIETY. SPECIAL PAPERS.
1978 3. See Michigan Basin Geological Society. 377.

MICHIGAN GEOLOGICAL SOCIETY. See MICHIGAN BASIN GEOLOGICAL SOCIETY. (377)

(379) MICHIGAN TECHNOLOGICAL UNIVERSITY.
1962 June See National Science Foundation. Summer Conference for College Teachers. 427.
1963 June See National Science Foundation. Summer Conference for College Teachers. 427.
1964 June See National Science Foundation. Summer Conference for College Teachers. 427.
1965 June See National Science Foundation. Summer Conference for College Teachers. 427.
1971 Sept. See Society of Economic Geologists. 583.

MICHIGAN. EASTERN MICHIGAN UNIVERSITY. See EASTERN MICHIGAN UNIVERSITY. (164)

MICHIGAN. UNIVERSITY. See UNIVERSITY OF MICHIGAN. MUSEUM OF PALEONTOLOGY. (697)

MICHIGAN. WAYNE STATE UNIVERSITY, DETROIT. See WAYNE STATE UNIVERSITY. [GEOLOGY DEPARTMENT] SUMMER CONFERENCE. (754)

MICHIGAN. WESTERN MICHIGAN UNIVERSITY. See WESTERN MICHIGAN UNIVERSITY. DEPARTMENT OF GEOLOGY. (768)

(380) MIDWEST GROUNDWATER CONFERENCE. GUIDEBOOK.
1960 5th Guidebook to the geology of the Rolla area emphasizing solution phenomena. (57, 397)

AzTeS	1960
CSt-ES	1960
CoDCh	1960
DI-GS	1960
IU	1960
InLP	1960
KLG	1960
MoSW	1960
NSyU	1960
ViBlbV	1960

(381) MIDWESTERN GEOLOGICAL SOCIETY.
1961 See Southwestern Association of Student Geological Societies (SASGS). 626.

(382) MILWAUKEE PUBLIC MUSEUM. LORE LEAVES.
1969 11. A trip on glacial geology in the North Kettle moraine area.
CoDGS	1969
WMMus	1969

(383) MINERALOGICAL ASSOCIATION OF CANADA.
1963 See Geological Association of Canada. 190.
1964 See American Association of Petroleum Geologists. 18.
1966 See Geological Association of Canada. 190.
1967 See Geological Association of Canada. 190.
1968 See Geological Association of Canada. 190.
1969 See Geological Association of Canada. 190.
1970 See Geological Association of Canada. 190.
1973 See Geological Association of Canada. 190.
1974 See Geological Association of Canada. 190.
1975 See Geological Association of Canada. 190.
1976 See Geological Association of Canada. 190.
1977 See Geological Association of Canada. 190.
1978 See Geological Society of America. 197.
1979 See Geological Association of Canada. 190.
1980 See Geological Association of Canada. 190.
1981 See Geological Association of Canada. 190.
1982 See Geological Association of Canada. 190.
1983 See Geological Association of Canada. 190.
1985 See Geological Association of Canada. 190.

(384) MINERALOGICAL SOCIETY OF AMERICA.
1965 See American Crystallographic Association. 33.
1978 See Geological Society of America. 197.
1979 See Geological Society of America. 197.
1981 See Geological Society of America. 197.
1982 See Geological Society of America. 197.
1983 See Geological Society of America. 197.
1984 See Geological Society of America. 197.
1985 See Geological Society of America. 197.

(385) MINERALOGICAL SOCIETY OF UTAH. FIELD TRIP.
1958 Apr. Gold Hill, Utah, Clifton district, Tooele County.
— May Birds Eye marble quarry, Birds Eye, Utah.
— Aug. Park City via Brighton.
DI-GS	1958

(386) Minnesota Academy of Science.

(386) MINNESOTA ACADEMY OF SCIENCE. GUIDEBOOK.
1964 Guidebook to the geology of the Duluth area.
1975 See Iowa Academy of Science. 330.

MnU	1964

(387) MINNESOTA GEOLOGICAL SURVEY. BULLETIN.
1925 No. 20. A guidebook to Minnesota Truck Highway No. 1.
1964 See Friends of the Pleistocene. Midwest Group. 183.

ATuGS	1925
CLU-G/G	1925
CMenUG	1925
CPT	1925
CSt-ES	1925
CU-EART	1925
CaOKQGS	1925
CaOON	1925
CaOWtU	1925
CoDGS	1925
CoG	1925
CoU	1925
DI-GS	1925
DLC	1925
ICF	1925
ICarbS	1925
IU	1925
IaU	1925
InLP	1925
InU	1925
KyU-Ge	1925
MNS	1925
MnU	1925
MoSW	1925
MoU	1925
MtBuM	1925
NSyU	1925
NcU	1925
NhD-K	1925
NjP	1925
NvU	1925
OCU-Geo	1925
OU	1925
OkT	1925
OrU-S	1925
PSt	1925
RPB	1925
SdRM	1925
TxDaDM	1925
TxDaM-SE	1925
TxHU	1925
TxLT	1925
UU	1925
ViBlbV	1925
WU	1925

(388) MINNESOTA GEOLOGICAL SURVEY. FIELD TRIP GUIDEBOOK. (TITLE VARIES)
1968 Geological field trip in the Rochester, Minnesota area.
1972 No. 2. See Geological Society of America. 197.
— No. 3. See Geological Society of America. 197.
— No. 4. See Geological Society of America. 197.
— No. 5. See Geological Society of America. 197.
— No. 6. See Geological Society of America. 197.
— No. 7. See Geological Society of America. 197.
— No. 8. See Geological Society of America. 197.
1979 No. 9. See Geological Society of America. North Central Section. 200. (1979(2))
— No. 10. See Geological Society of America. North Central Section. 200. (1979(8))
— No. 11. See Geological Society of America. North Central Section. 200. (1979(7))
— No. 12. Field trip guidebook for the Precambrian geology of east-central Minnesota.
— No. 13. See International Union of Geological Sciences (I.U.G.S.). Subcommission on Precambrian Stratigraphy. 325.
1982 No. 14. See Lunar and Planetary Institute. Topical Conference. 364.
1985 Pleistocene geology and evolution of the upper Mississippi Valley. (774)

CSt-ES	1968, 79(12)
CU-EART	1979(12)
DI-GS	1979(12)
IU	1979
IaU	1979(12), 85
InLP	1979(12)
MiKW	1968
MnU	1968, 79(12), 85
WU	1968, 79(12)

(389) MINNESOTA SPELEOLOGICAL SURVEY. MSS SPECIAL PUBLICATION.
1983 No. 1. A cave explorer's guide to Mystery Cave.
1984 No. 2. North country region; LaCrosse, Wisconsin.

DI-GS	1983-84
InTI	1983, 84
MA	1984
MBU	1984
MU	1984
MiD	1984
MiHM	1984
NGenoU	1984
NIC	1984
NNC	1984
NRU	1983, 84
NSbSU	1984
NhD-K	1984
OCU-Geo	1984
ODaWU	1984
OU	1984
PBU	1984
PPi	1984
PSt	1984

MINNESOTA. UNIVERSITY. See UNIVERSITY OF MINNESOTA. DULUTH BRANCH. (698)

(390) MISSISSIPPI GEOLOGICAL SOCIETY. GUIDEBOOK FOR THE FIELD TRIPS.
1940 1st Jackson to Recent.
— 2nd Claiborne and Wilcox.
— 3rd Upper Cretaceous of Mississippi and Alabama.
— 4th Northwest Alabama Paleozoics.
1943 Nov. Road log for Wilcox-Midway-Cretaceous portion of Mississippi Geological Society field trip.
1945 5th Eutaw-Tuscaloosa.
1948 6th Upper Eocene, Oligocene and lower Miocene of central Mississippi.
1949 7th Pre-Cambrian and Paleozoic rocks of northern Alabama and south-central Tennessee.
1950 8th Cretaceous of Mississippi and South Tennessee.
1952 9th Claiborne of western Alabama and eastern Mississippi.
1953 10th Wilcox and Midway groups, west-central Alabama.
1954 11th Paleozoic rocks, central Tennessee and Northwest Alabama.
1956 13th Covering outcrops of the Vicksburg, Jackson, Claiborne and Wilcox groups of central Mississippi.
1957 Wilcox and Midway groups, west-central Alabama. Supplement; road log.
1959 14th Upper Cretaceous outcrops, Northeast Mississippi and west-central Alabama.
1960 15th Cenozoic of southeastern Mississippi and southwestern Alabama.
1962 16th Paleozoics of Northwest Arkansas: Magnet Cove, Arkansas Valley, Ouachita Mountains, Ozark Highlands.
1978 17th Mississippian rocks of the Black Warrior Basin.

ATuGS	1940(2-4), 45, 48-50, 52-54, 56, 59-60, 62, 78
AzTeS	1948-49, 52, 54, 59-60, 62
CLU-G/G	1945, 48-50, 52-54, 59-60, 62
CLhC	1945, 48-49, 52-54
CMenUG	1945, 48-50, 52-54, 59-60, 62, 78
CPT	1945, 48-50, 52, 54, 59-60, 62
CSdS	1948-49, 52, 54, 59-60, 62
CSt-ES	1945, 48-50, 52-54, 56, 59-60, 62
CU-EART	1948-49, 53-54, 59, 62
CaACM	1949, 52-54
CaACU	1959-60, 62, 78

CaBVaU	1962
CaOHM	1948-49, 52, 54, 59-60, 62
CaOKQGS	1948-49, 54, 59-60, 62
CaOWtU	1940(3), 59-60, 62, 78
CoDCh	1945, 49, 50, 54
CoDGS	1945, 48-50, 52-53, 59-60, 62
CoG	1940, 45, 48-50, 52-54, 59-60, 62, 78
DFPC	1960
DI-GS	1940, 43, 45, 48-50, 52-54, 56, 59-60, 62, 78
DLC	1945, 48-50, 52-54
I-GS	1949
ICF	1948-49, 52, 54, 59-60, 62
ICarbS	1940, 45, 48-50, 52-54
IEN	1945, 48-49, 60, 62
IU	1945, 48-49, 52-54, 59-60, 62, 78
IaU	1978
InLP	1949, 52, 54, 59-60, 62, 78
InU	1945, 48-50, 52-54, 59-60, 62
KyU-Ge	1945, 48-50, 52-54, 59-60, 62, 78
LNU	1945, 48-50, 52-54, 59-60, 62
LU	1940, 45, 48-50, 52-54, 59-60, 62
MNS	1945
MiHM	1962
MiKW	1945
MiU	1945, 48-49, 52, 54, 59-60, 62
MnU	1945, 48-50, 52-54, 59-60, 62, 78
MoSW	1945, 48-50, 52-53, 60, 62
MoU	1945, 48, 54, 62
Ms-GS	1948-50, 52, 54, 56, 59-60, 62
MsJG	1948-50, 52, 54, 56, 59-60, 62
MtBuM	1945, 48
MtU	1948-49, 52, 54, 59, 62
NBiSU	1948-49, 52, 54, 59-60, 62
NNC	1940(2), 48-50, 52-54, 59-60, 62
NSbSU	1978
NcU	1948-50, 52, 54, 59-60, 62
NhD-K	1948-50, 52-54, 56, 59-60, 62
NjP	1940, 48-50, 52-54, 56, 59-60, 62
OU	1940(4th), 45, 48-50, 52-54, 59-60, 62
OkOkCGS	1940, 45, 48, 50, 52
OkT	1940, 45, 48-49, 53, 56, 59, 62
OkU	1940, 43, 45, 48-50, 52-54, 59-60, 62
OrU-S	1948-49, 62
PSt	1952-54, 59-60, 62
RPB	1945
TMM-E	1948-49, 52, 54, 59-60, 62
TU	1949, 54
TxBeaL	1945, 48-49, 52-54, 59-60, 62
TxDaAR-T	1940(2-4), 45, 48-50, 52-54, 60, 62
TxDaDM	1945, 48-50, 52-54, 59-60, 62, 78
TxDaGI	1940, 45, 48-49, 52-53
TxDaM-SE	1948-50, 52, 54, 59-60, 62
TxDaSM	1948, 52-54, 60, 62, 78
TxHSD	1948, 50, 52, 54, 62
TxHU	1940(1-2, 4), 45, 48-50, 52-54, 59-60, 62
TxLT	1940(4), 45, 48-50, 52-53
TxMM	1940(1, 3-4), 45, 48, 49, 52, 54, 59, 60, 62
TxU	1940, 45, 48-50, 52-54, 59-60, 62
TxU-Da	1940(2-4), 53
TxWB	1940, 45, 48-49, 52-54, 56, 59-60, 62, 78
UU	1948-49, 54
ViBlbV	1948-49, 52, 54, 59-60, 62
WU	1945, 48-49, 52-54, 59-60, 62, 78

(391) **MISSISSIPPI GEOLOGICAL, ECONOMIC AND TOPOGRAPHICAL SURVEY. FIELD TRIP.**
1961 Mar. Paleozoic field trip; N. Alabama & SW Tennessee.
    DI-GS   1961

(392) **MISSISSIPPI VALLEY FIELD CONFERENCE. GUIDEBOOK.**
1949 Natchez, Mississippi to Montgomery, Louisiana.
    Loess in the southern Mississippi Valley.
    IaU   1949

MISSISSIPPI. UNIVERSITY. See UNIVERSITY OF MISSISSIPPI. GEOLOGICAL SOCIETY. (699)

(393) **MISSOURI ASSOCIATION OF PROFESSIONAL SOIL SCIENTISTS. FIELD TRIP GUIDE.**
1982 Nov. Meet the Missouri Pleistocene field trip guide.
    IU   1982

(394) **MISSOURI SPELEOLOGY.**
1968 V. 10. No. 3. Guidebook to selected caves in southwestern Missouri. (435([9]))
    DI-GS   1968

MISSOURI. DIVISION OF GEOLOGICAL SURVEY AND WATER RESOURCES. See MISSOURI. DIVISION OF GEOLOGY AND LAND SURVEY. (397) and MISSOURI. DIVISION OF GEOLOGY AND LAND SURVEY. (395) and MISSOURI. DIVISION OF GEOLOGY AND LAND SURVEY. (396) and MISSOURI. DIVISION OF GEOLOGY AND LAND SURVEY. (399) and MISSOURI. DIVISION OF GEOLOGY AND LAND SURVEY. (398)

(395) **MISSOURI. DIVISION OF GEOLOGY AND LAND SURVEY. CONTRIBUTION TO PRECAMBRIAN GEOLOGY.**
1976 No. 6. See Association of Missouri Geologists. 87.
1981 No. 9. See Geological Society of America. 197.(15)

(396) **MISSOURI. DIVISION OF GEOLOGY AND LAND SURVEY. EDUCATIONAL SERIES.**
1978 4. Geologic wonders and curiosities of Missouri.
CMenUG	1978
CU-EART	1978
ICarbS	1978
InLP	1978
OkT	1978
WU	1978

(397) **MISSOURI. DIVISION OF GEOLOGY AND LAND SURVEY. [MISCELLANEOUS PUBLICATION]**
1960 See Midwest Groundwater Conference. 380.

(398) **MISSOURI. DIVISION OF GEOLOGY AND LAND SURVEY. OPEN FILE REPORT SERIES.**
1982 OFR-82-16-MR. Guidebook A: Field trip to the St. Francois Mountains and the historic Bonne Terre Mine.
DI-GS	1982
WU	1982

(399) **MISSOURI. DIVISION OF GEOLOGY AND LAND SURVEY. REPORT OF INVESTIGATIONS.**
1952 No. 13. See Kansas Geological Society. 341.
1954 No. 17. See Kansas Geological Society. 341.
1955 No. 20. See Association of Missouri Geologists. 87.
1958 No. 25. See Geological Society of America. 197.
1961 No. 26. See Association of Missouri Geologists. 87.
— No. 27. See Kansas Geological Society. 341.
1965 No. 30. See Geological Society of America. 197.
— No. 31. See Geological Society of America. 197.
1966 No. 34. See American Association of Petroleum Geologists. 18.
1967 No. 37. See National Speleological Society. Mississippi Valley-Ozark Regional Association. 434.
1975 No. 58. See Symposium on the Geology and Ore Deposits of the Viburnum Trend, Missouri. 643.
1976 No. 61. See Association of Missouri Geologists. 87.
1977 No. 62. Guidebook to the geology along Interstate-55 in Missouri.
1981 No. 67. See Geological Society of America. 197. (No. 15)
CPT	1977
CSt-ES	1977
CU-A	1977
CU-EART	1977
CaOWtU	1977, 81
CoU	1977
DI-GS	1977
DLC	1977
ICF	1977
ICarbS	1977

(400) Montana Bureau of Mines and Geology.

IU	1977
IaU	1977
InLP	1977
KyU-Ge	1977
MnU	1977
MtBuM	1977
NIC	1981
NNC	1981
NRU	1981
NhD-K	1977, 81
NvU	1977
OCU-Geo	1977
OCl	1981
OU	1977, 81
OkT	1977
OkU	1977
PBL	1977
SdRM	1977
TxBeaL	1977
TxU	1977
UU	1977
WU	1977

MISSOURI. GEOLOGICAL SURVEY. See ASSOCIATION OF MISSOURI GEOLOGISTS. (87) and MISSOURI. DIVISION OF GEOLOGY AND LAND SURVEY. (396) and MISSOURI. DIVISION OF GEOLOGY AND LAND SURVEY. (399) and MISSOURI. DIVISION OF GEOLOGY AND LAND SURVEY. (397) and MISSOURI. DIVISION OF GEOLOGY AND LAND SURVEY. (395) and MISSOURI. DIVISION OF GEOLOGY AND LAND SURVEY. (398)

MISSOURI. UNIVERSITY. See UNIVERSITY OF MISSOURI. GEOLOGY CLUB. (700)

(400) MONTANA BUREAU OF MINES AND GEOLOGY.
1937 See Rocky Mountain Association of Geologists. 549.

(401) MONTANA BUREAU OF MINES AND GEOLOGY. SPECIAL PUBLICATION.
1976 73. See Tobacco Root Geological Society. 658.
1980 82. See Tobacco Root Geological Society. 658.
1983 86. Guidebook of the Fold and Thrust Belt, west-central Montana.
   No. 1. Bozeman to Helena via Battle Ridge Pass, White Sulphur Springs, Townsend and Toston.
   No. 2. Helena to Bozeman via MacDonald Pass, Garrison, Deer Lodge, Butte, Homestake Pass, Jefferson Canyon and Three Forks.
1984 89. Profiles of Montana geology; a layman's guide to the Treasure State.
1985 92. See International Geological Correlation Programme (IGCP). Project 161. 300.

CLU-G/G	1983-84
CMenUG	1983-84
CPT	1983, 84
CSfCSM	1983-84
CSt-ES	1983-84
CU-EART	1983, 84
CU-SB	1983-84
CaACU	1983-84
CaOOG	1983-84
CaOTDM	1983
CaOWtU	1983, 84
CaQQERM	1983-84
CoDBM	1983
CoDCh	1983
CoDGS	1983-84
CoG	1983-84
CoU-DA	1983
DI-GS	1983-84
F-GS	1983
FBG	1983
ICarbS	1983, 84
IaU	1983
IdU	1983-84
InLP	1983-84
InU	1984
KLG	1983-84
KU	1983-84
KyU-Ge	1983
MiHM	1983-84
MnU	1983-84
MtBuM	1983, 84
NIC	1983
NNC	1983-84
NRU	1983, 84
NSbSU	1983
NcU	1983
NdU	1983-84
NhD-K	1983-84
NmSoI	1983-84
NmU	1983
NvU	1983-84
OCU-Geo	1983, 84
OkU	1983-84
PHarER-T	1983-84
PSt	1976
ScU	1984
TxDaM-SE	1983-84
TxMM	1983-84
TxWB	1983-84
WU	1983

(402) MONTANA GEOLOGICAL SOCIETY. GUIDEBOOK FOR THE ANNUAL FIELD CONFERENCE. (TITLE VARIES)
1950 1st   Trip through western Montana, covering the stratigraphic sections from the Cambrian to the Tertiary.
1951 2nd Rocks of Mississippian age (Big Snowy Group) and the overlying unconformity, central Montana.
1952 3rd   Black Hills [and] Williston Basin.
1953 4th   The Little Rocky Mountains-Montana [and] southwestern Saskatchewan.
1954 5th   Pryor Mountains-northern Bighorn Basin, Montana.
1955 6th   Sweetgrass arch-disturbed belt, Montana.
1956 7th   Judith Mountains, central Montana.
1957 8th   Crazy Mountain Basin.
1958 9th   Beartooth Uplift and Sunlight Basin. (784)
1959 10th Sawtooth-disturbed belt area.
1960 11th West Yellowstone-earthquake area.
1961 12th Float trip; Bighorn River from Kane, Wyoming to Yellowtail Dam site.
1962 13th Three Forks-Belt Mountains area and symposium; the Devonian system of Montana and adjacent areas.
1963 14th See Wyoming Geological Association. 781.
1964 15th See International Williston Basin Symposium. 328.
1965 16th Geology of Flint Creek Range, Montana.
1966 17th [Symposium] Jurassic and Cretaceous stratigraphic traps, Sweetgrass Arch.
1967 18th Centennial Basin of Southwest Montana.
1968 See Saskatchewan Geological Society. 559.
1969 20th Eastern Montana Symposium; the economic geology of eastern Montana and adjacent areas. (No road log)
1972 21st Crazy Mountains Basin.
1975 22nd Energy resources of Montana. No field trip.
1977 [23rd] See Wyoming Geological Association. 781.
1978 24th Williston Basin Symposium. The economic geology of the Williston Basin. (No road log) (328)
1979 30th Sun River Canyon-Teton Canyon Montana disturbed belt.
1981 Southwest Montana. (Cover title: Guidebook, Southwest Montana)
   First day. Billings to Big Timber.
   Big Timber to Livingston.
   Livingston to Three Forks.
   Three Forks to Twin Bridges.
   Twin Bridges to Dillon.
   Third day. Dillon to Wisdom to Dillon.
   Fourth day. Dillon to Three Forks.
1984 Northwest Montana and adjacent Canada.
   Map of the route of the Montana Geological Society, 1984 field conference: northwest Montana, southwest Alberta, southeast British Columbia.

AzTeS	1950, 52-60, 62, 67, 69, 72, 75
AzU	1950-61, 65-67, 69, 72
CChiS	1958

CLU-G/G	1950-62, 65-67, 69, 72, 75, 78-79, 81, 84
CLhC	1951-62, 65, 67, 69, 81
CMenUG	1950-63, 65-67, 69, 72, 75, 77-79, 81
CPT	1950-63, 65-67, 69, 72, 75, 78, 79, 81, 84
CSdS	1969, 72, 75, 78
CSfCSM	1960
CSt-ES	1950-63, 65-66, 69, 72, 81
CU-A	1959
CU-EART	1950-62, 65-67, 69, 72, 78-79
CU-SB	1967, 81
CaAC	1952, 55, 58, 62
CaACAM	1950-58, 60, 62, 65-67, 69, 72, 75, 81
CaACI	1950, 52-58, 62, 65-66, 69
CaACM	1950-60, 62, 65-67, 69, 78
CaACU	1950-60, 63, 65-75, 78-79, 81, 84
CaBVaU	1950-54, 57-60, 62
CaOHM	1953-62, 65-67
CaOKQGS	1952-53, 55, 57-60, 62, 65-67
CaOOG	1950, 52-62, 65-67, 69, 72, 75, 81, 84
CaOONM	1952-53
CaOTRM	1955-56
CaOWtU	1962, 72, 75
CoDCh	1950-52, 54-61, 65-67, 72, 79, 81, 84
CoDGS	1950-63, 65-67, 69, 72, 75, 78, 79, 81, 84
CoDU	1952-57, 59, 61-62, 66
CoG	1950-62, 65-67, 69, 72, 75, 78, 79, 81, 84
CoGrU	1967
CoLiM	1981
CoU	1950-62, 65-67, 69, 72
DI-GS	1950-62, 65-67, 69, 72, 81
DLC	1950-61, 66
ICF	1952-62, 66-67
ICarbS	1953-60, 62, 65
IEN	1950-60, 62, 65-67, 69, 72, 75
INS	1981
IU	1950-62, 65-67, 69, 72, 75
IaU	1950-62, 65-67, 69, 72, 81
IdBB	1952-60, 62, 66-67, 81
IdPI	1952-61
InLP	1950-62, 65-67, 69, 72, 75, 78, 79, 81, 84
InRE	1950-56
InU	1950-62, 65-67, 69, 72
KyU-Ge	1950-62, 64-67, 69
LU	1950-62, 65-67, 69, 72, 75
MCM	1981
MNS	1950-59
MiDW-S	1950-62, 66-67, 69
MiHM	1959, 65
MiKW	1954-56, 62
MnU	1950-62, 65-67, 69, 72, 75, 78-79, 81, 84
MoRM	1950-62
MoSW	1950-60
MtBC	1950-62, 65-67, 69, 72, 75
MtBuM	1950-62, 65-67, 69, 72, 75, 78-79, 81, 84
MtU	1950, 52-62, 65-67, 69
NIC	1981
NNC	1951-62, 65-67, 69, 72, 78, 81
NRU	1952-62
NSyU	1952-62, 65-67, 69, 72, 81
NbU	1950-62, 65-67, 69, 72
NcU	1950-62, 65-67, 69, 72, 75, 81
NdU	1950-58, 60, 62, 65-67, 69, 72, 75, 81
NhD-K	1950-62, 65-67, 69, 72, 81, 84
NjP	1950-62, 65-67, 69, 72
NvU	1953-62, 65-67, 69, 72, 81
OCU-Geo	1950-60, 62, 66-67, 69, 72, 78
OU	1950-62, 65-67, 69, 72, 78, 79, 81, 84
OkOkCGS	1950-60, 66
OkT	1950-60, 62, 78
OkTA	1981
OkU	1950-62, 65-67, 69, 72, 75, 81
OrCS	1978
OrU-S	1950-62, 65-67, 69, 72
PBL	1950-62, 65-67, 69, 72
PBU	1981
PSt	1950-62, 65-67, 69, 72, 75
RPB	1950-60, 62, 65-67, 69
SdRM	1950-62, 67, 69, 72
SdU	1950, 52-60, 62, 65-67, 69
TU	1953-60, 62, 65-66
TxDaAR-T	1950-62, 65-67, 69, 72
TxDaDM	1950-62, 65-67, 69, 75, 78
TxDaGI	1950-55, 57-60, 62, 65-67, 72, 75
TxDaM-SE	1951-62, 65, 81
TxDaSM	1950-60, 62, 66, 69
TxHSD	1951-62, 65-67, 69, 72, 75, 78
TxHU	1950-62, 65-67, 69, 72
TxLT	1950-60, 62
TxMM	1967, 69, 72, 81, 84
TxU	1950-62, 65-67, 69, 72, 75, 81
TxU-Da	1950-62, 65-66, 69
TxWB	1950-55, 58, 61, 65, 67, 69, 72, 75, 81
UU	1950-62, 66, 69, 72, 75
ViBlbV	1950-62, 69
WU	1950, 52-54, 57-62, 65-67, 69, 72, 75, 78-79, 81, 84
WyU	1950-62, 65-67, 81

(403) MONTANA GEOLOGICAL SOCIETY. SPECIAL PUBLICATIONS. PAPER.
1975 20. Three Forks to Livingston: a geologic road log along interstate 90.
1976 21. Billings, Montana to Belle Fourche, South Dakota; a geologic road log along interstate 90 and U.S. 212.
1978 23. Interstate 90 from the junction with U.S. 212 near Crow Agency, Montana to Sheridan, Wyoming; a geological road log.

IU	1975-76, 78
IdU	1978
InLP	1976, 78
MnU	1975-76, 78
MtBuM	1975-76, 78
NvU	1978
WU	1978

(404) MONTANA STATE UNIVERSITY. DEPARTMENT OF EARTH SCIENCES. PUBLICATION.
1982 10. See Geological Society of America. Rocky Mountain Section. 204.(1982(3))
— 11. See Geological Society of America. Rocky Mountain Section. 204.(1982(4))
— 12. See Geological Society of America. Rocky Mountain Section. 204.(1982(1))

MONTANA. UNIVERSITY. See UNIVERSITY OF MONTANA. GEOLOGY DEPARTMENT. (701)

(405) MOUNTAIN GEOLOGIST.
1964 V. 1. No. 3. See Rocky Mountain Association of Geologists. 549.
1965 V. 2. No. 3. See Rocky Mountain Association of Geologists. 549.
1968 V. 5. No. 3. See Rocky Mountain Association of Geologists. 549.
1969 V. 6. No. 3. See Rocky Mountain Association of Geologists. 549.
1970 V. 7. No. 3. See Rocky Mountain Association of Geologists. 549.
1972 V. 9. No. 2/3. See American Association of Petroleum Geologists. 18.
1973 V. 10. No. 3. See Rocky Mountain Association of Geologists. 549.
1977 V. 14. No. 3/4. See North American Paleontological Convention. 474.

(406) MURRAY STATE UNIVERSITY. DEPARTMENT OF GEOSCIENCES.
1980 See American Association of Petroleum Geologists. Eastern Section. 25.

NAGT See NATIONAL ASSOCIATION OF GEOLOGY TEACHERS. SOUTHWEST SECTION. (417) and NATIONAL ASSOCIATION OF GEOLOGY TEACHERS. TEXAS SECTION. (418) and NATIONAL ASSOCIATION OF GEOLOGY TEACHERS. NEW ENGLAND SECTION. (415) and NATIONAL ASSOCIATION OF GEOLOGY TEACHERS. NORTH CENTRAL SECTION. (416) and NATIONAL ASSOCIATION OF GEOLOGY TEACHERS. HAWAII SECTION. (413) and NATIONAL ASSOCIATION OF GEOLOGY TEACHERS. EAST CENTRAL SECTION. (410) and NATIONAL ASSOCIATION OF GEOLOGY TEACHERS. EASTERN SECTION. (411) and NATIONAL ASSOCIATION OF GEOLOGY TEACHERS. (408) and NATIONAL ASSOCIATION OF GEOLOGY TEACHERS. CENTRAL SECTION. ANNUAL MEETING. (409) and NATIONAL ASSOCIATION OF GEOLOGY TEACHERS. NAGT NEWSLETTER. (414) and NATION-

AL ASSOCIATION OF GEOLOGY TEACHERS. FAR WESTERN SECTION. (412)

(407) NATIONAL ACADEMY OF SCIENCES. PUBLICATION.
1955 See Clay Minerals Conference. 132.
1956 See Clay Minerals Conference. 132.

(408) NATIONAL ASSOCIATION OF GEOLOGY TEACHERS. FIELD TRIP.
1967 See American Association of Petroleum Geologists. 18.
1975 Mar. See National Science Teachers Association. Annual Meeting. 429.
1976 See National Science Teachers Association. Annual Meeting. 429.
1978 See Geological Society of America. 197.
1979 See Geological Society of America. 197.(No. 4)
1981 See Geological Society of America. 197.
1982 See Geological Society of America. 197.
1983 See Geological Society of America. 197.
1984 See Geological Society of America. 197.
1985 See Geological Society of America. 197.

(409) NATIONAL ASSOCIATION OF GEOLOGY TEACHERS. CENTRAL SECTION. ANNUAL MEETING. FIELD TRIP GUIDEBOOK.
1976 Apr. Field trip guidebook for the Precambrian geology of the St. Cloud granite district; east-central Minnesota.
1978 Apr. 1. Pleistocene deposits of McLean & Woodford counties, Illinois.
1980 April A field trip guidebook to the geology of the Fox River Valley and east central Wisconsin.
1983 Geology of the circumferential highway system.

CSt-ES	1976
DI-GS	1980
IU	1978, 83
IaU	1978, 80
KyU-Ge	1976
MnU	1976
OU	1980
WU	1980

(410) NATIONAL ASSOCIATION OF GEOLOGY TEACHERS. EAST CENTRAL SECTION. ANNUAL MEETING AND FIELD TRIP.
1962 No title, but it is about the bedrock and glacial geology of the Bellefontaine outlier.
1964 The geology of southeastern Indiana, mostly glacial.
1966 Apr. Some geological aspects of the Carboniferous of southern Indiana.
— Fall Southeastern New York.
1967 Middle coastal plain, Virginia.
1974 Energy reserves, Canton-Cadiz area.
1978 See Geological Society of America. North Central Section. 200.
1979 Physical setting and processes along the central Ohio shore of Lake Erie (Marblehead peninsula to Lorain). (493)
1982 See Geological Society of America. North Central Section. 200.
1984 See Geological Society of America. Southeastern Section. 206.
1985 See Geological Society of America. North Central Section. 200.

CaOWtU	1966, 68, 72-73, 77
DI-GS	1966-67, 74
IU	1964
InLP	1966(Apr.)
InU	1966
OU	1962, 79

(411) NATIONAL ASSOCIATION OF GEOLOGY TEACHERS. EASTERN SECTION. FIELD GUIDEBOOK. (TITLE VARIES)
1966 Field trip no. 1. Glacial geology and geomorphology between Cortland and Syracuse.
Field trip no. 2. Paleontology and stratigraphy of the Cortland-Syracuse-Ithaca area.
1968 Field trip A. The Silurian-Ordovician angular unconformity, southeastern New York.
Field trip B. The Rosendale readvance in the lower Wallkill Valley, New York.
1972 Field trip no. 1. Geology of the Ramapo fault system.
Field trip no. 2. Geomorphology of northern New Jersey and part of eastern Pennsylvania.
Field trip no. 3. Sedimentology and general structure of the northern portion of the Newark Basin.
Field trip no. 4. Mineralogy-petrology trip to northwestern New Jersey.
Field trip no. 5. Cretaceous and Tertiary greensands and their fauna, New Jersey coastal plain.
1973 Field trip no. 1. The geology of Chestnut Ridge Anticline in the vicinity of Laurel Caverns, Fayette County, Pennsylvania.
Field trip no. 2. Engineering geology at two sites on Interstate 279 and Interstate 79 northwest of Pittsburgh, Pennsylvania.
Field trip no. 3. Nuclear power.
1976 Guidebook to field trips at Skidmore College, Saratoga Springs, New York.
A. Geology of the southeastern Adirondacks.
B. Museums, geology, and education: From old to new in the New York State Museum.
C. Devonian alluvial and tidal lithofacies between Palenville and Gilboa, New York.
D. Minerals and mining.
1977 Guidebook for field trips in Maryland and the national capital area.
1979 Guidebook for field trips in Virginia.
[1] Geology of the Richmond-Petersburg area.
[2] Geology, soils and land use in central Virginia.
[3] Sedimentary paleoenvironments and the Paleozoic evolution of the Appalachian geosyncline in the northern Shenandoah Valley.
[4] Using coastal plain stratigraphy near Fredericksburg, Virginia, as a teaching tool.
[5] The Culpeper Basin of the Triassic lowlands of northern Virginia.
1980 May Geologic hazards of Pittsburgh.
1981 Spring The geology of the Genesee Valley area of western New York.
Trip A. General geology of the Genesee Valley region in Monroe and Genesee counties, New York.
Trip B. Stratigraphy, paleontology and paleoecology of the Upper Hamilton Group (Middle Devonian) in the Genesee Valley, Livingston County, New York.
Trip C. Glacial geology of the Genesee Valley-Dansville-Naples region.
1982 Guidebook to the Late Cenozoic geology of the lower York-James Peninsula, Virginia. (136(No.3). (Adapted in part from guidebook 2) Revised in 1982.)
1983 Spring Trip A. Deep-water gravity; displaced deposits marginal to the shelf edge of the now-vanished proto-Atlantic (Iapetus) Ocean.
Trip B. Mohawk Valley episodic discharges; the geomorphic and glacial sedimentary record.
— March See Geological Society of America. Northeastern Section. 201.
1984 Oct. F1: Physiography, geology and land use; Toronto to Madoc.
1985 May See Canadian Geoscience Council. EdGEO Conference. 119.

CSt-ES	1973, 77
CU-EART	1977
CU-SB	1977, 80
CaOWtU	1966, 68, 72-73, 77
DI-GS	1966, 72-73, 76-77, 79-81, 83
IU	1977, 82
IaU	1976-77, 79
InRE	1972
KyU-Ge	1973, 76(A-D), 77
NcU	1977
PHarER-T	1972-73, 80
TxBeaL	1977
ViBlbV	1977, 79
WU	1977

(412) NATIONAL ASSOCIATION OF GEOLOGY TEACHERS. FAR WESTERN SECTION. FIELD TRIP GUIDEBOOK. (TITLE VARIES)
1966 Berkeley Hills.
1967 Apr. See American Association of Petroleum Geologists. 18.
1971 Oct. Camino Cielo field trip guidebook (to southern Santa Barbara County).
1972 Apr. Groundwater geology of northern Sacramento County.

— Oct. [1] Geologic guidebook to the northern peninsular ranges, Orange and Riverside counties, California. (609)
[2] Four local geologic sites; a teacher's guide.
1973 Apr. Field trip to areas of active faulting and shallow subsidence in the southern San Joaquin Valley, California.
Some geologic hazards and environmental impact of development in the San Diego area. (Reprinted 1972 guidebook to field trip T-10, Southwestern Regional Meeting, National Science Teachers Association) (430)
1974 [Apr.] Oceanographic field trip, San Francisco Bay.
— [June] Geologic sites in Ventura County (teacher's guide).
— Oct. Northern California field trip.
1975 Mar. See National Science Teachers Association. Annual Meeting. 429.
— Oct. The Sierran superjacent and bedrock series in southwestern Placer County.
Trip 1: [No title given, but it is about local sites to visit on field trips]
Trip 2: Significant geology in southwestern Placer County and the Auburn Dam site.
1976 Oct. Geology field trip guidebook, Pt. Lobos State Reserve, Carmel, California. (413)
1977 Mar. The geology of the Crafton Hills and lower Mill Creek Canyon.
— Oct. No. 3. Mining history of the southern Mother Lode.
— Dec. Visitor's guide to the geology of Mount Diablo.
1978 Apr. Field guide for two trips in the Imperial Valley, California.
1. Eolian and lacustrine geomorphology of the western part of Salton Sink.
2. Features of San Jacinto fault zone and Borrego Mountain earthquake.
1979 Mar. [Coachella Valley, California]
1. Aeolian features of northern Coachella Valley and landforms and tectonic features of the San Andreas fault zone in the Indio Hills.
2. Guide to the geology and structure of Painted Canyon, Mecca Hills.
3. Martinez Mountain landslide and shoreline features of prehistoric Lake Cahuilla, southwestern margin of Coachella Valley.
— Oct. Franciscan rock types.
A. Geology and rock types of the Franciscan.
B. Geologic tour of the Geysers Geothermal Area.
1980 Apr. Field trip guide to the anorthosite-syenite terrain of the western San Gabriel Mountains, Los Angeles County, California, with emphasis on the origin of the layered gabbroic rocks.
— Sept. Guidebook to the glacial geology, volcanoes & earthquakes of Mammoth.
Reprinted 1973 guidebook to field trip T-10. (430)
1981 Apr. Tour and field guide to petroleum research and production facilities, the geology of the eastern Puente Hills, northeastern Los Angeles Basin, and the coastal geomorphology of southwestern Orange County, California.
— Fall Redding to Lassen Park.
1982 Mar. Field trip guide to the Sespe Creek area, northern Ventura County, California.
Late Cenozoic tectonics of the Ventura-Oak View area, Ventura County, California.
Guidebook to the Recent, Quaternary, Plio-Pleistocene and Franciscan geology of western Humboldt County. (Condensed version of Friends of the Pleistocene, Pacific Cell, field trip guidebook, 1981)
Field guide to Anacapa Island.
— Oct. Quaternary stratigraphy, soil geomorphology, chronology and tectonics of the Ventura, Ojai, and Santa Paula areas, western Transverse Ranges, California. (184)
1983 Mar. Death Valley region field guide.
— Fall Guidebook [to] selected field trips, Coalinga, Calif. area.
1984 Mar. Geology of Santa Catalina Island and nearby basins.
1. Santa Catalina Island field trip.
2. Geology of the San Pedro Basin; Los Angeles Harbor to Santa Catalina Island.
3. San Pedro Harbor to Catalina Island; a deep sea field trip.
4. The nearshore marine biota; Santa Catalina Island.
— Oct. Field trip A. Environmental geology of the Sacramento-Folsom-Auburn area, a brief history of the development of flood control, hydroelectric power, the Auburn Dam Project and placer gold mining.
Field trip B. Environmental geology and mining in the Mother Lode southeast of Sacramento, Michigan Bar, Ione, Penn Mine, Pardee Reservoir, Jackson and vicinity.
1985 Mar. Geology and geothermal energy of the Salton Trough.
— Oct. Northern Sierra Nevada. Self-guiding photo tour of geologic features of the Reno and Lake Tahoe areas, Nevada and California. (Cover title: Guidebook to the northern Sierra Nevada & Reno; Lake Tahoe areas)

CLU-G/G	1971, 72(Oct.), 73(1980 reprint), 74, 75(Oct.), 77-79, 81, 82(Apr., Oct.), 84, 85(Mar.)
CLhC	1972(Pt.1, Oct.), 79(Mar.), 81, 82(Mar., Oct.) 83(Mar., Fall), 84(Mar.)
CMenUG	1972, 73(Apr.), 74(Apr., Oct.), 75(Mar., Oct.), 77-85
CPT	1972(Apr., Oct.(1)), 73, 74(June), 77-79, 80-84, 85(Oct.)
CSdS	1972(Oct.1), 73(Fall), 78
CSfCSM	1974(Oct.), 81-84, 85(Mar,)
CSt-ES	1966, 72(Oct.), 73(Apr.), 77(Mar.), 78(Apr.), 79(Mar.), 80, 81, 82, 83, 84(Mar., Oct.), 85
CU-A	1966, 73, 74(June, Oct.), 75(Oct.), 77(Mar., Oct.), 78, 79, 80(Sept.), 81(Fall), 82, 83(Fall), 84(Mar., Oct.), 85(Oct.)
CU-EART	1972(Apr., Oct.1), 73-74, 75(Oct.), 77-78, 79(Oct.), 80, 81, 82, 83(Mar.), 84, 85(Mar.)
CU-SB	1974-75, 77, 80-84
DI-GS	1972(Apr., Oct.1), 73-74, 75(Oct.), 77-79, 81-85
IU	1972(Oct.(1)), 75(Oct.), 76(Oct.), 78
IaU	1972(Apr., Oct.(1)), 73-74, 75(Oct.), 77(Mar., Oct.), 78, 79(Mar.), 80-83, 84(Mar., Oct. A, B)
InLP	1971, 72(Oct.(1)), 73-79
InU	1980, 82
KyU-Ge	1971, 72(Oct.(1)), 73-74, 75(Oct.), 76, 77(Mar.), 78-83, 84(Oct.)
LNU	1982(Fall)
MnU	1972(Oct.1), 83-84
NcU	1977(Mar.), 79
NvU	1972(Oct.), 75(Oct.), 77(Oct.), 80(Apr.) 83(Fall), 84(Oct.), 85(Oct.)
OU	1972(Oct.(1)), 80(Apr.)
TxU	1977(Mar.)
ViBlbV	1971, 73, 74(Apr., Oct.), 75(Oct.), 77(Mar., Oct.), 78

(413) NATIONAL ASSOCIATION OF GEOLOGY TEACHERS. HAWAII SECTION.
1976 Oct. See National Association of Geology Teachers. Far Western Section. 412.

(414) NATIONAL ASSOCIATION OF GEOLOGY TEACHERS. NAGT NEWSLETTER.
1984 Sept Geologic notes on the Charleston Terre Haute and Paris area, Illinois and Indiana.

IU	1984

(415) NATIONAL ASSOCIATION OF GEOLOGY TEACHERS. NEW ENGLAND SECTION.
1979 Apr. Field trip guide to the Narragansett Basin, Massachusetts and Rhode Island.

KyU-Ge	1979

(416) NATIONAL ASSOCIATION OF GEOLOGY TEACHERS. NORTH CENTRAL SECTION. GUIDEBOOK.
1969 Apr. Geology of northeastern North Dakota. (480)

CLU-G/G	1969
CPT	1969
CSdS	1969
CSt-ES	1969
CU-EART	1969
CaOLU	1969
CaOWtU	1969
CoG	1969
DI-GS	1969
ICF	1969
ICarbS	1969
IEN	1969
IU	1969
IaU	1969
InLP	1969
InU	1969
KyU-Ge	1969

(417) National Association of Geology Teachers. Southwest Section.

MiDW-S	1969
MiHM	1969
MiU	1969
MnDuU	1969
MnU	1969
MoSW	1969
MoU	1969
MtBuM	1969
NBiSU	1969
NSyU	1969
NbU	1969
NcU	1969
NdU	1969
NhD-K	1969
NjP	1969
NvU	1969
OCU-Geo	1969
OU	1969
OkU	1969
OrU-S	1969
PBL	1969
PSt	1969
SdRM	1969
TxBeaL	1969
TxDaAR-T	1969
TxLT	1969
TxU	1969
ViBlbV	1969
WU	1969

(417) NATIONAL ASSOCIATION OF GEOLOGY TEACHERS. SOUTHWEST SECTION. [GUIDEBOOK] (TITLE VARIES.)
1970 Four Corners, Colorado Plateau, Central Rocky Mountain section.
1971 Guidebook to eastern Basin and Range, Colorado Plateau, southern Rocky Mountains.

AzFU	1970
AzTeS	1970-71
AzU	1970-71
CChiS	1970
CMenUG	1970-71
CSt-ES	1970
CaOLU	1971
CaOWtU	1970
CoDuF	1970
CoG	1970
DI-GS	1970-71
ICarbS	1970
IaU	1970
InLP	1970-71
InU	1970
KyU-Ge	1970
NdU	1970
PSt	1970
TxDaAR-T	1970-71
TxDaDM	1970
TxMM	1970-71
WaPS	1970

(418) NATIONAL ASSOCIATION OF GEOLOGY TEACHERS. TEXAS SECTION. FIELD TRIP.
1964 Dec. See Texas Academy of Science. 654.
1972 Environmental geology field trip.
1981 See Geological Society of America. South Central Section. 205.
1982 See Geological Society of America. South Central Section. 205.
1983 See Geological Society of America. South Central Section. 205.
1984 Mar. See Geological Society of America. South Central Section. 205.
    TxU   1972

NATIONAL CLAY CONFERENCE. See CLAY MINERALS CONFERENCE. (132)

(419) NATIONAL CONFERENCE ON CLAYS AND CLAY MINERALS. PROCEEDINGS. (SUPERSEDED BY THE PERIODICAL, CLAYS AND CLAY MINERALS)
1953 2nd See Clay Minerals Conference. 132.
1955 4th See Clay Minerals Conference. 132.
1956 5th See Clay Minerals Conference. 132.
1957 6th See Clay Minerals Conference. 132.
1958 See Clay Minerals Conference. 132.
1959 8th See Clay Minerals Conference. 132.
1960 9th See Clay Minerals Conference. 132.
1962 11th See Clay Minerals Conference. 132.
1963 12th See Clay Minerals Conference. 132.

(420) NATIONAL EARTHQUAKE PREDICTION AND HAZARD PROGRAMS. PROCEEDINGS OF WORKSHOP.
1984 28th On the Borah Peak, Idaho, earthquake. (220)
    Fault scarps, landslides and other features associated with the Borah Peak earthquake of October 28, 1983, central Idaho; a field trip guide.

CPT	1984
CSt-ES	1984
IaU	1984

(421) NATIONAL PARKS AND MONUMENTS SERIES. [GUIDEBOOK]
1981 Arches, National Park; an illustrated guide and history.

AzTeS	1981
CMenUG	1981
CU-EART	1981
CoDGS	1981
CoPS	1981
CoPs	1981
DI-GS	1981
NRU	1981
NcU	1981
UU	1981

(422) NATIONAL RESEARCH COUNCIL. COMMITTEE ON CLAY MINERALS. GUIDEBOOK TO THE FIELD EXCURSION.
1958 See Clay Minerals Conference. 132.

(423) NATIONAL SCIENCE FOUNDATION. COLLEGE TEACHERS CONFERENCE IN GEOLOGY. FIELD GUIDEBOOK.
1959 Geologic trips along Oregon highways. (517)
  1. Corvallis to Newport via Depoe Bay.
  2. Eugene to Coos Bay via Reedsport.
  3. Corvallis to Prineville via Bend and Newberry Crater.
  4. Prineville to John Day via Mitchell.
  5. John Day to upper Bear Valley.
  6. Logdell to Pine Creek (Jurassic).
  7. Picture Gorge to Portland via Arlington.

AzU	1959
CLU-G/G	1959
CPT	1959
CSt-ES	1959
CU-EART	1959
CaACU	1959
CaBVaU	1959
CaOOG	1959
CaOTDM	1959
CoDCh	1959
CoG	1959
DI-GS	1959
DLC	1959
I-GS	1959
IEN	1959
IU	1959
InU	1959
MH-GS	1959
MnU	1959
MoSW	1959
MtBuM	1959
NdU	1959
NjP	1959
NvU	1959
OU	1959
OrU-S	1959
PBL	1959
SdRM	1959
TxHSD	1959
TxMM	1959

TxU	1959
UU	1959
WU	1959

(424) NATIONAL SCIENCE FOUNDATION. IN-SERVICE INSTITUTE. EARTH SCIENCE FIELD TRIP GUIDEBOOK.
1963 Apr. Southeastern Pennsylvania.
1978 A collection of field trips in central Pennsylvania prepared for the National Science Foundation Earth Science Teachers Workshop at Edinboro State College.
  1. Laurel Creek-Mifflin field trip.
  2. Allegheny Plateau field trip.
  3. Sinking Valley field trip.
  4. Jacks Mountain field trip.
  5. Gettysburg field trip.
  6. Picture taking flights over parts of the field trip routes & specific outcrops for birds eye view of structure, morphology, and surface mine operations.

PHarER-T	1963, 78

(425) NATIONAL SCIENCE FOUNDATION. SCIENCE TEACHERS INSTITUTE.
1958 See Kansas State Teachers College of Emporia. Science Teachers Institute. Title varies. 342.
1960 See Kansas State Teachers College of Emporia. Science Teachers Institute. Title varies. 342.
1961 See Kansas State Teachers College of Emporia. Science Teachers Institute. Title varies. 342.
1962 See Kansas State Teachers College of Emporia. Science Teachers Institute. Title varies. 342.

(426) NATIONAL SCIENCE FOUNDATION. SUMMER CONFERENCE (TITLE VARIES.)
1963 See Wayne State University. [Geology Department] Summer Conference. 754.
1964 See Wayne State University. [Geology Department] Summer Conference. 754.
1965 See Wayne State University. [Geology Department] Summer Conference. 754.
1968 See Wayne State University. [Geology Department] Summer Conference. 754.
1970 See Wayne State University. [Geology Department] Summer Conference. 754.
1973 See Wayne State University. [Geology Department] Summer Conference. 754.

(427) NATIONAL SCIENCE FOUNDATION. SUMMER CONFERENCE FOR COLLEGE TEACHERS. [GUIDEBOOK]
1962 1st Geology of the Lake Superior region. (379)
  [1] Keweenaw Peninsula field trip.
  [2] Lake Superior excursion.
1963 2nd Geology of the Lake Superior region. (379)
  [1] Lake Superior excursion.
  [2] Keweenaw Peninsula field trip.
1964 3rd Geology of the Lake Superior region. (379)
  [1] Keweenaw Peninsula field trip.
  [2] Lake Superior excursion.
1965 4th Geology of the Lake Superior region. (379)
  [1] Keweenaw Peninsula field trip.
  [2] Lake Superior excursion.

CaOTDM	1962, 64
CaOWtU	1965
DLC	1963
IU	1963
MiHM	1962-64
NSyU	1965
TxU	1965

(428) NATIONAL SCIENCE FOUNDATION. SUMMER CONFERENCE ON FIELD GEOLOGY FOR SECONDARY SCHOOL TEACHERS. FIELD TRIP GUIDEBOOK.
1971 Geology of western Montana. 2v. (701)
  1. SW Montana, SE Idaho and Yellowstone National Park.
  2. NW Montana and Glacier National Park.

CaACU	1971
DI-GS	1971
InLP	1971
MtBuM	1971

(429) NATIONAL SCIENCE TEACHERS ASSOCIATION. ANNUAL MEETING. FIELD GUIDE.
1975 Mar. Field guide for three trips in Southern California. (408, 412)
  1. San Andreas-San Jacinto faults.
  2. Collecting minerals from pegmatites in the Southern California Batholith.
  3. Ecology of a coastal community.
1976 Mar. Guidebook to the coastal zone and coastal plain of southern New Jersey. (408)

CLU-G/G	1975
CPT	1975
CSfCSM	1975
CU-A	1975
CU-EART	1975
DI-GS	1975-76
IU	1975
IaU	1975
InLP	1975
KyU-Ge	1975-76
NvU	1975
OkU	1976
ViBlbV	1975

(430) NATIONAL SCIENCE TEACHERS ASSOCIATION. SOUTHWESTERN REGIONAL MEETING. GUIDEBOOK TO FIELD TRIP.
1972 T-10. Some geologic hazards and environmental impact of development in the San Diego area. (Reprinted, 1973, under the auspices of the Far Western Section, National Association of Geology Teachers. [Also] reprinted 1980) (412)

CLU-G/G	1972

(431) NATIONAL SOCIETY OF PROFESSIONAL ENGINEERS (NSPE). PROFESSIONAL ENGINEERS OF COLORADO. UTE CHAPTER. [FIELD TRIP GUIDEBOOK]
1985 Apr. See Oil Shale Symposium. 501.

(432) NATIONAL SPELEOLOGICAL SOCIETY. CHATTANOOGA GROTTO.
1979 See National Speleological Society. Southeastern Regional Association. (S.E.R.A.) 437.

(433) NATIONAL SPELEOLOGICAL SOCIETY. MID-APPALACHIAN REGION.
1976 See Pennsylvania. Bureau of Topographic and Geologic Survey. 536.

(434) NATIONAL SPELEOLOGICAL SOCIETY. MISSISSIPPI VALLEY-OZARK REGIONAL ASSOCIATION. GUIDEBOOK.
1967 Apr. Guidebook to the geology between Springfield and Branson, Missouri, emphasizing stratigraphy and cavern development. (399)

CLU-G/G	1967
CPT	1967
CSt-ES	1967
CU-EART	1967
CaOOG	1967
CoDCh	1967
CoG	1967
CoU	1967
DI-GS	1967
DLC	1967
I-GS	1967
ICF	1967
ICIU-S	1967
ICarbS	1967
IEN	1967

(435) National Speleological Society. NSS Convention.

IU	1967
IaU	1967
InLP	1967
InRE	1967
InU	1967
KyU-Ge	1967
LU	1967
MNS	1967
MiDW-S	1967
MnU	1967
MoSW	1967
MoU	1967
MtBuM	1967
NSyU	1967
NbU	1967
NdU	1967
NhD-K	1967
NjP	1967
NvU	1967
OCU-Geo	1967
OU	1967
OkT	1967
OkU	1967
OrU-S	1967
PSt	1967
RPB	1967
SdRM	1967
TxDaDM	1967
TxHU	1967
TxLT	1967
TxU	1967
ViBlbV	1967
WU	1967

(435) NATIONAL SPELEOLOGICAL SOCIETY. NSS CONVENTION. GUIDEBOOK SERIES. (TITLE VARIES)

- 1960 1. Carlsbad Caverns National Park.
- 1961 2. Field trip. Caves in Tennessee, Alabama, Georgia.
- 1962 3. Caves of the Black Hills, North Dakota.
- 1963 4. Major caves in the vicinity of Mountain Lake, Virginia.
- 1964 5. Caves of Texas.
- 1965 6. 1965 convention, Bloomington, Indiana.
- 1966 7. Caves of the Sequoia region, California.
- 1967 8. Caves of Alabama.
- 1968 [9] See Missouri Speleology. 394.
- 1971 [12] See National Speleological Society. Virginia Region. Region Record. 440.
- 1972 [13] Selected caves of the Pacific Northwest; with particular reference to the volcano-speleology of the state of Washington.
- 1973 [14] [Caves of Indiana]
- 1974 [15] Upper Mississippi Valley cave region.
- 1975 [16] [Description of caves in the Great Basin area: California-Nevada-Oregon]
- 1976 [17] Morgantown, West Virginia.
- 1977 [18] The Lakeshore convention; official 1977 guidebook, Alpena, Michigan.
- 1978 19. An introduction to the caves of Texas.
  — [19A] Caves and karst hydrogeology of the southeastern Edwards Plateau, Texas: guidebook, geology field excursion.
- 1979 20. An introduction to caves of the northeast.
  — [20A] Karst hydrogeology and geomorphology of eastern New York; a guidebook to the geology field trip.
- 1980 21. An introduction to caves of Minnesota, Iowa, and Wisconsin.
- 1982 [22] An introduction to caves of the Bend area.
  — June Guidebook of the geology and biology field trip; caves and other volcanic landforms of central Oregon.
- 1983 23. An introduction to the caves of east-central West Virginia.
- 1984 [24] Guidebook of the 1984 NSS convention [Caves of the Wild West; Bighorn-Sheridan area].
- 1985 25. The caves of south eastern Kentucky.

AzU	1964
CMenUG	1960, 62, 66-67, 72-75
CSt-ES	1966-67, 72-74
CU-EART	1962, 66-67, 72-78, 80, 82-85
CU-SB	1982
CaOHM	1966
DI-GS	1960-67, 72-77, 78(19), 79(20), 80, 82(22), 83-84
DLC	1973
IU	1960, 62-63, 67, 76-77, 79(20), 80
IaU	1974-80
InLP	1965-67, 72-80, 83, 85
InU	1965
KyU-Ge	1967, 74-78, 79(20), 80, 83-85
MnU	1976-77, 80
NcU	1967
NhD-K	1976
OCU-Geo	1960, 72-73
TxCaW	1960
TxU	1964, 80
WU	1980, 82(June)

(436) NATIONAL SPELEOLOGICAL SOCIETY. OREGON GROTTO. THE SPELEOGRAPH.
- 1982 18. 11. Traversing Oregon and looping into Washington; post '82 convention trip.

DI-GS	1982

(437) NATIONAL SPELEOLOGICAL SOCIETY. SOUTHEASTERN REGIONAL ASSOCIATION. (S.E.R.A.) [GUIDEBOOK]
- 1979 Cave carnival at Ketner's Mill, Tennessee. (432)

InLP	1979

(438) NATIONAL SPELEOLOGICAL SOCIETY. SOUTHERN MISSISSIPPI GROTTO. [GUIDEBOOK]
- 1974 Caves of Mississippi.

DLC	1974
InLP	1974

(439) NATIONAL SPELEOLOGICAL SOCIETY. VIRGINIA REGION.
- 1964 Caves of Virginia.

DLC	1964
MdBJ	1964
NcU	1964

(440) NATIONAL SPELEOLOGICAL SOCIETY. VIRGINIA REGION. REGION RECORD.
- 1971 V. 1. No. 4. 30th anniversary guidebook. (435)
  Section 3. Convention special tours. (435)
  [1] Karstlands excursion to southwest Virginia.
  [2] Geology field trip.
  [3] Biological tour of Greenbrier Caverns.
  [4] Blacksburg area scenic diversions.

CU-EART	1971
DI-GS	1971

(441) NATO ADVANCED STUDIES INSTITUTE ON METALLOGENY AND PLATE TECTONICS. [GUIDEBOOK]
- 1974 May Plate tectonic setting of Newfoundland mineral occurrences.
  (Cover title: Metallogeny and plate tectonics)

AzTeS	1974
CSt-ES	1974
CU-EART	1974
CU-SB	1974
CaAEU	1974
CaBVaU	1974
CaOWtU	1974
CoG	1974
DLC	1974
IU	1974
InLP	1974
MiHM	1974
MnU	1974
NhD-K	1974
NvU	1974
WU	1974

(442) NATO ADVANCED STUDY INSTITUTE.
1982 Aug. See International Geological Correlation Programme (IGCP). Project 27: The Caledonide Orogen. 302.

(443) NATURAL AREAS CONFERENCE. FIELD TRIP.
1974 May Appalachian plateaus. (763)
TxBeaL 1974

(444) NATURE CONSERVANCY. MAINE CHAPTER.
1977 Guidebook to geologic and beach features of the Rachel Carson Salt Pond area, New Harbor, Maine. (369)
IaU 1977

NBQUA See NEW BRUNSWICK QUATERNARY ASSOCIATION. (455)

(445) NEBRASKA ACADEMY OF SCIENCE. [GUIDEBOOK FOR FIELD CONFERENCE]
1965 See International Association for Quaternary Research (INQUA). 277.

(446) NEBRASKA GEOLOGICAL SOCIETY. FIELD CONFERENCE GUIDEBOOK.
1970 2nd Re-exploring the Missouri; an excursion aboard Corps of Engineers inspection boat, "The Sergeant Floyd", from Omaha to Rulo, Nebraska.
1974 Mineral aggregate industries in southeast Nebraska.
DI-GS 1970
NbU 1974
NhD-K 1974

(447) NEBRASKA GEOLOGICAL SURVEY. EDUCATIONAL CIRCULAR.
1979 EC-3 Geologic history of Scotts Bluff National Monument.
IaU 1979

(448) NEBRASKA GEOLOGICAL SURVEY. FIELD GUIDE.
1969 FG-1. Otoe County - Unadilla.
— FG-2. Cass County - Weeping Water.
— FG-3. Sarpy County - Gretna State Fish Hatchery area.
— FG-4. Gage County - Odell - Krider area.
— FG-5. Thayer County - Alexandria and Gilead areas.
— FG-6. Pawnee County - Table Rock area.
— FG-7. Jefferson County - Fairbury area.
CPT 1969
InLP 1969
MtBuM 1969(FG-1 thru FG-7)
WU 1969(FG-3)

(449) NEBRASKA GEOLOGICAL SURVEY. GUIDEBOOKS.
1966 See Friends of the Pleistocene. Midwest Group. 183.
1967 See Association of American State Geologists. 74.
1970 Guidebook to the geology along the Missouri River bluffs of southeastern Nebraska and adjacent areas.
1971 See Geological Society of America. North Central Section. 200.
CPT 1970
CSt-ES 1970
CaACU 1970
CaAEU 1970
CaOWtU 1970
CoDCh 1970
CoU 1970
DI-GS 1970
I-GS 1970
InLP 1970
InU 1970
KLG 1970
KyU-Ge 1970
MiKW 1970
MnU 1970
MoSW 1970
MtBuM 1970
NdU 1970
NvU 1970
OCU-Geo 1970
OU 1970
OkU 1970
TxU 1970
UU 1970
WU 1970

(450) NEBRASKA STATE MUSEUM.
1941 See Society of Vertebrate Paleontology. 604.

NEBRASKA. UNIVERSITY. See UNIVERSITY OF NEBRASKA--LINCOLN. DEPARTMENT OF GEOLOGY. (702)

(451) NEVADA BUREAU OF MINES AND GEOLOGY. OPEN-FILE REPORT.
1984 See Society of Economic Geologists. 583.

(452) NEVADA BUREAU OF MINES AND GEOLOGY. REPORT.
1974 No. 19. See Geological Society of America. Cordilleran Section. 199.
— No. 20. See Geological Society of America. Cordilleran Section. 199.
1975 No. 22. See Geological Society of America. Cordilleran Section. 199.
1976 No. 26. See American Institute of Mining, Metallurgical and Petroleum Engineers. Annual Meeting. 41.
1978 No. 32. See International Association on the Genesis of Ore Deposits. 280.

(453) NEVADA GEOLOGICAL SOCIETY.
1962 See Geological Society of Sacramento. 215. and Geological Society of Nevada. 211.

(454) NEVADA. UNIVERSITY. MACKAY SCHOOL OF MINES. DEPARTMENT OF GEOLOGY AND GEOGRAPHY.
1984 See Geological Society of America. 197.

(455) NEW BRUNSWICK QUATERNARY ASSOCIATION. NBQUA FIELD TRIP GUIDEBOOK.
1982 Aug. Quaternary studies in the upper St. John River Basin: Maine and New Brunswick.
CaOOG 1982

NEW ENGLAND GEOLOGICAL ASSOCIATION. See NEW ENGLAND INTERCOLLEGIATE GEOLOGICAL CONFERENCE. (456)

(456) NEW ENGLAND INTERCOLLEGIATE GEOLOGICAL CONFERENCE. ANNUAL MEETING GUIDEBOOK.
1937 33rd New York to Bear Mt. Park.
1938 34th Central Vermont marble belt.
1947 Rhode Island, glacial and shoreland features.
1948 Trip no. 1. Roxbury and Waterbury areas, Vermont.
Trip no. 2. Bedrock geology of the Burlington area.
Trip no. 3. Glacial geology.
1951 Purgatory Chasm in Sutton, Massachusetts.
1952 Geology of the Bennington Quadrangle [and four other trips].
1953 Triassic sedimentary rocks of central Connecticut, their petrology, petrography, stratigraphy and structure.
Trip A. From Summit Street, Hartford...to Durham and Highway 77.
Trip B. Hartland Formation and Nonewaug Granite trip.
Trip C. Surficial geology of the Hartford, Connecticut area.
Trip D. Crystallines of the eastern highlands quadrangles, Manchester and Rockville.
Trip E. Problems of the crystalline rocks west of New Haven.
1954 46th Glacial geology of the Hanover region [and five other trips].
1957 49th Geology of northern part; Connecticut Valley.
1959 51st Stratigraphy and structure of west-central Vermont and adjacent New York.
1960 52nd West-central Maine.
1961 53rd Vermont geologic map centennial, Montpelier.
1962 54th Area around Montreal.
1963 55th Geology of Rhode Island.

1964 56th Boston area and vicinity.
1965 57th Field trips in southern Maine.
1966 58th Mt. Katahdin region, Maine.
1967 59th Connecticut Valley of Massachusetts.
1968 60th Guidebook for field trips in Connecticut. (632)
Trip A-1. Post-glacial geology of the Connecticut shoreline.
Trip B-1. Two-till problem in Naugatuck-Torrington area, western Connecticut.
Trip B-2. Periglacial features and pre-Wisconsin weathered rock in the Oxford-Waterbury-Thomaston area, western Connecticut.
Trip B-3. Hydrogeology of southwestern Connecticut.
Trip B-4. Engineering geology as applied to highway construction.
Trip C-1. Sedimentology of Triassic rocks in the lower Connecticut Valley.
Trip C-2. General geology of the Triassic rocks of central and southern Connecticut.
Trip C-3. Geology of Dinosaur Park, Rocky Hill, Connecticut.
Trip C-4. Stratigraphy and structure of the Triassic strata of the Gaillard Graben, south-central Connecticut.
Trip C-5. Late Triassic volcanism in the Connecticut Valley and related structure.
Bedrock geology of western Connecticut.
Trip D-1. Progressive metamorphism of pelitic, carbonate, and basic rocks in south-central Connecticut.
Trip D-2. Multiple folding in western Connecticut; a reinterpretation of structure in the Naugatuck-New Haven-Westport area.
Trip D-4. Metamorphic geology of the Collinsville area.
Trip D-5. The bedrock geology of the Waterbury and Thomaston quadrangles.
Trip D-6. Geology of the Glenville area, southwesternmost Connecticut and southeastern New York.
Trip E-1. Animal-sediment relationships and early sediment diagenesis in Long Island Sound.
Bedrock geology of eastern Connecticut.
Trip F-1. The Honey Hill and Lake Char faults.
Trip F-3. Stratigraphy and structure of the metamorphic rocks of the Stony Creek Antiform (a "folded fold") and related structural features, southwestern side of the Killingworth Dome.
Trip F-4. A structural and stratigraphic cross-section traverse across eastern Connecticut.
Trip F-5. The Brimfield(?) and Paxton(?) formations in northeastern Connecticut.
Trip F-6. Mineral deposits of the central Connecticut pegmatite district.
1969 61st New York, Massachusetts and Vermont.
1970 62nd Guidebook for field trips in the Rangeley Lakes-Dead River basin, western Maine.
1971 63rd Guidebook for field trips in central New Hampshire and contiguous areas.
Trip A-1. Glacial features of the Winnipesaukee Wolfeboro area.
Trip A-2. Peterborough Quadrangle.
Trip A-3. The Cardigan Pluton of the Kinsman quartz monzonite.
Trip A-4. Geology of the Macoma mantled gneiss dome near Hanover, New Hampshire.
Trip A-5. Southwest side of the Ossipee Mountains, New Hampshire.
Trip A-6. Recumbent and reclined folds of the Mt. Cube area, New Hampshire-Vermont.
Trip A-7. The Hillsboro plutonic series in southeastern New Hampshire; field criteria in support of a partial melting petrogenetic model.
Trip A-8. Bedrock geology of the Ossipee Lake area.
Trip B-1. Jackson estuarine laboratory; sedimentation on Great Bay estuarine system; solid waste disposal in Gulf of Maine.
Guidebook supplement. Origin and distribution of gravel on Broad Cave Beach, Appledore Island, Maine.
Trip B-2. Geology of the Holderness Quadrangle.
Trip B-3. Geologic review of the Belknap Mountain complex.
Trip B-4. Surficial geology of the Merrimack River valley between Manchester and Nashua, New Hampshire.
Trip B-5. Igneous rocks of the Seabrook, New Hampshire-Newbury, Massachusetts area.
Trip B-6. Geology of the Concord Quadrangle.
1972 64th Guidebook for field trips in Vermont.

Bedrock geology trips.
Trip B-1. Stratigraphy of the east flank of the Green Mountain Anticlinorium, southern Vermont.
Trip B-2. Major structural features of the taconic allochthon in the Hoosick Falls area, New York-Vermont.
Trip B-3. Excursions at the north end of the taconic allochthon and the Middlebury Synclinorium, west-central Vermont, with emphasis on the structure of the Sudbury nappe and associated parautochthonous elements.
Trip B-4. The Champlain Thrust and related features near Middlebury, Vermont.
Trip B-5. Analysis and chronology of structures along the Champlain Thrust west of the Hinesburg Synclinorium.
Trip B-6. Sedimentary characteristics and tectonic deformation of Middle and Upper Ordovician shales of northwestern Vermont north of Mallets Bay.
Trip B-7. Rotated garnets and tectonism in Southeast Vermont.
Trip B-8. Stratigraphic and structural relationships across the Green Mountain Anticlinorium in north-central Vermont.
Trip B-9. Superposed folds and structural chronology along the southeastern part of the Hinesburg Synclinorium.
Trip B-10. Lower Paleozoic rocks flanking the Green Mountain Anticlinorium.
Trip B-11. Geology of the Guilford Dome area, southeastern Vermont.
Trip B-12. Stratigraphic and structural problems of the southern part of the Green Mountain Anticlinorium, Bennington-Wilmington, Vermont.
Trip B-13. Polymetamorphism in the Richmond area, Vermont.
Environmental geology trips.
Trip EG-1. Mount Mansfield trail erosion.
Trip EG-2. Feasibility and design studies; Champlain Valley sanitary landfill.
Glacial geology trips.
Trip G-1. Glacial history of central Vermont.
Trip G-2. Ice margins and water levels in northwestern Vermont.
Proglacial lakes in the Lamoille Valley, Vermont.
Trip G-3. Strandline features and late Pleistocene chronology of Northwest Vermont.
Trip G-5. Till studies, Shelburne, Vermont.
Trip G-6. Woodfordian glacial history of the Champlain Lowland, Burlington to Brandon, Vermont.
Lake studies trips.
Trip LS-1. The sludge bed at Fort Ticonderoga, New York.
Trip LS-2, LS-3. Sedimentological and limnological studies of Lake Champlain.
Paleontology trips.
Trip P-1. Ordovician paleontology and stratigraphy of the Champlain Islands.
Paleontology and stratigraphy of the Chazy Group (Middle Ordovician), Champlain Islands, Vermont.
Paleoecology of Chazy reef-mounds.
Trip P-2. Cambrian fossil localities in northwestern Vermont.
1973 Geology of New Brunswick, field guide to excursions.
Trip A-1. Zeolite mineral assemblage, Grand Manan Island, New Brunswick.
Trip A-2. The Variscan front in southern New Brunswick.
Trip A-3, B-6. The granitic rocks of southwestern New Brunswick.
Trip A-4. Carboniferous stratigraphy and sedimentology of the Chignecto Bay area, southern New Brunswick.
Trip A-5. Structural geology of the Bathurst-Newcastle District.
Trip A-6. The Bathurst mining camp.
Trip A-7. Pointe Verte to tide head, Chaleur Bay area, New Brunswick.
Trip A-8, B-7. Acadian orogeny in coastal southern New Brunswick.
Trip A-9. Tectonic evolution and mineral deposits of the northern Appalachians in southern New Brunswick.
Trip A-10, B-8. Post-Carboniferous and post-Triassic structures in southern New Brunswick.
Trip A-11. Minto coal fields.
Trip A-12. Molybdenum, tungsten, and bismuth mineralization at Brunswick Tin Mines, Ltd.
Trip A-13. Saint John area.
Trip A-14, B-11. The Harvey volcanic area.
Trip B-1. Tungsten mineralization at Burnt Hill tungsten mine.

Trip B-2. Silurian rocks of the Fredericton area.
Trip B-4. Vertebrate sites of northern New Brunswick and the Gaspe.
1974 66th Geology of east-central and north-central Maine.
Trip A-1. Metamorphism in the Belfast area, Maine.
Trip A-2. Recession of the late Wisconsin Laurentide ice sheet in eastern Maine.
Trip A-3. Sedimentary and slump structures of central Maine.
Trip A-4. The geology of the Camden-Rockland area.
Trip A-5. General bedrock geology of northeastern Maine.
Trip A-6. Precambrian rocks of Seven Hundred Acre Island and development of cleavage in the Islesboro Formation.
Trip A-7. Igneous petrology of some plutons in the northern part of the Penobscott Bay area.
Trip A-8. The paleontology of the present; littoral environments on a submerged crystalline coast, Gouldsboro, Maine.
Trip B-1. Late Wisconsin and Holocene geological, botanical, and archaeological history of the Orono, Maine region.
Trip B-2. Bedrock geology of Mount Desert Island.
Trip B-3. Stratigraphy and structure of central Maine.
Trip B-4. Economic deposits at Blue Hill.
Trip B-5. The concentrically zoned Tunk Lake Pluton; Devonian melting-anomaly activity?
Trip B-6. Buchan-type metamorphism of the Waterville Pelite, south-central Maine.
Trip B-7. Bedrock geology of northern Penobscott Bay area.
Trip B-8. The paleontology of the present: littoral environments on a submerged crystalline coast, Gouldsboro, Maine.
1975 67th Guidebook for field trips in western Massachusetts, northern Connecticut and adjacent areas of New York.
Trip A-1. Some basement rocks from Bear Mountain to the Housatonic Highlands.
Trip A-2. The Hudson Estuary.
Trip A-3. Structural and stratigraphic chronology of the Taconide and Acadian polydeformational belt of the central Taconics of New York State and Massachusetts.
Trip B-1. Selected localities in the Taconics and the implications for the place tectonic origin of the Taconic region.
Trip B-2. Fold-thrust tectonism in the southern Berkshire Massif, Connecticut and Massachusetts.
Trip B-3. Proposed Silurian-Devonian correlations east of the Berkshire Massif in western Massachusetts and Connecticut.
Trip B-4. Stratigraphic and structural relationships along the east side of the Berkshire Massif, Massachusetts.
Trip B-5. The Cambrian-Precambrian contact in northwestern Connecticut and west-central Massachusetts.
Trip B-6. Cross section of the Berkshire Massif at 42#DGN; profile of a basement reactivation zone.
Trip B-7. The late Quaternary geology of the Housatonic River basin in southwestern Massachusetts and adjacent Connecticut.
Trip B-8. The glacial geology of the Housatonic River region in northwestern Connecticut.
Trip B-9. Basement-cover rock relationships in the Pittsfield East Quadrangle, Massachusetts.
Trip B-10. General geology of the Stockbridge Valley marble belt.
Trip C-1. Repeat of B-1.
Trip C-2. Repeat of B-2.
Trip C-4. Repeat of B-4.
Trip C-5. Repeat of B-5.
Trip C-6. Repeat of B-6.
Trip C-7. Repeat of B-7.
Trip C-8. Boulder trains in western Massachusetts.
Trip C-9. Repeat of A-1.
Trip C-10. Stratigraphy and structural geology in the Amenia-Pawling Valley, Dutchess County, New York.
Trip C-11. Polyphase deformation in the metamorphosed Paleozoic rocks east of the Berkshire Massif.
1976 68th Geology of southeastern New England; a guidebook for field trips to the Boston area and vicinity.
Field trip: Boston Basin bedrock geology.
Trip no. A-1, B-1. The Boston Bay Group: The boulder bed problem.

Trip no. A-2, B-2. Geology of the Squantum 'Tillite.'
Trip no. F-1. Geology of Squaw Head, Squantum, Massachusetts.
Trip no. A-5. New evidence for glaciation during deposition of the Boston Bay Group.
Trip no. A-3, B-3. The Blue Hills igneous complex, Boston area, Massachusetts.
Trip no. A-4, B-4. Geologic relationships of the southern portion of the Boston Basin from the Blue Hills eastward.
Field trip: Surficial and Pleistocene geology.
Trip no. A-6, B-6. Surficial geology of Boston Basin.
Trip no. B-7. The effects of geology, topography, and disturbance on the soils and vegetation of the Middlesex Fells, Massachusetts.
Trip no. F-5, B-5. Stratigraphy and field analysis of glacial deposits near Worcester, Massachusetts.
Trip no. A-7. Glacial geology of southeastern Massachusetts.
Field trip: Coastal geology.
Trip no. A-8, B-8. Sedimentary and geomorphic origin and development of Plum Island, Massachusetts: An example of a barrier island system.
Trip no. A-9, B-9. Dynamics of sedimentation and coastal geology from Boston to Plymouth.
Trip no. A-10, B-10. Coastal geology and geomorphology of Cape Cod - an aerial and ground view.
Field trip: bedrock geology north, west and south of Boston.
North of Boston.
Trip no. A-11, B-11. Plutonic series in the Cape Ann area.
Trip no. B-12. Granite at Marblehead - igneous or metamorphic.
Trip no. A-17. Pre-Silurian stratified rocks southeast of the Bloody Bluff Fault.
Trip no. A-16, B-16. Stratigraphy and structural setting of the Newbury volcanic complex, northeastern Massachusetts.
West of Boston.
Trip no. F-3, A-13. Faults and related deformation in the Clinton-Newbury-Bloody Bluff fault complex of eastern Massachusetts.
Trip no. A-14, B-14. The Pre-Silurian Eugeosynclinal sequence bounded by the Bloody Bluff and Clinton-Newbury faults, Concord, Billerica, and Westford quadrangles, Massachusetts.
Trip no. A-15, B-15. Cataclastic and Plutonic rocks within and west of the Clinton-Newbury fault zone, east-central Massachusetts.
Trip no. F-4. Geologic setting of the Harvard Conglomerate, Harvard Massachusetts.
Trip no. B-13. Stratigraphy of the Webster-Worcester region, Massachusetts.
Trip no. F-2. Lower paleozoic rocks west of the Clinton-Newbury fault zone, Worcester area, Massachusetts.
Trip no. A-12. Pennsylvanian rocks of east-central Massachusetts.
South of Boston.
Trip no. B-18. Coal stratigraphy and flora of the northwestern Narragansett Basin.
Trip no. B-17. Pre-Pennsylvanian rocks of Aquidneck and Conanicut Islands, Rhode Island.
Trip no. F-6. Alleghenian deformation, sedimentation, and metamorphism in southeastern Massachusetts and Rhode Island.
Trip no. A-18. Mechanisms of Alleghenian deformation in the Pennsylvanian of Rhode Island.
Field trip: Environmental, engineering, and economic geology.
Trip no. B-19. The effect of urbanization on water quality.
Trip no. A-19. Engineering geology of the Charles River.
Trip no. F-7. Aspects of the economic geology of southeastern Massachusetts.
1977 69th Guidebook for field trips in the Quebec city area. (Cover title: Guide book of excursions in the province of Quebec)
1978 70th Guidebook for field trips in southeastern Maine and southwestern New Brunswick.
Trip no. A-l. The Silurian-Lower Devonian marine volcanic rocks of the Eastport Quadrangle, Maine.
Trip no. A-2. Geology of the Red Beach Granite.
Trip no. A-3. Geology of the Lower Devonian rocks of Passamaquoddy Bay, Southwest New Brunswick.
Trip no. A-4. Folding and cleavage in Triassic rocks of Southwest New Brunswick.

Trip no. A-5. Stratigraphy, structure, and progressive metamorphism of Lower Paleozoic rocks in the Calais area, southeastern Maine.
Trip no. A-6. Deglacial events in southeastern Maine.
Trip no. A-7. Precambrian basement in southwestern New Brunswick.
Trip no. A-8. Bedrock geology of the Wesley Quadrangle.
Trip no. B-1. Repeat of A-l.
Trip no. B-2. Field guide to Lower Paleozoic sedimentary and volcanic rocks of southwestern New Brunswick.
Trip no. B-3. Stratigraphy and structure of Silurian and pre-Silurian rocks in the Brookton-Princeton area, eastern Maine.
Trip no. B-4. The post-Acadian instrusives of southwestern New Brunswick.
Trip no. B-5. Stratigraphy and structure of the Silurian and Lower Devonian rocks of the St. George-Mascarene area, New Brunswick.
Trip no. B-6. The carboniferous deformed rocks west of St. John, New Brunswick.
Trip no. B-7. Multiple origins of the redbeds in the Eastport Formation (Devonian), Eastport Quadrangle, Maine.
Trip no. B-8. Repeat of A-8.
Trip no. B-9. The nature of the Early Ordovician/Late Silurian contact on Cookson Island, Oak Bay, New Brunswick.
Trip no. B-10. Copper mineralization in the coastal volcanic belt.

1979 71st Guidebook for field trips: Troy, New York. [Preface and acknowledgements page derived title: Classical sites of eastern New York state and adjoining New England] (469)
Trip no. A-1. Early and medial Devonian stratigraphy and paleoenvironments in east-central New York.
Trip no. A-2. Sedimentary environments and their products: shelf, slope and rise of proto-Atlantic (Iapetus) Ocean, Cambrian and Ordovician periods, eastern New York State.
Trip no. A-3. Sedimentary environments in glacial Lake Albany and its successors on the Albany, Delmar, Niskayuna, and Voorheesville 7 1/2 minute quadrangles.
Trip no. A-4. Structural framework of the southern Adirondacks.
Trip no. A-5. Microstructure of a Vermont slate, an Adirondack gneiss, and some laboratory specimens.
Trip no. A-6. Studies of cleavage and strain in the shaly rocks of the Cossayuna-Salem area, Washington County, New York.
Trip no. A-7. Thrust sheets of the central taconic region.
Trip no. A-8. Detailed stratigraphic and structural features of the Giddings Brook slice of the taconic allochthon in the Granville area.
Trip no. A-9. Sedimentology of a transgressive clastic wedge within the Marcellus Formation (Middle Devonian) in southeastern New York.
Trip no. A-10. Faults, stratigraphy, and the mineral waters of Saratoga: implications for Neogene rifting.
Trip no. B-1. Economic geology of the Hudson River Valley.
Trip nos. B-2, B-4 were cancelled.
Trip no. B-5a. Geology in state service.
Trip no. B-5b. The building stones of the Nelson A. Rockefeller Empire State Plaza.
Trip no. B-6. Stratigraphy of glacial Lakes Albany, Quaker Springs, Coveville, and relationships to late Woodfordian Mohawk and Hoosick River discharge history.
Trip no. B-7. Stratigraphy and depositional history of the Onondaga Limestone in eastern New York.
Trip no. B-8. Repeat of A-4.
Trip no. B-9. Repeat of A-5.
Trip no. B-10. Repeat of A-7.
Trip no. B-11. Recent structural investigations in Northwest Massachusetts for the new bedrock geologic map of Massachusetts.
Trip no. B-12. Precambrian structure and stratigraphy of the southeastern Adirondack uplands.
Trip no. B-13. Late Wisconsinan - recent geology of the lower Rondout Creek Valley, Ulster County, southeastern New York.
Trip no. B-14. Repeat of A-10.

1980 72nd The geology of northeastern Maine and neighboring New Brunswick. (100, 486, 692, 367)
Trip A-1. Geology of the Bottle Lake complex, Maine.
Trips A-2 and B-1. Geology and petrology of igneous bodies within the Katahdin Pluton.
Trips A-3 and B-2. Alpine glaciation of Mt. Katahdin.
Trip B-3. The Traveler rhyolite and its Devonian setting, Traveler Mountain area, Maine.
Trip B-4. The core of the Weeksboro-Lunksoos Lake Anticline, and the Ordovician, Silurian, and Devonian rocks on its northwest flank.
Trip B-5. Plant fossils (psilophytes) from the Devonian rocks on its northwest flank.
Trip B-6. Ordovician and Silurian stratigraphy of the Ashland Synclinorium and adjacent terraine.
Trip B-7 and C-2. Bedrock geology of the Presque Isle area.
Trip B-8. Wisconsinan glaciation of northern Aroostook County, Maine.
Trip B-9. Deglaciation of the Edmundston area and reappraisal of glacial Lake Madawaska interpretation.
Trip B-10. The geology and deformation history of the southern part of the Matapedia Zone and its relationship fo the Miramichi Zone and Canterbury Basin, N. Rast.
Trip C-1. Stratigraphy and sedimentology of the Siegas Formation (Early Llandovery) of northwestern New Brunswick.
Trip C-3. Biostratigraphic trip across northeastern Maine.
Trip C-4. Late-glacial and Holocene geology of the middle St. John River Valley.
Trip C-5. Sedimentology of Silurian flysch, Ashland Synclinorium.
Trip C-6. Wisconsinan glaciation of eastern Aroostook County, Maine.
Trip C-7. Structure and sedimentology of Siluro-Devonian between Edmunston and Grand Falls, New Brunswick.
Trip D-1. Stratigraphic and structural relations in the turbidite sequence of south-central Maine.

1981 73rd Guidebook to geologic field studies in Rhode Island and adjacent areas.
Trip no. A-1. Distribution and structural significance of the Oakdale Formation in northeastern Connecticut.
Trip no. A-2 & B-11. Pleistocene geology of Block Island.
Trip no. B-1. The diagenetic to metamorphic transition in the Narragansett and Norfolk basins, Massachusetts and Rhode Island.
Trip no. B-2. The geology of Precambrian rocks of Newport and Middletown, Rhode Island.
Trip no. B-3. The Blackstone series: evidence for an Avalonian plate margin in northern Rhode Island.
Trip no. B-4. Igneous rocks of northern Rhode Island.
Trip no. B-5. Contact relationships of the late Paleozoic Narragansett pier granite and country rock.
Trip no. B-6. Field guide to coastal environmental geology of Rhode Island's barrier beach coastline.
Trip no. B-7. The geologic setting of coal and carbonaceous material, Narragansett Basin, southeastern New England.
Trip no. B-8. Selected mineral collecting sites in northeastern Rhode Island.
Trip no. B-9. Sedimentation in microtidal coastal lagoons, southern Rhode Island.
Trip no. B-10 & C-9. Glacial geology in southern Rhode Island.
Trip no. C-1. The geology of Cambrian rocks of Conanicut Island, Jamestown, Rhode Island.
Trip no. C-2. Alleghenian deformation and metamorphism of southern Narrangansett Basin.
Trip no. C-3. Mafic dikes of northeastern Massachusetts.
Trip no. C-4. Felsic volcanic units in the Boston area, Massachusetts.
Trip no. C-5. Zircon geochronology and petrology of plutonic rocks in Rhode Island.
Trip no. C-6. The Boston Bay Group, Quincy, Massachusetts.
Trip no. C-7. Coastal zone management problems: RI coastal lagoons and barriers.
Trip no. C-8. Interpretation of primary sedimentary structures.

1982 74th Storrs, Connecticut.
Guidebook for fieldtrips in Connecticut and south central Massachusetts. (632)
Q1. The surficial geologic maps of Connecticut illustrated by a field trip in central Connecticut.
Q2. Anatomy of the Chicopee readvance, Massachusetts.
Q3. Mode of deglaciation of Shetucket River Basin.
Q4. Sedimentation in a proglacial lake: glacial Lake Hitchcock.

M1. Jurassic redbeds of the Connecticut Valley: (1) Brownstones of the Portland Formation; (2) Playa-Playa Lake-Oligomictic Lake model for parts of the East Berlin, Shuttle Meadow and Portland formations.
M2. Paleontology of the Mesozoic rocks of the Connecticut Valley.
M3. Mesozoic volcanism in north central Connecticut.
M4. Copper occurrences in the Hartford Basin of northern Connecticut.
P1. An investigation of the stratigraphy and tectonics of the Kent area, western Connecticut.
P1A. Chronology of metamorphism in western Connecticut: Rb-Sr ages.
P2. The Bonemill Brook fault zone, eastern Connecticut.
P3. High grade Acadian regional metamorphism in south-central Massachusetts.
P4. Stratigraphy and structure of the Ware-Barre area, central Massachusetts.
P5. Lake Char fault in the Webster, Massachusetts area: evidence for west-down motion.
P6. Structural relations at the junction of the Merrimack Province, Nashoba thrust belt and the Southeast New England platform in the Webster-Oxford area, Massachusetts, Connecticut, and Rhode Island.
P7. The structural geology of the Moodus seismic area, south-central Connecticut.
P8. Multi-stage deformation of the Preston Gabbro, eastern Connecticut.
P9. Structure and petrology of the Willimantic Dome and the Willimantic Fault, eastern Connecticut.
1983 75th The Greenville-Millinocket regions, north central Maine.
Trip A-1. Glacial traverse across the southern side of an ice cap.
Trip B-1. The Hurricane Mountain Formation melange and unconformably overlying Lower to Middle Ordovician volcanics, Brassua Lake and Moosehead Lake quadrangles.
Trip B-2. Geology of rocks along the east branch Pleasant River, White Cap Range, central Maine.
Trip B-3. Features of Upper Silurian and Lower Devonian sedimentary rock in the Caucomgomoc Lake area, northwestern Maine.
Trip B-4. The timing of the alpine glaciation of Mt. Katahdin.
Trip B-5. The northwest boundary fault of the Boundary Mountain anticlinorium.
Trip C-1. Petrology of the central portion of the Moxie pluton.
Trip C-2. Stratigraphy, metamorphism and geomorphology in the Greenville-Rockwood area, Maine.
Trip C-3. Stratigraphy and sedimentation in Silurian flysch east of Millinocket, Maine.
Trip C-4. The Seboomook Lake area became ice free; but how?
Trip C-5. Ophiolite and melange terrane, Caucomgomoc Lake area, northwestern Maine.
Trip D-1. Repeat of C-1.
Trip D-2. Repeat of C-2.
Trip D-3. Multiple-till localities in central Maine.
1984 76th Geology of the coastal lowlands, Boston to Kennebunk, Maine. (Cover title: Geology of the coastal lowlands, Boston, MA to Kennebunk, ME) (76)
Trip A1. Coastal geology of Winthrop and Yirrell beaches, Winthrop, MA and Thompson Island, Boston Harbor.
Trip A2. Engineering geology of the Boston Basin.
Trip A3. Hazardous waste problem sites.
Trip A4. Sedimentology and multiple deformation of the Kittery Formation in southwestern Maine and southeastern New Hampshire.
Trip A5. Igneous rocks of the Nashoba Block, eastern Massachusetts.
Trip A6. Mafic dikes from Boston to Cape Ann.
Trip B1. A geologic traverse across the Nashoba Block, Eastern MA.
Trip B2. Boston Basin restudied.
Trip B3. A Precambrian continent margin sequence (slope deposits & olistostromes): Boston north quadrangle, MA.
Trip B4. Ductile and brittle structures within the Rye Formation, of coastal Maine and New Hampshire.
Trip B5. Geologic framework of the Massabesic Anticlinorium and the Merrimack Trough, southeastern New Hampshire.
Trip B6. The geology of the Saddleback Mountain area, Northwood Quadrangle, southeastern New Hampshire.
Trip B7. Mount Pawtuckaway ring-dike complex.
Trip C1. Cambrian rocks of East Point, Nahant.
Trip C2. The Marlboro Formation in its type area and associated rocks just west of the Bloody Bluff fault zone, Marlborough area, Massachusetts.
Trip C3. Geology, petrology and origin of the Precambrian igneous rocks located in the area north of Boston.
Trip C4. The Bloody Bluff fault system.
Trip C5. Silurian and Devonian rocks in the Alton and Berwick quadrangles, New Hampshire and Maine.
Trip A7. Glaciomarine sediments and facies associations, southern York County, Maine.
Trip A8. Coastal geomorphology of Laudholm, Drakes's Island, Wells Inlet and Wells Beach, southeastern Maine.
Trip B8. Branch Brook in York County Maine: formation and maintenance of a drainage network by ground-water sapping.
Trip B9. Glacial lake history of the Merrimack Valley, southern New Hampshire.
Trip C7. A trip down the Alton Bay flow line.
Trip C8. Surficial geology and archaeology on Thompson Island, Boston Harbor, Massachusetts.
Trip C9. Sedimentary environments of the Plum Island and Parker River area, Newburyport, MA.
Trip C10. Rock lithology and glacial transport southeast of Boston.
1985 77th No. 6. Guidebook for fieldtrips in Connecticut and adjacent areas of New York and Rhode Island. (632(No.6))
A1. The geology of the Waterbury Dome.
A2. Ordovician ductile deformation zones in the Hudson highlands and their relationship to metamorphic zonation in the cover rocks of Dutchess County.
A3. Age and structural relations of granites, Stony Creek area, Connecticut.
A4. Bedrock geology of the Deep River area, Connecticut.
A5. Recessional moraines, southeastern Connecticut.
A6. Mesoscopic and microscopic structure of the Lake Char - Honey Hill mylonite zone.
B1. Geology of southern Connecticut, east-west transect.
B2. Stratigraphy and structural geology in the Bethel area southwestern Connecticut.
B3. The timing and nature of the Paleozoic deformation in the northern part of the Manhattan prong, southeast New York.
B4. The Hope Valley shear zone - a major late Paleozoic ductile shear zone in southeastern New England.
B5. The middle Haddam area, Connecticut, revisited.
B6. Deglaciation of the Middletown basin and the question of the Middletown readvance.
B7. The sedimentology, stratigraphy and paleontology of the lower Jurassic Portland formation, Hartford basin, central Connecticut.
C1. Geology of southern Connecticut, north south transect.
C2. Geology in the vicinity of the Hodges complex and the Tyler Lake granite, West Torrington, Connecticut.
C3. Geology of the Mt. Prospect region, western Connecticut.
C4. Honey Hill fault and Hunts Brook syncline.
C5. Pegmatites of the Middletown district, Connecticut.
C6. Late Quaternary deposits of the southern Quinnipiac-Farmington lowland and Long Island Sound basin: their place in a regional stratigraphic framework.
C7. Morphology of coastal marshes, southern Connecticut.

AzTeS	1965
AzU	1976
CLU-G/G	1959, 61-76, 79-85
CMenUG	1965-67, 69-71, 75-77, 80-83
CPT	1968
CSfCSM	1982
CSt-ES	1938, 57, 59, 62, 65, 68, 71-72, 76, 79, 81-85
CU-A	1962, 72, 76, 85
CU-EART	1938, 57, 62-64, 66-69, 72, 76, 78-79, 81-82
CU-SB	1981-82
CaACU	1962, 68, 76, 79
CaBVaU	1963-64
CaNSHD	1982
CaOHM	1963
CaOKQGS	1965-66, 68-69
CaOLU	1962, 68, 72
CaOOG	1959, 61-62, 67-68, 70-74, 76-77, 79-80, 82

(457) New Mexico Geological Society.

CaOTDM	1976
CaOTRM	1968
CaOWtU	1968, 72-76, 78-85
CoDCh	1979, 81
CoDGS	1938, 52, 59-61, 63-67, 69, 71-75, 78
CoG	1964, 68-69, 79
CoU	1967-69
CtNbC	1982, 85
DI-GS	1938, 47-48, 51-54, 59-77, 79-81, 83-84
DLC	1973, 76
I-GS	1968, 82
ICF	1968
ICtU-S	1967-68
ICarbS	1968, 82
IEN	1938, 63, 68, 72
IU	1938, 57, 59, 61-79, 80, 84-85
IaU	1938, 57, 63, 67-68, 76, 79-82
IdU	1982
InLP	1963, 67, 68, 72, 76, 79, 82, 85
InRE	1967
InU	1938, 60, 62, 65-68, 72-73, 76, 78
KyU-Ge	1937, 63, 66-68, 73, 75-77, 79-80, 82, 85
LNU	1965
M(Weston)	1980, 82
MA	1982, 85
MBU	1980, 81, 83-85
MCM	1980, 81, 83-84
ME-GS	1983-84
MH-GS	1938, 54, 57, 59-61, 63-76, 80-83, 85
MNS	1957, 62, 67-68, 71, 76-77, 81
MU	1980, 81-82
MWesB	1968, 80, 82-85
MWestonR	1983-85
Me	1980, 81
MeGS	1983-84
MiDW-S	1963, 68
MiHM	1968, 76
MiKW	1968
MnU	1938, 57, 59-73, 75-76, 79-82
MoRM	1968
MoSW	1938, 57, 62, 68, 82
MoU	1963, 68
MtBuM	1968
NBiSU	1959, 61, 63-66, 68-73
NN	1982
NNC	1938, 52, 57, 59, 61-69
NOneoU	1967
NRU	1938, 59, 61, 68, 71, 80-82
NSbSU	1963, 68, 69, 72, 76, 79, 82, 85
NSyU	1938, 57, 59, 62, 63, 65-66, 68, 70, 72, 73, 76-77, 80, 81
NbU	1959, 60, 68
NcLcU	1982
NcU	1938, 57, 59, 66-70, 72, 75-76, 79
NdU	1962-63, 68-69, 82
NhD-K	1954, 57, 59-72, 74-76, 79, 80-83, 85
NjP	1938, 52, 57, 59, 61-70, 75
NvU	1968
OCU-Geo	1961-62, 64, 68, 70, 72, 82, 85
OU	1938, 54, 68, 82
OkOkU	1967
OkU	1938, 59, 62, 67-69, 71, 76, 81-82
OrU-S	1968
PBL	1959
PBU	1982
PBm	1962, 67-68, 75
PSt	1963, 68-76, 78-79, 83-85
RPB	1938, 57, 63, 68
ScU	1982
TxBeaL	1938, 67, 72, 76, 79
TxDaAR-T	1967-68, 70-71, 73
TxDaDM	1957, 59-65, 68
TxDaM-SE	1959, 62, 82
TxHU	1962-63, 72
TxLT	1938
TxU	1938, 57, 59-68, 79, 82
TxU-Da	1963-66
UU	1968, 82
ViBlbV	1938, 67-69, 72, 76
Vt	1982
VtCasT	1982
WU	1968, 76, 79-83, 85
WaU	1979
WyU	1982

NEW ENGLAND PLEISTOCENE GEOLOGISTS. See FRIENDS OF THE PLEISTOCENE. EASTERN GROUP. (182)

(457) NEW MEXICO GEOLOGICAL SOCIETY. ANNUAL MEETING. GUIDEBOOK.

1949 [3rd] See Geological Society of America. 197.
1956 [10th] See Geological Society of America. Rocky Mountain Section. 204.
1961 [15th] Last Chance Canyon, Guadalupe Mountains, road log, Roswell to Sitting Bull Falls.
1967 21st Road log; Socorro-Magdalena-Rio Salado-Ladron Mountains.
1971 25th Road log of field trip from Roswell to Bottomless Lakes State Park. (550)
1978 32nd Field guide to selected cauldrons and mining districts of the Datil-Mogollon volcanic field, New Mexico. (459)

AzFU	1961
AzTeS	1961, 78
AzU	1978
CChiS	1961
CLU-G/G	1978
CSt-ES	1978
CU-A	1961
CU-EART	1967, 78
CaABU	1978
CaOWtU	1961, 78
CoDU	1961
CoDuF	1961
CoGrU	1961, 67, 71, 78
CoU	1961
DI-GS	1978
IU	1972
IaU	1978
InLP	1961, 71, 78
InU	1978
KyU-Ge	1961, 78
MiDW-S	1961
MiU	1961
MnU	1978
MoU	1961
NOneoU	1961
NSbSU	1978
NSyU	1978, 73
NcU	1978
NhD-K	1978
NjP	1961
NmPE	1961, 78
NmU	1961
NvU	1961, 78
OU	1961
OkU	1961, 78
PBm	1961
PSt	1961
SdRM	1961
TxBeaL	1978
TxDaGI	1961
TxHU	1961
TxLT	1961
TxMM	1961, 78
TxU	1961, 67, 78
TxU-Da	1961
TxWB	1961
UU	1961, 78
ViBlbV	1978
WU	1961, 78
WyU	1961

(458) NEW MEXICO GEOLOGICAL SOCIETY. GUIDEBOOK FOR THE ANNUAL FIELD CONFERENCE. (TITLE VARIES)

1950 1st Guidebook of the San Juan Basin, New Mexico and Colorado.
1951 2nd Guidebook of the south and west sides of the San Juan Basin, New Mexico and Arizona.
1952 3rd Rio Grande country, central New Mexico.

1953 4th Southwestern New Mexico.
1954 5th Southeastern New Mexico. Road logs.
— (1) Alamogordo to Alamo Canyon.
— (2) Alamogordo to Cloudcroft.
— (3) Cloudcroft to Carlsbad.
— (4) Guadalupe Mountains area, New Mexico and Texas.
— (5) Carlsbad to International Minerals and Chemical Corporation Potash Mine.
— (6) Carlsbad to Carlsbad Caverns.
— (7) Northern part of West Side Road, Sacramento Mountains.
— (8) Road log from Cloudcroft to Pinon.
— (9) Road log from Dunken to Pinon.
— (10) Road log from Pinon to Texas Hill.
— (11) Road log from Texas Hill Anticline to Hope, New Mexico.
— (12) Road log from Junction Highway 83 and State Road 24 to Bluewater Anticline.
— (13) Road log from Carlsbad to mouth of Dark Canyon.
— (14) Road log from Highway 62-180 to mouth of Slaughter Canyon.
— (15) Road log from Highway 62-180 to mouth of McKittrick Canyon.
1955 6th Guidebook of south-central New Mexico.
1956 7th Southeastern Sangre de Cristo Mountains.
1957 8th Southwestern San Juan Mountains, Colorado.
1958 9th Black Mesa Basin, northeastern Arizona. (65)
1959 10th West-central New Mexico.
1960 11th Rio Chama country, northern New Mexico.
1961 12th Albuquerque country, New Mexico.
1962 13th Mogollon Rim region, east-central Arizona. (65)
1963 14th Socorro region, New Mexico.
1964 15th Ruidoso country.
1965 16th Southwestern New Mexico, II.
1966 17th Taos-Raton-Spanish Peaks country, New Mexico and Colorado.
1967 18th Defiance-Zuni-Mt. Taylor region, Arizona and New Mexico.
1968 19th San Juan-San Miguel-La Plata region, New Mexico and Colorado. (254)
1969 20th Guidebook of the border region. [Chihuahua, Mexico and the United States]
1970 21st Guidebook of the Tyrone-Big Hatchet Mountains-Florida Mountains region. [New Mexico]
1971 22nd Guidebook of the San Luis Basin, Colorado.
1st day; Alamosa to the eastern San Juan Mountains, via Alamosa River, Jasper, Summitville, South Fork, and return.
2nd day; Alamosa to the Great Sand Dunes National Monument, Poncha Pass, Salida, Howard and return via Saguache and Monte Vista.
3rd day; Rail log from Antonito, Colorado to Chama, New Mexico. Supplemental logs.
1. Villa Grove to Bonanza.
2. Del Norte to Summer Coon volcanic area and return.
3. Fort Garland to Romeo via San Luis, San Acacio and Manassa.
4. Chama, New Mexico to Antonito, Colorado.
1972 23rd [1] Guidebook of the east-central New Mexico. Road logs.
1st day; Tucumcari, Mosquero and San Juan country.
2nd day; Tucumcari, Canadian Escarpment, and Santa Rosa country.
3rd day; Santa Rosa, Clines Corners, Encino, Duran, and Vaughn country.
[2] Subsurface geology of east-central New Mexico. [Supplement] (459)
1973 24th Guidebook of Monument Valley and vicinity, Arizona and Utah. Road logs:
1st day; Farmington, New Mexico to Kayenta, Arizona, via Shiprock, Four Corners, Aneth, Bluff, Cedar Mesa, Goosenecks and Mexican Hat.
2nd day; Kayenta, Arizona to Black Mesa and Navajo National Monument.
3rd day; Kayenta, Arizona to Gallup, New Mexico, via Dinnehotso, Rock Point, Round Rock, Many Farms, Chinle, Canyon de Chelly, Ganado, St. Michaels, Hunters Point and Lupton.
1974 25th Ghost Ranch, central-northern New Mexico. Road logs:
1st day: Ghost Ranch to Cuba and Nacimiento Mine and return.
2nd day; Coyote Junction to U.S. 84 and New Mexico 96 to Abiquiu, El Rito, Petaca, Tres Piedras, Hopewell Lake, Chama Basin, and return to Ghost Ranch.
3rd day; Junction of U.S. 84 and El Rito Turnoff to Espanola, Valle Grande, San Ysidro and Bernalillo; with optional trip beginning at mile 99.6 to examine Mississippian and Pennsylvanian rocks at Guadalupe Box.

3rd day: Optional trip; beginning at mile 111.6 to examine Cretaceous and Tertiary at south end of Nacimiento Uplift.
1975 26th Guidebook of the Las Cruces country (central-southern New Mexico). Road logs.
1st day; Las Cruces to southern San Andres Mountains and return.
2nd day; Las Cruces to the Sierra de las Uvas and Aden volcanic area and return.
3rd day; Las Cruces to North Mesilla Valley, Cedar Hills, San Diego Mountain, and Rincon area.
Exit A(North); Rincon area to Derry Hills via Interstate 25 North.
Exit B(West); Hatch and Deming via New Mexico 26.
Exit C(South); Upham Interchange to Anthony, New Mexico-Texas via Interstates 25 South and 10 East.
Exit D(East); Mesilla Valley to Tularosa Basin via U.S. 70 East.
1976 27th Guidebook of Vermejo Park, northeastern New Mexico. Road logs.
1st day; Las Vegas to Raton via Montezuma, Sapello, La Cueva (and vicinity), Ocate, Wagon Mound and Springer.
2nd day; Raton to Underwood Lakes, through the Raton coal field via the York Canyon mine, Vermejo Park and Gold Creek, with a discussion of timber types and site factors.
3rd day; Raton to Adams and Bartlett Lakes, Vermejo Park, New Mexico, through Trinidad coal field and Tercio Anticline, Colorado; return via Van Bremmer Canyon and Colfax, New Mexico.
1977 28th Guidebook of the San Juan Basin III, northwestern New Mexico.
1978 29th Land of Cochise, southeastern Arizona. (65)
1979 30th
[2] Archaeology and history of Santa Fe country. (459)
1980 31st Trans-Pecos region: southeastern New Mexico and West Texas.
1981 32nd Western slope, Colorado; western Colorado and eastern Utah.
1982 33rd Albuquerque Country II.
1983 34th Socorro Region II.
1984 35th Rio Grande Rift: northern New Mexico.
1985 36th Santa Rosa-Tucumcari region.

AzFU	1950-54, 56-73
AzTeS	1950-78, 79(2), 80, 83-84
AzU	1950-78, 79(1)
BEG	1980, 83-85
CChiS	1950-51, 53-54, 56-72
CLU-G/G	1950-85
CLhC	1952-85
CMenUG	1950-85
CPT	1950-85
CSdS	1950-54, 56-79
CSt-ES	1950-85
CU-A	1950-79, 81
CU-EART	1950-85
CU-SB	1950, 72, 80-85
CU-SC	1980-85
CaACU	1950-81, 84
CaAEU	1950-55, 57-82, 84
CaBVaU	1956, 60, 65, 69, 71, 77-78, 83
CaOKQGS	1953-54, 56-59, 62-63, 65
CaOOG	1964, 69, 72(1), 78-82, 84
CaOWtU	1950-78, 79(1), 80-85
CoDBM	1970-72
CoDCh	1950, 52, 54-57, 59-72(pt.1), 73-79(pt.1), 80-81, 83-84
CoDGS	1980-85
CoDU	1950-73
CoDuF	1950-62, 67-68, 71, 73
CoFS	1959-61, 63
CoG	1950-78, 80-81, 83-84
CoLiM	1981-84
CoU	1950-77, 79
DFPC	1953-58
DI-GS	1950-85
DLC	1955, 58-68, 75
I-GS	1952, 69
ICF	1952-54, 56-59, 62-64, 67
ICIU-S	1951, 53-54, 56-78
ICarbS	1978, 80, 84
INS	1982, 84
IU	1950-78, 79(1), 80
IaU	1950-85
IdBB	1958, 73
IdPI	1973

(459) New Mexico Geological Society.

IdU	1957-58
InLP	1950-85
InU	1950-79
KyU-Ge	1950-85
LNU	1950-51, 53-55, 57-60, 62-74, 76-78, 84
LRuL	1980
LU	1950-73
MH-GS	1951-52, 58, 69, 71-72, 74, 80-82, 84
MNS	1951, 58
MU	1981-83, 85
MiDW-S	1951-64
MiHM	1953, 66-67, 74
MiKW	1950-51, 57-59, 61, 66, 70, 73
MiU	1951, 53-67
MnSSM	1969
MnU	1950-84
MoSW	1950, 57, 74-75, 77, 80-81
MoU	1951, 53-54, 56-67
MtBuM	1962
MtU	1957
NIC	1980-85
NNC	1950-78, 80, 81, 83-85
NOneoU	1950-75
NRU	1953-54, 59-60, 62, 65-66, 69, 72-73, 76, 80-81, 85
NSbSU	1950-79, 81, 85
NSyU	1959, 79(2)
NbU	1952-73, 75
NcU	1950-78, 79(2)
NdU	1950-61, 63-71, 72(1), 73-78, 79(2), 80-84
NhD-K	1950-71, 72(1), 73-75, 78-79, 80-85
NjP	1950-71, 72(1), 73-74
NmLcU	1980-85
NmPE	1950-51, 53-79
NmSoI	1980-85
NmU	1950-56, 59-61, 63-68, 70, 80-85
NvLN	1971, 74
NvU	1950-66, 68-78, 79(1), 80-85
OCU-Geo	1950-71, 72-85
OU	1950-85
OkOkCGS	1950, 54
OkT	1950-65, 77, 79(1), 80-85
OkTA	1980-85
OkU	1950-78, 80-85
OrCS	1951-71, 72(1), 73-75, 78, 79(1), 81
OrU-S	1953-73, 80
PBU	1980-85
PBm	1950-78, 79(1)
PSt	1950-79, 80, 81
SdRM	1958, 61-62, 64-65
TxBeaL	1950-79
TxCaW	1959-73, 77, 80-82
TxDaAR-T	1951-74
TxDaDM	1951-77
TxDaGI	1950-74, 77
TxDaM-SE	1950-65, 80-81
TxDaSM	1950-71, 72(1), 73-77, 79
TxHMP	1980, 82-83
TxHSD	1950-66, 68-79
TxHU	1950-54, 56-71, 73
TxLT	1950-79
TxMM	1950-71, 72(1), 73-85
TxU	1950-71, 72(1), 73-85
TxU-Da	1950-74
TxWB	1950-69, 72, 74-75, 77-78, 80-84
UU	1951-77, 80-85
ViBlbV	1950-79
WBB	1980-82, 84
WU	1952-85
WaU	1950-52, 54-59, 61-66, 68-69, 71, 74-76
WyU	1953-54, 56-62, 68, 80-84

(459) NEW MEXICO GEOLOGICAL SOCIETY. SPECIAL PUBLICATION.
1972 4. See New Mexico Geological Society. 458.
1978 7. See New Mexico Geological Society. 457.
1979 8. See New Mexico Geological Society. 458.

(460) NEW MEXICO INSTITUTE OF MINING AND TECHNOLOGY.
1940 Nov. Sierra and Socorro counties, New Mexico. (462)
1941 Nov. Near Socorro, New Mexico. (462)
1945 Oct. Los Pinos Mountains--Chupadera Mesa--Manzano Mountains. (462)
1946 Sept. San Juan Basin. (462)

DI-GS	1945
InU	1946
KyU-Ge	1945
OkU	1945
TxDaAR-T	1940-41, 46
TxDaGI	1945-46
TxLT	1945-46
TxMM	1945-46
TxU	1946
TxU-Da	1945

(461) NEW MEXICO. BUREAU OF MINES AND MINERAL RESOURCES. CIRCULAR.
1976 149. See Geological Society of America. South Central Section. 205.
— 153. See Geological Society of America. Rocky Mountain Section. 204.(3)
— 154. See Geological Society of America. 197.
1978 163. See International Symposium on the Rio Grande Rift. 320.
1979 164. Model for beach shoreline in Gallup Sandstone (Upper Cretaceous) of northwestern New Mexico.

AzTeS	1979
CPT	1979
CSt-ES	1979
CU-EART	1979
InLP	1979

(462) NEW MEXICO. BUREAU OF MINES AND MINERAL RESOURCES. GUIDEBOOK.
1940 Nov. See New Mexico Institute of Mining and Technology. 460.
1941 Nov. See New Mexico Institute of Mining and Technology. 460.
1945 Oct. See New Mexico Institute of Mining and Technology. 460.
1946 Sept. See New Mexico Institute of Mining and Technology. 460.
1983 Apr. See Roswell Geological Society. 550.

(463) NEW MEXICO. BUREAU OF MINES AND MINERAL RESOURCES. MEMOIR.
1963 15. See Society of Economic Geologists. 583.
1976 31. See Geological Society of America. South Central Section. 205. (1977)
1981 No. 39. Soils and geomorphology in the Basin and Range area of southern New Mexico; guidebook to the desert project.

BEG	1981
CLU-G/G	1981
CMenUG	1981
CPT	1981
CSfCSM	1981
CSt-ES	1981
CU-A	1981
CU-SB	1981
CaOOG	1981
CoDGS	1981
CtY-KS	1981
IaCfT	1981
IaU	1981
InLP	1981
KLG	1981
KU	1981
MnU	1981
NNC	1981
NcU	1981
NdU	1981
NmLcU	1981
NmSoI	1981
NmU	1981
NvU	1981
OCl	1981
OU	1981
OkTA	1981
OkU	1981
TxCaW	1981
TxDaM-SE	1981
TxMM	1981

TxU	1981
UU	1981
WyU	1981

(464) NEW MEXICO. BUREAU OF MINES AND MINERAL RESOURCES. SCENIC TRIPS TO THE GEOLOGIC PAST.
- 1955 No. 1. Santa Fe, New Mexico.
- 1956 No. 2. Taos-Red River-Eagle Nest, New Mexico circle drive.
- 1958 No. 3. Roswell-Capitan-Ruidoso and Bottomless Lakes Park, New Mexico.
- — No. 4. Southern Zuni Mountains.
- 1959 No. 5. Silver City-Santa Rita-Hurley, New Mexico.
- 1960 No. 2. Taos-Red River-Eagle Nest, New Mexico circle drive. 3d ed.
- — No. 6. Trail guide to the Upper Pecos.
- 1961 No. 7. High Plains, northeastern New Mexico; Raton-Capulin Mountain-Clayton.
- 1964 No. 8. Mosaic of New Mexico's scenery, rocks and history; a brief guide for visitors.
- 1965 No. 7. High Plains, northeastern New Mexico, Raton-Capulin Mountain-Clayton. 2d ed.
- 1967 No. 3. Roswell-Capitan-Ruidoso and Bottomless Lakes State Park, New Mexico. 2d ed.
- — No. 5. Silver City-Santa Rita-Hurley, New Mexico. 2d ed.
- — No. 6. Trail guide to the Upper Pecos. 2d ed. (Rev. in 1974)
- — No. 7. High Plains, northeastern New Mexico-Raton-Capulin Mountain-Clayton. 3d ed.
- — No. 8. Mosaic of New Mexico's scenery, rocks, and history; a brief guide. 2d ed.
- 1968 No. 1. Santa Fe, New Mexico. 2d ed.
- — No. 2. Taos-Red River-Eagle Nest, New Mexico. 4th ed.
- — No. 4. Southern Zuni Mountains. 2d ed.
- 1969 No. 9. Albuquerque: its mountains, valley, water, and volcanoes.
- 1971 No. 4. Southern Zuni Mountains. Zuni-Cibola Trail. Revised 1971.
- — No. 10. Southwestern New Mexico: Lordsburg, Silver City, Deming, Las Cruces.
- 1972 No. 11. Cumbres and Toltec scenic railroad.
- 1974 No. 9. Albuquerque: its mountains, valley, water, and volcanoes. Revised 1974.
- — No. 11. Cumbres and Toltec scenic railroad. [Revised 1974]
- — No. 12. The story of mining in New Mexico.
- 1975 No. 6. Trail guide to the geology of the Upper Pecos. 3d ed., revised.
- 1980 No. 10. Southwestern New Mexico. 2d ed., revised.
- 1981 No. 3. Roswell-Ruidoso-Valley of Fires, including trips to Lincoln, Tularosa, and Bottomless Lakes State Park. 3d ed.
- 1982 No. 9. Albuquerque: its mountains, valley, water, and volcanoes. 3d ed.
- — No. 13. Espanola-Chama-Taos; a climb through time.

AzTeS	1955-56, 58-59, 60(6), 61, 64, 69, 72, 74(11-12)
AzU	1967-68, 71-72, 74-75, 82(13)
BEG	1981
CLU-G/G	1955-56, 58-59, 60(6), 61, 64, 67(3, 5), 68(4), 71(10), 72, 74(9, 12), 82(13)
CMenUG	1955-56, 58-59, 60(6), 61, 64, 68-69, 71, 72(11), 74, 80(10), 82(9, 13)
CPT	1955, 58-60, 67(8), 68(1), 71-72, 74(12), 75(6), 81(3), 82
CSfCSM	1981, 82(13)
CSt-ES	1955-56, 58(4), 59, 60(2), 61, 67(8), 69, 71(10), 72, 74(12), 75(6), 80, 82
CU-A	1955-56, 58-59, 60(6), 61, 71(10), 72, 74(9, 12)
CU-EART	1955, 58-59, 60(2), 61, 65, 67(6, 8), 69, 71, 75, 81, 82(13)
CaOOG	1960(2), 69, 81-82
CaOWtU	1967(7-8), 68(4), 72(11), 71(4)
CoDBM	1981-82
CoDCh	1956, 58-59, 61, 82
CoDGS	1982
CoDU	1959
CoG	1955-56, 58-59, 60(6), 61, 64, 69, 71-72, 74(12)
CoU	1955-56, 58-59, 60(6), 61, 64-65, 67-68, 71-72, 74(9, 12)
DI-GS	1955-56, 58-59, 60(6), 61, 64, 67(8), 71-72, 74(9, 12), 81, 82(13)
DLC	1955-56, 58-61, 64, 69, 71, 72, 74(9, 12)
F-GS	1982
FBG	1982
ICarbS	1955-56, 58-59, 60(6), 61, 64, 69, 71-72, 82
IU	1960(2), 69
IaU	1955-56, 58-59, 60(6), 61, 64, 68(1), 71-72, 74(12), 75, 81-82
IdBB	1968(1)
IdPI	1955-56, 58(3), 59, 60(6), 61, 64, 71-74
IdU	1955-56, 58-61, 64-65, 67-69, 71-72, 74-75, 80-82
InLP	1955-56, 58-61, 64, 67(3, 5, 7-8), 68-69, 71-72, 74(12), 75
InU	1955-56, 58-59, 60(6), 65, 71-72, 82
KU	1981-82
KyU-Ge	1955-56, 58-59, 60(6), 61, 64, 69, 71-72, 74(12)
MH-GS	1960
MNS	1955-56, 58-59, 60(6), 61, 71
MiHM	1955-56, 58, 60(6), 61, 64, 67-69, 71(4), 72
MnU	1955-56, 58-59, 60(6), 61, 64-65, 67-69, 71-72, 81-82
MoSW	1955-56, 58-59, 60(6), 61, 64-65, 67-68, 71-72, 74(9, 12), 75
MoU	1955-56, 58, 60(6), 61, 64
MtBuM	1955-56, 58-59, 60(6), 61, 64, 67(8), 69, 71-72, 74(9, 12), 80
NIC	1981
NMSoI	1980(10), 1982(9)
NbU	1955-56, 58-59, 60(6), 61, 64-65, 67-68, 71-72
NcU	1955-56, 58-59, 60(6), 61, 64, 69, 71-72, 74(9, 12), 81
NdU	1955-56, 58-59, 60(6), 61, 64-65, 67-68, 71-72, 81-82
NhD-K	1960(6), 72, 75
NjP	1967(3), 68, 71-72, 74(9)
NmLcU	1981-82
NmPE	1956, 58-61, 64-65, 67-69, 71-72, 74(9, 12), 75
NmSoI	1981-82
NmU	1981-82
NvLN	1971-72, 74(9, 12)
NvU	1955-56, 58-59, 60(6), 61, 64, 71(10), 72, 74(9, 12), 81, 82(13)
OCU-Geo	1955-56, 58-59, 60(6), 61, 64-65, 67-68, 71-72, 74(9, 12), 75, 82
OU	1955-56, 58-59, 60(6), 61, 64-65, 67-69, 71-72, 74(9, 12), 80-82
OkT	1955-56, 58-59, 60(6), 61, 64-65, 71(10), 72
OkU	1955-56, 58-59, 60(6), 61, 64-65, 67-68, 71-72, 74(9, 12), 81, 82(13)
PHarER-T	1981-82
PSt	1955-56, 58-59, 60(6), 61, 67(8), 69, 71-72
SdRM	1955-56, 58-59, 60, 61, 64-65, 67, 69, 71-72, 74-75, 80-81, 82(9)
TxBeaL	1955-56, 58-59, 60(6), 64
TxCaW	1961, 65, 67(3, 5-6, 8), 68, 71(10), 80(10), 81-82
TxDaAR-T	1955-56, 58-59, 61, 64, 71-72
TxDaGI	1955-56, 58-59, 60(6), 61, 64, 69, 71(10), 72(11)
TxDaM-SE	1981-82
TxHSD	1955-56, 58-59, 60(6), 64, 71(10)
TxHU	1955-56, 58-59, 60(6), 61, 64-65, 67-68, 71-72
TxLT	1955-56, 58-59, 60(6), 61, 64-65, 67
TxU	1955-56, 58-59, 60(6), 61, 64-65, 67-68, 71-72, 74(9, 12), 81-82
TxU-Da	1955
TxWB	1955-56, 58-61
UU	1955-56, 58-59, 60(6), 61, 64-65, 67(6-7), 68(1-2), 71-72, 74(9, 12), 75, 81
ViBlbV	1955-56, 58-59, 60(6), 61, 64-65, 67-68, 71-72, 74(9, 12)
WU	1955-56, 58-59, 60(6), 61, 64-65, 67-69, 71-72, 74(211), 75, 81-82
WyU	1981-82

(465) NEW MEXICO. HEALTH AND ENVIRONMENT DEPARTMENT. ENVIRONMENTAL IMPROVEMENT DIVISION. ENVIRONMENTAL EVALUATION GROUP. [FIELD TRIP]
- 1980 June WIPP site and vicinity; geological field trip; a report of a field trip to the proposed Waste Isolation Pilot project in southeastern New Mexico, June 16 to 18, 1980.

DLC	1980

NEW MEXICO. STATE SCHOOL OF MINES, SOCORRO. See NEW MEXICO INSTITUTE OF MINING AND TECHNOLOGY. (460)

(466) NEW ORLEANS GEOLOGICAL SOCIETY. GUIDEBOOK.
- 1961 Apr. Jefferson Island salt dome.
- 1962 Feb. Peninsula of Yucatan.
- 1965 See American Association of Petroleum Geologists. 18.
- 1966 Field trip to southeast pass of Mississippi River. (No guidebook issued)
- 1967 Nov. See Geological Society of America. 197.
- 1968 May Depositional environments; a comparison of Eocene and recent sedimentary deposits of the northern Gulf Coast.
- 1970 A study of the lower Mississippi River delta, its processes, sediments, and structures.
- 1971 May The Lafourche Delta and the Grand Isle Barrier Island; destruction of an ancient delta of the Mississippi River.
  Part 1. Flight over Bayou Lafourche, an ancient delta of the Mississippi.

(467) New Western Illinois University. Geology Club.

Part 2. Grand Isle Barrier Island in the Gulf of Mexico.
— Oct. See Gulf Coast Association of Geological Societies. 247.
1972 Spring Guidebook, Louisiana salt domes and the Mississippi deltaic plain, with a visit to Morton Salt Company Mine, "Weeks Island," Louisiana.
1973 Spring Geology of the Mississippi-Alabama coastal area and near-shore zone.
— Fall The Mississippi River: Vicksburg to Bonnet Carre.
1974 Nov. 12. See Geological Society of America. 197.
— Nov. 16. West-central Louisiana field trip.
1975 Fall Modern carbonate environments of the upper Florida Keys. (Cover title: Florida Keys field trip)
1976 See American Association of Petroleum Geologists. 18.
1978 See Gulf Coast Association of Geological Societies. 247.
— July A study of the lower Mississippi River delta, Garden Island Bay subdelta crevasse, and mud lumps of South Pass area.
1980 Feb. Geology of greater New Orleans: its relationship to land subsidence and flooding, with a geologic walking tour of downtown New Orleans.
1982 A tour guide to the building stones of New Orleans.
— June Coastal geology of Mississippi, Alabama and adjacent Louisiana areas.
— Oct. See Geological Society of America. 197.(10)
1985 Geology and hydrogeology of the Yucatan and Quaternary geology of northeastern Yucatan Peninsula ..., with a part on the history of northern Quintana Roo. (Also used for Gulf Coast Association of Geological Societies. 1985)
— Mar. See American Association of Petroleum Geologists. 18.

AzU	1968, 71
CChiS	1968
CLU-G/G	1961, 68, 71, 72-73, 74(Nov.16-17), 75, 85
CLhC	1961, 73(Spring)
CSt-ES	1962, 67, 68, 71, 72-73, 76, 85
CU-A	1962, 73(Spring)
CU-EART	1961-62, 68, 71, 72-73, 74(Nov.16-17), 75
CaACAM	1961, 71, 73
CaACU	1968, 71, 73(Spring), 74(Nov.16-17)
CaOHM	1968
CaOWtU	1973(Spring)
CoG	1961-62, 68, 70, 71, 72
CoLiM	1982(June)
CoU	1962
DI-GS	1961-62, 68, 71, 72-73, 74(Nov.16-17), 75, 82(June)
DLC	1968
IU	1961-62, 66, 68, 70, 71, 72, 73(Spring), 74(Nov.16-17), 75
IaU	1968, 71, 73(Fall), 75
InLP	1968, 72-75, 78
InRE	1973(Spring)
InU	1962, 68, 70, 71, 73, 74(Nov.16-17), 75, 80, 82
KyU-Ge	1968, 71-72, 73(Fall), 74-75, 82(June), 85
LNU	1961-62, 68, 71, 73(Fall), 74(Nov. 16-17), 75, 82
LU	1961-62
MnU	1961-62, 68, 71, 78, 80, 82(June)
MoU	1961
Ms-GS	1973(Spring)
MsJG	1973(Spring)
NBiSU	1973(Spring)
NNC	1961-62, 68, 70
NSyU	1968, 73(Spring)
NbU	1961-62
NcU	1962, 68, 71, 72, 73(Spring)
NdU	1968, 71, 72, 73(Spring), 74(Nov.16-17)
NhD-K	1961, 68, 70, 71, 72-73
NjP	1968, 73(Spring)
OCU-Geo	1962, 68, 80
OU	1968, 71(May), 72
OkT	1962
OkU	1961-62, 68, 71, 73-75, 85
TMM-E	1973(Spring)
TxBeaL	1961-62, 68, 71, 72, 73(Spring)
TxDaAR-T	1962, 68, 71, 72, 73(Spring)
TxDaDM	1961, 68, 70, 71, 72, 73(Spring)
TxDaGI	1968
TxDaM-SE	1961-62, 82(June)
TxDaSM	1961-62, 73, 75
TxHSD	1961-62, 70, 71, 74(Nov.16-17)
TxHU	1962
TxMM	1961-62, 65, 73-74, 82(June)
TxU	1961-62, 68, 70, 71, 72, 73(Spring), 82(June)
TxWB	1961, 67-68, 73-74
ViBlbV	1973(Fall)
WU	1973(Spring)
WyU	1982(June)

(467) NEW WESTERN ILLINOIS UNIVERSITY. GEOLOGY CLUB.
[N.D.] Pennsylvanian sedimentology lecture and field conference.
[n.d.] Pennsylvanian sedimentology lecture and field conference.

(468) NEW YORK ACADEMY OF SCIENCE.
1965 See Lunar Geological Field Conference. 366.

(469) NEW YORK STATE GEOLOGICAL ASSOCIATION. ANNUAL MEETING. FIELD TRIP GUIDEBOOK. (TITLE VARIES)
1926 2nd [Syracuse, New York, area]
1927 3rd [Poughkeepsie, New York, area]
1928 4th [Ithaca, New York, area]
1929 5th [Gouverneur, New York, area]
1930 6th [Schenectady, New York, area]
1931 7th [Ithaca, New York, area]
1932 8th [Rochester, New York, area]
1933 9th Excursion of the New York State Geological Association in the New York City region.
1934 10th [Hamilton, New York, area]
1935 11th [Mohawk Valley, New York, area]
1936 12th [Sterling, Pennsylvania, area] (531)
1937 13th [Syracuse, New York, area] (Syracuse University. Dept. of Geology and Geography. Bulletin v. 5, no. 3, May 1937)
1938 14th [Buffalo, New York, area] (Hobbies: The magazine of the Buffalo Museum of Science, v. 18, April 1938)
1939 15th [Canton, New York, area]
1940 16th Catskill, New York.
1941 17th [Rochester, New York, area]
1946 18th [Poughkeepsie, New York, area]
1947 19th [New York, New York, area]
1948 20th [Clinton, New York, area]
1949 21st Geology of the Cayuga Lake region.
1950 22nd [Geology around the Syracuse region]
1951 23rd [Plattsburg, New York, area]
  [1] Chazyan stratigraphy, Plattsburg.
  [2] Supplementary guidebook to Chazyan stratigraphy.
1952 24th [Buffalo, New York, area]
  1. Middle Devonian stratigraphy and paleontology.
  2. Niagara Gorge and Niagara Falls.
1953 25th [Canton, New York, area]
1954 26th [Poughkeepsie, New York, area]
1955 27th [Hamilton, New York, area]
1956 28th Western New York.
1957 29th Wellsville, New York.
  Geology of the southwestern tier oil and gas fields, oil refining field trips.
  Trip A. Geological outcrops in the Wellsville area.
  Trip B. Harrison Valley-Oriskany gas storage field.
  Trip C. Sinclair oil refinery (no log).
  Trip D. Oil fields and well shooting.
  Trip E. Stratigraphy of southwestern New York.
1958 30th Peekskill, New York.
1959 31st Geology of the Cayuga Lake basin.
1960 32nd Clinton, New York (region around Utica and southern Adirondack Mountains).
1961 33rd Troy, New York.
1962 34th Port Jervis, New York. [Southern taconics]
  Trip A. The Onondaga Limestone and Schoharie Formation.
  Trip B. Geology of the ... southern part of the Monroe Quadrangle.
  Trip C. Structure and stratigraphy of the Port Jervis-South Otisville quadrangles.
1963 35th Geology of south-central New York.
1964 36th South-central Adirondack highlands.
1965 37th The Schenectady area.

Trip A. Mohawk Valley strata and structure, Saratoga to Canajoharie.
Trip B. Geologic excursion from Albany to the Glen via Lake George.
Trip C. Glacial lake sequences in the eastern Mohawk-northern Hudson region.
Trip D. Geologic phenomena in the Schenectady area.
1966 38th Geology of western New York: Silurian, Devonian, and Pleistocene of western New York and adjacent Ontario.
1967 39th Field trips; mid-Hudson Valley region.
Trip A. Pleistocene geology of the Wallkill Valley.
Trip B. The economic geology of the mid-Hudson Valley region.
Trip C. Middle and Upper Devonian clastics of the Catskill front, New York.
Trip D. Upper Silurian-Lower Devonian stratigraphic sequence, western mid-Hudson Valley region, Kingston vicinity to Accord, Ulster County, New York.
Trip E. Geologic structure of the Kingston Arc of the Appalachian fold belt.
Trip F. Structure and petrology of the Pre-Cambrian allochthon and Paleozoic sediments of the Monroe area, New York.
Trip G. [No title available]
Trip H. [No title available]
1968 40th [New York City area]
Trip A. Bedrock geology in the vicinity of White Plains, New York.
Trip B. Cretaceous deltas in the northern New Jersey coastal plain.
Trip C. The Triassic rocks of the northern Newark Basin. Road log.
Trip D. Sterling and Franklin area in the highlands of New Jersey.
Trip E. Taconian islands and the shores of Appalachia.
Trip F. Pleistocene geology of the Montauk Peninsula.
Trip G. Structure and petrology of Pelham Bay Park.
Trip H. Stratigraphic and structural relations along the western border of the Cortlandt intrusives.
Trip I. Deep-well injection of treated waste water; an experiment in re-use of groundwater in western Long Island.
Trip J. Geology, geomorphology and late-glacial environments of western Long Island, New York.
1969 41st [Plattsburgh, New York]
Trip A. Sedimentary characteristics and tectonic deformation of Middle Ordovician shales in northwestern Vermont north of Malletts Bay.
Trip B. Recent sedimentation and water properties of Lake Champlain.
Trip C. Bedrock geology of the southern portion of the Hinesburg Synclinorium.
Trip D. Surficial geology of the Champlain Valley, Vermont.
Trip E. Stratigraphy of the Chazy Group (Middle Ordovician).
Trip F. The paleoecology of Chazyan (Lower Middle Ordovician) "reefs" or "mounds".
Trip G. Evidence of late Pleistocene local glaciation in the high peaks region, Adirondack Mountains, I.
Trip H. Adirondack meta-anorthosite.
Trip I. Deglacial history of the Lake Champlain-Lake George lowland.
Trip J. Evidence of late Pleistocene local glaciation in the high peaks region, Adirondack Mountains, II.
1970 42nd [Cortland, New York]
Trip A. Benthic communities of the Genesee Group (Upper Devonian).
Trip B. Upper Devonian deltaic environments.
Trip C. Transitional sedimentary facies of the Catskill deltaic system in eastern New York.
Trip D, H. Stratigraphy, paleontology and paleoecology of the Ludlowville and Moscow formations (upper Hamilton group), central New York.
Trip E, I. Mineral industries in parts of Onondaga, Cortland and Tompkins counties.
Trip F. Deglaciation of the eastern Finger Lakes region.
Trip G. Paleontology of the Cortland area.
Trip H. See Trip D.
Trip I. See Trip E.
Trip J. Proglacial lake sequence in the Tully Valley, Onondaga County.
Trip K. Glacial history of the Fall Creek Valley at Ithaca, New York.
1971 43rd Geological studies of the Northwest Adirondacks region.
Trip A. Some aspects of Grenville geology and the Precambrian/Paleozoic unconformity, northwest Adirondacks, New York.
Trip B. Precambrian and lower Paleozoic stratigraphy, Northwest St. Lawrence and north Jefferson counties.

Trip C. Some aspects of engineering geology in the St. Lawrence Valley and northwest Adirondack Lowlands.
Trip D. Economic geology of International Talc and Benson Iron Mines.
Trip E. Some Pleistocene features of St. Lawrence County, New York.
Trip F. Mineral collecting in St. Lawrence County.
1972 44th Hamilton and Utica, New York.
[A] The Clinton Group of east-central New York.
[B] Sedimentation and stratigraphy of the Salina Group (Upper Silurian) in east-central New York.
[C] Stratigraphy of the marine limestones and shales of the Ordovician Trenton Group in central New York.
[D] Glacial geology of the northern Chenango River valley.
[E] Stratigraphy and structure of the Canada Lake nappe, southernmost Adirondacks.
[F] Paleontological problems of the Hamilton Group (Middle Devonian).
[G] Paleoecology of a black limestone, Cherry Valley Limestone; Devonian of central New York.
[H] Sedimentary environments of biostratigraphy of the transgressive early Trentonian Sea (Middle Ordovician) in central and northwestern New York.
[I] Syracuse channels; evidence for a catastrophic flood.
[J] Half-day trip to Herkimer diamond grounds in Middleville, New York.
1973 45th Rochester, New York area.
Trip A. Glacial geology of the western Finger Lakes region.
Trip B. A comparison of environments, the Middle Devonian Hamilton Group in the Genesee Valley.
Trip C. Lower Upper Devonian stratigraphy from the Batavia-Warsaw meridian to the Genesee Valley; goniatite sequence and correlations.
Trip D. Eurypterid horizons and the stratigraphy of the Upper Silurian and Lower Devonian of western New York State.
Trip E. Late glacial and postglacial geology of the Genesee Valley in Livingston County, New York; a preliminary report.
Trip F. The Pinnacle Hills and the Mendon Kame area; contrasting morainal deposits.
Trip G. Pleistocene and Holocene sediments at Hamlin Beach State Park, New York.
Trip H. Mineral collecting at Penfield Quarry.
Trip I. Stratigraphy of the Genesee Gorge at Rochester.
1974 46th Geology of western New York State.
Trip A. Lockport (Middle Silurian) and Onondaga (Middle Devonian) patch reefs in western New York.
Trip B. Upper Devonian stratigraphy of Chautauqua County, New York.
Trip C. Late Middle and early Upper Devonian disconformities and paleoecology of the Moscow Formation in western Erie County, New York.
Trip D. From Lake Erie to the glacial limits and beyond.
Trip E. Environmental geology of the Fredonia-Dunkirk area.
Trip F. Glacial geology and buried topography in the vicinity of Fredonia, Gowanda and Zoar Valley, New York.
Trip G. A selected Middle Devonian (Hamilton) fossil locality reference section.
Trip H. The exploration, discovery, and production of natural gas in western New York.
1975 47th [Hempstead, New York]
Trip A-1. Stratigraphy, structure and petrology of the New York City Group.
Trip A-2. Structure and form of the Triassic basalts in north-central New Jersey.
Trip A-3. Placer mining and concentration processes of ilmenite sand deposits near Lakehurst, New Jersey.
Trip A-4. Shinnecock Inlet tidal flood delta and problems of coastal stabilization.
Trip A-5(AM). Barrier island accretion features, Democrat Point, Fire Island.
Trip A-5(PM). Environmental geology of the Jones Beach barrier island.
Trip A-6(AM). Geological oceanography of a segment of Long Island Sound.
Trip A-6(PM). Sedimentary dynamics of a coastal pond; Flax Pond, Old Field, Long Island.
Trip A-7. Quaternary geology of the Montauk peninsula.
Trip A-8(AM). Environmental engineering aspects and tour at Shoreham of Long Island Lighting Company Nuclear Reactor Site at Shoreham, Long Island.

Trip A-8(PM). Foreshore and backshore natural environments of a barrier island, Fire Island.
Trip B-1. Lower Paleozoic metamorphic stratigraphy of Mamaroneck area, New York.
Trip B-2. Geological aspects of Staten Island, New York.
Trip B-3(AM). Natural and man-made erosional and depositional features associated with stabilization of migrating barrier islands, Fire Island inlet, New York.
Trip B-3(PM). A major beach erosional cycle at Robert Moses State Park, Fire Island, during the storm of December 1-2, 1974.
Trip B-4. Jamaica Bay, Borough of Queens, New York City; a case study of geo-environmental stress.
Trip B-5. Wisconsinan glacial stratigraphy and structure of northwestern Long Island.
Trip B-6. Geological oceanography of a segment of Long Island Sound. (See Trip A-6(AM))

1976 48th [Poughkeepsie, New York]
Trip A-1. The Hudson River guide; a geological and historical guide to the lower mid-Hudson Valley region, as viewed from the river.
Trip B-1. Structure, petrology and geochronology of the Precambrian rocks of the Hudson Highlands.
Trip B-2. Structural geology of the taconic unconformity.
Trip B-3. Stratigraphy and paleontology of the Binnewater Sandstone from Accord to Wilbur, New York.
Trip B-4. Stratigraphy and structure of Silurian and Devonian rocks in the vicinity of Kingston, New York.
Trip B-5. Pleistocene history of the Millbrook, New York, region.
Trip B-6. Stratigraphic and structural geology in western Dutchess County, New York.
Trip B-7. Progressive metamorphism in Dutchess County, New York.
Trip B-8. Alluvial and tidal facies of the Catskill deltaic system.
Trip B-9. Environmental geology of the Lloyd Nuclear Power Plant site: a history of site study.
Trip B-10. Engineering and environmental geology of the Hudson Valley power sites.
Trip C-1. Repeat of B-1.
Trip C-2. Repeat of B-2.
Trip C-3. Repeat of B-3.
Trip C-4. Repeat of B-4.
Trip C-5. Repeat of B-5.
Trip C-6. Walking trip of historic Fishkill, New York.
Trip C-7. Stratigraphy and structural geology in the Harlem Valley, S.E. Dutchess County, New York.

1977 49th Oneonta, New York.
Trip A-1. Carbonate and terrigenous sedimentary facies of tidal origin, eastern New York.
Trip A-2. Ichnofossils of the Tully clastic correlatives in eastern New York state.
Trip A-3. Paleoenvironments of the Marcellus and lower Skaneateles formations of the Otsego County region (Middle Devonian).
Trip A-4. Geologic setting of upper Susquehanna and adjacent Mohawk region.
Trip A-5. Glacial morphology of upper Susquehanna drainage.
Trip A-6. Preliminary geological investigation of Otsego Lake.
Trip A-7. Physical stratigraphy and sedimentology of the Upper Devonian stream deposits of southeastern New York.
Trip A-8. Paleoecology and stratigraphy of the Ordovician Black River Group limestones: Central Mohawk Valley.
Trip A-9. Paleontology of the lower Trenton group of central New York State.
Trip A-10. The structural framework and stratigraphy of the southeastern Adirondacks.
Trip A-11. Mineralogy and geology of the Newcomb and Sanford Lake area.
Trips B-1 through B-6. Repeats of A-1 through A-6.
Trip B-7. Karst geomorphology of the Cobleskill area.
Trip B-8. Sedimentology and paleontology of portions of the Hamilton Group in central New York.
Trip B-9. Physical and bio-stratigraphy of the Onondaga Limestone in Otsego County, New York.
Trip B-10. The Panther Mountain circular structure: a possible buried meteorite crater.
Trip B-11. Geological contexts of archeological sites on the Susquehanna River flood plain.
Trip B-12. Wedge-shaped structures in bedrock and drift, central New York State.

1978 50th Syracuse, New York.
Trip A-1. Structures in lower Paleozoic rocks, Northwest Adirondacks.
Trip A-2. Structure and petrology of the central Adirondacks.
Trip A-3. Lower Hamilton Group paleoecology.
Trip A-4. Geomorphology of the Southeast Tughill area.
Trip A-5. Upper Hamilton Group paleoecology.
Trip A-6. Paleoenvironments of the Potsdam Sandstone and Theresa Formation, northwestern New York.
Trip A-7. Benthic communities in the Ordovician clastics, Tug Hill region.
Trip A-8. Punctuated aggradational cycles in the Black River-Trenton and Helderberg-Onondaga: a general model of deposition.
Trip A-9. Eurypterid horizons and stratigraphy of the upper Silurian and lower Devonian rocks of central-eastern New York.
Trip A-10. An examination of bottom sediments in Seneca Lake.
Trip A-11. Nine Mile Point nuclear site.
Trip A-12. Geology of the Tully Limestone.
Trip B-1(A-3). Lower Hamilton Group paleoecology.
Trip B-2. Syracuse channels.
Trip B-3. Valley Heads moraine in the Syracuse vicinity.
Trip B-4(A-8). Punctuated aggradational cycles in the Black River-Trenton and Helderberg-Onondaga: a general model of deposition.
Trip B-5(A-10). An examination of bottom sediments in Seneca Lake.
Trip B-6. Moss Island (an examination and history of the newest National Landmark in the U.S.).
Trip B-7. Upper Devonian biostratigraphy near Cortland.
Trip B-8(A-12). Geology of the Tully Limestone.

1979 51st See New England Intercollegiate Geological Conference. 456.
1980 52nd [Newark, New Jersey]
Field studies of New Jersey geology and guide to field trips: 52nd annual meeting of the New York State Geological Association.
1981 53rd Binghamton, New York.
Guidetook for field trips in south-central New York.
1982 54th (Reprinted for Clay Mineral Society 1983.) See American Association of Petroleum Geologists. Eastern Section. 25.
1983 Potsdam, New York. (638)
Field trip guidebook [Adirondacks - northern New York].
Trip 1. Seismology, tectonics, and engineering geology in the St. Lawrence Valley and northwest Adirondacks.
Trip 2. Geologic traverse from Potsdam to the Thousand Islands.
Trip 3. Selected mineral occurrences in St. Lawrence and Jefferson counties, New York.
Trip 4. Stratigraphy, structure and geochemistry of Grenvillian rocks in northern New York.
Trip 5. The Trenton Group of the Black River valley.
Trip 6. A few of the best outcrops in the North Country.
Trip 7. General geology of the Adirondacks.
Trip 8. Glacial geology and soils of the St. Lawrence-Adirondack Lowlands and the Adirondack Highlands.
Trip 9. Lower Ordovician stratigraphy and sedimentology, southwestern St. Lawrence Lowlands.

1984 56th Clinton, New York.
A-1. All day field trip to Black River and Mohawk River valleys for earth science teachers.
AB-2. Precambrian geology in the central Adirondacks: a two day field trip led by three authors.
Cross-section of the Loon Pond Syncline, Tupper Lake Quadrangle.
Bedrock geology of the Grampus Lake area, Long Lake Quadrangle, New York.
A traverse across the southern contact of the Marcy Anorthosite Massif, Santanoni quadrangle, New York.
BC-3. Foreland basin sedimentation in the Trenton Group, central New York.
BC-4. Animal-sediment relationships in Middle Ordovician habitats.
BC-5. Sedimentary structures and paleoenvironmental analysis of the Bertie Formation (Upper Silurian, Cayugan Series) of central New York State.
BC-6. The Sterling Mine, Antwerp, New York; a new look at an old locality.

B-7. Pleistocene geology, groundwater, and land use of the Tug Hill and Bridgewater Flats aquifers, Oneida County, New York.
B-8. Deglaciation and correlation of ice margins, Appalachian Plateau, New York.
B-8. Two till sequence at Dugway Road exposure southwest of Clinton, New York.
BC-9. Sedimentology and faunal assemblages in the Hamilton Group of central New York.
C-10. The late Wisconsin glaciation of the West Canada Creek Valley.
C-11. Structure and rock fabric within the central and southern Adirondacks.
C-12. Groundwater contamination from hazardous waste landfill; investigations and remedies.
C-13. Seismicity in the central Adirondacks with emphasis on the Goodnow, October 7, 1984 epicentral zone and its geology.

1985 57th Saratoga Springs, New York.
A-1, B-1. A trip to the Taconic problem and back and the nature of the eastern Taconic contact.
A-2, B-2. Thrusts, melanges, folded thrusts and duplexes in the Taconic Foreland.
A-3, B-3. Tri-corn geology; the geology-history-and environmental problems of the upper Hudson Champlain Valley.
A-4, B-4. The mineralogy of Saratoga County, New York.
A-5, B-5. Correlation of punctuated aggradational cycles, Helderberg Group, between Schoharie and Thacher Park.
A-6, B-6. Structure and rock fabric within the central and southern Adirondacks.
A-7. Cambro-Ordovician shoaling and tidal deposits marginal to Iapetus Ocean and Middle to Upper Devonian peritidal deposits of the Catskill Fan-Deltaic Complex.
A-8. Rocks and problems of the southeastern Adirondacks.
A-9. Cambrian and Ordovician platform sedimentation; southern Lake Champlain valley.
A-10. Glacial geology and history of the northern Hudson Basin, New York and Vermont.
B-7. Roberts Hill and Albrights Reefs; faunal and sedimentary evidence for an eastern Onondaga sea-level fluctuation.
B-8. Deglaciation of the middle Mohawk and Sacandaga valleys, or a tale of two tongues.

CLU-G/G	1957-71, 80
CSt-ES	1957-78, 79, 80-81, 82, 83-85
CU-EART	1933, 57-75, 76(B-1 thru C-7), 77-78, 80-81, 83-85
CaAC	1968
CaACU	1957-85
CaAEU	1965(A-B)
CaBVaU	1966
CaOKQGS	1980-81, 83-84
CaOLU	1961
CaOOG	1957-75, 80-81, 84-85
CaOTDM	1957, 67, 80-81, 83
CaOWtU	1958-66, 72-75, 76(B-1 thru C-1), 77-78, 80-84
CoG	1978, 80, 84
CoLiM	1980-81
CoU	1958-68, 71
DI-GS	1940, 56-78, 80-81, 83-85
I-GS	1958
IU	1955-78, 85
IaCfT	1980
IaU	1958-63, 65-66, 68, 70-72, 74-78, 80-81, 83-85
InLP	1957-78, 80-81, 83-85
InU	1958-60, 62, 64-65, 68, 76(A-1), 78
KyU-Ge	1957-78, 80, 85
LNU	1964, 68
M(Weston)	1980-84
MCM	1980
MH-GS	1957, 59, 61, 63-74
MWesB	1956, 58-79, 80-85
MWestonR	1980-85
MiHM	1957-69
MiU	1957, 59-67
MnU	1955, 57-75, 76(B-1 thru C-7), 77-78, 80-81
MoSW	1965, 71
MtBC	1968
NAIU	1980, 82-84
NBiSU	1957-72, 74
NFredU	1980-85
NIC	1980-82, 84
NIH	1980-85
NNC	1949-51, 55, 57-59, 61-75, 76(B-1 thru C-7), 77-78, 80, 81, 83
NOneoU	1965
NRU	1956-63, 66-74, 77, 80-84
NSbSU	1958, 60-62, 64, 67-69, 71, 75-77, 80, 81, 83-85
NSyU	1926-41, 46-78, 80-83, 85
NbU	1959
NcU	1957-75
NdU	1971
NhD-K	1958-74, 76-78, 80-84
NjP	1955-74
OCU-Geo	1956-75, 76(B-1 thru C-7), 77-85
OU	1957-75
OkU	1958-75, 76(B-1 thru C-7), 80
PBU	1980-84
PBm	1956(microfiche), 58, 59-61(all microfiche), 62-65, 66(microfiche), 68, 70-75(all microfiche), 76(B-1 thru C-7)(microfiche), 77(microfiche)
PHarER-T	1980-82
PSt	1985
TMM-E	1957, 65, 67
TxBeaL	1974-75, 77-78
TxDaAR-T	1957, 67-69, 71
TxDaDM	1957-66
TxDaM-SE	1957-62, 80
TxDaSM	1959, 62-63
TxHSD	1933, 50-52
TxMM	1956
TxU	1957-75, 76(B-1 thru C-7), 77-78, 80-81, 83-84
TxU-Da	1958
ViBlbV	1949, 57-74
WU	1956-64, 66-85

(470) NEW YORK STATE MUSEUM AND SCIENCE SERVICE. EDUCATIONAL LEAFLET.

1962 12. Field guide to the central portion of the southern Adirondacks.
1965 18. Guidebook field trips. Mohawk Valley strata and structure, from Albany to Glen via Lake George.

CSt-ES	1962, 65
CaAEU	1962, 65
CaOHM	1962
CaOLU	1962
CaOTDM	1962, 65
CaOWtU	1962, 65
CoFS	1962
CoG	1962, 65
DI-GS	1962, 65
DLC	1962, 65
I-GS	1962, 65
ICF	1962, 65
IaU	1962, 65
InLP	1962, 65
InU	1962, 65
LNU	1962
LU	1962
MH-GS	1962, 65
MNS	1962
MiHM	1965
MoSW	1962, 65
NOneoU	1962
NcU	1962
NdU	1962, 65
NvU	1965
OU	1962, 65
OkU	1962, 65
PBL	1962, 65
PBm	1962
TMM-E	1962
TxBeaL	1962
TxHSD	1965
TxU	1962, 65
ViBlbV	1962, 65
WU	1962, 65

(471) NEW YORK STATE MUSEUM AND SCIENCE SERVICE. HANDBOOK.
1927 No. 1.   A popular guide to the geology and physiography of Alleghany State Park.
1933 No. 14.   Guide to the geology of John Boyd Thacher Park (Indian Ladder region) and vicinity.
1942 No. 19.   Guide to the geology of the Lake George region, New York.

AzTeS	1942
CLU-G/G	1927, 33, 42
CaAEU	1942
CaBVaU	1942
CaOWtU	1927, 42
CoG	1927, 33, 42
DI-GS	1927, 33, 42
DLC	1927, 33, 42
ICF	1927, 33, 42
IEN	1927
IaU	1942
InLP	1927, 33, 42
MNS	1942
MiU	1927
MoU	1942
MtBuM	1927, 33, 42
NBiSU	1927, 33, 42
NOneoU	1942
NdU	1927, 33, 42
NhD-K	1942
OU	1942
OkU	1927, 33, 42
PBL	1942
PSt	1942
TxU	1942
ViBlbV	1927, 33, 42

(472) NEWFOUNDLAND AND LABRADOR. MINERAL DEVELOPMENT DIVISION. REPORT.
1979 794.   Geomorphology of the Avalon Peninsula, Newfoundland. A guidebook comprising itineraries and explanatory notes for six field trips followed by a systematic summary.
   A. Signal Hill-Cuckold Head area (Map A).
   B. Flat Rock area (Map B).
   C. St. Philipps and St. Thomas areas (Map C).
   D. Topsail and Manuels areas (Maps D and D#SU1#BS).
   E. Area traversed by the Trans-Canada Highways and Salmonier Line (Map E).
   F. Areas at the head of St. Mary's Bay (Maps F and F#SU1#BS).
1984 84-3.   Mineral deposits of Newfoundland; a 1984 perspective.

CU-EART	1979
CaOOG	1979, 84
CaOWtU	1979
CoDGS	1984
DI-GS	1979, 84
OkU	1984

(473) NORTH AMERICAN CLAY MINERALS CONFERENCE. GUIDEBOOK.
1964 13th   See Clay Minerals Conference. 132.
1965 14th   See Clay Minerals Conference. 132.

(474) NORTH AMERICAN PALEONTOLOGICAL CONVENTION. GUIDEBOOK FOR FIELD TRIP.
1969 [1st]   1. Minnesota and Wisconsin. Shallow-water Precambrian and Paleozoic communities. (522)
   2. The middle Paleozoic paleontology and biostratigraphy of eastern Iowa. (522)
   3. Devonian-Mississippian biostratigraphy of northeastern Missouri and western Illinois.
1977 2nd   [4] Field conference on Late Cenozoic biostratigraphy of the Texas Panhandle and adjacent Oklahoma. (761)
   5. Upper Chesterian-Morrowan stratigraphy and the Mississippian-Pennsylvanian boundary in northeastern Oklahoma and northwestern Arkansas. (Cover title: Mississippian-Pennsylvanian boundary in northeastern Oklahoma and northwestern Arkansas) (508, 522)
   [7] Cretaceous facies, faunas and paleoenvironments across the Western Interior basin. (405)
1982 Aug.   Montreal, Canada. Field trips guidebook.
   Field trip A. Cambrian and Ordovician stratigraphy and paleontology of the Champlain Valley.
   Field trip B. Fossil communities, paleoecology and tectonic history of the Trenton Group of central New York State.
   Field trip C. Ordovician stratigraphy and paleoecology of St. Lawrence lowlands and the frontal Appalachians near Quebec City.
   Field trip D. Middle and late Ordovician fossiliferous rocks of the Montreal area.

CLU-G/G	1977(5, 7)
CLhC	1977(5)
CMenUG	1982
CPT	1977(5, 7)
CSt-ES	1977(5, 7)
CU-A	1977(5)
CU-EART	1977(4-5, 7), 82
CaACU	1977(5, 7), 82
CaAEU	1977(7)
CaOKQGS	1982
CaOOG	1977(5)
CaOWtU	1977(5)
CoDCh	1977(5, 7)
CoDGS	1977, 82
CoG	1977(4-5)
CoU	1977(5, 7)
DI-GS	1969(3), 77(4-5, 7), 82
DLC	1977(7)
ICF	1969(1-2)
ICIU-S	1977(5)
ICarbS	1977(7)
IU	1969(1-3), 77(4-5, 7)
IaU	1969(1-2), 77(4-5, 7), 82
InLP	1977(5, 7)
InU	1977(5, 7)
KLG	1977(5)
KyU-Ge	1977(5, 7), 82
MiHM	1977(5)
MnU	1969(1-2), 77(5, 7), 82
MoU	1969(1-2)
MtBuM	1977(5)
NSbSU	1977(5)
NSyU	1982
NcU	1977(5)
NdU	1977(7), 82
NhD-K	1977(5, 7)
NvU	1977(5)
OU	1969, 77(4-5, 7)
OkU	1977(4-5, 7)
PSt	1977(5)
TxBeaL	1977(5, 7)
TxCaW	1977(4)
TxDaDM	1977(7)
TxU	1969(1, 3), 77(5, 7)
UU	1977(5)
ViBlbV	1977(7)
WU	1977(7), 82

NORTH ATLANTIC TREATY ORGANIZATION. See NATO ADVANCED STUDIES INSTITUTE ON METALLOGENY AND PLATE TECTONICS. (441) and NATO ADVANCED STUDY INSTITUTE. (442)

(475) NORTH CAROLINA ACADEMY OF SCIENCE. SCIENCE EDUCATION PROJECT COMMITTEE.
1967   Guide for geologic field trip in ... County for earth science.
   1. Alamance
   2. Anson
   3. Ashe
   4. Avery
   5. Buncombe
   6. Burke
   7. Cabarrus
   8. Caldwell
   9. Carteret-Craven
   10. Caswell

11. Chatham-Lee-Moore
12. Cherokee
13. Clay
14. Cleveland
15. Craven-Carteret-Onslow-Jones
16. Cumberland
17. Davidson
18. Davie
19. Durham
20. Forsyth
21. Franklin
22. Gaston
23. Granville
24. Guilford
25. Halifax
26. Hartnett
27. Haywood
28. Henderson
29. Iredell(pt.1-2)
30. Jackson
31. Johnson
32. Jones-Carteret
33. Lee-Moore-Chatham
34. Lincoln
35. MacDowell
36. Macon
37. Madison
38. Mecklenburg-Union
39. Mitchell
40. Montgomery
41. Moore-Lee-Chatham
42. Onslow-Carteret
43. Orange
44. Person
45. Polk
46. Randolph
47. Richmond
48. Rockingham
49. Rowan
50. Rutherford
51. Surrey
52. Stokes-Forsyth
53. Swain
54. Transylvania
55. Union-Mecklenburg
56. Vance
57. Wake
58. Warren
59. Watauga
60. Wayne(4pts.)
61. Yadkin
62. Yancey

DLC	1967(60)
InU	1967(60)
NcU	1967(1-62)
NjP	1967(60)
OU	1967(60)

AzTeS	1952, 54-55, 66
CLU-G/G	1952, 54-55
CPT	1952, 54-55
CSdS	1952, 54
CSt-ES	1952, 54-55
CU-EART	1952, 54-55
CaAC	1952, 54-55
CaACAM	1954
CaACI	1952, 54-55
CaACM	1952
CaACU	1952, 55, 66
CaOHM	1955
CaOOG	1952, 54-55
CaOWtU	1954
CoDCh	1952, 54-55
CoG	1952, 54-55
CoU	1952, 54-55
DI-GS	1952, 54-55
DLC	1952, 54-55
ICF	1952, 54-55
ICIU-S	1955
IEN	1954
IU	1952, 54-55, 66
IaU	1952, 54-55, 66
IdPI	1955
InU	1952, 54-55
KyU-Ge	1952, 54-55, 66
LU	1952, 54-55
MiDW-S	1954
MiU	1952, 54-55
MnU	1952, 54-55, 66
MoSW	1954
MoU	1954
MtBuM	1952, 55
MtU	1952, 54-55
NbU	1952, 54-55
NcU	1952, 54-55
NdU	1952, 54-55, 66
NjP	1954-55
NmU	1952, 54
OCU-Geo	1952, 54-55
OU	1952, 54-55
OkOkCGS	1952, 55
OkT	1952, 54-55
OkU	1952, 54-55
PBL	1954
PBm	1954
PSt	1952, 54-55
SdRM	1955
SdU	1952, 54-55, 66
TxDaAR-T	1952, 54-55, 66
TxDaDM	1952, 54-55, 66
TxDaGI	1952
TxDaM-SE	1952, 54-55
TxDaSM	1952, 54-55
TxHSD	1955
TxMM	1954-55
TxU	1952, 54-55, 66
TxU-Da	1952, 54-55
TxWB	1952, 54-55
UU	1955
ViBlbV	1952, 54-55, 66
WU	1954, 55, 66

(476) NORTH DAKOTA GEOLOGICAL SOCIETY. FIELD TRIP GUIDE.
1982   See International Williston Basin Symposium. 328.

(477) NORTH DAKOTA GEOLOGICAL SOCIETY. GUIDE BOOK FOR THE ANNUAL FIELD CONFERENCE.
1952 1st   Southern Manitoba and the Interlake area, Province of Manitoba.
1954 2nd   Southwestern North Dakota field conference.
1955 3rd   South Dakota Black Hills field conference.
1966 4th   Black Hills field conference including an informal study of adjacent areas in southwest North Dakota and northwest South Dakota.

(478) NORTH DAKOTA GEOLOGICAL SURVEY. BULLETIN.
1956 No. 30.   Guide for geologic field trip in northeastern North Dakota.

CMenUG	1956
CPT	1956
CSdS	1956
CSt-ES	1956
CaOLU	1956
CaOWtU	1956
CoDCh	1956
DI-GS	1956
DLC	1956
ICF	1956
ICarbS	1956
IU	1956
InLP	1956
MiHM	1956

(479) North Dakota Geological Survey.

MnDuU	1956
MnU	1956
MoSW	1956
MtBuM	1956
NcU	1956
NdU	1956
NhD-K	1956
NvU	1956
SdRM	1956
TxBeaL	1956
TxU	1956
UU	1956
WU	1956

(479) NORTH DAKOTA GEOLOGICAL SURVEY. EDUCATIONAL SERIES.
1972 No. 1. Geology along North Dakota Interstate Highway 94.
— No. 2. Guide to the geology of northeastern North Dakota, including Cavalier, Grand Forks, Nelson, Pembina, and Walsh counties.
— No. 3. Guide to the geology of southeastern North Dakota, including Barnes, Cass, Griggs, Ransom, Richland, Sargent, Steele, and Traill counties.
1973 No. 4. Geology along the South Loop Road, Theodore Roosevelt National Memorial Park.
— No. 6. Guide to the geology of south-central North Dakota, including Burleigh, Dickey, Emmons, Kidder, LaMoure, Logan, McIntosh and Stutsman counties.
1974 No. 7. Guide to the geology of north-central North Dakota, including Benson, Bottineau, Eddy, Foster, McHenry, Pierce, Ramsey, Rolette, Sheridan, Towner, and Wells counties.
1975 No. 8. Guide to the geology of Northwest North Dakota, including Burke, Divide, McLean, Mountrail, Renville, Ward, and Williams counties.
— No. 9. Guide to the geology of southwestern North Dakota, including Adams, Billings, Bowman, Dunn, Golden Valley, Grant, Hettinger, McKenzie, Mercer, Morton, Oliver, Sioux, Slope, and Stark counties.
1983 No. 16. Geology along North Dakota; Interstate highway 94.

AzTeS	1983
CLU-G/G	1983
CMenUG	1972-75, 83
CPT	1972(1, 2), 73(6), 74-75, 83
CSdS	1972-75
CSfCSM	1983
CSt-ES	1972-75, 83
CU-A	1975(9)
CU-EART	1972-74, 75(8)
CU-SB	1983
CaACU	1972(1, 3), 73-75, 83
CaOOG	1983
CaOWtU	1972, 73(6), 74-75
CoDCh	1972(2, 3), 73(6), 74(7), 75, 83
CoDGS	1983
CoG	1983
CoU	1972(2-3), 73-75
DI-GS	1972(2-3), 73-75, 83
DLC	1972(2-3), 73(6), 74, 75(9)
I-GS	1973(6)
ICF	1972-75
ICarbS	1972(2-3), 73-74, 75(8, 9)
IEN	1972-74, 75(8)
IU	1972-75
IaU	1972(2-3), 73-74, 83
IdU	1983
InLP	1972
InRE	1972(3), 73-74, 75(8)
InU	1972(2-3), 73-74, 75(8)
KLG	1983
KU	1983
KyU-Ge	1972(2-3), 73(6), 74-75
MH-GS	1975(8)
MiHM	1972(2-3), 73-75, 83
MiU	1972(3)
MnDuU	1975(8)
MnU	1972-75, 83
MoSW	1972-75
MtBuM	1972-75, 83
NSyU	1972(2-3), 73-74
NcU	1972(2-3), 73(4), 83
NdFA	1983
NdU	1972-74, 83
OCU-Geo	1972-75
OCl	1983
OU	1972-75
OkT	1972(3), 73-74, 83
OkU	1972(2-3), 73, 75
PBm	1983
PHarER-T	1983
PSt	1972(3), 73(6), 74
SdRM	1972-74, 75(9)
TxBeaL	1972-74, 75(9)
TxHSD	1972(3), 73(4)
TxLT	1972(2), 74
TxU	1972(2-3), 73, 75
UU	1972(2-3), 73, 75
ViBlbV	1972(2-3), 73-75
WU	1972-74, 75(9), 80-85
WyU	1975(3), 83

(480) NORTH DAKOTA GEOLOGICAL SURVEY. MISCELLANEOUS SERIES.
1957 No. 1. See Boy Scouts of America. 102.
— No. 2. See Boy Scouts of America. 102.
— No. 3. See Boy Scouts of America. 102.
— No. 4. See Boy Scouts of America. 102.
— No. 5. See Boy Scouts of America. 102.
— No. 6. See Boy Scouts of America. 102.
— No. 7. See Boy Scouts of America. 102.
— No. 8. See Boy Scouts of America. 102.
— No. 91. See Boy Scouts of America. 102.
1958 No. 10. See Friends of the Pleistocene. Midwest Group. 183.
1967 No. 30. See Friends of the Pleistocene. Midwest Group. 183.
1969 No. 39. See National Association of Geology Teachers. North Central Section. 416.
— No. 40. Geologic field trip from Grand Forks, North Dakota to Kenora, Ontario.
1970 No. 42. Guide to the geology of Burleigh County, North Dakota.
Trip no. 1. Southwest Burleigh County-Bismarck to McKenzie.
Trip no. 2. Southeast Burleigh County-McKenzie to Driscoll.
Trip no. 3. West central Burleigh County-Bismarck to Baldwin.
Trip no. 4. Northern Burleigh County-Wilton to Wing.
1972 No. 50 See Geological Society of America. 197.

AzTeS	1969(40), 70
CLU-G/G	1969(40), 70
CMenUG	1969(40), 70
CPT	1969(40), 70
CSdS	1969(40), 70
CSt-ES	1969(40), 70
CU-EART	1969(40), 70
CaACU	1969(40)
CaOLU	1969(40), 70
CaOOG	1969(40), 70
CaOWtU	1969(40), 70
CoDCh	1970
CoG	1969(40), 70
DI-GS	1969(40), 70
DLC	1970
ICF	1969(40), 70
ICarbS	1969-70
IEN	1969(40), 70
IU	1970
IaU	1969(40), 70
InLP	1969(40), 70
InU	1970
KyU-Ge	1969, 70
MNS	1970
MiDW-S	1969(40), 70
MiHM	1969(40), 70
MnDuU	1969(40), 70
MnU	1969(40), 70
MoSW	1969(40), 70
MtBuM	1969(40), 70
NBiSU	1969(40), 70
NSyU	1969(40), 70
NcU	1969(40), 70

NdU	1969(40), 70
NhD-K	1969(40), 70
NjP	1969(40), 70
NmSoI	1970
NvU	1969(40), 70
OCU-Geo	1958, 69(40), 70
OU	1969(40), 70
OkU	1969(40), 70
OrU-S	1969(40), 70
PBL	1969(40), 70
PSt	1970
SdRM	1969(40), 70
TxBeaL	1969(40), 70
TxDaAR-T	1969(40), 70
TxLT	1970
TxU	1969(40), 70
UU	1970
ViBlbV	1969(40), 70
WU	1969(40), 70

NORTH DAKOTA. UNIVERSITY. See UNIVERSITY OF NORTH DAKOTA. GEOLOGY DEPARTMENT. (703)

(481) NORTH TEXAS GEM AND MINERAL SOCIETY.
1957 Mid-winter field trip to Wichita Mountain area.
TxU   1957

(482) NORTH TEXAS GEOLOGICAL SOCIETY. ANNUAL FIELD TRIP.
1939 To study the Pease River Group (approximately equivalent to the Texas Blaine) and the Custer in North Texas, the Texas Panhandle, and in southwestern Oklahoma.
1940 Strawn and Canyon series of the Brazos and Trinity River valleys.
1947 Cambrian and Ordovician rocks of the Wichita Mountains.
1956 Facies study of the Canyon-Cisco series in the Brazos River area, north-central Texas.
1958 See Southwestern Federation of Geological Societies. 627.
1959 Guide to the Upper Permian and Quaternary of north-central Texas.

CLU-G/G	1939
CU-A	1939
CU-SB	1956
CaACU	1956, 59
CaOKQGS	1959
CoG	1956, 59
CoU	1959
DI-GS	1940, 47, 56, 59
ICF	1959
InLP	1956, 59
MnU	1956, 59
NNC	1956, 59
NbU	1959
NcU	1956, 59
NhD-K	1959
OU	1939, 59
OkU	1947, 56, 59
TxBeaL	1959
TxCaW	1956, 59
TxDaDM	1959
TxDaGl	1956, 59
TxDaM-SE	1959
TxDaSM	1956
TxHSD	1939-40, 47, 59
TxHU	1956, 59
TxLT	1947, 56
TxMM	1940, 47, 56, 59
TxU	1947, 56, 59
TxU-Da	1956
ViBlbV	1956, 59
WU	1956, 59

(483) NORTHEAST AMERICA SOCIETY OF AGRONOMY MEETING. GUIDEBOOK.
1980 June Footnote #1 Agronomy Series #64. See Northeast Cooperative Soil Survey Conference. 484.

(484) NORTHEAST COOPERATIVE SOIL SURVEY CONFERENCE. GUIDEBOOK.
1980 June University Park, Pennsylvania.
  Soils and geology of Nittany Valley. (532, 483)
  DI-GS   1980

(485) NORTHEAST LOUISIANA GEOLOGICAL SOCIETY.
1966 May Ouachita Mountains.
  TxWB   1966

(486) NORTHEASTERN UNIVERSITY. DEPARTMENT OF EARTH SCIENCES.
1980 Oct. See New England Intercollegiate Geological Conference. 456.

(487) NORTHERN CALIFORNIA GEOLOGICAL SOCIETY. [GUIDEBOOK]
1954 See American Association of Petroleum Geologists. Pacific Section. 28.
1962 See American Association of Petroleum Geologists. 18.
1968 Field trip to the Geysers, Sonoma County, California.
1969 Mount Diablo-Camp Parks (Northern California).
1970 San Andreas Fault and Point Reyes Peninsula.
1979 Spring Geology and engineering in the Livermore-Hayward region, California.
1981 May See American Association of Petroleum Geologists. 18.

CMenUG	1970
CSfCSM	1968
CU-A	1979
CU-EART	1970
DLC	1979
InU	1968
TxDaAR-T	1968
TxDaDM	1968
TxU	1968, 79

(488) NORTHERN OHIO GEOLOGICAL SOCIETY. [GUIDEBOOK]
1970 Guide to the geology of northeastern Ohio.
1978 Cuyahoga Valley National Recreation Area.

CSt-ES	1970
CaBVaU	1970
CaOWtU	1970
DI-GS	1970
IU	1970
InRE	1970
KyU-Ge	1970
MiU	1970
OCU-Geo	1970
OU	1970, 78
PSt	1970
TxDaAR-T	1970
TxU	1970

(489) NORTHWEST DISCOVERY.
1982 Fremont in Oregon; Deschutes River to Klamath Marsh.

CLU-G/G	1982
CMenUG	1982
CoDGS	1982
DI-GS	1982

(490) NORTHWEST GEOLOGY.
1982 V. 11. See Tobacco Root Geological Society. 658.(1981)

NORTHWEST SCIENTIFIC ASSOCIATION. COLUMBIA RIVER BASALT SYMPOSIUM. FIELD TRIP. See COLUMBIA RIVER BASALT SYMPOSIUM. (145)

(491) NORTHWEST SCIENTIFIC ASSOCIATION. GEOLOGY-GEOGRAPHY SECTION. FIELD TRIP.
1969 See Columbia River Basalt Symposium. 145.

(492) NOVA SCOTIA. DEPARTMENT OF MINES. GUIDEBOOK.
1948 The mineral province of eastern Canada; mineral and geological guidebook.

(493) Ohio Academy of Science. Section of Geology.

1954 Mineral and geological guide book.

CLU-G/G	1948
CSt-ES	1948
CaACU	1954
CaAEU	1948, 54
CaOWtU	1954
DI-GS	1948
DLC	1954
MtBuM	1948
TxHU	1954
TxU	1954

(493) OHIO ACADEMY OF SCIENCE. SECTION OF GEOLOGY. GUIDE TO THE ANNUAL FIELD CONFERENCE.

1948 23rd A study of the geology of Lucas County and the lime-dolomite belt.
1949 24th A study of the geology of Perry County.
1950 25th Spring. Glacial geology of west-central Ohio.
  Fall. Study of stratigraphy and sedimentation of Sharon Conglomerate northeastern Ohio.
1951 26th No title available; [southwestern Ohio].
1952 27th Glacial deposits of northeastern Ohio.
1953 28th No title available; [Columbus, Ohio area].
1954 29th Some geologic features of Athens County.
1955 30th Geology of the Bellefontaine outlier.
1956 31st The natural environment of the Springfield area.
1957 32nd Geology of the central lake plains area.
1958 33rd Geology of the Akron-Cleveland area.
1959 34th Geology of the Columbus-Galena-Gahanna area.
1960 35th Geology of the Yellow Springs region.
1961 36th Geology of the Cincinnati region.
1962 37th Geology of the Toledo area.
1963 38th Geology of the Highland-Adams County area.
1964 39th Upper Paleozoic stratigraphy of Lake and Geauga counties.
1965 40th Drainage history of a part of the Hocking River valley.
1966 41st Industrial minerals in northeastern Perry and southwestern Muskingum counties [Development, utilization and stratigraphy of].
1967 Silurian geology of western Ohio, Dayton.
1968 43rd Structures and fabrics in some Middle and Upper Silurian dolostones, northwestern Ohio.
1969 44th Guide to the Devonian Stratigraphy of central Ohio.
1970 45th Cincinnatian strata from Oregonia to the Ohio River with notes on Pleistocene geology along the route (Warren and Clermont counties).
1971 46th Geology and suburban-urban land use in portions of Summit, Portage and Stark counties, Ohio.
1973 48th Aspects of the engineering-environmental geology in the lower Cuyahoga River valley, Ohio.
1974 [49th] Geology field trip guidebook: Bedrock and Pleistocene geology of Wayne County, Ohio. ("Presented in conjunction with the annual meeting of The Ohio Academy of Sciences at The College of Wooster, Wooster, Ohio, April 27, 1974")
1975 [50th] Bedrock and Pleistocene geology of eastern Licking County, Ohio. (Cover title: Denison Geology; centennial field trip, 1965-1975. "This field trip guidebook is presented in conjunction with the 84th annual meeting of The Ohio Academy of Sciences at Denison University, Granville, Ohio, April 24-26, 1975")
1977 52nd Franklin and Fairfield counties.
1979 See National Association of Geology Teachers. East Central Section. 410.
1985 Upper Ordovician stratigraphy, sedimentology, and paleontology along Backbone Creek, Clermont County, Ohio.

CLU-G/G	1949, 71
CMenUG	1961
DI-GS	1949, 54-55, 57-66
I-GS	1955
ICF	1960
IU	1961-63, 65-67, 71, 74
InU	1967
KyU-Ge	1967
MH-GS	1958
MnU	1961, 70
NNC	1961
OCU-Geo	1952, 55-58, 60, 63, 65, 67, 70
OU	1948-67, 71, 74-75, 77, 85
TxDaAR-T	1959
TxDaDM	1949, 54-57, 68
TxU	1959-60, 62, 68, 70, 73

(494) OHIO GEOLOGICAL SOCIETY. FIELD TRIP. [GUIDEBOOK]

1965 Cambrian and Ordovician formations in the vicinity of the Cumberland overthrust block of Tennessee and Virginia.
1967 Guide to the annual field conference, northeastern Ohio.
1968 See Geological Society of Kentucky. 209.
1969 Spring A field guide to Allegheny deltaic deposits in the upper Ohio Valley with a commentary on deltaic aspects of Carboniferous rocks in the northern Appalachian Plateau. (541)
1970 May Guidebook to the Middle Devonian rocks of north-central Ohio.
1972 See American Association of Petroleum Geologists. Eastern Section. 25.
1983 Cuyahoga and Logan formations of central and eastern Licking County, Ohio.

AzTeS	1970
CLU-G/G	1965, 69
CPT	1970
CU-SB	1970, 72
CaACU	1969-70
CaOKQGS	1965, 70
CaOOG	1969
CaOWtU	1965, 69-70
CoG	1967
CoU	1969-70
DI-GS	1965, 69-70
I-GS	1969-70
IU	1969-70
IaU	1969
InLP	1965, 69
InU	1969-70
KyU-Ge	1965, 67, 69-70
MH-GS	1970
MnU	1965, 69-70
MoSW	1970
NBiSU	1969
NNC	1969
NcU	1969-70
NhD-K	1970
NjP	1969
OCU-Geo	1967, 69-70
OU	1965, 67, 69-70, 83
OkOkU	1969
OkU	1969
PBm	1969
PHarER-T	1969
PSt	1969
TxDaAR-T	1965, 67, 69-70
TxDaDM	1965, 69-70
TxDaGI	1969
TxHSD	1969
TxU	1969-70
ViBlbV	1969, 70
WU	1969, 70

(495) OHIO HISTORICAL SOCIETY. NATURAL HISTORY INFORMATION SERIES.

1980 No. 3. Rock and mineral collecting sites in Ohio.

DI-GS	1980

(496) OHIO INTERCOLLEGIATE FIELD CONFERENCE IN GEOLOGY. FIELD TRIP GUIDEBOOK. (TITLE VARIES)

1950 1st Cuyahoga Gorge and Chippewa Creek sections, Northeastern, Ohio.
1951 2nd Not held.
1952 3rd [Van Buren Lake]
1953 4th [Bedford Glens]
1954 5th Southeastern Stark County.
1955 6th [Delaware County]
1956 7th Geologic setting of Granville and Newark in Licking County, Ohio.
1957 8th Some Pennsylvania cyclothems in Athens County, Ohio.

1958	9th	[Oxford area]
1959	10th	Judy Gap trip. (See 16th)
1960	11th	[Northwestern Ohio]
1961	12th	[Delaware County]
1962	13th	Fairport Mine, Painesville, Ohio. (346)
1963	14th	Pennsylvanian in Muskingum and Coshocton (Ohio).
1964	15th	[Holmesville, Wooster] (Ohio)
1965	16th	Judy Gap trip. (Same as 10th)
1966	17th	[Southwestern Ohio]
1967	18th	[Vicinity Alliance, Ohio]
1968	19th	[Granville-Newark area]
1969	20th	Type localities of selected Mississippian and Pennsylvanian strata in northwestern Pennsylvania and northeastern Ohio. (345)
1974	25th	Some geological features in Pendleton County, West Virginia and Highland County, Virginia.
1975		No guidebooks.
1976		No guidebooks.
1979	30th	[Williamsport area] (786, 787)

CLU-G/G	1950, 52-69
CMenUG	1950, 53-54
CaOKQGS	1950, 52-69
CaOOG	1950, 52-69
CaOWtU	1950, 52-69
DI-GS	1950, 52-69, 74
IaU	1950, 52-69, 79
InLP	1950, 52-69
InRE	1950, 52-69
InU	1950, 52-69
KyU-Ge	1950, 53-69
LNU	1950, 52-69
NcU	1950, 52-69
NdU	1950, 52-69
OCU-Geo	1950, 52-69
OU	1950, 52-69, 74
PHarER-T	1969
PSt	1950, 52-69
TMM-E	1950, 52-69
TxDaDM	1950, 52-69
TxU	1950, 52-69
ViBlbV	1950, 52-69

(497) OHIO SEDIMENTOLOGY. (TITLE VARIES)
1979	1st	Shawnee State Park, Friendship, Ohio.
1980	2nd	Hueston Woods State Park, Oxford, Ohio.
1981	3rd	Cuyahoga and Logan formations of central and eastern Licken County, Ohio.
1983	5th	Aspects of the sedimentary geology along the southwestern shore of Lake Erie.
1984	6th	Limeville, Greenup County, Kentucky.
1985	7th	No field trip.
OU		1978, 80, 81, 83, 84, 85

(498) OHIO STATE UNIVERSITY. DEPARTMENT OF GEOLOGY AND MINERALOGY.
1985 Field guide to the geology of parts of the Appalachian Highlands and adjacent interior plains.

| CU-A | 1985 |
| KyU-Ge | 1985 |

(499) OHIO STATE UNIVERSITY. GEOLOGY CLUB.
1958	[No title given, but it is about the Appalachian region]
1960	The Southern Appalachians and the Great Smoky Mountains.
1962	Field trip to Missouri.
1963	Central Appalachian seminar and field trip.
1965	[Great Smoky Mountains]
1966	Appalachians highlands of Pennsylvania.
1968	Appalachian spring field trip.
1969	Southern Illinois fluorspar and southeast Missouri lead-zinc districts.
OU	1958, 60, 62-63, 65-66, 68-69
TxDaDM	1965

(500) OHIO STATE UNIVERSITY. INSTITUTE OF POLAR STUDIES. MISCELLANEOUS PUBLICATION.
1985	No. 225.	See Gondwana Symposium. 244.
	— No. 226.	See Gondwana Symposium. 244.
	— No. 227.	See Gondwana Symposium. 244.
	— No. 228.	See Gondwana Symposium. 244.
	— No. 229.	See Gondwana Symposium. 244.

OHIO. DIVISION OF GEOLOGICAL SURVEY. See GEOLOGICAL SURVEY OF OHIO. (228) and GEOLOGICAL SURVEY OF OHIO. (227) and GEOLOGICAL SURVEY OF OHIO. (229)

(501) OIL SHALE SYMPOSIUM. [FIELD TRIP GUIDEBOOK]
1985 18th   (1) Parachute Creek field trip. (27, 246, 603, 663, 431, 770.)
(2) Utah synfuels field trip. (27, 246)

| CU-SB | 1985 |
| OkTA | 1985 |

(502) OKLAHOMA ACADEMY OF SCIENCE.
1947	May 3	Braggs Mountain section.
	— May 4	Muskogee to Prague, Oklahoma; a geological road guide.
1952		Road log, geological field trip in eastern part of the Ouachita Mountains in Oklahoma. (576)
1953		Boiling Springs State Park, Woodward, Oklahoma.
1955		Road log, field trip from Dwight Mission to Tahlequah.
1959		Camp Egan area.

DI-GS	1952
OkU	1947, 52-53, 55, 59
TxLT	1947, 52-53

(503) OKLAHOMA CITY GEOLOGICAL SOCIETY. CONTINUING EDUCATION SHORT COURSE.
1983 Feb.   Cretaceous wave-dominated delta systems: Book Cliffs, east central Utah.

CoDCh	1983
IU	1983
OkU	1983

(504) OKLAHOMA CITY GEOLOGICAL SOCIETY. GUIDEBOOK FOR THE FIELD CONFERENCE.
1930	Stratigraphic section from Neva Ls thru Elgin Ss.
1932	See American Association of Petroleum Geologists. 18.
1936	Oct. Study of the Simpson Formation, Sections 24 and 25-T, 2S-R1E.
	—Nov. Field trip, study of the Simpson Formation, Sections 5, 8 and 17-T, 2S-R1W.
1937	Simpson Formation, Section 2 and 12-T1N-R6E.
1939	See American Association of Petroleum Geologists. 18.
1940	Field trip; structural and stratigraphic features of Wichita Mountains.
1941	Mesozoic rocks of the Oklahoma Panhandle, including an area in northeastern New Mexico.
1946	Lower Permian and Upper Pennsylvanian, north-central Oklahoma.
1949	Precambrian, Cambrian and Ordovician rocks of the Wichita Mountain area.
1950	Eastern part of the Ouachita Mountains in Oklahoma; with special reference to the pre-Pennsylvanian and Lower Pennsylvanian rocks.
1953	Pre-Atoka rock in western part of the Ozark Uplift, northeastern Oklahoma. (508)
1954	May On Desmoinesian rocks of northeastern Oklahoma. (508)
1955	Highway geology of Oklahoma. Road logs of the major highways of the state.
1956	Apr. Geology of the Turner Turnpike; road log, geologic profile, route map. (508, 662)
	— Sept. Panhandle of Oklahoma, northeastern New Mexico, south-central Colorado.
1964	Variations in limestone deposits, Desmoinesian and Missourian rocks in Northeast Oklahoma.
1968	See American Association of Petroleum Geologists. 18.
1970	The Bahamas and southern Florida. (564)
1972	A guidebook to the genesis and geometry of sandstones.
1978	See American Association of Petroleum Geologists. 18.

(505) Oklahoma Geological Survey.

1983 Cretaceous wave-dominated delta systems: Book Cliffs, east central Utah 1982, a field guide.

ATuGS	1940, 53-54, 56(Apr.)
AzFU	1955
AzTeS	1953-55, 56(Apr.), 64, 70
AzU	1953-54, 56(Apr.)
CLU-G/G	1949-50, 53-55, 56(Apr.), 64, 70, 72
CLhC	1953-54, 56(Apr.), 72
CPT	1953-54, 56
CSt-ES	1949, 53-55, 56(Apr.), 64, 70
CU-EART	1954-56
CU-SB	1968, 78
CaACM	1950, 55-56
CaACU	1955, 56(Apr.), 70, 72
CaAEU	1953-54, 56(Apr.)
CaBVaU	1955
CaOHM	1954-55, 56(Apr.), 64
CaOKQGS	1956(Apr.), 64
CaOLU	1954, 56(Apr.), 72
CaOOG	1953-54, 56(Apr.)
CaOTRM	1955
CaOWtU	1956(Apr.)
CoDCh	1946, 50, 53-54, 56, 64
CoDU	1954
CoFS	1956(Apr.)
CoG	1953-54, 56, 64
CoLiM	1983
CoU	1953-56, 64
DI-GS	1946, 49-50, 53-56, 64
DLC	1953-55, 56(Apr.)
I-GS	1941, 56(Apr.)
ICF	1953-54, 64
ICIU-S	1956(Apr.)
ICarbS	1953-55, 56(Apr.), 72
IEN	1949-50, 53-55, 56(Apr.)
IU	1946, 50, 53-54, 56(Apr.), 64
IaU	1954-55, 64
InLP	1953-55, 56(Oct.), 64, 70
InU	1949, 53-54, 56
KyU-Ge	1953-56
LNU	1955, 72
LU	1953-54, 56(Apr.)
MNS	1953-56
MiDW-S	1953-54, 56(Apr.)
MiU	1953-55, 56(Apr.)
MnDuU	1953-54
MnU	1953-56, 64
MoRM	1946
MoSW	1949-50, 53-55, 56(Apr.)
MoU	1953-54, 56(Apr.), 64
MtBuM	1953-54, 56(Apr.)
NNC	1949, 53-54, 56
NSyU	1953-54, 56
NbU	1950
NcU	1953-54, 56(Apr.), 64
NdU	1953-54, 56(Apr.)
NhD-K	1953-54, 56(Apr.)
NjP	1950, 53-54, 56(Apr.)
NmPE	1954-55, 72
NmU	1956
NvU	1953-55, 56(Apr.)
OU	1949, 53-54, 56, 64
OkOkCGS	1946, 50, 53-54, 64
OkOkU	1953-54
OkT	1950, 53-54, 56, 64, 72
OkU	1930, 36-37, 40-41, 46, 49-50, 53-56, 64, 70, 83
OrU-S	1953-54, 56(Apr.)
PBm	1953-54, 56(Apr.)
PSt	1953-54, 56(Apr.)
RPB	1953-54, 56
TMM-E	1956(Apr.)
TU	1953-54, 56(Apr.)
TxBeaL	1953-54, 56(Sept.), 64
TxCaW	1955
TxDaAR-T	1949-50, 53-54, 56, 70
TxDaDM	1936, 40, 50, 53-56, 64
TxDaGI	1950, 56(Apr.)
TxDaM-SE	1940, 49-50, 53-54, 56, 64
TxDaSM	1956(Sept.), 64, 70, 72
TxHSD	1950, 53-54, 56(Sept.), 72
TxHU	1940, 46, 50, 53-56
TxLT	1936-37, 40-41, 49-50, 53-55, 56(Apr.)
TxMM	1936, 40-41, 46, 49-50, 53-54, 56, 64, 72
TxU	1941, 50, 53-56, 64
TxU-Da	1953-54, 56
TxWB	1956
UU	1953-54, 56(Apr.)
ViBlbV	1953-56, 64, 70
WU	1946, 53-55, 56(Apr.)

(505) OKLAHOMA GEOLOGICAL SURVEY.
1964 Oct. See Kansas Geological Society. 341.
1973 Nov. See Geological Society of America. 197.
1978 Apr. See American Association of Petroleum Geologists. 18.

(506) OKLAHOMA GEOLOGICAL SURVEY. EDUCATIONAL PUBLICATION.
1971 No. 2. Guidebook for geologic field trips in Oklahoma. Book 1: Introduction, guidelines, and geologic history of Oklahoma.
1972 No. 3. Guidebook for geologic field trips in Oklahoma.
1981 No. 4. Guidebook for geologic field trips in north-central Oklahoma.

AzTeS	1971, 72, 81
CLU-G/G	1981
CPT	1971, 72, 81
CSt-ES	1971, 72, 81
CU-EART	1981
CaACU	1971-72
CaOKQGS	1981
CaOOG	1981
CaOWtU	1971-72, 81
CoDCh	1971-72
CoDGS	1981
DI-GS	1972, 81
DLC	1972
I-GS	1972
IU	1972
IaU	1972, 81
InLP	1971-72, 81
InU	1972
KyU-Ge	1971(2), 72, 81
MoSW	1972, 81
NdU	1972
NhD-K	1972, 81
NvU	1972
OCU-Geo	1971-72
OkT	1971-72, 81
OkU	1972
PHarER-T	1981
TxBeaL	1972
TxDaM-SE	1981
TxU	1972, 81
WU	1971, 72, 81

(507) OKLAHOMA GEOLOGICAL SURVEY. FIELD CONFERENCE. GUIDEBOOK
1927 10th Arbuckle Mountains and the Ardmore Basin.

AzU	1927
IU	1927
KyU-Ge	1927
OkU	1927
TxU	1927

(508) OKLAHOMA GEOLOGICAL SURVEY. GUIDEBOOK.
1953 1. See Oklahoma City Geological Society. 504.
1954 2. See Oklahoma City Geological Society. 504.
1955 3. See Ardmore Geological Society. 62.
1956 4. See Oklahoma City Geological Society. 504.
1957 5. See Panhandle Geological Society. 528.
1958 7. Guide to Robbers Cave State Park and Camp Tom Hale, Latimer County, Oklahoma.
1959 9. Guide to Roman Nose State Park, Blaine County, Oklahoma.
1963 11. Guide to Beavers Bend State Park.
— 12. Parks and scenic areas in the Oklahoma Ozarks.
— 13. Sample descriptions and correlations for wells on a cross section from Barber County, Kansas, to Cado County, Oklahoma. (Cover title: Well-sample descriptions, Anadarko Basin.
1966 16. Late Paleozoic conodonts from the Ouachita and Arbuckle Mountains of Oklahoma.

1968 See American Association of Petroleum Geologists. 18.
1969 15. Guide to Alabaster Cavern and Woodward County. (Revised in 1982)
— 17. Regional geology of the Arbuckle Mountains.
1977 18. See North American Paleontological Convention. 474.
1979 19. See International Congress of Carboniferous Stratigraphy and Geology. 290.(No.11)
1982 20. See Geological Society of America. South Central Section. 205.(2)
— 21. See Geological Society of America. South Central Section. 205.(1)

AzTeS	1958-59, 63, 66
AzU	1958-59, 63, 66, 69
CLU-G/G	1958-59, 63, 66, 69
CLhC	1966
CMenUG	1958-59, 63(12), 66, 69, 77
CPT	1958-59, 63, 66
CSt-ES	1958-59, 63, 66, 69
CU-EART	1958, 59, 63, 66, 69
CaACU	1958-59, 63, 66, 69, 77, 79
CaAEU	1958-59, 63, 66, 69
CaBVaU	1966, 69(17)
CaOHM	1958-59, 69(15)
CaOKQGS	1959, 63, 66, 69(15)
CaOLU	1959, 63, 66, 69(15)
CaOOG	1958-59, 63, 66, 69
CaOWtU	1959, 66, 69(15, 17)
CoDU	1958-59, 63, 66, 69(15)
CoFS	1958-59, 63, 69
CoG	1958-59, 63, 66, 69
CoU	1958-59, 63, 66, 69
DI-GS	1958-59, 63, 66, 69
DLC	1958-59, 63, 66, 69
I-GS	1958-59
ICF	1959
ICIU-S	1958-59, 63, 66, 69
ICarbS	1958-59, 63, 66(16), 69
IEN	1958-59, 63, 66, 69(15)
IU	1958-59, 63, 66, 69
IaU	1958-59, 63, 66, 69, 82(1-2)
IdBB	1969(15)
IdU	1958-59, 63, 66, 69(15)
InLP	1958-59, 63, 66, 69
InU	1958-59, 69,
KyU-Ge	1958-60, 63, 64, 66, 69
LNU	1959, 63, 66, 69
LU	1958-59, 63, 66, 69(15)
MNS	1958-59, 63, 66, 69
MiHM	1969
MiU	1958-59, 63, 66, 69(15)
MnU	1958-59, 63, 66, 69
MoSW	1958-59, 63, 66, 69(15)
MoU	1958-59, 63, 66, 69
MtBuM	1958-59, 66, 69
NNC	1958-59, 63, 66, 69
NSyU	1958-59, 63, 69(15)
NcU	1958-59, 63, 66, 69(17)
NdU	1958-59, 63, 66, 69(15)
NhD-K	1958-59, 63, 66, 69
NjP	1958-59, 63, 66, 69(15)
NmPE	1969
NvU	1958-59, 63, 66, 69
OCU-Geo	1963(12), 66, 69
OU	1958-59, 63, 66, 69
OkOkCGS	1963
OkOkU	1969(15)
OkT	1958-59, 63, 66, 69
OkU	1958-59, 63, 66, 69(15)
OrU-S	1958-59, 63, 66
PBm	1958-59, 63, 69(17)
PSt	1958-59, 63, 66, 69
TMM-E	1958-59, 63, 66, 69
TxBeaL	1959, 63, 66, 69(15)
TxCaW	1969(17)
TxDaDM	1969(17)
TxHSD	1958, 69(17)
TxHU	1969(15)
TxMM	1958-59, 63, 66, 69
TxU	1958-59, 63, 66, 69
UU	1958-59, 63, 66, 69(15)
ViBlbV	1958-59, 69
WU	1958-59, 63, 66, 69

(509) OKLAHOMA GEOLOGICAL SURVEY. SPECIAL PUBLICATION.
1984 84-1. See UNITAR Conference on Development of Shallow Oil and Gas Resources. 668.

(510) OKLAHOMA INDUSTRIAL AND MINERAL INDUSTRIES CONFERENCE.
1947 Nov. Industrial tour; manufacturing districts of Tulsa and Sand Springs. (511)

OkU	1947

(511) OKLAHOMA MINERAL INDUSTRIES CONFERENCE.
1945 Nov. Mineral resources field trip, Ada District.
1946 May Mineral resources field trip, Wichita Mountain District.
— Nov. Mineral resources field trip, Ada District.
1947 See Oklahoma Industrial and Mineral Industries Conference. 510.

CSt-ES	1945, 46(Nov.)
OkU	1946(May)
TxDaM-SE	1946(May)

OKLAHOMA. UNIVERSITY. SCHOOL OF GEOLOGY. See UNIVERSITY OF OKLAHOMA. SCHOOL OF GEOLOGY AND GEOPHYSICS. (704)

(512) ONTARIO ASSOCIATION FOR GEOGRAPHIC AND ENVIRONMENTAL EDUCATION (OAGEE). SPRING CONFERENCE. FIELD TRIP GUIDE.
1985 May See Canadian Geoscience Council. EdGEO Conference. 119.

ONTARIO DEPARTMENT OF MINES. See ONTARIO GEOLOGICAL SURVEY. (515) and LAMAR UNIVERSITY. DEPARTMENT OF GEOLOGY. (353) and ONTARIO GEOLOGICAL SURVEY. (513) and LAMAR UNIVERSITY. DEPARTMENT OF GEOLOGY. (353)

ONTARIO DIVISION OF MINES. See ONTARIO GEOLOGICAL SURVEY. (514) and ONTARIO GEOLOGICAL SURVEY. (515) and ONTARIO GEOLOGICAL SURVEY. (513)

(513) ONTARIO GEOLOGICAL SURVEY. GEOLOGICAL CIRCULAR.
1962 10 Geology and scenery along the north shore of Lake Superior.

OCU-Geo	1962

(514) ONTARIO GEOLOGICAL SURVEY. GUIDEBOOK. (TITLE VARIES)
1968 No. 1. Geology and scenery of Rainy Lake and east to Lake Superior.
1969 No. 2. Geology and scenery of north shore of Lake Superior.
— No. 3. Geology and scenery of Peterborough, Bancroft and Madoc area.
1972 No. 4. Geology and scenery, north shore of Lake Huron region.
1976 No. 5. Amethyst deposits of Ontario.
1979 No. 7. Geology and fossils, Craigleith area, Ontario.
1982 No. 6. Geology and scenery; Kilarney Provincial Park area, Ontario.

AzTeS	1976
CLU-G/G	1968-69, 72, 76, 79
CMenUG	1968-69, 76, 79, 82
CPT	1972, 79
CSt-ES	1968-69, 72, 76, 82
CU-EART	1968-69, 72, 76, 79, 82
CaACU	1968-69, 72, 76, 79, 82
CaBVaU	1969(3)
CaOHM	1968-69, 72
CaOKQGS	1968-69
CaOLU	1969(3)
CaOOG	1968-69, 76, 79
CaOONM	1968-69, 72
CaOTDM	1968-69, 72, 76, 79
CaOWtU	1968, 69(2), 72, 76, 79, 82
CoG	1972
CoU	1972
DI-GS	1968-69, 72
DLC	1972, 79, 82
ICF	1968-69, 72
ICIU-S	1968-69, 72

(515) Ontario Geological Survey.

IEN	1972
IU	1968-69, 72, 82
IaU	1976
KyU-Ge	1968-69, 72, 76, 79, 82
MH-GS	1968-69, 72
MiHM	1968-69, 72
MnU	1968-69, 72
NcU	1972, 79, 80
NhD-K	1972, 82
NjP	1968-69, 72
OCU-Geo	1968-69, 72, 76
OkU	1968, 69(3), 72
TxU	1968, 69(3)
UU	1968-69, 72
WU	1968-69, 72, 76, 79

(515) ONTARIO GEOLOGICAL SURVEY. MISCELLANEOUS PAPER.
1969 No. 29. A geological guide to Highway 60, Algonquin Provincial Park.
1983 No. 111. See Forum on the Geology of Industrial Minerals. 179.
1984 No. 118. Field trip guidebook to the Hemlo area.

CLU-G/G	1984
CMenUG	1969
CPT	1969, 84
CSt-ES	1969, 84
CaACU	1969
CaNSHD	1984
CaOKQGS	1984
CaOTDM	1969, 84
CaOWtU	1969, 83, 84
CoDGS	1984
CoG	1984
DI-GS	1969, 84
DLC	1969
ICF	1969
IaU	1984
KLG	1984
KU	1984
KyU-Ge	1969
MiHM	1969
MtBuM	1969
NIC	1983, 84
NNC	1983, 84
NRU	1983, 84
NcU	1984
NdU	1969, 84
NhD-K	1969, 83-84
OCU-Geo	1969
OkU	1969
TxU	1984
UU	1969
ViBlbV	1969
WU	1969, 83, 84

(516) OREGON HISTORICAL SOCIETY.
1985 See Geological Society of the Oregon Country. 216.

(517) OREGON. DEPARTMENT OF GEOLOGY AND MINERAL INDUSTRIES. BULLETIN.
1959 No. 50. See National Science Foundation. College Teachers Conference in Geology. 423.
1965 No. 57. See Lunar Geological Field Conference. 366.
1968 No. 62. See International Upper Mantle Committee. 327.
1973 No. 77. See Geological Society of America. Cordilleran Section. 199.
1980 No. 101. See Geological Society of America. Cordilleran Section. 199.

(518) OREGON. DEPARTMENT OF GEOLOGY AND MINERAL INDUSTRIES. OREGON GEOLOGY. (FORMERLY: THE ORE BIN)
1974 V. 36. No. 12. The Columbia River Gorge; the story of the river and the rocks.
1980 V. 42. No. 2. See Geological Society of America. Cordilleran Section. 199.
1982 V. 44. No. 8. Mount St. Helens road log: A visit to the lower Toutle and Cowlitz River drainages to observe the sedimentology, transported trees, and other effects of mudflows resulting from the eruptions of May 18, 1980, and March 19, 1982.
1983 V. 45. No. 7/8. Paleozoic and Triassic terranes of the Blue Mountains, Northeast Oregon: discussion and field trip guide, part II: road log and commentary.
No. 11. See Geothermal Resources Council. 243.
No. 12. See Geothermal Resources Council. 243.
1984 V. 46. No. 8. Exploring the Neogene history of the Columbia River: discussion and geologic field trip guide to the Columbia River Gorge. Part I. Discussion.
No. 9. Exploring the Neogene history of the Columbia River: discussion and geologic field trip to the Columbia River Gorge. Part II. Road log and comments.
1985 V. 47 No. 12. North Santaim mining area, Western Cascades - relations between alteration and volcanic stratigraphy: Discussion and field guide.

AzTeS	1982-84
CLU-G/G	1974, 82, 84, 85
CMenUG	1974, 82-83, 84(9)
CPT	1982-84
CSfCSM	1982-84
CSt-ES	1974, 82-84, 85
CU-EART	1974, 82-84
CU-SB	1982-84
CaACU	1974
CaAEU	1982-84
CaOOG	1982-84
CoDGS	1982-83, 84(8)
CoU	1974
DI-GS	1974, 82, 84
F-GS	1982-84
FBG	1982-84
IU	1974
IaU	1982-84
IdBB	1982-84
IdU	1982-84
InLP	1974, 82-84
InU	1982
MnU	1980, 82-84
MoSW	1974
MtBuM	1974
NIC	1982-84
NNC	1983, 84
NRU	1982-84
NcU	1974, 82-84
NdU	1982-84
NvU	1974, 83-84
OU	1982
OkU	1974
OrU-S	1982-84
PHarER-T	1982-84
TxDaM-SE	1982-84
TxU	1974, 82-84
UU	1982-84
WU	1974, 82-84
WyU	1982-84

(519) OREGON. DEPARTMENT OF GEOLOGY AND MINERAL INDUSTRIES. SPECIAL PAPER.
1978 No. 2. Field geology of S.W. Broken Top Quadrangle, Oregon.

CMenUG	1978
CPT	1978
CSt-ES	1978
CU-EART	1978
DI-GS	1978
IaU	1978
InLP	1978
OrCS	1978
ViBlbV	1978

OREGON. UNIVERSITY. See UNIVERSITY OF OREGON. MUSEUM OF NATURAL HISTORY. (706) and UNIVERSITY OF OREGON. DEPARTMENT OF GEOLOGY. (705)

(520) PACIFIC COAST PALEOGEOGRAPHY FIELD GUIDE.
1976 No. 1. See Society of Economic Paleontologists and Mineralogists. Pacific Section. 593.
1977 No. 2. See Society of Economic Paleontologists and Mineralogists. Pacific Section. 593.

1978 No. 3. See Society of Economic Paleontologists and Mineralogists. Pacific Section. 593.
1979 No. 4. See Society of Economic Paleontologists and Mineralogists. Pacific Section. 593.

(521) PALEOINDIAN LIFEWAYS, A SYMPOSIUM.
1983 See Friends of the Pleistocene. South-Central Cell. 186.

(522) PALEONTOLOGICAL SOCIETY.
1966 Precambrian-Cambrian succession; White-Inyo Mountains, California.
1969 See North American Paleontological Convention. 474.
1977 See North American Paleontological Convention. 474.
See Geological Society of America. 197.
1978 See Geological Society of America. 197.
1979 See Geological Society of America. 197.
1981 See Geological Society of America. 197.
1982 See Geological Society of America. 197.
1983 See Geological Society of America. 197.
1984 See Geological Society of America. 197.

CLU-G/G	1966
CU-EART	1966
CaOOG	1966

(523) PALEONTOLOGICAL SOCIETY. NORTH-CENTRAL SECTION. FIELD TRIP GUIDE BOOK.
1982 See Geological Society of America. North Central Section. 200.
1983 See Geological Society of America. North Central Section. 200.
1984 See Geological Society of America. Southeastern Section. 206.
1985 See Geological Society of America. North Central Section. 200.

(524) PALEONTOLOGICAL SOCIETY. PACIFIC COAST SECTION.
1973 Mar. See Geological Society of America. Cordilleran Section. 199.
1978 See Geological Society of America. Cordilleran Section. 199.
1982 See Geological Society of America. Cordilleran Section. 199.
1983 See Geological Society of America. Cordilleran Section. 199.
1984 Mar. See Geological Society of America. Cordilleran Section. 199.
1985 See Geological Society of America. Cordilleran Section. 199.

(525) PALEONTOLOGICAL SOCIETY. SOUTHEASTERN SECTION.
1980 See Geological Society of America. 197.(no.10)

PAN AMERICAN GEOLOGICAL SOCIETY. See SOUTHWESTERN ASSOCIATION OF STUDENT GEOLOGICAL SOCIETIES (SASGS). (626)

(526) PAN AMERICAN UNIVERSITY. GEOLOGICAL SOCIETY.
1964 See Southwestern Association of Student Geological Societies (SASGS). 626.
1970 See Southwestern Association of Student Geological Societies (SASGS). 626.
1974 See Southwestern Association of Student Geological Societies (SASGS). 626.

(527) PANDER SOCIETY. ANNUAL MEETING. GUIDEBOOK.
1972 See Geological Society of America. North Central Section. 200.
1973 Apr. See Geological Society of America. North Central Section. 200.
1974 See Geological Society of America. North Central Section. 200.
1978 Apr. See Geological Society of America. Rocky Mountain Section. 204.
1981 See Geological Society of America. North Central Section. 200.
1982 See Geological Society of America. North Central Section. 200.
1983 Apr.-May See Geological Society of America. North Central Section. 200.
1984 See Geological Society of America. Southeastern Section. 206.
1985 See Geological Society of America. Southeastern Section. 206.

(528) PANHANDLE GEOLOGICAL SOCIETY. FIELD TRIP GUIDEBOOK. (TITLE VARIES)
1938 Southern Colorado and northern New Mexico.
1939 Sacramento Mts. - White Sands - Siera Blanca region, New Mexico.
1946 Dry Cimarron River valley, Panhandle of Oklahoma and adjoining area; Front Range of Rocky Mountains and southeastern Colorado.
1949 Fossil and early man sites in the Texas Panhandle.
1951 May 5 Antelope Creek Pueblo and Triassic fossil site, Hutchinson and Potter counties, Texas.
— May 17 Dry Cimarron River valley, Panhandle of Oklahoma and adjoining area; Front Range of the Rocky Mountains and southeastern Colorado.
1953 Raton Basin region and the Sangre de Cristo Mountains of New Mexico.
1954 Fossil and early man sites in the Texas Panhandle.
1955 May Field trip of the Dry Cimarron River valley, the Panhandle of Oklahoma, northeastern New Mexico, lower Front Range of the Rocky Mountains, and southeastern Colorado.
1957 May Geology of the Wichita Mountain region of southwestern Oklahoma. (508, 704)
1958 Saddleback Pueblo and Rotten Hill Triassic fossil site, Oldham County, Texas.
1959 Southern Sangre de Cristo Mountains, New Mexico.
1961 Oct. See American Association of Petroleum Geologists. Regional Meeting. 29.
1963 Sept. Alibates flint quarries, Alibates Indian ruin, Santa Fe Trail, Sanford Dam.
1965 Field trip of the Rocky Dell site.
1969 Pre-Pennsylvanian geology of the western Anadarko Basin.
1973 Sept. Guidebook of interpretation of depositional environments from selected exposures of Paleozoic and Mesozoic rocks in north-central New Mexico.

AzFU	1957
AzU	1957, 59
CLU-G/G	1951(May 17), 53-55, 57-59, 63, 65, 69
CLhC	1957
CMenUG	1953-55, 58-59
CPT	1969
CSt-ES	1949, 57, 63, 69
CU-A	1963
CU-EART	1951(May 17), 54-55, 57
CU-SB	1973
CaACU	1953, 57, 69
CaOKQGS	1957
CaOOG	1957, 69
CoDCh	1953, 56, 57, 63, 67
CoDuF	1957
CoG	1938, 51(May 17), 54-55, 57, 59, 63
CoU	1951(May 5), 55, 57, 59
DI-GS	1946, 49, 51, 53-55, 57-59, 63, 65
DLC	1957, 69
ICF	1949, 54-55, 57-59, 63
IEN	1957, 59
IU	1946, 49, 54-55, 57-59, 63, 73
IaU	1969
InLP	1973
InU	1949, 57, 59
KyU-Ge	1951, 53, 54, 57, 63, 65, 69
LU	1949, 51(May 17), 53-55, 57, 59, 63, 65
MNS	1957
MiDW-S	1957, 69
MiU	1954-55, 57, 59, 63
MnU	1951, 53-55, 57-59, 63, 65, 69, 73
MoSW	1949, 57
MoU	1969
NBiSU	1969
NNC	1951, 53-55, 57-59, 63
NSyU	1957
NbU	1953, 55, 57, 59
NcU	1957
NhD-K	1957, 63, 73
NjP	1957
NmPE	1963, 73
NmU	1955
NvU	1957
OCU-Geo	1957
OU	1939, 49, 51(May 17), 53-55, 57-59, 63, 69
OkOkU	1957
OkT	1953-55, 57, 59, 69
OkU	1946, 49, 51, 53-55, 57, 59, 63, 65, 69, 73

(529) Peninsula Geological Society.

OrU-S	1957
PSt	1957
RPB	1957
TU	1957
TxBeaL	1946, 51, 53-54, 59, 63, 69, 73
TxCaW	1969
TxDaAR-T	1951(May 17), 53, 55, 57, 59
TxDaDM	1951, 54-55, 57-59, 63, 65, 69, 73
TxDaGI	1946, 51, 53, 59
TxDaM-SE	1954-55, 57-59, 63, 65
TxDaSM	1951, 53, 57, 59
TxHSD	1946, 51, 53, 55(May 17), 59, 63, 69
TxHU	1946, 53-55, 57-59, 63, 65
TxLT	1951(May 17), 53
TxMM	1946, 51(May 17), 53-55, 57, 59, 63, 73
TxU	1949, 51, 53-55, 57-59, 63, 69
TxWB	1951, 53-55, 57, 59, 63, 69
UU	1957, 59
ViBlbV	1954-55, 57, 63, 73
WU	1951(May 17), 53, 55, 57

(529) PENINSULA GEOLOGICAL SOCIETY. ANNUAL FIELD TRIP.
1979 1st Alcatraz Island; the allure and mystery of Alcatraz. (Geologists hit the rock)
1982 Geologic transect of the northern Sierra Nevada along the North Fork of the Yuba River.

CMenUG	1978, 82
DI-GS	1979
NvU	1982

(530) PENNSYLVANIA COUNCIL FOR GEOGRAPHY EDUCATION. SPRING CONFERENCE.
1967 14th Geology and geography of the South Mountain and environs.
1969 April Field trip guide to Fayette County, Pennsylvania.
  1. Settlement patterns in rural Fayette County (Geographic).
  2. Structure of Fayette County (Geologic).

CMenUG	1969
DI-GS	1969
InU	1967
PHarER-T	1969
TxU	1969

(531) PENNSYLVANIA EARTH SCIENCE TEACHER'S SOCIETY. FIELD TRIP GUIDEBOOK.
1973 May Johnstown, Pennsylvania.

PHarER-T	1973

PENNSYLVANIA GEOLOGICAL SURVEY. See PENNSYLVANIA. BUREAU OF TOPOGRAPHIC AND GEOLOGIC SURVEY. (537) and PENNSYLVANIA. BUREAU OF TOPOGRAPHIC AND GEOLOGIC SURVEY. (536) and PENNSYLVANIA. BUREAU OF TOPOGRAPHIC AND GEOLOGIC SURVEY. (534) and PENNSYLVANIA. BUREAU OF TOPOGRAPHIC AND GEOLOGIC SURVEY. (535) and PENNSYLVANIA. BUREAU OF TOPOGRAPHIC AND GEOLOGIC SURVEY. (538)

(532) PENNSYLVANIA STATE UNIVERSITY. AGRONOMY DEPARTMENT AGRONOMY SERIES.
1978 No. 52. See Friends of the Pleistocene. Eastern Group. 182.
1980 No. 64. See Northeast Cooperative Soil Survey Conference. 484.

(533) PENNSYLVANIA STATE UNIVERSITY. COAL RESEARCH SECTION. SHORT COURSE.
1976 June A field guidebook to aid in the comparative study of the Okefenokee Swamp and the Everglades-Mangrove Swamp-Marsh complex of southern Florida; a short course presentation. (Cover title: Environments of coal formation; Okefenokee and the Everglades)

CSt-ES	1976
CU-EART	1976
CaACU	1976
InLP	1976
PHarER-T	1976
PSt	1976
TxBeaL	1976
TxHSD	1976
TxU	1976

(534) PENNSYLVANIA. BUREAU OF TOPOGRAPHIC AND GEOLOGIC SURVEY. EARTH SCIENCE FIELD TRIP GUIDE.
[n.d.] [1] Cornwall Iron Mine, Lebanon County.
— [2] Guide to a cave.
— [3] Swatara Gap, Lebanon County.

DI-GS	n.d.(1-3)
PHarER-T	n.d.(1-3)

(535) PENNSYLVANIA. BUREAU OF TOPOGRAPHIC AND GEOLOGIC SURVEY. ENVIRONMENTAL GEOLOGY REPORT.
1979 [No.] 7. Oustanding scenic geological features of Pennsylvania.

CPT	1979
CSt-ES	1979
CaOWtU	1979
DLC	1979

(536) PENNSYLVANIA. BUREAU OF TOPOGRAPHIC AND GEOLOGIC SURVEY. GENERAL GEOLOGY REPORT.
1938 8. Guidebook: A Paleozoic section in south-central Pennsylvania.
— 11. Guidebook: A Paleozoic section at Delaware Water Gap.
1939 12. Guidebook: Highway geology from Philadelphia to Pittsburgh.
— 13. Guide to the geology from Dauphin to Sunbury.
— 14. Guide to the geology of the upper Schuylkill Valley.
— 15. Guidebook to the geology near Reading, Pennsylvania.
— 16. Guidebook to places of geologic interest in the Lehigh Valley, Pennsylvania.
— 17. Guidebook to the geology about Pittsburgh.
1942 20. Guidebook to the geology of the Pennsylvania Turnpike.
1949 24. Guidebook to the geology of the Pennsylvania Turnpike from Carlisle to Irwin.
1957 30. The geology of the Hidden Valley Boy Scout Camp area, Perry County, Pennsylvania.
1958 29. Guide to the highway geology from Harrisburg to Bald Eagle Mountain.
1961 35. Guide to the geology of Cornwall, Pennsylvania.
1964 41. Guidebook to the geology of the Philadelphia area.
1965 50. Guide to the Horse Shoe Curve section between Altoona and Gallitzin, central Pennsylvania.
1970 59. Geology of the Pittsburgh area.
1971 30. The geology of the Hidden Valley Boy Scout Camp area, Perry County, Pennsylvania. 2d ed.
1972 62. Upper Devonian marine-nonmarine transition, southern Pennsylvania.
1974 64. Pleistocene beach ridges of northwestern Pennsylvania.
1976 67. Caves of western Pennsylvania. (433)
1983 74. Geology of the Appalachian trail in Pennsylvania.

AzU	1938-39, 42, 49, 57-58, 61, 64-65
CLU-G/G	1938-39, 42, 49, 58, 59, 61, 64-65, 70, 72, 74, 76, 83
CPT	1938-39, 42, 49, 57-58, 61, 64-65, 70-72, 74, 76, 83
CSfCSM	1983
CSt-ES	1939, 42, 49, 57-58, 61, 64-65, 70-72, 74, 76, 83
CU-EART	1938, 42, 49, 57-58, 61, 64-65, 70-72, 74, 76, 83
CU-SB	1983
CaACU	1949, 72
CaBVaU	1961, 64-65
CaOHM	1949, 64
CaOKQGS	1939(17), 49, 64, 83
CaOLU	1949, 64
CaOOG	1957, 70-72, 83
CaOTDM	1983
CaOTRM	1964
CaOWtU	1939(17), 49, 57-58, 61, 64-65, 71-72, 83
CaQQERM	1983
CoDGS	1983
CoG	1938-39, 42, 49, 57-58, 64-65, 71
DFPC	1939(15, 17)
DI-GS	1938-39, 42, 49, 58, 61, 64-65, 83
DLC	1939, 49, 58, 61, 64-65, 70
ICF	1938, 39(12-14), 42, 49, 57-58, 64-65, 70-72, 74
IEN	1939(15-17), 49, 64, 70
IU	1938-39, 42, 49, 57-58, 64-65, 70-72
IaU	1939, 42, 49, 58, 64-65, 70-72, 74, 83
InLP	1938(8), 39(13-17), 42, 49, 57-58, 61, 64-65, 70-72, 74, 76, 83
InU	1939(15-17), 49, 57-58, 64-65, 70-72
KLG	1983
KyU-Ge	1938-39, 42, 49, 57-58, 61, 64-65, 70-72

MNS	1957, 71-72
MiHM	1939, 49, 61, 64-65, 70-71, 83
MnU	1938-39, 42, 49, 57-58, 61, 64-65, 70-72, 83
MoSW	1938(8), 39(13, 17), 42, 49, 58, 61, 64-65
MoU	1939(15-17), 49, 64
MtBuM	1938, 39, 42, 49, 58, 61, 64-65
NIC	1983
NNC	1938-39, 42, 49, 57-58, 61, 64-65, 70-71, 83
NOneoU	1964
NRU	1983
NSbSU	1939(13-17)
NSyU	1938-39, 42, 49, 58, 61, 64
NcU	1938-39, 42, 49, 57-58, 61, 64-65, 70-72, 76
NdU	1939, 42, 49, 57-58, 64-65, 70-72, 83
NhD-K	1938-39, 57, 64, 70-72, 83
NjP	1939(15-17), 49, 58, 64-65
NvU	1938(11), 39, 42, 49, 57-58, 61, 64-65, 76
OCU-Geo	1938-39, 42, 49, 57-58, 61, 64-65, 70-72, 74, 76
OU	1938-39, 42, 49, 57-58, 61, 64-65, 70-71
OkT	1938, 39, 42, 49, 57-58, 61, 64-65, 70-72, 74, 83
OkU	1938-39, 42, 49, 58, 61, 64-65, 83
OrU-S	1939(13-17), 57, 65, 70
PBL	1939, 64
PBm	1939, 42, 49, 58, 61, 64-65, 70-72, 83
PHarER-T	1939, 42, 49, 57-58, 61, 64-65, 70-72, 83
PSt	1938-39, 42, 49, 57-58, 61, 64-65, 70-72
SdRM	1939(17), 49, 57, 61, 64-65, 70-72
TMM-E	1964
TxBeaL	1949, 58, 61, 64-65
TxDaGl	1939(17), 49
TxDaM-SE	1983
TxHSD	1949, 64
TxU	1938-39, 42, 49, 57-58, 61, 64-65, 70-72, 83
UU	1949, 58, 61, 64-65
ViBlbV	1938(8), 39, 42, 57-58, 61, 64-65, 70-72
WU	1938, 39, 42, 49, 57-58, 61, 64-65, 70-72, 74, 76, 83
WaU	1939(15-17)

(537) PENNSYLVANIA. BUREAU OF TOPOGRAPHIC AND GEOLOGIC SURVEY. GEOLOGIC FIELD TRIP GUIDE.

[n.d.] Interstate 81 from Harrisburg to Hazelton.

IU	n.d.
PHarER-T	n.d.
WU	n.d.

(538) PENNSYLVANIA. BUREAU OF TOPOGRAPHIC AND GEOLOGIC SURVEY. PARK GUIDE.

[n.d.] Pennsylvania trail of geology.
—No. 12. Worlds End State Park, Sullivan County; geologic features of interest.
—No. 13. Ricketts Glen State Park, Luzerne County; the rocks, the glens, and the falls.
—No. 14. Nockamixon State Park, Bucks County; rocks and joints.
—No. 15. Caledonia and Pine Grove Furnace state parks, Cumberland, Adams, and Franklin counties; geologic features and iron ore industry.
—No. 16. Swatara State Park, Lebanon and Schuylkill counties; geologic features and iron ore industry.
—No. 17. Samuel S. Lewis State Park, York County; Mt. Pisgah and the lower Susquehanna Valley.
—No. 18. Promised Land State Park, Pike County; ancient rivers and ages of ice.
—No. 19. Raymond B. Winter State Park, Union County; scenery, rocks, and springs in eastern Brush Valley.

CU-EART	[n.d.](12-16)
CaOWtU	[n.d.](12-14)
DI-GS	[n.d.]
IaU	[n.d.](12-16)
InLP	[n.d.]
PHarER-T	[n.d.] (12-19)

PENNSYLVANIA. GEOLOGICAL SURVEY. See PENNSYLVANIA. BUREAU OF TOPOGRAPHIC AND GEOLOGIC SURVEY. (534)

PENNSYLVANIA. TOPOGRAPHIC AND GEOLOGIC SURVEY. See PENNSYLVANIA. BUREAU OF TOPOGRAPHIC AND GEOLOGIC SURVEY. (536)

PENROSE RESEARCH CONFERENCE FIELD TRIP. See GEOLOGICAL SOCIETY OF AMERICA. PENROSE CONFERENCE. (203)

(539) PETROLEUM EXPLORATION SOCIETY OF NEW YORK. GUIDEBOOK.

1974 [1st] Guidebook to field trip in Rockland County, New York.
1975 2nd Upper Cretaceous and lower Tertiary stratigraphy, New Jersey coastal plain.
1976 3rd Guidebook to the stratigraphy of the Atlantic Coastal Plain in Delaware.
1981 See Atlantic Margin Energy Symposium. 90. and American Association of Petroleum Geologists. National Energy Minerals Division. (Title varies) 27.

CLU-G/G	1975-76
CSt-ES	1974-75
CU-EART	1974
CaACU	1974
CaOOG	1975-76
CoDCh	1974-76
DI-GS	1974-76
DLC	1974
IaU	1974-76
InLP	1974-75
InU	1975
KyU-Ge	1975-76
MnU	1974
NRU	1974
NSyU	1974
NcU	1974
NhD-K	1974-76
OU	1974
OkU	1974-76
TxDaSM	1974
TxHSD	1975
TxU	1974-75
ViBlbV	1975-76

(540) PITTSBURG STATE UNIVERSITY (KANSAS). NATIONAL SCIENCE FOUNDATION CLASS.

1960 Pittsburg, Bourbon, Crawford and Cherokee counties. [Kansas]

IU	1960
KLG	1960

(541) PITTSBURGH GEOLOGICAL SOCIETY. GUIDEBOOK FOR THE FIELD TRIP.

1948 Oct. See American Association of Petroleum Geologists. Regional Meeting. 29.
1955 Mar. See American Association of Petroleum Geologists. 18.
1957 Oct. See Appalachian Geological Society. 58.
1959 Oct. See Appalachian Geological Society. 58.
1961 Oct. See Appalachian Geological Society. 58.
1963 Sept. See Appalachian Geological Society. 58.
1964 Oct. See Appalachian Geological Society. 58.
1969 See Ohio Geological Society. 494.
1974 Apr. See American Association of Petroleum Geologists. Eastern Section. 25.
1980 Oct. See Field Conference of Pennsylvania Geologists. 172.
1985 May [Title from Introduction: General geology of the Johnstown area]

PHarER-T	1985

(542) PLANETARY GEOLOGY FIELD CONFERENCE. GUIDEBOOK.

1974 Hawaiian planetology conference. Mars geologic mapping meeting.
1977 Aug. Volcanism of the eastern Snake River plain, Idaho: a comparative planetary geology guidebook.
1978 Jan. Aeolian features of Southern California: a comparative planetary geology guidebook. (66)
— June The Channeled Scabland: a guide to the geomorphology of the Columbia Basin, Washington.

AzU	1978(Jan.)
CLU-G/G	1974, 77-78
CPT	1974, 77, 78
CSt-ES	1974, 77-78
CU-A	1974
CU-EART	1974, 77-78

(543) Pleistocene Field Conference.

CaACU	1974, 78(June)	OU	1969
CaAEU	1974	OkU	1969
CaBVaU	1974		

(546) QUEBEC, UNIVERSITE, TROIS-RIVIERES. COLLECTION PALEO-QUEBEC.
1979 See Association Quebecoise pour l'Etude du Quaternaire. 88.
1980 No. 11. See Association Quebecoise pour l'Etude du Quaternaire. 88.

DI-GS	1977-78
DLC	1978
IU	1978(Jan.)
IaU	1977-78
InLP	1978
InU	1978(Jan.)
MH-GS	1974
MiHM	1974
MtBuM	1974
NcU	1978(Jan.)
NhD-K	1974, 77, 78(Jan.)
NvU	1974
OU	1974
OkU	1974, 77, 78(Jan.)
PSt	1974
TxU	1978(Jan.)
ViBlbV	1978
WU	1974, 77-78

(547) RIO GRANDE VALLEY INTERNATIONAL GEOLOGICAL SOCIETY.
1965 Mar. Field trip and road log: McAllen, Rio Grande City, Roma, Mier, Cerralvo.

IU	1965
OkU	1965
TxU	1965

(548) RIVER FALLS GEOLOGICAL SOCIETY
1978 See University of Wisconsin. 723.

(543) PLEISTOCENE FIELD CONFERENCE. GUIDEBOOK.
1947 1st State geologist's conference on the loess deposits, Illinois, Iowa, South Dakota, Nebraska.
1949 2nd Late Cenozoic geology of the Mississippi Valley; southeastern Iowa to central Louisiana.
1951 3rd Middle to late Pleistocene stratigraphy of the central Great Plains.
Part 1. Road log, post-Kansan Pleistocene deposits of central Great Plains.
Part 2. Road log, northeastern Kansas, southwestern Nebraska, western Kansas.
1953 4th Basis of subdivisions of Wisconsin glacial stage in northeastern Illinois...and Wisconsin stratigraphy of the Wabash Valley and west-central Indiana.
1955 5th [1] Wisconsin stratigraphy of northern and eastern Indiana.
[2] Pleistocene chronology of southwestern Ohio.

CLU-G/G	1955
CMenUG	1947, 49, 51, 55
CSt-ES	1955
DI-GS	1947, 49, 51, 53, 55
DLC	1955
I-GS	1947, 49, 51, 53, 55
ICarbS	1953
IU	1949, 51, 55
IaU	1949, 51, 53, 55
InU	1949, 51, 53, 55
KLG	1951
MH-GS	1949, 55
OCU-Geo	1955
OU	1953, 55

(544) PURDUE UNIVERSITY. DEPARTMENT OF GEOSCIENCES. [FIELD TRIP]
1982 May See Indiana University-Purdue University at Indianapolis. Geology Department. 269.
No. 1. See Geological Society of America. North Central Section. 200.
No. 2. See Geological Society of America. North Central Section. 200.
No. 5. See Geological Society of America. North Central Section. 200.

(545) QUEBEC (PROVINCE). GEOLOGICAL EXPLORATION SERVICE. FIELD GUIDE.
1969 Field guide to the Oka area, description and itinerary.

CPT	1969
CU-EART	1969
CaACU	1969
CaBVaU	1969
CaOKQGS	1969
CaOOG	1969
CaOWtU	1969
DLC	1969
ICarbS	1969
InU	1969
KyU-Ge	1969
MH-GS	1969
MiHM	1969
NhD-K	1969
NjP	1969

(549) ROCKY MOUNTAIN ASSOCIATION OF GEOLOGISTS. GUIDEBOOK FOR THE FIELD CONFERENCE.
1937 Bighorn Basin-Yellowstone Valley tectonics. (400, 665, 230, 784)
1938 Along the Front Range of the Rocky Mountains, Colorado. (341)
1947 Central Colorado.
1948 See American Association of Petroleum Geologists. 18.
1953 Northwestern Colorado.
1954 May Trip 1. Denver to Colorado Springs and return.
— Oct. Trip 2. Denver to Canon City and return.
1955 May Geology of the Front Range foothills west of Denver; Deer Creek to Ralston Creek, Jefferson County, Colorado.
— Aug. Geology of Northwest Colorado. (276)
1956 Geology of the Raton Basin, Colorado.
1957 Geology of north and middle Parks Basin, Colorado.
1958 10th Symposium on Pennsylvanian rocks of Colorado and adjacent areas.
1959 11th Washakie, Sand Wash, and Piceance basins. Symposium on Cretaceous rocks of Colorado and adjacent areas.
1960 [Oct.] [1] See Geological Society of America. 197.
[2] Geological road logs of Colorado.
Road log no. S1. Trinidad to Colorado-Kansas line.
Road log no. S2. Trinidad to Colorado-New Mexico line.
Road log no. S3. Pagosa Springs to Colorado-New Mexico state line.
Road log no. S4. Mineral-Archuleta county line to Durango.
Road log no. S5. Bayfield to Durango.
Road log no. S6. Intersection of U.S. 550 and U.S. 160 near Durango to Colorado-New Mexico line.
Road log no. S7. Durango to Colorado-Utah state line.
Road log no. S8. Cortez to Colorado-New Mexico state line.
Road log no. S9. Cortez to Whitewater.
Road log no. S10. Durango to Montrose.
Road log no. S11. Montrose to Grand Junction.
Road log no. S12. Newcastle to Meeker.
Road log no. S13. Rifle to Craig.
Road log no. S14. Junction U.S. 40 and Colorado 387 to Rio Blanco.
Road log no. S15. Craig to Utah state line.
Road log no. S16. Craig to Maybell.
Road log no. S17. Steamboat Springs to Craig.
Road log no. S18. Steamboat Springs to Hayden.
Road log no. S19. Granby to the Wyoming line.
Road log no. S20. Junction Colorado 125 and 127 to Wyoming line.
Road log no. S21. From Junction 0.2 mile east of Rand to Walden.
Road log no. S22. Kremmling to Dillon.
Road log no. S23. Glenwood Springs to Intersection of Colorado 82 and U.S. 24.
Road log no. S24. Rangely to Grand Junction, Colorado.
Road log no. S25. Poncha Springs to Montrose.
Road log no. S26. Trinidad to Walsenburg.
1961 12th South-central Colorado. Symposium on lower and middle Paleozoic rocks of Colorado.
1962 13th Exploration for oil and gas in northwestern Colorado.
1963 Geology of North Denver Basin and adjacent uplift.

1964 16th [Central Rockies, Colorado; tectonics and associated sedimentation] (405)
1965 Piceance and eastern Uinta basins, Colorado and northeast Utah. (405)
1966 No field trip.
1967 No field trip.
1968 Southeastern Colorado. (405)
1969 Raton Basin, Colorado and New Mexico. (405)
1970 Dakota and related rocks of the Front Range. (405)
1971 No field trip.
1972 See American Association of Petroleum Geologists. 18.
1973 Cretaceous stratigraphy, central Front Range. (405)
1974 25th Energy resources of the Piceance Creek Basin, Colorado.
1975 Deep drilling frontiers of the Central Rocky Mountains.
1976 Geology of the Cordilleran hingeline; hingeline sediments of the overthrust belt.
    Field trip route-segment no. 1. Mouth of Parleys Canyon to Wanship, Utah.
    Field trip route-segment no. 2. Wanship, Utah to Smith and Morehouse Reservoir to Coalville, Utah.
    Field trip route-segment no. 3. Coalville, Utah to Pineview oil field to Echo Reservoir and Echo Junction.
    Field trip route-segment no. 4. Echo Junction to mouth of Weber Canyon and south to Olympus Hills Shopping Center.
    Alternate field trip. Morgan section, Morgan, Utah.
1977 Exploration frontiers of the Central and Southern Rockies.
1978 Energy resources of the Denver Basin.
1979 Basin and Range symposium and Great Basin field conference.
1981 Geology of the Paradox Basin.
1982 Denver, Colorado.
  Geologic studies of the Cordilleran Thrust Belt.
  Cordilleran Overthrust Belt, Texas to Arizona; field trip, symposium, and articles on the styles of deformation between El Paso, Texas, and Wickenburg, Arizona.
1983 Rocky Mountain foreland basins and uplifts.
1984 Hydrocarbon source rocks of the greater Rocky Mountain region.
  (1) First day ... Lakewood to Pueblo.
  (2) Second day ... Pueblo to northern Raton Basin.
1985 See American Association of Petroleum Geologists. Rocky Mountain Section. 30.

AzFU	1955(Aug.), 56, 58-59, 60(2), 61-63, 65, 68, 70
AzTeS	1954-55, 57-58, 60(2), 61-63, 65, 68-70, 74, 76, 79
AzU	1947, 54-59, 60(2), 61-65, 68-70, 73-77, 79
CLU-G/G	1937-38, 47, 54-59, 60(2), 61-65, 68-70, 73-77, 79, 81-83
CLhC	1937, 47, 54-59, 60(2), 61-65, 68-70, 73-79, 81-83
CMenUG	1981-84
CPT	1938, 47, 54-65, 68-70, 73-84
CSdS	1955(Aug.), 64-65, 68-70, 73, 75
CSfCSM	1955, 58, 79
CSt-ES	1947, 54-59, 60(2), 61-64, 74-79, 81-83
CU-A	1957-58, 61, 63-65, 68-70, 73-79
CU-EART	1954-59, 60(2), 61-65, 68-70, 73-79, 81-83
CU-SB	1955, 56, 63, 72, 74, 81
CaAC	1977
CaACAM	1937, 76-78, 83
CaACI	1955, 60(2), 69-70
CaACM	1975-79
CaACU	1947, 60, 64-65, 68-70, 73-83
CaAEU	1963, 77, 79
CaBVaU	1958, 60(2), 68-70, 73, 75-77, 79
CaOHM	1954, 56-57, 60(2), 63, 68-70
CaOKQGS	1954, 56, 58-59, 60(2), 61, 63, 82
CaOLU	1960(2)
CaOOG	1955(Aug.), 58-59, 60(2), 61-65, 68, 75-78, 81-82
CaOWtU	1954, 58, 61, 68-70, 73-78
CoD	1981-83
CoDBM	1982
CoDCh	1937-38, 47, 53-54(May), 55(Aug.), 56-59, 60(2), 61-62, 74, 77-78, 81-83
CoDGS	1981-83
CoDU	1959, 60(2), 63
CoDuF	1955(Aug.), 60(2), 74
CoFS	1947, 62
CoG	1937-38, 47, 53-59, 60(2), 61-65, 68-70, 73-77, 79, 81-83
CoLiM	1983
CoU	1938, 47, 54-59, 60(2), 61-65, 68-70, 73-78
CoU-DA	1983
DI-GS	1937-38, 47, 53-59, 60(2), 61-63, 68-70, 73-79, 81-83
DLC	1938, 54-55, 57, 59, 60(2), 61, 64-65, 70, 73-76
I-GS	1938
ICF	1954-59, 60(2), 61, 63-65, 68-70, 73-79
ICarbS	1955(Aug.), 58-59, 60(2), 61-63, 69-70, 73, 75-77, 84
IEN	1938, 47, 54-59, 60(2), 61-64, 74
IU	1938, 47, 54, 55(Aug.), 56, 58-59, 60(2), 61-65, 68-70, 73-79
IaCfT	1981-82
IaU	1937-38, 54-59, 60(2), 61-65, 68-70, 73-79, 81-84
IdBB	1955-56, 58, 60(2), 62, 74
IdPI	1956-59, 61, 68-70, 72-73
IdU	1964-65, 68-70, 73, 81-83
InLP	1955-59, 61-65, 68-70, 73-84
InU	1937-38, 47, 53-59, 60(2), 61, 63-65, 68-70, 74, 76-79, 82
KyU-Ge	1938, 54-59, 60(2), 61-65, 70, 74-79, 81-84
LNU	1938, 54, 56, 58, 74-76
LU	1938, 47, 54-59, 60(2), 61-64
MH-GS	1937, 60(2)
MNS	1938, 54-56, 60(2), 63
MU	1981
MiDW-S	1938, 54, 55(Aug.), 56-57, 60(2)
MiKW	1947
MiU	1938, 47, 54-59, 61-65
MnStclS	1982-83
MnU	1937, 54-59, 60(2), 61-65, 68-70, 73-79, 81-84
MoSW	1938, 47, 54, 55(Aug.), 75
MoU	1938, 54-59, 60(2), 61-63
MtBC	1954-59, 60(2), 61-64, 74
MtBuM	1937, 47, 54, 55(Aug.), 58, 60(2), 61, 64-65, 68-69, 73-79, 82(v.2), 83-84
MtU	1955
NBiSU	1954, 56, 58-59, 60(2), 61-63
NIC	1982-83
NNC	1937-38, 47, 54-59, 60(2), 61-65, 75-79, 81, 83, 85
NRU	1956, 58-59, 60(2), 81-83
NSbSU	1974-79, 81-82, 84
NSyU	1955-58, 64-65, 68-70, 73-74, 76-79
NbU	1938, 47, 54-59, 60(2), 61-64
NcU	1938, 54, 55(Aug.), 56, 58-59, 74-76, 79, 81-83
NdU	1937, 56-57, 59, 60(2), 64-65, 68-70, 73-79
NhD-K	1937-38, 55(Aug.), 60(2), 63-65, 68-70, 73, 75-79, 81-82
NjP	1937-38, 54-59, 60(2), 62, 64-65, 68-70, 73-75
NmLcU	1981-82
NmPE	1958, 60(2), 76-77
NmSoI	1981, 83
NmU	1981-84
NvU	1955-59, 60(2), 61-64, 74-79, 81-84
OCU-Geo	1937, 55-59, 75-76, 82
OU	1937-38, 47, 54-59, 61-65, 68-70, 73-76, 81, 83
OkOkCGS	1952, 55(Aug.)
OkT	1937-38, 55(Aug.), 56-59, 61-62, 74-75, 77-78, 84
OkTA	1981-83
OkU	1937-38, 47, 54-59, 60(2), 61-65, 68-70, 73-74, 77-78, 81-85
OrCS	1947, 54, 56, 58-59, 61-65, 68, 70, 73-74, 76-79, 81
OrU-S	1955(Aug.), 57-59, 61-65, 68-70, 73, 76-77, 79
PBL	1937-38
PBU	1981-83
PBm	1982-83
PSt	1954-57, 60(2), 64-65, 68-70, 73, 75-79, 81, 84
RPB	1937, 55
SdRM	1953, 55-59
SdU	1954-55, 58-59, 60(2)
TU	1955(Aug.)
TxBeaL	1958, 61, 63, 65, 68-70, 73, 75-77
TxDaAR-T	1938, 47, 54-59, 60(2), 61-65, 68-70, 74
TxDaDM	1947, 54-59, 60(2), 61-65, 68-70, 73, 75-77, 79, 81
TxDaGI	1937, 47, 54, 55(Aug.), 56-58, 60(2), 62-64, 69, 74
TxDaM-SE	1937-38, 47, 54-59, 60(2), 61-65, 81-83
TxDaSM	1937, 54-59, 60(2), 61-63, 74-78
TxHMP	1981-82
TxHSD	1955(Aug.), 56-59, 62, 64, 74-79
TxHU	1938, 47, 54-59, 61-63, 74
TxLT	1954, 55(Aug.), 56-59, 60(2), 61-64
TxMM	1938, 47, 54-59, 60(2), 61, 63, 74, 81-83
TxU	1937-38, 47, 54-59, 60(2), 61-65, 68-70, 73-74, 76-79, 82-83
TxU-Da	1955(Aug.), 57, 60(2)
TxWB	1937, 47, 55-57, 65, 81-83
UU	1947, 54, 55(Aug.), 56, 58-59, 60(2), 74-79, 81-83
ViBlbV	1954, 58-59, 61, 74-79
WU	1938, 54-59, 60(2), 61-65, 68-70, 73-79, 81-84
WaU	1955(Aug.), 56-57, 79, 83
WyU	1937, 47, 53, 55-59, 60(2), 61-64, 81-83

(550) ROSWELL GEOLOGICAL SOCIETY. GUIDEBOOK OF THE FIELD CONFERENCE.

- 1950 2nd Pre-Permian stratigraphy of the Sacramento Mountains, New Mexico.
- 1951 4th Permian stratigraphy of the Capitan Reef area of the southern Guadalupe Mountains, New Mexico.
- — 5th Capitan-Carrizozo-Chupadera Mesa region, Lincoln and Socorro counties, New Mexico.
- 1952 6th Surface structures of the foothill region of the Sacramento and Guadalupe Mountains, Chaves, Eddy, Lincoln and Otero counties, New Mexico.
- — 7th The Pedernal positive element and the Estancia Basin, Torrance and northern Lincoln counties, New Mexico.
- 1953 8th Stratigraphy of the West Front and Sacramento Mountains, Otero and Lincoln counties, New Mexico.
- 1956 9th The Stauber Copper Mine and the Santa Rosa asphalt deposit, Guadalupe County, New Mexico.
- 1957 10th Slaughter Canyon, New Cave and Capitan reef exposure, Carlsbad Caverns National Park, New Mexico.
- 1958 11th Hatchet Mountains and the Cooks Range-Florida Mountain areas, Grant, Hidalgo and Luna counties, southwestern New Mexico.
- 1959 See Society of Economic Paleontologists and Mineralogists. Permian Basin Section. 594.
- 1960 12th Northern Franklin Mountains, southern San Andres Mountains with emphasis on Pennsylvanian stratigraphy.
- 1961 See Southwestern Federation of Geological Societies. 627.
- 1962 13th See West Texas Geological Society. 756.
- 1964 14th Geology of the Capitan Reef complex of the Guadalupe Mountains, Culberson County, Texas, and Eddy County, New Mexico.
- 1966 Feb. 1965-66 One Day Field Trip Series. (Plan was for a series of three one day field trips scheduled by the Roswell Geological Society) Field trip no.1. IMC Potash Mine field trip, Carlsbad, New Mexico.
- 1968 Roswell artesian basin.
- 1971 See West Texas Geological Society. 756.
- 1983 Apr. Guidebook for field trip to the Abo Red Beds (Permian), central and south-central New Mexico. (462)
- 1985 Geology of the backreef sediments equivalent to the Capitan Reef complex through Dark Canyon to Sitting Bull Falls, Eddy, County, New Mexico.

AzFU	1966
AzTeS	1958, 60, 64, 66, 68
AzU	1950-52, 57-58, 60, 64
CChiS	1958
CLU-G/G	1951(5), 53, 56-57, 60, 64, 66, 68, 83
CLhC	1958
CPT	1952(6th), 53, 56-68, 60, 64, 66, 83
CSdS	1951
CSt-ES	1956-58, 60, 64, 66, 68
CU-A	1983
CU-EART	1957, 60, 64, 66, 68, 83
CU-SB	1958, 60, 62, 64, 66, 71
CaACM	1964
CaACU	1951(5), 52(7), 60, 64, 66, 68, 71
CaOHM	1960, 64, 66
CaOKQGS	1968
CaOWtU	1960, 64
CoDCh	1950-56, 58, 60, 64, 83
CoDU	1964
CoG	1956-58, 60, 64
CoLiM	1983, 85
CoU	1952-53
DI-GS	1951(5), 52-53, 56-58, 60, 64, 66, 83
DLC	1968
ICF	1957-58, 60, 64
ICIU-S	1960, 64
IEN	1958
IU	1957-58, 60, 64, 66
IaU	1958, 64, 83
IdU	1983
InLP	1957, 60, 64, 66, 68
InU	1958, 60, 64, 66, 68
KyU-Ge	1960, 64, 66
LNU	1958
MiKW	1960
MiU	1960, 64, 66
MnU	1952(7), 56-58, 60, 64, 66, 85
MoSW	1958, 60, 64
MoU	1960, 64
MtU	1957-58, 60, 64
NNC	1951(5), 52-53, 57-58, 60, 64, 66, 68
NOneoU	1964
NbU	1960
NdU	1964
NhD-K	1960, 64
NjP	1960, 64, 66, 68
NmPE	1957, 60, 64, 66, 68
NmU	1956, 58, 60
OU	1951-53, 57-58, 64, 66
OkT	1958, 64
OkU	1950-52, 56-58, 60, 64, 66, 83
TxBeaL	1956-58, 60, 64
TxCaW	1960
TxDaAR-T	1951(5), 52-53, 56-58, 60, 64, 66
TxDaDM	1951(5), 52(6), 53, 56-58, 60, 64, 66
TxDaGI	1953, 58, 64, 66
TxDaM-SE	1956-58, 60, 64, 66
TxDaSM	1951(5), 53, 56-58, 64
TxHSD	1950, 56-57, 60, 64
TxHU	1956-58, 60, 64
TxLT	1956-58, 64, 66
TxMM	1950-51, 52(7), 53, 56-58, 60, 64, 83
TxU	1950, 52-53, 56-58, 60, 64, 66, 68
TxU-Da	1958, 66
TxWB	1952-53, 56-58, 60, 64-65
UU	1957-58, 60, 64
ViBlbV	1957, 60, 64, 68
WU	1964
WaU	1960
WyU	1964

(551) SABER SOCIETY.

- 1974 See Society to Adapt Building to Environment Reasonably. 605.

(552) SAFFORD CENTENNIAL SOCIETY. FIELD TRIP.

- 1974 Spring See Tennessee Academy of Science. Geology-Geography Section. 649.

(553) SAN ANGELO DESK AND DERRICK CLUB. [GUIDEBOOK]

- 1959 Geological field trip featuring plains and canyons.

TxU	1959

(554) SAN ANGELO GEOLOGICAL SOCIETY. GUIDEBOOK TO THE BIENNIAL FIELD TRIP.

- 1954 1st See West Texas Geological Society. 756.
- 1956 2nd Four provinces: central mineral region, Balcones fault zone, Kerr Basin, and Rio Grande Embayment.
- 1958 Apr. The Base of the Permian; a century of controversy.
- 1961 See West Texas Geological Society. 756.

AzTeS	1956, 58
AzU	1956, 58
CChiS	1958
CLU-G/G	1956, 58
CLhC	1956, 58
CMenUG	1956, 58
CPT	1956, 58
CSt-ES	1956, 58
CU-EART	1956, 58
CU-SB	1956, 58
CaBVaU	1956, 58
CaOOG	1956, 58
CoDCh	1956, 58
CoG	1956, 58
CoU	1956, 58
DI-GS	1956, 58
DLC	1956
ICF	1956, 58
IEN	1956, 58
IU	1956, 58
IaU	1958
InLP	1956, 58
InU	1956, 58
KyU-Ge	1956, 58
LNU	1956
LU	1956, 58

MNS	1958
MiDW-S	1958
MnU	1956, 58
MoU	1956, 58
NNC	1956, 58
NSyU	1956, 58
NbU	1956, 58
NcU	1956, 58
NdU	1956, 58
NhD-K	1956, 58
NjP	1956, 58
NvU	1956, 58
OCU-Geo	1956, 58
OU	1956, 58
OkT	1956, 58
OkU	1956, 58
PBL	1956, 58
PSt	1956, 58
RPB	1958
SdRM	1956
TU	1956, 58
TxBeaL	1956, 58
TxDaAR-T	1956, 58
TxDaDM	1956, 58
TxDaGI	1956
TxDaM-SE	1956, 58
TxDaSM	1956, 58
TxHSD	1956, 58
TxHSW	1956
TxHU	1956, 58
TxLT	1956, 58
TxMM	1956, 58
TxU	1956, 58
TxU-Da	1956, 58
TxWB	1956, 68
ViBlbV	1956, 58
WU	1954, 56

SAN ANTONIO GEOLOGICAL SOCIETY. See SOUTH TEXAS GEOLOGICAL SOCIETY. (612)

(555) SAN ANTONIO GEOLOGICAL SOCIETY. [FIELD TRIP GUIDEBOOK]
- 1981 Apr. See American Association of Petroleum Geologists. Southwest Section. 31.
- 1984 May See American Association of Petroleum Geologists. 18.

(556) SAN DIEGO ASSOCIATION OF GEOLOGISTS. GUIDEBOOK. (TITLE VARIES)
- 1973 May Studies on the geology and geologic hazards of the greater San Diego area, California. (84)
- 1975 Studies on the geology of Camp Pendleton and western San Diego County, California.
- 1977 Geology of southwestern San Diego County, California and northwestern Baja California.
- 1978 Natural history of the Coronado Islands, Baja California, Mexico.
- 1979 Nov. See Geological Society of America. 197.
- 1981 Apr. Geologic investigations of the coastal plain, San Diego County, California.
- 1982 Apr. Geologic studies in San Diego.
- 1985 Apr. On the manner of deposition of the Eocene strata in northern San Diego County.

CLU-G/G	1973, 75, 77, 85
CMenUG	1982
CPT	1977-78, 85
CSdS	1981-82, 85
CSfCSM	1973, 77, 81-82
CSt-ES	1973, 77-78, 85
CU-A	1973, 78
CU-EART	1977-78, 81-82, 85
CU-SB	1978
CaACU	1973, 77
CaOWtU	1973
CoLiM	1981-82
DI-GS	1973, 78, 81-82, 85
IU	1978
IaU	1977-78
KyU-Ge	1973, 78, 81, 82
NSyU	1973
NdU	1973
NhD-K	1977
NjP	1973
OU	1978
PSt	1978
SDSNH	1981-82, 85
TxBeaL	1978
ViBlbV	1973, 77-78
WU	1978

(557) SAN JOAQUIN GEOLOGICAL SOCIETY. GUIDEBOOK FOR THE FIELD TRIP. (TITLE VARIES)
- 1951 May See American Association of Petroleum Geologists. Pacific Section. 28.
- 1955 May See American Association of Petroleum Geologists. Pacific Section. 28.
- 1958 [May] Round Mountain area, Kern County, California.
- 1959 May Chico-Martinez Creek area, California.
- 1961 May See American Association of Petroleum Geologists. Pacific Section. 28.
- 1962 Oct. See American Association of Petroleum Geologists. Pacific Section. 28.
- 1963 June See American Association of Petroleum Geologists. Pacific Section. 28.
- 1964 Oct. See American Association of Petroleum Geologists. Pacific Section. 28.
- 1965 Apr. See Society of Economic Paleontologists and Mineralogists. Pacific Section. 593.
- 1977 Apr. See American Association of Petroleum Geologists. Pacific Section. 28.

AzTeS	1958-59
CChiS	1958
CLU-G/G	1958-59
CLhC	1958-59
CMenUG	1958-59
CPT	1958-59
CSfCSM	1958-59
CSt-ES	1958-59
CU-A	1958-59
CU-EART	1958-59
CU-SB	1958-59
CoG	1958-59
DI-GS	1958-59
ICF	1959
IU	1958-59
KyU-Ge	1958-59
MnU	1958-59
NNC	1958-59
NbU	1958-59
NvU	1958-59
OCU-Geo	1959
OU	1958-59
TxDaDM	1959
TxDaM-SE	1958-59
TxMM	1959
TxU	1958-59
TxU-Da	1958-59
UU	1958
ViBlbV	1958-59

(558) SAN JOSE STATE UNIVERSITY. DEPARTMENT OF GEOLOGY.
- 1974 See Society to Adapt Building to Environment Reasonably. 605.

SASGS. See SOUTHWESTERN ASSOCIATION OF STUDENT GEOLOGICAL SOCIETIES (SASGS). (626)

(559) SASKATCHEWAN GEOLOGICAL SOCIETY. FIELD TRIP.
- 1956 Field trip. Report of the Mississippian Names and Correlations Committee.
- 1958 Field trip. Report of the lower Palaeozoic Names and Correlations Committee.
- 1965 Geological road log of the Winnipeg and Interlake area, Manitoba.
- 1966 Geological road log of the Hanson Lake road, Flin-Flon area, Saskatchewan-Manitoba.

(560) Saskatchewan Geological Society.

1967 Geological road logs of the Lakes Winnepegosis and Manitoba areas, Manitoba.
1968 Field trip guidebook. (402)
1969 Field trip. Geological road log of the Cypress Hills area, southwestern Saskatchewan.
1970 Geological road log of the Montreal Lake-Lac La Ronge area, Saskatchewan.
1971 Field trip. Geological road log of the Cypress Hills-Milk River area southeastern Alberta.
1972 Field trip. Guidebook on geology and its application to engineering practice in the Qu'appelle Valley area.
1973 See Geological Association of Canada. 190.

CLU-G/G	1966-67, 69-72
CaACAM	1956, 58, 66-67, 69-72
CaACM	1967
CaACU	1965-71
CaOKQGS	1965
CaOOG	1966
CoG	1965-66
DI-GS	1965-67, 69-72
IU	1966-67, 69-72
IaU	1965
InLP	1966-67
MiU	1966-67
MnU	1965-67, 69, 71
MtU	1966
NdU	1965
OkU	1967, 69
TxDaDM	1965, 69
TxDaM-SE	1965-66
TxDaSM	1965
TxU	1965
WU	1969-71

(560) SASKATCHEWAN GEOLOGICAL SOCIETY. SPECIAL PUBLICATION.
1973 No. 1. See Geological Association of Canada. 190.
1982 No. 6. See International Williston Basin Symposium. 328.

(561) SCENIC TRIPS TO THE NORTHWEST'S GEOLOGIC PAST. [GUIDEBOOK]
1979 No. 1. The magnificent gateway; a layman's guide to the geology of the Columbia River Gorge.

CU-A	1979
CU-EART	1979
IdU	1979
KyU-Ge	1979
NcU	1979
WaPS	1979

SE-GSA (SOUTHEASTERN SECTION OF THE GEOLOGICAL SOCIETY OF AMERICA). FIELD TRIP. See GEOLOGICAL SOCIETY OF AMERICA. SOUTHEASTERN SECTION. (206)

SEDIMENTA. See UNIVERSITY OF MIAMI. ROSENSTIEL SCHOOL OF MARINE & ATMOSPHERIC SCIENCE. DIVISION OF MARINE GEOLOGY AND GEOPHYSICS. THE COMPARATIVE SEDIMENTOLOGY LABORATORY. (696)

SEG - SOCIETY OF ECONOMIC GEOLOGISTS. See SOCIETY OF ECONOMIC GEOLOGISTS. (583) and SOCIETY OF ECONOMIC GEOLOGISTS. PACIFIC SECTION. (584)

SEG - SOCIETY OF EXPLORATION GEOPHYSICISTS. See SOCIETY OF EXPLORATION GEOPHYSICISTS. (599) and SOCIETY OF EXPLORATION GEOPHYSICISTS. (602) and SOCIETY OF EXPLORATION GEOPHYSICISTS. (600) and SOCIETY OF EXPLORATION GEOPHYSICISTS. (601)

(562) SEISMOLOGICAL SOCIETY OF AMERICA. ANNUAL MEETING. FIELD TRIP.
1978 April Western Basin and Range Active Faulting Field Trip.
    NvU   1978

(563) SEISMOLOGICAL SOCIETY OF AMERICA. SOUTHEASTERN SECTION.
1969 [Appalachian field trip] (749)

KyU-Ge	1969
OU	1969
ViBlbV	1969
WU	1969

SEMINAIRE CANADIEN DE PALEONTOLOGIE ET BIOSTRATIGRAPHIE. GUIDE D'EXCURSIONS GEOLOGIQUES. See CANADIAN PALEONTOLOGY AND BIOSTRATIGRAPHY SEMINAR. (122)

SEPM See SOCIETY OF ECONOMIC PALEONTOLOGISTS AND MINERALOGISTS. (585) and SOCIETY OF ECONOMIC PALEONTOLOGISTS AND MINERALOGISTS. EASTERN SECTION. (586) and SOCIETY OF ECONOMIC PALEONTOLOGISTS AND MINERALOGISTS. GREAT LAKES SECTION. (587) and SOCIETY OF ECONOMIC PALEONTOLOGISTS AND MINERALOGISTS. GULF COAST SECTION. (588) and SOCIETY OF ECONOMIC PALEONTOLOGISTS AND MINERALOGISTS. SEPM GUIDEBOOK. (597) and SOCIETY OF ECONOMIC PALEONTOLOGISTS AND MINERALOGISTS. MID-CONTINENT SECTION. (589) and SOCIETY OF ECONOMIC PALEONTOLOGISTS AND MINERALOGISTS. MIDYEAR MEETING. (590) and SOCIETY OF ECONOMIC PALEONTOLOGISTS AND MINERALOGISTS. NORTH AMERICAN MICROPALEONTOLOGY SECTION. (591) and SOCIETY OF ECONOMIC PALEONTOLOGISTS AND MINERALOGISTS. NORTHEASTERN SECTION. (592) and SOCIETY OF ECONOMIC PALEONTOLOGISTS AND MINERALOGISTS. PACIFIC SECTION. (593) and SOCIETY OF ECONOMIC PALEONTOLOGISTS AND MINERALOGISTS. PERMIAN BASIN SECTION. (594) and SOCIETY OF ECONOMIC PALEONTOLOGISTS AND MINERALOGISTS. PERMIAN BASIN SECTION. (595) and SOCIETY OF ECONOMIC PALEONTOLOGISTS AND MINERALOGISTS. ROCKY MOUNTAIN SECTION. (596) and SOCIETY OF ECONOMIC PALEONTOLOGISTS AND MINERALOGISTS. SEPM RESEARCH CONFERENCE. (598)

(564) SHALE SHAKER
1970 V. 21. No. 1. See Oklahoma City Geological Society. 504.

(565) SHAWNEE GEOLOGICAL SOCIETY. FIELD CONFERENCE. (TITLE VARIES)
1938 [1] A study of surface rocks from Calvin Sandstone to Permian through Township 6 north to Township 10 north, Hughes, Seminole, and Pottawatomie counties, Oklahoma.
— [2] A study of surface rocks from McAlister Shale to Thurman Sandstone through Townships 4, 5, and 6 north, Ranges 12-17 east, Pittsburg County, Oklahoma.

DI-GS	1938(2)
IU	1938
MoSW	1938(2)
OU	1938(2)
OkU	1938
TxDaSM	1938(2)
TxLT	1938
TxMM	1938
TxU	1938(2)

(566) SHREVEPORT GEOLOGICAL SOCIETY. ANNUAL FIELD TRIP. GUIDEBOOK.
1923 [1st] Upper Cretaceous formation of South Arkansas.
1924 [2nd] Hope, Arkansas: Wilcox outcrop, Arkadelphia Formation and Midway outcrop.
1925 [3rd] Saratoga section, Rocky Comfort section, Cerro Gordo section, Basal section, Dequeen section, Dierks section, Basal Trinity section.
1926 4th [Title unknown]
1927 [5th] Wilcox and Clairborne outcrops of East Texas.
1928 [6th] Northeast Texas and Southeast Oklahoma.
1929 7th Tennessee-Muscle Shoals, Alabama-Tupelo, Mississippi.
1932 9th Tertiary formations of Mississippi and Alabama.

1933 10th Oligocene and Eocene Jackson formations of Caldwell and Catahoula parishes, Louisiana.
1934 11th Stratigraphy and paleontological notes on the Eocene (Jackson Group), Oligocene and lower Miocene of Clarke and Wayne counties, Mississippi.
1939 14th Upper and Lower Cretaceous of Southwest Arkansas, supplemented by contributions to the subsurface stratigraphy of South Arkansas and North Louisiana.
1947 15th Upper and Lower Cretaceous of southwestern Arkansas.
1948 16th Ouachita Mountains; Cambrian through Pennsylvanian Magnet Cove.
1949 17th Cretaceous of Austin, Texas area.
1951 18th Guidebook to the Paleozoic rocks of Northwest Arkansas. (67)
1953 19th Upper and Lower Cretaceous of southwestern Arkansas, Cambrian-Pennsylvanian of the Ouachita Mountains, and Magnet Cove.
1960 20th Interior salt domes and Tertiary stratigraphy of North Louisiana.
1961 21st Cretaceous of Southwest Arkansas and Southeast Oklahoma.
1965 22nd Cretaceous of Southwest Arkansas.
1966 Spring Iron ore deposits and surface geology (Tertiary) of North Lousiana. (358)
1967 Oct. Natchitoches Parish, Louisiana.
1969 23rd Comanchean stratigraphy of the Fort Worth-Waco-Belton area, Texas.
1970 24th Southeast Oklahoma and Northeast Texas.
1971 [25th] Geology of the Llano region and Austin area, Texas. (Modified in 1972 and published as Texas ... Bureau of Economic Geology guidebook No. 13. See University of Texas at Austin. Bureau of Economic Geology. 713.)
1973 [26th] A study of Paleozoic rocks in Arbuckle and western Ouachita Mountains of southern Oklahoma.
1975 See Louisiana State University. School of Geoscience. See Louisiana State University. School of Geoscience. 360.(1973)
1976 Oct. See Gulf Coast Association of Geological Societies. 247.
1979 May Rio Grande embayment, Coahuila Platform and Sierra Madre Oriental.

ATuGS	1934, 60-61, 65
CLU-G/G	1939, 49, 51, 53, 60, 66, 71, 73
CLhC	1947-49, 51, 53, 60
CMenUG	1951, 53, 60-61, 65, 69-70
CPT	1951
CSdS	1960
CSt-ES	1934, 39, 48-49, 51, 60-61, 64, 66, 68-71, 73
CU-EART	1949, 53, 60-61, 67, 69-70, 73
CU-SB	1932-34, 39, 60-61, 69-70
CaACU	1960-61, 69-70
CaOHM	1960-61, 65, 69
CaOKQGS	1960, 65
CaOWtU	1960-61, 69-70
CoG	1947, 49, 53, 60-61, 65-66, 69-71
CoU	1939, 51, 66
DI-GS	1932-34, 39, 47-49, 51, 53, 60-61, 65-66, 69-71, 73
DLC	1971
I-GS	1951
ICF	1951, 60
ICarbS	1951
IEN	1948-49, 51, 60
IU	1939, 47-49, 51, 53, 60-61, 65-66, 70, 73
IaU	1934, 39, 49, 51
InLP	1951, 53, 60-61, 65, 66, 69-71, 79
InU	1948-49, 51, 60
KyU-Ge	1949, 51, 53, 60-61, 65-66, 69-70
LNU	1934, 49, 53, 60-61, 65, 69-70, 79
LU	1932-34, 39, 49, 51, 53, 60-61, 65-66, 69-70
MNS	1939
MdBJ	1961, 66, 69-70
MiDW-S	1960
MiU	1939, 49, 53, 60-61, 65
MnU	1949, 51, 53, 60-61, 65-66, 69-71, 73
MoSW	1934, 47-49, 51, 53, 60
MoU	1939, 47-49, 51, 53, 60, 63, 73
Ms-GS	1934
MsJG	1934
NNC	1939, 48-49, 51, 53, 60-61
NSyU	1951, 66
NbU	1939, 49, 60-61
NcU	1949, 51, 60-61, 65, 69-70, 73
NdU	1951, 66
NhD-K	1960, 66
NjP	1934, 39, 53, 61, 65, 69-71, 73
NvU	1960
OCU-Geo	1939
OU	1933, 39, 48-49, 51, 53, 60-61
OkOkCGS	1939, 53
OkT	1939, 49, 51, 60-61, 71
OkU	1923-25, 27-29, 34, 39, 47-49, 51, 53, 60-61, 69-70, 73
OrU-S	1960
PSt	1971
TMM-E	1960-61, 71
TxBeaL	1934, 49, 60-61, 65, 69-71, 73
TxDaAR-T	1934, 48-49, 51, 61, 69, 71, 73
TxDaDM	1949, 51, 53, 60-61, 65-66, 69-70
TxDaGl	1932-34, 39, 47-49, 51
TxDaM-SE	1934, 39, 48-49, 51, 53, 60-61, 65
TxDaSM	1932-34, 39, 49, 51, 53, 61, 65, 71, 73
TxHSD	1947-49, 53, 60-61, 69, 73
TxHU	1932-34, 39, 47-49, 51, 53, 60-61, 65-67, 69-70
TxLT	1934, 39, 48-49, 51, 60
TxMM	1939, 49, 61
TxU	1925, 29, 32-34, 39, 47-49, 51, 53, 60-61, 65-66, 69-71, 73
TxU-Da	1933-34, 39, 47, 51
TxWB	1933-34, 39, 47, 49, 53, 60-61, 69-70
UU	1960
ViBlbV	1949, 53, 60-61, 69-70
WU	1948-49, 51, 53

(567) SIERRA CLUB. MOTHER LODE CHAPTER, YAHI GROUP. [GUIDEBOOK]
1966 A guide to the geology of the Yahi Trail, Bidwell Park, Chico, California.
CChiS 1966

(568) SIERRA CLUB. THE SIERRA CLUB GUIDE TO THE NATURAL AREAS ...
1981 Oregon and Washington.
1984 California.
1985 Middle Atlantic coast.

CMenUG	1984
CSdS	1984
CSt-ES	1981
CU-SB	1981, 84
DI-GS	1981, 84
IdU	1981
NN	1981
OCl	1984
OkT	1981, 84, 85
UU	1981, 84
WyU	1985

(569) SIERRA CLUB. THE SIERRA CLUB GUIDES TO THE NATIONAL PARKS.
1984 [1] Desert southwest.
   [2] Rocky Mountains and the Great Plains.
   [3] Pacific southwest and Hawaii.
1985 Pacific northwest and Alaska.

AzTeS	1984-85
CMenUG	1984(1), 85
CSt-ES	1984(1, 3)
CU-SB	1984(1-2), 85
CaACU	1984(2)
DI-GS	1984
IdU	1984-85
MnU	1984(2)
NN	1984
NmSoI	1984(1)
OCl	1985
OkT	1984-85
TxCaW	1984
UU	1984(1-2), 85
WyU	1984-85

(570) SIGMA GAMMA EPSILON.
[n.d.] SGE presents a field trip to Llano.
TxHU [n.d.]

(571) SIGMA GAMMA EPSILON. NATIONAL CONFERENCE. [FIELD TRIP GUIDEBOOK] (TITLE VARIES)
1937 11th [1] Luling oil fields and San Antonio.
    [2] Points of geologic interest in the vicinity of Austin, Texas.
1947 Criner Hills field trip.
1951 [1] Summary of Arkansas geology and field trip itineraries, Hot Springs.
    [2] Field trip to Magnet Cove and Potash Sulphur Springs area.
    [3] Bus trip to Little Rock, with stop enroute to inspect nepheline syenite quarries.
1968 [Assorted trips in Cincinnati, Ohio area]
1973 The Cretaceous of Parker and Tarrant counties, Texas.

CLU-G/G	1937(2)
CoG	1951
CoU	1951
DI-GS	1951
DLC	1951
IU	1947, 51
InU	1951
MoRM	1951
OCU-Geo	1951, 68
OU	1951
OkU	1951, 73
PSt	1951
TMM-E	1968
TxHU	1937(1)
TxMM	1937
TxU	1937

(572) SIGMA GAMMA EPSILON. ALPHA CHAPTER, UNIVERSITY OF KANSAS. GUIDEBOOK.
1949 Study of Marmaton-Pleasanton, Kansas City and Lansing Pennsylvanian between Lexington, Missouri and Lawrence, Kansas. (700)
1951 Joint field trip, University of Missouri--University of Kansas.
1958 2nd Upper Pennsylvanian rocks of eastern Kansas.
1965 Cyclothems of Kansas.

InU	1965
KLG	1951, 58, 65
MnDuU	1949
MoU	1949
NjP	1949
OkU	1949

(573) SIGMA GAMMA EPSILON. ALPHA NU CHAPTER, KANSAS STATE UNIVERSITY. ANNUAL SPRING FIELD TRIP GUIDEBOOK.
1969 Southeast Kansas, central Arkansas, southeast Missouri.

TxDaDM	1969

(574) SIGMA GAMMA EPSILON. BETA ZETA CHAPTER, UNIVERSITY OF NORTH DAKOTA.
1960 See University of North Dakota. Geology Department. 703.
1961 See University of North Dakota. Geology Department. 703.

(575) SIGMA GAMMA EPSILON. GAMMA ALPHA CHAPTER, UNIVERSITY OF WISCONSIN-- SUPERIOR. [GUIDEBOOK]
1974 6th Twin Ports area.

IU	1974

(576) SIGMA GAMMA EPSILON. GAMMA CHAPTER, UNIVERSITY OF OKLAHOMA. [GUIDEBOOK]
1936 Conference on the Pennsylvanian of Oklahoma, Kansas and North Texas.
1937 Conference on Permian of Oklahoma and southern Kansas.
1940 Conference on the Pennsylvanian of Texas, Oklahoma and Kansas.
1941 Muskogee area.
1947 Criner Hills field trip.
1952 See Oklahoma Academy of Science. 502.
1953 Pt. 1. Cambrian stratigraphy of the Wichita Mountains.
    Pt. 2. Stratigraphy and structure of the Criner Hills.
1959 Wichita Mountains field trip.

OkU	1941, 47, 53, 59
TxLT	1936-37, 40
TxMM	1937

(577) SIGMA GAMMA EPSILON. GAMMA ZETA CHAPTER, KENT STATE UNIVERSITY.
1975 See Geological Society of America. 197.

(578) SIGMA GAMMA EPSILON. RHO CHAPTER, INDIANA UNIVERSITY.
1965 A survey of Indiana geology with road logs for two field trips. (270)
1966 A survey of Indiana geology with road logs for two field trips. Rev. ed. (270)

AzTeS	1966
CLhC	1966
CPT	1966
CSt-ES	1966
CU-EART	1966
CaOOG	1966
CoG	1966
CoU	1966
DI-GS	1966
ICF	1966
IU	1966
IaU	1966
InLP	1966
InU	1965-66
KyU-Ge	1966
LU	1966
MiHM	1966
MoSW	1966
NBiSU	1966
NcU	1966
NdU	1966
NjP	1966
OCU-Geo	1966
OU	1966
OkU	1966
PSt	1966
TU	1966
TxBeaL	1966
TxDaAR-T	1966
TxU	1966
TxWB	1965-66
WU	1966

(579) SIGMA GAMMA EPSILON. XI CHAPTER, WASHINGTON STATE UNIVERSITY. GUIDEBOOK.
1962 Washington State University, Columbia Basin field trip guidebook. (The Xi Transactions of the Sigma Gamma Epsilon, Xi Chapter, v. 1(1) 1963)

OU	1962
WaPS	1962

(580) SIGMA GAMMA EPSILON. ZETA CHAPTER, THE UNIVERSITY OF TEXAS AT AUSTIN. [FIELD TRIP GUIDEBOOK] (TRIPS ARE NUMBERED FOR SCHOOL YEAR FROM FALL THROUGH SPRING)
1966 Jan. Northern Chihuahua field trip.
  —Nov. No. 1. Problems in Precambrian igneous and metamorphic rocks, Llano Uplift, central Texas.
  —Dec. No. 2. Tertiary sediments, central Texas.
1967 Jan. No. 3, pt. 1. Carbonate depositional environments, central Texas.
  Pt. 2. Carbonate depositional environments, central Texas.
  Pt. 3. Carbonate depositional environments, central Texas: Marble Falls and Glen Rose.
  No. 4. Problems in engineering geology in Austin, Texas.
  No. 5. Upper Cambrian succession, southeast Llano Uplift, Blanco County.
  —Apr. No. 6. West Texas field trip.
  No. 7. Problems in geomorphology, central Texas.
1968 Jan. No. 3. Early Cretaceous depositional environments.
  —May No. 4. Indio Lagoon system, Wilcox Group (Eocene), South Texas.
  —Nov. No. 1. Metamorphic rocks, Llano Uplift; sedimentary rocks, Marble Falls area.
  —Dec. No. 2. [Stratigraphic units of the Jackson Group and adjacent formations, south-central Texas]
1969 Trip 1. Environmental and engineering geology, Austin area.
1970 Trip 2. Austin-San Antonio-Eagle Pass-Piedras Negras-Nuevo Rosita-Monclova-Saltillo, Mexico.

1972 Mar.   Spring Break field trip. Parras Basin and vicinity.
1973 Oct.   Llano field trip.
1976 Nov.   Geomorphic features of Pedernales Falls State Park and Enchanted Rock, central Texas.
  DI-GS   1966(1), 67(5, 7), 68(1-3)
  TxU   1966-70, 72-73, 76

(581) SOCIEDAD GEOLOGICA MEXICANA. CONVENCION NACIONAL. GUIA DE LA EXCURSION GEOLOGICA.
1970 Libro-guia de la excursion Mexico-Oaxaca. Itineria geologica.
  Pt. 1. Mexico, D. F.-Cuautla, Mor.-Izucar de Matamoros, Pur.-Huajuapan de Leon, Oax.
  Pt. 2. Huajuapan de Leon, Oax.-Oaxaca, Oax.
1972 2nd   Mazatlan, Sinaloa.
  [1] Excursiones geologicas.
  [2] Torreon-Durango.
  [3] Cerro de Mercado, Durango, Dgo. (No text)
  [4] Durango-Mazatlan.
  [5] Distrito minero de Tayoltita, Dgo. (No text)
1974 3rd   Zacatecas, Zac.-Guanajuato, Gto.
1978 4th   Libro-guia de la excursion geologica a Tierra Caliente, estados de Guerrero y Mexico.
1982 Excursion geologica a la cuenca Mesozoica del centro de Mexico.
  [1] Libro-guia de la excursion geologica a la region de Zimpan y areas circumdantes, estados de Hidalgo y Queretaro.
  [2] Libreto-guia de la excursion las Calizas Lito - graficas de la Cantera Tlayuaen Tepexi de Rodriquez, Puebla.
  [3] Libro-guia de la excursion geologica a la parte central de la cuenca del alto del Rio Balsas, estados de Guerrero y Puebla.
  [4] Libro-guia de la excursion geologica a borde - noroeste de la paleo peninsula de Oaxaca (sureste del estado de Puebla).
  CSt-ES   1970
  DLC   1970, 72
  OCU-Geo   1970, 72
  OU   1972, 74, 78
  TxU   1970

(582) SOCIETY FOR THE STUDY OF EVOLUTION. FIELD EXCURSION.
1955 From Austin to Palmetto State Park via Bastrop, Red Rock, and Luling. San Marcos, Wimberley, and Dripping Springs.
  TxU   1955

(583) SOCIETY OF ECONOMIC GEOLOGISTS. GUIDEBOOK.
1963 Geology and technology of the Grants uranium region. Prepared for Uranium Field Conference. (No roadlog) (463)
1969 1. Papers on the stratigraphy and mine geology of the Kingsport and Mascot formations (Lower Ordovician) of East Tennessee. (651)
      2. Field conference on Wyoming uranium deposits. (726)
1970 Lead-zinc deposits in the Kootenay Arc, northeastern Washington and adjacent British Columbia. (No roadlog)
1971 June   Field conference on Wyoming trona deposits. (726)
— Sept.   Michigan copper district. (379)
1973 Apr.   See Geological Society of America. South Central Section. 205.
— Aug.   Guidebook for the Butte field meeting. (revised in 1978)
1974 Feb.   Arkansas-Texas economic geology field trip. (67(1975))
— Nov.   Field conference on kaolin and fuller's earth. (237)
1975 Guidebook to the Bingham mining district.
1977 Apr.   See Geological Association of Canada. 190.
1978 Feb.   Guidebook for the Butte Field Meeting, 1973. Second printing, rev. and enl.
— Oct.   See Geological Society of America. 197.
1979 See Geological Society of America. 197.
1980 See Geological Society of America. 197.(No.8,19)
1981 See Geological Society of America. 197.
1982 Sept. Relationship of mineralization to volcanic evolution of the San Juan Mountains, Colorado.
— Oct.   Economic geology of central Dominican Republic. (197)
1983 See Geological Society of America. 197.(No.10,16)
1984 Oct.   Sediment hosted precious metal deposits. (197, 451)
  Guidebook geology of the Pinson Mine, Humboldt County, Nevada. (451)
1985 See Geological Society of America. 197.(No.6)

  AzFU   1963
  AzTeS   1963, 69(1)
  AzU   1963, 69, 75
  CLU-G/G   1963, 69(1), 74(Nov.), 82
  CLhC   1969(2)
  CMenUG   1973, 75, 84
  CPT   1963, 69-70, 71(June), 74(Nov.), 75
  CSt-ES   1963, 69-71, 74(Nov.), 75, 84
  CU-A   1970
  CU-EART   1963, 69(1), 75, 78, 84
  CU-SB   1973
  CaACU   1969(2), 71(June), 78
  CaBVaU   1969-70, 71(June)
  CaOKQGS   1971(Sept./Oct.)
  CaOLU   1969(1), 70
  CaOOG   1963, 70, 73(Aug.), 82
  CaOWtU   1963, 69-71, 75
  CoDGS   1969(2), 71(June)
  CoG   1963, 69(1), 71(Sept./Oct.), 73(Oct.)
  CoU   1970
  DI-GS   1969, 71, 73(Aug.), 74(Nov.), 75
  DLC   1963, 69
  GU   1974(Nov.)
  ICF   1969-70
  ICIU-S   1969(1)
  ICarbS   1963, 69, 71
  IU   1969, 71, 74(Nov.), 75
  IaU   1963, 69(1), 75
  IdU   1969-70
  InLP   1963, 69-70, 71(June), 74(Nov.), 75, 78, 84(2)
  InU   1963, 69(1), 71
  KyU-Ge   1969(2), 71, 73(Aug.), 74-75, 84
  LNU   1963, 74
  MdBJ   1969
  MiHM   1963, 69, 71
  MnU   1963, 69-71, 74(Nov.), 75
  MoSW   1963, 69(1), 74(Nov.)
  MoU   1973(Aug.)
  MtBuM   1969(1), 73, 78
  NSyU   1969(1)
  NbU   1969(1)
  NcU   1963, 69, 71(June), 74(Nov.), 75
  NdU   1963, 69(1), 75
  NhD-K   1963, 69(1), 70, 71(Sept./Oct.), 75
  NjP   1969(1), 75
  NvU   1963, 69-70, 71(June), 75, 84
  OCU-Geo   1963, 69(1), 74(Nov.), 75, 84(Oct.)
  OU   1963, 69(1), 74
  OkU   1961, 63, 69-70, 73, 74(Nov.)
  OrU-S   1971(June)
  PBm   1974(Nov.)
  PSt   1963, 69(1), 73(Aug.), 75
  SdRM   1963, 69(1), 70
  TMM-E   1969(1)
  TxBeaL   1963, 69, 71(Sept.), 75
  TxDaAR-T   1969(2), 74(Nov.)
  TxDaDM   1969, 71(June), 74
  TxDaGI   1963
  TxHSD   1969(1)
  TxU   1963, 69
  UU   1963, 69, 71(June), 74(Nov.)
  ViBlbV   1963, 69-71, 75
  WU   1963, 69-70, 71(June), 74

(584) SOCIETY OF ECONOMIC GEOLOGISTS. PACIFIC SECTION. GUIDEBOOK.
1980 See American Association of Petroleum Geologists. Pacific Section. 28.

(585) SOCIETY OF ECONOMIC PALEONTOLOGISTS AND MINERALOGISTS. ANNUAL MEETING. GUIDEBOOK FOR ANNUAL FIELD TRIP.
1929 See American Association of Petroleum Geologists. 18.
1930 Oct.   Arbuckle Mountains, Oklahoma.
1932 See American Association of Petroleum Geologists. 18.
1933 See American Association of Petroleum Geologists. 18.
1936 See American Association of Petroleum Geologists. 18.
1937 See American Association of Petroleum Geologists. 18.
1939 See American Association of Petroleum Geologists. 18.
1940 See American Association of Petroleum Geologists. 18.

(586) Society of Economic Paleontologists and Mineralogists. Eastern Section.

1941 See American Association of Petroleum Geologists. 18.
1947 See American Association of Petroleum Geologists. 18.
1948 See American Association of Petroleum Geologists. 18.
1949 See American Association of Petroleum Geologists. 18.
1950 See American Association of Petroleum Geologists. 18.
1951 See American Association of Petroleum Geologists. 18.
1952 See American Association of Petroleum Geologists. 18.
1953 Mar. See American Association of Petroleum Geologists. 18.
1954 See American Association of Petroleum Geologists. 18.
1955 Mar. See American Association of Petroleum Geologists. 18.
1956 See American Association of Petroleum Geologists. 18.
1957 See American Association of Petroleum Geologists. 18.
1958 Mar. See American Association of Petroleum Geologists. 18.
1959 See American Association of Petroleum Geologists. 18.
1960 See American Association of Petroleum Geologists. 18.
1962 See American Association of Petroleum Geologists. 18.
1963 [36th] See American Association of Petroleum Geologists. 18.
1964 Mar. See American Association of Petroleum Geologists. 18.
1965 See American Association of Petroleum Geologists. 18.
1966 See American Association of Petroleum Geologists. 18.
1967 See American Association of Petroleum Geologists. 18.
1968 See American Association of Petroleum Geologists. 18.
1969 See American Association of Petroleum Geologists. 18.
1970 See American Association of Petroleum Geologists. 18.
1971 See American Association of Petroleum Geologists. 18.
1972 See American Association of Petroleum Geologists. 18.
1973 See American Association of Petroleum Geologists. 18.
1974 See American Association of Petroleum Geologists. 18.
1975 See American Association of Petroleum Geologists. 18.
1976 See American Association of Petroleum Geologists. 18.
1977 See American Association of Petroleum Geologists. 18.
1978 See American Association of Petroleum Geologists. 18.
1979 See American Association of Petroleum Geologists. 18.
1980 June SEPM trips 1-3, 5. See American Association of Petroleum Geologists. 18.
1981 May SEPM Trips 1-6. See American Association of Petroleum Geologists. 18.
1983 Apr. See American Association of Petroleum Geologists. 18.
1984 May See American Association of Petroleum Geologists. 18.

ICF	1930
IU	1930
TxDaGI	1930
TxLT	1930
TxMM	1930
TxU	1930

(586) SOCIETY OF ECONOMIC PALEONTOLOGISTS AND MINERALOGISTS. EASTERN SECTION. FIELD TRIP GUIDEBOOK.

1969 Field trip guidebook; coastal environments of northeastern Massachusetts and New Hampshire. (694)
1970 Sedimentology and origin of Upper Ordovician clastic rocks, central Pennsylvania.
1971 Interpretation of calcarenite paleoenvironments.
1972 Sedimentary facies: products of sedimentary environments in Catskill Mountains, Mohawk Valley, and Taconic sequence, eastern New York State.
1974 See American Association of Petroleum Geologists. Eastern Section. 25.
1975 June A. Ancient sediments of Nova Scotia.
    B. [Paleozoic and Early Mesozoic sedimentary rocks of Nova Scotia] (Reprint of Ancient sediments of Nova Scotia in "Maritime Sediments", v. 11, p. 9-46, 55-140)
1977 June Coastal processes and resulting forms of sediment accumulations, Currituck Spit, Virginia-North Carolina. (748)
1978 Apr. Guide to the redbeds of central Connecticut. (693)
1979 Environmental geologic guide to Cape Cod National Seashore.
1980 May Helderberg PACs. Application of the PAC hypothesis to limestones of the Helderberg Group.
1981 Oct. See American Association of Petroleum Geologists. National Energy Minerals Division. (Title varies) 27.
1982 March See Geological Society of America. Northeastern Section. 201.
— May Tuscarora Formation of Pennsylvania.
1983 Field trip guide to lower Paleozoic carbonate rocks, Roanoke region, Virginia.
1985 Oct. The Taconian clastic sequence in northern Virginia and West Virginia.

AzTeS	1969
AzU	1969
CLU-G/G	1969, 82
CLhC	1969
CSt-ES	1970, 77-78, 82
CU-A	1977
CU-EART	1975(B), 77
CU-SB	1977, 80
CaACU	1975(A)
CaAEU	1977
CaBVaU	1975(A)
CaOOG	1969, 72, 75(A), 77, 80
CaOWtU	1969-72, 77-79
CoDGS	1970
CoU	1969
DI-GS	1969-70, 72, 75(A), 77-80, 82-83, 85
ICarbS	1969
IU	1969, 77-79, 85
IaU	1977-78, 80, 82
InLP	1972, 77-80, 82-83, 85
InU	1969, 75(A), 77-78, 82
KyU-Ge	1969, 77-78, 82-83, 85
MH-GS	1969
MiU	1969
MnU	1977-78, 80, 82
MoSW	1969
NBiSU	1969
NNC	1980
NRU	1980
NSbSU	1972, 77, 79-80, 82
NSyU	1970
NcU	1969, 77-79, 83, 85
NdU	1969
NhD-K	1969, 78, 80, 82
NjP	1969
NmSoI	1980
OCU-Geo	1969, 70, 77, 79, 80, 82
OkU	1969, 82
OrCS	1977
PBm	1972, 77, 82
PHarER-T	1982
PSt	1971
TxBeaL	1969, 78
TxDaAR-T	1969
TxDaDM	1975(B)
TxDaSM	1969, 77
TxHSD	1969
TxU	1969, 72, 80, 82-83
TxWB	1983
ViBlbV	1969, 77
WU	1975(A), 77, 78, 82, 83, 85

(587) SOCIETY OF ECONOMIC PALEONTOLOGISTS AND MINERALOGISTS. GREAT LAKES SECTION. ANNUAL FIELD CONFERENCE. GUIDEBOOK.

1971 1st See Geological Society of America. 197.
1972 [2nd] Devonian strata of Alpena and Presque Isle counties, Michigan. (200(1970))
1973 3rd Field conference on Borden Group and overlying limestone units, south-central Indiana. (270)
1974 4th Coastal geology, sedimentology, and management, Chicago and the Northshore. (265(No.12))
    [1] Chicago River to Northshore and return aboard the M.V. Trinidad.
    [2] Illinois Beach State Park to Lake Bluff and return.
1975 5th Silurian reef and interreef environments with emphasis on interreef petrology, paleontology, stratigraphy, and sedimentation. (3rd printing in 1981)
1976 6th Variability within a high constructive lobate delta: the Pounds Sandstone of southeastern Illinois.
1977 7th Field guidebook to the biostratigraphy and paleoenvironments of the Cincinnatian series of southeastern Indiana.

1978 8th Lithostratigraphy, petrology, and sedimentology of Late Cambrian-Early Ordovician rocks near Madison, Wisconsin, with special papers. (777)
1979 Sept. The sedimentological and paleontological features of late Pennsylvanian and early Permian rocks of southeastern Ohio and western West Virginia.
1980 10th Middle and Late Pennsylvanian strata on margin of Illinois Basin, Vermilion County, Illinois, Vermillion and Parke counties, Indiana.
 Sediments of the modern meander belt of the Vermilion River near Eugene, Indiana and sedimentology of the lake Wisconsinan terrace deposits along the Bend area of the Wabash River.
1981 11th Field guidebook to the sedimentary environments and regional tectonic setting of the Huronian Supergroup, north shore of Lake Huron, Ontario, Canada.
 Silurian reef and interreef environments with emphasis on interreef petrology, paleontology, stratigraphy, and sedimentation. (5th Annual Field Conference, October 4 and 5, 1975, 3rd printing 1981)
— April-May See Geological Society of America. North Central Section. 200.
1982 12th Field guidebook to the paleoenvironments and biostratigraphy of the Newman and Breathitt formations (Mississippian-Pennsylvanian) in northwestern Kentucky.
1983 13th Ordovician Galena Group of the Upper Mississippi Valley; deposition, diagenesis, and paleoecology.
— Apr./May See Geological Society of America. North Central Section. 200.
— Sept. A compilation of informal and ad hoc field trip guides to the geology of the Great Lakes area.
1984 14th Bowling Green, Ohio.
 Field guidebook to the geomorphology and sedimentology of late Quaternary lake deposits, southwestern Lake Erie and glaciolacustrine deltaic deposits in the Oak Opening Sands of northwestern Ohio.
— Apr. See Geological Society of America. Southeastern Section. 206.
1985 15th Rock Island, Illinois.
 Devonian and Pennsylvanian stratigraphy of the quad cities region, Illinois-Iowa.
— Apr. See Geological Society of America. North Central Section. 200.

AzTeS	1977
AzU	1977
CLU-G/G	1974
CPT	1974, 77
CSt-ES	1973-74, 77-82
CU-A	1974, 80(2)
CU-EART	1973-74, 77, 79-80, 83
CU-SB	1974, 77, 82
CaACU	1977, 80
CaAEU	1974
CaBVaU	1980
CaOOG	1973, 80
CaOTRM	1974
CaOWtU	1973-74, 76-78
CoDGS	1973
DI-GS	1973-76, 78, 80-85
DLC	1973-74, 77
I-GS	1980, 83(Sept.)
ICF	1974
ICIU-S	1974
ICarbS	1974
IEN	1973
IU	1972-79, 80, 84
IaU	1974, 76-78, 80, 82-83, 85
InLP	1972-79, 80, 82-85
InU	1973-74, 76-78, 81-82
KyU-Ge	1973-79, 82
MiHM	1973-76, 78
MnU	1972-83
MtBuM	1974
NNC	1980, 81
NRU	1973, 80
NSbSU	1973-80, 82
NSyU	1974, 76-78
NcU	1973-75, 77, 80-81
NdU	1973
NhD-K	1974
NvU	1974
OCU-Geo	1973-76, 77-78, 80, 84
OU	1972-79, 80, 84-85
OkU	1973-74, 77, 79-83
PBL	1974
PBm	1974
PSt	1974-75, 77
SdRM	1974
TxDaM-SE	1980, 82
TxDaSM	1973
TxU	1977, 80, 82
TxWB	1982
UU	1974
ViBlbV	1973, 77
WU	1974-75, 77-78, 81, 83
WaU	1974

(588) SOCIETY OF ECONOMIC PALEONTOLOGISTS AND MINERALOGISTS. GULF COAST SECTION. GUIDEBOOK FOR THE ANNUAL FIELD TRIP.

1956 May A summary of the geology of Florida with emphasis on the Miocene deposits and a guidebook to the Miocene exposures.
— Nov. See Gulf Coast Association of Geological Societies. 247.
1957 Oct. Oligocene-Eocene of western Mississippi, central Louisiana and extreme eastern Texas.
1958 Oct. Upper and middle Tertiary of Brazos River valley, Texas. (255)
1959 May Lower Tertiary and Upper Cretaceous of Brazos River valley, Texas. (255)
— Nov. See Gulf Coast Association of Geological Societies. 247.
1960 May Jackson Group, Catahoula and Oakville formations and associated structures of northern Grimes County, Texas. (255)
1961 May Middle Eocene of Houston Co., Texas.
1962 Nov. See Gulf Coast Association of Geological Societies. 247.
1963 Mar. See American Association of Petroleum Geologists. 18.
1965 May Recent organisms and sediments, Galveston Heald Bank area, Gulf of Mexico.
1967 Sept. Selected Cretaceous and Tertiary depositional environments.
1969 Oct./Nov. See Gulf Coast Association of Geological Societies. 247.
1970 Oct. See Gulf Coast Association of Geological Societies. 247.
1971 Oct. See Gulf Coast Association of Geological Societies. 247.
1974 Oct. See Gulf Coast Association of Geological Societies. 247.
1977 Oct. See Gulf Coast Association of Geological Societies. 247.
1978 Oct. See Gulf Coast Association of Geological Societies. 247.
1979 Apr. See American Association of Petroleum Geologists. 18.
— Oct. See Gulf Coast Association of Geological Societies. 247.
1980 May Middle Eocene coastal plain and nearshore deposits of East Texas; a field guide to the Queen City Formation and related papers.
— Oct. See Gulf Coast Association of Geological Societies. 247.
1982 Apr. Deltaic sedimentation on the Louisiana coast.
1983 Apr. Upper Cretaceous lithostratigraphy and biostratigraphy in Northeast Mississippi, Southwest Tennessee and Northwest Alabama, shelf chalks and coastal clastics.
 (2) Stratigraphic and structural overview of Upper Cretaceous rocks exposed in the Dallas vicinity.
1984 May See American Association of Petroleum Geologists. 18.

ATuGS	1957, 61
AzTeS	1961, 80
AzU	1957-58, 59(May), 61
CLU-G/G	1957-58, 59(May), 60-61, 67, 80
CLhC	1960, 82, 83(May)
CMenUG	1983
CPT	1961, 80
CSdS	1957, 61, 83
CSt-ES	1957-58, 60, 82, 83
CU-EART	1957, 59(May), 61, 67, 82, 83
CU-SB	1957, 61, 67, 80, 82, 83
CaACU	1956(May), 82, 83
CaOHM	1957, 59(May), 61, 67
CaOKQGS	1957, 59(May)
CaOTRM	1958, 59(May)
CaOWtU	1980, 83
CoDCh	1959(May)
CoDGS	1957-63, 83
CoG	1957-58, 59(May), 60-61, 67
CoLiM	1983
CoU	1957-58, 59(May), 60, 67
CtY-KS	1982
DFPC	1961
DI-GS	1957-58, 60, 82-83

(589) Society of Economic Paleontologists and Mineralogists. Mid-Continent Section.

DLC	1957, 80
ICF	1957-58, 59(May), 60-61
ICarbS	1959(May), 60
IEN	1959(May), 60-61
IU	1957-58, 59(May), 61, 67, 80, 82-83
IaU	1957-58, 59(May), 60-61
InLP	1957, 59(May), 61, 80, 82-83
InU	1957-58, 59(May), 60, 67, 80
KyU-Ge	1957-58, 59(May), 61, 80, 82-83, 84
LNU	1957-58
LRuL	1980
LU	1957-58, 59(May), 60, 67
MH-GS	1984
MNS	1958, 60
MiKW	1958, 60
MiU	1957, 59(May), 60
MnU	1957-58, 59(May), 60-61, 65(May), 67, 80, 82, 83
MoSW	1957-58, 59(May), 61, 67
MoU	1956(May)
Ms-GS	1956(May)
MsJG	1956(May)
MtU	1957, 61
NNC	1957-58, 59(May), 60-61, 67, 82, 83, 84
NRU	1957
NSbSU	1961, 80, 82
NSyU	1958, 79(Oct.)
NbU	1957-58, 59(May), 60-61
NcU	1957, 61, 64, 65, 67, 68, 73, 79
NdU	1967
NjP	1957, 67
OCU-Geo	1957, 59(May), 60-61, 67
OU	1958, 59(May), 60-61, 67, 80
OkT	1958, 59(May), 60
OkTA	1980, 83
OkU	1958, 59(May), 60, 65(May), 67, 80, 83-84
PSt	1959(May), 60-61
TMM-E	1957, 61
TxBeaL	1957, 59(May), 61, 67
TxDaAR-T	1957-58, 59(May), 60, 67
TxDaDM	1957-58, 59(May), 60-61, 67
TxDaGI	1959(May)
TxDaM-SE	1957-58, 59(May), 60-61, 83
TxDaSM	1956(May), 57-58, 60, 67
TxHSD	1959(May), 60, 65(May), 67
TxHU	1957-58, 59(May), 61, 67
TxMM	1958, 59(May), 83
TxU	1957-58, 59(May), 60-61, 67, 80, 83
TxU-Da	1957-58, 59(May), 60
TxWB	1958-60
ViBlbV	1959(May)
WU	1957-58, 80, 82, 83, 84

(589) SOCIETY OF ECONOMIC PALEONTOLOGISTS AND MINERALOGISTS. MID-CONTINENT SECTION. ANNUAL MEETING AND FIELD CONFERENCE.

1983 V. 1. Tectonic-sedimentary evolution of the Arkoma Basin and guidebook to deltaic facies, Hartshorne Sandstone.

1984 2nd [Little Rock, Arkansas]
A guidebook to the post-St. Peter Ordovician and the Silurian and Devonian rocks of north-central Arkansas. (67)

1985 3rd Lawrence, Kansas.
Recent interpretations of late Paleozoic cyclothems.

AGC	1984
BEG	1984
CMenUG	1984
CSfCSM	1984
CU-EART	1983
CoDGS	1984
CoLiM	1984
DI-GS	1983-84
DLC	1983
IaU	1983, 85
InLP	1983-85
InU	1983
KLG	1985
KU	1984
KyU-Ge	1983-85
MnU	1984-85
OkT	1985
OkTA	1983, 85
OkU	1983-85
TxDaDM	1985
TxU	1983

(590) SOCIETY OF ECONOMIC PALEONTOLOGISTS AND MINERALOGISTS. MIDYEAR MEETING. FIELD TRIP.

1984 Aug. San Jose, California.
[Field trip no. 1.] Stratigraphic, tectonic, thermal, and diagenetic histories of the Monterey Formation, Pismo and Huasna Basin, California. (597)
Field trip no. 2. Latest Cretaceous and early Tertiary depositional systems of the northern Diablo Range, California.
Field trip no. 3. Depositional facies of sedimentary serpentinite: selected examples from the Coast Ranges, California.
Field trip no. 5. Pleistocene shoreline and shelf deposits at Fort Funston and their relation to sea-level changes.

1985 Aug. Field trip no. 9. Fine-grained deposits and biofacies of the Cretaceous western interior seaway; evidence of cyclic sedimentary processes. (597(No.3), 596)
No. 2. A field guidebook to the geology of Glacier National Park, Montana and vicinity.
No. 3. Environments of deposition of Cretaceous sandstones of the western interior.
No. 4. Pennsylvanian algal carbonates and associated facies, central Colorado.
No. 6. Sedimentology, dolomitization, mineralization and karstification of the Leadville Limestone (Mississippian), central Colorado.
No. 7. Canyonlands National Park float trip, Utah.
No. 8. Triassic-Jurassic fluvial systems, northeastern Arizona and southeastern Utah.
No. 10. Depositional facies of the Cretaceous Castlegate and Blackhawk formations, Book Cliffs, eastern Utah.
No. 11. Shelf sandstones: Shannon and Haystack mountains formations, Wyoming.
No. 13. Holocene braided rivers of eastern Colorado and the sedimentologic effects of the Lawn Lake Dam failure in the Rocky Mountain National Park.

CLU-G/G	1984(1-3), 85(3, 4, 6, 8, 10-13)
CLhC	1984(1), 85(9)
CMenUG	1984(1, 2), 85(9)
CSdS	1984
CSt-ES	1984, 85(9)
CU-EART	1984(1)
CU-SB	1985(2-5, 6-8, 10-11, 13)
CU-SC	1984, 85(9)
CaACU	1985(2-5, 6-8, 10-11, 13)
CaOHM	1984(1)
CaOOG	1984(1), 85(2)
CoDCh	1985
CoDGS	1985
DI-GS	1984(1), 85(2-5, 6-8, 10-11, 13)
IU	85(9)
IaU	1984(2-3, 5)
IdU	1984(1, 3), 85(9)
InLP	1984
InU	1985(9)
KyU-Ge	1984
MdBJ	1984(1)
MnU	1984
NcU	1984(1)
NmU	1985(2-5, 6-8, 10-11, 13)
NvU	1984, 85
ODaWU	1984
OkTA	1984
OkU	1984, 85(9)
PSt	1984(3), 85(9),
TxDaM-SE	1984
TxMM	1984
TxWB	1985
WU	1984, 85(2-4, 6, 8, 10, 11, 13)

(591) SOCIETY OF ECONOMIC PALEONTOLOGISTS AND MINERALOGISTS. NORTH AMERICAN MICROPALEONTOLOGY SECTION. GUIDEBOOK FOR THE FIELD TRIP.
1985 May   Stratigraphy and paleontology of the outcropping Tertiary beds in the Pamunkey River region, central Virginia coastal plain. (This guidebook is a repeat of the 1984 field trip sponsored by Atlantic Coastal Plain Geological Association). (89(1984))

CSt-ES	1985
CU-A	1985
CU-EART	1985
InLP	1985
MH-GS	1985
MnU	1985

(592) SOCIETY OF ECONOMIC PALEONTOLOGISTS AND MINERALOGISTS. NORTHEASTERN SECTION. FIELD TRIP GUIDEBOOK.
1968   Coastal sedimentary environments, Lewes-Rehoboth Beach, Delaware. (683)
1983   See Geological Society of America. Northeastern Section. 201.

CaOOG	1968
DI-GS	1968
InU	1968
OkU	1968
TxDaAR-T	1968
TxDaSM	1968

(593) SOCIETY OF ECONOMIC PALEONTOLOGISTS AND MINERALOGISTS. PACIFIC SECTION. GUIDEBOOK FOR THE FIELD TRIP.
1944 Nov.   See American Association of Petroleum Geologists. Pacific Section. 28.
1947 Sept.   See American Association of Petroleum Geologists. Pacific Section. 28.
1948 May   San Emigdio Creek, Kern County, California.
1949 Oct.   Hollister Ranch, El Bulito Canyon and vicinity of Refugian type section, Santa Barbara, California.
1950 May   See American Association of Petroleum Geologists. Pacific Section. 28.
1951 May   See American Association of Petroleum Geologists. Pacific Section. 28.
— Dec.   Road log. Death Valley to San Fernando.
1953 Apr.   See American Association of Petroleum Geologists. Pacific Section. 28.
— May   Santa Ana Mountains, California.
1954 May   See American Association of Petroleum Geologists. Pacific Section. 28.
1955 May   See American Association of Petroleum Geologists. Pacific Section. 28.
1956 Feb.   Tejon formation in Liveoak Canyon, [Tehachapi Mts.].
— May   Huasna Basin, San Luis Obispo County, California.
1957 May   La Jolla area, California.
1958 May   See American Association of Petroleum Geologists. Pacific Section. 28.
1959 Apr.   [Big Basin area, Santa Cruz Mountains, California]
1960 Apr.   Type Panoche-Panoche Hills area, Fresno County, California.
1961 May   See American Association of Petroleum Geologists. Pacific Section. 28.
1962 Mar.   See American Association of Petroleum Geologists. 18.
— Oct.   See American Association of Petroleum Geologists. Pacific Section. 28.
1963 May   (28)
— June   Hathaway Ranch Area. (557)
1964 Oct.   See American Association of Petroleum Geologists. Pacific Section. 28.
1965 Apr.   Geology of southeastern San Joaquin Valley, California; Kern River to Grapevine Canyon. (557, 601)
— Oct.   Western Santa Ynez Mountains, Santa Barbara, California. (134)
1966 Mar.   See American Association of Petroleum Geologists. Pacific Section. 28.
— June   See American Association of Petroleum Geologists. Pacific Section. 28.
1967 Apr.   See American Association of Petroleum Geologists. 18.
— Oct.   See American Association of Petroleum Geologists. Pacific Section. 28.
1968 Mar.   See American Association of Petroleum Geologists. Pacific Section. 28.
— Oct.   See American Association of Petroleum Geologists. Pacific Section. 28.
1969 Mar.   See American Association of Petroleum Geologists. Pacific Section. 28.
— Oct.   Geologic setting of upper Miocene gypsum and phosphorite deposits, upper Sespe Creek and Pine Mountain, Ventura County, California. (Cover title: Upper Sespe Creek)
1970 Mar.   See American Association of Petroleum Geologists. Pacific Section. 28.
— Nov.   See American Association of Petroleum Geologists. Pacific Section. 28.
1971 Oct.   Geologic guidebook; Newport Lagoon to San Clemente, California, coastal exposures of Miocene and early Pliocene rocks.
1972 Mar.   See American Association of Petroleum Geologists. Pacific Section. 28.
— June   See American Association of Petroleum Geologists. Pacific Section. 28.
— Oct.   Cretaceous of the Coalinga area.
1973 May   See American Association of Petroleum Geologists. 18.
— Oct.   Cretaceous stratigraphy of the Santa Monica Mountains and Simi Hills, Southern California.
1974 Apr.   See American Association of Petroleum Geologists. Pacific Section. 28.
— Oct.   The Paleogene of the Panoche Creek-Cantua Creek area, central California.
1975 Apr.   See American Association of Petroleum Geologists. Pacific Section. 28.
— Oct.   Geologic field guide of the eastern Santa Maria area.
1976 Apr.   See American Association of Petroleum Geologists. Pacific Section. 28.
— Oct.   Depositional environments of lower Paleozoic rocks in the White-Inyo Mountains, Inyo County, California. (520)
1977 Apr.   See American Association of Petroleum Geologists. Pacific Section. 28.
— Oct.   Cretaceous geology of the California coast ranges, west of the San Andreas Fault. (520)
1978 Apr.   See American Association of Petroleum Geologists. Pacific Section. 28.
— Oct.   Depositional environments of Tertiary rocks along Sespe Creek, Ventura County, California. (520)
1979 Mar.   See American Association of Petroleum Geologists. Pacific Section. 28.
— Oct.   Tertiary and Quaternary geology of the Salinas Valley and Santa Lucia Range, Monterey County, California. (520)
— Nov.   See Geological Society of America. 197.
1980   See American Association of Petroleum Geologists. Pacific Section. 28.
— Oct.   Neogene biostratigraphy of the northern La Panza Range, San Luis Obispo County, California.
1981 May   See American Association of Petroleum Geologists. 18.
— Oct.   Simi Hills Cretaceous turbidites, Southern California.
1982 Apr.   See American Association of Petroleum Geologists. Pacific Section. 28.
— Oct.   Vol. 25. Monterey Formation and associated coarse clastic rocks, central San Joaquin Basin, California.
1983 May   [1] Guidebook to the Stony Creek Formation, Great Valley Group, Sacramento Valley, California.
[2] Tectonic transect of Sierran Paleozoic through Jurassic accreted belts.
[3] Geology and sedimentology of the southwestern Sacramento Basin and East Bay Hills.
— Oct.   Cenozoic geology of the Simi Valley area, Southern California.
1984 Apr.   2. Upper Cretaceous depositional systems; Southern California-northern Baja California.
— June   See Los Angeles Basin Geological Society. 356.
— Sept.   Ashland, Oregon.
Geology of the Upper Cretaceous Hornbrook Formation, Oregon and California.
1985 May   Anchorage, Alaska. See American Association of Petroleum Geologists. Pacific Section. 28.

(594) Society of Economic Paleontologists and Mineralogists. Permian Basin Section.

— Oct. Coalinga, California.
Geology of the Temblor Formation, western San Joaquin Basin, California.

AzTeS	1951(Dec.), 56, 59-60, 72(Oct.), 73(Oct.), 78
AzU	1977-78
CLU-G/G	1953(May), 55, 56(May), 59-60, 69(Oct.), 71, 72(Oct.), 73(Oct.), 75(Oct.), 76(Oct.), 77-78, 80(Oct.), 81(Oct.), 82(Oct.), 83, 84(Apr., Sept.), 85
CLhC	1960, 65(Oct.), 69(Oct.), 71, 72(Oct.), 74(Oct.), 76(Oct.), 77-79, 80(Oct.), 81(Oct.), 82(Oct.), 83(May), 84
CMenUG	1965, 80(Oct.), 82(Oct.), 83(May, Oct.), 84, 85(Oct.)
CPT	1951, 56(May), 57, 60, 69(Oct.), 71, 72-74, 76-79, 80(Oct.), 81(Oct.), 83(May, Oct.), 84(Apr., Sept.), 85(Oct.)
CSdS	1969(Oct.), 71, 72(Oct.), 73(Oct.), 74(Oct.), 76(Oct.), 77(Oct.), 78(Oct.), 79(Oct.), 80(Oct.), 81(Oct.), 82(Oct.), 83(May, Oct.), 84(Apr.)
CSfCSM	1957, 59-60, 69(Oct.), 71, 72(Oct.), 74(Oct.), 76(Oct.), 77, 80(Oct.), 81(Oct.), 82(Oct.), 83, 84(Apr.), Sept.)
CSt-ES	1956(May), 59-60, 69(Oct.), 71, 73(Oct.), 74(Oct.), 76(Oct.), 77-79, 80(Oct.), 81(Oct.), 82, 83(May), 84-85
CU-A	1948, 51(Dec.), 53(May), 56-57, 60, 69(Oct.), 71, 73(Oct.), 74(Oct.), 80-82, 83(May), 84(Apr.)
CU-EART	1948-49, 51(Dec.), 53(May), 56-57, 59-60, 69(Oct.), 71, 72(Oct.), 73(Oct.), 74(Oct.), 76(Oct.), 77-79, 80(Oct.), 82(Oct.), 83, 84, 85(Oct)
CU-SB	1953, 59, 63, 69, 73-75, 77-80, 81(May), 82(Oct.), 83(May), 84(Apr.)
CU-SC	1982(Oct.), 83(May, Oct.), 84(Apr., Sept.)
CaACU	1972(Oct.), 73(Oct.), 76(Oct.), 77(Oct.), 78(Oct.), 79(Oct.), 81
CaOHM	1972(Oct.), 74(Oct.)
CaOKQGS	1965(Oct.)
CaOOG	1983-84
CaOTRM	1965(Oct.)
CaOWtU	1965(Oct.), 74(Oct.), 76(Oct.), 77-78, 80(Oct.)
CoDGS	1950-51, 54, 56, 60-66, 70, 74, 77-79, 80(Oct.), 81-83
CoFS	1965(Oct.)
CoG	1956(May), 60
CoLiM	1980(Oct.), 81(May), 83, 84(Apr., Sept.), 85(Oct.)
DI-GS	1948, 51(Dec.), 53(May), 56-57, 60, 69(Oct.), 71, 72(Oct.), 73(Oct.), 74(Oct.), 75(Oct.), 76(Oct.), 77-81, 83-84
DLC	1974(Oct.), 76-78
IU	1971, 72(Oct.), 73(Oct.), 75(Oct.), 77-79, 80
IaU	1965(Oct.), 69(Oct.), 71, 72(Oct.), 73(Oct.), 74(Oct.), 75(Oct.), 76(Oct.), 78, 79, 81, 83(May)
IdBB	1965(Oct.), 71
InLP	1974(Oct.), 76(Oct.), 77-84
InU	1959, 69(Oct.), 71, 72(Oct.), 77, 82
KyU-Ge	1951(Dec.), 56(May), 60, 69(Oct.), 71, 72(Oct.), 73(Oct.), 74(Oct.), 76(Oct.), 77-79, 80(Oct.), 81(May), 83, 84(Apr., Sept.)
MH-GS	1971
MU	1981
MiHM	1983, 84
MnU	1948-49, 51(Dec.), 53(May), 56-57, 69(Oct.), 71, 72(Oct.), 73(Oct.), 74(Oct.), 75(Oct.), 76(Oct.), 77-78, 80(Oct.), 81(Oct.), 83(May)
MoSW	1965(Oct.)
MoU	1965(Oct.), 69(Oct.)
NBiSU	1959
NIC	1981, 83
NNC	1959, 69(Oct.)
NSbSU	1976-82, 83-85
NSyU	1973(Oct.), 76(Oct.), 77
NbU	1959-60
NcU	1969(Oct.), 72(Oct.), 74(Oct.), 76(Oct.), 77-78, 81
NhD-K	1973(Oct.), 74(Oct.), 77(Oct.), 82
NjP	1965(Oct.), 72(Oct.), 74(Oct.)
NvU	1951(Dec.), 56(May), 60, 72(Oct.), 74(Oct.), 75(Oct.), 76(Oct.), 77-78, 79(Oct.), 80(Oct.), 81(Oct.), 82(Oct.), 83(May), 84(Apr.), Sept.)
OCU-Geo	1956, 60, 69(Oct.), 71, 76(Oct.), 77(Oct.), 79(Oct.)
OKentC	1981
OU	1956-57, 60, 69(Oct.), 71, 72(Oct.), 74(Oct.), 76(Oct.)
OkTA	1984(Apr.)
OkU	1965(Oct.), 71, 72(Oct.), 77, 80(Oct.), 83-84
SDSNH	1981(Oct.), 83(May)
ScU	1984
TxBeaL	1953(May), 69(Oct.)
TxDaAR-T	1969(Oct.), 71, 72(Oct.), 73(Oct.), 74(Oct.)
TxDaDM	1956, 59-60, 69(Oct.), 71, 72(Oct.), 73(Oct.)
TxDaM-SE	1981, 83-84
TxDaSM	1959-60, 71, 72(Oct.), 74(Oct.)
TxHSD	1965(Oct.), 69(Oct.), 73(Oct.), 74(Oct.)
TxMM	1960
TxU	1948-49, 53(May), 56-57, 59-60, 69(Oct.), 71, 72(Oct.), 74(Oct.), 80-81, 83-84
TxU-Da	1951(Dec.), 60, 71, 72(Oct.), 73(Oct.)
TxWB	1973, 77-79, 80(Oct.), 84(Apr.)
ViBlbV	1951(Dec.), 69(Oct.), 72(Oct.), 73(Oct.), 74(Oct.), 75(Oct.), 76(Oct.), 77
WU	1971, 76(Oct.), 77(Oct.), 78(Oct.), 79(Oct.), 80-83, 85
WaU	1983(May), 84(May[2])
WyU	1983(May)

(594) SOCIETY OF ECONOMIC PALEONTOLOGISTS AND MINERALOGISTS. PERMIAN BASIN SECTION. GUIDEBOOK FOR THE FIELD TRIP.

1955 Permian field conference to the Guadalupe Mountains.
1956 Symposium of the Fort Worth Basin area and field study of the Hill Creek beds of the lower Strawn, southwestern Parker County, Texas.
1957 Wolfcamp of the Glass Mountains and the Permian Basin.
1958 Cretaceous platform and geosyncline, Culberson and Hudspeth counties, Trans-Pecos, Texas.
1959 Sacramento Mountains of Otero County, New Mexico. (550)
1960 Apr. Traverse of Post-Avis Cisco rocks, Brazos Valley, north central Texas; A study of the Cisco facies near Breckenridge, Texas.
1962 Leonardian facies of the Sierra Diablo region, West Texas.
1964 Filling of the Marathon Geosyncline; symposium and guidebook.
1965 Amistad Dam field trip. (Guidebook used for this trip was the West Texas Geological Society Val Verde Basin guidebook, 1959)
1967 One day field trip; Bank to basin transition in Permian (Leonardian) carbonates, Guadalupe Mountains, Texas.
1968 Guadalupian facies, Apache Mountains area, west Texas.
1969 Ordovician symposium. (167)
1970 Apr. Geology of the southern Quitman Mountains area, Trans-Pecos, Texas.
1971 Robledo Mountains, New Mexico, Franklin Mountains, Texas.
1972 Capitan reef, New Mexico and Texas: facts and questions to aid interpretation and group discussion. (Publication 72-14) (595)
1974 Feb. See American Association of Petroleum Geologists. Southwest Section. 31.
 — Oct./Nov. Lower Cretaceous shelf, platform, reef, and basinal deposits, Southwest Texas and northern Coahuila. (Publication 74-64) (756, 595)
1975 Apr. Geology of the Eagle Mountains and vicinity, Trans-Pecos Texas. (Publication 75-15) (756, 595)
 — Oct./Nov. Permain exploration, boundaries, and stratigraphy. Symposium and field trip: Midland and El Paso, Texas. (Publication 75-65) (756)
1977 Upper Guadalupian facies, Permian reef complex, Guadalupe Mountains, New Mexico and West Texas. Volume 2: Road logs and locality guides. (Publication 77-16) (595)
1978 Tectonics and Paleozoic facies of the Marathon Geosyncline, West Texas. (Publication 78-17) (595)
1979 Guadalupian Delaware Mountain Group of West Texas and Southeast New Mexico. (Publication 79-18) (595)
1980 Geology of the Finlay Mountains, south central Hudspeth County, Texas. (Publication 80-19) (595)
1981 Apr. Marathon-Marfa region of West Texas; symposium and guidebook. (Publication 81-20) (595)
1982 Middle and upper Pennsylvanian system of north-central and West Texas (outcrop to subsurface). (Publication 82-21) (595)
1983 May Geology of the Sierra Diablo and southern Hueco Mountains, West Texas. (Publication 83-22) (595)
1984 Apr. Lower Permian to upper Pliocene carbonate and clastic facies, southern eastern shelf, west central Texas. (Publication 84-23) (595)

AzTeS	1957, 59-60, 62, 64, 68, 70-72, 74-75, 77-80, 83-84
AzU	1955-57, 59-60, 62, 68-72, 74, 75(Oct./Nov.), 77-79
BEG	1980-81
CChiS	1964, 68
CLU-G/G	1955, 57-60, 62, 64, 68, 70-72, 74-75, 77-84
CLhC	1959, 60, 64, 72, 82
CMenUG	1955-60, 62, 64, 67-68, 70-72, 74-75, 77-84
CPT	1959-60, 64, 68, 70-71, 74, 75(Oct/Nov), 77-84
CSdS	1960, 62, 68
CSt-ES	1955-60, 62, 64-65, 68, 70-71, 74-75, 77-84
CU-A	1956-60, 62, 64, 67-68, 70-71, 74-75, 77-79, 84

CU-EART	1956-60, 62, 64, 68, 70-71, 74-75, 77-84
CU-SB	1957-60, 62, 64, 68, 70, 75
CU-SC	1980-84
CaACM	1977
CaACU	1955-60, 62, 68, 70-72, 74(Oct/Nov), 75, 77
CaAEU	1972
CaBVaU	1959, 64, 68-70, 75(Oct./Nov.)
CaOKQGS	1980-83
CaOLU	1964
CaOOG	1957, 60, 62, 64, 68-72, 74-75, 77-84
CaOWtU	1959-60, 62, 68-72, 74(Oct/Nov), 75, 77-82, 84
CoDCh	1955-59, 62, 64, 68, 70, 77-78, 80-83
CoDGS	1955-60, 62, 68, 70-72, 75, 77-78, 80-83
CoDU	1959, 62, 64, 68, 70-71
CoG	1956-60, 62, 64-65, 68-71, 77
CoGrU	1977
CoLiM	1980-84
CoU	1956-60, 62, 64-65, 68-71, 75
CoU-DA	1982-84
DI-GS	1955-60, 62, 64, 67-72, 74-75, 77-84
DLC	1974-75, 77, 83
ICF	1959, 69, 71
ICarbS	1971-72, 75(Apr.), 77-84
IU	1955-60, 62, 64, 71-72, 74-75, 77-79, 80
IaU	1957, 59-60, 62, 64-65, 68-72, 74-75, 77-84
IdPI	1972
IdU	1975, 77-78, 80-83
InLP	1956-60, 62, 65, 68-72, 74, 75(Apr.), 77-84
InU	1959-60, 64, 71-72, 77-79, 82
KyU-Ge	1956-60, 62, 64, 68, 70-72, 74-75, 77-84
LNU	1960, 70-71, 74, 75(Apr.), 77, 82-84
LRuL	1980-81
LU	1964
MH-GS	1964, 68-69
MNS	1977
MdBJ	1972, 77, 79, 82-84
MdU	1955-60, 62, 64
MiDW-S	1971
MiHM	1955-60, 62, 74
MiKW	1957
MiU	1956-60, 62, 64
MnU	1956-60, 62, 64-65, 68-72, 74-75, 77-78, 81-82, 84
MoRM	1959, 62
MoSW	1969-70, 75(Oct./Nov.)
MoU	1969-72, 75, 77
MtBuM	1977
MtU	1955-60, 62
NBiSU	1959-60, 62, 68, 70-71
NMSoI	1967
NNC	1955-60, 62, 64, 68, 70-72, 74, 75(Apr.), 77-79, 80, 81, 83-84
NOneoU	1972
NRU	1967, 77
NSbSU	1960, 71, 75, 77, 80-82
NSyU	1975(Apr.), 77-78, 81-83
NbU	1957, 59-60, 62, 64, 67-68, 70-72, 74, 75(Oct./Nov.)
NcU	1958-60, 62, 64, 68, 70-72, 74-75, 77-81
NdU	1955-57, 59-60, 62, 64, 68, 70-72, 74, 75(Apr.), 77-82
NhD-K	1965, 70-72, 74-75, 77, 80-84
NmLcU	1980-81, 83
NmPE	1955, 57-60, 62, 64, 68, 70-71
NmU	1959, 80-83
NvU	1965, 70, 74, 75(Oct./Nov.)
OCU-Geo	1959, 62, 64, 67-72, 74-75, 77
ODaWU	1980, 83-84
OKentC	1984
OU	1955-60, 62, 64, 68, 70-71, 77-84
OkT	1958
OkTA	1980-84
OkU	1955-56, 59-60, 62, 64, 67-68, 70-72, 75, 77, 80-84
OrCS	1975(Oct./Nov.), 77-81
PBL	1967, 70
PBU	1980-82, 84
PSt	1974
RPB	1964, 68
SdRM	1959, 62
TMM-E	1959-60, 62, 68, 70-72
TxBeaL	1955-58, 60, 62, 64-65, 68, 70-72, 74, 75(Apr.), 77-79
TxCaW	1970, 74
TxDaAR-T	1956-60, 62, 64, 69-70, 74
TxDaDM	1960, 71, 74
TxDaGI	1955-56
TxDaM-SE	1959-60, 80-82
TxDaSM	1955-60, 64, 68-71, 74, 75(Oct./Nov.), 77-79
TxHMP	1980-84
TxHSD	1955, 57-59, 62, 64, 68, 71, 74-75, 77-78
TxHU	1955-60, 62, 64, 67-68, 70-71, 74
TxLT	1955-60, 62, 64, 74, 75(Oct./Nov.)
TxMM	1955-60, 62, 64, 68-72, 75(Oct./Nov.), 77, 79-84
TxU	1955-60, 62, 64-65, 67-71, 74-75, 77-84
TxU-Da	1960
TxWB	1955-60, 62, 64, 67-68, 70-72, 75-77, 79, 81-82, 84
UU	1970, 77
ViBlbV	1974, 75(Oct./Nov.)
WU	1956-60, 62, 64, 67-68, 70-72, 74-75, 77-84
WaPS	1977
WaU	1959
WyU	1955, 59, 80-82

(595) SOCIETY OF ECONOMIC PALEONTOLOGISTS AND MINERALOGISTS. PERMIAN BASIN SECTION. PUBLICATION.

1972 72-14. See Society of Economic Paleontologists and Mineralogists. Permian Basin Section. 594.
1974 See Society of Economic Paleontologists and Mineralogists. Permian Basin Section. 594.
1975 75-15. See Society of Economic Paleontologists and Mineralogists. Permian Basin Section. 594.
1977 77-16. See Society of Economic Paleontologists and Mineralogists. Permian Basin Section. 594.
1978 78-17. See Society of Economic Paleontologists and Mineralogists. Permian Basin Section. 594.
1979 79-18. See Society of Economic Paleontologists and Mineralogists. Permian Basin Section. 594.
1980 80-19. See Society of Economic Paleontologists and Mineralogists. Permian Basin Section. 594.
1981 81-20. See Society of Economic Paleontologists and Mineralogists. Permian Basin Section. 594.
1982 82-21. See Society of Economic Paleontologists and Mineralogists. Permian Basin Section. 594.
1983 83-22. See Society of Economic Paleontologists and Mineralogists. Permian Basin Section. 594.
1984 84-23. See Society of Economic Paleontologists and Mineralogists. Permian Basin Section. 594.

(596) SOCIETY OF ECONOMIC PALEONTOLOGISTS AND MINERALOGISTS. ROCKY MOUNTAIN SECTION. FIELD TRIP.

1974 See American Association of Petroleum Geologists. Rocky Mountain Section. 30.
1979 June See American Association of Petroleum Geologists. National Energy Minerals Division. (Title varies) 27.
— Sept. Cretaceous of the Rock Springs Uplift, Wyoming.
1981 Fall Depositional systems Upper Cretaceous Mancos Shale and Mesaverde Group, northwestern Colorado.
1982 Fall Past and present Eolian sediments.
Day 1: Nebraska Sand Hills.
Day 2: Leo Sandstones.
Day 3: Unkpapa Sandstones, Inyan Kara Group, and Sundance Formation.
1983 Mid-Cretaceous Codell Sandstone Member of Carlile Shale, eastern Colorado.
1984 Spring Sedimentology of the Fountain Fan Delta complex, near Manitou Springs and Canon City, Colorado.
1985 See American Association of Petroleum Geologists. Rocky Mountain Section. 30. and Society of Economic Paleontologists and Mineralogists. Midyear Meeting. 590.

CPT	1984
CU-EART	1979(Sept.), 81
CU-SB	1982
CaACM	1980
CaACU	1981
CoDCh	1981-84
CoDGS	1981
CoLiM	1983-84
DI-GS	1979(Sept.), 81-84
IaU	1981, 84
InLP	1981
InU	1982
MnU	1981-82
MtBuM	1981

(597) Society of Economic Paleontologists and Mineralogists. SEPM Guidebook.

NmPE	1981
NvU	1982
OkTA	1982-83
OkU	1981
TxDaM-SE	1982-83
TxU	1981, 84
TxWB	1981, 83
UU	1981
WU	1982-84
WyU	1982

(597) SOCIETY OF ECONOMIC PALEONTOLOGISTS AND MINERALOGISTS. SEPM GUIDEBOOK.
1984 No. 2. See Society of Economic Paleontologists and Mineralogists. Midyear Meeting. 590.([1])

(598) SOCIETY OF ECONOMIC PALEONTOLOGISTS AND MINERALOGISTS. SEPM RESEARCH CONFERENCE. [FIELD TRIPS]
1980 Aug. Modern shelf and ancient cratonic sedimentation; the orthoquartzite-carbonate suite revisited.

CSt-ES	1980
IaU	1980
MnU	1980
WU	1980

(599) SOCIETY OF EXPLORATION GEOPHYSICISTS. ANNUAL MEETING.
1932 See American Association of Petroleum Geologists. 18.
1933 See American Association of Petroleum Geologists. 18.
1936 See American Association of Petroleum Geologists. 18.
1937 See American Association of Petroleum Geologists. 18.
1939 See American Association of Petroleum Geologists. 18.
1940 See American Association of Petroleum Geologists. 18.
1941 See American Association of Petroleum Geologists. 18.
1947 See American Association of Petroleum Geologists. 18.
1948 See American Association of Petroleum Geologists. 18.
1949 See American Association of Petroleum Geologists. 18.
1950 See American Association of Petroleum Geologists. 18.
1951 See American Association of Petroleum Geologists. 18.
1952 See American Association of Petroleum Geologists. 18.
1953 Mar. See American Association of Petroleum Geologists. 18.
1954 See American Association of Petroleum Geologists. 18.
1955 Mar. See American Association of Petroleum Geologists. 18.
1969 39th Calgary, Alberta, Canada.
Field trip: Banff-Lake Louise area.

CaACAM	1969
MnU	1969

(600) SOCIETY OF EXPLORATION GEOPHYSICISTS. MIDWESTERN SECTION.
1963 Mar. See American Association of Petroleum Geologists. 18.

(601) SOCIETY OF EXPLORATION GEOPHYSICISTS. PACIFIC SECTION.
1961 See American Association of Petroleum Geologists. Pacific Section. 28.
1962 See American Association of Petroleum Geologists. 18.
1965 See Society of Economic Paleontologists and Mineralogists. Pacific Section. 593.
1966 See American Association of Petroleum Geologists. Pacific Section. 28.
1967 See American Association of Petroleum Geologists. 18.
1968 See American Association of Petroleum Geologists. Pacific Section. 28.
1969 See American Association of Petroleum Geologists. Pacific Section. 28.
1970 See American Association of Petroleum Geologists. Pacific Section. 28.
1972 See American Association of Petroleum Geologists. Pacific Section. 28.
1973 See American Association of Petroleum Geologists. 18.
1974 See American Association of Petroleum Geologists. Pacific Section. 28.
1975 See American Association of Petroleum Geologists. Pacific Section. 28.
1976 See American Association of Petroleum Geologists. Pacific Section. 28.
1977 See American Association of Petroleum Geologists. Pacific Section. 28.
1978 See American Association of Petroleum Geologists. Pacific Section. 28.
1979 See American Association of Petroleum Geologists. Pacific Section. 28.
1982 Apr. See American Association of Petroleum Geologists. Pacific Section. 28.
1985 See American Association of Petroleum Geologists. Pacific Section. 28.

(602) SOCIETY OF EXPLORATION GEOPHYSICISTS. REGIONAL MEETING.
1950 See Geological Association of Canada. 190.

(603) SOCIETY OF MINING ENGINEERS OF AIME. COLORADO PLATEAU CHAPTER. [FIELD TRIP GUIDEBOOK]
1985 Apr. See Oil Shale Symposium. 501.

(604) SOCIETY OF VERTEBRATE PALEONTOLOGY. GUIDEBOOK.
1941 1st Tertiary and Pleistocene of Nebraska. (Reprinted in Plateau, v. 20, no. 1, July, 1947) (450)
1947 2nd Continental Triassic of northern Arizona. (Reprinted in Plateau, v. 20, no. 1, July, 1947)
1948 3rd Southeastern Wyoming.
1950 4th Northwestern New Mexico.
1951 5th Western South Dakota.
1953 6th Northeastern Utah.
1954 Oct. The late Eocene beds of the Sespe Formation on Pearson Ranch, Simi Valley (now known as the Wharton Ranch) and the Tick and Mint Canyon faunas in the Soledad Basin.
1956 7th Miocene Texas Gulf Coastal Plain.
1958 8th Western Montana.
1961 9th Tertiary and Pleistocene of western Nebraska.
1964 Nov. Central Florida.
1966 Nov. 12 [A] Southern California field trip.
Field trip [1] Anza-Borrego desert and Barstow areas.
— Nov. 19 [B] Berkeley area.
Trip no. 1. Fort Funston ... Thornton Beach Pliocene and Pleistocene.
— Nov. 20 Trip no. 2. The beginning of continental deposition in the Mt. Diablo area.
1972 Aug. Field conference on Tertiary biostratigraphy of southern and western Wyoming.
1978 See Geological Society of America. 197.
1982 Aug. [40th] University of Michigan.
Early Cenozoic paleontology and stratigraphy of the Bighorn Basin, Wyoming.
— 42nd Mexico City.
[Trip 1]. Cenozoic Vertebrate localities in northeast and central Mexico.
[Trip 2]. Lithographic limestone like deposits in Tepexi de Rodriguez, Puebla, Mexico.
1985 45th Fossiliferous Cenozoic deposits of western South Dakota and northwestern Nebraska. (155)

CLU-G/G	1958
CMenUG	1941, 48, 50-51, 53-58
CPT	1941, 48, 58, 82
CSdS	1961
CSt-ES	1941, 48, 50-51, 66(B1-2), 72
CU-EART	1941, 47-48, 50-51, 53-54, 56, 58, 61, 66(B1-2), 72, 85
CU-SB	1948
CaAEU	1958
CaBVaU	1951, 58
CoDCh	1941
CoDGS	1941, 48, 50, 53, 58, 61
CoG	1948, 50-51
CoU	1950-51
CtY-KS	1982
DI-GS	1941, 48, 50-51, 53, 58, 61, 72, 85
DLC	1948, 61
I-GS	1941
ICF	1947-48, 61
IEN	1958
IU	1948, 50-51, 61
IaU	1941, 58
InLP	1958
InU	1948, 50-51, 53, 58, 61
KyU-Ge	1941, 47-48, 50-51, 53, 58
LU	1941, 47-48, 50-51, 53, 58, 61
MNS	1958
MdBJ	1982
MiDW-S	1958, 61

MiHM	1951, 58
MiU	1950, 58, 72
MnU	1948, 50-51, 53, 58, 1982
MoRM	1958
MoSW	1958
MoU	1941, 58, 64
MtBuM	1958
NN	1982
NNC	1941, 50-51, 53, 58, 61, 64
NRU	1958
NbU	1941, 47, 61
NcU	1941, 47-48, 50-51, 53-54, 56, 58
NdU	1951, 58
NhD-K	1972
NjP	1941, 50, 58, 64
NmU	1950
OCU-Geo	1958
OU	1948, 50, 58, 64, 72
OkU	1941, 47-48, 50-51, 53-54, 56, 58, 61, 64
OrCS	1941, 48, 50-51, 58, 61
OrU-S	1958
PBL	1958
PBm	1948, 58
PSt	1958
SDSNH	1985
SdRM	1950-51
TxCaW	1972
TxDaAR-T	1950
TxDaDM	1951
TxDaM-SE	1941, 47
TxMM	1948, 50-51
TxU	1941, 47-48, 50-51, 53, 58, 64, 66, 72
UU	1958
WU	1958, 64, 85
WyU	1941, 48, 64, 72

(605) SOCIETY TO ADAPT BUILDING TO ENVIRONMENT REASONABLY.
1974 Physical hazards & land use. (551, 558)
CSt-ES	1974

(606) SOIL SCIENCE SOCIETY OF AMERICA.
1970 Distribution and genesis of soils and geomorphic surfaces in a desert region of southern New Mexico.
1977 Nov. See American Society of Agronomy. 56.
1978 Dec. Soil geomorphology of northeastern Illinois.
CLU-G/G	1970
IU	1978
KyU-Ge	1978
NdU	1970
TxU	1970

(607) SOUTH CAROLINA. DIVISION OF GEOLOGY. BULLETIN.
1959 B-24 See Carolina Geological Society. 129.

(608) SOUTH CAROLINA. DIVISION OF GEOLOGY. GEOLOGIC NOTES.
1961 V. 5.   No. 5. See Carolina Geological Society. 129.
1963 V. 7.   No. 5. See Carolina Geological Society. 129.
1965 V. 9.   No. 2. See Carolina Geological Society. 129.
1968 V. 12. No. 4. See Carolina Geological Society. 129.
1969 V. 13. No. 1. See Carolina Geological Society. 129.

SOUTH CAROLINA. UNIVERSITY. See UNIVERSITY OF SOUTH CAROLINA. DEPARTMENT OF GEOLOGY. (707)

(609) SOUTH COAST GEOLOGICAL SOCIETY. FIELD TRIP GUIDEBOOK.
1971 See Los Angeles Basin Geological Society. 356.
1972 Oct. See National Association of Geology Teachers. Far Western Section. 412.
1973 Oct. Guidebook to the Tertiary geology of eastern Orange and Los Angeles counties, California.
1974 Guidebook to selected features of the Palos Verdes Peninsula and Long Beach, California.
1976 Oct. Geologic guidebook to the southwestern Mojave Desert Region, California.
1977 Oct. Geological guidebook to the Ridge Basin, northwestern Los Angeles County, California.
1978 Oct. Geologic guidebook to the Santa Ana River Basin, Southern California.
1979 Oct. Guidebook to selected geologic features, coastal areas of southern Orange and northern San Diego counties, California.
1980 Oct. Geology and mineral wealth of the California desert; Dibblee volume.
1981 Oct. No. 9. Geology of the San Jacinto Mountains.
1982 Oct. Geology and mineral wealth of the California Transverse Ranges; Mason Hill volume.
1983 Guidebook to selected geologic features, coastal area of southern San Diego County.
1985 Nov. No. 13. Geology of Santa Catalina Island.
AzTeS	1977
CLU-G/G	1973-74, 76-82
CLhC	1980-82
CMenUG	1973-74, 76-82, 80, 82, 85
CPT	1973-74, 76-82
CSfCSM	1974, 76-78, 80, 82, 85
CSt-ES	1973-74, 76-83
CU-A	1974, 77-78
CU-EART	1973-74, 76-82
CU-SB	1980-81
CoDGS	1980, 82, 85
CoG	1980
CoLiM	1981(Oct.), 83
DI-GS	1973-74, 77-82, 85
IU	1980, 81
IaU	1974, 77-78, 80-82
InU	1981, 82
MnU	1973-74, 76-77, 80
NcU	1980
NvU	1974, 78, 80, 82
OkU	1981
WU	1980, 82

(610) SOUTH DAKOTA GEOLOGICAL SURVEY. GUIDEBOOK.
1965 1st Upper Mississippi Valley. (Reprint of South Dakota part of International Association for Quaternary Research, 7th Congress, with supplemental data)
1969 2nd Guidebook to the Major Cenozoic deposits of southwestern South Dakota.
AzTeS	1969
AzU	1969
CLU-G/G	1965, 69
CPT	1969
CSt-ES	1969
CU-EART	1969
CaACU	1969
CoDGS	1965, 69
CoG	1965, 69
CoU	1965, 69
DI-GS	1965, 69
DLC	1969
I-GS	1965
IEN	1965, 69
IaU	1965, 69
IdBB	1965, 69
InLP	1965, 69
InRE	1965, 69
InU	1965, 69
KyU-Ge	1965, 69
MiDW-S	1965, 69
MiHM	1969
MiKW	1969
MiU	1965
MoSW	1965, 69
MtBuM	1965, 69
MtU	1969
NBiSU	1965, 69
NMSoI	1969
NSyU	1969
NcU	1969
NdU	1969
NhD-K	1965, 69

(611) South Dakota State University. Remote Sensing Institute.

NjP	1969
NvLN	1965, 69
NvU	1965, 69
OU	1969
OkU	1965, 69
OrU-S	1969
PBm	1969
PSt	1969
RPB	1969
SdRM	1965, 69
TxBeaL	1965, 69
TxDaAR-T	1965, 69
TxLT	1965, 69
TxU	1965, 69
UU	1969
ViBlbV	1965, 69
WU	1965, 69

(611) SOUTH DAKOTA STATE UNIVERSITY. REMOTE SENSING INSTITUTE. GUIDEBOOK.
1979 SDSU-RSI-79-07. See William T. Pecora Symposium. 773.

(612) SOUTH TEXAS GEOLOGICAL SOCIETY. GUIDEBOOK FOR THE FIELD TRIP.
1930 1st Southern Uvalde and northern Zavala counties.
1931 Mar. Darst Creek, Salt Flat and Luling oil fields; data on Edward Lime oil fields; paleontological field trip, Bexar and Medina counties; road log, San Antonio to Laredo; brief road log, Laredo to Monterrey; road log, Laredo to Corpus Christi; brief notes, Corpus Christi to San Antonio.
1932 3rd San Antonio to Corpus Christi.
— 4th Laredo to Roma, Texas; Miel, General Trevino, Agualeguas, Sabinas Hidalgo, Monterrey, and Saltillo, Mexico; Monterrey to Laredo; Laredo to Corpus Christi; Laredo to San Antonio.
1933 5th Victoria to Corpus Christi and George West to Casa Blanca.
1934 6th San Antonio to Laredo via Campbellton and Government Wells. Road log in Tamaulipas and Nuevo Leon, Mexico portion of trip.
1935 7th Laredo to Mexico City. Geologic map.
1936 8th Oct. 16-18. no. 1. Pleasanton to Pearsall (Texas).
No. 2. Pearsall to Loma Vista (Texas).
Nov. 18. no. 3. Loma Vista to Eagle Pass (Texas).
No. 4. Carrizo Springs to north of Eagle Pass (Texas).
1937 June Rock asphalt deposits, Uvalde County.
— July Bee and Live Oak counties.
— Nov. Cretaceous of Bexar and Medina counties.
1938 Gasoline extraction plants; McCampbell, Plymouth and Saxet fields.
1939 Oct. No. 1. Rio Grande Valley; Laredo to Mission and Brownsville. (628)
No. 2. [Rio Grande delta and Quaternary formations]
— Nov. 1. Points of geologic interest in the vicinity of Austin, Texas. (628, 715, 716)
2. Marshall Ford Dam. (628, 715, 716)
1940 12th Lower Tertiary: Austin-Manor-Elgin-McDade-Page-Bastrop-Smithville.
1941 13th Road log from Monterrey, Mexico to Laredo, Texas.
1947 14th See American Association of Petroleum Geologists. Regional Meeting. 29.
1948 15th A spectacular circle tour: Mexico-Tuxpan-Mexico.
1949 16th San Antonio-Uvalde-Bracketville-Del Rio-Langtry.
1950 17th Monterrey to Huasteca Canyon.
1951 18th See American Association of Petroleum Geologists. Regional Meeting. 29.
1953 19th Geological section; Taylor to Glen Rose: San Antonio-Bandera-Leakey-Campwood-Brackettville-Del Rio-Rock Springs-Kerrville.
1954 20th Fault line: Seguin, Darst Creek, and Luling fields.
— 21st Cook Mountain to Jurassic: San Antonio-Monterrey-Saltillo-Eagle Pass-San Antonio.
1956 Nov. See Gulf Coast Association of Geological Societies. 247.
1958 24th Eocene-Miocene, oil-uranium: Falls City, Tordilla Hill and Fashing areas, Wilson, Karnes, Atascosa counties, Texas.
1959 25th Mesozoic stratigraphy and structure, Saltillo-Galeana areas, Coahuila and Nuevo Leon, Mexico.
1960 26th Geological section; Taylor to Glen Rose.
1961 See Gulf Coast Association of Geological Societies. 247.
1964 Dec. Mesozoic stratigraphy and structure, Monterrey-Saltillo-Monclava areas, Coahuila and Nuevo Leon, Mexico.
1967 Oct. See Gulf Coast Association of Geological Societies. 247.
1974 See American Association of Petroleum Geologists. 18.
1976 Oct. Economic geology of south central Texas. (Reprinted for the Houston Geological Society Field Trip. 1980.) See Houston Geological Society. 255.
1979 Oct. See Gulf Coast Association of Geological Societies. 247.
1981 See Geological Society of America. South Central Section. 205.
1984 May See American Association of Petroleum Geologists. 18.
1985 May Selected Cretaceous exposures of southwest Texas: Culebra Arch, Knippa basalt quarry, Maverick Basin margin.

AzTeS	1930-34, 36-38, 39(Oct.), 40, 49-50, 53-54, 58-59
AzU	1953-54, 58-59
CLhC	1959
CMenUG	1930-41, 49-50, 53-54, 58-59, 85
CSt-ES	1930-40, 50, 53-54, 58-59, 76
CU-A	1930-36, 37(June, Nov.), 39(Oct.), 40, 49-50, 53-54, 58-60
CU-EART	1953-54, 58-59
CU-SB	1976
CaACU	1953, 58-59
CaOHM	1953-54, 58-59
CaOWtU	1958, 81, 84
CoDGS	1981(1), 85
CoG	1930-41, 50, 53-54, 58-59
DI-GS	1936-38, 39(Oct.), 40-41, 48-50, 53-54, 58-60, 81, 84-85
ICF	1959
IU	1953-54, 58-59, 76(Oct.)
IaU	1976
InU	1960
KyU-Ge	1953, 58-59, 76(Oct.) 81
LU	1930-38, 39(Oct.(1)), 40-41, 49, 53-54, 58-59
MH-GS	1935, 84
MNS	1941, 53-54
MnU	1930-34, 39(Oct.), 40-41, 49-50, 53-54, 58-60, 76, 81
MoSW	1958
MoU	1953-54, 58-59
MtBuM	1935
NNC	1930-31, 32(4), 33-38, 39(Oct.), 40, 58-59
NSbSU	1981,
NbU	1940-41, 48-50, 53-54
NcU	1947, 51-54, 56, 58-59, 61, 79
NmPE	1953, 58-59
OCU-Geo	1953
OU	1940-41, 58-59
OkU	1930-34, 36(Oct.), 37-38, 39(Oct.), 40-41, 49, 53-54, 58-59
OrCS	1954, 59
RPB	1953-54, 56
TU	1953-54, 56(Apr.)
TxBeaL	1930-38, 39(Oct.), 40, 49, 53, 58-59
TxDaAR-T	1940, 53, 58-59
TxDaDM	1930, 32(4), 36-38, 41, 48, 53, 54(20), 58-59
TxDaGI	1940-41, 53, 54(21), 60
TxDaM-SE	1930-31, 32(4), 33-38, 39(Oct.), 40-41, 50, 53, 54(20), 58, 81(1)
TxDaSM	1930-34, 36-38, 39(Oct.), 40-41, 49, 53, 54(20), 58
TxHMP	1981
TxHSD	1931-32, 34-38, 40-41, 48, 54(21), 59-60, 76
TxHU	1931, 39(Oct.), 40-41, 53, 54(21), 59-60
TxLT	1931, 58
TxMM	1939(Oct.(2)), 40-41, 53, 54(20), 58-60, 76, 81, 85
TxU	1930-41, 48-50, 53-54, 58-60, 64, 76, 81(Apr.), 85
TxWB	1939-41, 49, 53-54, 58-60, 85
ViBlbV	1953, 58-59
WU	1981

(613) SOUTHEAST MISSOURI STATE UNIVERSITY. EARTH SCIENCE DEPARTMENT. GUIDEBOOK.
1975 Nov. See Big Rivers Area Geological Society. 99.

(614) SOUTHEASTERN GEOLOGICAL SOCIETY. FIELD CONFERENCE. GUIDEBOOK.
1944 1st Southwestern Alabama.
— 2nd Southwestern Georgia.
1945 3rd Western Florida.
1946 4th Southeastern Alabama.
1947 5th Portion of central Florida; Gulf Coast.
1948 6th Cretaceous of east-central Alabama.

1951 7th	Geology of the crystalline rocks and of the Paleozoic area of Northwest Georgia.
1954 8th	Carbonate deposits in South Florida.
1960 9th	Late Cenozoic stratigraphy and sedimentation of central Florida.
1963 10th	Summary of Paleocene and Eocene stratigraphy and economic geology of southeastern Alabama.
1965 11th	Some highlights of the Cretaceous and crystalline terranes of Georgia.
1966 12th	Geology of the Miocene and Pliocene series in the North Florida-South Georgia area. (89, 237)
1967 13th	Miocene-Pliocene problems of peninsular Florida.
1970 14th	Geology and geohydrology of the Cross-Florida Barge Canal area, Ocala, Florida.
1971 15th	Geological review of some North Florida mineral resources.
1972 16th	Space age geology; terrestrial applications, techniques and training.
1973 Oct.	Carbonate rock environments, Florida Keys and western Bahamas.
1975 19th	Hydrogeology of west-central Florida.
1976 20th	Tertiary carbonates; Citrus, Levy, Marion counties, west central Florida. (Cover title: Mid-Tertiary carbonates; Citrus, Levy, Marion counties, west central Florida) (147)
1977 21st	Environment of the central Florida phosphate district.
1978 22nd	Hydrogeology of south-central Florida. (615)
1979 [23rd]	Guide to sedimentation for the Dry Tortugas. (615)
1980	Holocene geology and man in Pinellas and Hillsborough counties, Florida. (615)
1981 Oct.	Karst hydrogeology and Miocene geology of the upper Suwannee River Basin, Hamilton County, Florida. (615)
1982 Sept.	Cenozoic vertebrate and invertebrate paleontology of North Florida. (206(1983 no. 4.), 615)
1983 Oct.	Cenozoic geology of the Apalachicola River area, Northwest Florida. (197(1985-No.12), 615)
1984 Dec.	Karst hydrogeology and geomorphology of the Dougherty Plain, southwest Georgia. (176, 615)
ATuGS	1946-48, 51, 54, 60, 63
AzU	1970-72
CLU-G/G	1946-47, 51, 54, 60, 63, 65-67
CLhC	1946-48
CMenUG	1945-46, 48, 51, 54, 60, 63, 65-67, 70-71, 75-84
CPT	1946-48, 51, 54, 60, 63, 76
CSt-ES	1946-47, 51, 54, 60, 63, 65-67, 70-72, 75-78, 80-84
CU-EART	1946-47, 51, 60, 63, 65-67, 70-72, 75-76, 83
CU-SB	1970-72, 75-84
CaACU	1976
CaAEU	1970
CaOHM	1970
CaOKQGS	1951, 60, 63
CaOOG	1984
CaOWtU	1951, 60, 65-67, 70-72, 75-79, 82
CoDCh	1954
CoDGS	1942-46, 51, 54, 60, 63, 65, 67, 75-84
CoG	1945-48, 51, 54, 60, 63, 65-67, 70-72, 75-78
CoLiM	1981-84
CoU	1951, 54, 60, 63, 65-67
DI-GS	1944-48, 51, 54, 60, 63, 65-67, 70-72, 75-84
DLC	1944(1), 47-48
F-GS	1980-81, 83-84
FBG	1980-81, 83-84
I-GS	1970
ICF	1951, 54, 60, 63, 66
ICIU-S	1966, 70
ICarbS	1951, 54, 60, 63, 65-67, 70-72, 75-76, 78-84
IEN	1951, 54, 60
IU	1944-48, 51, 54, 60, 63, 65-67, 70, 76-78
IaU	1946-47, 51, 54, 60, 63, 65-67, 70-72, 75-77, 79-84
InLP	1951, 54, 60, 63, 65-67, 76
InRE	1951, 54, 60
InU	1944-46, 51, 54, 60, 63, 65-67
KyU-Ge	1944-47, 51, 54, 60, 63, 65-67, 70-72, 75-76
LNU	1946-48, 51, 60, 63, 65-67, 70-72, 75-76
LU	1944(2), 45-48, 51, 54, 60, 63, 65-67
MNS	1946, 48
MiDW-S	1966, 70
MiU	1966
MnU	1946-47, 51, 60, 63, 65-67, 70-72, 75-82
MoSW	1946-47, 51, 54, 60, 63, 65-67, 70
MoU	1944(2), 45-48, 51, 54, 60, 63, 65-67, 70, 72
Ms-GS	1945, 48, 51, 54, 60, 63, 65, 67
MsJG	1945, 48, 51, 54, 60, 63, 65, 67
MtU	1951, 54, 60, 63
NBiSU	1951, 60, 63, 65-67, 70-72
NIC	1980, 82-84
NNC	1946-48, 51, 54, 60, 63, 65-67
NSbSU	1970
NSyU	1944-46, 76
NbU	1947-48, 51, 54, 60, 63, 70
NcU	1946, 51, 54, 65-67, 70-72, 75-77
NdU	1951, 60, 65-67
NhD-K	1951, 54, 60, 63, 65-67, 70-72, 75-77, 80-84
NjP	1944-48, 51, 54, 60, 63, 65-67, 70-72
NvU	1951, 54, 60, 63, 66
OCU-Geo	1944(1), 46-48, 51, 54, 60, 63, 65-67, 70
OU	1944-48, 51, 54, 60, 63, 65-67, 70-72, 75-77, 80-82
OkT	1944-48, 51, 54
OkTA	1981
OkU	1971-72, 75-77, 80, 83, 84
OrU-S	1970
PBU	1984
PSt	1951, 54, 60, 63, 66-67, 70-71, 75-81
RPB	1944(1)
TU	1951
TxBeaL	1944, 51, 54, 60, 63, 65-67, 70
TxDaAR-T	1944(2), 46-48, 51, 54, 60, 63, 65-66, 73
TxDaDM	1946-48, 51, 54, 60, 63, 65-67, 70-71
TxDaGI	1944, 46-48, 60
TxDaM-SE	1946-48, 51, 54, 60, 63, 81, 84
TxDaSM	1944(1), 46, 51, 54, 77
TxHSD	1946-48, 60
TxHU	1944-48, 51, 54, 60, 65-66, 70
TxLT	1951, 60, 63, 65-67
TxMM	1944, 54
TxU	1944-48, 51, 54, 60, 63, 65-67, 70, 76, 78-80, 84
TxU-Da	1944, 46, 48, 54
TxWB	1944-45. 47-48, 51, 54, 66
UU	1970
ViBlbV	1951, 54, 60, 63, 65, 67, 70-72, 75
WU	1945-46, 48, 51, 54, 60, 63, 65-67, 70-72, 75-84

(615) SOUTHEASTERN GEOLOGICAL SOCIETY. PUBLICATION.
1978 20. See Southeastern Geological Society. 614.
1979 21. See Southeastern Geological Society. 614.
1980 No. 22. See Southeastern Geological Society. 614.
1981 No. 23. See Southeastern Geological Society. 614.
1982 No. 24. See Southeastern Geological Society. 614.
1983 No. 25. See Southeastern Geological Society. 614.
1984 No. 26. See Southeastern Geological Society. 614.

(616) SOUTHEASTERN GEOLOGY.
1963 V. 4. No. 4. See Geological Society of America. Southeastern Section. 206.

(617) SOUTHEASTERN GEOLOGY. SPECIAL PUBLICATION.
1968 No. 1. See Geological Society of America. Southeastern Section. 206.

SOUTHERN ASSOCIATION OF STUDENT GEOLOGIC SOCIETIES (SASGS). FIELD TRIP. See SOUTHWESTERN ASSOCIATION OF STUDENT GEOLOGICAL SOCIETIES (SASGS). (626)

(618) SOUTHERN CALIFORNIA PALEONTOLOGICAL SOCIETY. BULLETIN.
1984 V. 16. Nos. 11 & 12. Field trip: North Santa Ana Mountains.
Field trip: Costa Mesa, check list of Late Pleistocene macrofossils.
CLU-G/G	1984
CSt-ES	1984
CU-EART	1984
DI-GS	1984

(619) SOUTHERN CALIFORNIA PALEONTOLOGICAL SOCIETY. [MEETING]
1980 Nov. Paleontological tour of the Mojave Desert, California- Nevada. (116)

CLU-G/G	1980	
CMenUG	1980	
CU-A	1980	
CU-EART	1980	
CU-SB	1980	
DI-GS	1980	

(620) SOUTHERN CALIFORNIA PALEONTOLOGICAL SOCIETY. SPECIAL PUBLICATIONS.
1980 No. 2. See Southern California Paleontological Society. 619.

(621) SOUTHERN GEOLOGICAL SOCIETY, SEGSA. PUBLICATION.
1981 No. 2. See Geological Society of America. Southeastern Section. 206.
1984 No. 3. See Association of Engineering Geologists. Lower Mississippi Valley Section. 79.

(622) SOUTHERN ILLINOIS UNIVERSITY AT CARBONDALE. DEPARTMENT OF GEOLOGY. [GUIDEBOOK]
1965 Devonian of Jackson and Union counties, Illinois.
1977 See Geological Society of America. North Central Section. 200.
1983 See American Association of Petroleum Geologists. Eastern Section. 25.

CU-SB	1965
ICF	1965
ICarbS	1965
KyU-Ge	1965

(623) SOUTHERN ILLINOIS UNIVERSITY AT EDWARDSVILLE. DEPARTMENT OF GEOLOGY. GUIDEBOOK.
1970 Guidebook to engineering geologic features and land use relationships in the St. Louis metropolitan area.

ICarbS	1970
NNC	1970

(624) SOUTHERN METHODIST UNIVERSITY. FONDREN SCIENCE SERIES.
1951 No. 4. See East Texas Geological Society. 162.
1955 No. 5. See Dallas Geological Society. 156.

(625) SOUTHWEST MISSOURI STATE UNIVERSITY. GEOSCIENCE SERIES.
1978 5. See Geological Society of America. Rocky Mountain Section. 204.

(626) SOUTHWESTERN ASSOCIATION OF STUDENT GEOLOGICAL SOCIETIES (SASGS). FIELD CONFERENCE GUIDEBOOK. (TITLE VARIES)
1960 1st Cretaceous stratigraphy of the Grand and Black prairies, east-central Texas. (94)
1961 2nd Structure and stratigraphy of the Arbuckle Anticline and the Criner Hills area, Oklahoma. (381)
1962 3rd Field conference on the Llano Uplift, Llano and Burnet counties, Texas. (716)
1963 4th Geology of west-central Texas [Pennsylvanian-Permian-Cretaceous sequences of west-central Texas] (94, 249)
1964 5th Monterrey, Mexico. (Partial translation of Guidebook C-5, 20th International Geological Congress, 1956) (526)
1965 6th Recent and Pleistocene sediments and geomorphology of southwestern Louisiana and southeastern Texas. (353)
1966 7th Geology of the Palo Duro Canyon State Park and the Panhandle of Texas. (759)
1967 8th Geology of south-central Texas. (629)
1968 9th Geology of the Claiborne Group of central Texas. (653)
1969 10th Geology of the Pennsylvanian-Permian Quaternary sequences of west-central Texas. (249)
1970 11th Geology of northeastern Mexico. (526)
— Fall Thee guidebook. First fall field trip. [Field trip through the classic type lower Comanchean section of North America, vicinity of Glen Rose, Texas] (94)
1971 [Spring] Tertiary of central Louisiana and Mississippi. (708)
1972 Spring Paleozoic geology of the Arbuckle Mountains, Oklahoma. (163)
— [Fall] Urban geology, guidebook for field trip I-35 growth corridor, central Texas. (94)
1973 Spring Caves, ground water, and karst features of the Cretaceous in south central Texas. (629)
— Fall Eocene of central Louisiana and East Texas. (640, 708)
1974 Spring Economic geology of the central mineral region of Texas. (716)
— [Fall?] Geology of Huizachal-Peregrina anticlinorium. (526)
1975 Spring Geologic field guide to Mariscal Canyon. (716)
— Fall Structural geology of central Texas. (94)
1976 Spring Recent and Pleistocene sediments of the southeastern Gulf Coast. (352)
— Fall Geologic guidebook to the Permian reef complex of the Guadalupe Mountain region. (716)
1977 Fall Geology of the western Ouachita Mountains. (163)
1979 Spring The nature of the Cretaceous-Precretaceous contact, central Texas. (94)
— Fall Comanchean sedimentation of central Texas. (640)
1980 Apr. Holocene depositional environments of the Southeast Texas-Southwest Louisiana Gulf Coast. (351)
1981 Geology and urban growth, central Texas. (94)
1983 Spring Guidebook to the Cretaceous and Tertiary sediments of the Austin area. (715)

AzU	1970(Fall)
CChiS	1960
CLU-G/G	1965, 68, 70, 72
CLhC	1970
CPT	1970, 72(Spring)
CSt-ES	1964, 70, 72, 73(Fall)
CU-EART	1970, 72
CU-SB	1977
CaACU	1972(Fall)
CaBVaU	1972(Spring)
CaOWtU	1972(Spring), 76(Spring), 80
CoLiM	1980
CoU	1963
DI-GS	1960-63, 65-72, 73(Spring), 74-75, 79-81
ICF	1960, 70(11)
IU	1960-66, 70(Fall), 72(Fall), 73-76, 77(Fall), 79(Fall), 80
IaU	1970, 80
InLP	1972, 79(Spring)
InU	1960, 70(11), 71-72
KyU-Ge	1970(11), 72, 80-81
LNU	1962-63
LU	1960
MH-GS	1970
MiU	1961
MnU	1960, 66, 70(11)
MoU	1972(Spring)
NNC	1960-64
NSbSU	1972, 80
NSyU	1972(Fall)
NdU	1968, 72
NjP	1970
OCU-Geo	1975(Fall), 80
OU	1960, 70(11), 72(Spring)
OkT	1960
OkU	1960, 63-65, 68, 72(Spring)
PBm	1972(Spring)
PSt	1960
TMM-E	1972(Spring)
TxBeaL	1964, 66, 70(11), 72(Fall)
TxCaW	1966, 72(Spring)
TxDaAR-T	1960, 63, 65, 70(11), 72(Spring)
TxDaDM	1960-61
TxDaGI	1964
TxDaM-SE	1960-61, 63
TxDaSM	1960, 70, 72(Spring)
TxHSD	1972(Spring)
TxHU	1960-61, 63-65, 67, 70(11)
TxLT	1960, 63
TxU	1960-70, 72, 74, 75(Spring), 76(Fall), 77(Fall), 80, 83
TxU-Da	1960
TxWB	1960-74, 76-83, 85
WU	1960

(627) SOUTHWESTERN FEDERATION OF GEOLOGICAL SOCIETIES. GUIDEBOOK FOR REGIONAL MEETING FIELD TRIP.

1958 1st Guide to the Strawn and Canyon series of the Pennsylvanian system in Palo Pinto County, Texas. (29, 482)
1961 4th Guidebook for the southern Franklin Mountains. (68, 550)
1962 5th Geology of the type area, Canyon Group, north-central Texas. (Journal of the Graduate Research Center, v. 30, no. 3, 1962)
1965 7th Fredericksburg facies, Austin area.

AzTeS	1958
AzU	1958
CLU-G/G	1958, 61
CPT	1958
CSt-ES	1958
CaACU	1958
CoG	1958, 61-62
DI-GS	1958, 61
ICF	1958, 62
IU	1961
IaU	1958
InLP	1958
InU	1958
LNU	1961
MNS	1958
MnU	1958, 61
NNC	1958, 61
NbU	1961-62
NcU	1958
NvU	1958
OU	1958
OkU	1956, 58
PSt	1958
TxBeaL	1958
TxCaW	1958
TxDaAR-T	1961
TxDaGI	1958
TxDaM-SE	1962
TxHSD	1958
TxHU	1961
TxU	1958, 61-62, 65
TxWB	1958
UU	1958
ViBlbV	1958
WU	1958

(628) SOUTHWESTERN SOCIETY. FIELD TRIP.

1937 [1] Lower Cretaceous.
 [2] Lower Tertiary.
 [3] Llano Uplift.
1939 See South Texas Geological Society. 612.

TxHSD	1937
TxMM	1937
TxU	1937

(629) ST. MARY'S GEOLOGICAL SOCIETY.

1967 See Southwestern Association of Student Geological Societies (SASGS). 626.
1973 Spring See Southwestern Association of Student Geological Societies (SASGS). 626.

(630) STANFORD UNIVERSITY PUBLICATIONS. GEOLOGICAL SCIENCES.

1967 V. 13. See Conference on Geologic Problems of the San Andreas Fault System, Stanford University. 150.
1983 V. 18. See Circum-Pacific Terrane Conference. Proceedings. 131.

STANFORD UNIVERSITY, CALIFORNIA. CONFERENCE ON GEOLOGIC PROBLEMS OF THE SAN ANDREAS FAULT SYSTEM. See CONFERENCE ON GEOLOGIC PROBLEMS OF THE SAN ANDREAS FAULT SYSTEM, STANFORD UNIVERSITY. (150)

(631) STATE GEOLOGICAL AND NATURAL HISTORY SURVEY OF CONNECTICUT. BULLETIN.

1927 41. Guide to the geology of Middletown, Connecticut, and vicinity.

CMenUG	1927
CPT	1927
CSt-ES	1927
CU-EART	1927
CaACU	1927
CoDGS	1927
DLC	1927
InLP	1927
MdBJ	1927
OCU-Geo	1927
WU	1927

(632) STATE GEOLOGICAL AND NATURAL HISTORY SURVEY OF CONNECTICUT. GUIDEBOOK.

1965 No. 1. Postglacial stratigraphy and morphology of coastal Connecticut. (Originally issued as Geological Society of America. Annual meeting. Guidebook for the field trips, 1963, no. 5)
1968 No. 2. See New England Intercollegiate Geological Conference. 456.
1970 No. 3. Stratigraphy and structure of the Triassic strata of the Gaillard Graben, south-central Connecticut.
1978 No. 4. Guide to the Mesozoic redbeds of central Connecticut. (Reprint of "Guide to the redbeds of central Connecticut: 1978 field trip, Eastern Section of the Society of Economic Mineralogists and Paleontologists.")
1982 No. 5. See New England Intercollegiate Geological Conference. 456.
1985 No. 6. See New England Intercollegiate Geological Conference. 456.

ATuGS	1965
AzTeS	1965
CMenUG	1965, 70, 78
CPT	1965
CSt-ES	1965, 70, 78
CU-EART	1965, 70, 78
CaACU	1965, 70
CaOLU	1965
CaOOG	1978
CaOTRM	1970, 78
CaOWtU	1965, 70, 78
CoDCh	1970, 78
CoDGS	1965, 70, 78
CoG	1965
CoU	1965, 70
DI-GS	1965, 70, 78
I-GS	1965, 70
ICF	1965, 70
ICIU-S	1965, 70
ICarbS	1965, 70, 78
IEN	1965
IU	1965, 70, 78
IaU	1965, 70, 78
InLP	1965, 70, 78
InU	1965, 70, 78
KyU-Ge	1965, 70, 78
MH-GS	1970
MWesB	1978
MdBJ	1965, 70
MiDW-S	1965, 70
MiHM	1965, 70
MiKW	1965
MnU	1965, 70, 78
MoSW	1965, 70
MoU	1965, 70
MtBuM	1965, 70, 78
NBiSU	1965, 70
NMSoI	1978
NRU	1965
NSyU	1965, 70
NbU	1970
NcU	1965, 70, 78
NdU	1965, 70, 78
NhD-K	1965, 70, 78
NjP	1965, 70
NvU	1965, 70, 78
OCU-Geo	1970, 78
OU	1965, 70, 78
OkU	1965, 70, 78
OrU-S	1965, 70
PBm	1965, 70, 78
PSt	1965, 70, 78
RPB	1965
TxBeaL	1978
TxDaM-SE	1965

(633) State Geological Survey of Kansas.

TxU	1965, 70, 78
UU	1965, 70
ViBlbV	1978
WU	1965, 70, 78

(633) STATE GEOLOGICAL SURVEY OF KANSAS.
1956 See University of Kansas. Science and Mathematics Camp. 689.
1958 See University of Kansas. Science and Mathematics Camp. 689.
1959 See University of Kansas. Science and Mathematics Camp. 689.
1960 See University of Kansas. Science and Mathematics Camp. 689.

(634) STATE GEOLOGICAL SURVEY OF KANSAS. FIELD TRIP GUIDE.
1956 See Geologic Field Conference for Kansas School Teachers. 189.
1959 Oct. See American Association of Petroleum Geologists. Regional Meeting. 29.
1960 May Geologic field trip in south Osage Cuesta region; vicinity of Neodesha, Wilson and Montgomery counties, Kansas. (For science students, Neodesha High School)
— July A geologic tour in the Turkey Creek camp of south-central Kansas. (For Girl Scouts of Turkey Creek camp)

IU	1960
KLG	1960(July)

(635) STATE GEOLOGICAL SURVEY OF KANSAS. GUIDEBOOK SERIES.
1976 1. See Friends of the Pleistocene. Midwest Group. 183.
1978 2. See American Association of Petroleum Geologists. 18.
— 3. See American Association of Petroleum Geologists. 18.
1979 4. See International Congress of Carboniferous Stratigraphy and Geology. 290.

(636) STATE GEOLOGICAL SURVEY OF KANSAS. SPECIAL DISTRIBUTION PUBLICATION.
1971 53. See Friends of the Pleistocene. Midwest Group. 183.

(637) STATE GEOLOGICAL SURVEY OF KANSAS. OPEN-FILE REPORT.
1985 85-20. See Association of Earth Science Editors. 75.

STATE GEOLOGISTS FIELD TRIP. See PLEISTOCENE FIELD CONFERENCE. (543)

(638) STATE UNIVERSITY COLLEGE OF ARTS AND SCIENCE, POTSDAM.
1983 See New York State Geological Association. 469.

(639) STATE UNIVERSITY OF NEW YORK AT BINGHAMTON. PUBLICATIONS IN GEOMORPHOLOGY. CONTRIBUTION.
1973 3. Glacial geology of the Binghamton-western Catskill region.

AzTeS	1973
CSt-ES	1973
ICarbS	1973
IaU	1973
MnU	1973
NSyU	1973

(640) STEPHEN F. AUSTIN STATE UNIVERSITY. DEPARTMENT OF GEOLOGY. FIELD TRIP.
1973 Fall See Southwestern Association of Student Geological Societies (SASGS). 626.
1979 Fall See Southwestern Association of Student Geological Societies (SASGS). 626.

SUI GEOLOGY CLUB, STATE UNIVERSITY OF IOWA. See UNIVERSITY OF IOWA GEOLOGY CLUB. (687)

(641) SUL ROSS STATE UNIVERSITY. GEOLOGICAL SOCIETY.
1969 A field guide to Permian, Cretaceous, and Tertiary rocks in the Pinto Canyon, Presidio County, Texas, with auxiliary field guides to Paleozoic rocks of the Marathon Basin and Tertiary volcanics in the vicinity of Alpine, Texas.

IaU	1969
TxU	1969

(642) SYMPOSIUM ON COASTAL SEDIMENTOLOGY. PROCEEDINGS.
1983 6th Near-shore sedimentology.

CMenUG	1983
DI-GS	1983
WU	1983

(643) SYMPOSIUM ON THE GEOLOGY AND ORE DEPOSITS OF THE VIBURNUM TREND, MISSOURI. GUIDEBOOK.
1975 The geology and ore deposits of selected mines [in the] Viburnum Trend, Missouri. (399)

CLU-G/G	1975
CPT	1975
CSt-ES	1975
CU-EART	1975
CaOOG	1975
CaOWtU	1975
CoU	1975
DI-GS	1975
DLC	1975
I-GS	1975
ICF	1975
ICarbS	1975
IEN	1975
IU	1975
IaU	1975
InLP	1975
KLG	1975
KyU-Ge	1975
MnU	1975
MoSW	1975
MoU	1975
MtBuM	1975
NNC	1975
NbU	1975
NcU	1975
NdU	1975
NvU	1975
OCU-Geo	1975
OU	1975
OkU	1975
SdRM	1975
TxBeaL	1975
TxU	1975
UU	1975
ViBlbV	1975
WU	1975

(644) SYMPOSIUM ON THE GEOLOGY OF ROCKY MOUNTAIN COAL (ROMOCOAL).
1980 4th Romocoal field trip.
  1. Road log from Denver south through Castle Rock, Colorado Springs, Pueblo, and Trinidad, Colorado to Raton, New Mexico.
  2. Road log from Raton to Vermejo Park through the Raton Coal Field via the York Canyon Mine.
  3. Road log from Raton to Cokedale over Raton Pass through the eastern part of the Trinidad Coal Field, Colorado.
1982 5th Rocky Mountain Coal Symposium field trip.
  1. Ferron Sandstone Member. (730)
  2. Northern Wasatch coal field of central Utah. (730)

CSt-ES	1980, 82
KyU-Ge	1980
NN	1980
NvU	1980

(645) SYMPOSIUM ON THE GEOLOGY OF THE BAHAMAS. PROCEEDINGS.
1984 2nd Symposium on the Geology of the Bahamas (Cover title: Proceedings of the Second Symposium on the Geology of the Bahamas) (135)

Field guide to the Cockburn Town Fossil Coral Reef, San Salvador, Bahamas. Second symposium on the geology of the Bahamas field trip to Pigeon Creek. Geology of the great Exuma Island. (135)

CSt-ES	1984
CU-A	1984
CU-EART	1984
CtY-KS	1984
IU	1984
KyU-Ge	1984
MdBJ	1984
MnU	1984
NN	1984
NSbSU	1984
OU	1984
OkU	1984

(646) SYMPOSIUM ON THE PALEOENVIRONMENTAL SETTING AND DISTRIBUTION OF THE WAULSORTIAN FACIES. [GUIDEBOOK]

1982 1st Waulsortian facies, Sacramento Mountains, New Mexico: guide for an international field seminar, March 2-6, 1982. (167, 718)

CSt-ES	1982
CoDGS	1982
IU	1982
InLP	1982
OkU	1982

(647) SYMPOSIUM ON THE QUATERNARY OF VIRGINIA. [GUIDEBOOK]

1984 A geologic travelog from Charlottesville to Saltville, Virginia, for the Symposium on the Quaternary of Virginia.

CU-EART	1984
DI-GS	1984
InLP	1984
VDMR	1984
Vi-MR	1984

(648) TECTONOPHYSICS.

1978 V. 61. No. 1-3. See Lunar and Planetary Institute. Topical Conference. 364.

(649) TENNESSEE ACADEMY OF SCIENCE. GEOLOGY-GEOGRAPHY SECTION. FIELD TRIP.

1954 Cumberland Plateau.
1960 Mississippian stratigraphy of the northwestern highland rim.
1972 Spring [Mineral industry in Scott County, northern Cumberland Plateau]
1974 Spring Geology along Interstate 40 through Pigeon River Gorge, Tennessee-North Carolina. (552)

ATuGS	1954, 60
DI-GS	1972, 74
IU	1972
IaU	1974
InLP	1974
KyU-Ge	1954, 74
TMM-E	1972

(650) TENNESSEE. DIVISION OF GEOLOGY. BULLETIN.

1973 No. 70. See Geological Society of America. Southeastern Section. 206.

(651) TENNESSEE. DIVISION OF GEOLOGY. REPORT OF INVESTIGATIONS.

1958 No. 5. Guidebook to geology along Tennessee highways.
1961 No. 12. See Geological Society of America. Southeastern Section. 206.
— No. 13. See Geological Society of America. Southeastern Section. 206.
1969 No. 23. See Society of Economic Geologists. 583.
1972 No. 33. See Geological Society of America. Southeastern Section. 206.
1975 No. 36. See Geological Society of America. Southeastern Section. 206.
1978 No. 37. See Geological Society of America. Southeastern Section. 206.
1981 No. 38. Mississippian and Pennsylvanian section on Interstate 75, south of Jellico, Campbell County, Tennessee.
— No. 39. Guide to the geology along the interstate highways in Tennessee.

AzTeS	1958
CLU-G/G	1958, 81(39)
CMenUG	1981(39)
CPT	1958, 81(39)
CSfCSM	1981(39)
CSt-ES	1958, 81
CU-EART	1958, 81(39)
CaACU	1958, 81
CaOOG	1981(39)
CaOWtU	1958, 81
CoDGS	1958
CoG	1958, 81(39)
DI-GS	1958, 81(39)
DLC	1958
ICF	1958
ICIU-S	1958
ICarbS	1958
IU	1958
IaU	1981(39)
IdU	1958
InLP	1958
InU	1958
KLG	1981(39)
KU	1981(39)
KyU-Ge	1958, 81
LNU	1981(39)
MdBJ	1958
MnU	1958, 81(39)
MoSW	1958
MoU	1958
NSbSU	1958
NSyU	1958
NbU	1958
NcU	1958, 81(39)
NdU	1958, 81(39)
NhD-K	1981
NvU	1958
OCU-Geo	1958
OCl	1981
OU	1958, 81
OkT	1958
OkU	1958, 81(39)
OrU-S	1981(39)
PBm	1981(39)
PPi	1981
PSt	1958
SdRM	1958, 81
TMM-E	1958
TxBeaL	1958
TxDaGl	1958
TxDaM-SE	1981(39)
TxLT	1958
TxU	1958, 81(39)
TxWB	1981(39)
UU	1958
ViBlbV	1958
WBB	1981(39)
WU	1958, 81(38, 39)
WyU	1981(39)

TENNESSEE. UNIVERSITY. See UNIVERSITY OF TENNESSEE, KNOXVILLE. DEPARTMENT OF GEOLOGY AND GEOGRAPHY. (710) and UNIVERSITY OF TENNESSEE, KNOXVILLE. DEPARTMENT OF GEOLOGICAL SCIENCES. (709) and UNIVERSITY OF TENNESSEE, KNOXVILLE. LOWER DIVISION CURRICULUM COMMITTEE. (711)

(652) TEXAS A & M UNIVERSITY. DEPARTMENT OF GEOLOGY. FIELD TRIP. (TITLE VARIES)

1940 Road log of a field trip from Hempstead to College Station over Highway 6.
1948 Geologic excursion: [North Texas trip].
[n.d.] Geological field trip: Houston, Columbus, La Grange, Smithville, Bastrop, Austin, central mineral region.

CSt-ES	n.d., 1948
TxU	1940

(653) TEXAS A & M UNIVERSITY. GEOLOGICAL SOCIETY.
  1968 See Southwestern Association of Student Geological Societies (SASGS). 626.

(654) TEXAS ACADEMY OF SCIENCE.
  1939 Fall Geologic excursion; Paleozoic section of the Llano Uplift. (178, 756)
  1957 The Fredericksburg Group in the valley of the Trinity River, Texas.
  1964 Dec. Geology and the city of Waco, a guide to urban problems. (94, 418)

DI-GS	1939, 64
IU	1964
LU	1939
MnU	1964
NNC	1964
NbU	1939
OCU-Geo	1964
TxBeaL	1957
TxHU	1939, 64
TxLT	1939, 64
TxMM	1939
TxU	1939, 57, 64
TxU-Da	1939, 57

(655) TEXAS TECH UNIVERSITY.
  1970 Ogallala aquifer symposium; field trip guidebook, Lubbock-Lake Ransom Canyon.

IEN	1970
TxU	1970

(656) TEXAS TECH UNIVERSITY. INTERNATIONAL CENTER FOR ARID AND SEMI-ARID LAND STUDIES. FIELD TRIP.
  1983 See Friends of the Pleistocene. South-Central Cell. 186.

(657) TEXAS TECH UNIVERSITY. MUSEUM. FIELD TRIP.
  1983 See Friends of the Pleistocene. South-Central Cell. 186.

TEXAS. AGRICULTURAL AND MECHANICAL UNIVERSITY. See TEXAS A & M UNIVERSITY. DEPARTMENT OF GEOLOGY. (652) and TEXAS A & M UNIVERSITY. GEOLOGICAL SOCIETY. (653)

TEXAS. BUREAU OF ECONOMIC GEOLOGY. See UNIVERSITY OF TEXAS AT AUSTIN. BUREAU OF ECONOMIC GEOLOGY. (713) and UNIVERSITY OF TEXAS AT AUSTIN. BUREAU OF ECONOMIC GEOLOGY. (714) and UNIVERSITY OF TEXAS AT AUSTIN. BUREAU OF ECONOMIC GEOLOGY. (712)

TEXAS. LAMAR STATE COLLEGE OF TECHNOLOGY, BEAUMONT. See LAMAR UNIVERSITY. DEPARTMENT OF GEOLOGY. (353) and LAMAR UNIVERSITY GEOLOGY CLUB. (352) and LAMAR UNIVERSITY GEOLOGICAL SOCIETY. (351)

TEXAS. PAN AMERICAN UNIVERSITY, EDINBURG. See PAN AMERICAN UNIVERSITY. GEOLOGICAL SOCIETY. (526)

TEXAS. SUL ROSS STATE UNIVERSITY, ALPINE. See SUL ROSS STATE UNIVERSITY. GEOLOGICAL SOCIETY. (641)

TEXAS. UNIVERSITY AT DALLAS ... See UNIVERSITY OF TEXAS AT DALLAS. (717)

TEXAS. UNIVERSITY AT EL PASO. See UNIVERSITY OF TEXAS AT EL PASO. (718) and UNIVERSITY OF TEXAS AT EL PASO. (719)

TEXAS. UNIVERSITY. See UNIVERSITY OF TEXAS AT AUSTIN. DEPARTMENT OF GEOLOGICAL SCIENCES. (FORMERLY: DEPARTMENT OF GEOLOGY) (715) and UNIVERSITY OF TEXAS AT AUSTIN. UNIVERSITY STUDENT GEOLOGICAL SOCIETY (716)

TEXAS. WEST TEXAS STATE UNIVERSITY, CANYON. See WEST TEXAS STATE UNIVERSITY GEOLOGICAL SOCIETY. (759) and WEST TEXAS STATE UNIVERSITY. DEPARTMENT OF GEOLOGY AND ANTHROPOLOGY. (760) and WEST TEXAS STATE UNIVERSITY. DEPARTMENT OF GEOLOGY AND ANTHROPOLOGY. (761)

(658) TOBACCO ROOT GEOLOGICAL SOCIETY. FIELD CONFERENCE. GUIDEBOOK.
  1976 1st Guidebook; [Southwestern Montana] the Tobacco Root Society 1976 field conference. (401(No.73))
  1978 [3rd] Guidebook of the Drummond-Elkhorn areas, west-central Montana. (401)
  1980 Guidebook of the Drummond-Elkhorn areas, west-central Montana. (Prepared in cooperation with the Tobacco Root Geological Society Field Conference, 1978) (401(No.82))
  Road log no. 1. Missoula to Flint Creek via upper Clark Fork valley and Drummond.
  Road log no. 2. Missoula to florule locales near Drummond and Lincoln.
  Road log no. 3. Elkhorn mining district.
  1981 Sept. Pocatello, Idaho. (490)
  1982 7th Dillon, Montana.
  The overthrust province in the vicinity of Dillon, Montana, and how this structural framework has influenced mineral and energy resources accumulation. (This 1982 ... road logs [were] originally presented at the Tobacco Root Geological Society Field Conference in Pocatello, Idaho, September 8-13, 1981).
  1984 9th Boise, Idaho.
  Geology, tectonics, and mineral resources of western and southern Idaho.
  1985 10th Geology and mineral resources of the Tobacco Root Mountains and adjacent region: "A homecoming to the Tobacco Roots".
  Trip no. 1. Archean geology of the Spanish Peaks area, southwestern Montana.
  Trip no. 2. Structural and stratigraphic geology of the central Bridger Range, Montana.
  Trip no. 3. Golden sunlight and Butte mining districts.
  Trip no. 4. Nature of deformation in foreland anticlines and impinging thrust belt: Tobacco Root and southern Highland Mountains, Montana.
  Trip no. 5. Field guide to the Quaternary geology and biogeography of the east flank of the central Bridger Range, Gallatin County, Montana.

AzU	1976
CLU-G/G	1980-82, 84-85
CMenUG	1980-82, 84, 85
CPT	1976, 78, 81
CSfCSM	1976, 78
CSt-ES	1976, 78, 81-82, 84
CU-A	1980
CU-EART	1978, 80-82, 84-85
CU-SB	1981-82
CaACU	1976, 78
CaAEU	1981
CaOOG	1976, 78, 81-82
CaOTDM	1976
CaOWtU	1976, 78, 81
CoDBM	1976, 80
CoDCh	1976, 78, 80
CoDGS	1981
CoU	1976
DI-GS	1976, 78, 80-82, 84
DLC	1976
ICF	1978
ICIU-S	1976
ICarbS	1976, 78
IU	1976
IaU	1976, 78, 80-81
IdBB	1984
IdU	1980-82, 84
InLP	1976, 78, 80, 82, 84-85
InU	1976, 78, 80, 82
KLG	1981
KU	1981
KyU-Ge	1976, 78, 80-82, 84
MiHM	1976, 80
MnU	1976, 78, 80-82
MoSW	1976
MtBuM	1976, 78, 80, 82, 84
NIC	1980-82

NNC	1981
NRU	1976, 82
NSbSU	1981, 85
NdU	1981
NhD-K	1976, 78, 80
NvU	1976, 78, 82, 84
OCU-Geo	1976, 80, 81
OU	1976, 78, 80, 81
OkTA	1982
OkU	1976, 80-82
OrU-S	1980-82
PSt	1976
SdRM	1976, 80, 82
TxDaDM	1976, 84
TxDaM-SE	1982, 84
TxHSD	1976
TxLT	1978
TxMM	1982
TxU	1981-82
UU	1976, 81-82
ViBlbV	1976
WU	1976, 78, 81-82
WyU	1980-82

(659) TRI-STATE GEOLOGICAL FIELD CONFERENCE. GUIDEBOOK FOR THE ANNUAL FIELD CONFERENCE.

- 1933 1st Upper Mississippi Valley.
- 1934 2nd Southern Wisconsin.
- 1935 3rd Clinton, Jackson and Dubuque counties, Iowa.
- 1936 4th Calhoun and Jersey counties, Illinois.
- 1937 5th Southeastern Wisconsin.
- 1938 6th Madison, Dallas, Guthrie, and Polk counties, Iowa.
- 1939 7th Western Illinois.
- 1940 8th North-central Wisconsin, igneous rocks.
- 1941 9th Southeast Iowa: Montpelier, Muscatine County, Iowa, Burlington-Keokum area.
- 1946 10th Northeastern Illinois and adjacent Indiana.
- 1947 11th Eastern Wisconsin.
- 1948 12th Northeast Iowa.
- 1949 13th LaSalle Anticline, Northeast Illinois.
- 1950 14th Central plain of Wisconsin.
- 1951 15th Devonian of north-central Iowa.
- 1952 16th Central northern Illinois. (265(No.2))
- 1953 17th Northern Wisconsin.
- 1954 18th Northeastern Iowa.
- 1955 19th West-central Illinois.
- 1956 [20th] Upper Mississippi zinc-lead district.
- 1957 21st Southeast Iowa.
- 1958 22nd Southern Illinois fluorspar district.
- 1959 23rd Southwestern Wisconsin.
- 1960 24th Stratigraphic sequence of north-central Iowa.
- 1961 25th Southern Illinois; Marion to Dongola, Illinois on Interstate 57.
- 1962 26th Northeastern Wisconsin, McCaslin Syncline-Tigerton Anorthosite.
- 1963 27th Bioherms and biostromes in the Silurian-Devonian of eastern Iowa. (Cover title: Silurian-Devonian of eastern Iowa) (265(No.5))
- 1964 28th Western Illinois. (265(No.6))
- 1965 29th Cambro-Ordovician stratigraphy of Southwest Wisconsin. (778)
- 1966 30th Devonian of northern Iowa Cedar Valley, Shell Rock and Lime Creek.
- 1967 31st The Mississippi River arch.
- 1968 32nd A greenstone belt in central Wisconsin.
- 1969 33rd The many faces of geology. (334)
- 1970 34th Trip 1. Cambro-Ordovician stratigraphy and structure of north-central Illinois, plus underground gas storage.
- Trip 2. Stratigraphy of Pleistocene deposits in northeastern Illinois.
- 1971 35th Geology of the Twin Ports area, Superior-Duluth.
- 1972 36th General geology in the vicinity of northern Iowa.
- 1973 37th Depositional environments of selected Lower Pennsylvanian and Upper Mississippian sequences of southern Illinois.
- 1974 38th Field trip no. 1. Precambrian rocks of the Chippewa region.
- Field trip no. 2. Paleozoic rocks of the Eau Claire area.
- Field trip no. 3. Quaternary.
- 1975 39th Field trip No. 1. Devonian limestone facies; Cedar Valley and State Quarry limestones in the Iowa City region.
- Field trip no. 2. Strip mine reclamation in south-central Iowa.
- Field trip no. 3. Ordovician structure and mineralization in northeastern Iowa.
- 1976 40th Field trip guidebook to the geology of west central Illinois.
- 1977 41st Geology of southeastern Wisconsin.
- Field trip no. 1. Silurian and Devonian stratigraphy.
- Field trip no. 2. Pleistocene geology of Southeast Wisconsin.
- Field trip no. 3. The upper Ordovician Neda Formation and lower Silurian Mayville dolomite in Dodge County, Wisconsin.
- Field trip no. 4. Environmental field trip.
- 1978 42nd Geology of east-central Iowa. (332)
- Trip no. 1. A field guide to the Plum River fault zone in east central Iowa.
- Trip no. 2. The Iowan erosion surface: an old story, an important lesson, and some new wrinkles.
- Trip no. 3. Applied geology problems in the Cedar Rapids area.
- Trip no. 4. Geomorphology and basal Maquoketa stratigraphy in the vicinity of Dubuque.
- Trip no. 5. Geology and history of Stone City.
- 1979 43rd Geology of western Illinois. (265(No.14))
- Trip 1. Structure and Paleozoic stratigraphy of the Cap au Gres faulted flexure in western Illinois.
- Trip 2. Stratigraphy of Wisconsinan and older loesses in southwestern Illinois.
- Trip 3. Pre-Illinoian till stratigraphy in the Quincy, Illinois area.
- 1980 44th Geology of eastern and northeastern Wisconsin.
- Trip 1. Precambrian basement complex of northeastern Wisconsin.
- Trip 2. Paleozoic and Late Wisconsinan stratigraphy of eastern Wisconsin.
- Trip 3. Environmental geology of the Green Bay area.
- Trip 4. Green Bay metropolitan wastewater treatment plant tour.
- Trip 5. Structure and landform evolution in the Green Bay, Wisconsin area.
- 1981 45th Trip 1. Cherokee Sandstone and Fort Dodge gypsum beds.
- Trip 2. [No title, but it is about Pint's Quarry]
- Field trip guide no. 3. Rockford Pit, Floyd County, Iowa.
- 1982 46th Geology of the Mississippi River arch in the quad-cities area. (Cover title: The Mississippi River arch quad-cities region)
- 1983 47th Three billion years of geology: a field trip through the Archean, Proterozoic, Paleozoic and Pleistocene geology of the Black River Falls area of Wisconsin; discussion, geological stop descriptions and roadlog. (Cover title: Three billion years of geology, Archean, Proterozoic, Paleozoic and Pleistocene geology of the Black River Valley) (777)
- 1984 48th General geology of north-central Iowa.
- 1985 49th Geology of the Chicago area.
- Trip 1. Late Quaternary glacial and glacial lake history of the Chicago region.
- Trip 2. Classic Silurian reefs of the Chicago area.

CLU-G/G	1965, 68, 79
CPT	1948, 51, 55-56, 79
CSt-ES	1948-49, 52, 63-65, 69-70, 72-73, 77, 79, 83
CU-A	1984
CU-EART	1969, 79
CU-SB	1979, 83
CaACU	1964, 79
CaAEU	1979
CaBVaU	1969
CaOKQGS	1966, 70
CaOOG	1968-79, 82-83
CaOTRM	1979
CaOWtU	1965, 68, 79
CoDCh	1948-61, 79
CoDGS	1948, 57-73, 78-79
CoFS	1964
CoG	1960, 65, 69, 79
CoU	1965, 69
DI-GS	1933-41, 46-55, 57-62, 64-73, 78-80, 82-84
DLC	1965, 69
I-GS	1933-41, 46-62, 64-75, 80
ICF	1938, 41, 46, 48-49, 54-62, 65, 67-68, 73, 77, 79
ICIU-S	1965, 79
ICarbS	1952, 61, 64-65, 73, 79
IEN	1948-49, 52, 64-65, 69-70
IU	1934-41, 46-62, 64-79
IaCfT	1982-85
IaU	1933-41, 46-73, 76-80, 82-85
InLP	1933-48, 50-53, 56, 58-59, 62-70, 72, 74-75, 77-80, 82-84

InTI	1979
InU	1951, 53, 55, 60, 64-66, 68-70, 79
KLG	1951-52, 55
KyU-Ge	1955, 65, 70-71, 73-74, 76, 78-80, 82-84
LNU	1962, 64
LU	1964
MH-GS	1969
MNS	1964-65
MiHM	1965, 68, 71, 79
MnDuU	1938, 40-41, 62, 68, 75(1)
MnU	1948-49, 51-52, 54, 59-60, 62, 65, 68-74, 76-80, 83-84
MoSW	1951, 55, 64-65, 69
MoU	1938, 66, 69
MtBuM	1965, 79
MtU	1969
NBiSU	1966, 69
NNC	1948-55
NRU	1973
NSbSU	1979
NSyU	1963, 64, 79
NbU	1964
NcU	1965, 74, 79
NdU	1965, 69-70, 73
NhD-K	1973, 77(2), 79
NjP	1938, 48-49, 65, 67-69
NvU	1965
OCU-Geo	1955, 65, 67, 73, 79
OU	1935, 38, 49, 51-52, 55, 61, 65, 67, 69, 79, 82
OkU	1940-41, 48-49, 55, 61, 64-65, 69-70, 78-79
PBL	1965
PBm	1979
PSt	1965, 69, 79
RPB	1964
SdRM	1966, 69, 79
TxBeaL	1973, 79
TxDaAR-T	1951, 54, 65, 68, 70-74
TxDaDM	1948, 58, 60-61, 64-68, 79
TxDaGI	1941, 52
TxDaM-SE	1965, 83
TxDaSM	1954-55, 60, 73
TxHSD	1965, 73
TxHU	1965
TxLT	1969
TxU	1936, 38, 41, 51-52, 54-55, 57, 59-62, 64-70, 73, 76
UU	1965, 69, 79
UWGB	1980
WGrU	1980
WU	1937-38, 41, 52, 55, 57, 59, 61-71, 74-83, 85
WaU	1952, 63-64, 79

TRI-STATE GEOLOGICAL SOCIETY. See TRI-STATE GEOLOGICAL FIELD CONFERENCE. (659)

(660) TRIARTHRUS CLUB, AUSTIN, TEXAS.

1940 Feb. 4 Field trip through the Cretaceous section near Austin, Texas.
TxU   1940

(661) TULSA GEOLOGICAL SOCIETY. DIGEST.

1972 37. See Geological Society of America. South Central Section. 205.(1978)

(662) TULSA GEOLOGICAL SOCIETY. GUIDEBOOK FOR THE FIELD CONFERENCE. (TITLE VARIES)

1936 See American Association of Petroleum Geologists. 18.
1941 Oct. Tulsa to Choteau and Grand River area and return.
— Nov. Pennsylvanian stratigraphy of Tulsa County.
1946 Pennsylvanian and Mississippian rocks of eastern Oklahoma.
1947 Western part of the Ouachita Mountains in Oklahoma.
1950 Oct. To the mines and mill of the Eagle-Picher Mining and Smelting Company.
— Dec. Coody's Bluff to Burbank, Oklahoma.
1951 Apr. Tulsa to Spavinaw, Oklahoma.
— May Kansas, Oklahoma to Marble City, Oklahoma.
1954 Tulsa-Woolaroc-Bartlesville.
1956 Apr. See Oklahoma City Geological Society. 504.
1957 Oct./Nov. See American Association of Petroleum Geologists. Regional Meeting. 29.
1961 Arkoma Basin and north-central Ouachita Mountains of Oklahoma. (177)
1968 Oct. Guidebook and roadlog; geology of the Tulsa metropolitan area.
1972 Apr. Sandstone environments, Keystone Reservoir area.
1973 Apr. Guide to sandstone environments, Keystone Reservoir area, Oklahoma.
— Oct. "The big lime" southern margin of the Oologah Limestone banks.
1975 May Basinward facies changes in the Wapanuck Limestone (Lower Pennsylvanian) Indian Nations Turnpike, Ouachita Mountains, Oklahoma.
1976 Apr. Coal and oil potential of the Tri-State area.
1977 Apr. A guidebook to the geology of the Arkansas Paleozoic area (Ozark Mountains, Arkansas Valley, and Ouachita Mountains).(revised in 1980.) (67(1980))
1979 May Lithostratrigraphy of selected Devonian and Carboniferous units, Northwest Arkansas.
— Oct. See American Association of Petroleum Geologists. Mid-continent Section. 26.
1981 Nov. Geologic features along the flank of the Ozark Uplift, N.E. Oklahoma; a survey of structural, stratigraphic, mining, and human factors, geology of the area.
1983 Mar. See American Institute of Professional Geologists (AIPG). 49.
1984 May Upper Pennsylvanian source beds of northeastern Oklahoma and adjacent Kansas.
(2) North of Tulsa.
(3) Tulsa area and south of city.

AzTeS	1950(Oct.)
CLU-G/G	1950-51, 54, 61, 68
CPT	1950-51, 54
CSt-ES	1950-51, 54, 73(Oct.)
CU-EART	1961, 77
CU-SB	1977
CaACI	1961
CaACU	1976, 79
CaOHM	1968
CaOTRM	1961
CoDCh	1973, 76-77
CoDGS	1947, 50-51, 54, 56, 61, 68, 77
CoG	1950-51, 61
CoLiM	1984
CoU	1954, 68
DFPC	1961
DI-GS	1941(Oct.), 46-47, 50-51, 54, 61, 68, 73, 75, 77, 84
DLC	1947, 50-51
I-GS	1954, 72
ICF	1961
ICIU-S	1968
IEN	1947
IU	1947, 50-51, 54, 61, 68, 73(Oct.), 77, 79
IaU	1961, 77, 79(May)
InLP	1950-51, 61, 77
InU	1947, 50-51, 54, 61, 68
KyU-Ge	1947, 50-51, 68, 73(Oct.), 77, 79(May)
LU	1947, 50-51, 54, 61
MH-GS	1947
MNS	1947, 50-51, 54
MiHM	1950-51
MnU	1950-51, 54, 61, 68, 73(Oct.), 77, 79(May)
MoRM	1947, 50
MoSW	1947, 50-51
NNC	1947, 54, 61, 68
NSbSU	1977
NbU	1947, 61
NcU	1968, 73
NdU	1950-51, 54, 73
NhD-K	1947, 73(Oct.)
NjP	1950-51, 54, 61
NmPE	1961, 68
NvU	1977
OCU-Geo	1947, 54, 77
OU	1947, 50-51, 54, 68, 73(Oct.)
OkOkCGS	1947, 50-51
OkT	1941(Oct.), 47, 50-51, 54, 61, 68, 73(Oct.), 84
OkTA	1981, 84
OkU	1941, 46-47, 50-51, 54, 61, 68, 72-73, 76-77, 84

RPB	1947
TxBeaL	1950-51, 61, 77
TxDaAR-T	1947, 50-51, 54, 61
TxDaDM	1961, 68
TxDaGI	1947, 50-51, 54, 61
TxDaM-SE	1947, 50-51, 54, 61, 76
TxDaSM	1950-51, 54, 61, 76
TxHSD	1947, 61, 77, 79(May)
TxHU	1941, 46-47, 61
TxLT	1941(Oct.), 46-47, 50-51
TxMM	1941(Oct.), 46-47, 50-51, 61
TxU	1941, 46-47, 50-51, 54, 61, 68, 77
TxU-Da	1947, 50-51, 54
TxWB	1941, 47, 54, 68
UU	1950-51, 54
ViBlbV	1968, 73(Oct.)
WU	1950-51, 77, 84

(663) U.S. DEPARTMENT OF ENERGY. [FIELD TRIP GUIDEBOOK]
1985 Apr.   See Oil Shale Symposium. 501.

(664) U.S. DEPARTMENT OF ENERGY. EASTERN GAS SHALES PROJECT. GUIDEBOOK.
1979 Sept.   Middle and Upper Devonian clastics, central and western New York state.

CSt-ES	1979
DI-GS	1979
IU	1979
IaU	1979
InLP	1979
KyU-Ge	1979
MnU	1979
NcU	1979
NhD-K	1979
OCU-Geo	1979
WU	1979

U.S. GEOLOGICAL SURVEY. See GEOLOGICAL SURVEY (U.S.). (218) and GEOLOGICAL SURVEY (U.S.) (219) and GEOLOGICAL SURVEY (U.S.). (217) and GEOLOGICAL SURVEY (U.S.) (220) and GEOLOGICAL SURVEY (U.S.). WRD DISTRICT CHIEFS' CONFERENCE. (221)

(665) U.S. NATIONAL PARK SERVICE.
1937   See Rocky Mountain Association of Geologists. 549.

(666) U.S. SYMPOSIUM ON ROCK MECHANICS. FIELD TRIP.
1984 25th   Longwall coal mining in Illinois.

I-GS	1984

(667) U.S. WORKING GROUP OF THE INTERNATIONAL GEOLOGICAL CORRELATION PROGRAM. ARCHEAN GEOCHEMISTRY FIELD CONFERENCE.
1981 19th   Upper Peninsula & northern Wisconsin.
[1] Wakefield, Michigan, area.
[2] Marenisco-Watersmeet area.
[3] The northern complex.
[4] Southern complex stops.
[5] Gneissic rocks of northeastern Wisconsin.
[6] Archean and proterozoic rocks of central Wisconsin.
[7] Marathon County area.

DI-GS	1981
WU	1981

(668) UNITAR CONFERENCE ON DEVELOPMENT OF SHALLOW OIL AND GAS RESOURCES. GUIDEBOOK.
1984   Guidebook for Arbuckle Mountain field trip, southern Oklahoma. (509)

BEG	1984
CSfCSM	1984
CSt-ES	1984
CaOOG	1984
CaOWtU	1984
DI-GS	1984
IU	1984
IaU	1984
KLG	1984
KU	1984
MnU	1984
MoSW	1984
NdU	1984
NvU	1984
OCl	1984
OkT	1984
PHarER-T	1984
TxDaM-SE	1984
WU	1984

UNITAR/UNDP INFORMATION CENTRE FOR HEAVY CRUDE AND TAR SANDS. See AMERICAN ASSOCIATION OF PETROLEUM GEOLOGISTS. AAPG RESEARCH COMMITTEE CONFERENCE. (TITLE VARIES) (19)

UNITED NATIONS INSTITUTE FOR TRAINING AND RESEARCH (UNITAR). CONFERENCE ON THE DEVELOPMENT OF SHALLOW OIL AND GAS RESOURCES. See UNITAR CONFERENCE ON DEVELOPMENT OF SHALLOW OIL AND GAS RESOURCES. (668)

UNITED NATIONS INSTITUTE FOR TRAINING AND RESEARCH/UNITED NATIONS DEVELOPMENT PROGRAMME. See AMERICAN ASSOCIATION OF PETROLEUM GEOLOGISTS. AAPG RESEARCH COMMITTEE CONFERENCE. (TITLE VARIES) (19)

(669) UNITED STATES-JAPAN CONFERENCE ON RESEARCH RELATED TO EARTHQUAKE PREDICTION.
1966 2nd   Guidebook for Nevada earthquake sites.

CU-SB	1958
CoDGS	1966
NvU	1966

(670) UNIVERSIDAD NACIONAL AUTONOMA DE MEXICO. INSTITUTO DE GEOLOGIA.
1946   Guia geologica de Oaxaca, Mexico.
1981 Mar.   See Geological Society of America. Cordilleran Section. 199.

CLU-G/G	1946
DI-GS	1946
ICarbS	1946
InU	1946
NhD-K	1946

(671) UNIVERSITY OF ALABAMA GEOLOGY CLUB. GUIDEBOOK FOR THE ANNUAL FIELD TRIP.
1968 2nd   A field guide to petrographic variations of the Pinckneyville Granite complex and other related rocks, Coosa County, Alabama.
1973   Selected geology of Tuscaloosa, Jefferson, Shelby and Bibb counties, Alabama with special emphasis on environmental geology.
1978   Geologic profiles of the Cahaba River in central Bibb County, Alabama.
[n.d.] Spring   1. Field trip; [Northern Alabama].

ATuGS	1973, 78
Ar-GS	1968
DI-GS	1968, 73
OkU	[n.d.], 1968

(672) UNIVERSITY OF ARIZONA. BUREAU OF GEOLOGY AND MINERAL TECHNOLOGY. BULLETIN.
1965 174.   Guidebook 1. Highways of Arizona. U.S. Highway 666.
1967 176.   Guidebook 2. Highways of Arizona. Arizona Highways 77 and 177.
1971 183.   Guidebook 3. Highways of Arizona. Arizona Highways 85, 86 and 386.
— 184.   Guidebook 4. Highways of Arizona. Arizona Highways 87, 88 and 188.

AzTeS	1965, 67
AzU	1965, 67, 71
CLU-G/G	1965, 67, 71
CLhC	1971(184)
CPT	1965, 67, 71
CSdS	1965, 67, 71
CSt-ES	1965, 67, 71
CU-EART	1965, 67, 71
CaOKQGS	1965, 67

(673) University of Arizona. Bureau of Geology and Mineral Technology.

CaOWtU	1971
CoDGS	1965, 67, 71
CoG	1965, 67
CoU	1965, 67, 71
DI-GS	1965, 67, 71
DLC	1965, 67, 71
ICF	1965, 67, 71
IEN	1965, 67, 71
IU	1965, 67, 71
IaU	1965, 67, 71
InLP	1965, 67, 71
InU	1965, 67, 71
KyU-Ge	1965, 67, 71
LNU	1967, 71
MH-GS	1967
MiHM	1965, 67, 71
MnU	1965, 67, 71
MoSW	1965, 67, 71
MoU	1965, 67
MtBuM	1965, 67, 71
NSyU	1967
NbU	1965, 67
NcU	1971
NdU	1965, 67, 71
NhD-K	1965, 67, 71
NjP	1965, 71
NvLN	1965, 67
NvU	1965, 67, 71
OCU-Geo	1965, 67, 71
OU	1965, 67
OkT	1965, 67, 71
OkU	1965, 67
OrU-S	1965, 67
PBL	1965, 67
PSt	1965, 67, 71
TxDaAR-T	1971
TxHSD	1965, 67, 71
TxU	1965, 67, 71
UU	1965, 67, 71
ViBIbV	1965, 71
WU	1965, 67, 71

(673) UNIVERSITY OF ARIZONA. BUREAU OF GEOLOGY AND MINERAL TECHNOLOGY. SPECIAL PAPER.
1978   2.   See Geological Society of America. Cordilleran Section. 199.

(674) UNIVERSITY OF ARIZONA. BUREAU OF GEOLOGY AND MINERAL TECHNOLOGY. OPEN-FILE REPORT.
1979   79-4.   See Geological Society of America. 197.(15)
1983   83-24   See Arizona Geological Society. 65.

(675) UNIVERSITY OF BRITISH COLUMBIA REPORT.
1968   No. 6.   See Geological Association of Canada. 190.(A)
—   No. 7.   See Geological Association of Canada. 190.(B)

(676) UNIVERSITY OF BRITISH COLUMBIA. DEPARTMENT OF GEOLOGICAL SCIENCES. GUIDEBOOK FOR GEOLOGICAL FIELD TRIPS.
1962   San Juan Island, Washington.
1965   San Juan Island, Washington.
1968   See Geological Association of Canada. 190.
1977   [1]   Guidebook for geological field trips in Burnaby, Ioco, Port Moody, Pitt Meadows, Maple Ridge, Port Coquitlam, and Vancouver, B.C. (3(No.17))
—   [2]   Guidebook for geologic field trips in the Lynn Canyon-Seymour area of North Vancouver, B.C. (3(No.22))

CaACAM	1977(1)
CaAEU	1962
CaBVaU	1962, 65, 77
NNC	1962
NbU	1962
TxU	1962

(677) UNIVERSITY OF CALGARY, DEPARTMENT OF GEOLOGY AND GEOPHYSICS. CSPG-U OF C SHORT COURSE [NOTES].
1983 May   See Canadian Society of Petroleum Geologists. 124.

UNIVERSITY OF CALIFORNIA, LOS ANGELES. DEPARTMENT OF EARTH AND SPACE SCIENCES. ESSSO GUIDEBOOK. See EARTH AND SPACE SCIENCE STUDENT ORGANIZATION, U.C.L.A. (161)

(678) UNIVERSITY OF CALIFORNIA, RIVERSIDE. CAMPUS MUSEUM CONTRIBUTIONS.
1971   No. 1.   See Geological Society of America. Cordilleran Section. 199.
1979   No. 5.   See Geological Society of America. 197.

(679) UNIVERSITY OF CALIFORNIA, SAN DIEGO, GEOLOGY DEPARTMENT. GUIDEBOOK FOR FIELD TRIPS.
1961   See Geological Society of America. Cordilleran Section. 199.

(680) UNIVERSITY OF CALIFORNIA, SANTA BARBARA. DEPARTMENT OF GEOLOGICAL SCIENCES. FIELD TRIP.
1958   Field trip.
1979   See Geological Society of America. 197.

UNIVERSITY OF CENTRAL FLORIDA. COLLEGE OF ENGINEERING. FLORIDA SINKHOLE RESEARCH INSTITUTE. FSRI REPORT. See FLORIDA SINKHOLE RESEARCH INSTITUTE. (176)

(681) UNIVERSITY OF CINCINNATI. DEPARTMENT OF GEOLOGY. GUIDEBOOK.
1982 Oct.   Geology of eastern midcontinent.
1984   Geology of the eastern midcontinent; the Blue Ridge of East Tennessee and western North Carolina.
1985 Nov.   Geology of the eastern midcontinent; Mammoth Cave area.
   OCU-Geo   1982-85

(682) UNIVERSITY OF COLORADO. MUSEUM.
1968 Aug.   Field Conference for the high altitude and mountain basin deposits of Miocene age in Wyoming and Colorado.

CoDGS	1968
CoDuF	1968
CoU	1968
DI-GS	1968

(683) UNIVERSITY OF DELAWARE. DEPARTMENT OF GEOLOGY.
1968   See Society of Economic Paleontologists and Mineralogists. Northeastern Section. 592.

(684) UNIVERSITY OF GEORGIA. MARINE INSTITUTE, SAPELO ISLAND. CONTRIBUTION.
1966   No. 105.   See Geological Society of America. Southeastern Section. 206.

(685) UNIVERSITY OF HOUSTON. DEPARTMENT OF GEOLOGY. FIELD TRIP.
1965   Historical geology field trip.
1966   Big Bend field trip.
1967   Big Bend field trip.
1968   Coastal geology field trip.
1970   Houston to Uvalde, via U.S. 90A and 90 to Big Bend region.
[n.d.]   Physical geology trip to central Texas.

IU	[n.d.], 1965-67
OkU	1968
TxHU	1965, 70

(686) UNIVERSITY OF ILLINOIS AT URBANA-CHAMPAIGN. DEPARTMENT OF GEOLOGY. GEOLOGIC GUIDEBOOK.
1967 Mar.   1. Guide to the geology of the Cagle's Mill Spillway, Turkey Run State Park and the Pennsylvanian sequence at Montezuma, Indiana. Corrected printing, March, 1967.
   IU   1967

(687) UNIVERSITY OF IOWA GEOLOGY CLUB. FIELD TRIP.
1959   1st   Devil's Lake State Park, Wisconsin.
   IaU   1959

(688) UNIVERSITY OF IOWA. DEPARTMENT OF GEOLOGY. [GUIDEBOOK]
1961 Lake Superior region petrology-economic geology field trip.
1962 Central and northeast Wisconsin petrology-economic geology field trip.
1963 Southeast Missouri and South Illinois stratigraphy-economic geology, Precambrian.
1964 Missouri-Illinois-Arkansas-Oklahoma.
1968 May See Geological Society of America. North Central Section. 200.
— Spring The Grand Canyon.
1970 Big Bend area, Texas, Gomez field trip.
1971 Spring The Southern Appalachians.
1972 Spring Florida.
1973 Spring The Grand Canyon and vicinity; a guidebook by participants.
1974 Big Bend field trip.
1978 South Florida carbonate sediments. (1972 reprint of the Guidebook for Field Trip 1, Geological Society of America Annual Meeting, 1964) (696, 197(1964))
1979 [No title given but is on the Southern Appalachians]
1984 Geology spring field trip guide to South Florida.

AzU	1968(Spring), 70
CSt-ES	1978
CU-EART	1978
DI-GS	1968(Spring)
IU	1978
IaU	1961-64, 70, 78-79, 84
InLP	1962
InU	1978
KyU-Ge	1978
MnU	1978
NSyU	1978
NcU	1978
NhD-K	1978
TxBeaL	1963
TxU	1968(Spring)
ViBlbV	1978
WU	1978

(689) UNIVERSITY OF KANSAS. SCIENCE AND MATHEMATICS CAMP. GEOLOGIC FIELD CONFERENCE.
1956 1st Northeastern Kansas. (633)
1957 [2nd] Geologic field conference in northeastern Kansas between Lawrence and Wyandotte County Park.
1958 3rd Northeastern Kansas in Douglas, Johnson, and Wyandotte counties between Lawrence and Wyandotte County Park. (633)
1959 Northeastern Kansas between Lawrence and Wyandotte County Park, Douglas, Johnson, Wyandotte, and Leavenworth counties. (633)
1960 Northeastern Kansas between Lawrence and Wyandotte County Park; Douglas, Johnson, Wyandotte, and Leavenworth counties. 8
1961 [6th] Same title as 1957.

CoDCh	1957
IU	1961
KLG	1956, 57, 58-60, 61
OU	1957

(690) UNIVERSITY OF KENTUCKY. GEOLOGY DEPARTMENT.
1938 No.1. Elk Lick Falls.
— No.2. Clay's Ferry.
— No.3. Boone's Cave.
— No.4. Faircloth Flurite Vein.
— No.5. Burdett Knob.
— No.6. Herrington Lake.
— No.7. Natural Bridge.

KyU-Ge	1938

(691) UNIVERSITY OF LOUISVILLE. GEOLOGY CLUB. ANNUAL FIELD TRIP GUIDEBOOK.
1958 1st Annual field trip.
1959 2nd Annual field trip.

KyU-Ge	1958

(692) UNIVERSITY OF MAINE, PRESQUE ISLE.
1980 Oct. See New England Intercollegiate Geological Conference. 456.

(693) UNIVERSITY OF MASSACHUSETTS AT AMHERST. DEPARTMENT OF GEOLOGY AND GEOGRAPHY. CONTRIBUTION.
1978 No.32. See Society of Economic Paleontologists and Mineralogists. Eastern Section. 586.
1983 See Friends of the Grenville. [Grenville Province of Quebec] 181.

(694) UNIVERSITY OF MASSACHUSETTS. COASTAL RESEARCH GROUP.
1969 1. See Society of Economic Paleontologists and Mineralogists. Eastern Section. 586.

(695) UNIVERSITY OF MIAMI. DEPARTMENT OF GEOLOGY. FIELD TRIPS.
1977 See Miami Geological Society. 374.

(696) UNIVERSITY OF MIAMI. ROSENSTIEL SCHOOL OF MARINE & ATMOSPHERIC SCIENCE. DIVISION OF MARINE GEOLOGY AND GEOPHYSICS. THE COMPARATIVE SEDIMENTOLOGY LABORATORY. SEDIMENTA.
1978 2. See University of Iowa. Department of Geology. 688.
1983 9. Oligocene reef tract development, southwestern Puerto Rico. Part III. Field guide to representative exposures and modern analog.
1984 Modern and ancient carbonate environments of Jamaica.

CMenUG	1983
CSt-ES	1983
CaOKQGS	1983
CoDGS	1983
DI-GS	1983
IaU	1983, 84
NcU	1983
TxU	1983
WU	1983

(697) UNIVERSITY OF MICHIGAN. MUSEUM OF PALEONTOLOGY. PAPERS ON PALEONTOLOGY.
1974 No. 7 See Michigan Basin Geological Society. 377.(1976)
1980 No. 24. See Society of Vertebrate Paleontology. 604.(1980)

(698) UNIVERSITY OF MINNESOTA. DULUTH BRANCH.
1959 No title but the subject is the Pleistocene glacial deposits and granitic outcrops from Duluth to Burntside Lake, Minnesota.

OkU	1959

(699) UNIVERSITY OF MISSISSIPPI. GEOLOGICAL SOCIETY.
1972 The classical Tertiary of southwest Alabama.

Ms-GS	1972
MsJG	1972

(700) UNIVERSITY OF MISSOURI. GEOLOGY CLUB. FIELD TRIP GUIDEBOOK.
1949 See Sigma Gamma Epsilon. Alpha Chapter, University of Kansas. 572.
1950 [Ordovician, Devonian, Mississippian, and Pennsylvanian, central Missouri]
1951 Kansas River valley, Northeast Kansas, Upper Pennsylvanian and Lower Permian.
1955 A study of the Ordovician, Devonian, Mississippian and Pennsylvanian.

I-GS	1951
MnDuU	1950
MoU	1950, 55
NjP	1950
TxHU	1950
TxLT	1950, 55
TxU	1955

(701) UNIVERSITY OF MONTANA. GEOLOGY DEPARTMENT. GEOLOGICAL SERIES PUBLICATION.
1971 Summer See National Science Foundation. Summer Conference on Field Geology for Secondary School Teachers. 428.
— Sept. See Friends of the Pleistocene. Pacific Cell (Formerly: Pacific Coast Group). 184.
1977 See Geological Society of America. Rocky Mountain Section. 204.

(702) UNIVERSITY OF NEBRASKA--LINCOLN. DEPARTMENT OF GEOLOGY. ANNUAL SPRING FIELD TRIP.
1951 3rd   Geology of West Texas.
1954 Apr. West Texas.
1959   West Texas.
n.d. 2nd   The Paleozoic of the central United States.
 NbU   n.d., 1951, 54, 59
 WU   n.d., 1951

(703) UNIVERSITY OF NORTH DAKOTA. GEOLOGY DEPARTMENT. GUIDEBOOK.
1960 [1] Geology of the Black Hills and route between Grand Forks and Rapid City. (574)
1961 2.   The geology of southeastern Minnesota and northwestern Iowa and routes between Grand Forks, North Dakota and Dubuque, Iowa. (574)
1972 3.   See Geological Society of America. 197.
 CU-EART   1960-61
 CoDCh   1960
 MnU   1960-61
 NdU   1960-61
 TxBeaL   1960-61
 TxDaSM   1960-61

(704) UNIVERSITY OF OKLAHOMA. SCHOOL OF GEOLOGY AND GEOPHYSICS.
1957 May   See Panhandle Geological Society. 528.
1964 Oct. See Kansas Geological Society. 341.

(705) UNIVERSITY OF OREGON. DEPARTMENT OF GEOLOGY.
1965 See Lunar Geological Field Conference. 366.

(706) UNIVERSITY OF OREGON. MUSEUM OF NATURAL HISTORY. BULLETIN.
1973 No. 21. Guide to the geology of the Owyhee region of Oregon.
1977 No. 22. Guide to the geology and lore of the wild reach of the Rogue River, Oregon.
 AzTeS   1973, 77
 CSt-ES   1973
 CU-EART   1973, 77
 DI-GS   1973, 77
 DLC   1973, 77
 ICF   1973
 ICarbS   1973, 77
 IU   1973
 IaU   1977
 IdBB   1973
 IdU   1973, 77
 NcU   1973, 77
 NmPE   1973, 77
 NvLN   1973
 NvU   1973
 OrCS   1977
 TxLT   1973
 TxU   1973
 UU   1973, 77

(707) UNIVERSITY OF SOUTH CAROLINA. DEPARTMENT OF GEOLOGY.
1978 Apr. See American Association of Petroleum Geologists. Department of Education. 23.
1979   See Geological Society of America. Southeastern Section. 206.(No. 4)

(708) UNIVERSITY OF SOUTHWESTERN LOUSIANA GEOLOGICAL SOCIETY.
1971 [Spring] See Southwestern Association of Student Geological Societies (SASGS). 626.
1973 Fall   See Southwestern Association of Student Geological Societies (SASGS). 626.

(709) UNIVERSITY OF TENNESSEE, KNOXVILLE. DEPARTMENT OF GEOLOGICAL SCIENCES. STUDIES IN GEOLOGY.
1977 [No.] 1. See International Symposium on the Ordovician System. 319.
1980 No. 4.   See Geological Society of America. 197.

1985 No. 9.   See Geological Society of America. Southeastern Section. 206.
— No. 10   See Geological Society of America. Southeastern Section. 206.

(710) UNIVERSITY OF TENNESSEE, KNOXVILLE. DEPARTMENT OF GEOLOGY AND GEOGRAPHY. REGIONAL STUDIES IN ECONOMIC GEOLOGY. FIELD TRIP.
1949 2nd   Arkansas, Oklahoma, and Missouri.
1950 3rd   Southeastern states.
1964   Georgia-Florida.
1965   Tennessee, Virginia, Pennsylvania, New Jersey.
1966   Alabama, Mississippi, Tennessee, Arkansas, Oklahoma, Missouri.
1967   North Carolina, South Carolina, Georgia and north Florida.
 TU   1949-50, 64-67

(711) UNIVERSITY OF TENNESSEE, KNOXVILLE. LOWER DIVISION CURRICULUM COMMITTEE. FIELD GUIDE.
[n.d.] [1] Geoscience I field trip: The University of Tennessee.
— [2]   Geoscience II field trip: The University of Tennessee.
 IaU   n.d.(1-2)

(712) UNIVERSITY OF TEXAS AT AUSTIN. BUREAU OF ECONOMIC GEOLOGY. FIELD CONFERENCE ON THE ELLENBERGER GROUP.
1945 June Instructions and road log for a field conference on the Ellenburger Group of the Llano region, Texas.
 TxU   1945

(713) UNIVERSITY OF TEXAS AT AUSTIN. BUREAU OF ECONOMIC GEOLOGY. GUIDEBOOK SERIES.
1958 No. 1. See Association of American State Geologists. 74.
1960 No. 2. Texas fossils; an amateur collector's handbook.
1961 No. 3. See Clay Minerals Conference. 132.
1962   See Geological Society of America. 197.(1)
1963 No. 4. The geologic story of Longhorn Cavern.
— No. 5.   Geology of Llano region and Austin area. (Reprint (with minor modifications) of Field trip no. 1 from GEOLOGY OF THE GULF COAST AND CENTRAL TEXAS, published by the Houston Geological Society for the 1962 Annual Meeting of The Geological Society of America and Associated Societies) (revised in 1972.)
1964 No. 6. Texas rocks and minerals; an amateur's guide.
1968 No. 7. The Big Bend of the Rio Grande, a guide to the rocks, landscape, geologic history and settlers of the area of Big Bend National Park.
1969 No. 8. The geologic story of Palo Duro Canyon.
— No. 9.   See Clay Minerals Conference. 132. and Geological Society of America. 197.(1973-No.11)
1970 No. 10. Geologic and historic guide to the state parks of Texas.
— No. 11.   Recent sediments of Southeast Texas; a field guide to the Brazos alluvial and deltaic plains and Galveston Barrier Island complex. (Reprint of FIELD GUIDE prepared by Shell Development Company)(Reprinted in 1973)
1971 No. 12. See American Association of Petroleum Geologists. 18.
1972 No. 13. Geology of the Llano region and Austin area, field excursion. (Bureau of Economic Geology Guidebook no. 5, 1963, updated and combined with Shreveport Geological Society's Geology of the Llano region and Austin area, Texas, 1971) (566, 205(1975(3)))
1973   Reprint of No. 11, 1970. See Geological Society of America. South Central Section. (205(1976(3)))
— No. 14.   See Geological Society of America. 197.
— No. 15.   See Geological Society of America. 197.
1976 No. 16. Guide to points of geologic interest in Austin.
1979 No. 18. See American Association of Petroleum Geologists. 18.
— No. 19.   Cenozoic geology of the Trans-Pecos volcanic field of Texas. (149(1978))
1980 No. 17. Padre Island National Seashore; a guide to the geology, natural environment, and history of a Texas barrier island.
— No. 20.   Modern depositional environments of the Texas coast.
1984 No. 21. Geology of Monahans Sandhills State Park, Texas.
 ATuGS   1963(5)
 AzTeS   1960, 63-64, 68, 69(8), 70, 76, 79, 80, 84
 AzU   1963-64, 68, 69(8), 70, 72
 BEG   1980, 84

CLU-G/G	1960, 63-64, 68, 69(8), 70, 72, 76, 79-80, 84
CLhC	1963(4), 70(11), 72, 76, 79-80
CMenUG	1980(20), 84
CPT	1968, 84
CSdS	1970(11)
CSfCSM	1980(20), 84
CSt-ES	1960, 63-64, 68, 70, 72, 76, 79-80, 84
CU-A	1960, 63(4), 70(11), 72, 76, 79
CU-EART	1960, 63-64, 68, 69(8), 70, 72, 76, 79-80, 84
CU-SC	1980(20), 84
CaACM	1970(11)
CaACU	1960, 63, 64, 68, 69(8), 70, 72, 76, 79-80
CaAEU	1970(11)
CaBVaU	1968
CaOHM	1970, 72
CaOKQGS	1963(5), 70
CaOLU	1970, 72
CaOOG	1960, 63-64, 68, 69(8), 70, 72, 76, 79-80, 84
CaOWtU	1963(4), 64, 68-70, 72, 76, 79-80, 84
CoDGS	1960, 63-64, 68-70, 72-73, 76, 79-80, 84
CoG	1960, 63, 68, 69(8), 70(11)
CoU	1963-64, 68, 69(8), 70, 72
DI-GS	1960, 63-64, 68, 69(8), 70, 72, 76, 79-80, 84
DLC	1960, 63-64, 68, 69(8), 70, 72, 76, 79
F-GS	1980, 84
FBG	1980, 84
I-GS	1963(4), 64, 68, 69(8), 70(11), 72
ICF	1960, 63-64, 68, 69(8), 70, 72, 76, 79
ICarbS	1968, 69(8), 70, 72, 76, 79, 80, 84
IEN	1960, 63-64, 68, 69(8), 70, 72
IU	1960, 63-64, 68, 69(8), 70, 72, 76, 79, 80
IaU	1960, 64, 68, 69(8), 70, 72, 76, 79-80, 84
IdBB	1970
IdPI	1970
IdU	1960, 64, 68-70
InLP	1960, 63-64, 68, 69(8), 70, 72, 76, 79, 80
InRE	1968, 72
InU	1960, 63(4), 64, 68, 69(8), 70, 79
KLG	1972, 80, 84
KyU-Ge	1960, 63-64, 68, 69(8), 70, 72, 76, 79-80, 84
LNU	1960, 63-64, 70(11), 72, 76, 79-80
LU	1963
MNS	1960, 63-64, 68, 69(8)
MU	1980
MiDW-S	1963-64, 68, 69(8), 70, 72
MiHM	1963(5), 64, 68, 69(8), 70, 72
MiU	1960, 63-64, 68, 69(8), 70
MnU	1960, 63-64, 68, 69(8), 70, 72, 76, 79-80, 84
MoSW	1960, 63-64, 68, 69(8), 70, 72
MoU	1960, 63(4), 64, 68, 69(8), 70, 72
MtBuM	1960, 63-64, 68, 69(8), 70, 72
NBiSU	1960, 63-64, 68, 69(8), 70, 72
NIC	1980, 84
NNC	1960, 63-64, 68, 69(8), 70, 72, 80
NSbSU	1970(11), 78, 79(19), 80
NSyU	1960, 63-64, 68, 69(8), 70, 72, 76
NbU	1963
NcU	1960, 63(5), 64, 68, 69(8), 70, 76, 79-80
NdU	1963(4), 64, 68, 69(8), 70, 80, 84
NhD-K	1963, 68, 70, 72, 76, 79, 80
NjP	1960, 63-64
NmU	1980, 84
NvLN	1972
NvU	1960, 63-64, 68, 69(8), 70
OCU-Geo	1960, 63(4), 64, 68, 69(8), 70, 72, 76, 79, 80
ODaWU	1980
OU	1960, 63-64, 68, 69(8), 70, 72, 76, 79, 80
OkT	1968, 69(8), 70(10), 72, 76
OkTA	1980
OkU	1960, 63-64, 68, 69(8), 70, 72, 76, 79, 80
OrCS	1976, 79-80
OrU-S	1960, 63-64, 68, 69(8), 70, 72
PBL	1960, 70(11)
PBm	1980(20)
PSt	1960, 63-64, 68, 69(8), 70, 76, 79, 80(17)
RPB	1963
SdRM	1960, 64
TMM-E	1960, 63-64, 68, 69(8), 70(11), 72
TxBeaL	1960, 63-64, 68, 69(8), 70, 72, 76
TxCaW	1963-64, 68, 70, 72, 76, 79-80, 84
TxDaAR-T	1960, 63, 68, 69(8), 70, 72
TxDaGI	1963(4), 64, 69(8), 70(10)
TxDaM-SE	1963, 68, 80, 84
TxDaSM	1960, 63-64, 68, 69(8), 70(11), 72
TxHMP	1980
TxHSD	1960, 63-64, 68, 69(8), 70, 72, 79
TxHU	1960, 63-64, 68, 69(8), 70, 72
TxLT	1963, 69(8), 70(11), 72
TxMM	1963(5), 70(11), 80, 84
TxU	1960, 63-64, 68, 69(8), 70, 72, 76, 79-80, 84
TxU-Da	1960, 63-64, 68, 69(8), 70
TxWB	1960, 63-64, 68, 69(8), 70, 72, 76, 79-80, 84
ViBlbV	1960, 63-64, 68, 69(8), 70, 72, 76, 79
WU	1960, 63-64, 68, 69(8), 70, 72, 76, 79-80, 84
WaU	1960, 63, 64, 69-70, 72, 76, 79-80

(714) UNIVERSITY OF TEXAS AT AUSTIN. BUREAU OF ECONOMIC GEOLOGY. RESEARCH NOTE.

1977 7. See Uranium in Situ Symposium. 728.
— 10. See Gulf Coast Association of Geological Societies. 247.

(715) UNIVERSITY OF TEXAS AT AUSTIN. DEPARTMENT OF GEOLOGICAL SCIENCES. (FORMERLY: DEPARTMENT OF GEOLOGY) [GUIDEBOOK]

1937 No. 1. Lower Tertiary. Route: Austin, Manor, Elgin, McDade, Paige, Bastrop, Smithville, Bastrop, Austin.
No. 2. Lower Cretaceous. Route: Austin, Cedar Park, Leander, Travis Peak, Marshall Ford Dam, Austin, Buda, San Marcos, Austin.
No. 3. Llano uplift. Route: Austin, Fredericksburg, Llano, Burnet, Austin.
1939 Nov. See South Texas Geological Society. 612.
1967 Cooperative soils; geology field study [by] University of Texas, Department of Geology, USDA Soil Conservation Service and Texas A & M University Soil and Crop Sciences Department.
1982 Nov. Arbuckle-Ouachita field trip.
1983 See Southwestern Association of Student Geological Societies (SASGS). 626.

TxBeaL	1937(2)
TxDaGI	1937
TxU	1937, 67, 82

(716) UNIVERSITY OF TEXAS AT AUSTIN. UNIVERSITY STUDENT GEOLOGICAL SOCIETY GUIDEBOOK.

1939 See South Texas Geological Society. 612.
1949 See Fault Finders Geological Society. 171.
1950 Cretaceous field trip.
1951 Mar. Tertiary field trip.
— Apr. Geology of Pilot Knob.
1952 Igneous metamorphic field trip. [Central Texas]
1953 Tertiary field trip. [Bastrop and Fayette counties]
1954 Tertiary field trip.
1955 Tertiary field trip: structural geology, sedimentary structures, petrographic stratigraphy of the Texas Gulf Coast Tertiary, soils, vegetation. Contains road log of trip. [Bastrop and Fayette counties]
1956 [1] Paleozoic field trip; Cambrian stratigraphy, central Texas.
— Nov. Pre-Cambrian field trip; igneous and metamorphic rocks of central Texas.
1957 [Carboniferous strata of San Saba County]
1958 Apr. Pre-Cambrian field trip; igneous and metamorphic rocks of central Texas.
— Dec. Cretaceous field trip. [Travis County]
1959 Tertiary field trip; petrographic stratigraphy of the Texas Gulf Coast Tertiary. [Bastrop and Fayette counties]
1960 Feb. Igneous and metamorphic rocks of the southeastern part of the Llano Uplift, central Texas.
— Dec. Tertiary field trip. Same as field trip of Oct. 1955, except no conventional road log is included.
1961 May 13 Llano Uplift field trip.
1962 See Southwestern Association of Student Geological Societies (SASGS). 626.
1963 Nov. Interpretation of Eocene depositional environments, Little Brazos River Valley, Texas. (94)

(717) University of Texas at Dallas.

1974 Spring See Southwestern Association of Student Geological Societies (SASGS). 626.
1975 Spring See Southwestern Association of Student Geological Societies (SASGS). 626.
1976 Fall See Southwestern Association of Student Geological Societies (SASGS). 626.
1983 See Southwestern Association of Student Geological Societies (SASGS). 626.
    CLU-G/G   1951-53, 55-60
    DI-GS   1954-55, 56(Nov.), 58, 60(Dec.), 63
    IU   1955
    LNU   1961
    TxBeaL   1953
    TxDaAR-T   1963
    TxMM   1953
    TxU   1950-61, 63

(717) UNIVERSITY OF TEXAS AT DALLAS. PROGRAMS IN GEOSCIENCES.
1983 Structural styles of the Ouachita Mountains, southeastern Oklahoma.
    CLU-G/G   1983
    DI-GS   1983
    OkTA   1983
    OkU   1983
    TxDaM-SE   1983
    TxHMP   1983
    TxU   1983
    TxWB   1983
    WU   1983

(718) UNIVERSITY OF TEXAS AT EL PASO.
1982 Mar. See Symposium on the Paleoenvironmental Setting and Distribution of the Waulsortian Facies. 646.

(719) UNIVERSITY OF TEXAS AT EL PASO. SCIENCE SERIES.
1980 No. 7. El Paso's geologic past.
    DLC   1980
    NmPE   1980

UNIVERSITY OF WISCONSIN - SUPERIOR. SIGMA GAMMA EPSILON. See UNIVERSITY OF WISCONSIN. (723)

(720) UNIVERSITY OF WISCONSIN--EAU CLAIRE. DEPARTMENT OF GEOLOGY. GEOLOGICAL FIELD CONFERENCE. (TITLE VARIES)
1971 3rd Geology of the Eau Claire region.
— Oct. Wausau field trip.
1977 9th Eau Claire [area].
1980 See Institute on Lake Superior Geology. 273.
    LNU   1971(Oct.)
    WU   1971(3rd), 77

(721) UNIVERSITY OF WISCONSIN--MADISON. DEPARTMENT OF GEOLOGY AND GEOPHYSICS. GUIDEBOOK FOR ANNUAL SPRING FIELD TRIP.
1946 1st Southern Missouri, Arkansas, and eastern Oklahoma.
1947 2nd Rocky Mountain Front Range.
1948 3rd Southern Appalachians.
1951 Field trip to the Lake Superior region.
1979 Guidebook Middle and Upper Devonian Clastics, central and western New York state.
    CSt-ES   1951
    TxBeaL   1948
    WU   1946-48

(722) UNIVERSITY OF WISCONSIN--MADISON. DEPARTMENT OF GEOLOGY AND GEOPHYSICS. GUIDEBOOK: FIELD TRIP IN ECONOMIC GEOLOGY.
1970 Southern Illinois and southeastern Missouri.
1971 Central and eastern Tennessee; western North Carolina.
1972 The Colorado mineral belt.
1973 Mineral deposits of northern Ontario and northwestern Quebec.
1974 Mineral deposits of the central Mississippi Valley.
1977 Economic geology field trip [Superior Province; Mesabi Iron Range, Duluth complex; Manitouwadge, Ont.; Michipicoten District, Ont.; Sudbury District, Ont.; Blind River area, Ont.; White Pine, Mich.; Marquette Iron Range, Mich.]
1979 [Southern Canadian Shield]
1981 May See International Proterozoic Symposium. 306.
    WU   1970-74, 77, 79

(723) UNIVERSITY OF WISCONSIN. ANNUAL WISCONSIN STATE UNIVERSITY FIELD CONFERENCE. (CONFERENCES ARE HELD ON A ROTATING BASIS BY THE PARTICIPATING UNIVERSITIES WITHIN THE STATE UNIVERSITY NETWORK. TITLE VARIES)
1969 [1st] Superior.
1970 2nd UW-Platteville.
    The Upper Mississippi Valley zinc lead district.
1975 7th UW-Platteville.
    The Upper Mississippi Valley zinc lead district.
1978 10th UW-River Falls
    Geology of the St. Croix Valley, Wisconsin and Minnesota. (548)
1981 13th UW-Platteville.
    Field trip guidebook for the Ordovician geology of southwestern Wisconsin.
1983 15th UW-Parkside.
    Mid-Paleozoic and Quaternary geology of southeastern Wisconsin.
    MiHM   1981
    WPlaU   1970, 75, 81
    WU   1983

(724) UNIVERSITY OF WYOMING. GEOLOGY CLUB FIELD TRIP. GUIDEBOOK.
1976 6th Geologic road log, Boise to Challis via Idaho 21 and U.S. 93.
1977 Apr. Guidebook to the Bighorn Basin fieldtrip.
— Sept. Guidebook to the Wyoming Thrustbelt.
1978 Apr. Guidebook to the Black Hills.
— Sept. Guidebook to Northern Colorado.
1979 Apr. Guidebook to the western Wind River Basin.
— Sept. Guidebook to Yellowstone National Park.
1981 Sept. Wyoming overthrust belt.
    CoDCh   1977(Sept.), 78(Sept.)
    IdU   1976
    WyU   1977, 78, 79, 81

(725) UNIVERSITY OF WYOMING. DEPARTMENT OF GEOLOGY AND GEOPHYSICS. GEOLOGY 301G HONORS SECTION. FIELD TRIP GUIDE. [TITLE VARIES]
1967 Road logs [Laramie - Medicine Bow]
1968 Road logs [Laramie - Newcastle]
1969 Bighorn Basin, Wyoming.
1970 Bighorn Basin, Wyoming.
1971 Canyonlands area of the Colorado Plateau, Grand and San Juan counties, Utah.
1973 Uinkaret Plateau of the western Grand Canyon District.
1974 The Great Basin/Colorado Plateau.
    WyU   1967-71, 73, 74

(726) UNIVERSITY OF WYOMING. DEPARTMENT OF GEOLOGY. CONTRIBUTIONS TO GEOLOGY. (TITLE VARIES)
1969 V. 8:2. See Society of Economic Geologists. 583.
1971 V. 10:1. See Society of Economic Geologists. 583.
1972 V. 11:2. See Geological Society of America. Rocky Mountain Section. 204.

(727) UNIVERSITY OF WYOMING. GEOMORPHOLOGY SEMINAR.
1985 A Wyoming geologic tour; Laramie via Lander to Jackson to Rock Springs.
    WyU   1985

(728) **URANIUM IN SITU SYMPOSIUM.**
1977 Sept. Guide to modern barrier environments of Mustang and north Padre islands and Jackson (Eocene) Barrier/Lagoon facies of the South Texas Uranium District. (714, 44)
CU-EART	1977
IaU	1977
InLP	1977

(729) **UTAH FIELD HOUSE OF NATURAL HISTORY.**
1948 No. 1. The Uinta Mountains and vicinity, a field guide to the geology.
CLU-G/G	1948
DI-GS	1948
InLP	1948
UU	1948

(730) **UTAH GEOLOGICAL AND MINERALOGICAL SURVEY. BULLETIN.**
1966 No. 80. See Geological Society of America. 197.
1969 No. 82. See Geological Society of America. Rocky Mountain Section. 204.
1982 No. 118. See Symposium on the Geology of Rocky Mountain Coal (ROMOCOAL). 644.

(731) **UTAH GEOLOGICAL AND MINERALOGICAL SURVEY. EARTH SCIENCE EDUCATION SERIES.**
1966 No. 1. Field guide to the geology of the Uinta Mountains and adjacent synclinal basins. (Revision of Reprint 43. 1954) (732)
CLU-G/G	1966
CU-EART	1966
CoDGS	1966
CoU	1966
DI-GS	1966
ICIU-S	1966
IaU	1966
InLP	1966
MiHM	1966
MoSW	1966
NvU	1966
OU	1966
UU	1966

(732) **UTAH GEOLOGICAL AND MINERALOGICAL SURVEY. REPRINT.**
1954 No. 43. The Uinta Mountains and vicinity; a field guide to the geology. (Reprint from Paper No. 1 of the Utah Field House of Natural History. 1948) (Revised in 1966 See Utah Geological and Mineralogical Survey. 731.)
CaBVaU	1954
CoDGS	1954
DI-GS	1954
InLP	1954
UU	1954

(733) **UTAH GEOLOGICAL AND MINERALOGICAL SURVEY. SPECIAL STUDIES.**
1983 59. See Geological Society of America. Rocky Mountain Section. 204.
— 60. See Geological Society of America. Rocky Mountain Section. 204.
— 61. See Geological Society of America. Rocky Mountain Section. 204.
— 62. See Geological Society of America. Rocky Mountain Section. 204.

(734) **UTAH GEOLOGICAL ASSOCIATION. FIELD CONFERENCE. GUIDEBOOK.**
1971 1st Environmental geology tour of the Wasatch Front; [road logs]. (735)
UGA field trip no.1. Farmington to Draper.
UGA field trip no.2. Nephi to Draper.
UGA field trip no.3. Brigham City to Farmington.
1972 Plateau-Basin and Range transition zone, central Utah. (735)
1973 Geology of the Milford area. (735)
1974 Sept. Energy resources of the Uinta Basin, Utah. Road logs: reference data bibliography. (735)
Trip no.1. Salt Lake City-Duchesne-Vernal.
Trip no.2. Vernal-Bonanza-Vernal.
Trip no.3. Vernal-Sunnyside-Salt Lake City.
1976 Sept. Geology of the Oquirrh Mountains and regional setting of the Bingham mining district, Utah. Road log and abstracts of talks. (735)
1977 See Wyoming Geological Association. 781.
1980 Henry Mountains symposium. (735(No.8))
1981 Central Wasatch geology. (735(No.9))
1982 Overthrust belt of Utah. (735(No.10))
1983 Sept. Energy resources and geologic overview of the Uinta Basin. (735(No.11))
1984 Geology of Northwest Utah, southern Idaho and Northeast Nevada. (735(No.13))
1985 Geology and energy resources, Uinta Basin of Utah. (735)
Orogenic patterns and stratigraphy of north-central Utah and southeastern Idaho. (735(No.12))

AzTeS	1972
AzU	1971-74
CLU-G/G	1971-74, 76, 80-85
CLhC	1982, 83
CMenUG	1980-85
CPT	1972-74, 76, 80-85
CSdS	1972
CSt-ES	1971-73, 76, 80-85
CU-A	1971-73, 76
CU-EART	1971-74, 76, 80, 82-85
CU-SB	1976, 80-82, 84
CU-SC	1980
CaACU	1972, 74, 76, 82
CaBVaU	1972-74, 76, 80, 82, 84-85
CaOOG	1980-82, 84-85
CaOWtU	1971, 76
CoDCh	1971-74, 82, 84-85
CoDGS	1971-74, 76, 80-85
CoG	1971-74, 80, 82
CoLiM	1980
CoU	1971-73, 76
CoU-DA	1980, 84
DI-GS	1971-74, 76, 81-85
DLC	1971
ICarbS	1971-74, 76, 80-85
IU	1971-74, 76
IaU	1971-74, 76, 80-85
IdU	1971-74, 76, 80-85
InLP	1971-74, 80, 83-85
InU	1972, 74
KyU-Ge	1971-74, 76, 80-85
MH-GS	1980-83, 85
MU	1980, 82
MnU	1971-74, 76, 80-85
NNC	1980, 81, 85
NSbSU	1980, 81, 83-85
NSyU	1972, 76
NcU	1971, 80-82
NdU	1971-74
NhD-K	1971-74, 76, 80, 81
NmU	1980
NvU	1971-74, 76, 81-85
OCU-Geo	1971-74, 76, 80-85
OKentC	1980
OU	1971-74, 80-82
OkTA	1980
OkU	1972-74, 76, 81, 83, 85
PSt	1971-74, 76, 85
TxBeaL	1971
TxDaAR-T	1971-73
TxDaDM	1974, 82
TxDaM-SE	1980-82, 84
TxDaSM	1972
TxHSD	1972
TxMM	1972, 80-82
TxU	1971-74, 76, 80-84
UU	1971-74, 76, 80-85
ViBlbV	1971-74, 76
WU	1972-74, 76, 80-85
WyU	1980-85

(735) **UTAH GEOLOGICAL ASSOCIATION. PUBLICATION.**
1971 No. 1. See Utah Geological Association. 734.

(736) Utah Geological Society.

1972 No. 2.   See Utah Geological Association. 734.
1973 No. 3.   See Utah Geological Association. 734.
1974 No. 4.   See Utah Geological Association. 734.
1976 No. 6.   See Utah Geological Association. 734.
1978 No. 7.   See International Association on the Genesis of Ore Deposits. 280.
1980 No. 8.   See Utah Geological Association. 734.
1981 No. 9.   See Utah Geological Association. 734.
1982 No. 10. See Utah Geological Association. 734.
1983 No. 11. See Utah Geological Association. 734.
1984 No. 13. See Utah Geological Association. 734.
1985 No. 12. See Utah Geological Association. 734.

(736) UTAH GEOLOGICAL SOCIETY. FIELD CONFERENCE. ROAD LOG.
1955 See Utah Geological Society. 737.

(737) UTAH GEOLOGICAL SOCIETY. GUIDEBOOK TO THE GEOLOGY OF UTAH.
1946 1st   The geology and geography of the Henry Mountain region.
1947 2nd   Some structural features of the intrusions in the Iron Springs District. (No road log)
1948 3rd   Geology of the Utah-Colorado salt dome region with emphasis on Gypsum Valley, Colorado.
1949 4th   The transition between the Colorado Plateau and the Great Basin in central Utah.
1950 5th   See Intermountain Association of (Petroleum) Geologists. 276.
1951 6th   See Intermountain Association of (Petroleum) Geologists. 276.
1952 7th   See Intermountain Association of (Petroleum) Geologists. 276.
1953 8th   Geology of the central Wasatch Mountains.
1954 9th   Uranium deposits and general geology of southeastern Utah.
1955 10th Tertiary and Quaternary geology of the eastern Bonneville Basin. Utah Geological Survey field trip; Tertiary and Quaternary stratigraphic features of the Ogden Valley area, Utah. Supplemental road log.
1956 11th Geology of parts of northwestern Utah. (No road log)
1957 12th Geology of the East Tintic Mountains and ore deposits of the Tintic mining districts.
1958 13th Geology of the Stansbury Mountains, Tooele County, Utah.
1959 14th Geology of the southern Oquirrh Mountains and Five Mile Pass, northern Boulter Mountain area, Tooele and Utah counties, Utah.
1960 15th Geology of the Silver Island Mountains; Box Elder and Tooele Counties, Utah and Elko County, Nevada.
1961 16th Geology of the Bingham mining district and northern Oquirrh Mountains.
1963 17th Beryllium and uranium mineralization in western Juab County, Utah. (No road log)
1964 18th The Wasatch fault zone in north central Utah. (No road log)
1965 [19th] See Intermountain Association of (Petroleum) Geologists. 276.
1966 20th The Great Salt Lake. (No road log)
1967 21st Uranium districts of southeastern Utah.
1968 22nd [Geology of the] Park City [mining] district, Utah.
1970 23rd Western Grand Canyon district.
1972 See Virginia Geology Field Conference. 747.
1973 See Virginia Geology Field Conference. 747.
1978 See Virginia Geology Field Conference. 747.

AzFU	1953, 54, 58, 61, 63, 66, 68, 70
AzTeS	1948-49, 53-61, 63-64, 66-68, 70
AzU	1946-49, 53-61, 63-64, 66-68, 70
CLU-G/G	1946-49, 53-61, 63-64, 66-68, 70
CLhC	1947-49, 53-61, 63-64, 66-68, 70
CPT	1953-55, 57-61, 63-64, 66
CSdS	1948-49, 53-55, 58, 64, 70
CSt-ES	1946-49, 54-61, 67-68, 70
CU-A	1952, 61
CU-EART	1946-49, 53-61, 63-64, 66-68, 70
CU-SB	1947, 49, 53-61, 63-64, 66-68, 70
CaACU	1946, 49, 53-56, 58-61, 64, 66, 68, 70
CaAEU	1956
CaBVaU	1948-49, 53-61, 66-68, 70
CaOHM	1953, 55
CaOKQGS	1953, 54-61, 63-64, 66-68, 70
CaOLU	1954, 57, 61, 70
CaOOG	1946-49, 53, 56
CaOWtU	1955, 64, 66, 67, 70
CoDCh	1946, 56, 59-60, 66-68
CoDGS	1946-49, 53-64, 66-68, 70
CoDU	1948-49, 53-61, 63-64, 66-68, 70
CoFS	1948, 61
CoG	1946-49, 53-61, 63-64, 66-68, 70
CoU	1946-49, 53-61, 63-64, 66-68, 70
DI-GS	1946-49, 53-61, 63-64, 66-68, 70
DLC	1946-49, 53-61, 70
ICF	1946, 48-49, 53-61, 63-64, 67, 70
ICIU-S	1953-61, 63-64, 66-68, 70
ICarbS	1948-49, 53-61, 63-64, 66-68, 70
IEN	1946-49, 53-61, 63-64, 66-68, 70
IU	1946-49, 53-61, 63-64, 66, 68
IaU	1946-49, 53-61, 63-64, 66-68, 70
IdBB	1953-59, 63, 66-67
IdPI	1956-57, 60
IdU	1946-49, 53-61, 63-64, 66-68
InLP	1947-49, 53-61, 63-64, 66-68, 70
InRE	1946-49, 53-56, 66, 70
InU	1946-49, 53-61, 63-64, 66-68, 70
KyU-Ge	1946-49, 53-61, 63-64, 66-68, 70
LU	1948-49, 53-61, 63-64, 66-68, 70
MH-GS	1946-48, 54, 57, 66
MNS	1949, 53, 58, 64
MiDW-S	1946, 48-49, 53-61, 63-64, 66-68, 70
MiHM	1946-48, 54, 61, 66, 68
MiKW	1953-61, 63-64, 66-68, 70
MiU	1948-49, 53-61, 63-64, 66-68, 70
MnU	1946-49, 53-61, 63-64, 66-68, 70
MoRM	1946-49, 53-59, 61, 63-64, 66-68, 70
MoSW	1947-49, 54-61, 63-64, 66-68, 70
MoU	1946-49, 53-61, 63-64, 66-68, 70
MtBuM	1946-49, 53-58, 61
NBiSU	1953-61
NNC	1946-49, 53-61, 63-64, 66-68, 70
NOneoU	1970
NSyU	1948-49, 53-61, 63-64, 66-68
NbU	1946-49, 53-61, 63-64, 66-68, 70
NdU	1946-48, 53-55, 57-61, 63-64, 66-68, 70
NhD-K	1946-49, 53-61, 63-64, 66-68, 70
NjP	1946-49, 53-61, 63-64, 66-68, 70
NvLN	1967
NvU	1947-49, 53-61, 63-64, 66-68, 70
OCU-Geo	1946-48, 53-61, 63-64, 66-68, 70
OU	1946-49, 53-61, 63-64, 66-68, 70
OkT	1946, 48-49, 53-60, 64
OkU	1946-49, 53-61, 63-64, 66-68, 70
OrCS	1948-49, 53-61, 63-64, 66-68, 70
OrU-S	1946-49, 53-61, 63-64, 66-68, 70
PBL	1946-49
PBm	1946-47, 61, 70
PSt	1946-49, 53-61, 63-64, 66-68, 70
RPB	1946-49, 53-61, 63-64, 66-67
SdRM	1946-49, 53-61, 63-64, 66
TMM-E	1970
TU	1966
TxBeaL	1970
TxDaAR-T	1946, 48-49, 53-61, 63-64, 67-68, 70
TxDaDM	1946, 48-49, 53-61, 63-64, 66-67, 70
TxDaGI	1946-49, 53, 56, 66-67
TxDaM-SE	1946-49, 53-61, 63-64, 66-67
TxDaSM	1948-49, 53-57, 59, 61, 67, 70
TxHSD	1947-49, 53-61, 63, 67
TxHU	1946-49, 53-61, 64, 66-68, 70
TxMM	1948-49, 54
TxU	1946-49, 53-61, 63-64, 66-68, 70
TxU-Da	1953
TxWB	1948, 54-56, 58-59, 63-64
UPB	1946-49, 53-61, 63-64, 66-68
UU	1946-49, 53-61, 63-64, 66-68, 70
ViBlbV	1946-49, 53-61, 63-64, 66-68, 70
WU	1946-49, 53-61, 63-64, 66-68, 70
WaU	1948-49, 53-61, 63-64, 66-68
WyU	1946-49, 53-61, 63-64, 66-68

(738) VERMONT GEOLOGICAL SOCIETY. GUIDEBOOK.
1985 Guidebook [No. 1] See Vermont Geology. 739.

(739) **VERMONT GEOLOGY. GUIDEBOOK SERIES.**
1985 Guidebook [No.] 1. "The first of several such volumes which the Vermont Geological Society plans to publish in the 'Vermont Geology' series, contains guides of field trips the Society has sponsored in its 11 year history." - Foreword. (738)
A1-A31. Thetford Mines area, P.Q.
B1-B9. Geology of the Guildford Dome area, Brattleboro Quadrangle, southeastern Vermont.
D1-D14. The Crown Point section, New York.
E1-E21. The Cambrian platform in northwestern Vermont.

CSt-ES	1985
IaU	1985
KyU-Ge	1985
MH-GS	1985
OkU	1985
PSt	1985

(740) **VIRGINIA ACADEMY OF SCIENCE. GEOLOGY SECTION. FIELD TRIP GUIDEBOOK.**
1953 [Log only]
1956 Pleistocene terraces south of the James River, Virginia.
1958 [Log only]
1959 [Log only]
1960 [Log only]
1962 40th Norfolk meeting.
Trip A. Tour of the Langeley Research Center of the National Aeronautics and Space Administration.
Trip B. Tour of the U.S. Naval Station, Norfolk, Virginia.
Trip C. Tour of the Norfolk Botanical Gardens.
Trip D. Tour of the Chesapeake Bay Bridge-Tunnel Project.
Part 1. Field trip log.
Part 2. Pleistocene record in the subsurface of the Norfolk area, Virginia.
Part 3. Fact file on the Chesapeake Bay Bridge Tunnel.
Trip E. Excursion to Seashore State Park.
Trip F. Excursion to the Back Bay National Wildlife Refuge.
1965 [Log only]
1968 Structure and Paleozoic history of the Salem Synclinorium, southwestern Virginia. (749)
1972 See Virginia Geology Field Conference. 747.
1973 See Virginia Geology Field Conference. 747.
1978 See Virginia Geology Field Conference. 747.

CaACU	1962
DI-GS	1953, 56, 58-60, 62, 65, 68
I-GS	1962
IEN	1968
IU	1962
InLP	1968
KyU-Ge	1968
MH-GS	1962
NcU	1962, 68
NjP	1968
ViBlbV	1956, 60, 62, 68
WU	1968

(741) **VIRGINIA DIVISION OF MINERAL RESOURCES. BULLETIN.**
1976 No. 86. Geology of the Shenandoah National Park, Virginia. Appendix: Roadlog along Skyline Drive.

CPT	1976
CSt-ES	1976
CU-EART	1976
CaACU	1976
CaOWtU	1976
CoDGS	1976
CoU	1976
DI-GS	1976
ICF	1976
ICarbS	1976
IU	1976
IaU	1976
InLP	1976
InU	1976
MiHM	1976
MnU	1976
MoSW	1976
MtBuM	1976
NcU	1976
NhD-K	1976
NvU	1976
OU	1976
TxU	1976
ViBlbV	1976
WU	1976

(742) **VIRGINIA DIVISION OF MINERAL RESOURCES. GUIDE LEAFLET.**
1938 No. 1. See Field Conference of Pennsylvania Geologists. 172.

(743) **VIRGINIA DIVISION OF MINERAL RESOURCES. INFORMATION CIRCULAR.**
1962 No. 6. See Atlantic Coastal Plain Geological Association. 89.
1971 No. 16. See Geological Society of America. 197.

(744) **VIRGINIA DIVISION OF MINERAL RESOURCES. PUBLICATION.**
1977 No. 2. Geology of the Blairs, Mount Hermon, Danville, and Ringgold quadrangles, Virginia.
—No. 3. Geology of the Waynesboro East and Waynesboro West quadrangles, Virginia.
—No. 4. Geology of the Greenfield and Sherando quadrangles, Virginia.
1980 No. 23. Geologic structure and hydrocarbon potential along the Saltville and Pulaski thrusts in southwestern Virginia and northeastern Tennessee.
Part A. Regional structure and hydrocarbon potential.
Part B. Deformation in the hanging wall of the Pulaski thrust sheet near Ironto, Montgomery County, Virginia.
Part C. Saltville Fault footwall structure at Stone Mountain, Hawkins County, Tennessee.
1981 No. 29. See American Institute of Mining, Metallurgical, and Petroleum Engineers. Virginia Section. 48.

AzU	1977(2-4)
CLU-G/G	1977, 80
CMenUG	1980
CPT	1977, 80
CSfCSM	1980
CSt-ES	1977(2-4), 80
CU-EART	1977(2-4), 80
CaACU	1980
CaBVaU	1977(2-4)
CaOOG	1977(2-4)
DI-GS	1980
IU	1977(2-4), 80
IaU	1977(2-4), 80
InLP	1977(2-4), 80
InU	1977(2-4)
KyU-Ge	1977(2-4)
MdBJ	1977
MiHM	1977(2, 4)
MnU	1977(2-4), 80
NNC	1980
NSbSU	1980
NhD-K	1977(2-4), 80
OU	1977(2-4), 80
OkU	1980
PHarER-T	1980
PSt	1977(2-4)
ScU	1977(2)
SdRM	1977
TxU	1977(2), 80
ViBlbV	1977(2-4)
WU	1977(2-4), 80

(745) **VIRGINIA DIVISION OF MINERAL RESOURCES. REPORT OF INVESTIGATIONS. (TITLE MODIFIED WITH THIS EDITION.)**
1974 No. 35. Geology of Woodstock, Wolf Gap, Conicville and Edinburg quadrangles, Virginia.
1975 No. 40. Geology of the Front Royal quadrangle, Virginia.

CPT	1974-75
CSt-ES	1974-75
CU-EART	1974-75
CaBVaU	1974
CoDGS	1974-75
DI-GS	1974-75
ICF	1974-75
ICarbS	1974-75
IU	1974-75
InLP	1974-75
MdBJ	1974-75
MoSW	1974-75
MtBuM	1974-75
NcU	1974-75
NhD-K	1974-75
OCU-Geo	1974-75
OkT	1974-75
SdRM	1974-75
ViBlbV	1974-75
WU	1974-75

(746) VIRGINIA DIVISION OF MINERAL RESOURCES. VIRGINIA MINERALS. ROAD LOGS. (TITLE VARIES)

1969 Road logs: Staunton, Churchville, Greenville, and Stuarts Draft quadrangles.
1970 Road log of storm-damaged areas in central Virginia.
1971 Road log of the geology of Frederick County, Virginia.
1973 See Virginia Geology Field Conference. 747.
1975 [1] Road log to some abandoned gold mines of the gold-pyrite belt, northeastern Virginia.
 [2] Road log of the geology from Madison to Cumberland counties in the Piedmont, central Virginia.
1976 Road log of the geology in the northern Appalachian Valley of Virginia.
1980 26. No. 1. Road log to the geology of the Abingdon and Shady Valley quadrangles.
1981 27. No. 3. Field guide to selected Paleozoic rocks, Valley-Ridge Province, Virginia.
1982 28. No. 3. Geology and mineral resources of the Farmville Triassic Basin, Virginia.

CSt-ES	1980-81
CU-EART	1969-71, 75-76, 80-82
CaOWtU	1969-71, 75-76
CoDGS	1969-71, 75-76, 80-82
DI-GS	1969-71, 75-76, 81
ICF	1969-71, 75-76
IU	1969-71, 75-76
IaU	1981
InLP	1969-71, 75-76, 80-81
MoSW	1969-71, 75(1)
NNC	1981
NRU	1981
NcU	1969-71, 75-76
NdU	1975
NvU	1969-71, 75-76, 80-81
OU	1981
OkT	1981
OkU	1969-71, 75-76
PHarER-T	1981
TxDaM-SE	1981
ViBlbV	1969-71, 75-76
WU	1969-71, 75-76, 81
WyU	1981

(747) VIRGINIA GEOLOGY FIELD CONFERENCE. GUIDEBOOK.

1969 1st See Atlantic Coastal Plain Geological Association. 89.
1970 2nd Central Valley and Ridge.
1971 3rd Triassic Basin-Culpeper.
1972 Geologic features of the Bristol and Wallace quadrangles, Washington County, Virginia and anatomy of the Lower Mississippian Delta in southwestern Virginia. (168)
1973 5th Field trip across the Blue Ridge Anticlinorium, Smith River Allochthon, and Sauratown Mountains Anticlinorium near Martinsville, Virginia. (740, 746)
1974 6th See Atlantic Coastal Plain Geological Association. 89.
1975 7th Geology of Southwest Virginia coal fields and adjacent areas.
1976 8th Weathering processes and natural hazards in the Blue Ridge and Piedmont, northwestern Virginia.
1977 9th Geology of Little North Mountain and the central Shenandoah Valley.
1978 10th The faulted coastal plain margin at Fredericksburg, Virginia. (740)
1979 11th Nature of thrusting along the Allegheny Front near Pearisburg and of overthrusting in the Blacksburg-Radford area of Virginia. (749)
1980 12th See Atlantic Coastal Plain Geological Association. 89.
1981 13th Stratigraphic relationships between rocks of the Blue Ridge Anticlinorium and the Smith River Allochthon in the southwestern Virginia Piedmont.
1982 14th Regional Geology along the Leesburg Turnpike; history and application.
1983 15th Sedimentology, diagenesis and stratigraphy of Pleistocene coastal deposits in southeastern Virginia.
1984 16th Stratigraphy and structure in the thermal springs area of the western anticlines.
1985 17th Day One. Geology of the crystalline portion of the Richmond #DG1 x #DG2 quadrangle; a progress report.
Day Two. Field guide to the metamorphic stratigraphy of western Hanover County, Virginia.

CLhC	1975
CSt-ES	1971, 76-78, 82-83
CU-EART	1973, 75, 79
CU-SB	1970
CaACU	1975
CaOWtU	1973
CoDGS	1973
DI-GS	1972, 75-78, 81-84
ICF	1973
IU	1972-73, 75-789
IaU	1971-73, 75-78
InLP	1971, 75-84
KyU-Gc	1973, 75-79, 81, 82
MnU	1970-71, 73, 75-79, 81-82, 84
MoSW	1973
NcU	1969, 72, 73, 75-77
NvU	1973
OCU-Geo	1971, 73, 75-77
OkU	1973
TxBeaL	1973
TxDaAR-T	1971, 73
TxDaDM	1975
TxU	1975-77
TxWB	1981
VDMR	1985
Vi-MR	1985
ViBlbV	1972, 73, 75-79
WU	1971, 73, 75-79, 80-82

(748) VIRGINIA INSTITUTE OF MARINE SCIENCE. SPECIAL REPORT IN APPLIED MARINE SCIENCE AND OCEAN ENGINEERING (SRAMSOE).

1977 No. 143. See Society of Economic Paleontologists and Mineralogists. Eastern Section. 586.

(749) VIRGINIA POLYTECHNIC INSTITUTE AND STATE UNIVERSITY. DEPARTMENT OF GEOLOGICAL SCIENCES. GEOLOGICAL GUIDEBOOK.

1961 No. 1. See Geological Society of America. 197.
1963 No. 2. See Geological Society of America. Southeastern Section. 206.
1968 No. 3. See Virginia Academy of Science. Geology Section. 740.
1969 No. 4. See Seismological Society of America. Southeastern Section. 563.
1971 No. 5. See Geological Society of America. Southeastern Section. 206.
— No. 6. See Geological Society of America. 197.
1979 No. 7. See Geological Society of America. Southeastern Section. 206.
— No. 8. See Virginia Geology Field Conference. 747.

VIRGINIA POLYTECHNIC INSTITUTE, BLACKSBURG. ENGINEERING EXTENSION DIVISION. ENGINEERING EXTENSION SERIES. See VIRGINIA POLYTECHNIC INSTITUTE AND STATE UNIVERSITY. DEPARTMENT OF GEOLOGICAL SCIENCES. (749)

(750) VIRGINIA. DEPARTMENT OF EDUCATION. DIVISION OF SCIENCES AND ELEMENTARY ADMINISTRATION. [FIELD GUIDE]
1981 June   Earth science field guide; region I.
    CoDGS   1981
    DI-GS   1981
    IaU   1981
    WU   1981

VIRGINIA. DIVISION OF MINERAL RESOURCES. See VIRGINIA DIVISION OF MINERAL RESOURCES. (743) and VIRGINIA DIVISION OF MINERAL RESOURCES. (742) and VIRGINIA DIVISION OF MINERAL RESOURCES. (744) and VIRGINIA DIVISION OF MINERAL RESOURCES. (741) and VIRGINIA DIVISION OF MINERAL RESOURCES. (745) and VIRGINIA DIVISION OF MINERAL RESOURCES. (746)

VIRGINIA. GEOLOGICAL SURVEY. See VIRGINIA DIVISION OF MINERAL RESOURCES. (743) and VIRGINIA DIVISION OF MINERAL RESOURCES. (741) and VIRGINIA DIVISION OF MINERAL RESOURCES. (742) and VIRGINIA DIVISION OF MINERAL RESOURCES. (744) and VIRGINIA DIVISION OF MINERAL RESOURCES. (745) and VIRGINIA DIVISION OF MINERAL RESOURCES. (746)

WASHINGTON (STATE). DIVISION OF MINES AND GEOLOGY. See WASHINGTON. DIVISION OF GEOLOGY AND EARTH RESOURCES. (752) and WASHINGTON. DIVISION OF GEOLOGY AND EARTH RESOURCES. (753)

(751) WASHINGTON GEOLOGICAL SOCIETY.
1977   See American Association of Petroleum Geologists. 18.

(752) WASHINGTON. DIVISION OF GEOLOGY AND EARTH RESOURCES. BULLETIN.
1970 No. 61.   See Society of Economic Geologists. 583.
1980 No. 72.   Washington coastal geology between the Hoh and Quillayute rivers. Part 1. Rock formations, geologic processes, and events. Part 2. Geologic observations and interpretations along the coast. A review of geologic processes and events as revealed by the rock formations and deposits of the Washington coast, with historical notes and hiking information.
    CPT   1980
    CSt-ES   1980
    CU-EART   1980
    DLC   1980
    SdRM   1980
    WaU   1980

(753) WASHINGTON. DIVISION OF GEOLOGY AND EARTH RESOURCES. INFORMATION CIRCULAR. (TITLE VARIES)
1963 No. 38.   A geologic trip along Snoqualmie, Swauk, and Stevens Pass highways.
1975 No. 54.   A geologic road log over Chinook, White Pass, and Ellensburg to Yakima highways.
    CLU-G/G   1963, 75
    CPT   1963, 75
    CSt-ES   1963, 75
    CU-A   1975
    CU-EART   1963, 75
    CaACU   1963
    CaOOG   1963
    CaOWtU   1975
    CoDGS   1963, 75
    CoU   1975
    DI-GS   1963, 75
    DLC   1975
    ICF   1963, 75
    ICarbS   1963
    IEN   1963
    IU   1963, 75
    IaU   1963, 75
    IdU   1963, 75
    InLP   1963, 75
    InU   1963, 75
    KyU-Ge   1963, 75
    MnU   1963, 75
    MoSW   1963, 75
    MoU   1963
    MtBuM   1963, 75
    NcU   1963, 75
    NdU   1963, 75
    NhD-K   1963, 75
    NvU   1963
    OCU-Geo   1975
    OU   1975
    PBL   1975
    SdRM   1963, 75
    TxU   1963
    UU   1975
    ViBlbV   1963, 75
    WU   1963, 75
    WaPS   1975

(754) WAYNE STATE UNIVERSITY. [GEOLOGY DEPARTMENT] SUMMER CONFERENCE. [GUIDEBOOK]
1963   Structures and origin of volcanic rocks, Montana-Wyoming-Idaho. (A summer conference sponsored by Wayne State University through a grant from the National Science Foundation)
1964 June   Structures and origin of volcanic rocks, Montana-Wyoming-Idaho. (A summer conference sponsored by Wayne State University through a grant from the National Science Foundation)
1968   Structures and origin of volcanic rocks, Montana-Wyoming-Idaho. (A summer conference sponsored by Wayne State University through a grant from the National Science Foundation)
1970   Structures and origin of volcanic rocks, Montana-Wyoming-Idaho.
1973 July   Structures and origin of volcanic rocks, Montana-Wyoming-Idaho. (A summer field course sponsored by Wayne State University through a grant from the National Science Foundation)
    CU-EART   1964
    CoU   1963
    DI-GS   1968
    IEN   1963
    IdBB   1963, 68
    InU   1963, 70
    LNU   1963
    MiHM   1973
    MoSW   1968
    MtBuM   1963
    NdU   1963
    NjP   1968
    NvU   1963, 68
    OCU-Geo   1968
    OkU   1968
    TxCaW   1970
    TxU   1963

(755) WEST SLOPE INTERCOLLEGIATE GEOLOGY FIELD CONFERENCE. GEOLOGIC GUIDES AND ROAD LOGS.
1982 Sept.   Boise, Idaho.
[Pt. 1] Boise to Lowman via Idaho State Highway 21.
[Pt. 2] Lowman to Galena Summit via Stanley.
[Pt. 3] Craters of the Moon National Monument.
    IdBB   1982

(756) WEST TEXAS GEOLOGICAL SOCIETY. GUIDEBOOK FOR THE FIELD TRIP. (TITLE VARIES)
1933   San Saba to Mason, central Texas mineral region; road log. Contains composite section showing formations of Ellenberger Group.
1936   Marathon Basin area.
1937   Hueco and Franklin Mountains.
1938 Spring   East side outcrops (of Permian Basin).
— Sept.   See American Association of Petroleum Geologists. Regional Meeting. 29.
1939 Spring   Van Horn area, Culberson and Hudspeth counties, Texas; road log.
— Fall   See Texas Academy of Science. 654.
1940 Spring   Sacramento Mountains, New Mexico.

(756) West Texas Geological Society.

— Fall Eddy County, New Mexico.
1941 Spring Fort Worth to Midland, Texas on Highway 80 along the Texas and Pacific. Additional notes on El Reno and Whitehorse Group.
— Fall Big Bend Park area, Brewster County, Texas.
1942 Fall El Paso to Carlsbad along U.S. Highway 62.
1946 Spring Stratigraphy of the Hueco and Franklin Mountains.
— Fall Glass Mountain-Marathon Basin.
1947 Guadalupe Mountains of New Mexico and Texas.
1948 Green Valley and Paradise Valley; Wire Gap and Solitario; Limpia Canyon and Barrilla Mountains.
1949 Nov. See Geological Society of America. 197.
1950 Sierra Blanca region, Franklin Mountains, Texas.
1951 Spring Pennsylvanian of Brazos River and Colorado River valleys, north-central Texas.
— Fall Apache Mountains of Trans-Pecos Texas.
1952 Spring Marathon Basin, Brewster and Pecos counties, Trans-Pecos Texas.
1953 Spring Chinati Mountains, Presidio County, Texas.
— Fall Sierra Diablo, Guadalupe and Hueco areas of Trans-Pecos Texas.
1954 Mar. Cambrian field trip--Llano area. (554)
1955 Spring Big Bend National Park, Texas.
1956 Spring Eastern Llano Estacado and adjoining Osage plains. (362)
1957 Fall Glass Mountains.
1958 Fall Franklin and Hueco Mountains.
1959 Nov. Geology of the Val Verde Basin.
1960 Sept. Geology of the Delaware Basin.
1961 Oct. Upper Permian to Pliocene; San Angelo area. (554)
1962 Permian of the central Guadalupe Mountains, Eddy County, New Mexico. (254, 550)
1964 Oct. Geology of Mina Plomosas-Placer de Guadalupe area, Chihuahua, Mexico.
1965 Oct. Geology of Big Bend area, Texas; with road log and papers on natural history of the area. (Publication 65-51)(Reprinted in 1972) (757)
1966 Geology of the Val Verde Basin. (Note: 1959 guidebook used, with supplement)
1967 No field trip.
1968 Oct./Nov. Delaware Basin exploration; Guadalupe Mountains, Hueco Mountains, Franklin Mountains, geology of the Carlsbad Caverns. (Publication 68-55) (757)
1969 Nov. Supplement to Delaware Basin exploration; Guadalupe Mts., Hueco Mts., Franklin Mts., Geology of the Carlsbad Caverns. (Publication 68-55a) (757)
1970 See Society of Economic Paleontologists and Mineralogists. Permian Basin Section. 594.
1971 Oct. Field trip guidebook and Memoir 24; Stratigraphy and structure of Pecos country, southeastern New Mexico. (Publication 71-58) (462, 550, 757)
1972 Nov. Geology of the Big Bend area, Texas. Field trip guidebook with road log and papers on natural history of the area. (Publication 72-59; Reprint of Publication 65-51 (October 1965), with additions) (757)
1974 Apr. Guidebook to the Mesozoic and Cenozoic geology of the southern Llano Estacado. (362)
— Oct. Geologic field trip guidebook through the states of Chihuahua and Sinaloa, Mexico. (Publication 74-63) (757)
— Oct./Nov. See Society of Economic Paleontologists and Mineralogists. Permian Basin Section. 594.
1975 Apr. See Society of Economic Paleontologists and Mineralogists. Permian Basin Section. 594.
— Oct./Nov. See Society of Economic Paleontologists and Mineralogists. Permian Basin Section. 594.
1977 Oct. Geology of the Sacramento Mountains, Otero County, New Mexico. (Publication no. 77-68) (757)
1979 Sept. Geology of Cancun, Quintana Roo, Mexico. (Publication 79-72) (757)
1980 Oct. Geology of the Llano region, central Texas. (757)
1981 Nov. Lower Cretaceous stratigraphy and structure, northern Mexico. (757)
1982 Oct. Delaware Basin field trip guidebook. (757)
1983 Oct. Structure and stratigraphy of the Val Verde Basin-Devils River uplift, Texas. (757)
1984 Oct. Geology and petroleum potential of Chihuahua, Mexico. (757)
1985 Oct. Structure and tectonics of Trans-Pecos Texas. (757)

ATuGS	1941, 46(Fall), 47, 50, 52-53, 56
AzFU	1955-56, 58-62, 64, 68, 71-72
AzTeS	1950-52, 55-57, 58-62, 64-65, 68, 71-72, 74(Oct.), 77, 79-85
AzU	1950, 51(Spring), 52-62, 64, 68-69, 71, 74(Oct.), 77, 79
BEG	1980-81, 83-85
CLU-G/G	1936, 39(Spring), 50-62, 64-65, 68-69, 71-72, 74(Oct.), 77, 79-85
CLhC	1946(Spring), 47-48, 50-62, 64, 68, 80-82, 85
CMenUG	1980-85
CPT	1950-52, 53(Spring), 54-56, 58-65, 68-69, 71-72, 74(Oct.), 77, 79-85
CSdS	1951-52, 53(Spring), 55, 64
CSt-ES	1950-62, 64-65, 69, 71-72, 74(Oct.), 77, 79-85
CU-A	1941(Spring), 46-47, 53(Fall), 57-62, 64-65, 68, 72, 74(Oct.), 77, 79-80, 83, 84
CU-EART	1951-62, 64-65, 68-69, 71-72, 74, 77, 79-85
CU-SB	1951, 54, 61, 62, 71
CaAC	1957, 59
CaACAM	1977
CaACM	1968
CaACU	1952, 54, 56-62, 64, 69, 71-72, 74, 77, 80, 83
CaAEU	1968
CaBVaU	1956-62, 64-65, 68, 83-84
CaOHM	1956-59, 61-62, 64, 66, 68-69, 71-72
CaOKQGS	1952, 56-62, 64-65, 68, 80-85
CaOLU	1955, 57, 60, 62, 64, 68
CaOOG	1950, 53(Spring), 59-62, 64-65, 68-69, 79-83
CaOTRM	1956-62
CaOWtU	1957, 61-62, 64, 68, 72, 77, 79
CoDCh	1954, 64, 81-84
CoDGS	1974, 80-85
CoDU	1962, 68, 71
CoFS	1956-59, 61-62, 64-65, 68
CoG	1940(Fall), 41, 47-48, 50-62, 64-65, 68, 71-72, 74, 77, 83
CoLiM	1980-82, 84-85
CoU	1950, 52-62, 64-66, 68-69, 71-72, 77
CoU-DA	1981-82
DI-GS	1940(Fall), 41, 46-48, 50-62, 64-65, 68-69, 71-72, 74, 77, 79-83, 85
DLC	1952-58, 60
I-GS	1950, 52-57, 59
ICF	1952, 55-62, 64
ICIU-S	1956, 58-62, 64, 68, 71-72, 74(Oct.), 77, 79
ICarbS	1940(Fall), 52, 53(Fall), 57, 85
IEN	1950-53, 56-59, 61-62, 64-65, 68, 71-72, 74(Oct.)
IU	1946(Spring), 47-48, 50-62, 64-66, 69, 71-72, 74(Apr.), 77, 79, 80
IaCfT	1980-85
IaU	1941(Fall), 52-53, 56-62, 64, 68, 71, 74(Oct.), 77, 79-85
IdPI	1972
IdU	1959, 80-83
InLP	1954-55, 57-59, 61-62, 64-65, 68, 71-72, 74(Oct.), 77, 79, 80-85
InRE	1952
InU	1950, 51(Fall), 52-62, 68, 71, 82
KLG	1980-85
KU	1980-85
KyU-Ge	1940(Fall), 41(Spring), 46(Spring), 47, 50-62, 64-65, 68-69, 71-72, 74(Oct.), 77, 80-85
LNU	1946(Spring), 56-59, 61-62, 64, 68, 72, 74(Oct.), 80-83
LU	1939(Spring), 50-62, 64-65, 68, 71
MH-GS	1980-83
MNS	1941, 53-54, 58
MdU	1980, 81
MiDW-S	1956-62, 64-66, 68, 71-72
MiHM	1952, 54-62, 64-65, 68, 71-72, 74(Oct.), 77, 79, 80
MiKW	1956-57
MiU	1956-62, 64-65, 68, 71-72, 74(Oct.)
MnU	1948, 50-51, 53-62, 64-66, 68-69, 71-72, 74(Oct.), 77, 79-85
MoRM	1948, 52-53, 56-58
MoSW	1946(Spring), 47-48, 52-59, 61-62, 64-65, 68
MoU	1950-62, 64, 68
MtBuM	1956
MtU	1950-52, 54, 56-62, 64
NBiSU	1956, 58-62, 64, 68
NNC	1954, 58-62, 64-66, 68-69, 71-72, 74, 77, 79, 80, 81, 83, 85
NOneoU	1965
NRU	1956-58, 61-62
NSbSU	1962
NSyU	1968, 79, 80, 81
NbU	1941(Spring), 47, 50-62, 64-65, 68, 71-72

NcU	1949, 52, 56-62, 64, 69, 71, 74(Oct.), 75, 77, 79-85
NdU	1952, 54, 56-62, 64-65, 68, 71-72
NhD-K	1952, 54, 56-57, 59-60, 62, 64, 68-69, 71-72, 74(Oct.), 77, 79, 80, 81
NjP	1951-56, 58-59, 61-62, 64-65, 68-69, 71-72
NmLcU	1982
NmPE	1956, 58, 60, 62, 64-65, 68, 71-72
NmU	1962, 71, 80-84
NvU	1941, 53-62, 64-65, 68-69, 71-72, 77, 79-81
OCU-Geo	1956, 58-62, 64, 68, 71-72, 74, 77, 79-85
ODaWU	1980, 81
OU	1950-62, 64-65, 68-69, 71-72, 74(Oct.), 77, 79, 80-83, 85
OkT	1940(Fall), 46-47, 50, 51(Spring), 52-54, 56, 59, 61-62, 64
OkTA	1981-85
OkU	1941, 46-48, 50-62, 64-66, 68-69, 71-72, 74(Oct.), 77, 79-85
OrCS	1952, 53(Fall), 54, 56-58, 60-62, 64-65, 69, 71-72, 74(Oct.), 77, 79-80
OrU-S	1952-62, 64-65, 68-69, 71-72, 80-84
PBL	1959, 62, 68
PBU	1980-85
PSt	1954, 56-62, 64-66, 68, 71-72, 74(Oct.), 77 85
SdRM	1959, 61-62, 64, 68
TU	1952-53, 55-60, 62, 64-65
TxBeaL	1951-62, 64-65, 68, 71-72, 77
TxCaW	1952, 56-62, 64-65, 68, 72, 74(Oct.), 77
TxDaAR-T	1937, 40(Fall), 41(Spring), 48, 52, 53(Spring), 54-62, 64-65, 68, 71-72
TxDaDM	1948, 52-62, 64-65, 68-69, 71-72
TxDaGI	1936, 41-42, 46-48, 51-54, 57, 59, 60-62, 65, 74(Oct.), 77
TxDaM-SE	1948, 52-62, 64-65, 68, 81-83
TxDaSM	1941(Spring), 47-48, 50-57, 59-62, 64-65, 68, 71, 74(Oct.)
TxHMP	1980-85
TxHSD	1941, 46-47, 51-62, 64-65, 71-72, 74(Oct.), 77, 79
TxHU	1940(Fall), 41, 48, 52, 53(Fall), 54-62, 64-65, 68-69, 71-72
TxLT	1938(Spring), 48, 50-62, 64-65, 68-69, 71-72, 74, 77
TxMM	1936-37, 38(Spring), 39(Spring), 40-42, 46-48, 50-62, 64-65, 68, 71-72, 74(Apr.), 77, 79-85
TxU	1933, 36-37, 38(Spring), 39(Spring), 40-42, 46-48, 50-62, 64-66, 68-69, 71-72, 74, 77, 79-85
TxU-Da	1946(Spring), 47-48, 50, 54-55, 57-58
TxWB	1939-40, 47, 49-55, 57-62, 66, 68-72, 74-75, 78-85
UU	1953(Spring), 56-62, 64-65
ViBlbV	1950-51, 54, 56-62, 64-65, 68-69, 71-72, 74(Oct.), 77, 79
WU	1947, 50-54, 56-62, 64-65, 68-69, 71-72, 77, 79-85
WyU	1941, 46-48, 50, 52-53, 55-62, 64-65, 68, 80-83

(757) WEST TEXAS GEOLOGICAL SOCIETY. PUBLICATION.
- 1965 No. 65-51. See West Texas Geological Society. 756.
- 1968 No. 68-55. See West Texas Geological Society. 756.
- 1969 No. 68-55a. See West Texas Geological Society. 756.
- 1971 No. 71-58. See West Texas Geological Society. 756.
- 1972 No. 72-59. See West Texas Geological Society. 756.
- 1974 No. 74-63. See West Texas Geological Society. 756.
- 1977 No. 77-68. See West Texas Geological Society. 756.
- 1979 No. 79-72. See West Texas Geological Society. 756.
- 1980 No. 80-73. See West Texas Geological Society. 756.
- 1981 No. 81-74. See West Texas Geological Society. 756.
- 1982 No. 82-76. See West Texas Geological Society. 756.
- 1983 No. 83-77. See West Texas Geological Society. 756.
- 1984 No. 84-80. See West Texas Geological Society. 756.
- 1985 No. 85-81. See West Texas Geological Society. 756.

(758) WEST TEXAS GEOLOGICAL SOCIETY. [SPECIAL PUBLICATION]

1958 Geological road log along U.S. Highways 90 and 80 between Del Rio and El Paso, Texas. (Cover title: Road log, Del Rio-El Paso)

AzU	1958
CSt-ES	1958
CU-A	1958
CU-EART	1958
CaACM	1958
CaACU	1958
CaBVaU	1958
CaOHM	1958
CaOKQGS	1958
CaOTRM	1958
CoFS	1958
CoG	1958
CoU	1958
DI-GS	1958
DLC	1958
IU	1958
IaU	1958
IdPI	1958
InLP	1958
InU	1958
KyU-Ge	1958
MNS	1958
MiHM	1958
MiU	1958
MnU	1958
MoU	1958
MtU	1958
NBiSU	1958
NRU	1958
NcU	1958
NdU	1958
NhD-K	1958
NmPE	1958
NvU	1958
OCU-Geo	1958
OU	1958
OkU	1958
OrCS	1958
TxBeaL	1958
TxCaW	1958
TxDaAR-T	1958
TxDaDM	1958
TxDaM-SE	1958
TxHSD	1958
TxHU	1958
TxLT	1958
TxU	1958
TxU-Da	1958
UU	1958
ViBlbV	1958
WU	1958
WyU	1958

WEST TEXAS STATE COLLEGE. GEOLOGICAL SOCIETY. See **WEST TEXAS STATE UNIVERSITY GEOLOGICAL SOCIETY. (759)**

(759) WEST TEXAS STATE UNIVERSITY GEOLOGICAL SOCIETY. FIELD TRIP GUIDEBOOK.
- 1959 Claude-Silverton field trip.
- 1960 2nd Palo Duro.
- 1961 Apr. Palo Duro.
- — Nov. Claude-Silverton. (Same as 1959, except has 6 stops instead of 8.)
- 1962 Palo Duro, 4th edition.
- 1964 Palo Duro, 5th edition.
- 1966 See Southwestern Association of Student Geological Societies (SASGS). 626.
- 1967 Palo Duro.
- 1971 Palo Duro.
- 1972 Palo Duro.
- 1973 Sedimentology of the upper Triassic sandstones of the Texas high plains. (760)
- 1975 Palo Duro Canyon.

CU-SB	1962, 71, 75
IU	1960, 61
NSbSU	1975
PSt	1971
TxCaW	1960, 62, 64, 67, 71-73, 75
TxDaAR-T	1973
TxDaSM	1973
TxHSD	1973
TxU	1959-62, 72

(760) WEST TEXAS STATE UNIVERSITY. DEPARTMENT OF GEOLOGY AND ANTHROPOLOGY.
- 1973 See West Texas State University Geological Society. 759.

(761) WEST TEXAS STATE UNIVERSITY. DEPARTMENT OF GEOLOGY AND ANTHROPOLOGY. SPECIAL PUBLICATION.
1977 No. 1.   See North American Paleontological Convention. 474.

(762) WEST VIRGINIA GEOLOGICAL AND ECONOMIC SURVEY. BULLETIN.
1979 37-2.   See American Association of Petroleum Geologists. Eastern Section. 25.

(763) WEST VIRGINIA GEOLOGICAL AND ECONOMIC SURVEY. [FIELD TRIP GUIDEBOOK]
1950 Sept.   Field guide for the special field conference on the stratigraphy, sedimentation and nomenclature of the Upper Pennsylvanian and Lower Permian strata (Monongahela, Washington, and Green series) in the northern portion of the Dunkard Basin of Ohio, West Virginia, and Pennsylvania.
1953   See Appalachian Geological Society. 58.
1955   See Appalachian Geological Society. 58.
1957   See Appalachian Geological Society. 58.
1958   Conservation of non-renewable resources, log of field trip.
1961   See Appalachian Geological Society. 58.
1963   See Association of American State Geologists. 74.
1964   See Appalachian Geological Society. 58.
1969   See Geological Society of America. 197.
1972   I.C. White memorial symposium field trip.
1974   See Natural Areas Conference. 443.

CSt-ES	1972
CU-EART	1972
CaAEU	1972
CaOKQGS	1972
CaOWtU	1972
DI-GS	1950, 58, 72
I-GS	1950, 72
ICF	1950, 72
IU	1950, 72
IaU	1950, 72
InLP	1972
InU	1950, 72
KyU-Ge	1972
NBiSU	1972
NSyU	1972
NbU	1972
NcU	1972
NhD-K	1972
OCU-Geo	1972
OU	1950, 72
OkU	1950, 72
PBm	1972
PHarER-T	1950
PSt	1972
UU	1972
WU	1972
WaPS	1972

(764) WEST VIRGINIA GEOLOGICAL AND ECONOMIC SURVEY. PUBLICATION MAP.
1985   Geology along I-79 Harrison County, West Virginia.

IU	1985
OCU-Geo	1985

(765) WEST VIRGINIA GEOLOGICAL AND ECONOMIC SURVEY. PUBLICATION WV. (TITLE VARIES)
1981 13.   Geology along I-79, Monongalia County, West Virginia.
— 14.   Geology along the West Virginia portion of U.S. Route 48.
— 15.   Geology along I-79, Marion County, West Virginia.
1985 18.   Geology along I-79, Harrison County, West Virginia.
— 19.   Geology along U.S. Route 33; Canfield to Bowden, Randolph County, West Virginia.

CoDGS	1981, 85
DI-GS	1981
IU	1985
OCU-Geo	1981

(766) WEST VIRGINIA GEOLOGICAL AND ECONOMIC SURVEY. STATE PARK SERIES BULLETIN.
1958 No. 6.   Blackwater Falls State Park and Canaan Valley State Park; resources, geology and recreation. Includes road log of "Little Switzerland Tour." Rev. 1958.
1971 No. 6.   Blackwater Falls State Park and Canaan Valley State Park; resources, geology and recreation, Rev. 1971.

CSt-ES	1958, 71
CU-EART	1971
CaOWtU	1971
CoDGS	1958, 71
DI-GS	1958
DLC	1971
ICarbS	1971
IU	1958
InLP	1958, 71
LNU	1958
MdBJ	1958, 71
MnU	1958
MoSW	1958
NhD-K	1958
OCU-Geo	1958, 71
OU	1958
OkU	1958
TxU	1958
UU	1958
ViBlbV	1958
WU	1958

(767) WEST VIRGINIA UNIVERSITY. DEPARTMENT OF GEOLOGY.
1957   See Geological Society of America. Southeastern Section. 206.
1979 Oct.   See American Association of Petroleum Geologists. Eastern Section. 25.

(768) WESTERN MICHIGAN UNIVERSITY. DEPARTMENT OF GEOLOGY. FIELD GUIDE.
1970   Studies in geology.
   No. 1. Coastal sedimentation of southeastern Lake Michigan; field trip guidebook.
1972   A field guide to the geology of southwestern Michigan. (Publication ES-1)
1973   Source book and field guide to the geology of the west-central lower peninsula of Michigan. (Publication ES-2)
1976 May   A field guide and sourcebook to geology of the eastern upper peninsula of Michigan intended primarily for secondary earth science teachers. (Publication ES-3)
See Geological Society of America. North Central Section. 200.

CU-A	1976
CU-EART	1976
CU-SB	1976
CaOLU	1972
CaOOG	1973
CaOWtU	1970
DI-GS	1970, 72-73, 76
IU	1972-73, 76
IaU	1976
InLP	1972-73, 76
InRE	1972
InU	1972, 76
KyU-Ge	1972, 73, 76
LNU	1972
MiHM	1972, 76
NRU	1976
NSbSU	1976
NcU	1976
NhD-K	1976
NjP	1972
OCU-Geo	1973, 76
OU	1970, 73
TMM-E	1972
TxDaSM	1970
TxU	1970
ViBlbV	1976

(769) WESTERN RESERVE GEOLOGICAL SOCIETY. ANNUAL FIELD TRIP. GUIDEBOOK.
    1953   Euclid Creek reservation, Bedford Glens.
    DI-GS   1953

(770) WESTERN SYNFUELS SYMPOSIUM. [FIELD TRIP GUIDEBOOK]
    1985 Apr.   See Oil Shale Symposium. 501.

(771) WICHITA FALLS GEOLOGICAL SOCIETY. FIELD TRIP.
    1946   Wichita Group-Red Bed phase, southeast Baylor County, southwest Archer County; non-red phase, Throckmorton County, Shackelford County.
    DI-GS   1946
    TxHSD   1946
    TxU   1946

WICHITA STATE UNIVERSITY. DEPARTMENT OF GEOLOGY

(772) WICHITA STRATIGRAPHIC SOCIETY. INFORMAL FIELD CONFERENCE.
    1936   Mississippi stratigraphy, southwest Missouri.
    IU   1936

(773) WILLIAM T. PECORA SYMPOSIUM. FIELD TRIP GUIDEBOOK.
    1979 5th   Remote sensing of hydrology in eastern South Dakota. (611)
    DI-GS   1979

(774) WINONA STATE UNIVERSITY

(775) WISCONSIN ACADEMY OF SCIENCE. GUIDEBOOK.
    1975   See Iowa Academy of Science. 330.

(776) WISCONSIN CLAY MINERALS COMMITTEE.
    1964   See Clay Minerals Conference. 132.

(777) WISCONSIN GEOLOGICAL AND NATURAL HISTORY SURVEY. FIELD TRIP GUIDEBOOK.
    1978 No. 1.   See Institute on Lake Superior Geology. 273.
    — No. 2.   See Institute on Lake Superior Geology. 273.
    — No. 3.   See Society of Economic Paleontologists and Mineralogists. Great Lakes Section. 587.
    1979 No. 4.   See Geological Society of America. North Central Section. 200.
    1982 No. 5.   See Friends of the Pleistocene. Midwest Group. 183.
    1983 No. 6.   A guide to the glacial landscapes of Dane County, Wisconsin.
    — No. 7.   See Geological Society of America. North Central Section. 200.
    — No. 8.   See Geological Society of America. North Central Section. 200.
    — No. 9.   See Tri-State Geological Field Conference. 659.
    1984 No. 10.   A voyageur's guide to the lower Wisconsin River.
    No. 11. See Friends of the Pleistocene. Midwest Group. 183.
    CSt-ES   1984
    CU-SB   1983
    CaOOG   1983
    DI-GS   1983-84
    IaU   1983-84
    InLP   1983-84
    KyU-Ge   1983-84
    MnU   1983-84
    MoSW   1983
    OkU   1983-84
    TxDaM-SE   1983-84
    WKenU   1983-84
    WU   1983-84

(778) WISCONSIN GEOLOGICAL AND NATURAL HISTORY SURVEY. INFORMATION CIRCULAR.
    1965 No. 6.   See Tri-State Geological Field Conference. 659.
    1966 No. 7.   See Michigan Basin Geological Society. 377.
    1970 No. 11.   See Geological Society of America. 197.
    — No. 13.   See Geological Society of America. 197.
    — No. 14.   See Geological Society of America. 197.
    — No. 15.   See Geological Society of America. 197.
    — No. 16.   See Geological Society of America. 197.
    — No. 17.   See Geological Society of America. 197.

WISCONSIN. UNIVERSITY - RIVER FALLS See UNIVERSITY OF WISCONSIN. (723)

WISCONSIN. UNIVERSITY - SUPERIOR. See UNIVERSITY OF WISCONSIN. (723)

WISCONSIN. UNIVERSITY--EAU CLAIRE. See UNIVERSITY OF WISCONSIN. (723)

WISCONSIN. UNIVERSITY--MADISON. See UNIVERSITY OF WISCONSIN--MADISON. DEPARTMENT OF GEOLOGY AND GEOPHYSICS. (721) and UNIVERSITY OF WISCONSIN--MADISON. DEPARTMENT OF GEOLOGY AND GEOPHYSICS. (722)

WISCONSIN. UNIVERSITY. See UNIVERSITY OF WISCONSIN. (723)

(780) WYOMING FIELD SCIENCE FOUNDATION.
    1974   A field guide to the Alcova area.
    1978   A field guide to the Casper Mountain area. (233)

(781) WYOMING GEOLOGICAL ASSOCIATION. ANNUAL FIELD CONFERENCE. GUIDEBOOK.
    1946 1st   Central and southeastern Wyoming.
    1947 2nd   Bighorn Basin. (784)
    1948 3rd   Wind River basin, Wyoming.
    1949 4th   Powder River basin.
    1950 5th   Southwest Wyoming.
    1951 6th   South-central Wyoming.
    1952 7th   Southern Bighorn Basin, Wyoming.
    1953 8th   Laramie Basin, Wyoming, and North Park, Colorado.
    1954 9th   Casper area, Wyoming.
    1955 10th   Green River basin.
    1956 11th   Jackson Hole.
    1957 12th   Southwest Wind River basin.
    1958 13th   Powder River basin.
    1959 14th   Bighorn Basin.
    1960 15th   Overthrust belt of southwestern Wyoming and adjacent areas.
    1961 16th   Symposium on Late Cretaceous rocks, Wyoming and adjacent areas.
    1962 17th   Symposium on Early Cretaceous rocks of Wyoming and adjacent areas.
    1963 18th   Northern Power River basin, Wyoming and Montana, guidebook. (402)
    1965 19th   Sedimentation of Late Cretaceous and Tertiary outcrops, Rock Springs Uplift.
    1966   Symposium on recently developed geologic principles and sedimentation of the Permo-Pennsylvanian of the Rocky Mountains. Road log: Casper-Casper Mountain-Bates Hole and Alcova via Circle Drive and Highway 220; Bates Creek to Casper, Wyoming; Wyoming; Alcova Store to Fremont Canyon.
    1968 20th   Black Hills area, South Dakota, Montana, Wyoming.
    1969 21st   Symposium on Tertiary rocks of Wyoming.
    1970 22nd   Symposium on Wyoming sandstones.
    1971 23rd   Symposium on Wyoming tectonics and their economic significance.
    1972 24th   South-central Wyoming; stratigraphy, tectonics and economics.
    1973 25th   [1] Symposium and Core Seminar on the geology and mineral resources of the greater Green River basin, Casper, Wyoming. [2] Core book.
    1974 26th   Muddy Sandstone-Wind River basin. (782)
    1975 27th   Geology and mineral resources of the Bighorn Basin, Cody, Wyoming.
    1976 28th   Geology and energy resources of the Powder River.

(781) Wyoming Geological Association.

1977 29th Rocky Mountain thrust belt geology and resources. (402, 736, 734)
   Trip 1. Front of the thrust belt to the foreland.
   Trip 2. Geology of the Snake River and adjacent areas.
   Trip 3. Snake River float trip log.
1978 30th Resources of the Wind River basin.
1980 31st Jackson Hole, Wyoming.
   Stratigraphy of Wyoming.
1981 32nd Jackson Hole, Wyoming.
   Energy resources of Wyoming.
   Trip 1. Front of the thrust belt to foreland.
   Trip 2. Geology of the Snake River range and adjacent areas.
   Trip 3. Snake River float trip lot; Elbow Campground to Sheep Gulch.
   Trip 4. Alpine, Wyoming to Little Greys River Anticline.
1982 33rd Mammoth Hot Springs, Wyoming.
   Geology of Yellowstone Park area.
   West Yellowstone to Hebgen Lake area.
   West Yellowstone to Madison Canyon Slide.
   West Yellowstone to Canyon Junction.
1983 34th Billings, Montana.
   Geology of the Bighorn Basin.
   Road log 1. Billings to Bridger.
   Road log 2. Bridger to junction of Wyoming routes 120 and 292 (turnoff to Clarks Fork Canyon).
   Road log 3. Clarks Fork Canyon and Dead Indian Hill to Cody.
   Road log 4. Cody to Buffalo Bill Dam and Rattlesnake Mountain.
   Road log 5. Return Buffalo Bill Dam, via Horse Center Anticline, to Cody.
   Road log 6. Cody to Lovell.
   Road log 7. Lovell to Goose Egg Anticline.
   Road log 8. Lovell to new U.S. 14-A roadcut and Five Springs thrust.
   Road log 9. U.S. Highway 14-A to Devils Canyon overlook in the Bighorn Canyon recreational area.
   Road log 10. Lovell, Wyoming to Bridger, Montana.
   Road log [11] Jackson to Dinwoody and return. (232(No. 20))
1984 35th Casper, Wyoming.
   The Permian and Pennsylvanian geology of Wyoming.
   Day 1. Casper to north fork Crazy Woman Creek & return.
   Day 3. Casper to Guernsey Reservoir to Newcastle.
   Day 4. Newcastle to lower Hell Canyon and return to Casper via Reno Junction.
   Remainder - Reno Junction to Midwest, Wyoming.
1985 36th Casper, Wyoming.
   The Cretaceous geology of Wyoming.
   Road log [1] Casper-Pathfinder-Hat Six road.
   Road log [2] Casper-Kaycee-Redwall.

AzFU	1956-58, 60-63, 68-70
AzTeS	1956-58, 60-63, 65, 68-71, 73-78, 80-84
AzU	1947, 49-63, 65, 68-71, 73, 75
BEG	1980-81
CChiS	1953, 55-58, 60-63, 65, 68-73
CLU-G/G	1947, 49-58, 60-63, 65, 68-71, 73-78, 80-85
CLhC	1947-63, 65, 68-73, 75-78, 80
CMenUG	1980-85
CPT	1947-62, 65, 68-78, 80-85
CSdS	1948-56, 58, 60-63, 65, 68-71, 73
CSfCSM	1950, 52-56
CSt-ES	1947-63, 65, 68-78, 80-84
CU-A	1952, 56, 60, 62-63, 65, 72-78, 80, 85
CU-EART	1947-54, 56-63, 65, 69-78, 80-84
CU-SB	1947-51, 54, 56-58, 60-63, 65, 68-78, 80-85
CU-SC	1980, 82
CaAC	1963
CaACAM	1947, 49, 58, 60, 63, 70-71, 73
CaACI	1947, 49, 57-58, 60-62, 65
CaACM	1955, 57, 68-69, 77-78
CaACU	1949-63, 68-78, 80-84
CaAEU	1980
CaBVaU	1954, 56-61, 77, 82
CaOHM	1956-63, 65
CaOKQGS	1947, 53-58, 60-61, 63, 69-70, 80-84
CaOLU	1954, 56-58, 60-62, 68
CaOOG	1950, 56-58, 60-61, 63, 71, 80-85
CaOWtU	1970, 77, 83(11)
CaQQERM	1983(1)
CoDCh	1946-48, 50, 52-56, 59-60, 69, 71-73, 75, 80-84
CoDGS	1946-85
CoDuF	1963
CoFS	1960
CoG	1946-63, 65, 68, 71, 73, 75-77, 80-85
CoGrU	1963, 82
CoLiM	1982-83
CoU	1947-63, 65, 68-76
CoU-DA	1980-83
DI-GS	1946-63, 65, 68-78, 80-85
DLC	1952-53, 56-58, 61-62, 65, 68, 73, 75-76, 78
I-GS	1947-48
ICF	1947, 50-51, 54-58, 60-63, 69
ICIU-S	1948-49, 57-58, 61, 68-76
ICarbS	1954, 56-58, 60-63, 65, 68-70, 74, 85
IEN	1946-47, 49-63, 65, 68-73
INS	1981, 82, 84
IU	1947-63, 65, 68, 71, 73-78, 80
IaU	1947-58, 60-63, 65, 68-70, 72-78, 80-85
IdBB	1956, 58, 60-61, 63, 68-69
IdPI	1953-62, 73
IdU	1950, 53-58, 68-69, 71, 74-75, 77, 80-84
InLP	1949, 50-54, 56-58, 60-63, 65, 68-71, 73-78, 80, 83-85
InRE	1954, 56, 58-61
InU	1946-63, 65, 68-69, 71, 73, 75-78, 82, 85
KLG	1983(11)
KyU-Ge	1946-63, 65, 68-78, 80-85
LNU	1947-53, 56, 63, 68-69, 71, 73, 75
LU	1948-63, 65, 68-71
MH-GS	1947-48, 50, 56, 81
MNS	1951, 53-58
MWesB	1948-53
MiDW-S	1949-50, 53-63, 68, 70
MiKW	1968
MiU	1947, 49-50, 60-63, 65, 68-70
MnU	1946, 48-63, 65, 68-78, 80-83
MoRM	1947-48, 53-58, 60-61
MoSW	1948-50, 53-58, 60-62, 65, 75, 84
MoU	1947-48, 52-58, 61-63, 68, 70
MtBC	1947, 52, 54-63, 65, 68-69, 71, 73
MtBuM	1946-58, 60-63, 65, 68-71, 73, 75-78, 82-85
MtU	1948-54, 56-63, 65, 68-71, 73
NBiSU	1956-57, 60-63
NIC	1980-82
NNC	1946-58, 60-63, 65, 68-70, 73, 77-78, 80, 81, 83-85
NOneoU	1956, 61, 68-70
NRU	1947, 53-54, 56-58, 60, 63, 65, 68-69, 71, 73-78, 80-85
NSbSU	1980-82
NSyU	1954, 56-57, 59, 63, 68
NbU	1947-58, 60-63
NcU	1948, 50-55, 57-58, 60-63, 65, 68-78, 80-81, 83, 85
NdFA	1983(11)
NdU	1949-51, 53-58, 60-63, 65, 68-71, 73-74, 77-78, 80-85
NhD-K	1949-52, 55, 57-63, 65, 68-70, 76-78, 80, 81-83
NjP	1947-63, 65, 68-74
NmPE	1958, 63, 68, 71, 73
NmSoI	1980, 83(11)
NmU	1980, 82, 84
NvLN	1956-57, 60
NvU	1949-50, 53-63, 68-69, 71, 73, 75-78, 80-84
OCU-Geo	1947, 50-58, 60-63, 65, 68-76, 80, 82, 84
ODaWU	1984
OU	1947-49, 51-58, 60-63, 65, 68-71, 74, 77, 81, 82
OkOkCGS	1948-50, 52, 54, 58, 62-63, 68
OkT	1947-58, 60-61, 65, 70-71, 75-76
OkTA	1980-84
OkU	1946-58, 60-63, 65, 68-74, 78, 80-84
OrCS	1947, 49-63, 65, 68-73, 76-78
OrU-S	1949, 51-63, 65, 68-73, 80-84
PBL	1954, 56-63, 65, 69-70
PBU	1980-85
PBm	1947, 54, 56-58, 60, 73
PSt	1947, 51, 53-63, 68-78, 81
RPB	1953-63, 65, 68
ScU	1983(11)
SdRM	1947-63, 65, 68-71, 75-78
TMM-E	1956, 60-61, 65, 68-69
TU	1956-57, 60-63, 65
TxBeaL	1949-51, 53-61, 63, 73-76
TxDaAR-T	1946-58, 60-63, 65, 68-73

TxDaDM	1947-50, 52-63, 65, 68-77, 80, 84-85
TxDaGI	1947-54, 56-63
TxDaM-SE	1947-63, 65, 80, 82-83
TxDaSM	1947, 49-58, 60-63, 65, 68-71, 73, 75-78
TxHSD	1947-52, 54, 56-58, 60-63, 65, 68-71, 73, 76-78
TxHU	1948, 50-63, 65, 68-69
TxLT	1963, 68, 71, 73, 75-76
TxMM	1947-63, 65, 68-71, 73, 77, 80-85
TxU	1947-63, 65, 68-73, 75-77, 80-85
TxU-Da	1947, 49, 51-63, 65, 69
TxWB	1947-63, 65-70, 75-83
UU	1948-58, 60-63, 65, 68-78, 80-85
ViBlbV	1953-58, 60-61, 63, 68-69, 71, 73-78
WU	1950-63, 65, 68-71, 73-78, 80-85
WyU	1946-63, 65, 68-69, 80-84

(782) WYOMING GEOLOGICAL ASSOCIATION. EARTH SCIENCE BULLETIN.
1973 V. 6. No. 4. See American Association of Petroleum Geologists. Rocky Mountain Section. 30.(1974)
1974 V. 7. No. 1. See Wyoming Geological Association. 781.
1979 V. 12. No. 4. See American Association of Petroleum Geologists. National Energy Minerals Division. (Title varies) 27.
1981 V. 13. No. 4. See Wyoming Geological Association. 783.(1964)
1983 V. 14. See Wyoming Geological Association. 783.(1964)

(783) WYOMING GEOLOGICAL ASSOCIATION. [GUIDEBOOK]
1964 Highway geology of Wyoming; road logs of highways, including points of geologic, economic, historic and scenic interest. (Parts reprinted in 1981 and in 1983 issues of Wyoming Geological Association. Earth Sciences Bulletin. (V.13, 14)
1979 Laramie, Wyoming.
Field trip guidebook for the Laramie - Elk Mountain - Hanna area, Albany and Carbon counties, Wyoming. (230)

AzTeS	1964
AzU	1964
CChiS	1964
CLU-G/G	1964
CLhC	1964
CSt-ES	1964
CU-EART	1964
CaBVaU	1964
CaOHM	1964
CaOWtU	1964
CoDGS	1964
CoG	1964
CoU	1964
DI-GS	1964
DLC	1964
ICF	1964
ICarbS	1964
IEN	1964
IaU	1964
InLP	1964
LNU	1964
MiDW-S	1964
MiU	1964
MtBuM	1964
MtU	1964
NNC	1964
NRU	1964
NcU	1964
NdU	1964
NhD-K	1964
NjP	1964
NvU	1964
OCU-Geo	1964
OU	1964
OkU	1964
OrU-S	1964
PSt	1964
RPB	1964
SdRM	1964
TxDaDM	1964
TxDaGI	1964
TxDaM-SE	1964
TxHU	1964

TxU	1964
UU	1964
WU	1964
WyU	1964, 79

WYOMING. GEOLOGICAL SURVEY. See GEOLOGICAL SURVEY OF WYOMING. (230) and GEOLOGICAL SURVEY OF WYOMING. (231) and GEOLOGICAL SURVEY OF WYOMING. (232)

WYOMING. UNIVERSITY. See UNIVERSITY OF WYOMING. DEPARTMENT OF GEOLOGY. (726) and UNIVERSITY OF WYOMING. (724)

(784) YELLOWSTONE-BIG HORN RESEARCH ASSOCIATION.
1937 See Rocky Mountain Association of Geologists. 549.
1947 See Wyoming Geological Association. 781.
1958 Aug. See Montana Geological Society. 402.

(785) YORK ROCK AND MINERAL CLUB, INC. SPECIAL PUBLICATION.
1980 No. 2. Geologic guide of York County, Pennsylvania.
PHarER-T 1980

(786) YOUNGSTOWN STATE GEOLOGICAL SOCIETY.
1979 See Ohio Intercollegiate Field Conference in Geology. 496.

(787) YOUNGSTOWN STATE UNIVERSITY.
1979 See Ohio Intercollegiate Field Conference in Geology. 496.

GEOGRAPHIC INDEX

Bahamas
General 197—1974(3); 504—1970; 645—1984
West 197—1974(12); 247—1969(1); 374—1969(1), 71; 614—1973; 272—86
Bimini 374—1970
Exuma Island 645—1984
Grand Cayman Island 197—1982(5)
Great Bahama Bank 197—1964(2); 341—1977
San Salvador 135—1981, 83, 84; 645—1984
San Salvador Island 197—1985(2)

Barbados
General 310—1977(1)

Belize (formerly British Honduras)
General 310—1977(2)
Shelf, South 247—1971

Bermuda
General 98—1970; 190—1980(1)

CANADA

Regional
General 296—1933(28,29), 72(A31/C31,C23,A30,C22)
East 296—1972(A59,A51a,C51b); 302—1982
West 289—1971(5)
Appalachian Mountains 190—1985(1); 272—120
Atlantic Provinces 190—1966(1-8), 80(3); 85(6,8)
Avalon Zone 302—1982
Bay of Fundy 190—1980(23); 279—1982(6A)
Canadian Shield 296—1972(A31/C31,A33/C33,A35/C35,A36/C36,A51a,A53/C53,A56/C5 6,A65,C38,B1,B6,B18,B21,B23-B27)
Canadian Shield, South 722—1979
Cordillera 190—1976(C11), 77(8); 199—1985; 204—1971(3); 296—1913(8), 33(28), 72(Various); 318—1967
Cordillera, West 190—1983(4)
Great Lakes, Lower 279—1982(9B)
Grenville Province 190—1984(9A,10A)
Gulf of St. Lawrence, South 279—1982(6A)
Highland Valley Camp 193—1985
Interior Plains 296—1972(A21,A25/C25,A26,A68,C18,C22,C23)
Lake Superior region 190—1982(4)
Maritime Provinces 190—1980(24); 296—1913(1), 72(A57/C57,A60,A61/C61,A63/C63); 492—1948
Meguma Zone 302—1982(Aug.)
Pacific Coast 296—1972(A06/C06)
Rocky Mountains 296—1913(8), 72(A08/C08,A10,A19,A26,C17,C18,C22,A20,CA15/C15,A16,A25/C25)
Rocky Mountains, Central 125—1984(1,1A); 190—1976(C7,10)
Rocky Mountains, South 125—1954, 60(Map1), 68; 190—1981(10,14); 317—1981(2)
Trans-Canada Highway 204—1971(1b,2a)

Alberta
General 18—1982(1,7); 166—1969, 70; 183—1969; 190—1976(A6); 204—1971(1,1a); 296—1972(A12,A16,A21,A25/C25,A26,C17); 12—1981; 321—1980
Central 296—1913(9), 72(A12,A25/C25,C18)
North 272—139
South 18—1970(1), 82(2,3,5,6,8,11-15); 125—1953, 78(June 4), 83(12); 185—1976; 190—1976(A-5,C-11), 77(8) 81(4,5,11,12); 296—1913(8), 72((A16,A21,A25/C25,C17,C22); 279—1982(20B,21A,28B); 301—1982; 272—94; 305—1984; 126—1979
Southeast 190—1973(7,8)
Southwest 125—1952, 62(1,2), 71, 75(May,Sept.), 80(June, 83(3); 190—1950(2); 402—1984; 272—26
West-central 190—1976(A3); 296—1972(A10,A16)
Alberta Caper 185—1976
Alberta Foothills, Central 190—1976(A5)
Alberta Plains 321—1980
Athabasca Oil Sands 279—1982(22); 272—139
Athabasca Oil Sands area 125—1973; 190—1976(C9)
Athabasca River 272—32
Banff 117—1974; 125—1960(Map 2); 148—1957(1); 204—1971(1b,2a,5,6)
Banff National Park 225—1960, 67, 77
Banff area 190—1950(1); 226—1971; 599—1969
Big Hill 125—1978(Sept.)
Bow Valley 18—1982(10); 125—1956, 83(7)
Burnt Timber Creek 125—1977(Oct.)
Cadomin 166—1959, 66
Calgary 125—1953; 199—1985; 204—1971(1b,2a,5,6)
Canyon Creek 190—1981(9)
Coalspur area 125—1983(9)
Crowsnest Pass area 125—1953, 83(4); 190—1981(1)
Cypress Hills 296—1972(C22)
Cypress Hills Plateau 125—1965, 78(Sept 2)
Cypress Hills area 559—1971
David Thompson Highway 166—1965
Dinosaur Provincial Park 125—1983(1), 84(2)
Drumheller 18—1982(6,15); 125—1959, 83(2); 190—1981(8)
Drumheller area 190—1976(A5)
Edmonton 183—1961
Edmonton area 190—1976(A4,C8)
Ghost River area 125—1963
Glacier National Park 125—1977(Sept.)
Grande Cache area 125—1983(9)
Grassi Lakes 125—1979(1), 80(May)
Highwood Pass 125—1983(11)
Jasper National Park 125—1955; 166—1961, 64, 69; 225—1963; 279—1982(27A)
Jura Creek 125—1977(May), 81(Sept.)
Kananaskis Valley 125—1952
Kananaskis area 125—1961, 83(11)
Kicking Horse Pass 125—1977(July), 83(10)
Lake Louise 148—1957(1)
Lake Louise area 599—1969
Lake Minnewanka 125—1977(June), 78(Aug.), 84(1)
Lundbreck 125—1980(June)
Maligne Lake 166—1964
Medicine Lake 166—1964
Middle Sand Hills 296—1972(C22)
Milk River 559—1971
Moose Mountain 125—1959, 78(May); 190—1981(9)
Mount Wilson 125—1980
Nordegg 125—1958
Peace River 166—1962, 70
Plateau Mountain 18—1982(9); 125—1978(Sept. 1), 79(2), 81(Aug.)
Red Deer River 296—1913(8), 72(C18)
Red Deer River badlands 18—1970(2)
Rock Lake 166—1960
Rocky Mountains 18—1982(3,4); 125—1978(June 2), 80(Sept.); 148—1957(1); 296—1972(A20,A24/C24,A68,C17,C18,C22); 287—1978(6); 317—1981(2); 279—1982(28B); 321—1980
Rocky Mountains, Moose Dome 125—1974

Seebe 125—1981(Sept.)
Southeast, Milk River 559—1971
Southwest, Crowsnest Pass area 125—1953, 83(4); 190—1981(1)
Southwest, Ghost River area 125—1963
Southwest, Jura Creek 125—1977(May), 81(Sept.)
Southwest, Kicking Horse Pass 125—1977(July), 83(10)
Southwest, Lundbreck 125—1980(June)
Southwest, Seebe 125—1981(Sept.)
Southwest, Trap Creek 125—1980(June)
Sunwapta Pass 166—1963
Trap Creek 125—1980(June)
Waterton 18—1982(2); 125—1957, 77(Sept.)
Waterton Lakes National Park 225—1964(10); 292—1974(7); 272—8
Waterton National Park 296—1972(A25/C25)
Whiteman Gap 125—1979(1), 80(May)

British Columbia
General 190—1977(3,4,5,9,13,16); 204—1971(1); 296—1972(A12,A25/C25); 279—1982(21A); 281—1959(1C)
Central 199—1985(14); 296—1913(10), 72(A09/C09,A12)
Northeast 166—1969; 296—1972(A10)
Northwest 296—1913(10)
North-central 190—1977(2), 83(2)
South 76—1977(1-A); 190—1976(C-11), 77(6,8,9), 81(12), 83(10); 191—1982; 296—1913(8-9), 72(A09/C09); 301—1982
Southeast 125—1962(1,2), 71; 190—1968(A1,A2), 77(1), 83(13); 199—1985(11-12); 402—1984; 272—26
Southwest 190—1983(6); 199—1960, 85(9,15); 676—1977(1); 279—1982(30A)
South-central 190—1983(1)
West-central 190—1983(14)
Alaska Highway 226—1973
Arrow Lakes 296—1913(9)
B. C. Molybdenum Mine 296—1972(A06/C06)
Boundary District 296—1913(9)
Cascade Range 296—1972(A05/C05)
Clayburn 296—1913(8)
Coast Mountains 199—1960(1), 85(4)
Coast Mountains, South 190—1983(15)
Coast Range 296—1913(8,10), 72(A04/C04,A05/C05,A08/C08)
Copper Mountain 191—1982
Crowsnest Pass area 125—1983(4); 190—1981(1,2,12,15)
Dawson Creek 226—1973
Flathead Valley 125—1964; 204—1959(3), 77(4)
Fraser Lowland 190—1977(10,14), 83(15); 197—1977(6); 199—1985(15)
Fraser River Canyon 296—1913(8)
Fraser River delta 190—1977(12)
Garibaldi Lake area 191—1975
Glacier National Park 225—1965
Granduc 296—1972(A06/C06)
Interior Plateaus 296—1913(8)
Island Copper Mine 296—1972(A06/C06)
Kamloops 199—1960(3)
Kimberly 296—1972(A24/C24,A25/C25)
Kootenay Arc 583—1970
Kootenay Lake 296—1913(9)
Kootenay National Park 225—1964(9)
Kootenay, East 323—1967(1)
Long Beach Segment 225—1975
Merritt 296—1913(9)
Monashee Mountains 296—1913(9)
Mount Revelstoke National Park 225—1965
North Vancouver 676—1977(2)
Okanagan region 190—1981(3)
Pacific Coast 296—1913(10), 72(A04/C04)
Pacific Rim National Park 225—1975
Phoenix 296—1913(9)
Phoenix Mining District 191—1982

Purcell Mountains 97—1973
Queen Charlotte Islands 190—1983(8); 296—1972(A06/C06)
Revelstoke 204—1971(1b,3)
Rocky Mountains 18—1982(4); 317—1981(2)
Rossland 296—1913(9)
Silkameen District 296—1913(9)
Skeena River District 296—1913(10)
Slocan District 296—1913(9)
South-central, Okanagan region 190—1981(3)
Southern Interior 190—1968(8)
Sullivan Mine 296—1972(A24/C24)
Tasu Mine 296—1972(A06/C06)
Texada Island 296—1972(A06/C06)
Vancouver 190—1977(11,14); 296—1972(A05/C05)
Vancouver Island 190—1977(7), 83(9); 199—1985(1,7,8); 296—1913(8), 72(A05/C05,A06/C06)
Vancouver Island, Coast Mountains 190—1968(A1)
Vancouver Island, North 190—1968(A1)
Vancouver Island, South 190—1983(7,11)
Vancouver Island, Southeast 190—1983(5)
Vancouver Island, Southwest 190—1983(12)
Vancouver area 190—1977(11,14); 191—1973, 77; 676—1977(1)
Vancouver, North 199—1960(2)
Victoria 296—1913(8-9)
Yoho National Park 225—1962(4)

Labrador
General 287—1978(5)
South 279—1982(1A)

Manitoba
General 296—1913(8), 72(A26,A30,A31/C31,C23,A33/C33,C22); 477—1952
Central 190—1970(1)
North 190—1970(2)
South 296—1913(9), 72(A33/C33,C22); 477—1952
Southeast 190—1970(4,7), 82(9,15)
Southwest 190—1982(10)
West 190—1970(2), 73(3,11)
West-central 190—1982(6,14)
Dawson Bay area 190—1970(6)
Flin Flon 296—1972(A31/C31)
Flin-Flon area 559—1966
Fox Mine 190—1982(2)
Interlake area 190—1970(5), 73(1), 82(11); 477—1952; 559—1965
Keewatin District, North 287—1978(4)
Lake Manitoba area 559—1967
Lake Winnepegosis 296—1913(8)
Lake Winnepegosis area 559—1967
Lynn Lake 296—1972(A31/C31)
Manitoba Escarpment 190—1970(6)
Manitoba Nickel Belt 190—1970(1)
Nelson River 190—1982(5)
Pembina Mountain 190—1982(11)
Riding Mountain National Park 225—1974
Ruttan Mine 190—1982(2)
Thompson Mineral Deposits 296—1972(A31/C31)
Winnipeg 296—1913(8), 72(A33/C33,A35/C35,C22)
Winnipeg area 190—1973(1), 82(1,7); 559—1965

New Brunswick
General 190—1980(5), 85(13); 296—1913(1), 72(A57/C57,A58/C58,A59,A60,A61/C61); 456—1973(A-12,A-14,B-11), 80(B-10); 302—1982(C)
North 190—1966(2), 85(4); 456—1973(B-4)
Northeast 190—1985(2); 296—1972(A58/C58,A59,A63/C63)
Northwest 456—1980(C-1,C-7)
South 296—1972(A58/C58,A59,A60,A61/C61); 456—1973(A-2,A-8,B-7,A-10,B-8); 279—1982(4A)
Southeast 190—1985(5); 302—1982(D)

Southwest 190—1966(1), 80(17), 85(7,9); 456—1973(A-3,B-6), 78(A-2,A-4,A-7,B-2,B-4,B-5,B-10)
Appalachian Mountains 456—1973(A-9)
Bathurst Mineral Deposits 296—1972(A58/C58)
Bathurst area 190—1980(16); 303—1983; 456—1973(A-5,A-6); 302—1982(B)
Bay of Fundy 296—1972(A58/C58)
Burnt Hill Mine 456—1973(B-1)
Chaleur Bay area 456—1973(A-7)
Chignecto Bay area 456—1973(A-4)
Coast, South 302—1982(H)
Edmunston area 456—1980(B-9)
Fredericton area 456—1973(B-2)
Fundy National Park 225—1962(2)
Grand Manan Island 456—1973(A-1)
Harvey Station 456—1973(A-14,B-11)
Lake Madawaska 456—1980(B-9)
Mascarene area 456—1978(B-5)
Minto area 456—1973(A-11)
Newcastle Mineral Deposits 296—1972(A58/C58,A63/C63)
Newcastle area 456—1973(A-5)
Oak Bay, Cookson Island 456—1978(B-9)
South Passamaquoddy Bay 456—1978(A-3)
South, Chignecto Bay area 456—1973(A-4)
St. George area 456—1978(B-5)
St. John River basin, Upper 455—1982
St. John River valley 456—1980(C-4)
St. John area 456—1973(A-13), 78(B-6)

Newfoundland
General 146—1967; 190—1974(B-8,B-10), 80(14); 296—1972(A57/C57,A61/C61,A62/C62,A63/C63); 441—1974; 472—1984; 302—1982(F)
North 190—1974(B-3,B-3b,B-7)
Southeast 190—1974(A-5,B-6,S-2)
Southwest 190—1974(B-11); 296—1972(A61/C61,A62/C62)
West 190—1974(A-2,B-2,B-4), 80(13); 279—1982(2B)
Advocate Asbestos Mine 190—1974(B-3a)
Appalachian Mountains 190—1974(A-1)
Avalon Peninsula 190—1974(A-4), 80(12); 472—1979
Avalon Platform 190—1974(B-5,S-1); 296—1972(A62/C62,A63/C63)
Avalon Zone 302—1982
Avalon Zone, West 190—1980(2)
Baie Verte 146—1967(5)
Betts Cove 190—1974(B-3i)
Bonne Bay 146—1967(6)
Boyds Cove 146—1967(4)
Buchans area 190—1974(B-3h)
Cape Ray fault zone 190—1974(B-11)
Carmanville 146—1967(3)
Carol Lake Ore Deposits 296—1972(A55)
Central Mobile Belt 190—1974(B-3,B-7)
Change Islands 146—1967(3)
Churchill Falls 296—1972(A51a)
Consolidated Rambler Deposits 190—1974(B-3c)
Corner Brook 146—1967(7); 190—1966(8)
Cuckold Head area 472—1979(A)
Deer Lake 146—1967(6)
Flat Rock area 472—1979(B)
Fleur de Lys Belt 190—1974(B-7)
Gander 146—1967(2-5)
Gander Lake Belt 190—1974(B-7)
Harp Lake region 296—1972(A54)
Humber Arm Stephenville 146—1967(7)
Kiglapait Mountains 296—1972(A54)
Labrador 296—1972(A51a,A54,A55)
Manuels area 472—1979(D)
Michikamau Lake region 296—1972(A54)
Nain 296—1972(A54)

New World Island 146—1967(4)
Pilleys Island area 190—1974(B-3f,3g)
Port au Port Peninsula 146—1967(7); 190—1974(B-4)
Random Island 146—1967(2)
Scully Mine 296—1972(A55)
Signal Hill area 472—1979(A)
Snyder Island 296—1972(A54)
Springdale Peninsula 190—1974(B-3e)
St. Anthony 146—1967(6)
St. John's area 146—1967(1-2)
St. Lawrence 190—1974(B-9)
St. Mary's Bay 472—1979(F)
St. Philipps area 472—1979(C)
St. Thomas area 472—1979(C)
Terra Nova National Park 225—1966
Tilt Cove 190—1974(B-3d)
Topsail area 472—1979(D)
Wabush Lake area 296—1972(A55)
Western Plaform 296—1972(A62/C62)

Northwest Territories
General 296—1972(A27,A28,A29,A30)
Southwest 166—1969
Arctic Islands 296—1972(A66,A68)
Baker Lake 296—1972(A32a/A32b)
Dempster Highway 287—1983(3)
Great Bear 296—1972(A27,A66)
Great Slave Lake 296—1972(A28)
Great Slave Lake region 190—1976(A2)
Hudson Bay 296—1972(A30)
MacKenzie River delta 125—1978(June 1); 287—1983(3)
Mackenzie Mountains 296—1972(A66,A68)
Mackenzie River delta 296—1972(A30,A66)
Mackenzie River valley, Lower 287—1978(3)
Mackenzie River valley, Upper 287—1978(2)
Pine Point 166—1969
Pine Point Mineral District 296—1972(A24/C24)
Powell Creek 296—1972(A14)
Svedrup Basin 296—1972(A68)
Thelon Basin 296—1972(A32a-b)
Yellowknife 296—1972(A27,A28,A30); 323—1967(2)

Nova Scotia
General 121—1984; 190—1966(3,5), 80(3,6-7,18,20,22), 85(14); 296—1913(1), 72(A57/C57,A58/C58,A59,A60,A61/C61,A63/C63); 492—1948, 54; 586—1975(A,B); 279—1982(4A)
Central 190—1980(11)
East 190—1985(11)
South 190—1980(21)
Southwest 190—1980(9)
Avalon Terrain 190—1985(3)
Bay of Fundy region, North 190—1980(8)
Cape Breton Highlands 190—1985(10)
Cape Breton Highlands National Park 225—1962(5)
Cape Breton Island 190—1966(7); 296—1972(A61/C61)
Cobequid Highlands 190—1980(19)
Mabou Basin 197—1978(7)
Meguma Zone 279—1982(5B)
Minas Basin 190—1966(4)
North, Avalon Terrain 190—1985(3)
Pictou Field 197—1978(7)
Sable Island 190—1980(10)
Sydney Basin 197—1978(7)
Walton area 190—1966(6)

Ontario
General 190—1975; 273—1966(12th), 74(4), 85(4); 296—1913(2), 72(A33/C33,A39/A39b/C39,A35/C35,A36/C36,A40/C40,A42,A48, A47/C47,A65,C34); 377—1951; 514—1976

Central 18—1964; 197—1978(25); 296—1913(9); 72(A41/C41); 377—1965; 279—1982(10A)
East 190—1967(1-6), 75(A-2); 296—1913(2,3,6,7), 72(A48,C49); 514—1969(3)
North 43—1970; 121—1967; 722—1973
Northeast 272—120
Northwest 190—1982(3,8); 197—1978(5); 296—1972(C34)
South 25—1982(A-2,A-3); 179—1983; 182—1948; 190—1975(B-4,B-5); 197—1978(9,26); 296—1913(2,3,6), 72(A42,A45/C45,A48,C51b,A53/C53,C49); 279—1982(10A,11A,12A,16B)
Southeast 190—1967(6); 197—1978(10,23); 273—1967; 469—1966; 272—102
Southwest 190—1975(B-7,B-10), 84(1,2,16); 296—1913(4,5), 72(A42,C34); 377—1956; 514—1968
South-central 411—1985(1); 365—1983
West 190—1984(17)
Algoma District 273—1981
Algonquin Provincial Park 515—1969
Alliston area 197—1978(11)
Appalachian Basin 279—1982(17A)
Appalachian Mountains 279—1982(17A)
Atikokan 273—1970(D); 296—1972(A35/C35,C34)
Bancroft 190—1967(1-2), 75(1); 296—1972(A35/C35,A53/C53)
Bancroft area 197—1953(1,2), 78(8)
Batchawana 273—1974(1)
Blind River area 273—1966(3)
Brent 296—1972(A65)
Bruce Peninsula 197—1978(18)
Callander Bay 296—1972(A65)
Cameron Lake 273—1985(1)
Canadian Shield 197—1978(1,6)
Canadian Shield, Southern Province 197—1978(3)
Chandos Township area 181—1984
Cobalt 190—1967(1), 75(A-12); 197—1953(1); 296—1913(7), 72(A39/39b/C39)
Coldwell 296—1913(8), 72(C34)
Craigleith area 514—1979
Dryden 273—1985(5)
Elliot Lake 296—1972(C38,C67)
Elliot Lake Mining District 190—1984(3)
Elliot Lake area 121—1983; 197—1978(3)
Fort Frances 273—1982(1)
Georgian Bay North Shore 197—1978(25)
Great Lakes 296—1972(A42,A43,C34,C38,A48)
Great Lakes region 325—1979
Grenville 190—1967(4), 69(4)
Grenville Front Tectonic Zone 197—1978(6)
Grenville Province 190—1964(4), 75(A-1); 296—1972(A53/C53,A55,A65,C3g,B1,B2,B18)
Haliburton 296—1972(A53/C53)
Haliburton area 197—1978(8)
Hamilton 377—1972
Hemlo 121—1985
Hemlo area 515—1984
Iron Spur 296—1913(8)
Keewatin 296—1913(8)
Kenora 480—1969(40)
Kenora area 273—1985(3)
Kilarney Provincial Park area 514—1982
Kingston 296—1913(2), 72(A43,A48,C49,C51b)
Kingston area 190—1967(3,5)
Kipawa River 296—1972(A53/C53)
Kirkland Lake 190—1984(4); 296—1972(A39/39b/C39,A40/C40)
Kirkland Lake area 197—1953(8), 78(17)
Kitchner area 190—1984(8)
Lake Erie 296—1913(5), 72(A42,A43)
Lake Erie Islands 377—1971
Lake Erie shore, central 190—1984(12)
Lake Huron 296—1972(A36/C36,C38)

Lake Huron shore 514—1972
Lake Huron, North shore 273—1974(3); 587—1981
Lake Nipissing 296—1972(A53/C53,A65)53/C53,A65)
Lake Ontario 296—1913(4,5), 72(A42,A43,A48)
Lake Ontario Basin, West 197—1978(15)
Lake Ontario, North shore 197—1978(20)
Lake Simcoe 296—1913(5), 72(A39/39b/C39)
Lake Superior 309—1977; 513—1962
Lake Superior shore 514—1969(2)
Lake Timagami 296—1972(A39/39b/C39)
Lake Wanapitei 296—1972(A65)
Lake of the Woods 273—1985(2); 296—1913(8), 72(A33/C33)
Larder Lake area 197—1953(8)
London 182—1959
Madawaska Highlands 190—1975(A-2)
Madoc 296—1913(6)
Manitoulin Island 296—1913(5); 377—1954, 68, 78
Manitouwadge 121—1985
Manitouwadge Lake area 273—1966(2)
Marathon 273—1977(A)
Marathon area 296—1972(C34)
Michipicoten 273—1974(5)
Minden Township area 181—1984
Mine Centre 296—1913(8)
Mine Centre area 273—1982(1)
Mining Districts 43—1970
Muskoka 296—1913(6)
Niagara Escarpment 197—1978(18)
Niagara Falls 197—1953(6), 78(18); 296—1913(4), 72(A43,A45/C45,A48,C51b)
Niagara Falls Dam 197—1978(12)
Niagara Peninsula 25—1982(A-2); 190—1975(C-9); 197—1953(4-5); 377—1955; 411—1985(2,3)
Ottawa 75—1973; 225—1968; 296—1913(3), 72(A48,B24,B26)
Ottawa Lowlands 296—1972(B23,B26)
Ottawa River valley 197—1978(10)
Ottawa River valley area 296—1972(B25,B27)
Owen Sound 377—1946
Owen Sound area 197—1978(26)
Paris-Hamilton District 190—1975(C8)
Parry Sound 183—1982(Oct.)
Peterborough 296—1972(C49)
Pickerining Generating Station 197—1978(14)
Porcupine 296—1913(7)
Porcupine Mining District 197—1953(9)
Port Arthur 296—1913(8), 72(A33/C33)
Port Coldwell 273—1970(C)
Rainy Lake 296—1913(8)
Red Lake area 190—1982(8)
Sault Ste. Marie 296—1972(A35/C35,A36/C36,A40/C40,C38)
Sault Ste. Marie area 273—1966(1)
Sheffield Lake 296—1972(A53/C53)
Steeprock Lake 296—1913(8)
Sturgeon Lake 273—1977(C)
Sturgeon River 273—1970(B): 296—1972(C34)
Sudbury 148—1957(2); 190—1967(1); 273—1966(4); 296—1913(7), 72(A35/C35,A36/C36,A39/39b/C39,A43,A65,C38)
Sudbury Mining Area 190—1984(3)
Sudbury area 121—1983: 197—1953(1,7), 78(19)
Superior Province 197—1978(1); 722—1977
Superior Province, English River Belt 197—1978(4)
Superior Province, Uichi Subprovince 197—1978(5)
Thousand Islands 469—1983(2)
Thunder Bay 273—1970(A), 77(B)
Timmins 190—1984(14)
Timmins Mineral Deposits 296—1972(A39/39b/C39,A40/C40)
Timmins Mining Area 197—1978(2,21)
Timmins area 190—1975(A-3)
Toronto 18—1964; 296—1913(6,8-9), 72(A48,C51b); 315—1984
Toronto area 190—1975(B-6); 197—1953(3), 78(13,24,26)

Wabigoon Subprovince 190—1982(3)
Waubaushene 377—1946
Winston Lake 121—1984

Prince Edward Island
General 296—1972(A57/C57,A61/C61)
Prince Edward National Park 225—1962(3)

Prince Edwards Island
General 190—1985(14)

Quebec
General 296—
  1972(A46/C46,A51a,A47/C47,A53/C53,A56/C56,A65,C52);
  456—1977; 88—1984; 272—43
East 296—1913(1), 72(B9); 326—1981
North 296—1972(A55,A512); 287—1978(5)
Northwest 121—1967; 197—1953(10); 722—1973; 272—120
South 190—1963(1-10); 296—1913(2,3),
  72(A40/C40,A41/C41,A44/C44,A56/C56,A58/C58,C52,B4,B7,C49,
  B8,B21,A63/C63): 181—1985; 302—1982(A)
West 43—1970; 121—1980; 190—1967(7-8)
Abitibi 272—13
Abitibi area 190—1979(A-4,A-2)
Anticosti Island 326—1981
Appalachian Highlands 296—
  1972(A56/C56,A57/C57,A58/C58,A63/C63,B5,B18,B21)
Appalachian Mountains 190—1969(1), 79(A-14,B-6); 279—
  1982(7B)
Appalachian Mountains, West 190—1963(9)
Asbestos 190—1979(B-3,A-17); 296—1972(B8)
Bagot County 190—1979(A-16)
Brome Complex 296—1972(B13)
Brome Mountain 190—1969(6C)
Champlain Valley 474—1982(A)
Charlevoix area 296—1972(B6)
Chibouagamau 296—1972(A41/C41)
Chibougamau area 190—1979(B-1)
Covey Hill 296—1972(C52)
De-Mix Quarry 296—1972(B15)
Drummondville region 182—1956
Eastern townships 296—1972(C49,B7,B8,B13)
Gagnon Mineral Deposits 296—1972(A51a)
Gaspe Copper Mines 296—1972(A58/C58,A63/C63)
Gaspe Peninsula 190—1979(A-6,B-2); 296—
  1972(A56/C56,A57/C57,A58/C58,A59); 456—1973(B-4); 326—
  1981; 272—81; 281—1959(23)
Gatineau area 132—1962
Glen Almond 190—1963(4)
Grenville Province 190—1967(4), 69(4); 181—1985
Havre St. Pierre 296—1972(B9)
Hudson Bay 88—1980
James Bay 76—1982(4)
Jeffery Mine 296—1972(B8)
Kilmar Mineral Deposits 296—1972(C49)
Knob Lake Ore Deposits 296—1972(A55)
Lac St. Jean 121—1981(Sept./Oct.); 190—1969(5)
Lac Tio Ore Deposit 296—1972(B9)
Lake Jeannine 296—1972(A51a,A55)
Lake St. Jean 296—1972(A41/C41,A44/C44,A46/C46)
Magdalen Islands 190—1979(A-5)
Manicougan River Hydroelectric Complex 296—1972(A51a)
Matagomi 296—1972(A41/C41)
Mining Districts 43—1970
Monteregian Hills 190—1969(1-9); 296—1913(3),
  72(A65,B3,B4,B10,B11,B12,B14,B15,B17)
Montreal 138—1973; 190—1969(6A); 296—1913(3),
  72(C49,C51b,C52,B3,B12,B17,B18)
Montreal area 76—1982(1-3); 190—1979(A-18); 456—1962; 474—
  1982(D)

Mount Johnson 190—1963(8), 69(6B:2); 296—1972(B14)
Mount Rougemont 190—1969(6B:1)
Mount Royal 190—1963(6); 296—1972(B12)
Mount St. Hilaire 190—1963(8), 67(7), 69(9); 296—1972(B15)
Mount Wright 296—1972(A51a,A55)
Noranda 190—1984(14); 296—1972(A40/C40,A41/C41)
Noranda area 190—1979(A1)
Oka Hills area 296—1972(B10,B11)
Oka area 190—1963(1), 67(7), 69(6A); 545—1969
Outardes River Hydroelectric Complex 296—1972(A51a)
Portneuf County 190—1979(B-12)
Quebec City 296—1972(A56/C56,A63/C63,C52,B19,B20); 456—
  1977
Quebec City area 190—1979(A-9,A-11,B-11); 474—1982(C): 122—
  1985(1)
Riviere-du-Loup 182—1963
Rougemont 296—1972(B14)
Rouyn area 190—1979(A1)
Scheffesville Ore Deposits 296—1972(A55)
Seguenay River 190—1969(5)
Shawinigan 296—1972(B2)
Shawinigan area 190—1967(8), 69(4), 79(B-11); 88—1976
Shefford Complex 296—1972(B13)
Shefford Mountain 190—1969(6C)
Sherbrooke 182—1969
Sherbrooke area 190—1969(7), 79(A-12,B-4); 296—1972(B5)
Southwest, Temiscamingue 272—13
St. Jean Vianney 296—1972(A51a)
St. Lawrence Lowlands 190—1963(2,5), 69(8A); 296—
  1972(A56/C56,C49,C52,B4,B17,B18,B20,B21)
St. Lawrence River 296—
  1972(A46/C46,C52,A63/C63,C51b,C52,B4)
St. Lawrence area 272—120
St. Paulin 296—1972(B2)
St. Urbain 190—1979(A-8,B-8); 296—1972(B6)
Ste. Agathe des Monts 296—1972(B1)
Temiscamingue 272—13
Thetford Mines 190—1979(A-13,B-5,B-10)
Thetford Mines area 296—1972(A56/C56,A58/C58,B8,B21); 739—
  1985(A1-A31)
Trois Rivieres area 88—1976
Trois-Riviers area 296—1972(B21)
Val d'Or 296—1972(A41/C41,A46/C46)
Val d'Or-Amos Area 190—1979(A-4)
Weedon Mining District 296—1972(B7)

Saskatchewan
General 121—1981(Sept.); 183—1969; 190—1981(6,7); 296—
  1972(A26,C22,C23,C23); 559—1956, 58
Central 296—1913(9)
East 190—1973(3,11)
North 190—1973(10), 76(A1)
Northeast 190—1973(9)
South 190—1973(C), 81(4); 296—1913(8), 72(A21,C22)
Southwest 190—1973(8); 402—1953
South-central 190—1973(6)
Athabasca Basin 296—1972(A32a-b,C67)
Beaverlodge Uranium District 296—1972(C67)
Cypress Hills Plateau 125—1965, 78(Sept 2)
Cypress Hills area 559—1969
Estevan Coal Field 190—1973(5)
Great Sand Hills 296—1972(C22)
Hanson Lake area 559—1966
Lac La Ronge area 559—1970
Lake Manitoba 190—1973(2)
Lake Winnipegosis 190—1973(2)
Montreal Lake 190—1973(4)
Montreal Lake area 559—1970
Qu'appelle Valley area 559—1972
Rabbit Lake Uranium Deposits 296—1972(C67)

Guidebooks of North America

Wapawekka Hills area 190—1973(4)

Yukon Territory
Central 287—1978(1)
North 287—1983(3)
South 296—1913(10), 72(A11,A12)
West-central 296—1913(10), 72(A11)
Alaska Highway 226—1973
Anvil Mine 296—1972(A24/C24)
Dempster Highway 287—1983(3)
Klondike gold fields 296—1913(10)
Peel River 296—1972(A14)
Richardson Mountains 296—1972(A68)
Royal Creek 296—1972(A14)

Caribbean Islands
General 36—1970

Cayman Islands
Grand Cayman Island 310—1977(3)

Cuba
General 153—1938

Dominican Republic
Central 583—1982

Guatemala
General 18—1976(6); 197—1967(8); 247—1974(3); 374—1977

Haiti
General 197—1985(11); 374—1982

Jamaica
General 247—1969(3); 310—1977(4); 696—1984
Central 197—1974(11)
North 197—1974(11)
Lime Cay 197—1974(11)
Wagwater Belt 197—1974(11)

MEXICO

Regional
General 69—1982(14); 612—1935, 48; 316—1982(6)
Central 581—1982
East 296—1956(C7)
East-central 18—1983(1)
North 18—1974(SEPM 2:1); 296—1956(A13); 756—1981
Northeast 18—1984(2); 152—1979, 83; 205—1984(3); 626—1970, 74(Fall)
Northwest 152—1978
South 296—1956(C15/C15b)
Southeast 296—1956(C7)
South-central 197—1968(9)
West 296—1956(C7)
East Coast 296—1956(C16)
Gulf of Mexico 296—1956(C11)
Mexican Basin 197—1968(8)
North Central Plain 296—1956(A2/A5)
Sierra Madre Occidental 296—1956(A2/A5)
Sierra Madre Oriental 197—1968(3); 296—1956(A2/A5,A14/C6,C8,C10)
Sierra Madre Oriental, North 296—1956(C5)
Sierra Madre de Oaxaca 296—1956(A6)
Sierra Madre del Sur 296—1956(A6,C7)
Sierra de Guadalupe 296—1956(C9)
South Central Plain 296—1956(A6)
Yucatan Peninsula 197—1974(2); 296—1956(C7)

Baja California Norte
General 197—1979(13,26)
North 593—1984(2)
Northwest 28—1984(Apr.); 197—1979(23); 199—1961(5); 556—1977
Coronado Islands 556—1978
Mexican Highway 1 197—1979(12)
Pacific Slope 28—1970(Nov.), 84

Baja California Sur
General 296—1956(A1/C4,A7)
Boleo Copper District 296—1956(A1/C4)
Isla Cedros 197—1979(10)
Lucifer Manganese Mine 296—1956(A1/C4)
Mexican Highway 1 197—1979(12)
Santa Rosalia 296—1956(A1/C4)
Sierra de la Giganta 296—1956(A7)
Vizcaino Peninsula 197—1979(10)

Campeche
General 296—1956(C7)

Chiapas
General 296—1956(C15/C15b)
North 296—1956(C15,C15a)
Simojovel Basin 296—1956(C7)

Chichuahua
North 458—1969

Chihuahua
General 756—1964, 74(Oct.), 84
North 167—1983; 580—1966(Jan)
Aquiles Serdan Mineral District 296—1956(A2/A5)
Cerro de Cristo Rey Uplift 205—1977(3a)
Cerro de Muleros 296—1906(20)
Naica Mineral District 296—1956(A2/A5)
Parral 296—1906(21,22), 56(A13)
Parral Mineral District 296—1956(A2/A5)
San Francisco del Oro 296—1956(A2/A5)
Santa Barbara 296—1956(A2/A5)
Sierra de Juarez 31—1974; 167—1972, 80

Coahuila
General 296—1906(23,27,28), 56(C5); 612—1959, 64
East 580—1970
North 594—1974
Coahuila Fold Belt 152—1978
Coahuila Platform 566—1979
Monclava Mineral District 296—1956(C3)
Nueva Rosita Mineral District 296—1956(C3)
Parras Basin 580—1972
Sabinas 296—1956(C5)
Sabinas Coal Basin 197—1968(1)
Saltillo 296—1906(29); 612—1964
Sierra Madre Oriental 566—1979
Sierra de Parras 296—1956(C5)
Torreon 581—1972(2)

Colima
Volcan de Colima 296—1906(13)

Durango
General 581—1972(2,3,4,5)
Mapimi 296—1906(18)
Sierra de Banderas 296—1906(19)

Federal District
General 581—1970(1)
Basilica de Guadalupe 296—1956(C9)

La Caldera 296—1956(C9)
Mexican Basin 296—1956(A14/C6,C10)
Mexico City 18—1974(4); 612—1935, 48
Mexico City area 197—1968(4)
Penon de Los Banos 296—1956(C9)
Ticoman 296—1956(C9)
Zacatenco 296—1956(C9)

Guanajuato
General 296—1906(15); 581—1974
Central 604—1982(1)
Northeast 604—1982(1)
Guanajuato Mineral District 296—1956(A2/A5,C3)
Valle de Santiago 296—1906(14)

Guerrero
General 197—1968(9); 296—1956(A9/C12,A12,C9); 581—1978, 82(3)
Taxco Mineral District 296—1956(A4/C2)

Hidalgo
General 296—1956(A10/C13); 581—1982(1)
Pachuca Mineral District 296—1956(A3/C1)
Pachuca Mining District 197—1968(2)
Pathe 197—1968(6)
Real del Monte 296—1956(A3/C1)
Real del Monte Mining District 197—1968(2)
Zimapan 296—1956(A3/C1,C10)

Jalisco
Atemajac Valley 296—1956(A16)
Guadalajara Basin 296—1956(A16)
Tesistan Valley 296—1956(A16)
Tuxpan 612—1948

Mexico
General 296—1956(A10/C13); 581—1970(1), 78
Ixtacihuatl 197—1968(8)
Mexican Basin 296—1956(A14/C6,C10)
Volcan Nevado de Toluca 296—1906(9)

Michoacan
General 296—1956(A15)
Patzcuaro 296—1906(8)
Volcan Paricutin 296—1956(A15)
Volcan de Jorullo 296—1906(11)
Volcan de San Andreas 296—1906(10)

Morales
Popocatepetl 197—1968(8)

Morelos
General 197—1968(9); 296—1956(C9); 581—1970(1)
Cacahuamilpa Caverns 296—1956(C14)
Morelos Basin 197—1968(5)

Nayarit
Ixtlan 296—1906(12)

Nuevo Leon
General 152—1952, 56, 61; 296—1956(C5); 612—1931, 32(4th), 34, 54(21st), 59, 64
Cortinas Canyon 152—1952
Hidalgo Canyon 152—1970
Huasteca Canyon 152—1952, 79; 612—1950
La Popa Canyon 152—1970
Monterrey 152—1952, 56, 61; 296—1906(29), 56(A13); 612—1941, 50; 626—1964
Portrero Garcia 152—1979
Villa de Garcia 296—1956(C5)

Oaxaca
General 296—1956(A12); 581—1970(1,2); 670—1946
Northeast 296—1956(A6)
Southeast 296—1956(C15/C15b)
Canon de Tomellin 296—1906(5)
Istmo de Tehuantepec 296—1906(31), 56(C7,C15b)
Mitla 296—1906(6)
Natividad Mineral District 296—1956(A6)
Salina Basin 296—1956(C7)
Tezoatlan area 197—1968(7)
Tlaxico 296—1956(A12)

Puebla
General 296—1956(A12,C8); 581—1970(1), 82(2,3)
Northwest 296—1956(A10/C13)
South 296—1956(A11)
Huachinango 296—1956(A10/C13,C8)
Nexaca Dam 296—1956(C8)
Petlalcingo 296—1956(A11)
Petlalcingo area 197—1968(7)
Puebla 296—1956(A11)
San Juan Raya 296—1956(A11)
Sierra de Santa Rosa 296—1956(A11)
Tehuacan 296—1906(4,7), 56(A11)
Tepexi de Rodriguez 604—1982(2)
Zapotitlan 296—1906(7)

Queretaro
General 581—1982(1)

Quintana Roo
Cancun 756—1979

San Luis Potosi
General 296—1906(30)
East 296—1956(C16)
Cuidad Valles 296—1956(C10)

Sinaloa
General 756—1974(Oct.)
Central 581—1972(4)

Sonora
General 199—1959(6); 296—1956(A1/C4)
North 197—1979(27)
Northwest 199—1981
Altar 296—1956(A8)
Caboraca 296—1956(A8)
Cananea Mining District 296—1956(A1/C4)
Pinacate Mountains 199—1959(6)
Sonoran Desert 285—1969

Tabasco
General 296—1956(C7)
Jalapa 296—1906(1)
Petroleum Fields 296—1956(C7)

Tamaulipas
General 296—1956(A14/C6,C10); 612—1931, 32(4th), 34, 41, 54(21st)
North 152—1952, 61
Ciudad Victoria 626—1974(Fall)
Coast 296—1956(C16)
Novilla Canyon 247—1981(2)
Peregrina Canyon 152—1963; 247—1981(2)
Reynosa 152—1952, 61
Sedimentary Basin 296—1956(C16)
Sierra de El Abra 152—1963
Tampico 296—1906(30), 56(C16)

Veracruz
   General 296—1956(C7,C8)
   Angostura Fields 296—1956(C7)
   Casa Blanca Fields 296—1956(C7)
   Chavarillo 296—1906(2)
   Coast 296—1956(C16)
   Coatzacoalcos 296—1956(C11)
   Esperanza 296—1906(3)
   Huayacocotla 296—1956(C8)
   Istmo de Tehuantepec 296—1906(31), 56(C7,C15b)
   Jaltipan Sulfur Dome 296—1956(C7)
   Misantla 296—1956(C16)
   North Coast 296—1956(A10/C13)
   Orizaba 296—1906(2)
   Petroleum Fields 296—1956(C7)
   Poza Rica Petroleum District 296—1956(A10/C13,C10)
   Rio Papaloapan Basin 296—1956(C7)
   San Andreas Tuxtla 296—1956(C7)
   Santa Maria Tatetla 296—1906(2)
   Sedimentary Basin 296—1956(C16)
   Veracruz 296—1906(2), 56(C11)
   Veracruz Basin 296—1956(C7)

Yucatan
   General 197—1982(10)
   Northeast 247—1978(2)
   Coast, Northeast 247—1973(3); 255—1972
   Yucatan Peninsula 18—1965(4), 1976(5); 197—1967(7); 466—1962, 85

Zacatecas
   General 296—1906(16,17); 581—1974
   Aranzazu 296—1906(25)
   Avalos Mineral District 296—1956(C3)
   Concepcion del Oro 296—1906(24)
   Conception del Oro Mineral District 296—1956(C3)
   Fresnillo Mineral District 296—1956(A2/A5)
   Mazapil Mineral District 296—1956(C3)
   San Martin Mineral District 296—1956(A2/A5)
   Sierra de Mazapil 296—1906(26)
   Sombrerete Mineral District 296—1956(A2/A5)
   Zacatecas Mineral District 296—1956(A2/A5)

Netherlands Antilles
   General 190—1984(13)

Panama
   San Blas Islands 310—1977(5)

Puerto Rico
   General 76—1970(6); 197—1964(9); 214—1968
   Central 127—1959
   Southwest 696—1983
   West 127—1959
   Guayanes River Dam 76—1968(6)

UNITED STATES

Regional
   General 296—1933(28,29); 272—73, 77, 78, 79, 55, 56
   Central 702—n.d.
   East-central 710—1965; 498—1985
   Northeast 435—1979
   South 217—1933
   Southeast 206—1972(B3-4), 74; 710—1950, 67; 169—1981(1,2); 299—1985
   Southwest 217—1915(613); 277—1965(H); 296—1933(14); 417—1970, 71; 569—1984(1); 235—1985
   South-central 710—1949, 66
   West 217—1915, 22, 33; 289—1971(5)
   Adirondack Mountains 296—1933(1)
   Allegheny Front 172—1972
   Appalachian Basin 29—1948; 58—1970; 290—1979(4)
   Appalachian Basin, North 25—1982
   Appalachian Basin, South 197—1981(3,4)
   Appalachian Highlands 498—1985
   Appalachian Mountains 23—1982; 25—1985(No.3); 197—1971(10-11), 85; 270—1968(Apr.), 72(May); 290—1979(2,6); 296—1933(10,28); 443—1974; 499—1958, 68; 272—83
   Appalachian Mountains, Central 18—1977(4); 197—1971(10,11); 201—1982(6,7); 203—1978; 296—1933(7); 499—1963; 279—1982(19B); 291—1981; 54—1976
   Appalachian Mountains, East 319—1977(2)
   Appalachian Mountains, Northeast 288—1976; 302—1979(1)
   Appalachian Mountains, Piedmont 129—1978
   Appalachian Mountains, South 5—1965, 69, 71, 74, 84, 85; 76—1985(Oct.5-6); 197—1955(1), 80(7); 201—1982(5); 203—1978; 206—1972(3), 78(1); 296—1933(3); 319—1977(3); 499—1960; 688—1971, 79; 721—1948; 302—1979(2); 244—1985(2)
   Appalachian Mountains, Southwest 206—1971, 85(4)
   Appalachian Plateau 206—1969
   Appalachian Plateau, North 494—1969
   Appalachian region 206—1979(4)
   Atlantic Coast 568—1985
   Atlantic Coastal Plain 277—1965(B-1); 296—1933(10)
   Atlantic Coastal Plain, Central 197—1957(6)
   Atlantic Coastal Plain, South 76—1985(Oct.11-13)
   Basin and Range 549—1979
   Basin and Range, West 562—1978
   Black Hills 270—1968(June), 70(June)
   Blue Ridge Mountains 203—1978
   Central States 341—1932, 33, 34; 543—1947; 688—1964; 272—68
   Central States, North 74—1959
   Chesapeake Bay region 296—1933(5)
   Colorado Plateau 197—1979(3); 725—1971, 74
   Colorado Plateau, Northwest 197—1975(10)
   Colorado Plateau, South 272—114
   Cordillera 296—1933(28)
   Cordilleran Thrust Belt 549—1982
   East Coast 27—1981(5,Trip2); 55—1980
   East Coastal Plain, South 76—1985(Oct.11-13)
   Eastern Overthrust Belt 58—1982, 84
   Eastern States 586—1980
   Four Corners area 417—1970, 71
   Great Basin 277—1965(I); 280—1978(C-1); 549—1979; 317—1981(1); 725—1974
   Great Basin area 435—1975
   Great Lakes 244—1985(7)
   Great Lakes region 277—1965(G); 325—1979; 493—1957; 587—1975, 81, 83(Sept.)
   Great Lakes, South 272—38
   Great Plains 569—1984(2)
   Great Plains region 296—1933(28)
   Great Plains, Central 277—1965(D); 543—1951(1,2)
   Great Plains, West 272—28
   Great Smoky Mountains 499—1960, 65
   Great Smoky Mountains National Park 294—1973(1)
   Gulf Coast 18—1941; 206—1979(4); 247—1956(1-2), 59; 277—1965(B-3)
   Gulf Coast region 588—1967
   Gulf Coast, North 466—1968
   Gulf Coast, Southeast 626—1976(Spring)
   Gulf States, Coast 18—1985(2)
   Illinois Basin 290—1979(4,9)
   Interstate 40 272—160
   Lake Michigan Basin 183—1973
   Lake Michigan shore, South 76—1979; 200—1982(1)
   Lake Superior 200—1979(5); 273—1971(A), 75; 296—1891(C), 1933(27); 309—1977; 306—1981

Lake Superior region 270—1970(Aug.); 688—1961; 721—1951
Lake Superior, South 273—1984(2)
Midcontinent, East 681—1982, 85
Midwest 270—1949(June), 62(June)
Midwestern States, East 681—1982
Mississippi River basin 296—1933(26)
Mississippi River delta 277—1965(B-3)
Mississippi River valley 32—1968; 74—1949; 290—1979(8); 543—1949
Mississippi River valley Zinc-Lead District 723—1970, 75
Mississippi River valley, Central 722—1974
Mississippi River valley, Lower 197—1967(2); 392—1949
Mississippi River valley, Upper 197—1956(2); 277—1965(C); 435—1974; 587—1983; 610—1965; 659—1933, 56, 67, 82; 723—1970, 75
Missouri River headwaters 446—1970; 449—1970
New England 76—1976(3); 182—1935; 277—1965(A); 290—1979(5); 456—1948(3), 52, 54; 469—1985(A-1,B-1,A-2,B-2); 272—66
New England area 296—1933(1)
New England, Central 456—1971(A-3)
New England, Connecticut Valley 456—1957
New England, East 456—1984(A-3,B-7)
New England, East, Alton Bay 456—1984(C-7)
New England, Southeast 456—1976, 81(B-5,C-8), 82(Q4,P6)
New England, Southeast, Hope Valley Shear Zone 456—1985(B-4)
New England, Southeast, Narragansett Basin 456—1981(B-7,C-2)
New England, West-central 303—1983
North Central States 296—1933(26)
Northeastern States 296—1933(30)
Ohio River valley 277—1965(G)
Ouachita Mountains 205—1985(Apr.4); 270—1969, 73(May); 290—1979(11); 485—1966
Ozark Mountains 270—1935, 51(1), 69(Mar./Apr.), 73(May); 290—1979(11)
Pacific Coast 16—1915
Pacific Northwest 76—1981; 217—1915(611); 277—1965(J); 435—1972; 569—1985; 272—109
Pacific Southwest 569—1984(3)
Rocky Mountains 30—1974; 179—1979; 270—1968(June), 70(June); 277—1965(E); 290—1979(15); 292—1974(6); 296—1891(B); 549—1983, 84; 781—1977; 569—1984(2); 272—28, 112
Rocky Mountains, Central 549—1975, 77; 272—98, 76
Rocky Mountains, Front Range 528—1946, 51(May 17), 55; 549—1970; 721—1947
Rocky Mountains, Front Range, Central 549—1973
Rocky Mountains, North 270—1980; 272—7
Rocky Mountains, South 18—1972(A1,2); 417—1970, 71; 549—1977; 272—69, 76
Theodore Roosevelt National Memorial Park 569—1984(2)
Transcontinental 34—1912; 296—1933(28-30)
Valley and Ridge Province 203—1978
Western Coastal Ranges 296—1933(28)
Western Interior 590—1985(3)
Western Interior Basin 474—1977(7)
Western States 272—68
Yellowstone National Park 217—1915(612); 296—1933(24)

Alabama
General 5—1964(2nd), 66, 70, 85; 74—1948; 390—1940(3rd); 435—1961, 67; 566—1932
Central 5—1968, 85; 206—1972(B2), 80(1); 296—1933(2); 671—1973
East 5—1968, 77, 81, 83, 85; 206—1974
East-central 614—1948
North 5—1967, 75; 197—1967(5); 206—1972(A); 391—1961; 390—1949; 671—[n.d.]
Northwest 206—1980(2); 390—1940(4th), 54; 566—1929; 588—1983(Apr.)
South 222—1973; 247—1962
Southeast 614—1946, 63
Southwest 5—1976; 197—1955(2); 206—1981(1); 390—1960; 614—1944(1st); 699—1972
West 390—1945, 52
West-central 197—1955(3); 247—1968, 83; 390—1943, 53, 57, 59
Appalachian Mountains 5—1965(3rd), 69, 71, 74, 84; 197—1980(16,21,23), 82(13)
Bibb County, Central 671—1978
Birmingham 5—1978; 206—1958
Black Warrior Basin 5—1982; 390—1978
Blount County 5—1964
Chattahoochee River valley 197—1980(20)
Chilton County 206—1972(B1)
Coast 5—1972; 466—1982(June)
Coastal Plain 74—1968; 197—1967(1); 206—1958; 222—1968
Colbert County 206—1980(4)
Coosa County 671—1968
Eutaw 390—1945
Hillabee 5—1979
Jefferson County 5—1964, 78
Kelley Mountain 5—1985
Mining Districts 43—1971
Montgomery area 223—1960
Piedmont 5—1964(2nd), 70, 75, 81
Piedmont, North 5—1975
Pike County 4—1962
Pine Mountain Window 5—1981; 206—1980(3)
Russellville 4—1965
South-Central 296—1913(10)
Talladega 5—1973, 85
Talladega Slate Belt 197—1980(23)
Tuscaloosa 390—1945
Warrior Basin 18—1976(4)

Alaska
General 287—1978(1); 569—1985
Central 277—1965(F), 77
East-central 28—1985(1)
North 287—1983(4)
South 6—1963, 64, 70, 73
Southeast 199—1984
South-central 277—1965(F), 77; 7—1981; 287—1983(1)
Alaska Railroad 287—1983(6)
Alaska Range 287—1983(1)
Anchorage 6—1963-64, 73; 28—1985; 199—1984(4)
Birch Creek Valley 197—1977(15)
Caribou Creek 6—1964
Castle Mountain Fault 199—1984(6)
Chugach Mountains 199—1984(2)
Coast, Southeast 296—1913(10)
Colville River delta 287—1983(2)
Cook Inlet Basin 6—1970; 199—1984(3)
Copper River basin 287—1983(1)
Dalton Highway 287—1983(4)
Denali National Park 272—46
Elliott Highway 287—1983(4)
Fairbanks 28—1985(1); 199—1984(9)
Glacier Bay 199—1984(1); 296—1913(10)
Glenn Highway 287—1983(1)
Kenai Peninsula 28—1985(2)
Malaspina Glacier 296—1913(10)
Matanuska Glacier 287—1983(1)
Matanuska Valley 287—1983(1)
Mount McKinley National Park 272—46
Prudhoe Bay 287—1983(5)
Richardson Highway 287—1983(1)
River delta River area 287—1983(1)
Seward Highway 7—1981
Sutton 6—1963-64
Tanana River valley, Middle 287—1983(1)

Turnagain Arm 199—1984(5)
Willow Creek 199—1984(6)
Yakutat Bay 296—1913(10)

Arizona
General 28—1982(1); 199—1952(4), 59(1,2); 272—29
Central 199—1978
East 672—1965, 67, 71(183)
East-central 32—1978; 40—1974; 458—1962
North 204—1974; 296—1933(18); 327—1970; 604—1947; 269—1983; 272—114, 101
Northeast 180—1979; 458—1958, 67; 590—1985(8)
North-central 204—1974(2)
South 199—1952, 59, 68(6), 81
Southeast 185—1965; 199—1959(5), 68(1), 81; 458—1978
Southwest 65—1979
South-central 32—1971([1]); 672—1971(183)
West 197—1979(16)
Apache County 180—1955, 63
Basin and Range 197—1979(15)
Benson 199—1968(3)
Bisbee Mining District 296—1933(14)
Black Mesa 197—1979(15); 204—1974(7)
Black Mesa Basin 180—1955; 458—1958
Coconino County 185—1970
Coconino County, San Francisco Volcanic Field 65—1982
Colorado Plateau 197—1979(15); 296—1933(18)
Colorado River, Marble Gorge 272—130
Empire Mountains 199—1968(2)
Gila River, Upper 185—1965
Graham County 185—1964
Grand Canyon 104—1969; 161—1979(Mar.); 180—1969; 217—1915(613); 290—1979(13); 327—1970; 688—1968(spring), 73; 269—1983; 725—1973
Grand Canyon National Park 296—1933(18); 569—1984(1)
Grand Canyon National Park, East 103—1968(5); 272—12, 105, 104
Grand Canyon National Park, West 103—1969(2)
Grand Canyon, Colorado River 272—130, 155
Grand Canyon, South 204—1974(1)
Grand Canyon, West 737—1970
Granite Wash Mountains 65—1983
Greenlee County 185—1964
Hackberry Mountain 204—1974(5)
Harquahala Mountains, West 65—1983
Hopi Buttes 204—1974(6)
Jerome District 204—1974(8)
Lake Mead National Recreation Area 272—18
Little Colorado River 204—1974(7)
Marine Corps Gunnery Range 65—1979
Meteor Crater 32—1971(1); 373—1979
Monument Valley 327—1970
Monument Valley area 458—1973
Navaho County 180—1963
Navajo Buttes 204—1974(6)
Petrified Forest National Park 32—1971(1); 569—1984(1)
Pima 199—1968(4)
Pima County, Santa Rita Mountains 65—1966
Pinnacles National Park 272—80
Plomosa Mountains, North 65—1983
Queen Creek 199—1952(1)
Salt River Canyon 64—1981
San Francisco Mountains 204—1974(4)
San Francisco Peaks 55—1976; 185—1970; 204—1974(3); 272—103
San Juan Basin 458—1951
San Pedro Valley 185—1968; 199—1968(6)
Santa Catalina Mountains 199—1952(3), 59(3)
Santa Cruz County, Santa Rita Mountains 65—1966
Santa Rita Mountains 65—1966; 199—1968(2)
Silver Bell Mountains 199—1959(2)
Silver Mining District 65—1979
Superstition Wilderness 272—128
Tucson 199—1968(3)
Tucson Mountains 199—1952(2), 59(4), 68(5)
Uinkaret Plateau 725—1973
Verde Valley 204—1974(5)
Waterman Mountains 199—1959(2)

Arkansas
General 67—1975; 583—1974(Feb.); 662—1976; 721—1946
Central 74—1979(2); 177—1982; 197—1967(3); 205—1973(4), 85(Apr.[3]); 296—1933(2,16); 573—1969
North 360—1973
Northeast 205—1973(3)
Northwest 67—1955; 205—1978(1); 341—1956(20th); 390—1962; 474—1977(5); 566—1951; 662—1979
North-central 205—1979(1); 589—1984
South 566—1923, 39
Southwest 67—1980(80-1,80-2), 82(80-2); 205—1985(Apr.[3]); 566—1925, 47, 53, 61, 65
Arkansas Valley 55—1984; 67—1980(77-1); 662—1977
Arkansas Valley Basin, West 177—1959
Arkansas Valley area 197—1982(3)
Arkansas Valley, Southeast 177—1963
Batesville 205—1973(3)
Bauxite 91—1954
Bauxite Deposits 296—1933(2)
Boston Mountains, South 205—1985(1)
Buffalo National River 205—1979(2)
DeGray Dam site 349—1963
Hope 566—1924
Hot Springs 571—1951(1)
Lake Ouachita 67—1980(80-1); 74—1979(1); 205—1973(2)
Little Rock 205—1973(3); 571—1951(3)
Magnet Cove 154—1977; 341—1956(20th); 566—1953; 571—1951(2)
Ouachita Mountains 18—1959, 68(2), 75(SEPM 1), 78(3), 78(SEPM 1); 67—1980(77-1), 83, 84(80-2); 91—1954; 177—1963, 64; 197—1982(3,6); 205—1973(1); 341—1931, 66; 566—1948, 53; 662—1977; 715—1982; 589—1984; 485—1966
Ouachita Mountains, East 255—1982
Ouachita Mountains, Magnet Cove 177—1964, 82
Ozark Mountains 67—1980(77-1); 262—1973; 341—1928; 662—1977
Ozark Mountains region, South 205—1985(2)
Ozark area 197—1982(6)
Potash Sulphur Springs 571—1951(2)
Stone County, Blanchard Springs Cavern 205—1979(3)

California
General 18—1947, 52; 76—1969(1); 197—1979(3); 215—1955; 277—1965(I); 314—1976; 412—1972(Oct.)(2), 79(Oct.A,B); 593—1984(Apr.); 568—1984; 590—1985
Central 132—1965(1); 593—1983(2); 272—115
East-central 215—1962
North 18—1962, 81; 39—1979(7); 111—1964; 109—1968; 197—1984(16); 199—1977(9); 412—1974(Oct.); 272—9, 52, 53
Northeast 272—119
Northwest 199—1979(12); 215—1960; 37—1985
North-central 107—1961(1,2); 593—1984(Sept.)
South 18—1967(B1,B4), 73(1,SEG 1, SEPM 2); 28—1979(Mar.), 82(1); 45—1956; 84—1975, 76; 161—1973, 78(Fall); 199—1971, 82(3-4,14); 296—1933(15); 338—1979; 429—1975(3); 542—1978(Jan.); 593—1951(Dec.), 84(2); 604—1966(A); 282—1985(4); 272—71, 72, 122, 123, 125
Southeast 197—1979(16,22)
West-central 215—1978
Alameda County 199—1963
Alta California 28—1970(Nov.)
American River, South fork 272—157

Anacapa Island 412—1982(Oct.)
Antelope Valley 28—1964(Oct.)
Anza-Borrego Desert 604—1966(A)
Auburn area 412—1984(Oct.)(A)
Baldwin Hills 18—1967(A6)
Barstow area 604—1966(A)
Basin and Range 199—1975
Berkeley Hills 199—1963(1); 296—1933(16); 412—1966
Berkeley area 604—1966(B)
Big Mountain 28—1967(June)
Bodega Bay 18—1981(3)
Borrego Mountain 412—1978(2)
Burney Falls area 215—1969
Cajon Pass 28—1962(Oct.); 296—1933(15); 356—1984
Calaveras County 412—1984(Oct.)(B)
Calaveras Fault 199—1970(5), 79(2,5)
Calaveras fault zone 76—1969(2A,2B); 199—1979(5)
Caliente Range 28—1951
California Aquaduct 28—1968(June)
California Desert 609—1980
Cantua Creek 593—1974(Oct.)
Capay Valley 28—1954(May 7-8)
Carmel 412—1976
Castaic 28—1973(June)
Castaic Dam 76—1973(8)
Castle Steam Field 28—1978
Central, Cantua Creek 593—1974(Oct.)
Central, Panoche Creek 593—1974
Chico 567—1966
Chico area 557—1959
Clear Lake 199—1977(2)
Coachella Valley 412—1979(Mar. 1,2,3), 85(Mar.)
Coalinga 412—1983(Fall); 593—1972(Oct.)
Coalinga area 272—40
Coast Ranges 298—1978; 590—1984(3); 19—1984
Coast Ranges, Central 18—1981(1); 199—1970(1); 215—1959
Coast Ranges, North 197—1966(B5); 215—1960, 71, 72, 78
Coast Ranges, South 18—1973(SEPM 2); 161—1971(Fall), 79(Fall); 593—1977(Oct.)
Coast, Central 18—1981(4,A); 114—1969
Coast, South 18—1981(B); 429—1975(3); 593—1971; 590—1984(1)
Coast, Southwest 197—1979(1)
Contra Costa County 28—1950; 199—1963
Crestmore Quarry 199—1971(5)
Cucamonga fault zone 199—1982(12)
Cuyama 28—1951
Death Valley 41—1976; 161—1978(Dec.), 80(Dec.); 197—1954(1); 199—1974(1); 412—1983(Mar.); 593—1951(Dec.)
Death Valley National Monument, Titus Canyon 108—1985
Death Valley area 215—1970
Devil's Den 28—1955
Diablo Canyon 199—1970(4)
Diablo Range 18—1981(2); 199—1979(1)
Diablo Range, North 590—1984
Diablo Range, South 28—1955
East-central, Truckee 284—1983
Eel River basin 215—1960
El Dorado County 215—1973
Feather River, South Fork 107—1961(2)
Fillmore 23—1977
Fresno County 199—1965(2-3)
Fresno County, Panoche Hills 593—1960
Gabilan Range 28—1967(Oct.)
Geysers Geothermal Area 32—1983; 199—1977(2); 412—1979(B); 82—1976
Geysers Geothermal Field 18—1981(AAPG 2); 53—1970, 72
Great Basin 197—1984(7)
Great Valley, Central 215—1963
Hall Canyon 18—1967(B3)

Hayward Fault 199—1970(5), 79(2)
Hayward area 487—1979
Hayward fault zone 38—1976; 76—1969(2A,2B)
Huasna Basin 593—1956 (May); 590—1984(1)
Humboldt County, West 412—1982(Oct.)
Imperial Valley 18—1973(2); 28—1958; 243—1978; 412—1978(Mar.), 85(Mar.)
Interstate 15, East 272—111
Inyo County, White-Inyo Mountains 522—1966; 593—1976(Oct.)
Inyo Mountains, Papoose Flats 161—1971(Spring)
Inyo Mountains, South 199—1983(7)
Kern County 28—1961(May), 81(2); 322—1963
Kern County, Round Mountain 557—1958
Kern County, San Emigdio Creek 593—1948
Kings Canyon Highway 199—1979(11)
Kings Canyon National Park 32—1973(2)
Kingvale 197—1984(A)
Klamath Mountains 199—1977(6)
Klamath Mountains, South 215—1974
Klamath River 272—53
La Jolla area 593—1957
Lake Cahuilla 412—1979(3)
Lake Casitas 28—1979(June)
Lake County 53—1970, 72
Lake Shasta area 215—1969
Lake Tahoe 151—1964(1); 197—1984(23); 412—1985(Oct.)
Lake Tahoe area 215—1968
Lassen National Park 412—1981(Fall)
Lassen Peak area 215—1969
Lassen Volcanic National Park 272—119
Livermore area 199—1963(1); 487—1979
Long Beach 609—1974
Long Beach Harbor 76—1973(5)
Long Beach offshore 355—1967
Long Valley 161—1976
Los Angeles 18—1958, 67(A4,A5,A6,B1)
Los Angeles Basin 18—1937, 67(A5), 73(SEG 1,SEPM 1); 28—1966(Mar.), 69(Mar.), 70(Mar.), 75(Apr.); 197—1954(3-4); 296—1933(15)
Los Angeles Basin, Northeast 412—1981(Apr.)
Los Angeles Coastal Plain 76—1973(7)
Los Angeles County 28—1944, 65(June); 199—1982(5); 609—1973
Los Angeles County, Northwest 609—1977
Los Angeles Harbor 76—1973(5)
Los Angeles area 77—1978
Los Angeles area, La Brea Tar Pits 76—1973(4)
Los Angeles area, Point Fermin 76—1973(2)
Los Angeles area, Portuguese Bend 76—1973(2)
Los Angeles area, Sierra Madre Fault 76—1973(3)
Madera County 199—1965(3)
Malibu Coast 76—1973(1)
Mammoth 412—1980(Sept.)
Mammoth Lakes 184—1971; 272—134
Marin County 50—1977
Martinez Creek area 557—1959
Martinez Mountain 412—1979(3)
McClure Valley 28—1955
Mecca Hills 412—1979(2)
Medicine Lake 39—1979(7)
Mendocino County, Laytonville Quarry 197—1966(B7)
Merced Canyon 18—1962(4)
Modoc Plateau 215—1980
Mojave Desert 161—1980(Spring); 197—1954(1); 199—1971(3), 82(1,2,9,13); 619—1980
Mojave Desert, East 197—1984(14)
Mojave Desert, Northwest 197—1984(7)
Mojave Desert, Southwest 609—1976
Mono Craters 161—1976; 197—1984(11)
Mono Lake 272—42
Monterey Bay 199—1979(10)

Monterey County, Salinas Valley 593—1979
Monterey County, Santa Lucia Range 593—1979
Mother Lode 412—1977(Oct.), 84(Oct.)(B)
Mother Lode area 112—1948; 199—1977(3); 215—1958; 296—1933(16); 272—24
Mount Diablo 18—1962(6); 28—1950; 199—1955(2); 215—1964; 412—1977(Dec.); 487—1969; 604—1966(B); 272—19, 20
Newport Lagoon 593—1971
North, Crafton Hills 412—1977(Mar.)
North, Mill Creek Canyon 412—1977(Mar.)
Oakland 199—1955(2)
Oceanside 199—1971(8)
Ojai Valley, West 28—1979(June)
Orange County 28—1970(Mar.); 76—1973(3); 412—1972(Oct.)(1)
Orange County, Coast, South 609—1979
Orange County, Costa Mesa 618—1984
Orange County, East 609—1973
Orange County, Southwest 412—1981(Apr.)
Owens Valley 161—1980(Fall)
Painted Canyon 412—1979(2)
Palos Verdes Hills 18—1967(A6)
Palos Verdes Peninsula 199—1982(10); 609—1974
Panamint Valley 184—1978
Panoche Creek 593—1974
Panoche Pass 199—1965(1)
Peninsular Ranges 28—1974(Apr.); 161—1972(Spring); 197—1979(21); 199—1971(4), 75(2)
Peninsular Ranges, North 197—1954(5), 79(5,20-21,24); 412—1972(Oct.)(1)
Peninsular Ranges, South 161—1972(Spring)
Petaluma 199—1955(1)
Pinnacles National Monument 272—80
Piru 23—1977
Pismo Basin 590—1984(1)
Placer County, South 412—1975(Oct.)(1,2)
Plumas County 215—1956, 65
Point Lobos 412—1976
Point Reyes 18—1962(3); 197—1966(B1); 487—1970
Point Reyes area 215—1971
Point Sal 199—1975(5)
Puente Hills, East 412—1981(Apr.)
Pyramid Dam 76—1973(8)
Ridge Basin 28—1973(June), 82(3); 199—1982(1); 609—1977
Rincon Island 28—1961(June)
Riverside County 412—1972(Oct.)(1)
Riverside County, Central 197—1979(22)
Sacramento Basin, South 199—1979(9)
Sacramento County 412—1972(Apr.)
Sacramento County, East 215—1973
Sacramento County, North 215—1967
Sacramento River delta 28—1976 (Apr.)
Sacramento Valley 18—1962(1); 197—1966(B5); 199—1977(8); 215—1958; 322—1963; 593—1983(1)
Sacramento Valley, East-central 215—1961
Sacramento Valley, North 107—1961(1)
Sacramento Valley, Southwest 593—1983(3)
Sacramento Valley, West 28—1954(May 7-8)
Sacramento area 412—1984(Oct.)(A)
Salinas Valley 28—1963
Salton Basin 197—1979(4)
Salton Trough 197—1979(7,22); 199—1971(2); 412—1978(1), 85(Mar.); 20—1981
San Andreas fault zone 150—1967; 199—1970(5), 75(1), 82(4); 296—1933(15); 322—1963; 487—1970; 272—95
San Andreas fault zone, Antelope Valley area 28—1964(Oct.)
San Andreas fault zone, Cajon Pass 28—1962(Oct.), 84(June); 356—1984
San Andreas fault zone, Carrizo Plain 199—1979(4)
San Andreas fault zone, Central 197—1966(B3)
San Andreas fault zone, Gabilan Range area 28—1967(Oct.)

San Andreas fault zone, Indio Hills 412—1979(1)
San Andreas fault zone, Los Angeles area 197—1979(8)
San Andreas fault zone, North 18—1962(3); 197—1966(B1); 199—1977(5); 50—1977
San Andreas fault zone, Salinas area 28—1963
San Andreas fault zone, San Francisco 28—1971
San Andreas fault zone, South 199—1971(7), 75(1), 82(4,14); 429—1975(1)
San Bernardino Mountains 84—1976, 78; 199—1982(6,9), 71(9)
San Cayetano 28—1977(June)
San Clemente 593—1971
San Diego 24—1977; 184—1975; 199—1971(8); 412—1973(reprint of 1972 T-10); 556—1973, 82
San Diego County 197—1979(5,18-19,20-21,23); 199—1961(1-4)
San Diego County, Coast, North 609—1979
San Diego County, Coastal Plain 197—1979(23); 556—1981
San Diego County, La Jolla 32—1973(1); 197—1979(11)
San Diego County, North 556—1985
San Diego County, Point Loma 197—1979(11)
San Diego County, Point Sal area 197—1979(24)
San Diego County, South 609—1983
San Diego County, Southwest 556—1977
San Diego County, Torrey Pines State Preserve 32—1973(1)
San Diego County, West 556—1975
San Diego area 197—1979(11,18-19); 430—1972
San Fernando 76—1973(6); 356—1971; 593—1951(Dec.)
San Francisco 32—1983; 197—1966(C); 272—152
San Francisco Bay 76—1969(6)
San Francisco Bay area 38—1976, 83, 84; 76—1969(4,6); 112—1951; 132—1965(2); 184—1972; 199—1979(2); 412—1974(Apr.); 604—1966(B)
San Francisco Bay area, Alcatraz Island 529—1979
San Francisco Bay area, East 593—1983(3)
San Francisco Bay area, East Bay 197—1966(B4)
San Francisco Bay area, North 131—1983
San Francisco Peninsula 18—1962(5); 28—1971; 76—1969(3,5); 197—1966(B2); 296—1933(16)
San Francisco Peninsula, Fort Funston 590—1984(5)
San Gabriel Fault 28—1960
San Gabriel Mountains 199—1982(5)
San Gabriel Mountains, West 412—1980(Apr.)
San Jacinto Mountains 609—1981
San Jacinto fault zone 199—1971(6); 412—1978(2); 429—1975(1)
San Joaquin Basin, Central 593—1982(V.25)
San Joaquin Basin, West 593—1985
San Joaquin River delta 199—1977(10)
San Joaquin Valley 18—1937; 56—1977; 322—1963
San Joaquin Valley, East 107—1959
San Joaquin Valley, South 28—1961(May), 68(Mar.), 77(Apr.); 412—1973(Apr.)
San Joaquin Valley, Southeast 28—1965(Apr.)
San Joaquin Valley, West-central 28—1972(Mar.)
San Jose area 605—1974
San Luis Obispo County 199—1967(3), 70(4); 593—1980
San Luis Obispo County, La Panza Range 593—1980
San Luis Reservoir 199—1965(4)
San Mateo County 199—1979(8)
San Miguelito 28—1970(June)
San Miguelito Field 45—1956
San Nicolas Island 199—1975(3)
San Onofre 28—1979(Mar.); 76—1973(7)
San Pedro Basin 412—1984(Mar.)(2,3)
San Pedro Bay 28—1966(Mar.)
Santa Ana Mountains 28—1982(2); 199—1982(8); 593—1953(May)
Santa Ana Mountains, North 618—1984
Santa Ana River basin 609—1978
Santa Barbara Channel 18—1973(3)
Santa Barbara Coastal Plain 296—1933(15)
Santa Barbara County 28—1947, 54(May 15), 65(Oct.), 72(June); 199—1967(2); 220—1984; 412—1971; 593—1949, 65(Oct.)

Santa Catalina Island 18—1967(B2); 84—1977; 197—1954(6); 412—1984(Mar.)(1-4); 609—1985
Santa Clara County 199—1979(3)
Santa Clara Valley 199—1979(3)
Santa Clara Valley, North 197—1966(B4)
Santa Cruz County 199—1979(8)
Santa Cruz Mountains 18—1962(2); 199—1979(6,7)
Santa Cruz Mountains, Big Basin area 593—1959(Apr.)
Santa Maria area, East 593—1975(Oct.)
Santa Monica Mountains 18—1967(A7); 76—1973(1); 84—1979; 296—1933(15); 593—1973(Oct.)
Santa Monica Mountains, Central 28—1980
Santa Paula River basin 296—1933(15)
Santa Rosa Island 28—1968(Oct.)
Santa Susana Mountains 28—1966(June)
Santa Ynez Mountains 28—1972(June); 199—1967(2); 593—1965(Oct.); 680—1958
Searles Valley 184—1967
Sequoia National Park 32—1973(2)
Sequoia region 435—1966
Sespe Creek 593—1969(Oct.), 78(Oct.)
Shasta County 215—1957
Sierra National Forest, Fresno County 108—1984
Sierra Nevada 151—1964(1); 184—1971, 79, 84; 296—1933(16); 272—135
Sierra Nevada Province 199—1975, 77
Sierra Nevada, Central 76—1975; 197—1966(B6); 199—1979(11)
Sierra Nevada, East 161—1980(Fall); 184—1979
Sierra Nevada, East-central 185—1971; 215—1966
Sierra Nevada, Mammoth Lakes 272—134
Sierra Nevada, North 197—1984(5); 199—1977(1); 412—1985(Oct.)
Sierra Nevada, Southwest 199—1977(7)
Sierra Nevada, West 161—1982(Fall); 412—1977(Oct.)
Sierra Nevada, Yosemite area 272—2
Simi Hills 593—1973(Oct.), 81(Oct.)
Simi Valley 593—1983(Oct.); 604—1954
Soledad Basin 28—1965(June); 604—1954
Sonoma 199—1955(1)
Sonoma County 53—1970, 72; 487—1968; 82—1976
South Mountain 18—1967(A2)
South, Camp Pendleton 556—1975
South, Mecca Hills 20—1981
South, Palos Verdes Peninsula 272—110
South, coast 272—124
Southern California Batholith 429—1975(2)
Stanislaus River 215—1975
Tehachapi Mountains 28—1968(June)
Tehachapi Mountains, Liveoak Canyon 593—1956(Feb.)
Temblor Mountains 28—1964(Oct.)
Transverse Ranges 199—1982(11); 609—1982; 680—1958; 272—95
Transverse Ranges, West 184—1981
Truckee 151—1964(1); 284—1983
U. S. Interstate 80 272—49
Ventura 18—1958; 28—1970(June)
Ventura Avenue Field 45—1956
Ventura Basin 18—1967(A8,B3); 28—1962(June), 69(Mar.), 76(June); 74—1953(Feb.), 73; 197—1954(2); 199—1982(3)
Ventura County 18—1937, 67(A2); 23—1977; 28—1953, 61(June), 67(June), 69(June), 75(June), 79(June), 82(June); 199—1967(1); 412—1974(June), 82(Mar.)
Ventura County, Pine Mountain 593—1969(Oct.)
Ventura County, Sespe Creek 593—1969(Oct.), 78(Oct.)
Warner Mountains 215—1980
Wheeler Canyon 18—1967(B3)
White-Inyo Mountains 161—1980(Fall); 522—1966; 593—1976(Oct.)
Whittier 18—1967(A5)
Whittier Hills 28—1944
Whittier fault zone 28—1975(Apr.)
Wilmington Field 18—1967(A4)
Yosemite Valley 18—1962(4); 197—1966(B6); 296—1933(16); 272—3
Yuba County, Western World Massive Sulfide Deposit 85—1984(6)

Colorado

General 179—1979; 280—1978(C-3/C-4); 296—1933(19); 549—1958-59; 682—1968; 722—1972; 293—1985(4); 139—1972; 272—92, 91, 30, 100
Central 18—1948(1-3), 72(1,3); 32—1980; 185—1981; 197—1960(A-3), 76(16); 204—1958, 73(8); 549—1947, 54(1,2); 158—1982; 78—1981; 590—1985(4,6); 272—106
East 596—1983; 590—1985(13); 293—1985(5)
North 204—1965(2,4); 724—1978(Sept.)
Northeast 55—1984(2)
Northwest 197—1976(19); 549—1953, 55(Aug.), 60(S12,S13,S14,S15,S16,S17); 596—1981
North-central 204—1979(1B); 296—1933(19); 549—1960(S19,S20,S21,S22)
South 458—1968; 528—1938
Southeast 18—1980(SEPM 3); 528—1946, 51(May 17), 55; 549—1960(S1,S2), 68
Southwest 180—1955, 60; 246—1962; 549—1960(S4,S5,S6,S7,S8,S9,S10); 141—1985
South-central 197—1960(A-2), 76(1); 341—1930, 58; 458—1966, 71(3); 504—1956(Sept.); 549—1960(S3,S26), 61
West 458—1981; 282—1985(2)
West-central 27—1980; 197—1960(A-1); 246—1983(Sept.); 549—1960(S11,S23,S24,S25); 142—1985
Arapaho Cirque 55—1984([1],[2])
Arkansas Valley, Upper 55—1984
Bellvue 204—1979(1A)
Big Thompson Canyon 204—1965(6), 79(2)
Big Thompson River Project 197—1960(C-2)
Boulder 204—1968(3), 73(4,6); 277—1965(K); 272—117
Boulder area 197—1976(21)
Canon City area 197—1960(C-1); 596—1984
Central City Mining District 197—1976(7)
Clear Creek County 197—1960(C-3,C-4)
Climax 197—1960(B-3)
Climax Molybdenum District 296—1933(19)
Colorado Piedmont 55—1984(2)
Colorado Plateau 180—1972; 296—1933(19); 364—1978
Colorado Plateau, North 272—113
Colorado Springs area 197—1960(C-1); 272—39
Colorado-Wyoming State Line 204—1979(4)
Creede Mining District 158—1981
Cripple Creek Mining District 296—1933(19)
Custer County 185—1972
Deer Creek 204—1958
Denver 219—1967
Denver Basin 549—1978
Denver Basin, Northwest 204—1979(1)
Denver Basin, West 197—1976(18); 549—1963
Denver area 179—1979(appendix); 197—1976(5); 245—1969
Dinosaur National Monument 296—1933(17); 272—59, 148
Dunton area 246—1962
Durango 30—1964; 204—1984(4)
Estes Park 185—1952
Foothill region 296—1933(19)
Fort Collins 204—1965(4,5), 79
Fremont County 185—1972
Front Range 55—1984([1],[2],3,5,11); 190—1982(12); 197—1960(B-1,B-2,C-3,C-4,C-6), 76(A3,7-8,15,22); 204—1958, 67, 73(1,7), 79(1C,2-3); 341—1929
Front Range Foothills, Central 197—1960(C-4)
Front Range Foothills, North 197—1960(C-5)
Front Range, Central 197—1976(22)
Front Range, East-central 55—1984(8)

Front Range, North 197—1976(3,8,15)
Front Range, South 197—1976(1)
Garfield County, Parachute Creek 501—1985(1)
Gilpen County 197—1960(C-3,C-4)
Golden 204—1958, 67(4), 73(4); 296—1933(19)
Golden area 197—1976(12,21)
Grand Canyon District, Western 725—1973
Great Sand Dunes National Monument 458—1971(2)
Greeley 204—1973(10)
Green River 272—58
Gunnison County, South 158—1983
Gunnison Gold Belt 158—1983
Gunnison area 272—106
Gypsum Valley 737—1948
Henderson Mine 197—1976(20)
Idaho Springs Mining District 197—1976(7)
Jefferson County 197—1976(12); 549—1955(May)
Jefferson County, Schwartzwalder Uranium Mine 197—1976(9)
La Plata County 180—1972
Lake Devlin 204—1973(1)
Larimer County 204—1979(1A,1C)
Leadville Mining District 296—1933(19)
Livermore 204—1979(1A)
Loveland 204—1973(10)
Lyons 204—1973(4,5)
Manitou Springs area 596—1984
Mesa Verde National Park 569—1984(1)
Mining Districts 43—1969
Niwot Ridge 55—1984(4); 204—1973(7)
North Park 781—1953
Paradox Basin 18—1972(B2); 180—1955, 60; 246—1983(Oct.); 549—1981
Parks Basin 549—1957
Piceance Basin 246—1982; 549—1965, 74
Piceance Basin, West 246—1982
Pikes Peak Batholith 197—1976(15)
Pueblo 203—1983
Raton Basin 549—1956, 69
Raton Basin, South 197—1984(3)
Red Cliff Mining District 296—1933(19)
Red Mountain 197—1976(20)
Rico area 246—1962
Rio Grande Rift 197—1976(16); 320—1978; 463—1981
Rocky Mountain National Park 590—1985(13); 293—1985(5); 272—153, 31, 112
Rocky Mountains 341—1929; 549—1964, 84
Rocky Mountains, Front Range 18—1972(B1), 80(SEPM 1); 549—1938, 55
San Juan Basin 204—1984(5); 296—1933(19); 458—1950
San Juan Mountains 197—1976(17); 204—1984(1); 246—1978; 583—1982; 158—1981
San Juan Mountains, East 458—1971(1)
San Juan Mountains, North 246—1960
San Juan Mountains, Southwest 458—1957
San Juan Mountains, West 30—1964; 246—1961; 296—1933(19)
San Luis Basin 458—1971
San Luis Valley 197—1976(2,16)
Sand Hills 55—1984(2)
Silverton 30—1964
St. Vrain Drainage Basin, North 204—1979(3)
Twin Lakes 185—1953
Uncompahgre Plateau 141—1985
Uncompahgre Uplift 246—1983(Oct.)
Wellington 204—1965(4), 79(1D)

Connecticut
General 456—1968(B-4), 75(B-7,C-7), 85(C-4); 272—62
Central 296—1933(1); 456—1953(A-E), 68(C-2,F-6), 82(Q1); 586—1978; 632—1978
East 456—1968(F-4), 82(P8)
Northeast 456—1968(F-5), 81(A-1)
Northwest 456—1975(A-1,B-5,B-7,B-8,C-5,C-7)
North-central 456—1982(M-3)
South 456—1968(C-2), 85(B-1,C-1,C-6,C-7)
Southeast 197—1963(2); 456—1985(A-5)
Southwest 456—1968(B-3,D-6)
South-central 456—1968(C-4,D-1,F-3), 82(P7); 632—1970
West 456—1968(B-1,B-2,D-2), 82(P1A)
Berkshire Massif 456—1975(B-2,B-3,B-6,C-2,C-6,C-11)
Bethel area 456—1985(B-2)
Burlington area 456—1948(2)
Central, Hartford Basin 456—1985(B-7)
Coast 197—1963(5); 456—1968(A-1); 632—1965
Collinsville area 456—1968(D-4)
Connecticut River valley 51—1968(H); 456—1968(C-1,C-5), 82(M1,M2)
Deep River area 456—1985(A-4)
East, Bonemill Brook fault zone 456—1982(P2)
Gaillard Graben 456—1968(C-4); 632—1970
Haddam area 456—1985(B-5)
Hartford 74—1953(Sept.); 456—1953(A,C)
Hartford Basin 456—1985(B-7)
Honey Hill Fault 456—1968(F-1), 85(C-4)
Kent area 456—1982(P1)
Killingworth Dome 456—1968(F-3)
Lake Char Fault 456—1968(F-1)
Long Island Sound 456—1968(E-1)
Manchester 456—1953(D)
Middletown area 456—1985(C-5); 631—1927
Mount Prospect area 456—1985(C-3)
New Haven 456—1953(E)
North, Hartford Basin 456—1982(M-4)
Rockville 456—1953(D)
Rocky Hill, Dinosaur Park 456—1968(C-3)
Shetucket River basin 456—1982(Q3)
Southwest, Bethel area 456—1985(B-2)
Stony Creek area 456—1985(A-3)
Thomaston Quadrangle 456—1968(D-5)
Waterbury Dome 456—1985(A-1)
Waterbury Quadrangle 456—1968(D-5)
West Torrington 456—1985(C-2)
Willimantic Fault 456—1982(P9)
Willlimantic Dome 456—1982(P9)

Delaware
General 18—1977(3); 74—1977(4); 89—1967; 182—1976
North 74—1977(2,3); 132—1958
Coast 74—1977(5)
Coastal Plain 76—1976(1); 89—1961, 73; 197—1971(1); 539—1976
Lewes 592—1968
Piedmont 74—1977(1)
Rehoboth Beach 592—1968

District of Columbia
General 201—1982(8,9); 304—1962; 411—1977
Washington 296—1891(A), 1933(16th)
Washington, DC area 69—1968(A1-9,B)

Florida
General 74—1952; 132—1963; 175—1959, 64; 197—1985(9); 374—1969; 588—1956; 614—1967; 688—1972; 710—1964
Central 179—1971; 197—1985(3); 206—1983; 374—1981; 604—1964; 614—1960, 70, 77
North 197—1985(3); 614—1966, 71, 82
Northwest 197—1985(12); 614—1983
South 197—1964(1,7,10), 74(6,7); 374—1968, 83; 504—1970; 533—1976; 614—1954; 688—1978, 84
Southeast 197—1964(4)
Southwest 247—1969(4); 374—1979, 80

South-central 614—1978
West 206—1967(2); 614—1945
West-central 197—1964(6); 614—1975, 76
Alligator Point 206—1967(1)
Apalachicola River 614—1983
Apalachicola River area 197—1985(12)
Atlantic Coast 197—1974(5)
Atlantic Coast, Central 197—1985(4)
Barrier Islands 197—1974(5), 85(1)
Coast, Southwest 197—1985(13)
Dry Tortugas 614—1979
Everglades 55—1972(2); 533—1976
Everglades National Park 197—1964(10); 374—1967
Florida Keys 197—1964(3), 74(12); 247—1969(1); 374—1969(1), 71; 614—1973; 272—86
Florida Keys, Upper 466—1975
Grayton Beach State Park 174—1984
Gulf Coast, Central 197—1985(1); 614—1947
Hamilton County 614—1981
Hillsborough County 614—1980
Ichetucknee Springs State Park 174—1982
Manatee Springs State Park 174—1982
O'Lebi State Park 174—1982
Ocala 614—1970
Okefenokee Swamp 197—1974(6); 533—1976
Orlando, Winter Park Sinkhole 197—1985(8)
Panhandle 206—1956
Pinellas County 614—1980
Southwest, Coast 197—1985(13)
St. Andrews State Park 174—1984
St. George Island State Park 174—1984
St. Joseph Peninsula State Park 174—1984
Suwannee River State Park 174—1982
Suwannee River basin, Upper 614—1981
Windley's Key Quarry 55—1972(1)

Georgia
General 132—1963; 206—1962(1,2); 240—1969, 71(4), 79; 435—1961; 614—1965; 710—1964
East 240—1968, 80
East-central 197—1980(15); 206—1966(2)
North 169—1981[1]
Northeast 240—1974(2)
Northwest 197—1980(16); 240—1972; 614—1951; 169—1981[2]; 272—85
North-central 206—1962(3)
South 614—1966
Southwest 206—1983; 614—1944(2nd), 84
West 5—1981, 83; 206—1974, 80(1)
Atlanta 238—1980
Atlanta area 197—1980(2)
Bartow County 240—1977
Bartow County, Cartersville District 197—1980(19)
Blue Ridge Mountains 129—1976
Brasstown Antiform 240—1974(1)
Brevard fault zone 197—1980(12), 85(5); 240—1970
Cartersville 240—1966, 70
Cartersville District 240—1982; 296—1933(2)
Chattahoochee River valley 197—1980(11,20)
Chattahoochee River valley, Lower 240—1975
Cloudland Canyon State Park 236—1983
Coast 197—1980(4,14); 240—1971, 73
Coastal Plain 206—1974
Coastal Plain, Augusta area 197—1980(22)
Cumberland Island National Seashore 236—1982
Dougherty Plain 206—1983; 614—1984
Georgia Fall Line area 240—1971
Gordon area 240—1971
Jasper County, West 240—1971(1)
Lamar County 240—1967
Lithonia 206—1962(2)
Lookout Mountain area 197—1980(13)
Monroe County 240—1971(1)
Mount Arabia 197—1980(3)
Murphy Syncline 240—1978
Panola Mountain State Park 236—1977(2-3)
Piedmont 5—1981; 197—1980(18); 206—1980(3)
Piedmont, East 197—1980(5)
Piedmont, Southwest 197—1985
Pine Mountain Window 5—1981
Polk County 240—1977
Rocky Mountain Storage Project 197—1980(1)
Sapelo Island 206—1966(1)
Savannah River 206—1966(3); 240—1976
Stone Mountain 197—1980(3,18); 206—1962(2)
Stone Mountain Park 236—1980
Sweetwater Creek State Park 236—1977(1)

Hawaii
General 190—1983(16); 199—1972; 251—1979; 542—1974; 569—1984(3); 272—137, 138
Mauna Loa 272—70
Oahu 252—1939

Idaho
General 17—1979(Dec.); 253—1975; 259—1963; 276—1967; 754—1963-64, 68, 70, 73
Central 204—1964(3), 75(1); 420—1984; 755—1982(Pt. 2)
East 781—1977
North 185—1983
South 199—1983; 204—1980(1); 658—1984; 734—1984
Southeast 185—1961; 204—1957, 80(3); 276—1953; 428—1971(1); 734—1985
Southwest 185—1962; 256—1982; 755—1982(Pt. 1); 284—1983
West 658—1984
Bannock Overthrust 296—1933(17)
Bear Lake Valley 185—1961
Benewah County 97—1973(3)
Bitterroot Range 204—1959(4), 64(3)
Boise 204—1975(1,6), 85(5,7)
Boise foothills 256—1982
Borah Peak 204—1985(4); 420—1984
Brownlee Dam 204—1975(8)
Coeur d'Alene 74—1961
Coeur d'Alene District 253—1975
Conda Phosphate Fields 296—1933(17)
Craters of the Moon National Monument 755—1982(Pt. 3); 272—136
Fish Creek Reservoir 204—1985(5)
Hebgen Lake 292—1974(3)
Hells Canyon Dam 204—1975(8)
Idaho Batholith 17—1979(Dec.); 199—1976(4); 204—1977(1)
Idaho-Wyoming Thrust fault zone 204—1975(4)
Lewiston 39—1979(1); 216—1984
Lost River Range 204—1985(4)
Massacre Rocks State Park 272—142
Mountain Home 204—1985(5)
North Clark Fork area 97—1973(1)
Owyhee Mountains 204—1985(2)
Payette River 204—1975(5)
Pocatello 658—1981
Snake River 199—1976(5); 204—1975(8), 85(8)
Snake River Birds of Prey Area 204—1985(1,6)
Snake River Canyon 185—1962
Snake River Plain 199—1983(4,9); 204—1975(3), 85(7); 542—1977
Snake River Plain, East 197—1977(15)
St. Joe 97—1973(4)
St. Maries River 17—1979(June)
State Highway 21 755—1982
U.S. Highway 93 259—1963

Weiser 204—1975(7)
Whitebird Hill 204—1975(1)
Wood River 204—1985(5)
Yellowstone National Park 428—1971(1)

Illinois
General 132—1956; 262—1938, 46; 296—1933(26); 659—1969, 83(1,2); 666—1984
Central 183—1979
East-central 133—1981; 183—1972; 197—1983(3); 414—1984
North 74—1927; 183—1985
Northeast 200—1972(4), 76(3), 85(1); 262—1953; 543—1953; 606—1978; 659—1946, 49, 70(2)
Northwest 197—1970(8)
North-central 262—1939; 659—1952, 70(1)
South 18—1966(1); 25—1983(No.1); 200—1977(2,3,5-7); 262—1956, 82; 499—1969; 659—1958, 61, 73; 688—1963; 722—1970; 272—57
Southeast 262—1959; 296—1933(2); 587—1976
Southwest 18—1949; 262—1956; 341—1939; 659—1979(2)
West 183—1952; 197—1958(3); 262—1957; 341—1941, 61; 474—1969(3); 659—1939, 64, 79(1)
West-central 18—1954; 183—1963; 659—1955, 76; 414—1984; 261—1980
Alto Pass 264—1965(1)
Baldwin 25—1983(No.3)
Calhoun County 659—1936
Carbondale 200—1977
Carbondale County 290—1979(J)
Central, Bloomington 264—1964(1)
Central, Carlock 264—1972(1)
Central, Danvers 264—1982(C)
Central, Decatur 264—1982(A)
Central, Edinburg 264—1964(4)
Central, Eureka 264—Pre-1960
Central, Farmer City 264—1979(B)
Central, Farmington 264—1963(3)
Central, Greenville 264—1962(3), 84(A)
Central, Havana 264—1962(2)
Central, Mahomet 264—1969(3)
Central, McLean County 264—Pre-1960
Central, Middle Illinois Valley 264—1978(C)
Central, Monticello 264—1969(3)
Central, Normal 264—1982(C)
Central, Pana 264—1960(4)
Central, Pekin 264—Pre-1960, 85(C)
Central, Peoria 264—1946(F), 62(5)
Central, Petersburg 264—Pre-1960 67(5)
Central, Pontiac 264—Pre-1960, 63(7), 84(C)
Central, Princeville 264—Pre-1960
Central, Shelbyville 264—1950(B), 65(5)
Central, Springfield 264—Pre-1960, 78(A)
Central, St. Elmo 264—1968(4)
Central, Streator 264—Pre-1960, 84(C)
Champaign County 133—1981; 264—1969(3)
Chicago River 587—1974(1)
Chicago area 18—1940(1), 50; 76—1979; 659—1985(1,2)
Clark County 264—1979(A)
Coast, Northeast 587—1974(1,2)
Cook County 264—1979(C)
Cora 25—1983(No.3)
DeKalb County 264—Pre-1960
Des Plaines River valley 200—1985(4)
Dupage County 264—1978(B)
East, Danville 264—1948(B), 72(2,4), 81(A)
East-central, Casey 264—1949(A), 79(A)
East-central, Champaign County 264—Pre-1960, 77(1)
East-central, Charleston 264—Pre-1960, 61(1)
East-central, Fairbury 264—Pre-1960
East-central, Georgetown 264—1961(3)
East-central, Greenup 264—Pre-1960
East-central, Homer 264—Pre-1960
East-central, Hoopeston 264—Pre-1960
East-central, Marshall 264—Pre-1960
East-central, Neoga 264—1962(4)
East-central, Newton 264—Pre-1960
East-central, Oakwood 264—1967(4)
East-central, Paris 264—1951(B), 66(3)
East-central, Potomac 264—1972(4)
East-central, Urbana 264—Pre-1960, 77(1)
East-central, Watseka 264—Pre-1960
East-central, Westfield 264—1979(A)
Flourspar Deposits 296—1933(2)
Hancock County, West 264—1970(2), 71(2)
Hardin County 179—1973(1); 264—1982(D)
Herrin 290—1979(D)
Illinois Basin 25—1983(No.2)
Illinois Beach State Park 587—1974(2)
Illinois-Kentucky Mining District 179—1973(1)
Jackson County 622—1965
Jersey County 659—1936
Kankakee County 264—1975(4), 76(3)
Keensburg area 197—1983(9)
La Salle 18—1940(2); 264—1971(3), 72(3)
La Salle County 200—1972(3); 264—Pre-1960
Lake Bluff 587—1974(2)
Lake County 264—1981(C)
Lake Michigan shore 197—1983(6)
Livingston County, Northwest 264—Pre-1960
McHenry County 200—1972(2)
McLean County 264—1982(C); 409—1978
Mississippi Embayment, Upper 179—1973(1)
Mississippi River valley 197—1970(7); 200—1977(7)
Monroe County, General 264—1981(D)
North shore 587—1974(1)
North-central, Amboy 264—1962(1)
North-central, Belvidere 264—Pre-1960, 63(1)
North-central, Byron 264—Pre-1960, 66(1)
North-central, Capron 264—1982(B)
North-central, Dixon 264—Pre-1960, 68(2)
North-central, Marseilles 264—Pre-1960
North-central, Oregon 264—Pre-1960
North-central, Pecatonica 264—Pre-1960
North-central, Princeton 264—1968(3), 69(5)
North-central, Rochelle 264—1964(7)
North-central, Rockford 264—Pre-1960, 74(3), 82(B)
North-central, Rockton 264—1973(3), 74(4)
North-central, Starved Rock State Park 264—1962(7)
Northeast, Bourbonnais 264—1967(1)
Northeast, Chicago Heights 264—Pre-1960
Northeast, Crystal Lake 264—Pre-1960
Northeast, Des Plaines 264—Pre-1960
Northeast, Downers Grove 264—Pre-1960
Northeast, Dundee 264—1980(C)
Northeast, Elgin 264—Pre-1960, 61(2)
Northeast, Elmhurst 264—1978(B)
Northeast, Evergreen Park 264—1979(C)
Northeast, Joliet 264—1950(C)
Northeast, Kanakakee 264—Pre-1960
Northeast, Lake Bluff 264—1981(C)
Northeast, Lake County 264—Pre-1960
Northeast, Lake Region 264—Pre-1960
Northeast, Momenu 264—1975(4), 76(3)
Northeast, Morris 264—1961(6), 84(B)
Northeast, Naperville 264—Pre-1960, 78(B)
Northeast, North Shore 264—Pre-1960
Northeast, Palos Hills 264—1971(6)
Northeast, Palos Park 264—Pre-1960
Northeast, St. Anne 264—1975(4), 76(3)
Northeast, Thornton 264—1979(C)

Northeast, Waukegan 264—Pre-1960
Northeast, West Chicago 264—1947(D)
Northeast, Wheaton 264—1962(8)
Northeast, Wilmington 264—1946(E)
Northeast, Woodstock 264—1960(9)
Northeast, Yorkville 264—Pre-1960, 1965(6)
Northeast, Zion 264—1981(C)
Northwest, Apple River Canyon 264—1948(C)
Northwest, Elizabeth 264—Pre-1960, 85(B)
Northwest, Freeport 264—Pre-1960, 63(4), 66(2), 70(1)
Northwest, Fulton 264—Pre-1960
Northwest, Galena 264—Pre-1960, 65(4), 71(1)
Northwest, Hillsdale 264—1980(B)
Northwest, Kawanee 264—Pre-1960
Northwest, Lena 264—1961(5)
Northwest, Milan 264—1974(2), 75(3)
Northwest, Moline 264—Pre-1960
Northwest, Morrison 264—1964(6)
Northwest, Mt. Carroll 264—Pre-1960, 69(4), 70(4)
Northwest, Port Byron 264—Pre-1960
Northwest, Rock Island County 264—Pre-1960, 60(3)
Northwest, Savanna 264—1951(C), 63(8)
Northwest, Stockton 264—Pre-1960, 72(6), 73(4)
Northwest, Wyoming 264—Pre-1960
Ogle County 264—Pre-1960
Piatt County, General 264—1969(3)
Pine Hills 264—Pre-1960, 63(6)
Pope County 179—1973(1)
Quad Cities region 587—1985
Quincy area 659—1979(3)
Rock Island County 200—1985(3); 290—1979(G)
Rock Island area 263—1950
Saline County, Raleigh Field 25—1983(No.4)
Sheffield 200—1985(4)
South, Anna area 264—Pre-1960
South, Cairo 264—1948(F), 81(A)
South, Carbondale 264—Pre-1960
South, Metropolis 264—Pre-1960, 75(2)
South, Murphysboro 264—Pre-1960
South, Union County 264—Pre-1960
South-central, Benton 264—Pre-1960
South-central, Breese 264—1974(1)
South-central, Carrier Mills 264—1965(3)
South-central, Nashville 264—Pre-1960
South-central, Pinkneyville 264—Pre-1960
South-central, Salem 264—Pre-1960, 60(7), 85(A)
South-central, Vienna 264—Pre-1960, 66(6)
Southeast, Cave-in-Rock 264—Pre-1960, 67(3), 82(D)
Southeast, Eldorado 264—Pre-1960
Southeast, Elizabethtown 264—1967(3)
Southeast, Equality 264—1969(1), 80(A)
Southeast, Fairfield 264—Pre-1960
Southeast, Golconda 264—1962(2), 83(D)
Southeast, Harrisburg 264—1951(14), 60(2)
Southeast, Lawrenceville 264—Pre-1960, 85(D)
Southeast, Marion 264—Pre-1960, 63(5), 78(D)
Southeast, Mt. Carmel 264—1976(2), 77(3)
Southeast, Olney 264—Pre-1960
Southeast, Robinson 264—1973(2)
Southeast, Rosiclare 264—Pre-1960, 60(6), 82(D)
Southeast, Shawneetown 264—Pre-1960
Southwest, Chester 264—1950(A), 64(2)
Southwest, Dupo 264—1963(2), 70(3)
Southwest, Grand Tower 264—1947
Southwest, Jonesboro 264—1964(5)
Southwest, Makanda 264—1971(4)
Southwest, Millstadt 264—Pre-1960, 70(3)
Southwest, Red Bud 264—Pre-1960, 72(5)
Southwest, Sparta 264—1960(8), 61(7)
Southwest, Steeleville 264—1966(5)
Southwest, Thebes 264—Pre-1960, 68(5)
Southwest, Valmeyer 264—1961(8), 81(D)
Southwest, Waterloo 264—1950(F), 81(D)
St. Louis area 253—1969
Thornton 18—1956
Union County 622—1965
Vermilion County 133—1972(4), 81; 264—1972(4); 587—1980
West, Alton area 264—1947(F)
West, Dallas City 264—Pre-1960
West, Monmouth 264—Pre-1960, 77(2)
West, Mt. Sterling 264—Pre-1960, 71(5)
West-central, Barry 264—1968(1)
West-central, Beardstown 264—1965(2)
West-central, Canton 264—Pre-1960
West-central, Carlinville 264—Pre-1960, 79(D)
West-central, Carrollton 264—Pre-1960, 75(1), 76(1)
West-central, Colchester 264—1964(3)
West-central, Galesburg 264—1949(B)
West-central, Grafton 264—1947, 60(1)
West-central, Hamilton 264—1961(4), 70(2), 71(2)
West-central, Hardin 264—1949(2)
West-central, Jacksonville 264—Pre-1960
West-central, Knoxville 264—1973(1)
West-central, Macomb 264—Pre-1960
West-central, Pere Marquette 264—Pre-1960
West-central, Pere Marquette State Park 264—Pre-1960
West-central, Pittsfield 264—Pre-1960, 62(6)
West-central, Quincy 264—1950(E), 60(5), 66(4), 80(D)
West-central, Warsaw 264—1947(E)
West-central, Winchester 264—1970(5)
Woodford County 409—1978

Indiana
General 42—1952; 74—1966; 132—1960; 267—1962(Spring); 270—1965; 435—1973; 578—1966
East 543—1955(1)
North 18—1940(3); 197—1983(12); 200—1976(3a); 268—1949, 61; 543—1955(1)
Northwest 197—1983(5); 659—1946
North-central 197—1961(2)
South 197—1981(2), 83(1,2); 200—1980(4); 206—1984(3); 268—1948; 271—1940, 72, 75; 290—1979(7); 410—1966(Apr.); 272—33, 34, 35; 691—1958
Southeast 197—1961(9), 81(1), 83(15); 200—1967(2), 80(3); 268—1947, 53, 55; 410—1964; 587—1977
Southwest 183—1978; 197—1961(3,6), 83(9); 200—1967(4); 268—1957
South-central 183—1957; 197—1983(14); 200—1967(1), 80(5); 268—1954; 296—1933(2); 578—1966; 587—1973
West 268—1950
West-central 197—1983(3); 200—1967(3); 268—1951, 73(1); 543—1953
Bedford 200—1980(1)
Bloomington 200—1980(1); 435—1965
Borden River delta 197—1983(2)
Brazil 268—1973(1)
Brown County 267—1955
Carroll County 200—1982(5)
Cave River Valley Park 130—1962
Clark County, South 268—1973(5)
Clay County, North 268—1973(1)
Crawford County 209—1952
Crawford Upland 268—1965
Falls of the Ohio 197—1961(9)
Floyd County, New Albany area 268—1973(5)
Fort Wayne area 268—1973(4)
Fountain County 290—1979(B)
Grant County 200—1982(5)
Illinois Basin, North 197—1983(10)
Indianapolis area 197—1983(8)

Kentland 200—1975(1), 82(4)
Kentland Dome area 197—1983(5)
Kentland Uplift 18—1940(3)
Lafayette 40—1961
Lafayette area 253—1967
Lake Michigan 200—1976(4)
Lake Michigan shore 268—1973(2)
Lake Monroe area 197—1983(2)
Marshall County, Lake Maxinkuckee 267—1965
McCormick's Creek State Park 267—1959
Mitchell Plain 197—1983(7); 268—1965
Monroe County 200—1980(2)
Montezuma 686—1967
New Harmony 197—1983(11)
Parke County 197—1983(10); 290—1979(B); 587—1980
Perry County 209—1952
Richmond area 377—1953
Steuben County, North 267—1954
Terre Haute 268—1973(1)
Vermillion County 587—1980
Vigo County 268—1973(1)
Wabash River valley 183—1983; 197—1983(13); 200—1982(2)
Washington County 130—1962
Wyandotte Cave region 197—1983(14)

Iowa
General 179—1972(1-4); 183—1965; 208—1960(Apr.,Sept.), 68(June), 72(May), 81(36); 341—1927; 435—1980; 659—1969, 83(1,2); 703—1961; 272—156
Central 200—1981(4)
East 183—1952; 197—1970(3); 208—1963(May); 296—1933(26); 331—1967; 341—1935; 474—1969(2); 587—1984-85; 659—1963
East-central 331—1967(2-4); 659—1935, 78(1-5)
North 659—1966, 72
Northeast 208—1962(July), 65, 83(39,40); 331—1967(1,2); 659—1948, 54, 75(3)
Northwest 200—1981(3); 703—1961; 86—1982
North-central 208—1982(37); 331—1967(5-7); 659—1951, 60, 84
Southeast 200—1968(3); 208—1972, 77; 331—1967(8-9); 659—1941, 57; 333—1984, 85
Southwest 183—1955; 208—1964
South-central 331—1967(8); 659—1938, 75(2)
Algona Moraine 208—1981(36)
Allamakee County 208—1979
Allamakee County, Upper Iowa Valley 333—1983
Black Hawk County 208—1972(July), 84(42)
Buchanan County 208—1984(42)
Cedar County 200—1968(1)
Cedar Rapids area 208—1967(2); 659—1978(3)
Des Moines County 208—1967(June 3)
Des Moines County, Sperry Mine 179—1972(4); 200—1968(2)
Des Moines Lobe 200—1981(2)
Dubuque area 659—1978(4)
Fayette County 208—1983(39,40)
Floyd County, Rockford Pit 659—1981(3)
Fort Dodge area 208—1976
Guthrie County 208—1982(38)
Henry County 208—1959(3)
Iowa City 332—1984(7)
Iowa City area 659—1975(1)
Iowa City area, Conklin Quarry 208—1984(41)
Johnson County 200—1968(6); 272—82
Jones County, West 208—1981(35)
Lake Calvin 200—1968(5)
Lake Calvin Basin 333—1984
Linn County 200—1968(1)
Linn County, East 208—1981(35)
Linn County, Southeast 208—1962(May)
Little Sioux River valley 208—1980(34)
Madison County 208—1980(33)
Marion County 208—1970
Middle River 200—1968(4); 208—(1967 Fall), 68(Mar.), 80(33)
North Osage area 208—1961
North-central, Garner area 208—1963(July)
North-central, Mason City area 208—1963(July)
Oskaloosa area 333—1985
Pella area 333—1985
Plum River fault zone 659—1978(1)
Plymouth County 86—1982
Quad Cities region 587—1985
Raymond 208—1978
Saylorville Dam 208—1985
South, Douds 208—1959(2)
Southeast Selma 208—1959(1)
Sperry 208—1962(Mar.), 67(June 2)
Stone City 659—1978(5)
Upper Iowa Valley 333—1983
Washington County 200—1968(6)
Winneshiek County 208—1983(39); 330—1975
Winneshiek County, Upper Iowa Valley 333—1983

Kansas
General 74—1959; 76—1972(1); 102—1959; 290—1979(10); 343—1976; 572—1965; 576—1936, 40; 272—23
Central 205—1980(1)
East 18—1978(SEPM 2); 87—1957; 296—1933(20); 341—1949(June, Oct.), 56(19th), 57
East-central 572—1949
Northeast 76—1972(2); 183—1971; 253—1970; 341—1936, 59; 689—1956, 61; 589—1985
Northwest 75—1985
North-central 205—1972
South 576—1937
Southeast 26—1975; 296—1933(2); 341—1937, 46(Mar.,June), 47, 49(Apr.), 62, 82(1), 84; 540—1960; 573—1969; 634—1960(May); 662—1984(May)
Southwest 341—1955
South-central 29—1959(Oct.); 634—1960(July)
West 18—1978(SEPM 3); 197—1965(2)
Anadarko Basin 508—1963(13)
Anderson County 197—1965(1)
Camp Naish 189—1956
Chase County 342—1958; 188—1958
Chautauqua County 341—1983
Cherokee County 73—1984
Crawford County 197—1965(6); 73—1984
Doniphan County 205—1969
Flint Hills 342—1957, 60-62
Franklin County, West 340—1958
Kansas City 76—1972(3-5)
Kansas City area 75—1985; 197—1965(5); 253—1970
Kansas River valley 197—1965(3); 700—1951; 73—1984
Kansas River valley, Lower 197—1965(7)
Lawrence 341—1969; 589—1985
Lawrence area 340—1951
Lyon County 341—1951
Meade County 102—1967; 183—1976
Missouri River valley 183—1971
Montgomery County 197—1965(1)
Osage Cuesta region 634—1960(May)
Osage Cuestas 342—1960
Reno County 205—1980(2)
Sumner County, West 341—1982(2)
Tuttle Creek Dam 340—1953
Wilson County 197—1965(1)
Wyandotte County 102—1956; 189—1956

Kentucky
General 74—1956; 206—1970(Pt.1,Pt.2), 84(4); 209—1941, 62, 72; 681—1982

Central 197—1981(11); 206—1970(2), 84(6); 375—1965; 690—1938; 691—1959
East 23—1978; 197—1971(9), 81(14); 206—1984(2,7); 209—1953, 81; 244—1985 (6)
East-central 206—1984(5)
Northeast 25—1976; 209—1955, 68, 71, 77, 80
Northwest 197—1981(2); 206—1984(3); 290—1979(7); 587—1982
North-central 209—1984
Southeast 206—1970(1); 209—1942, 46(May); 435—1985
Southwest 209—1954, 56
South-central 206—1970(3); 209—1963
West 25—1980; 40—1970; 200—1977(3); 209—1966, 79
Appalachian Basin 197—1981(3,4)
Barkley Dam Site 209—1962
Bluegrass 206—1960(2), 84(4)
Bluegrass region 272—65
Boone's Cave 690—1938
Breckinridge County 209—1952
Burdett Knob 690—1938
Cincinnati area 197—1961(3)
Corbin 206—1984(8)
Crittenden County 179—1973(2)
Cumberland Gap area 209—1961
Elizabethtown area 209—1964
Elkhorn City area 209—1967
Elklick Falls 690—1938
Faircloth Flourite Vein 690—1938
Falls of the Ohio 197—1961(9)
Flourspar Deposits 296—1933(2)
Green River valley 197—1981(17)
Greenup County, Limeville 497—1984
Herrington Lake 690—1938
Hopkins County 209—1969
Interstate 64 206—1970(4)
Jackson Purchase region 209—1972
Jefferson County 272—33, 34, 35
Kentucky River Fault 209—1975
Lake Cumberland area, West 209—1978
Lexington 206—1960(1)
Lexington area 209—1965(1)
Litchfield 209—1941
Livingston County 179—1973(2)
Louisville area 197—1983(1,2); 209—1958
Mammoth Cave 206—1960(1); 291—1981
Mammoth Cave area 197—1981(8,17), 83(7); 209—1954, 64; 681—1985
Midland Trail 272—75, 74; 348—1930
Mississippi River 209—1984
Natural Bridge 690—1938
Nelson County 209—1959
Ohio River valley 197—1961(8,9)
Ohio River valley, Middle 209—1974
Pine Mountain 206—1984(1)
Pine Mountain Thrust Sheet 197—1981(19)
Pine Mountain front 209—1967
Turnhole Spring ground water basin 313—1976
Versailles area 209—1965(2)
Webster County 209—1969
West-central, Rough Creek Fault 271—1973

Louisiana
General 186—1985
Central 206—1964(2); 350—1962, 68; 588—1957; 626—1971, 73(Fall)
North 566—1939, 60, 66
Northeast 18—1985(3)
Northwest 162—1939
North-central 247—1970(1)
South 197—1944(4), 55
Southeast 247—1980(3)
Southwest 255—1978(Nov.); 626—1965, 80
West-central 197—1955(4); 466—1974(Nov.)
Avery Island Salt Mine 18—1976(2); 247—1974(1)
Barataria Basin 197—1982(4)
Baton Rouge fault zone 206—1964(3)
Bayou Lafourche 466—1971(1)
Belle Isle 247—1966(2)
Caldwell Parish 566—1933
Catahoula Parish 566—1933
Chenier Plain 247—1980(4); 255—1978(Nov.)
Coast 197—1962(9), 82(12); 247—1980(1-4); 466—1973(Spring), 82(June); 588—1982
Coastal Plain 197—1982(1,14)
Five Islands 206—1964(4)
Gladys McCall No. 1 Well Site 197—1982(14)
Grand Isle 466—1971(2)
Iberia Parish 197—1955(5)
Iberia Parish, Cote Blanche Island 247—1980(1)
Jefferson Island 466—1961
Jefferson Island Salt Dome 197—1982(11)
Lake Peigneur 197—1982(11)
Mississippi River delta 18—1965(1-3), 76(2); 197—1967(6,A), 82(7); 206—1964(4); 247—1980(3); 466—1970, 71(1), 72; 272—121
Mississippi River valley 392—1949(2)
Mississippi River valley, Lower 32—1968; 79—1984
Natchitoches Parish 566—1967
New Orleans 466—1980, 82
New Orleans area 197—1967(B)
Offshore 197—1955(6)
Rapides Parish, Cane River 247—1980(2)
Salt Domes 197—1967(6)
South, Avery Island Salt Mine 18—1976(2); 247—1974(1)
South, Belle Isle 247—1966(2)
Weeks Island 18—1976(2); 466—1972

Maine
General 456—1983(C-1,D-1)
Central 456—1974(A-3,B-3)
East 456—1974(A-2), 78(B-3)
East-central 210—1983(9)
North 456—1980(A-1,A-2,B-1,B-4,B-5,B-6,C-5)
Northeast 456—1974(A-5), 80(C-1)
Northwest 456—1983(B-3)
North-central 456—1983(A-1,B-1)
South 456—1965
Southeast 210—1977(1,2), 83(9); 456—1978(A-2,A-6,B-10), 84(A-8)
Southwest 182—1961; 456—1984(A-4)
South-central 456—1974(B-3,B-6), 80(D-1)
West 456—1970
West-central 456—1960, 83(C-4)
Acadia National Park 272—27
Appledore Island 456—1971(B-1
Aroostock County, East 456—1980(C-6)
Aroostock County, North 456—1980(B-8)
Aroostock County, Northwest 210—1980
Belfast area 456—1974(A-1)
Berwick Quadrangle 456—1984(C-5)
Blue Hill 456—1974(B-4)
Boundary Mountain 456—1983(B-5)
Brassua Lake Quadrangle 456—1983(B-1)
Brooks Quadrangle 210—1983(8)
Brookton area 456—1978(B-3)
Calais area 210—1977(2); 456—1978(A-5)
Camden area 456—1974(A-4)
Cancomgomoc Lake area 456—1983(B-3,C-5)
Coast 456—1984(B-4)
Coopers Mills area 210—1978(4)
Eastport Quadrangle 456—1978(A-1,B-1,B-7)

Eastport area 210—1977(1)
Gardiner Quadrangle 210—1983(6)
Gouldsboro 456—1974(A-8,B-8)
Greenville area 456—1983(C-2)
Gulf of Maine 456—1971(B-1)
Knox County 210—1983(7)
Leeds 210—1983(11)
Liberty area 210—1978(4)
Lincoln County 210—1983(7)
Machias area 210—1977(1)
Millinocket area 456—1983(C-3)
Moosehead Lake Quadrangle 456—1983(B-1)
Mount Desert Island 456—1974(B-2)
Mount Katahdin 456—1966, 80(A-3,B-2), 83(B-4)
New Harbor, Rachel Carson Salt Pond area 444—1977
Newbury Port 456—1984(C-9)
Norumbega fault zone 210—1983(8)
Orono area 456—1974(B-1)
Penobscot Bay area, North 456—1974(A-7,B-7)
Penobscot Bay area, West 456—1974(A-4)
Portland area 210—1978(3)
Presque Isle area 456—1980(B-7,C-2)
Princeton area 456—1978(B-3)
Rockland area 456—1974(A-4)
Rockwood area 456—1983(C-2)
Sebago Lake area 210—1983(12)
Seboomook Lake area 456—1983(C-4)
Seven Hundred Acre Island 456—1974(A-6)
St. John River area 210—1980
St. John River basin, Upper 455—1982
St. John River valley 456—1980(C-4)
Traveler Mountain area 456—1980(B-3)
Tunk Lake area 456—1974(B-5)
Waldoboro Moraine 210—1983(7)
Wayne 210—1983(11)
Wesley Quadrangle 456—1978(A-8,B-8)
Wesley area 210—1977(2)
White Cap Range 456—1983(B-2)
Wiscasset Quadrangle 210—1983(6)
York County 456—1984(B-8)
York County, South 456—1984(A-7)

Maryland
General 18—1960(3); 25—1979; 69—1968(A-8); 172—1958, 79; 197—1950(1); 201—1976(2,6); 296—1933(10); 411—1977
East 18—1977(3)
Northeast 132—1958
South 197—1950(3); 296—1933(12)
Southwest 32—1984
West 29—1937; 58—1952, 82
Allegheny County 172—1972
Allegheny Front 58—1984; 172—1972
Annapolis 253—1978
Appalachian Mountains 18—1960(4-5); 172—1958
Baltimore Harbor 69—1968(A-9)
Bear Island 197—1950(2), 71(3)
Chesapeake Bay 296—1933(5)
Chesapeake Bay region 201—1976(7a,7b), 82(1)
Chesapeake and Ohio Canal 76—1970(5)
Coastal Plain 69—1968(A-6); 89—1961, 68; 197—1950(3), 71(5)
Columbia 69—1968(A-5)
Great Falls 69—1968(A-2); 272—147
Greenbelt 69—1968(A-5)
Piedmont, East 172—1970; 197—1971(3)
Potomac River 69—1968(B)
Potomac River, Great Falls 69—1968(A-2); 272—147
Silver Springs 201—1982(9)
South Mountain 18—1960(4-5)
Wills Mountain 58—1982

Massachusetts
General 456—1969, 76(B-19), 82(Q2); 272—63
Central 296—1933(1); 456—1982(P4), 83(C-2,C-3,D-3)
East 456—1976(F-3,A-13)
East-central 456—1976(A-12,A-15,B-15)
Northeast 456—1976(A-16,B-16,A-14,B-14), 81(C-3), 84(C-3)
Northwest 296—1933(1); 456—1979(B-11)
Southeast 89—1983; 456—1976(A-7,F-6,F-7), 84(C-10)
Southwest 456—1975(A-1,B-7,C-7)
South-central 456—1982(P3)
West 456—1975(C-8)
West-central 456—1975(B-5,C-5)
Berkshire Massif 456—1975(B-2,B-3,B-4,B-6,C-2,C-4,C-6,C-11)
Bloody Bluff fault zone 456—1976(A-13,A-14,A-17), 84(C-2,C-4)
Boston area 456—1964, 76(A-1,B-1,A-2,A-5,A-3,B-3,A-4,B-4,A-6,B-6), 81(C-4), 84(A-2,B-2,B-3); 272—14, 131, 132, 16, 15
Boston area, Thompson Island 456—1984(A-1,C-8)
Boston area, Winthrop 456—1984(A-1)
Buzzards Bay area 197—1952(4)
Cape Anne area 456—1976(A-11/B-11)
Cape Cod 69—1971; 182—1968; 456—1976(A-10/B-10)
Cape Cod National Seashore 32—1972; 586—1979
Cape Cod, West 197—1952(4)
Charles River 456—1976(A-19)
Chelmsford area 197—1952(3)
Clinton-Newbury fault zone 456—1976(A-13,A-14,A-15,F-2,F-3,B-15)
Coast, Central 456—1976(A-9,B-9)
Coast, North 456—1984(A-6)
Connecticut River valley 456—1967, 82(M-1,M-2)
East, Nashoba Block 456—1984(A-5,B-1)
Harvard 456—1976(F-4)
Marblehead 456—1976(B-12)
Marlborough area 456—1984(C-2)
Martha's Vineyard 89—1983; 182—1964(27th)
Middlesex Falls 456—1976(B-7)
Nahant 456—1984(C-1)
Narragansett Basin 456—1976(B-18), 81(B-1); 415—1979
Nashoba Block 456—1984(A-5,B-1)
Nashua River valley 182—1981
Newbury area 456—1971(B-5)
Norfolk Basin 456—1981(B-1)
Northeast Coast 586—1969
Pittsfield East Quadrangle 456—1975(B-9)
Plum Island 456—1976(A-8/B-8)
Quincy 456—1981(C-4)
Squantum, Squaw Head 456—1976(F-1)
Stockbridge Valley 456—1975(B-10)
Sutton, Purgatory Chasm 456—1951
Taconic Mountains 456—1975(A-3,B-1,C-1)
Webster area 456—1976(B-13)
Webster area, Lake Char Fault 456—1982(P5)
Worchester area 456—1976(F-5,B-5,B-13,F-2)

Michigan
General 273—1972(A); 377—1947, 69, 74; 583—1971(Sept.)
North 377—1983
Northwest 273—1975(2,3)
South 200—1978(1)
Southeast 197—1951(2); 200—1978(2); 376—1935; 377—1952
Southwest 768—1972
Afton area 376—1940
Alpena 435—1977
Alpena County 200—1970(1); 587—1972
Charlevoix County 377—1976
Copper Range 296—1933(27)
Detroit area 197—1951(1)
Dickinson County 377—1958
Emmet County 377—1976
Gogebic Range 296—1933(27)

Grand Valley 200—1970(2)
Iron County 377—1958
Kalamazoo 200—1976(2)
Keweenaw Copper Range 74—1957
Keweenaw Peninsula 273—1983(1); 427—1962-65
Lake Erie Islands 377—1971
Lake Michigan shore, Southeast 200—1976(4); 768—1970(1)
Lake Michigan, South 377—1962
Lake Superior region 427—1962-65
Lower Peninsula, North 377—1949, 59, 73, 83
Lower Peninsula, Northeast 183—1981
Lower Peninsula, Northwest 183—1956
Lower Peninsula, West-Central 768—1973
Mackinac Island 376—1941; 377—1959
Mackinac Straits 377—1944
Marenisco-Watersmeet area 667—1981(2)
Marquette 200—1970(3); 273—1964, 75(4)
Marquette County 376—1939; 377—1977
Marquette County, Roper Gold Mine 273—1983(2)
Marquette Range 74—1957; 197—1970(5); 273—1972(B); 296—1933(27)
Menominee County 376—1939
Michigan Basin 200—1978(3)
Monroe County 376—1935
Onaway area 376—1940
Palmer, Empire Mine 273—1975(5,6)
Presque Isle County 200—1970(1); 587—1972
St. Ignace 376—1941
Thumb area 200—1978(1)
Twin Ports 659—1971
Upper Peninsula 91—1970; 377—1948, 57, 61, 67, 80; 722—1977; 667—1981; 272—48, 99
Upper Peninsula, East 768—1976
Upper Peninsula, Escanaba 377—1950
Upper Peninsula, Stonington 377—1950
Wakefield area 667—1981(1)
Washtenaw County 164—1959
Watersmeet area 667—1981(2)

Minnesota
General 435—1980; 474—1969(1); 703—1961; 272—93, 118
Central 183—1954
East 183—1964; 341—1935
East-central 197—1956(3); 200—1979(2); 388—1979(12); 409—1976
Northeast 197—1956(1); 200—1979(7-8); 698—1959
Southeast 183—1951; 197—1972(3); 200—1979(6); 703—1961; 389—1984
West 197—1972(6)
Cook County 197—1972(5); 273—1971(B)
Cuyuna Iron Ore District 296—1933(27)
Duluth area 197—1983(4); 200—1979(3); 386—1964
Ely 273—1968
International Falls 273—1982(2)
Kabetogama area 273—1982(2)
Lake Superior region 197—1983(4)
Lake Superior shore, North 197—1972(2)
Lake Superior, West 200—1979(5)
Mesabi Range 200—1979(4); 273—1971(C); 296—1933(27)
Minneapolis 273—1976(B)
Minneapolis area 197—1972(7)
Minnesota River 197—1972(4)
Minnesota River valley 364—1982
Mystery Cave 389—1983
Olmstead County 272—64
Rochester 388—1968
St. Cloud Granite District 409—1976
St. Croix Valley 723—1978
St. Paul 273—1965, 76(B); 272—67
St. Paul area 197—1972(7)

Trunk Highway No. 1 387—1925
Vermilion District, West 200—1979(7)
Vermilion Range 197—1972(1); 273—1971(D)

Mississippi
General 206—1964(1); 350—1965; 390—1940(1st-3rd), 50; 438—1974; 566—1932; 626—1971; 662—1976
Central 18—1985(1); 197—1955(2); 206—1981(4); 247—1983; 390—1948, 56
East 390—1952
East-central 206—1981(3,4); 247—1975(3)
Northeast 390—1959; 566—1929; 588—1983(Apr.)
South 206—1981
Southeast 390—1960
Southwest 18—1976(1), 85(3); 255—1938; 360—1961
West 588—1957
Black Warrior Basin 390—1978
Chickasawhay 206—1981(3)
Clarke County 566—1934
Coast 466—1973(Spring), 82(June)
Hattiesburg 206—1981(5)
Horn Island 247—1960(1)
Mississippi River 466—1973(Fall)
Mississippi River valley 392—1949(2)
Mississippi River valley, Lower 32—1968; 79—1984
Pascagoula Valley 247—1960(2)
Rankin County, Thomasville Field sour gas plant 247—1975(1)
South, Horn Island 247—1960(1)
South, Pascagoula Valley 247—1960(2)
Vicksburg 247—1975(2); 186—1985
Wayne County 566—1934

Missouri
General 87—1976; 102—1959; 197—1958(5); 396—1978; 499—1962; 643—1975; 700—1955; 393—1982
Central 87—1956, 58, 59, 67; 197—1965(4); 200—1973(5); 341—1941; 700—1950
East 341—1941, 61
East-central 87—1974, 77; 132—1953; 296—1933(2)
Northeast 87—1970; 474—1969(3)
Northwest 183—1971; 296—1933(20); 341—1936
North-central 200—1973(2)
South 341—1954; 721—1946
Southeast 18—1949, 51; 74—1970(2); 87—1954; 91—1953; 197—1958(1); 200—1977(4); 262—1965; 341—1939; 399—1977; 499—1969; 573—1969; 688—1963; 722—1970; 286—1982
Southwest 296—1933(2); 434—1967; 772—1936; 394—1968
South-central 87—1964
West 87—1955; 183—1975
West-central 87—1966; 341—1952; 572—1949
Bates County 87—1978
Bonne Terre 87—1969
Bonne Terre Mine 398—1982; 286—1982
Buick Mine 286—1982
Cape Girardeau 18—1951
Cape Girardeau area 87—1962, 81
Devils Elbow area 74—1970(1)
Fletcher Mine 200—1973(3); 286—1982
Frank R. Milliken Mine 286—1982
Interstate 55 399—1977
Iron Mountain 296—1933(2)
Joplin area 87—1963
Kansas City 75—1985; 272—44
Kansas City area 87—1971
Kaysinger Bluff Dam 197—1965(8)
Lead District 296—1933(2)
Magmont Mine 286—1982
Mineral District, Central 200—1973(4)
Mining Districts 43—1971
Missouri River valley 183—1971

North Union County 102—1968
Onondaga Cave 197—1958(4)
Ozark Mountains 183—1975; 341—1928
Pilot Knob 296—1933(2)
Rolla area 74—1970(1); 87—1972; 380—1960
Springfield area 87—1973
St. Charles County 18—1966(2)
St. Francis River 200—1977(4)
St. Francois Mountains 91—1953; 99—1975; 197—1961(10), 81(15); 398—1982; 286—1982
St. Francois Mountains area 87—1961, 76
St. Joseph area 87—1968
St. Louis 18—1957; 87—1960; 623—1970
St. Louis County 18—1966(2); 87—1960; 200—1977(1)
St. Louis region 197—1958(2)
Stockton Dam 197—1965(8)
Viburnum Trend 643—1975
Warrensburg area 87—1975

Montana
General 203—1979; 204—1959(1-4); 253—1975; 276—1967; 401—1984; 402—1975; 559—1968; 754—1963-64, 68, 70, 73; 282—1985(3)
Central 402—1951, 62
East 402—1969; 781—1984, 85
Northwest 97—1973(2); 190—1983(13); 402—1984; 428—1971(2)
North-central 328—1982
South 185—1963
Southeast 403—1976; 781—1968
Southwest 204—1953(2), 59(3,4), 68(3), 82(3); 401—1983(1,2); 402—1967, 81; 403—1975; 428—1971(2); 658—1976, 85(1-5); 781—1982
South-central 55—1970(C,D); 403—1978; 300—1985
West 204—1977(3), 82(2); 402—1950; 604—1958; 781—1977
West-central 204—1982(2); 658—1978, 80(1-3)
Absaroka Range 292—1974(2)
Avon 204—1977(2)
Bangtail Ridge 55—1970(C)
Beartooth Range 402—1958; 365—1981
Big Fork 204—1977(2)
Bighorn Basin 296—1933(24); 402—1954; 781—1983
Bighorn Basin, North 55—1984
Bitterroot Range 204—1959(4), 64(3), 77(1)
Bolder 204—1982(3)
Bridger Range 55—1970(C)
Bridger Range, Central 658—1985(2,5)
Butte 204—1953(2), 68(1); 292—1974(5); 296—1933(23); 583—1973(Aug.), 78
Butte Mining District 658—1985(3)
Central, Judith Mountains 402—1956
Crazy Mountains Basin 204—1968(2); 402—1957, 72
Dewey 204—1953(3)
Dillon area 658—1982
Divide 204—1953(3)
Drummond 204—1953(5), 59(2); 658—1978
Elkhorn Mining District 658—1980(3)
Flathead Valley 204—1959(3), 77(4)
Flint Creek Range 402—1965
Gallatin County 185—1974
Gallatin Valley 204—1982(6)
Glacier National Park 125—1977(Sept.); 292—1974(7); 428—1971(2); 569—1984(2); 590—1985(2); 272—8, 108
Golden Sunlight Mining District 658—1985(3)
Hebgen Lake 185—1974; 292—1974(3)
Helmville 204—1959(2)
Highland Mountains, South 658—1985(4)
Interstate 90 403—1975, 76, 78
Judith Mountains 402—1956
Lake Missoula area 197—1977(13:2)
Libby Dam 292—1974(6)
Little Rocky Mountains 402—1953; 328—1982
Madison River 185—1963
Missoula 204—1977(5)
Missouri River headwaters 204—1982(5)
Northeast, Little Rocky Mountains 402—1953
Northwest, Sun River Canyon 402—1979
Northwest, Teton Canyon 402—1979
Ovando 204—1959(2)
Phillipsburg 204—1953(5)
Platte Valley 204—1959(3), 77(4)
Powder River basin, North 781—1963
Pryor Mountains 402—1954
Red Lodge 296—1933(24)
Sawtooth Range 402—1959
Schmitt Chert Mine 204—1982(5)
Snake River Range 781—1977
South-central, Pryor Mountains 402—1954
Spanish Peak area 658—1985(1)
Stillwater County 204—1968(4)
Sun River Canyon 402—1979
Sweet Grass County 204—1968(4)
Sweetgrass Arch 402—1955, 66
Teton Canyon 402—1979
Three Forks 204—1953(1)
Tobacco Root Mountains 204—1982(3)
Tobacco Root Mountains region 658—1985(1-5)
U.S. 212 403—1976
U.S. Highway 191 185—1974
Williston Basin 402—1952
Yellowstone National Park 781—1982
Yellowstone River 185—1963
Yellowstone Valley 55—1970(D); 204—1968(5), 82(1)

Nebraska
General 200—1971(1); 290—1979(10); 596—1982; 604—1941; 272—141
East 183—1966; 200—1971(1-3)
Northeast 200—1981(3)
Northwest 335—1964; 604—1985
South 200—1971(4)
Southeast 74—1967; 446—1974; 449—1970; 272—25
West 604—1961
West-central 46—1965
Calamus River 293—1985(6)
Cass County, Weeping Water 448—1969(FG-2)
Gage County, Krider area 448—1969(FG-4)
Gage County, Odell area 448—1969(FG-4)
Jefferson County, Fairbury area 448—1969(FG-7)
Otoe County, Unadilla 448—1969(FG-1)
Pawnee County, Table Rock area 448—1969(FG-6)
Platte River valley 200—1971(3)
Sarpy County, Gretna area 448—1969(FG-3)
Scotts Bluff National Monument 447—1979
South-central, Republican River valley 272—97
Thayer County, Alexandria area 448—1969(FG-5)
Thayer County, Gilead area 448—1969(FG-5)
Weeping Water Valley 200—1971(3)

Nevada
General 203—1979; 669—1966
Central 197—1984(15); 199—1974(3); 272—115
East 103—1973
East-central 199—1983(6); 276—1960
North 199—1966
Northeast 734—1984
Northwest 284—1983
North-central 199—1966(6); 85—1984(7,9)
South 197—1984(2,19,20); 199—1974(4); 282—1985(4)
Southeast 276—1952
West 197—1984(22)

West-central 151—1964(2); 197—1984(13,18); 215—1962; 296—1933(16)
Alligator Ridge Mine 211—1985
Basin and Range, West 562—1978
Battle Mountain 93—1969
Black Mountain 199—1974(2)
Buckhorn Mine 211—1985
Caliente 204—1966
Carson River basin 197—1984(4)
Churchill County, Bell Mountain Mine 85—1984(4)
Clark County 93—1965
Dead Mountain 93—1965
Death Valley 199—1974(1)
Dixie Valley 199—1956, 66(1)
Eldorado Mountain Ranges 93—1965, 69
Elko County 737—1960
Esmeralda County, Goldfield District 85—1984(3,4)
Eureka County, Carlin Mine 85—1984(1)
Fairview Peak 151—1964(2); 199—1956, 66(1)
Fallon area 197—1984(13)
Florida Canyon Deposit 211—1984
Frenchman Mountain 93—1965, 69
Humboldt County, Pinson Mine 583—1984; 85—1984(1)
Humboldt County, Sulfur District 85—1984(8)
Humboldt Range 197—1984(9)
Lake Tahoe 197—1984(23)
Lake Tahoe area 151—1964(1); 215—1968; 412—1985(Oct.)
Las Vegas 41—1976; 93—1965, 69; 204—1966
Little Bald Mountain Mine 211—1985
Lyon County 85—1984(10)
Miller Mountain area 197—1984(7)
Mineral County 197—1984(6)
Mineral County, Borealis Mine 85—1984(3)
Mineral County, Candelaria Mine 85—1984(4)
Mining Districts 43—1969
Mojave Desert 619—1980
Nelson Mountains 93—1965
Nevada Test Site 197—1984(10,21); 220—1969
Newberry Mountain Ranges 93—1965, 69
Nye County, Nevada Moly Mine 85—1984(7)
Nye County, Northumberland Mine 85—1984(2)
Nye County, Sterling Mine 85—1984(3)
Nye County, Tonopah District 85—1984(3,4)
Pershing County 199—1966(5)
Pershing County, Sulfur District 85—1984(8)
Pinson Mine 211—1984
Relief Canyon Deposit 211—1984
Reno 93—1965; 296—1933(16)
Reno area 197—1984(18,22)
Rochester District 211—1984
Ruby Mountains 197—1984(9)
Schell Creek Range 197—1984(1)
Sevier Thrust Belt 197—1984(19)
Sierra Nevada 151—1964(1)
Slide Mountain 197—1984(12)
Snake Range 197—1984(1)
Spring Mountains 93—1969
Steamboat Springs 151—1964(1)
Storey County, Gooseberry Mine 85—1984(12)
Storey County, Virginia City Mining District 85—1984(11)
Truckee Canyon 199—1966(3-4)
Truckee River basin 197—1984(4)
U. S. Interstate 80 272—49
Virginia City area 151—1964(2)
Washoe County 199—1975(4)
White Pine County, Alligator Ridge Mine 85—1984(2)
White Pine County, Taylor Mine 85—1984(2)
Yerington District 197—1984(8)

New Hampshire
Central 182—1937
Southeast 456—1971(A-7), 84(A-4,B-5)
Alton Quadrangle 456—1984(C-5)
Appalachian Mountains 197—1952(1)
Belknap Mountain 456—1971(B-3)
Coast 456—1984(B-4); 586—1969
Concord Quadrangle 456—1971(B-6)
Great Bay Estuary 456—1971(B-1)
Hanover area 456—1954, 71(A-4)
Holderness Quadrangle 456—1971(B-2)
Merrimack River valley 456—1984(B-9)
Merrimack River valley, Lower 456—1971(B-4)
Mt. Cube area 456—1971(A-6)
Mt. Washington 182—1937
Northwook Quadrangle 456—1984(B-6)
Ossipee Lake area 456—1971(A-8)
Ossipee Mountains 456—1971(A-5)
Peterborough Quadrangle 456—1971(A-2)
Seabrook area 456—1971(B-5)
Southeast, Northwook Quadrangle 456—1984(B-6)
Winnipesankee area 456—1971(A-1)
Wolfeboro area 456—1971(A-1)

New Jersey
General 27—1981; 76—1970(2); 172—1956; 469—1980
East 197—1948(2,7)
North 27—1981(5); 51—1968(D); 197—1969(3); 411—1972(2)
Northwest 172—1977; 411—1972(4)
North-central 197—1969(4); 469—1975(A-2)
South 18—1977(3)
Appalachian Valley and Ridge 76—1976(3)
Bangor Slate region 296—1933(8)
Coast, North 469—1968(B)
Coastal Plain 27—1981(5,Trip1,3); 76—1976(1); 89—1960, 61; 197—1957(1), 69(2,6); 411—1972(5); 539—1975
Coastal Plain, North 18—1960(1)
Coastal Plain, South 429—1976; 195—1985
Delaware Valley 197—1969(1A)
Delaware Valley, Central 197—1957(2)
Delaware Water Gap 51—1968(G)
Delaware Water Gap area 197—1957(4)
Franklin Mine 197—1948(1)
Franklin Zinc Deposits 296—1933(8,9)
Franklin area 469—1968(D)
Great Valley 182—1985
Highlands 197—1957(3)
Lakehurst area 469—1975(A-3)
Lehigh Cement District 296—1933(8)
Morris County 197—1957(3)
Newark 25—1966
Newark Basin 27—1981(5,Trip3); 411—1972(3); 195—1984
Newark Basin, North 469—1968(C)
Passaic County, Paterson 197—1948(11)
Paterson 296—1933(9)
Stokes Forest 213—1959
Sussex County 172—1952; 197—1957(3)
Tock Island area 197—1969(5)
Trenton area 197—1969(3)
Yards Creek Hydro-electric Project 197—1969(6)

New Mexico
General 464—1964, 67(8), 72, 74(11,12)
Central 30—1975; 458—1952; 460—1945; 550—1983
East-central 458—1972(1,2), 85
North 341—1930; 464—1956, 60(2,6), 68(2), 82(13); 458—1960, 68, 84; 528—1938; 272—114
Northeast 464—1961, 65(7), 67(7); 458—1966, 76; 504—1941, 56(Sept.); 528—1955

Northwest 197—1976(B); 461—1979; 458—1967, 77; 604—1950; 37—1983
North-central 458—1971(3), 74; 528—1973
South 18—1975(SEPM 2); 167—1981, 84; 185—1966; 197—1949(5); 606—1970; 311—1983; 646—1982
Southeast 74—1955; 464—1958(3), 67(3), 81; 457—1961, 71; 458—1954, 64, 80; 594—1979; 756—1942; 465—1980
Southwest 185—1965; 199—1968(1); 296—1933(14); 464—1959, 67(5), 71(10), 80; 458—1953, 65, 69, 70; 550—1958
South-central 205—1977(1); 458—1955, 75; 550—1983
West-central 204—1956; 457—1967, 78; 458—1959
Alamogordo 29—1938(1)
Albuquerque 204—1956; 464—1969, 74(9), 82(9)
Albuquerque Basin 204—1976(3)
Albuquerque area 458—1961, 82
Carlsbad Caverns National Park 29—1938 (3); 154—1977; 296—1933(13); 435—1960; 458—1954(6); 550—1957, 66; 569—1984(1); 272—90
Cerro de Cristo Rey Uplift 200—1977(3a)
Chaco Canyon National Monument 197—1976(B)
Datil-Mogollon Vocanic field 457—1978
Dona Ana County 167—1970, 73
Eddy County 756—1940(Fall)
Estancia Basin 550—1952(7)
Florida Mountains 167—1974
Franklin Mountains 550—1960
Gila River, Upper 185—1965
Grants 204—1956; 583—1963
Guadalupe County 550—1956
Guadalupe Mountains 457—1961; 458—1954(4); 550—1951(4), 52(6), 64; 594—1955, 72, 77; 626—1976(Fall); 756—1947, 62
Hanover 296—1933(14)
Hidalgo County 185—1964
Jemez Mountains, North 221—1975
Las Cruces area 458—1975
Lincoln County 550—1951(5), 52(7)
McKinley County 180—1957
Pecos County 756—1971
Pecos River, Upper 464—1960(6), 67(6), 75(6)
Portillo Basalt Field 205—1977(1)
Raton Basin 197—1984(3); 528—1953; 549—1969
Rio Grande 185—1966
Rio Grande Rift 204—1976(3); 320—1978; 463—1981; 458—1984
Rio Grande Valley 167—1970
Rio Grande Valley, North 221—1975
Robledo Mountains 594—1971
Roswell 550—1968
Ruidosa area 458—1964
Sacramento Mountains 18—1975(SEPM 2); 458—1954(7); 550—1950, 52(6), 53; 594—1959; 756—1940(Spring), 77; 646—1982
San Andres Mountains 550—1960
San Juan Basin 180—1955, 57; 458—1950-51, 77; 460—1946
San Juan County 180—1955, 57, 63, 72
Sangre de Cristo Mountains 528—1953, 59
Sangre de Cristo Mountains, Southeast 458—1956
Santa Fe 464—1955, 68(1)
Santa Fe Trail 528—1963
Santa Fe area 458—1979(1,2)
Santa Rita 296—1933(14)
Sierra County 460—1940
Silver City 29—1938(5); 197—1949(3)
Socorro County 460—1940; 550—1951(5)
Socorro area 458—1963, 83; 460—1941
Torrance County 550—1952(7)
Valles Caldera 272—47
Vermejo Park 458—1976
West Front Mountains 550—1953
Zuni Mountains, South 464—1958(4), 68(4), 71(4)

New York
General 277—1965(A); 296—1933(1,4); 411—1972(1), 73(3), 76(B,D), 83(A); 456—1969; 312—1983; 324—1981; 272—116, 150
Central 469—1970, 72, 77(A-9,B-8,B-12), 84(BC-3,BC-5,BC-9); 474—1982(B); 586—1971; 664—1979; 269—1982(May)
East 197—1963(1); 296—1933(1); 411—1983(B); 435—1979(A); 456—1979(A-2,B-7); 469—1970(C), 77(A-1,A-2,B-1,B-2), 85; 586—1972
East-central 456—1975(A-1,C-9), 79(A-1,A-3,B-6); 469—1972(A,B), 78(A-9), 85(A-5,B-5); 272—140
North 469—1983
Northeast 456—1959
Northwest 25—1982(A-4); 469—1972(H), 73(G), 78(A-6)
North-central 411—1966(1-2)
Southeast 201—1983(1); 296—1933(9); 410—1966(Fall); 411—1968(A,B), 82; 456—1937, 68(D-6), 79(A-9); 469—1977(A-7)
Southwest 469—1957
South-central 469—1963, 81
West 25—1982(A-1,A-3,A-5,B-1,B-2,B-3,B-5,B-6); 182—1954; 377—1951; 411—1981; 469—1956, 66, 73(D), 74; 664—1979; 269—1982(May)
Adirondack Mountains 182—1979; 296—1933(1); 456—1979(A-5,B-9); 469—1969(G,H,J), 83(7,8); 181—1983; 272—158
Adirondack Mountains, Central 469—1977(A-11), 78(A-2), 84(AB-2,C-11,C-13), 85(A-6,B-6)
Adirondack Mountains, Northwest 469—1971, 78(A-1), 83(2,7,8)
Adirondack Mountains, South 456—1979(A-4,B-8); 469—1960, 64, 72(E), 77(A-10), 84(C-11), 85(A-6,B-6); 470—1962
Adirondack Mountains, Southeast 411—1976(A); 456—1979(B-12); 469—1985(A-8)
Albany 456—1979(B-5a,B-5b)
Alleghany Plateau 25—1982(A-4)
Alleghany State Park 471—1927
Antwerp area 469—1984(BC-6)
Appalachian Basin 25—1982; 279—1982(17A)
Appalachian Mountains 18—1955; 197—1952(1); 279—1982(17A)
Appalachian Plateau 469—1984(B-8)
Bear Mountain 197—1948(3)
Bedford 296—1933(9)
Binghamton 182—1973
Binghamton area 639—1973
Black River valley 469—1983(5), 84(A-1)
Buffalo area 469—1938, 52(1,2)
Buttermilk Valley 25—1982(B-4)
Canton area 469—1939, 53
Catskill Front 469—1967(C)
Catskill Mountains 182—1941; 296—1933(9a); 411—1976(C); 586—1972; 293—1985(1)
Catskill Mountains, West 639—1973
Catskill Plateau 27—1981(5,Trip3)
Catskill River delta 469—1970(C), 76(B-8), 85(A-7)
Catskill area 469—1940
Cattaraugus County 182—1960
Cayuga Lake area 469—1949, 59
Champlain Valley 469—1969(I), 85(A-3,B-3)
Champlain Valley, South 469—1985(A-9)
Chautauqua County 469—1974(B)
Chenango River valley, North 469—1972(D)
Clinton area 469—1948, 84(B-8)
Cobleskill area 469—1977(B-7)
Cortland 201—1983(1)
Cortland area 469—1970, 78(B-7)
Crown Point 739—1985(D1-D14)
Dunkirk area 469—1974(E)
Dutchess County 197—1948(10); 456—1985(A-5); 469—1976(B-7)
Dutchess County, Amenia-Pawling Valley 456—1975(C-10)
Dutchess County, Southeast 469—1976(C-7)
Dutchess County, West 469—1976(B-6)
East-central, Cobleskill area 469—1977(B-7)

East-central, Granville area 456—1979(A-8)
Erie County 469—1974(C)
Erie Lowland 25—1982(A-4)
Finger Lakes 182—1947
Finger Lakes region 296—1933(4)
Finger Lakes, East 469—1970(F)
Finger Lakes, West 469—1973(A)
Fire Island 469—1975(A-8(AM))
Fire Island, Democrat Point 469—1975(A-5(AM))
Fire Island, Robert Moses State Park 469—1975(B-3(PM))
Fishkill 469—1976
Fort Ticonderoga 456—1972(LS-1)
Fredonia area 469—1974(E,F)
Genesee Valley 411—1981(A,B,C); 469—1973(B-C,E)
Genessee Gorge 469—1973(I)
Gouverneur area 469—1929
Granville area 456—1979(A-8)
Great Falls 272—147
Hamilton County 469—1978(A-3,A-5)
Hamilton area 469—1934, 55
Hamlin Beach State Park 469—1973(G)
Hempstead area 469—1975
Hoosick Falls area 456—1972(B-2)
Hudson Basin, North 469—1985(A-10)
Hudson Estuary 456—1975(A-2)
Hudson Highlands 51—1968(F); 456—1985(A-2)
Hudson Highlands, Central 469—1976(B-1,C-1)
Hudson River 272—1
Hudson River valley 296—1933(1); 456—1979(B-1); 469—1976(B-9,10)
Hudson River valley, Central 27—1981(5); 201—1983(2); 469—1967, 76
Hudson River valley, North 469—1965(C), 85(A-3,B-3,A-10)
Ithaca 182—1950
Ithaca area 469—1928, 31
Ithaca, Fall Creek Valley 469—1970(K)
Jamesville 272—5
Jefferson County 469—1971(B), 83(3)
John Boyd Thacher Park 471—1933
Kingston Arc 469—1967(E)
Kingston area 469—1976(B-4,C-4)
Lake Albany 456—1979(A-3)
Lake Champlain 456—1972(LS-2,LS-3); 469—1969(B,I)
Lake Erie 469—1974(D)
Lake George 469—1969(I)
Lake George area 471—1942; 470—1965
Lake Ontario, West 25—1982(B-3)
Livingston County 469—1973(E)
Long Island 182—1964(28th); 197—1948(8), 63(7); 469—1975(A-6,B-6)
Long Island Sound 456—1968(E-1), 85(C-6)
Long Island, Jones Beach State Park 469—1975(A-5(PM))
Long Island, Montauk Peninsula 469—1968(F), 75(A-7)
Long Island, Northwest 469—1975(B-5)
Long Island, Shinnecock Inlet 469—1975(A-4)
Long Island, Shoreham 469—1975(A-8(AM))
Long Island, West 469—1968(I,J)
Malone area 182—1955
Manhattan Prong 456—1985(B-3)
Middleville 469—1972(J), 78(B-3)
Millbrook area 469—1976(B-5,C-5)
Mohawk Valley 411—1983(B); 469—1935, 65(A), 77(A-4), 81(A-1), 85(B-8); 586—1972
Mohawk Valley, Central 469—1977(A-8)
Mohawk Valley, East 469—1965(C); 470—1965
Monroe area 469—1967(F)
Moss Island 469—1978(B-6)
New York City 25—1956; 51—1968; 197—1948(4,12); 296—1933(9)
New York City area 469—1933, 47, 68

New York City area, Mamaroneck 469—1975(B-1)
New York City area, Pelham Bay Park 469—1968(G)
New York City, Bronx to Manhattan 51—1968(B)
New York City, Central Park 272—51
New York City, Fort Tryon Park 51—1968(A)
New York City, Indwood Hill Park 51—1968(A)
New York City, Palisade Ridge 51—1968(C)
New York City, Staten Island 51—1968(E)
New York City, Watchung Ridge 51—1968(C)
Newcomb Lake area 469—1977(A-11)
Niagara Falls 133—1983; 296—1933(4); 469—1952(2); 272—145
Niagara Falls Dam 197—1978(12)
Niagara Gorge 25—1982(B-1,B-2,B-3)
Nine Mile Point Area 469—1978(A-11)
North, Malone area 182—1955
North-central, Jamesville 272—5
Oneida County 469—1984(B-7)
Onondaga County 469—1970(J)
Otsego County 469—1977(A-3,B-3,B-9)
Otsego Lake 469—1977(A-6,B-6)
Panther Mountain 469—1977(B-10)
Passaic Valley 296—1933(9)
Peekskill 469—1958
Penfield Quarry 469—1973(H)
Plattsburg area 469—1951(1,2), 69
Port Jervis area 469—1962
Potomac River Gorge 272—147
Potsdam area 469—1983(2)
Poughkeepsie area 469—1927, 46, 54, 76
Queens, Jamaica Bay 469—1975(B-4)
Ramapo Fault 411—1972(1)
Rochester 296—1933(4)
Rochester area 469—1932, 41, 73
Rockland County 197—1969(4); 539—1974
Sacandaga Valley 469—1985(B-8)
Sanford Lake area 469—1977(A-11)
Saratoga County 469—1985(A-4,B-4)
Saratoga Springs 456—1979(A-10,B-14)
Schenectady area 469—1930, 65
Seneca Lake 469—1978(A-10,B-5)
St. Lawrence County 469—1971(E-F), 83(3)
St. Lawrence County, Northwest 469—1971(B)
St. Lawrence Valley 469—1971(C), 83(1,8,9)
Staten Island 197—1948(2); 469—1975(B-2)
Susquehanna River valley 469—1977(B-11)
Susquehanna River valley, Upper 469—1977(A-4,A-5,B-4,B-5)
Syracuse Channels 469—1972(I)
Syracuse area 469—1926, 37, 50
Syracuse area, Valley Head Moraine 469—1978(B-3)
Taconic Foreland 469—1985(A-2,B-2)
Taconic Mountains 456—1975(A-3,B-1,C-1)
Taconic Region, Central 456—1979(A-7,B10)
Taconic Region, South 197—1963(3)
Thousand Islands 469—1983(2)
Troy area 469—1961
Tughill area 469—1978(A-4,A-7)
Ulster County 456—1979(B-13); 469—1967(D)
Upstate 272—151, 149
Utica area 469—1960
Wallkill Valley 411—1968(B); 469—1967(A)
Washington County, South 456—1979(A-6)
Wellsville area 469—1957
West Canada Creek valley 469—1984(C-10)
West Valley 25—1982(B-4)
Westchester County 197—1948(3)
White Plains area 469—1968(A)

North Carolina
General 206—1971(2)
Northeast 89—1963

Northwest 129—1983
North-central 129—1964
South-central 129—1984
West 649—1974; 722—1971
Alamance County 475—1967(1)
Albemarle Quadrangle 129—1959
Anson County 475—1967(2)
Appalachian Mountains 197—1955(1)
Appalachian Mountains, South 206—1978(1)
Ashe County 475—1967(3)
Asheboro 206—1977(1)
Avery County 475—1967(4)
Blue Ridge Mountains 74—1975; 129—1975, 76, 83; 197—1981(10); 681—1984
Buncombe County 475—1967(5)
Burke County 475—1967(6)
Cabarrus County 129—1966; 475—1967(7)
Caldwell County 475—1967(8)
Cape Fear River 89—1964
Carolina Slate Belt 129—1984; 197—1980(8), 85(6)
Carteret County 475—1967(9,15,32,42)
Caswell County 475—1967(10)
Chatham County 475—1967(11,33,41)
Cherokee County 475—1967(12)
Clay County 475—1967(13)
Cleveland County 475—1967(14)
Coastal Plain 89—1971, 72, 76, 79, 82; 129—1955; 206—1955, 59(2)
Corundum Hill 33—1965(4)
Craven County 475—1967(9,15)
Cumberland County 475—1967(16)
Currituck Spit 586—1977
Dan River basin 129—1970
Davidson County 475—1967(17)
Davie County 475—1967(18)
Denton Quadrangle 129—1959
Durham County 475—1967(19)
Durham Triassic Basin 129—1977
Fayetteville 206—1959(2)
Forsyth County 475—1967(20,52)
Franklin County 475—1967(21)
Gaston County 475—1967(22)
Graingers Wrench Zone 89—1982
Grandfather Mountain 129—1960
Granville County 475—1967(23)
Great Smoky Mountains 129—1952; 272—36
Guilford County 475—1967(24)
Halifax County 475—1967(25)
Hartnett County 475—1967(26)
Haywood County 475—1967(27)
Henderson County 475—1967(28)
Iredell County 475—1967(29)
Jackson County 475—1967(30)
Johnson County 475—1967(31)
Jones County 475—1967(15,32)
Kings Mountain Belt 129—1981; 206—1977(3)
Lee County 475—1967(11,33,41)
Lincoln County 475—1967(34)
Lincolnton Quadrangle 129—1953
MacDowell County 475—1967(35)
Macon County 475—1967(36)
Madison County 475—1967(37)
Mecklenburg County 475—1967(38,55)
Mitchell County 475—1967(39)
Montgomery County 475—1967(40)
Moore County 129—1962; 475—1967(11,33,41)
Mount Rogers area 129—1967
Murphy Belt 129—1971
Murphy Syncline 240—1978
New Bern 89—1982

New River valley 72—1980
Onslow Bay 206—1968(1)
Onslow County 475—1967(15,42)
Orange County 475—1967(43)
Person County 475—1967(44)
Piedmont Province 197—1955(Addenda), 85(6)
Polk County 475—1967(45)
Raleigh 206—1959(1)
Randolph County 475—1967(46)
Richmond County 475—1967(47)
Rockingham County 475—1967(48)
Rowan County 475—1967(49)
Rutherford County 475—1967(50)
Sauratown Mountain 206—1968(2)
Shelby Quadrangle 129—1953
Spruce Pine District 197—1955(Addenda)
Spruce Pine Mining District 74—1975
Stokes County 475—1967(52)
Surrey County 475—1967(51)
Swain County 475—1967(53)
Transylvania County 475—1967(54)
Union County 475—1967(38,55)
Vance County 475—1967(56)
Wadesboro area 129—1974
Wake County 475—1967(57)
Warren County 475—1967(58)
Watauga County 475—1967(59)
Wayne County 475—1967(60)
Yadkin County 475—1967(61)
Yancey County 475—1967(62)

North Dakota
General 703—1960-61
East-central 183—1958
Northeast 102—1957(9); 416—1969; 478—1956; 479—1972(2)
Northwest 479—1975(8)
North-central 479—1974
South 479—1972(1), 83
Southeast 102—1957(8); 479—1972(3)
Southwest 197—1972(8); 477—1954, 66; 479—1975(9)
South-central 479—1973(6)
Bismarck-Mandan area 102—1957(4)
Black Hills 435—1962
Burleigh County 480—1970(42)
Coteau du Missouri 183—1967
Devils Lake area 102—1957(3)
Dickinson area 102—1957(5)
Grand Forks 480—1969(40)
Jamestown area 102—1957(7)
Minot area 102—1957(2)
Theodore Roosevelt National Park 479—1973(4)
Valley City area 102—1957(1)
Williston Basin 402—1952
Williston area 102—1957(6)

Ohio
General 25—1972(2); 160—1973; 206—1984(5); 410—1962; 493—1957; 495—1980
Central 183—1962; 197—1981(13); 229—1955; 493—1958; 244—1985(3,4,9)
East 197—1961(4)
North 493—1957
Northeast 197—1981(13); 200—1974(5); 410—1974; 488—1970; 493—1950(Fall), 52, 71; 494—1967; 496—1969; 272—146
Northwest 25—1972(1); 197—1951(2); 200—1978(2); 376—1935; 377—1952; 493—1968; 496—1960; 587—1984
North-central 410—1979; 494—1970
South 209—1968; 244—1985
Southeast 587—1979

Southwest 183—1962; 197—1981(1,9); 200—1969(3); 493—1951; 496—1966; 543—1955(2); 497—1980
South-central 200—1968(2); 227—1979; 497—1979
West 227—1982; 377—1963; 493—1967
West-central 197—1983(12); 493—1950(Spring)
Adams County 493—1963
Akron area 200—1974(2); 493—1958
Alliance area 496—1967
Athens County 493—1954; 496—1957
Bedford Glens 496—1953
Bellefontaine area 493—1955
Cadiz 200—1974(6)
Canton 200—1974(6)
Chippewa Creek 496—1950
Cincinnati 228—1982
Cincinnati area 197—1961(2,3,5,6,7), 81(12,18); 377—1953; 493—1961; 571—1968
Clermont County 493—1970
Cleveland area 200—1974(1); 493—1958
Columbus 182—1952; 200—1969(1)
Columbus area 493—1953
Coshocton County 496—1963
Cuyahoga Gorge 496—1950
Cuyahoga River valley 200—1974(3); 493—1973
Delaware County 496—1956, 61
Dunkard Basin 763—1950
Fairfield County 493—1977
Franklin County 493—1977
Geauga County 493—1964
Highland County 200—1969(2); 493—1963
Hocking Hills State Park 228—1975
Hocking River valley 493—1965
Holmesville 496—1964
Judy Gap 496—1959, 65
Lake County 493—1964
Lake Erie 228—1973
Lake Erie shore 200—1974(4); 410—1979; 497—1983
Lake Erie shore, South 197—1981(7)
Lake Erie, Southwest 587—1984
Licking County 200—1969(4); 493—1975; 496—1956; 497—1981
Lucas County 376—1935; 493—1948
Mad River valley 197—1961(6)
Marblehead Peninsula 410—1979
Maumee 200—1978(3)
Miami River valley 197—1961(6)
Mining Districts 43—1971
Muskingum County 200—1969(4); 493—1966; 496—1963
Newark 200—1969(4)
Newark area 496—1968
Northeast, Bedford Glens 496—1953; 769—1953
Northeast, Chippewa Creek 496—1950
Ohio River valley 197—1961(5); 494—1969
Oxford area 496—1958
Painesville 496—1962
Perry County 493—1949, 66
Scioto River 179—1974
Serpent Mound 197—1961(8), 81(6,16)
Springfield area 493—1956
Stark County, Southeast 496—1954
Toledo 228—1982
Toledo area 493—1962
Van Buren Lake 496—1952
Warren County 493—1970
Wayne County 493—1974
Williamsport area 496—1979
Wooster 496—1964
Yellow Springs area 493—1960

Oklahoma
General 18—1932, 68(1); 296—1933(2,6); 504—1930, 36(Oct., Nov.), 37, 55 72; 506—1971-72; 528—1963; 576—1936-37, 40; 662—1950(Oct.), 73(Oct.)
Central 565—1938(1)
East 18—1936; 29—1957; 662—1946; 721—1946
East-central 26—1979
North 205—1978(2)
Northeast 205—1974(2-3); 296—1933(2); 341—1937, 60, 64, 80; 474—1977(5); 504—1954, 56(Apr.), 64; 662—1941(Oct.), 51(Apr., May), 54, 81, 84(May)
North-central 504—1946; 506—1981; 662—1950(Dec.)
South 62—1937(Mar.), 57, 63; 94—1985; 317—1981(3); 668—1984; 272—133
Southeast 62—1956; 566—1928, 61, 70; 717—1983
Southwest 18—1978(4); 132—1959; 197—1973(6); 482—1939; 21—1983
South-central 62—1937(Mar.), 38, 63; 205—1982(2); 316—1982(2)
West 474—1977(4)
Ada District 511—1945, 46(Nov.)
Alabaster Cavern 508—1969(15)
Anadarko Basin 18—1939; 508—1963(13); 528—1969
Arbuckle Mountains 18—1932(2), 39(2), 68(3), 75(1), 78(1); 42—1950; 62—1946, 50, 52, 55, 69; 94—1985; 178—1969; 197—1931, 73(5); 205—1978(3), 84(4); 247—1976; 253—1971; 270—1973(May); 296—1933(6); 341—1931; 507—1927; 508—1966, 69(17); 566—1973; 585—1930; 626—1961, 72(Spring); 715—1982; 49—1983; 668—1984
Arbuckle Mountains, West 197—1932
Ardmore Basin 18—1939(2); 62—1936(Mar.,Apr.14,May), 37(Mar.), 48, 66; 74—1964; 197—1932; 296—1933(6); 507—1927
Ardmore area 62—1948; 63—1977; 253—1971
Arkoma Basin 18—1968(2), 75(SEPM 1), 78(2,3); 662—1961; 589—1983
Arkoma Basin, West 26—1981(No.2)
Atoka County 62—1952; 205—1978(1)
Baum area 62—1938
Beavers Bend State Park 508—1963(11)
Berwyn area 62—1938
Black Mesa area 102—1961
Blaine County 508—1969
Blaine County, Roman Nose State Park 508—1959
Blue Creek Canyon 21—1983
Boiling Springs State Park 502—1953
Braggs Mountain 502—1947(May 3)
Camp Egan area 502—1959
Carter County 62—1950, 69
Cimarron County, Black Mesa area 102—1968
Cimarron River valley 102—1961; 528—1946, 51(May 17), 55
Coffeyville 205—1978(2)
Criner Hills 62—1937(May), 46, 57; 576—1947, 53(2); 571—1947
Criner Hills area 626—1961
Dwight Mission 502—1955
Johnston County 62—1950, 52
Keystone Reservoir 662—1972, 73(Apr.)
Lake Murray area 62—1957
Latimer County, Camp Tom Hale 508—1958
Latimer County, Robbers Cave State Park 508—1958
Murray County 62—1950, 69
Muskogee 502—1947
Muskogee area 576—1941
Oil Fields 296—1933(6)
Oklahoma City 18—1932(1)
Oklahoma Coal Basin 62—1954
Okmulgee 173—1964
Osage County 341—1983

Guidebooks of North America

Ouachita Mountains 18—1959, 68(2), 75(SEPM 1), 78(3),
  78(SEPM 1); 62—1936(June), 52, 56; 205—1982(3), 84(4), 85;
  341—1931, 66; 502—1952; 504—1950; 508—1966; 566—1948,
  53, 73; 626—1977; 662—1947, 61, 75; 715—1982; 717—1983;
  485—1966
Ouachita Mountains, West 247—1976
Ozark Mountains 508—1963(12)
Ozark Uplift 504—1953; 662—1981
Panhandle 504—1941, 56(Sept.)
Pittsburg County 565—1938(2)
Prague 502—1947
Rocky Dell 528—1965
Sand Springs 510—1947
Slick Hills 21—1983
Stillwater 40—1960
Tahlequah 502—1955
Tulsa 32—1977, 205—1974(1), 78(4); 510—1947
Tulsa County 102—1963; 662—1941(Nov.)
Tulsa area 662—1968, 84(2,3)
Washita Valley 62—1940
Wichita Mountains 18—1978(4); 62—1936(Apr.18); 94—
  1962(Mar.); 197—1973(6); 205—1967, 76(2), 84(1); 341—1931;
  481—1957; 482—1947; 504—1940, 49; 528—1957; 576—
  1953(1), 59
Wichita Mountains area 132—1959; 511—1946(May)
Wichita Mountains, Southwest 205—1983(1)
Woodward 502—1953
Woodward County 508—1969(15)

Oregon
General 216—1971, 81; 423—1959; 436—1982; 568—1981; 272—
  10, 6
Central 39—1979(2-6); 216—1965(1), 68(1), 69(1-4), 70(1), 82;
  296—1933(21); 423—1959(3-6); 435—1982(22); 489—1982(1,3-
  4)
East 706—1973; 284—1983
North 199—1973; 423—1959(7)
Northeast 39—1979(1)
Northwest 199—1973(3)
North-central 199—1973(1)
Southwest 216—1970(1,2:1-2:4); 593—1984(Sept.)
South-central 489—1982
West-central 423—1959(1,2)
Applegate River 216—1970(2:3)
Belknap Crater area 366—1965(4)
Bend 199—1969(2); 435—1982
Blue Mountains 518—1983
Broken Top Quadrangle 519—1978
Broken Top area 366—1965(1)
Brothers fault zone 39—1979(5)
Cascade Range 132—1976(2); 327—1968(5); 519—1978
Cascade Range, Central 39—1979(3); 243—1983(A-1,2)
Cascade Range, North 216—1977; 253—1979
Christmas Lake Valley Basin 39—1979(6)
Coast 76—1971(2); 132—1976(1)
Coast Range 132—1976(1); 199—1973(2)
Coast Range, North 216—1964
Coast, North 272—4
Columbia River 216—1967(1); 518—1984(8,9)
Columbia River Gorge 76—1971(1); 199—1958, 73(4); 216—
  1965(2); 518—1974, 84(8,9); 561—1979(1)
Columbia River Gorge, Lower 253—1979
Coos Bay 199—1958
Crater Lake 132—1976(2)
Crater Lake area 327—1968(2); 366—1965(5)
Dayville 216—1969(1)
Deschutes Basin 243—1983(A-1)
Deschutes Canyon 216—1968(2)
Deschutes County 489—1982
Deschutes River 216—1967(2); 489—1982

Douglas County 216—1973
Eugene 199—1958, 69(2)
Fort Rock 39—1979(6); 366—1965(3)
Grant County 216—1985
Harney Basin 39—1979(5)
Harney County 216—1985
High Lava Plains 39—1979(5)
Hole in the Ground area 366—1965(3)
Idaho-Oregon Border 204—1975(8)
Illinois River 216—1970(2:2)
John Day River, South Fork 216—1969(1)
Jordan Valley 185—1969; 204—1985(3)
Klamath County, North 489—1982
Klamath Mountains 199—1969(1), 80
Lava Butte 366—1965(1)
McKenzie Pass area 327—1968(1)
Medford 216—1970(2:4)
Mitchell 216—1969(3)
Mount Hood 76—1971(3); 243—1983(A-1)
Mount Hood area 327—1968(4)
Mount Mazama 132—1976(2)
Mt. Ashland 216—1970(2:4)
Newberry Caldera area 327—1968(3)
Newberry Volcano 39—1979(4)
Newberry Volcano area 366—1965(2)
Newport 199—1969(3)
Ochoco Summit area 216—1969(2)
Owyhee River 204—1985(3)
Owyhee area 706—1973
Painted Hills 216—1969(4)
Portland 76—1971(4); 199—1958, 73(5)
Rogue River 216—1970(2,2:1)
Rogue River area 706—1977
Siskiyou River 216—1970(2)
The Dalles 199—1958
Twickenham area 216—1969(4)
Wallowa County 216—1972(1-6)
Willamette Valley 199—1958

Pennsylvania
General 18—1960(3); 74—1982; 160—1973; 172—1982, 83; 534—
  n.d.(2); 586—1982; 250—1985; 535—1979
Central 32—1969; 172—1931(II,IV), 46, 55(2),
  66(B1,B2,F1,S1,S2), 73, 85(1-4); 197—1959(1); 288—1976;
  296—1933(8); 424—1978; 536—1939(13), 65; 586—1970; 137—
  1976
East 27—1981(4); 172—1932, 53, 60, 67, 84; 296—1933(2,8);
  411—1972(2)
East-central 197—1963(4)
North 25—1982(A-6); 182—1954
Northeast 76—1970(2), 78(3); 172—1937, 63, 69, 75, 78; 197—
  1969(1C)
Northwest 172—1950(2), 59(B,C), 76; 197—1959(5); 496—1969;
  536—1974
South 296—1933(10); 536—1939(12), 42, 49, 72; 250—1982
Southeast 74—1960; 172—1948, 49, 58, 66(A), 74; 424—1963;
  537—n.d.
Southwest 29—1937; 172—1950(1,3), 62
South-central 172—1968, 79; 536—1938(8), 58; 250—1983; 272—
  129
West 32—1969; 172—1964, 65, 85(5); 197—1959(2); 329—1965;
  536—1976
Adams County, Caledonia State Park 538—15
Allegheny County 172—1980
Allegheny Front 32—1969; 172—1931(III), 41, 72
Allegheny Plateau 424—1978(2)
Allentown 201—1973
Anthracite Fields 296—1933(8)
Anthracite region, East 197—1963(4), 59(1)
Appalachian Mountains 18—1955; 197—1957(7); 499—1966

Appalachian Mountains, Central 58—1963
Appalachian Mountains, Piedmont 172—1960
Appalachian Plateau 32—1969
Appalachian Trail 536—1983
Appalachian Valley and Ridge 76—1976(2,3), 78(2)
Avella area 197—1978(11)
Bedford County 172—1972
Bellefonte 172—1966(1)
Bethlehem area 172—1947
Blair County 172—1931(III), 41
Bradford County 172—1937, 81
Bucks County, Nockamixon State Park 538—14
Caledonia State Park 538—15
Central, Nittany Valley 484—1980
Centre County, Bald Eagle Mountain 172—1966(B1,F1)
Centre County, Bald Eagle Valley 172—1931(II)
Centre County, Bellefonte area 172—1931(I), 55(1), 66(B1)
Clearfield County 172—1955(3), 66(B)
Conemaugh(Glenshaw) 25—1974
Cornwall 536—1961
Cornwall Iron Mines 296—1933(8)
Cornwall Mines 172—1948
Crawford County 172—1959(A)
Cumberland County, Pine Grove Furnace State Park 538—15
Cumberland Valley 172—1966(A)
Delaware Valley 197—1969(1A)
Delaware Water Gap 51—1968(G); 536—1938(11)
Delaware Water Gap area 197—1957(4)
Dunkard Basin 763—1950
Erie County 172—1959(A)
Fayette County 411—1973(1); 530—1969(1,2)
Franklin County, Caledonia State Park 538—15
Friedensville Mine 201—1973
Gettysburg 424—1978(5)
Harrisburg 74—1982
Interstate 80 272—21
Interstate 81 272—22
Jacks Mountain 424—1978(4)
Johnstown 531—1973; 541—1985
Lackawanna County 172—1971
Lebanon County 197—1959(4); 534—n.d.(1,3)
Lebanon County, Swatara State Park 538—16
Lebanon Valley 172—1966(A)
Lehigh County 172—1961; 197—1959(4)
Lehigh Valley 172—1961; 536—1939(16)
Lehighton area 197—1969(1B)
Lock Haven State College 354—1981
Luzerne County, Ricketts Glen State Park 538—13
Mifflin 424—1978(1)
Monroe County 197—1969(1B)
Nockamixon State Park 538—14
Northampton County 172—1961
Perry County 536—1957, 71
Philadelphia 172—1935, 51; 536—1964
Philadelphia area 172—1935, 51; 197—1957(5)
Piedmont 76—1976(2,3); 172—1974
Pike County, Promised Land State Park 538—18
Pine Grove Furnace State Park 538—15
Pittsburgh 411—1980; 536—1939(17), 70
Pittsburgh Mining District 296—1933(2)
Pittsburgh area 197—1959(6), 71(6); 411—1973(2)
Presque Isle Peninsula 25—1982(A-6)
Promised Land State Park 538—18
Raymond B. Winter State Park 538—19
Raystown Dam 76—1970(1:1)
Reading Hills 172—1961
Reading Prong 76—1976(2)
Reading area 536—1939(15)
Ricketts Glen State Park 538—13
Samuel S. Lewis State Park 538—17
Schuylkill County, Swatara State Park 538—16
Schuylkill Valley 536—1939(14)
Sinking Valley 424—1978(3)
South Mountain 172—1948, 58; 530—1967
South Mountain area 250—1982
Sterling area 469—1936
Sullivan County, Worlds End State Park 538—12
Susquehanna County 172—1971
Susquehanna River 250—1983
Susquehanna River valley 182—1978
Susquehanna River, Lower 76—1978(1)
Swatara State Park 538—16
Tioga County 172—1981
Union County, Raymond B. Winter State Park 538—19
Washington area 197—1959(3)
Worlds End State Park 538—12
Wyoming County 172—1971
York County 785—1980
York County, Samuel S. Lewis State Park 538—17
York-Hanover Valley 250—1984

Rhode Island
General 197—1963(2); 456—1947, 63, 76(A-18,F-6), 81(C-5)
North 456—1981(B-3,B-4)
Northeast 456—1981(B-8)
South 182—1962; 456—1981(B-9,B-10,C-9)
Aquidneck Island 456—1976(B-17)
Block Island 456—1981(A-2,B-11)
Coast 456—1981(B-6,C-7)
Conanicut Island 456—1976(B-17)
Conanicut Island, Jamestown 456—1981(C-1)
Middletown 456—1981(B-2)
Narragansett Basin 456—1981(B-1); 415—1979
Newport 456—1981(B-2)
Norfolk Basin 456—1981(B-1)

South Carolina
General 206—1969(1)
Central 89—1965; 129—1968, 78; 206—1969(2)
Northwest 129—1969
Aiken County 129—1982
Appalachian Mountains 197—1955(1)
Blue Ridge Mountains 129—1976
Carolina Slate Belt 129—1961, 84, 85; 197—1980(8), 85(6)
Charlotte Belt 129—1961; 206—1969(3)
Coast 197—1980(14)
Coastal Plain 89—1971, 76; 129—1957; 197—1980(9); 206—1954
Kings Mountain Belt 129—1981; 206—1977(3)
Lake Murray area 129—1958
McCormick County 169—1981[1]
Newberry County 129—1961
Oconee County 129—1963
Pageland area 129—1974
Pickens County 129—1963
Piedmont 129—1973; 197—1955(Addenda); 206—1954
Piedmont Province 197—1985(6)
Piedmont, East 197—1980(5)
York County 129—1965

South Dakota
General 204—1951; 703—1960
East 183—1960; 197—1972(6); 773—1979
Northwest 477—1966
Southeast 200—1981(3)
Southwest 335—1964; 610—1969
West 204—1960; 341—1940; 604—1951, 85; 725—1968
Badlands National Monument 204—1981(6); 569—1984(2)
Bell Fourche 204—1960
Big Badlands 204—1970(3)

Black Hills 74—1946; 190—1982(12); 204—1970(2), 81(1-5), 85(Revised); 253—1977(A,B); 341—1929; 402—1952; 477—1955, 66; 703—1960; 282—1985(1); 724—1978(Apr.); 272—37
Black Hills area 296—1933(25); 781—1968
Deadwood 204—1960
Homestake Mine 204—1970(1), 81(7)
Interstate 90 204—1970(1)
Mining Districts 43—1969
Pine Ridge 204—1981(6)
Rapid City 204—1960, 70(1,4,5)
Spearfish 204—1960
Sturgis 204—1960
White River Badlands 296—1933(25)
Wind Cave National Park 569—1984(2)

Tennessee
General 40—1970; 435—1961; 437—1979; 566—1929; 649—1960; 651—1958, 81; 662—1976; 711—n.d.(1-2); 681—1982
Central 206—1953, 65(3); 390—1954; 722—1971
East 129—1983; 206—1985(6,7); 296—1933(2); 494—1965; 583—1969(1); 722—1971; 272—36
Northeast 206—1985(3); 744—1980(A-C)
South 206—1972(A); 390—1950
Southeast 206—1962(3)
Southwest 99—1978; 206—1975(5); 391—1961; 588—1983(Apr.)
South-central 5—1967; 390—1949
West 206—1975(4)
Appalachian Basin 197—1981(3)
Appalachian Mountains 33—1965; 197—1955(1), 80(10)
Appalachian Mountains, South 206—1978(1)
Blue Ridge Mountains, East 681—1984
Campbell County 651—1981(38)
Chattanooga 206—1978(4)
Copper Ridge District 206—1985(1)
Cumberland Gap area 209—1961
Cumberland Plateau 206—1961, 65(1); 649—1954, 72
Cumberland Plateau, South 206—1978(5)
Ducktown Copper District 33—1965(2); 206—1985(5)
Ducktown Mining District 296—1933(2)
East Tennessee Marble District 33—1965(1)
Gatlinburg 33—1965
Great Smoky Mountains 129—1952; 272—36
Hawkins County 744—1980(C)
Jefferson City 206—1961
Jefferson City Zinc District 296—1933(2)
Ketner's Mill 437—1979
Knox County 206—1973(3)
Knoxville 206—1973(1-2)
Mascot 206—1961
Mascot-Jefferson City Zinc District 33—1965(3); 206—1961, 85(1)
Memphis 206—1975(2)
Mining Districts 43—1971
Mississippi Embayment 197—1982(8)
Mount Rogers area 129—1967
Newport 206—1984(8)
Ocoee River Gorge 294—1973(2)
Pigeon River Gorge 649—1974
Pine Mountain 206—1984(1)
Pine Mountain Thrust Sheet 197—1981(19)
Polk County 294—1973(2)
Raccoon Mountain area 197—1980(13)
Reelfoot Lake 206—1975(3)
Scott County 649—1972
Sequatchie Valley 206—1965(1)
Tennessee River 206—1975(1)
Walden Ridge 206—1985(2)
Wells Creek Basin 206—1965(2)

Texas
General 18—1953, 84(3); 67—1975; 152—1948; 171—1949; 255—1959(3); 275—1970; 296—1933(6); 418—1972; 435—1964, 78; 554—1956, 58; 576—1940; 580—1968(Jan.); 583—1974(Feb.); 594—1969; 626—1974(Spring); 628—1937(1-2); 685—1965; 713—1960, 64, 70(10); 715—1967; 716—1950, 51(Mar.), 54; 759—1973; 272—126, 159, 127
Central 2—1946, 48(Jan.), 49-51, 52([1]); 18—1971(4), 74(SEPM 1-2), 75(3); 29—1951; 62—1937(Apr.); 92—1974, 84(5), 85(8); 94—1958, 59, 61, 64, 66, 70(3), 73(1,2), 76, 78(2), 79(1), 83(Fall), n.d.; 95—1960, 61(5th,6th); 132—1961, 80; 152—1955, 66, 82; 162—1951, 59; 197—1973(1); 205—1975(1), 83(2); 247—1977(1); 580—1966(Nov.-Dec.), 67(Jan.3:1,3:2, Apr.7); 582—1955; 612—1950, 53, 60; 626—1968, 72(Fall), 74(Spring), 75(Fall), 79(Spring, Fall), 81; 652—n.d.; 685—n.d.; 715—1937(1-3); 716—1952, 56, 58(Apr.), 60(Feb.); 756—1933, 41(Spr.); 317—1981(3)
East 132—1969; 152—1954; 156—1955; 159—1951; 162—1939, 45, 84; 197—1973(11); 255—1952; 566—1927; 588—1957, 80; 626—1973(Fall)
East-central 18—1979(5); 94—1962(Feb.); 152—1953; 247—1977(2); 626—1960
North 482—1939; 576—1936; 652—1948
Northeast 2—1957; 566—1928, 70
Northwest 197—1940(8)
North-central 18—1969(1), 75(2); 31—1985; 197—1973(8); 482—1956, 59; 566—1969; 594—1960, 82; 627—1962; 771—1946; 316—1982(5)
South 18—1971(3), 1979(4); 29—1947; 152—1949, 50, 56, 57, 59, 62, 68, 75; 205—1981(1,2,3); 247—1958, 67(1-2), 81(1); 547—1965; 580—1968(May), 70; 612—1931-34, 36-40, 49, 53-54, 58, 60; 685—1970; 728—1977
Southeast 18—1971(2), 1979(2); 69—1982(2); 197—1962(2-4,10); 205—1976(3); 255—1978(Nov.); 626—1965, 80; 652—n.d.; 713—1970(11)
Southwest 18—1984(6); 152—1951; 167—1975; 197—1949(1); 594—1974(Oct/Nov); 612—1985; 758—1958
South-central 197—1940(3); 255—1980; 580—1968(Dec.); 612—1976, 80; 626—1967, 73(Spring)
West 152—1960; 167—1984; 296—1933(13); 458—1980; 580—1967(Jan.6); 594—1979; 702—1951, 54, 59; 756—1948; 311—1983
West-central 2—1948, 52(2), 55, 61; 18—1969(4); 31—1981; 594—1984; 626—1963, 69
Abilene 2—1946, 48-49, 51, 52(1)
Alpine area 641—1969
Amistad Dam 594—1965
Anacacho 18—1984(1)
Anadarko Basin 18—1939
Apache Mountains 594—1968
Atascosa County 612—1958
Austin 571—1937(2); 580—1967(Jan.4), 69
Austin area 92—1973, 79, 81, 82, 85(7); 152—1955; 197—1940(1-2), 62(1); 205—1975(2-3), 83(3); 566—1949, 71; 612—1939(Nov.1), 40; 626—1983; 627—1965; 660—1940; 713—1963(5), 72, 76; 715—1937(1-3); 316—1982(1)
Balcones Escarpment 69—1982(12,13)
Balcones fault zone 554—1956
Bastrop County 716—1953, 55, 59, 60(Dec.)
Bee County 612—1937(July)
Bell County 18—1979(SEPM 2), 84(SEPM 3); 92—1985(8); 197—1973(1)
Bell County, Moffatt Mound 247—1973(2)
Bexar County 612—1937(Nov.)
Big Bend 18—1974(1); 685—1956, 67, 70; 688—1970, 74
Big Bend National Park 69—1982(15); 152—1984(Spring); 154—1977; 713—1968; 756—1941(Fall), 55, 65, 72; 569—1984(1)
Big Bend region 197—1949(1)
Black Prairie 94—1974(1)
Blanco County 580—1967(Jan.5)

Bosque Basin 94—1970(1)
Bosque County 95—1961(5th); 132—1980
Bosque Watershed 94—1969(1), 74(2)
Brazoria County, Damon Mound 255—1978(Apr.)
Brazos County 205—1983(1)
Brazos River valley 2—1948, 54-55: 18—1979(2), 84(SEPM 5); 197—1962(2); 205—1970, 83(4); 482—1940; 588—1958, 59; 594—1960; 756—1951(Spring); 316—1982(3)
Brown County 62—1936(Dec.)
Brownwood District 2—1949
Bryan area 205—1983(1)
Burnet County 626—1962
Central, Marble Falls 580—1967(Jan.3:3), 68(Nov.)
Central, Pedernales Falls State Park 580—1976
Cherokee County 162—1960
Chittim Arch 152—1960
Coast 197—1982(12); 713—1980(20)
Coast, South-central 247—1964
Coastal Plain 18—1984(4); 197—1940(7), 62(3,4), 73(7); 247—1965
Coleman County 62—1936(Dec.)
College Station area 205—1983(1)
Colorado River valley 2—1948, 55, 61; 756—1951(Spring)
Colorado River, Lower 197—1940(6)
Comal County, Canyon Dam 197—1962(12)
Comanche County 171—1949
Cook Mountain 612—1954(21st)
Corpus Christi 612—1932(3rd)
Correll County, East 95—1962
Culberson County 167—1971; 594—1958; 756—1939
Dallas 69—1982(3); 76—1980; 94—1978(3)
Dallas area 18—1983(2); 205—1984(2); 588—1983(2)
Damon Mound 18—1979(3)
Delaware Basin 756—1960, 68-69, 82
Devils River Uplift 756—1983
Duval County 152—1975
East Texas Embayment, Palestine Dome 197—1962(6)
East, Lone Star area 247—1970(2)
Eastland County 62—1936(Dec.)
Ector County, Meteor Crater 197—1940(9)
Edwards Escarpment 152—1959
Edwards Plateau 18—1974(2); 435—1978(A)
Edwards Plateau, East 247—1979(1)
Edwards Plateau, South 247—1961
El Paso 205—1977; 296—1933(13); 594—1975(Oct./Nov.); 719—1980
Enchanted Rock 69—1982(5); 580—1976
Fairfield 197—1973(10)
Falls County 132—1980
Fayette County 716—1953, 55, 59, 60(Dec.)
Fort Bend County 18—1933
Fort Bend County, Boiling Field 255—1959(1)
Fort Worth Basin 2—1957; 594—1956
Fort Worth area 18—1929
Franklin Mountains 167—1967, 68; 594—1971; 756—1937, 46(Spring), 50, 58
Franklin Mountains, South 627—1961
Galveston Heald Bank area 588—1964
Galveston Island 205—1976(3); 713—1970(11)
Glass Mountains 29—1938(2); 594—1957; 756—1946(Fall), 57
Glen Rose 580—1967(Jan.3:3)
Glen Rose area 626—1970(Fall)
Grand Prairie 94—1974(1), 1979(2)
Great Saline Dome 197—1962(6)
Green Valley 29—1938(4)
Grimes County, North 588—1960
Guadalupe Mountains 458—1954(4); 550—1964; 594—1955, 67, 72, 77; 626—1976(Fall); 756—1947; 569—1984(1)
Guadalupe Mountains National Park 154—1977

Gulf Coast 152—1980; 255—1968; 604—1956; 685—1968; 716—1955, 59, 60(Dec.)
Gulf Coast, Houston-Galveston area 316—1982(4)
Gulf Coast, South-central 247—1964
Gulf Coastal Plain 197—1982(1)
Hays County 92—1984(6)
Hill Country 69—1982(6,10); 92—1974; 94—1972; 95—1963
Hog Creek Watershed 94—1968
Houston 18—1963, 79(1)
Houston County 588—1961
Houston area 197—1962(5); 205—1976(1); 255—1961, 64; 652—1940
Hudspeth County 167—1971; 594—1958; 756—1939
Hudspeth County, Finlay Mountains 594—1980
Hueco Mountains 594—1983; 756—1937, 46(Spring), 58
Hutchinson County 528—1951(May 5)
Karnes County 612—1958
Kerr Basin 554—1956
Lake Belton 18—1984(SEPM 3)
Lake Ransom Canyon 655—1970
Lampasas Cut Plain 94—1970(2)
Laredo 612—1935, 41
Leon River valley 94—1979(3)
Little Brazos River valley 716—1963
Live Oak County 612—1937(July)
Llano 570—n.d.; 580—1973
Llano Country, Southeast 94—1983
Llano County 626—1962
Llano Estacado 197—1949(2); 756—1956, 74; 759—1973; 186—1983
Llano Uplift 2—1948, 50, 57; 29—1951; 62—1937(Apr.); 92—1984(5); 94—1966; 152—1982; 580—1966(Nov.), 67(Jan.5), 68(Nov.); 626—1962; 628—1937(3); 654—1939; 715—1937(3); 716—1960(Feb.), 61
Llano Uplift, Northeast 2—1957
Llano region 74—1958; 197—1940(4,5), 62(1); 205—1975(1,3); 566—1971; 713—1963(5), 72; 712—1945; 756—1954, 80
Lone Star area 247—1970(2)
Longhorn Cavern 713—1963(4)
Lubbock 655—1970
Lubbock region 205—1971
Luling 571—1937
Malone Mountain 29—1938(4)
Marathon 594—1981
Marathon Basin 18—1969(5); 29—1938(2); 594—1964, 81; 641—1969; 756—1936, 46(Fall), 52
Marathon area 197—1949(1); 296—1933(13); 594—1981
Marble Falls 580—1967(Jan.3:3), 68(Nov.)
Marfa 594—1981
Mariscal Canyon 626—1975(Spring)
Marshall Ford Dam 716—1939(Nov.2)
Maverick Basin 18—1984(6)
McLennan County 132—1980; 197—1973(1)
McLennan County, Northwest 95—1961(6)
McLennan County, Southwest 95—1962
McLennan County, West-central 95—1960
Medina County 612—1937(Nov.)
Midland 594—1975(Oct./Nov.)
Mineral District 2—1949
Mineral District, Central 204—1983(5); 205—1983(5)
Monahans Sandhills State Park 713—1984
Mound Valley 94—1969(2)
Mustang Island 69—1982(1); 728—1977
New Braunfels area 69—1982(12)
Oil Fields 296—1933(6)
Oldham County 528—1958
Osage Plains 756—1956
Padre Island 69—1982(1)
Padre Island National Seashore 247—1972; 713—1980(17)
Padre Island, North 728—1977

Palestine 18—1924
Palo Duro 759—1960, 61(Apr.), 62, 64, 67, 71-72
Palo Duro Canyon 154—1977; 626—1966; 713—1969(8); 759—1975
Palo Duro Canyon State Park 29—1961; 626—1966
Palo Pinto County 62—1936(Dec.); 627—1958
Paluxy River 94—1973(3), 78(1)
Panhandle 341—1930; 474—1977(4); 482—1939; 528—1949, 54; 626—1966; 759—1959, 61(Nov.)
Parker County 62—1936(Dec.); 571—1973
Parker County, Southwest 594—1956
Pedernales Falls State Park 580—1976
Permian Basin 594—1957
Permian Basin, East 756—1938
Pilot Knob 205—1975(4); 716—1951(Apr.)
Potter County 528—1951(May 5)
Presidio County 641—1969
Presidio County, Chinati Mountains 756—1953(Spring)
Ranger District 2—1949
Rio Grande Embayment 152—1959, 65; 554—1956; 566—1979
Rio Grande River delta 612—1939(2)
Rio Grande Valley 197—1949(2)
Rio Grande Valley, East 612—1939(1)
Salt Domes 296—1933(6)
San Angelo 553—1959
San Angelo area 756—1961
San Antonio 18—1974(3); 69—1982(A,4,7-9,11); 205—1981(2); 571—1937(1); 612—1932(3rd)
San Antonio area 197—1962(11)
San Saba County 716—1957
Sierra Blanca area 756—1950
Sierra Diablo Mountains 594—1962, 83
Smith County 162—1960
Tarrant County 205—1968; 571—1973
Trans-Pecos Volcanic Field 149—1978
Trans-Pecos area 197—1949(4-5); 594—1958; 713—1979; 756—1942, 52, 53(Fall), 85
Trans-Pecos area, Apache Mountains 756—1951(Fall)
Trans-Pecos area, Eagle Mountains 594—1975(Apr.)
Trans-Pecos area, Finlay Mountains 594—1980
Trans-Pecos area, Quitman Mountains 594—1970
Travis County 92—1984(6), 85(8); 716—1958(Dec.)
Trinity River delta 255—1978(Nov.)
Trinity River valley 156—1955; 482—1940; 654—1957
U.S. Highway 75 255—1952
U.S. Highway 77 255—1952
U.S. Highway 80 255—1959(3)
U.S. Highway 90 255—1959(3)
Uvalde County 152—1982; 612—1937(June)
Uvalde County, South 612—1930
Val Verde Basin 756—1959, 66, 83
Valley of the Giants 94—1967, 73(3)
Van Horn District 296—1933(13)
Waco 162—1951; 654—1964
Waco area 94—1968(2)
Walnut Prairie 94—1971
Webb County 152—1975
West, Alpine area 641—1969
West, Amistad Dam 594—1965
Wharton County, Boling 197—1962(7)
Wharton County, Boling Field 255—1959(1)
Whitney Reservoir 94—1974(3)
Williamson County 92—1985(8)
Wilson County 612—1958
Zapata County 152—1975
Zavala County, North 612—1930

Utah
General 30—1984; 184—1969; 199—1983; 204—1978; 276—1967; 737—1948; 501—1985(Pt.2); 272—113, 101

Central 197—1966(A), 75(2,10); 199—1983; 276—1954; 734—1972; 737—1949; 282—1985(2); 644—1982
East 280—1978(C-3/C-4); 458—1981; 282—1985(2)
East-central 276—1956; 503—1983(Feb.)
North 185—1960, 61; 199—1983; 204—1969(1,2,4,6); 276—1953
Northeast 199—1983; 604—1953
Northwest 199—1983; 204—1980(1); 734—1984; 737—1956
North-central 204—1957; 385—1958(Aug.); 734—1985; 737—1964
South 296—1933(18); 327—1970
Southeast 737—1954, 67; 590—1985(8); 272—107
Southwest 197—1975(11), 84(2); 199—1983(4,9); 276—1952, 63; 280—1978(C-2)
South-central 105—1957, 59; 220—1982(850); 276—1954, 65
West 103—1973; 204—1978(4)
Antelope Island 204—1969(6)
Arches National Park 199—1983(8); 217—1974; 421—1981
Bannock Overthrust 296—1933(17)
Bear Lake Valley 185—1961
Beaver Basin 220—1982(850)
Big Cottonwood Canyon 204—1980(4B)
Bingham 204—1952, 69(2), 80(7)
Bingham Mining District 296—1933(17); 583—1975; 734—1976; 737—1961
Birds Eye 385—1958(May)
Bonneville Basin 737—1955
Bonneville Lake 103—1968(4)
Book Cliffs 590—1985(10); 503—1983(Feb.)
Boulter Mountains, North 737—1959
Bryce Canyon National Park 105—1957, 59; 161—1977; 569—1984(1)
Canyonlands 217—1974
Canyonlands National Park 103—1971; 180—1971, 75; 199—1983(8); 590—1985(7)
Carbon County 197—1966(A)
Cataract Canyon 180—1971
Cedar City 204—1966
Colorado Plateau 296—1933(18)
Colorado Plateau, North 199—1983(7)
Colorado Plateau, Northwest 197—1975(10)
Colorado River 180—1971; 272—88
Cottonwood region 296—1933(17)
Dinosaur National Monument 296—1933(17); 272—148
Disturbed Belt 204—1978(10)
East Tintic Mountains 737—1957
Farmington 204—1980(6B)
Grand County 180—1960, 75, 79; 725—1971
Great Basin 197—1984(1)
Great Salt Lake 204—1969(6); 737—1966
Great Salt Lake area 197—1975(13); 296—1933(17)
Green River 272—88
Green River, Desolation Canyon 272—87
Green River, Gray Canyon 272—87
Green River, North 272—58
Henry Mountains 734—1980; 737—1946
Hopi Buttes 327—1970
Iron Springs Mining District 197—1975(11); 737—1947
Juab County, West 737—1963
Kaysville 204—1980(4A)
Lake Bonneville 199—1983(10)
Little Mountain 204—1980(6A)
Little Valley area 197—1975(13)
Long Bench 204—1980(4C)
Milford area 734—1973
Millard County 276—1951
Mining Districts 43—1969
Monument Valley 327—1970
Monument Valley area 458—1973
Ogden 204—1980(4C)
Ogden Valley 737—1955
Oquirrh Mountains 734—1976

Oquirrh Mountains, North 737—1961
Oquirrh Mountains, South 737—1959
Overthrust Belt 549—1976(1-4); 734—1982
Paradox Basin 18—1972(B2); 180—1955, 60, 63; 246—1983(Oct.); 276—1958; 549—1981
Park City Mining District 296—1933(17)
Park City area 737—1968
Pine Grove Molybdenum Deposit 85—1984(7)
Salt Lake County 184—1969
San Juan County 180—1955, 60, 63, 71, 73, 75, 79; 725—1971
San Juan River 180—1973
Silver Island Mountains 737—1960
Stansbury Mountains 737—1958
Tintic Mining District 296—1933(17)
Tooele County 737—1958, 59
Tooele County, Clifton District, Gold Hill 385—1958(Apr.)
U. S. Interstate 80 272—49
Uinta Basin 197—1976(6); 199—1983(2); 276—1950, 57, 64; 549—1965; 734—1974, 83, 85
Uinta Mountains 276—1959, 69; 296—1933(17); 729—1948; 731—1966; 732—1954
Utah County 197—1966(A); 737—1959
Wasatch Front 103—1968(2); 204—1978(10), 80(4,6); 734—1971
Wasatch Front, Central 197—1975(4)
Wasatch Mountains 199—1983(2); 204—1962; 276—1959; 296—1933(17); 385—1958(Aug.)
Wasatch Mountains, Central 103—1980; 734—1981; 737—1953
Wasatch Mountains, South 103—1968(3)
Wasatch Plateau 197—1975(2)
Wasatch fault zone 737—1964
Wayne County 180—1975
Zion National Park 105—1957, 59; 161—1977; 569—1984(1); 272—50

Vermont
General 456—1961, 69, 72, 79(A-4,B-8)
Central 456—1938, 72(G-1)
Northwest 296—1933(1); 456—1972(B-5,B-6,B-9,G-2,G-3,P-2); 469—1969(A); 739—1985(E1-E21)
North-central 456—1972(B-8)
South 456—1972(B-1)
Southeast 456—1972(B-7,B-11); 739—1985(B1-B9)
West 296—1933(1)
West-central 456—1959, 72(B-3)
Appalachian Mountains 197—1952(1)
Battleboro Quadrangle, Gildford Dome area 739—1985(B1-B9)
Champlain Islands 456—1972(P-1)
Champlain Valley 456—1972(EG-2,G-6); 469—1969(D); 474—1982(A)
Green Mountains 296—1933(1); 456—1972(B-1,B-8,B-10)
Green Mountains, South 456—1972(B12)
Hinesburg Synclinorium 469—1969(C)
Hoosick Falls area 456—1972(B-2)
Hudson Basin, North 469—1985(A-10)
Lake Champlain 296—1933(1); 456—1972(LS-2,LS-3); 469—1969(B)
Lamoille Valley 456—1972(G-2)
Marble Belt 296—1933(1)
Middlebury area 456—1972(B-4)
Montpelier 456—1961
Mount Mansfield 456—1971(EG-1)
Mt. Cube area 456—1971(A-6)
Richmond area 456—1972(B-13)
Roxbury area 456—1948(1)
Shelburne 456—1972(G-5)
Waterbury area 456—1948(1)

Virginia
General 69—1968(A-8); 172—1938; 206—1971(3), 77(2); 290—1979(1); 303—1978; 439—1964
Central 201—1982(5); 296—1933(11); 411—1979(2); 746—1970
North 69—1968(A-7,B); 201—1976(5), 82(3,4); 296—1933(11); 411—1979(5); 586—1985
Northeast 32—1984; 746—1975(1)
Northwest 58—1953; 745—1974; 747—1976
Southeast 25—1985(No.1); 411—1979(1); 740—1956; 747—1983; 750—1981
Southwest 25—1984; 201—1976(3); 206—1963, 71(1,2,4), 79(2), 85(3); 209—1946(May), 50; 740—1968; 744—1980(A-C); 747—1972, 75, 79, 81; 647—1984; 440—1971(1)
South-central 744—1977(2)
West 18—1960(2); 29—1937; 494—1965; 747—1984
West-central 197—1971(8:2); 746—1969
Abingdon Quadrangle 746—1980
Allegheny County 747—1984
Andersonville Quadrangle 48—1981
Appalachian Basin 197—1981(4)
Appalachian Mountains 197—1971(2); 411—1979(3); 563—1969
Appalachian Valley 18—1960(2); 296—1933(3)
Appalachian Valley, North 746—1976
Atlantic Coastal Plain Margin 201—1982(2)
Back Bay National Wildlife Refuge 740—1962(E)
Blacksburg 206—1979
Blacksburg area 440—1971(4)
Blue Ridge Mountains 201—1976(5), 82(6); 206—1979(1)
Catoctin 201—1982(4)
Central, Madison County 746—1975(2)
Chesapeake 182—1966
Chesapeake Bay 296—1933(5)
Chesapeake Bay Bridge Tunnel 740—1962
Chesapeake Bay region 201—1976(7b), 82(1)
Chesapeake and Ohio Canal 197—1971(7)
Coastal Plain 410—1967
Coastal Plain, Central 25—1985(No.5); 89—1984; 591—1985
Coastal Plain, North 89—1962
Coastal Plain, South 89—1970(1)
Culpeper Basin 201—1982(3); 411—1979(5); 747—1971
Cumberland County 746—1975(2)
Cumberland Gap area 209—1961
Currituck Spit 586—1977
Delmarva Peninsula, South 25—1985(No.2)
Fairfax County 18—1977(6); 202—1976
Farmville Triassic Basin 746—1982
Frederick County 746—1971
Fredericksburg 201—1976(1,4); 411—1979(4); 747—1978
Front Royal Quadrangle 745—1975
Great Dismal Swamp 25—1985(No.1)
Great Falls 69—1968(A-2)
Greenbrier Caverns 440—1971(3)
Greenfield Quadrangle 744—1977(4)
Hanover County, West 747—1982(2)
Highland County 496—1974
James River 206—1979(1)
Lee County 25—1984
Leesburg area 747—1982
Little North Mountain 747—1977
Madison County 746—1975(2)
Martinsville area 747—1973
Midlothian area 89—1970(2)
Mining Districts 43—1971
Montgomery County 206—1971(4); 744—1980(B)
Mount Rogers area 129—1967
Mountain Lake 435—1963
Mt. Vernon 69—1968(A-3)
Narrows 206—1971(3)
New River valley 72—1980
Norfolk area 740—1962
Pamunkey River 25—1985(No.5); 591—1985
Piedmont region 296—1933(11)
Pine Mountain 206—1984(1)

Potomac River, Great Falls 272—147
Reston 69—1968(A-4)
Richmond 206—1979(1)
Richmond area 89—1974; 747—1985(1)
Roanoke 206—1963
Roanoke area 586—1983
Rockingham County, Harrisonburg area 58—1955
Roseland District 201—1982(5); 296—1933(11)
Seashore State Park 740—1962
Shady Valley Quadrangle 746—1980
Shenandoah National Park 741—1976
Shenandoah Valley 197—1971(8); 411—1979(3)
Shenandoah Valley, Central 747—1977
Sherando Quadrangle 744—1977(4)
Southeast, Back Bay National Wildlife Refuge 740—1962(E)
Southeast, Seashore State Park 740—1962
Valley and Ridge Province 201—1982(6); 296—1933(3); 746—1981
Valley and Ridge Province, Central 747—1970
Virginia Piedmont 206—1979(1)
Waynesboro Quadrangle 744—1977(3)
Willis Mountain Quadrangle 48—1981
York-James Peninsula 75—1982; 89—1969, 80

Washington
General 76—1968(1-5); 435—1972; 436—1982; 568—1981; 272—11
Central 197—1977(13:1); 753—1963, 75
East 145—1969; 185—1983; 199—1976(1)
Northwest 184—1966; 190—1983(12)
South 199—1973
Southeast 39—1979(1); 199—1976(2)
Southwest 197—1977(9); 199—1973(3-4)
West-central 197—1977(13:3)
Bacon Siphon 76—1977(1-C)
Bellingham Bay 199—1985(6)
Bridgeport Slide 76—1977(1-B)
Cascade Front 76—1977(3)
Cascade Range 292—1974(9)
Cascade Range, Central 197—1977(3,10)
Cascade Range, East-central 184—1969
Cascade Range, North 197—1977(1)
Cascade Range, South 197—1977(4)
Channeled Scablands 296—1933(22); 542—1978(June)
Channeled Scablands area 197—1977(13:2)
Chief Joseph Dam 76—1977(1-B)
Coast 752—1980
Columbia Basin 199—1976(3); 542—1978(June); 579—1962
Columbia Plateau 292—1974(8)
Columbia Plateau, West 197—1977(12)
Columbia River Gorge 561—1979(1)
Grand Coulee 76—1977(1-C); 296—1933(22)
Grand Coulee Dam 292—1974(8)
Hoh River 752—1980
Kootenay Arc 583—1970
Mount St. Helens 199—1973(7); 518—1982
Mt. Challenger Quadrangle 272—143
Mt. Rainier 292—1974(9)
Northeast, Kootenay Arc 583—1970
Olympic National Park 197—1977(2); 272—144
Olympic Peninsula 199—1964(1)
Olympic Peninsula, Coast 197—1977(2)
Olympic Peninsula, North 76—1977(5); 199—1964(2)
Pasco Basin 199—1973(6)
Quillayute River 752—1980
San Juan Islands 190—1983(5); 197—1977(11); 676—1962, 65
San Juan Islands, Orcas Island 197—1977(5)
Satsop 76—1977(4)
Seattle area 197—1977(14)
Snoqualmie 76—1977(3)
Teton Dam 76—1977(2)

Whitman County 97—1973(3)
Yakima River Valley 197—1977(13:3)

West Virginia
General 18—1977(2); 29—1937; 25—1979; 172—1939; 206—1971(3); 290—1979(1); 763—1958, 72
East 58—1952, 82; 586—1985
East-central 435—1983
North 25—1972(2); 58—1949; 197—1969(A); 60—1984
North-central 765—1981(14)
South 23—1978; 201—1976(3)
Southeast 58—1953
West 587—1979
Allegheny Front 58—1982, 84; 172—1972
Appalachian Basin 197—1981(4)
Appalachian Mountains 74—1963; 197—1961(1), 69(A)
Beckley 206—1971(3)
Berkeley Springs 58—1959; 76—1970
Blackwater Falls State Park 58—1957(Oct.), 61; 766—1958, 71
Cacapon State Park 58—1959
Canaan Valley State Park 766—1958, 71
Dunkard Basin 763—1950
Grant County 172—1972
Great Valley 58—1964
Greenbrier County, East 22—1984
Greer 206—1957(1)
Harrison County 765—1985(18)
Humphrey 206—1957(2)
Marion County 765—1981(15)
Middlesboro Basin 58—1957(Apr.)
Mineral County 172—1972
Mining Districts 43—1971
Monongalia County 765—1981(13)
Morgantown 206—1957(1,2); 435—1976
New River valley 72—1980
Pendleton County 496—1974
Pocahontas County, South 22—1984
Randolph County 765—1985(19)
Terra Alto gas field 58—1949
Warm Spring Ridge 18—1960(4-5)
West Virginia Turnpike 197—1961(1)
Wheeling area 197—1959(3)
White Sulphur Springs 52—1948
Wills Mountain 58—1982

Wisconsin
General 55—1974; 69—1975; 183—1950; 273—1972(A); 435—1980; 474—1969(1); 659—1969, 83(1-2); 272—99
Central 55—1974(1); 200—1983(5); 273—1969; 377—1960; 659—1950, 68; 688—1962; 667—1981(6,7)
East 200—1983(3); 659—1947, 80(2-5)
East-central 132—1964; 197—1970(9); 409—1980
North 200—1979(1); 273—1973(1); 659—1953; 667—1981
Northeast 55—1974(1); 183—1953; 273—1984(1); 659—1962, 80(1); 688—1962; 667—1981(5)
North-central 659—1940
South 132—1964; 197—1970(4); 262—1939; 296—1933(26); 659—1934; 723—1970, 75
Southeast 197—1970(6); 200—1983(7); 659—1937, 77(1-4); 723—1983
Southwest 55—1974(2); 200—1983(2); 659—1959, 65; 598—1980; 723—1981, 83
South-central 132—1964; 200—1983(4); 273—1978(3)
West 197—1970(1); 377—1966
West-central 183—1959, 84; 659—1974(1-3)
Baraboo District 197—1970(2)
Black River Falls 659—1983
Black River valley 273—1980(2)
Chippewa Valley 273—1980(1)
Chippewa region 659—1974(1)

Clark County 273—1973(2)
Dane County 777—1983
Dane County, Southwest 272—96
Devil's Lake State Park 687—1959
Dodge County 659—1977
Driftless area 55—1974(2); 183—1982(May)
Eau Claire area 659—1974(2); 720—1971(3), 77
Fox River valley 197—1970(6); 409—1980
Gogebic Range 296—1933(27)
Jackson County 273—1973(2)
Kickapoo River valley 183—1982(May)
Lake Superior region 190—1984(5); 272—84
Madison area 587—1978
Marathon County 273—1980(4); 667—1981(7)
McCaslin Syncline 659—1962
Milwaukee 273—1978(2)
Mississippi River valley 208—1969
Mississippi River valley area 197—1970(7)
North Kettle 382—1969
Rock River valley 197—1970(6)
St. Croix Valley 723—1978
Superior 723—1969
Twin Ports area 575—1974
Two Creeks Buried Forest 197—1970(9)
Upper Mississippi valley 273—1973(3), 78(1)
Wausau 273—1980(3), 84(3); 720—1971(Oct.)
Wisconsin River valley 183—1982(May)
Wisconsin River, Lower 777—1984

Wyoming
General 27—1979(No.2); 30—1974; 231—1971; 276—1967; 583—1969(2), 71(June): 682—1968; 754—1963-64, 68, 70, 73; 783—1964; 781—1961-62, 69-71, 80-81(1); 37—1982; 590—1985(11); 724—1981; 272—100; 725—1967; 727—1985
Central 30—1967, 74; 781—1946
East 341—1940; 725—1968
North 20—1983(Aug.)
Northwest 55—1970(D); 296—1933(24); 293—1985(2)
North-central 435—1984
South 204—1965(2,4); 604—1972
Southeast 204—1972(3); 604—1948; 783—1964; 781—1946; 232—1984(21)
Southwest 781—1950, 60
South-central 781—1951, 72
West 604—1972; 232—1983; 20—1983(Aug.)
West-central 781—1981(4)
Absaroka Range 292—1974(2)
Albany County 783—1979
Alcova area 780—1974
Beartooth Range 365—1981
Bighorn Basin 292—1974(1); 435—1984; 549—1937; 604—1982(40th); 783—1964; 781—1947, 59, 75, 83; 293—1985(2); 724—1977(Apr.): 725—1969, 70
Bighorn Basin, North 55—1984(1); 781—1983
Bighorn Basin, South 781—1952
Bighorn River 402—1961
Black Hills 74—1946; 204—1981(2), 85(Revised); 282—1985(1)
Black Hills area 781—1968
Carbon County 783—1979
Casper Mountain 27—1979(1); 30—1967
Casper Mountain area 780—1978
Casper area 781—1954, 66, 73, 84-85
Cheyenne 204—1965(4)
Cody 781—1975
Colorado-Wyoming State Line 204—1979(4)
Dinwoody 232—1983
Fremont County 232—1984(23)
Front Range 197—1976(3); 341—1929
Grand Teton National Park 292—1974(4)
Green River 204—1972(1)
Green River basin 781—1955, 73
Gros Ventre Valley 292—1974(4)
Hanna 204—1972(2)
Heart Mountain 292—1974(2)
Highland Uranium Deposit 30—1974(4)
Jackson 232—1983
Jackson Hole 185—1958; 292—1974(4); 781—1956, 80, 81
Laramie Basin 197—1976(3); 204—1972(4); 781—1953
Laramie Range 197—1960(B-2); 204—1965(1)
Mining Districts 43—1969
Overthrust Belt 724—1977(Sept.)
Parkman River delta 30—1974(1)
Powder River 781—1976
Powder River Coal Basin 232—1980
Powder River Uranium Mines 27—1979(3)
Powder River basin 27—1979(3,4); 783—1981; 781—1949, 58, 63; 282—1985(1)
Powder River basin, North 781—1963
Rock Springs Uplift 596—1979(Sept.); 781—1965
Rocky Mountains 341—1929
Sheridan 435—1984
Snake River 781—1981(3)
Snake River Range 781—1981(2)
Teton Pass 20—1983(Aug.)
Wind River basin 781—1948, 74, 78
Wind River basin, Southwest 781—1957
Wind River basin, West 724—1979(Apr.)
Wyoming Thrust Belt 20—1983(Aug.)
Yellowstone National Park 204—1968(6); 292—1974(3); 296—1933(24); 428—1971(1); 781—1982; 569—1984(2); 724—1979(Sept.); 272—60, 98, 41
Yellowstone National Park, West 402—1960
Yellowstone Valley 55—1970(D); 549—1937

Virgin Islands
General 197—1980(24)
Caribbean area 127—1971
St. Croix 127—1968(2); 170—1974; 197—1974(1); 220—1981; 310—1977(6)
St. John 127—1968(1)
St. Thomas 127—1968(1)

STRATIGRAPHIC INDEX

Accomack Barrier Complex 25—1985(2)
Albert Formation 199—1985(#6)
Anacacho Asphalt Deposit 18—1984(May)
Androscoggin Lake Igneous Complex 210—1983(#11)
Annona Chalk 162—1945
Arkadelphia Marl 566—1924
Athabasca Oil Sands 125—1973, 78(#3); 190—1976(C-9); 561—74(#100); 279—1982(#22); 308—1978(Jun 17,Jun 9)
Atoka Formation 204—1985(#1); 205—1978(#1)
Austin Chalk 92—1985(#7); 199—1983(#3); 205—1975(#4)
Banff Formation 190—1950(#2)
Bangor Limestone 5—1980(#17)
Barnwell Formation 240—1979
Bartlesville Sand 18—1968(#1)
Bearpaw Formation 18—1982(#11)
Belly River Formation 125—1984(Aug)
Belt Supergroup 97—1973(#1), 83(#1); 204—1959(#1), 77(#3,#5)
Berry River Formation 273—1985
Bertie Formation 469—1984(BC-5)
Big Snowy Group 402—1951
Binnewater Sandstone 469—1976(B3)
Bird River Greenstone Belt 190—1982(#9)
Black River Group 469—1977(A8), 78(A8,B4)
Blackhawk Formation 590—1985(#10)
Blackstone Series 456—1981(B3)
Blaine Formation 482—1939
Blood Reserve Sandstone 125—1983(#12)
Blue Hills Igneous Complex 456—1976(A3,B3)
Bluejacket Sandstone Member 18—1968(#1)
Bois Blanc Formation 25—1982(A-2)
Bone Valley Formation 197—1964(#6)
Borden Group 200—1980(#2); 206—1970(#3); 209—1980(Oct); 587—1973
Boston Bay Group 456—1976(A1,A5,B1), 81(C6)
Bottle Lake Comlex 456—1980(A1)
Boulder Batholith 204—1953(#3,#4), 68(#3), 82(#3)
Breathitt Formation 587—1982
Brimfield Schist 456—1968(F5)
Brome Igneous Complex 296—1972(B13)
Cairn Formation 125—1985
Calaveras Formation 199—1977(#4)
Calvin Sandstone 565—1938(#1)
Canyon Group 2—1954(Nov); 18—1975(#2); 482—1940, 56; 627—1958, 62
Capitan Formation 550—1951(#4), 57, 64, 85; 594—1972
Captain Creek Limestone Member 90—1978(SEPM 2)
Carbondale Formation 197—1970(#8); 290—1979(J)
Cardigan Gneiss 456—1971(A3)
Carlile Shale 596—1983
Carolina Slate Bed 129—1961, 84(Oct), 85(Nov); 197—1980(V1 #8), 85(#6); 206—1977(#1); 583—1985
Casco Bay Group 210—1978(#3,#4)
Casper Formation 197—1960(B-2)
Castlegate Sandstone Member 590—1985(#10)
Catahoula Formation 588—1960
Catskill Delta 469—1985(A7); 293—1985(#1)
Cayugan Series 469—1984(BC-5)
Cedar Valley Formation 208—1984(#42); 659—1975(#1)
Champlain Sea Clays 198—1978(#10)
Charlotte Belt 129—1961; 206—1969(#3)
Chattanooga Shale 197—1981(V2 #3)
Chazy Group 456—1972(P1); 469—1969(E,F)
Chelmsford Granite 197—1952(#3)
Cherokee Group 200—1981(#2); 290—1979(#10); 659—1981(#1)
Cherry Valley Limestone 469—1972(G)
Chesapeake Group 201—1976(7b), 82
Chesterian 209—1952; 262—1938, 46, 56, 82; 271—1940, 75
Chickamauga Group 240—1969(#4); 561—(#78), 74(#64)

Chuckanut Formation 199—1985(#6)
Cincinnatian Series 197—1981(V1 #1,V1 #12)
Cisco Group 482—1956; 594—1960(Apr)
Citadel Formation 122—1985(Sep #2)
Claiborne Group 18—1979(#2); 162—1939, 60; 205—1970; 247—1956(#1); 255—1938; 390—1940(2nd), 52, 56; 566—1927; 626—1968
Clays Ferry Formation 209—1965(#1)
Clear Fork Group 2—1946(Nov)
Clinton Group 25—1982(B-1); 469—1972(A)
Codell Sandstone Member 596—1983
Coffeyville Formation 205—1978(#2)
Coldwell Complex 273—1977(A)
Columbia River Basalt 39—1979(#1); 145—1969
Comanchean 18—1974(SEPM2:1)
Coosa Deformed Belt 5—1974(12th)
Cortlandt Complex 197—1948(#3); 201—1983(#1); 469—1968(H)
Crouse Limestone Member 205—1972
Custer Group 482—1939
Cuyahoga Formation 494—1983; 497—1981(3)
Dakota Formation 18—1980(SEPM 3); 197—1960(B-2); 204—1973(#9), 79(#1C)
Deer Creek Limestone Member 341—1984
Deese Formation 62—1937(Mar)
Delaware Mountain Group 594—1979
Desmoinesian 26—1979(Oct); 87—1955(2nd); 197—1958(#5); 90—1978(2)
Dolores Formation 204—1984
Duluth Complex 273—1968
Eagle City Beds 208—1960(Sep)
Eastport Formation 456—1978(B7)
Edwards Formation 18—1974(#2,#3), 75(#3); 92—1984(#6), 85(#8); 162—1959; 197—1962(#11); 247—1979(#1); 713—1973(#1,#4)
El Reno Formation 2—1946(Nov)
Ellenberger Limestone 712—1945(Jun); 756—1933
Elberton Granite 240—1980
Ervine Creek Limestone 341—1984
Eutaw Formation 5—1980(#17); 223—1960(#18); 390—1945
Fairholme Carbonate Complex 125—1978
Farmington Canyon Complex 204—1980(#6B)
Fernie Formation 301—1982
Flin Flon Volcanic Belt 190—1982(#6)
Fort Dodge Gypsum 659—1981(#1)
Fountain Formation 30—1984; 293—1985(#4)
Fox Hills Formation 18—1980(SEPM 2); 204—1979(#1D)
Francis Creek Shale 197—1970(#8)
Franciscan Formation 18—1981(#1,AAPG 2); 38—1983(Dec); 114—1969; 199—1979(#12); 356—1984(Nov); 412—1979(Oct A), 82(Oct)
Fredricksburg Group 627—1965; 654—1957
Frontier Formation 30—1974(#3)
Galena Dolomite 200—1983; 330—1975
Gallup Sandstone 461—1979
Galveston Barrier Island Complex 205—1976(#3); 713—1970(#11)
Genesee Group 25—1982(A-1); 469—1970(A)
Gilmore City Limestone 208—1960(Sep)
Glen Rose Formation 32—1979
Gog Formation 125—1983(#10)
Gower Formation 208—1981(#35)
Great American Bank Facies 201—1982
Great Valley Sequence 18—1981(#1); 28—1978(Apr); 199—1977(#8); 593—1983(May)
Green River Formation 204—1972(#1)
Green Series 197—1959(#3); 763—1950(Sep)
Guadalupian 594—1977
Guichon Creek Batholith 190—1977(#3)
Hamburg Klippe 172—1984
Hamilton Group 411—1981(B); 469—1970(D), 72(F), 73(B), 74(G), 77(B8), 78(A3,A5,B1), 84(BC-9)

Hartland Formation 456—1953(B)
Hartselle Sandstone 5—1980(#17)
Hartshorne Sandstone 589—1983
Harvard Conglomerate 456—1976(F4)
Haystack Mountains Formation 590—1985(#11)
Helderberg Group 25—1984(Oct); 197—1963(#1); 469—1978(A8,B4), 85(A5,B5); 586—1980
Helena Formation 97—1983(#1,#7)
Herrin Coal Member 290—1979(D)
High Bridge Group 197—1981(V1 #11)
Hillabee Chlorite Schist 5—1979
Hillsboro Plutonic Series 456—1971(A7)
Hodges Mafic Complex 456—1985(C2)
Holdenville Formation 26—1981(#2)
Home Creek Limestone 2—1948(June)
Hornbrook Formation 356—1984(Sep)
Howard Limestone 341—1982(#1,#2)
Hoxbar Formation 62—1937(Mar)
Hummingbird Reef Complex 18—1982(#7); 125—1981
Huronian 273—1974(#4); 587—1981; 279—1982(#13B)
Hurricane Mountain Formation 456—1983(B1)
Idaho Batholith 17—1979(Dec); 199—1976(#4); 204—1959(#4), 64(#3), 77(#1)
Inyan Kara Group 596—1982(#3)
Iowa Falls Dolomite Member 208—1960(Sep)
Isleboro Formation 456—1974(A6)
Jackson Group 206—1964(#1); 255—1938; 390—1956; 566—1933, 34; 580—1968(#2); 588—1960
Jacobsville Sandstone 273—1975(#3)
John Day Formation 39—1979(#2)
Jokulhlaups 185—1983(Aug)
Judith River Formation 125—1983(#1), 84(#2)
Kansas City Group 197—1965(#5)
Katahdin Granite 456—1980(A2,B1)
Key Largo Limestone 55—1972(#1)
Kinderhookian 208—1967(Jun); 331—1967(#6)
Kings Mountain Belt 129—1981(Oct); 206—1977(#3)
Kingsport Formation 583—1969
Kinsman Quartz Monzonite 456—1971(A3)
Kittery Formation 456—1984(A4)
Knip Beds 127—1977(#5)
Knippa Basalt 612—1985
Kootenay Formation 125—1983(#4,#11); 190—1981
Kreyenhagen Shale 356—1984(Apr)
LaHood Formation 204—1982(#4)
Lac du Bonnet Batholith 190—1982(#15)
Lake Chatuge Sill 240—1974(#1)
Lamotte Sandstone 456—1977(#4)
Laramie Anorthosite Complex 197—1960(B-2); 204—1965(#1), 72(#3)
Laramie Formation 204—1979(#1D)
Leadville Formation 590—1985(#6)
Leo Sandstone 596—1982(#2)
Leonardian 594—1962
Levis Formation 122—1985(Sep #3)
Lexington Limestone 206—1970(#2); 209—1965(#1)
Lock-Haven Reef 172—1985(#1)
Lockport Formation 469—1974(A)
Logan Formation 494—1983; 497—1981(3)
Ludlowville Shale 469—1970(D)
Lynn Lake Greenstone Belt 190—1982(#2)
Lyons Sandstone 204—1973(#5)
Mancos Shale 197—1960(A-1); 596—1981(Fall)
Maquoketa Formation 208—1962(Jul); 659—1978(#4)
Marcellus Shale 456—1979(A9); 469—1977(A3)
Marcy Anorthosite 469—1984(AB-2)
Marlboro Formation 456—1984(C2)
Martinsburg Formation 172—1954, 82
Mascot Dolomite 583—1969
Mayville Limestone 659—1976(#3)
McAlister Shale 565—1938(#2)

McMurray Formation 18—1982(#1)
Meganos Formation 199—1979(#9)
Meguma Group 121—1984(Oct); 190—1980(#11); 199—1985(#11); 279—1982(#5B)
Meramec Group 208—1959(Apr)
Mesaverde Group 197—1960(A-1); 596—1981(Fall)
Miami Limestone 374—1983
Midway Group 390—1943, 53, 57; 566—1924
Miette Complex 279—1982(#27A)
Miette Group 125—1983(#10)
Missoula Group 97—1973(#2), 83(#1,#7)
Monongahela Group 197—1959(#3); 763—1950(Sep)
Monterey Formation 18—1981(AAPG 2), 84; 23—1977(OCT.); 28—1982(Oct); 220—1984; 356—1984(Apr); 593—1982(OCT); 590—1984(#2)
Morin Complex 296—1972(B1)
Morrison Formation 204—1979(#1B)
Moscow Formation 25—1982(A-1); 469—1970(D), 74(C)
Mount Arabia Migmatite 197—1980(V1 #3)
Moxie Pluton 456—1983(C1)
Muddy Sandstone 781—1974
Murphy Belt 129—1971
Muskox Intrusion 296—1972(A29)
Narragansett Pier Granite 456—1981(B5)
Neda Formation 200—1983(#3); 659—1976(#3)
New Albany Shale 197—1983(V2 #15)
New Virgilian 341—1956
New York City Group 469—1975(A1)
Newark Group 197—1969(#4)
Newbury Volcanic Complex 456—1976(A16,B16)
Newman Limestone 587—1982
Niagaran Bioherms 200—1968(#1)
Niagaran Reef 18—1950, 56
Nicola Group 190—1977(#5)
Nonewaug Granite 456—1953(B)
North Mountain Basalt 190—1980(#18)
North Shore Volcanics 197—1972(#2); 200—1979(#7); 273—1971(A)
Oak Hill Group 190—1979(A-14); 200—1979(A19,B6)
Oakdale Quartzite 456—1981(A1)
Oakville Sandstone 588—1960
Ocoee Series 206—1962(#3)
Ohio Shale 197—1981(V2 #3)
Oka Complex 190—1963(#1); 296—1972(B11)
Onondaga Limestone 25—1982(A-2); 197—1963(#1); 456—1979(B7); 469—1962(A), 74(A), 77(B9), 78(A8,B4)
Oologah Limestone 662—1973(Oct)
Oread Limestone 26—1983; 341—1983
Oriskany Formation 172—1931(II)
Osagian 179—1972(#3); 208—1967(Jun)
Pacific Rim Complex 199—1985(#7)
Palisade Diabase 197—1948(9)
Palliser Formation 190—1950(#2)
Patch Reef 200—1978(#3)
Paxton Quartz Schist 456—1968(F5)
Pease River Group 482—1939
Pecan Gap Chalk Member 162—1945
Peninsular Ranges Batholith 199—1986(#9); 408—1979(#5,#21)
Permian Reef Complex 594—1977; 626—1976(Fall); 311—1983
Phosphoria Formation 204—1980(#3)
Pierre Shale 204—1965(#5)
Pinckneyville Granite Complex 671—1968
Pocatello Formation 199—1983(Pt.2 #3)
Pocono Formation 172—1969
Point Sal Ophiolite 408—1979(#5,#24)
Port Coldwell Alkalic Complex 273—1970(C)
Portland Formation 456—1982(M1), 85(B7)
Potrillo Basalt/Volcanics 205—1977(#1)
Potsdam Sandstone 469—1978(A6)
Pottsville Group 5—1964(1st), 82
Pounds Sandstone 587—1976

Powderhorn Carbonitite 158—1983
Preston Gabbro 456—1982(B8)
Queen City Formation 588—1980
Ravalli Group 97—1983(#1,#7)
Red Beach Granite 456—1978(A2)
Rochester Formation 25—1982(B-1)
Rye Formation 456—1984(B4)
Sainte Genevieve Limestone 262—1938
Salem Limestone 197—1983(V2 #16); 200—1980(#1); 268—1954
Salina Group 469—1972(B)
San Gabriel Formation 161—1973(Spring); 199—1982(#5)
Schoharie Grit 469—1962(A)
Scotch Grove Formation 208—1981(#35)
Selma Group 5—1968; 223—1960(#18)
Seminole Formation 26—1981(#2)
Sespe Formation 604—1954(Oct)
Shady Dolomite 206—1977(#2)
Shannon Sandstone Member 590—1985(#11)
Sharon Conglomerate 200—1974(#2); 493—1950(Fall)
Shawnee Group 341—1984
Shefford Igneous Complex 296—1972(B13)
Shelt Sandstone 590—1985(#11)
Shuswap Complex 199—1985(#12)
Shuttle Meadow Formation 456—1982(M1)
Siegas Formation 456—1980(C1)
Sierra Nevada Batholith 197—1966(#6)
Silver Cliff Volcanics 158—1980
Simpson Group 504—1936, 37
Skaneateles Shale 469—1977(A3)
Southesk Formation 125—1985
Spergen Formation 208—1977(Oct)
Springfield Coal Member 197—1983(V2 #9)
Squantum Tillite Member 456—1976(A2,B2)
St. Cloud Red Granite 409—1976
St. Urbain Anorthosite 296—1972(B6)
State Quarry Limestone 659—1975(#1)
Stephensport Group 271—1975
Stettin Syenite Pluton 273—1980(#3)
Stillwater Complex 204—1968(#4); 365—1981; 300—1985
Stone Mountain Granite 197—1980(V1 #3)
Stony Creek Basalt 593—1983(May)
Strawn Series 2—1950(Nov), 54(Nov); 482—1940; 594—1956; 627—1958
Sundance Formation 596—1982(#3)
Sycamore Canyon Member 28—1944(Nov)
Talladega Front 5—1973(11th), 85; 197—1980(V2 #23)
Tejon Formation 593—1956(Feb)
Temblor Formation 593—1985(Oct)
Theresa Dolomite 469—1978(A6)
Thornton Reef 200—1976(#3B)
Thurman Sandstone 565—1938(#2)
Tigerton Anorthosite 659—1962
Tioga Bentonite 197—1971(#8)
Tobacco Root Batholith 204—1982(#3)
Topanga Formation 28—1980(#49)
Trafalgar Formation 197—1983(V1 #3)
Traveler Rhyolite 456—1980(B3)
Trenton Group 25—1984(Oct); 469—1972(C), 77(A9), 78(A8,B4), 83(#5), 84(BC-3); 474—1982(B); 122—1985(Sep #1)
Trinity Group 18—1974(SEPM2:1,SEPM2:2)
Tully Limestone 469—1977(A2), 78(A12,B8)
Tunk Lake Granite 456—1973(B5)
Tunk Lake Pluton 456—1974(B5)
Tuscaloosa Formation 390—1945
Tuscarora Formation 586—1982
Tyler Lake Granite 456—1985(C2)
Unkpapa Sandstone 596—1982(#3)
Vicksburg Group 206—1964(#1); 390—1956
Virgilian 341—1982(#1)
Virginia Dale Ring-Dike Complex 204—1973(#2)
Wabaunsee Group 341—1982(#1)

Wapanucka Limestone 662—1975(May)
Wapiabi Formation 125—1984(Aug)
Washikemba Formation 127—1977(#9)
Washington Series 197—1959(#3); 763—1950(Sep)
Washita Group 156—1955
Watchung Basalt 197—1948(#11)
Waterville Formation 456—1974(B6)
Waulsortian Facies 646—1982(1st)
Wausau Granite 273—1980(#3), 84(#3)
Wedron Formation 197—1983(V1 #3)
West Point Formation 190—1979(B-2)
Whitehorse Sandstone 2—1946(Nov)
Wilcox Group 162—1960; 255—1968(Oct); 390—1940(2nd), 43, 53, 56, 57; 566—1924, 27; 580—1968(#4)
Windsor Group 190—1980(#5,#22); 199—1985(#5)
Witchita Group 771—1946
Wolfe City Sand Member 162—1945
Woodbine Formation 162—1951; 205—1968
Yakima Basalt 197—1977(#12); 199—1976(#1)

Form for reporting new field trip guidebooks

(After a preliminary list of new guidebooks is sent to you use this form to report any not on the list.)

NUC Symbol _____ Date _____ Your initials _____

Please be sure to include your NUC library symbol.

Year of guidebook_____

Organization name_____

UL number_____ Volume number_____ Part number _____

Title (or general title)_____

_____

_____

Title of part_____

_____

_____

Secondary series (Co-sponsers)

UL number_____ Volume number_____ Part number_____

_____   _____   _____

Return to:
Union List Update
American Geological Institute
4220 King Street
Alexandria, VA 22302

YOU MAY PHOTOCOPY THIS PAGE

Please attach a copy of the title page and table of contents - This is very important.

Search retrivial printouts will be accepted in lieu of this form.